Lecture Notes in Computer Science 5257

Commenced Publication in 1973
Founding and Former Series Editors:
Gerhard Goos, Juris Hartmanis, and Jan van Leeuwen

T0224099

Masami Ito   Masafumi Toyama (Eds.)

Developments in Language Theory

12th International Conference, DLT 2008
Kyoto, Japan, September 16-19, 2008
Proceedings

 Springer

Volume Editors

Masami Ito
Kyoto Sangyo University
Faculty of Science, Kyoto 603-8555, Japan
E-mail: ito@cc.kyoto-su.ac.jp

Masafumi Toyama
Kyoto Sangyo University
Faculty of Computer Science and Engineering
Kyoto 603-8555, Japan
E-mail: toyama@cc.kyoto-su.ac.jp

Library of Congress Control Number: 2008934866

CR Subject Classification (1998): F.4.3, F.4.2, F.4, F.3, F.1, G.2

LNCS Sublibrary: SL 1 – Theoretical Computer Science and General Issues

ISSN 0302-9743
ISBN 978-3-540-85779-2 Springer Berlin Heidelberg New York

Springer is a part of Springer Science+Business Media

springer.com

© Springer-Verlag Berlin Heidelberg 2008

Typesetting: Camera-ready by author, data conversion by Scientific Publishing Services, Chennai, India
Printed on acid-free paper   SPIN: 12519972   06/3180   5 4 3 2 1 0

Preface

The 12th International Conference on Developments in Language Theory (DLT 2008) was held at Kyoto Sangyo University, Kyoto, September 16–19, 2008. This was the second time DLT took place in Kyoto. Previous DLTs were held in Turku (1993), Magdeburg (1995), Thessaloniki (1997), Aachen (1999), Vienna (2001), Kyoto (2002), Szeged (2003), Auckland (2004), Palermo (2005), Santa Barbara (2006) and Turku (2007). These will be followed by the next DLT in Stuttgart (2009).

The topics dealt with at the conference were: grammars, acceptors and transducers for words, trees and graphs; algebraic theories of automata; algorithmic, combinatorial and algebraic properties of words and languages; variable length codes; symbolic dynamics; cellular automata; polyominoes and multidimensional patterns; decidability questions; image manipulation and compression; efficient text algorithms; relationships to cryptography, concurrency, complexity theory and logic; bio-inspired computing; quantum computing.

The Program Committee selected 36 papers from 102 submitted papers. Each submitted paper was evaluated by at least three members of the Program Committee. All 36 selected papers are contained in this volume together with 6 invited papers. The editors thank the members of the Program Committee for the evaluation of the papers and the many referees who assisted the Program Committee members in this process. We are also grateful to the contributors to DLT 2008, in particular to the invited speakers for the realization of a successful conference. Moreover, we thank the Organizing Committee for its splended work, in particular Szabolcs Iván for his technical assistance. Without his assistance, it would have been impossible to perform the paper selecting process. Finally, we would like to express our gratitude to the Steering Committee for its support, in particular to Prof. Grzegorz Rozenberg for his valuable advice. This conference was dedicated to his 65th birthday.

The conference was supported by Kyoto Sangyo University and the Japanese Society for the Promotion of Science. The conference was held under the auspices of the European Association for Theoretical Computer Science. We are grateful to these institutions.

July 2008

Masami Ito
Masafumi Toyama

Organization

Steering Committee

Jean Berstel	Marne-la-Vallée, France
Cristian S. Calude	Auckland, New Zealand
Volker Diekert	Stuttgart, Germany
Juraj Hromkovic	Zurich, Switzerland
Oscar H. Ibarra	Santa Barbara, USA
Masami Ito	Kyoto, Japan
Werner Kuich	Vienna, Austria
Ghorge Paun	Bucharest, Rumania
Antonio Restivo	Palermo, Italy
Grzegorz Rozenberg, Chair	Leiden, Netherlands
Arto Salomaa	Turku, Finland
Sheng Yu	London, Canada

Program Committee

Symeon Bozapalidis	Thessaloniki, Greece
Olivier Carton	Paris, France
Alessandra Cherubini	Milan, Italy
Aldo de Luca	Naples, Italy
Zoltán Ésik	Szeged, Hungary
Volker Diekert	Stuttgart, Germany
Jozef Gruska	Brno, Czech Republic
Tero Harju	Turku, Finland
Hendrik Jan Hoogeboom	Leiden, Netherlands
Oscar H. Ibarra	Santa Barbara, USA
Masami Ito, Chair	Kyoto, Japan
Yuji Kobayashi	Funbashi, Japan
Werner Kuich	Vienna, Austria
Markus Lohrey	Leipzig, Germany
Victor Mitrana	Bucharest, Rumania
Friedrich Otto	Kassel, Germany
Jaques Sakarovitch	Paris, France
Kai Salomaa	Kingston, Canada
Pedro V. Silva	Porto, Portugal
Denis Thérien	Montreal, Canada
Sheng Yu	London, Canada
Mikhail Volkov	Ekaterinburg, Russia

Organizing Committee

Masami Ito, Co-chair   Kyoto, Japan
Szabolcs Ivan   Szeged, Hungary
Yoshiyuki Kunimochi   Fukuroi, Japan
P. Leupold   Kyoto, Japan
Jianqin Liu   Kobe, Japan
Masafumi Toyama, Co-chair   Kyoto, Japan
Kayoko Tsuji   Tenri, Japan

External Referees

Anil Ada   Omer Egecioglu
Cyril Allauzen   Laszlo Egri
Jean-Paul Allouche   Chiara Epifanio
Marco Almeida   Christiane Frougny
Dmitry Ananichev   Yuan Gao
Marcella Anselmo   Zsolt Gazdag
Jean Berstel   Viliam Geffert
Franziska Biegler   Silvio Ghilardi
Luc Boasson   Hugo Gimbert
Benedikt Bollig   Stefan Göler
Henning Bordihn   Archontia Grammatikopoulou
Robert Brijder   Erich Grädel
Srecko Brlek   Giovanna Guaiana
Veronique Bruyere   Irene Guessarian
Michelangelo Bucci   Stefan Gulan
Thierry Cachat   H. Peter Gumm
Cezar Campeanu   Klinkner Gwénaël
Arturo Carpi   Yo-Sub Han
Giuseppa Castiglione   Johanna Högberg
Arkadev Chattopadhyay   Markus Holzer
Christian Choffrut   Lucian Ilie
Marek Chrobak   Szabolcs Ivan
Alfredo Costa   Petr Jancar
Maxime Crochemore   Antonios Kalampakas
Sinisa Crvenkovic   Juhani Karhumaki
Elena Czeizler   Jarkko Kari
Eugen Czeizler   Efim Kinber
Juergen Dassow   Daniel Kirsten
Alessandro De Luca   Jetty Kleijn
Klaus Denecke   Walter Kosters
Michael Domaratzki   Mojmir Kretinsky
Pal Domosi   Manfred Kudlek
Jérôme Durand-Lose   Manfred Kufleitner

Michal Kunc
Yoshiyuki Kunimochi
Petr Kurka
Dietrich Kuske
Salvatore La Torre
Klaus-Jorn Lange
Eric Laugerotte
Peter Leupold
Kamal Lodaya
Christof Loeding
Sylvain Lombardy
Florin Manea
Sebastian Maneth
Christian Mathissen
Pierre McKenzie
Alexander Meduna
Ingmar Meinecke
Mark Mercer
Wolfgang Merkle
Hartmut Messerschmidt
Naoto Miyoshi
Frantisek Mraz
Emanuele Munarini
Lorand Muzamel
Mark-Jan Nederhof
Dirk Nowotka
Hitoshi Ohsaki
Alexander Okhotin
Gennaro Parlato
Andrei Paun
Dominique Perrin
Holger Petersen
Jean-Eric Pin
Brunetto Piochi
Martin Platek

Matteo Pradella
Panagiota Pournara
George Rahonis
Bala Ravikumar
Klaus Reinhardt
Antonio Restivo
Christophe Reutenauer
Kristina Richomme
Emanuele Rodaro
Jan Rutten
Nicolae Santean
Andrea Sattler-Klein
Shinnosuke Seki
Olivier Serre
Alexander Shen
Arseny Shur
Michael Sipser
Jean-Claude Spehner
Paola Spoletini
Ludwig Staiger
Heiko Stamer
Frank Stephan
Howard Straubing
Izumi Takeuti
Carolyn Talcott
Pascal Tesson
Wolfgang Thomas
Sandor Vagvolgyi
Birgit van Dalen
Stefano Varricchio
György Vaszil
Laurent Vuillon
Pascal Weil
Thomas Worsch
Hsu-Chun Yen

Sponsoring Institutions

Kyoto Sangyo University
Japan Society for the Promotion of Science

Table of Contents

Invited Talks

Contributed Papers

Iteration Semirings

Zoltán Ésik*

Research Group on Mathematical Linguistics
Rovira i Virgili University, Tarragona, Spain

Abstract. A Conway semiring is a semiring S equipped with a unary operation * : $S \to S$, called star, satisfying the sum star and product star equations. An iteration semiring is a Conway semiring satisfying Conway's group equations. In this extended abstract, we review the role of iteration semirings in the axiomatization of regular languages and rational power series, and in the axiomatization of the equational theory of continuous and complete semirings.

1 Introduction

One of the most fundamental algebraic structures introduced in Computer Science are the algebras of regular languages over an alphabet. These algebras are idempotent semirings equipped with a star operation, denoted * and called Kleene iteration. Other structures with similar operations include binary relations equipped with the operations of union, relation composition and reflexive-transitive closure which provide a theoretical basis for nondeterministic imperative programming. The reflexive-transitive closure operation on binary relations and the operation of Kleene iteration on languages can both be defined as least upper bounds of a chain of approximations, or by an infinite geometric sum. More generally, each ω-continuous or ω-complete semiring gives rise to a star operation. It is known that regular languages satisfy the very same equations as binary relations, or as idempotent ω-continuous semirings, or ω-complete semirings with an infinitary idempotency condition: the sum of every nonempty countable sum of any element a with itself is a. Moreover, the *-semirings of regular languages can be characterized as the free algebras in the class of all models of these equations.

The equational theory of regular languages has received a lot of attention since the late 1950's. Early results by Redko [40,41] and Conway [13] show that there is no finite basis of equations. On the other had, there are several finitely axiomatized first-order theories capturing the equational theory of regular languages. The first such first-order theory was described by Salomaa [42]. He proved that a finite set of equations, together with a guarded unique fixed point rule is sound and complete for the equational theory of regular languages. This result has been the model of several other complete axiomatizations involving power series, trees and tree languages, traces, bisimulation semantics, etc., see Morisaki and Sakai [38], Elgot [17],

* Supported in part by grant no. MTM 2007-63422 from the Ministry of Science and Education of Spain. On leave from University of Szeged, Hungary.

M. Ito and M. Toyama (Eds.): DLT 2008, LNCS 5257, pp. 1–20, 2008.

Elgot, Bloom, Tindell [18], Ito and Ando [28], Milner [36,37], Rabinovitch [39], Corradini, De Nicola, Labella [14], to mention a few examples. Salomaa's axiomatization has been further refined by Archangelsky and Gorshkov [1], Krob [34], Kozen [30,31] and Boffa [10,11]. See also [7]. In particular, the system found by Kozen relies on the least pre-fixed point rule for left and right linear fixed point equations. This result is important since the (guarded) unique fixed point rule is not sound in several important models such as binary relations over a nontrivial set.

Another direction of research has been trying to describe infinite nontrivial equational basis of regular languages, cf. [34,5]. A Conway semiring [6] is a *-semiring (i.e., a semiring equipped with a star operation) satisfying the well-known sum star and product star equations. In [13], Conway associated an equation with each finite group in any *-semiring. An iteration semiring is a Conway semiring which is a model of these group equations. Conway conjectured that the iteration semiring equations, together with a strengthened form of idempotency (the equation $1^* = 1$) are complete for regular languages. His conjecture was confirmed by Krob. (See also [19] for a proof from a more general perspective.) The completeness of this system readily implies the completeness of all known first-order axiomatizations, including the completeness of the systems given by Krob, Kozen, Boffa and others mentioned above. The equations associated with the finite groups are rather complex. For commutative (or even solvable) groups, they can be reduced to the classical "power equations", see Conway [13] and Krob [34], but the simplification of the general group equations is still open.

In this paper our aim is to review the role of Conway and iteration semirings in some recent axiomatization results extending those mentioned above. These recent results show that the iteration semiring equations, together with three simple additional equations, provide a complete account of the equations that hold in ω-continuous, or ω-complete semirings without any idempotency conditions. Moreover, the same equations provide a complete account of the valid equations of rational power series over the semiring $\mathbb{N}_\infty$, obtained by adding a point of infinity to the usual semiring $\mathbb{N}$ of nonnegative integers. We also provide finite first-order axiomatizations extending the results of Krob and Kozen. We will also include results regarding the equational theory of rational series over $\mathbb{N}$, or the semirings $\mathbf{k}$ on the set $\{0, 1, \cdots, k-1\}$ obtained from $\mathbb{N}_\infty$ by collapsing ∞ with the integers $\geq k-1$. For proofs we refer to [8] and [9].

2   Preliminaries

A *semiring* [24] is an algebra $S = (S, +, \cdot, 0, 1)$ such that $(S, +, 0)$ is a commutative monoid, $(S, \cdot, 1)$ is a monoid, moreover, the following hold for all $a, b, c \in S$:

$$0 \cdot a = 0$$
$$a \cdot 0 = 0$$
$$a(b + c) = ab + ac$$
$$(b + c)a = ba + ca$$

The operation $+$ is called *sum* or *addition*, and the operation $\cdot$ is called *product* or *multiplication*. A semiring S is called *idempotent* if

$$1 + 1 = 1, \quad \text{or equivalently}, \quad a + a = a$$

for all $a \in S$. A morphism of semirings preserves the sum and product operations and the constants 0 and 1. Since semirings are defined by equations, the class of all semirings is a variety (see e.g., [25]) as is the class of all idempotent semirings.

For any integer $k \geq 1$, we will also write k for the *term* which is the k-fold sum of 1 with itself: $1 + \cdots + 1$, where 1 appears k-times. Thus, a semiring S is idempotent if the equation $1 = 2$ holds in S. More generally, we will find occasion to deal with semirings satisfying the equation $k - 1 = k$, for some $k \geq 2$.

An important example of a semiring is the semiring $\mathbb{N} = (\mathbb{N}, +, \cdot, 0, 1)$ of nonnegative integers equipped with the usual sum and product operations, and an important example of an idempotent semiring is the *boolean semiring* $\mathbb{B}$ whose underlying set is $\{0, 1\}$ and whose sum and product operations are disjunction and conjunction. Actually $\mathbb{N}$ and $\mathbb{B}$ are respectively the initial semiring and the initial idempotent semiring. More generally, for every $k \geq 2$, we let $\mathbf{k}$ denote the semiring obtained from $\mathbb{N}$ by identifying all integers $\geq k - 1$. We may represent $\mathbf{k}$ as the semiring on the set $\{0, \cdots, k - 1\}$ where sum and product are the usual operations if the result of the operation is in this set, otherwise the operation returns $k - 1$. For each k, the semiring $\mathbf{k}$ is initial in the class of semirings satisfying $k - 1 = k$.

We end this section by describing two constructions on semirings. For more information on semirings, the reader is referred to Golan's book [24].

2.1   Polynomial Semirings and Power Series Semirings

Suppose that S is a semiring and A is a set. Let A^* denote the free monoid of all words over A including the empty word ϵ. A *formal power series*, or just *power series* over S in the (noncommuting) letters in A is a function $s : A^* \to S$. It is a common practice to represent a power series s as a formal sum $\sum_{w \in A^*}(s, w)w$, where the *coefficient* (s, w) is ws, the value of s on the word w. The *support* of a series s is the set $\mathrm{supp}(s) = \{w : (s, w) \neq 0\}$. When $\mathrm{supp}(s)$ is finite, s is called a *polynomial*. We let $S\langle\!\langle A^* \rangle\!\rangle$ and $S\langle A^* \rangle$ respectively denote the collection of all power series and polynomials over S in the letters A.

We define the sum $s + s'$ and product ss' of two series $s, s' \in S\langle\!\langle A^* \rangle\!\rangle$ as follows. For all $w \in A^*$,

$$(s + s', w) = (s, w) + (s', w)$$
$$(ss', w) = \sum_{uu' = w} (s, u)(s', u').$$

We may identify any element $s \in S$ with the series, in fact polynomial that maps ϵ to s and all other elements of A^* to 0. In particular, 0 and 1 may be viewed as polynomials. It is well-known that equipped with the above operations and constants, $S\langle\!\langle A^* \rangle\!\rangle$ is a semiring which contains $S\langle A^* \rangle$ as a subsemiring.

The semiring $S\langle A^*\rangle$ can be characterized by a universal property. Consider the natural embedding of A into $S\langle A^*\rangle$ such that each letter $a \in A$ is mapped to the polynomial whose support is $\{a\}$ which maps a to 1. By this embedding, we may view A as a subset of $S\langle A^*\rangle$. Recall also that each $s \in S$ is identified with a polynomial. The following fact is well-known.

Theorem 1. *Given any semiring S', any semiring morphism $h_S : S \to S'$ and any function $h : A \to S'$ such that*

$$(sh_S)(ah) = (ah)(sh_S) \tag{1}$$

for all $a \in A$ and $s \in S$, there is a unique semiring morphism $h^\sharp : S\langle A^*\rangle \to S'$ which extends both h_S and h.

The condition (1) means that for any $s \in S$ and letter $a \in A$, sh_S *commutes with* ah. In particular, since $\mathbb{N}$ is initial, and since when $S = \mathbb{N}$ the condition (1) holds automatically, we obtain that any map $A \to S'$ into a semiring S' extends to a unique semiring morphism $\mathbb{N}\langle A^*\rangle \to S'$, i.e., the polynomial semiring $\mathbb{N}\langle A^*\rangle$ is *freely generated by A in the class of semirings*. In the same way, for each $k \geq 2$, $\mathbf{k}\langle A^*\rangle$ is freely generated by A in the class of semirings satisfying the equation $k - 1 = k$.

Note that a series in $\mathbb{B}\langle\!\langle A^*\rangle\!\rangle = \mathbf{2}\langle\!\langle A^*\rangle\!\rangle$ may be identified with its support. Thus a series in $\mathbb{B}\langle\!\langle A^*\rangle\!\rangle$ corresponds to a language over A and a polynomial in $\mathbb{B}\langle A^*\rangle$ to a finite language. The sum operation corresponds to set union and the product operation to concatenation. The constants 0 and 1 are the empty set and the singleton set $\{\epsilon\}$.

The power series semirings $S\langle\!\langle A^*\rangle\!\rangle$ can be generalized in a straightforward way to semirings of series $S\langle\!\langle M\rangle\!\rangle$, where $M = (M, \cdot, 1)$ is a *locally finite partial monoid*.[1] Here, a partial monoid is a set M equipped with a partially defined product $(m, m') \mapsto mm'$ and a constant 1 such that for any m, m', m'' in M, $(mm')m''$ is defined iff $m(m'm'')$ is defined in which case they are equal. Moreover, the products $1m$ and $m1$ are always defined and equal to m. We say that a partial monoid is locally finite if each $m \in M$ can be written only a finite number of different ways as a product $m_1 \cdots m_k$ with $m_i \neq 1$ for all i. Clearly, every free monoid is locally finite. Another example is given by *data word* monoids. Suppose that A is an alphabet and D is set of *data values*. A *data word* over (A, D) is either the symbol 1 or a word in $D(AD)^+$. Let $(A, D)^*$ denote this set. We define a partial product operation on $(A, D)^*$ called *fusion* and denoted $\bullet$: For any $u = d_0 a_1 \cdots a_n d_n$ and $u' = d_0' a_1' \cdots a_m' d_m'$ in $D(AD)^+$ we define $u \bullet u' = d_0 a_1 \cdots a_n d_n' a_1' \cdots a_m' d_m'$ if $d_n = d_0'$, and leave $u \bullet u'$ undefined otherwise. Moreover, we define $1 \bullet u = u \bullet 1 = u$ for all $u \in (A, D)^*$. Let M be a locally finite partial monoid. We call an element $m \in M$ *irreducible* if $m \neq 1$ and m cannot be written as a nontrivial product of elements different from 1. It is clear that each $m \in M$ is a product of irreducibles (where the empty product is 1).

---

[1] This notion generalizes the locally finite monoids of Eilenberg [16].

2.2   Matrix Semirings

When S is a semiring, then for each $n \geq 0$ the set $S^{n \times n}$ of all $n \times n$ matrices over S is also a semiring denoted $S^{n \times n}$. The sum operation is defined pointwise and product is the usual matrix product. The constants are the matrix 0_{nn} all of whose entries are 0 (often denoted just 0), and the diagonal matrix E_n whose diagonal entries are all 1.

We can associate an $n \times n$ zero-one matrix with each function or relation ρ from the set $\{1, \cdots, n\}$ to the set $\{1, \cdots, m\}$, whose (i, j)th entry is 1 when $i\rho j$, and 0 otherwise. We usually identify ρ with the associated matrix, called a *functional* or *relational matrix*. A matrix associated with a bijective function is called a *permutation matrix*.

3   Conway Semirings

The definition of Conway semirings involves two important equations of regular languages. Conway semirings appear implicitly in Conway [13] and were defined explicitly in [6]. See also [35]. On the other hand, the applicability of Conway semirings is limited due to the fact that the star operation is total, whereas many important semirings only have a partially defined star operation. Moreover, it is not true that all such semirings can be embedded into a Conway semiring with a totally defined star operation. The following definition is taken from [9].

Definition 1. *A partial* *-semiring *is a semiring* S *equipped with a partially defined star operation* * $: S \to S$ *whose domain is an* ideal *of* S. *A* *-semiring *is a partial* *-semiring S such that* * *is defined on the whole semiring* S. *A morphism* $S \to S'$ *of (partial)* *-semirings is a semiring morphism* $h : S \to S'$ *such that for all* $s \in S$, *if* s^* *is defined then so is* $(sh)^*$ *and* $s^*h = (sh)^*$.

Thus, in a partial *-semiring S, 0^* is defined, and if a^* and b^* are defined then so is $(a + b)^*$, finally, if a^* or b^* is defined, then so is $(ab)^*$. When S is a partial *-semiring, we let $D(S)$ denote the domain of definition of the star operation.

Definition 2. *A partial Conway semiring* *is a partial* *-semiring S satisfying the following two axioms:*

1. Sum star equation*:*

$$(a + b)^* = a^*(ba^*)^*$$

   for all $a, b \in D(S)$.
2. Product star equation*:*

$$(ab)^* = 1 + a(ba)^*b,$$

   for all $a, b \in S$ *such that* $a \in D(S)$ *or* $b \in D(S)$.

A Conway semiring *is a partial Conway semiring S which is a* *-semiring (i.e., $D(S) = S$). A morphisms of (partial) Conway semirings is a (partial)* *-semiring morphism.*

Note that in any partial Conway semiring S,

$$aa^* + 1 = a^*$$
$$a^*a + 1 = a^*$$
$$0^* = 1$$

for all $a \in D(S)$. Moreover, if $a \in D(S)$ or $b \in D(S)$, then

$$(ab)^*a = a(ba)^*.$$

It follows that also

$$aa^* = a^*a$$
$$(a + b)^* = (a^*b)^*a^*$$

for all $a, b \in D(S)$. When $a \in D(S)$ we will denote $aa^* = a^*a$ by a^+ and call $^+$ the *plus* operation.

When S is a (partial) Conway semiring, each semiring $S^{n \times n}$ may be turned into a (partial) Conway semiring.

Definition 3. *Suppose that S is a partial Conway semiring with $D(S) = I$. We define a partial star operation on the semirings $S^{k \times k}$, $k \geq 0$, whose domain of definition is $I^{k \times k}$, the ideal of those $k \times k$ matrices all of whose entries are in I. When $k = 0$, $S^{k \times k}$ is trivial as is the definition of star. When $k = 1$, we use the star operation on S. Assuming that $k > 1$ we write $k = n + 1$. For a matrix $\begin{pmatrix} a & b \\ c & d \end{pmatrix}$ in $I^{k \times k}$, define*

$$\begin{pmatrix} a & b \\ c & d \end{pmatrix}^* = \begin{pmatrix} \alpha & \beta \\ \gamma & \delta \end{pmatrix} \tag{2}$$

where $a \in S^{n \times n}$, $b \in S^{n \times 1}$, $c \in S^{1 \times n}$ and $d \in S^{1 \times 1}$, and where

$$\alpha = (a + bd^*c)^* \qquad \beta = \alpha bd^*$$
$$\gamma = \delta ca^* \qquad \delta = (d + ca^*b)^*.$$

Theorem 2. (Conway[13], Krob [34,35], Bloom, Ésik [6], Bloom, Ésik, Kuich [9]) *Suppose that S is a partial Conway semiring with $D(S) = I$. Then, equipped with the above star operation, each semiring $S^{k \times k}$ is a partial Conway semiring with $D(S^{k \times k}) = I^{k \times k}$.*

Theorem 3. (Conway[13], Krob [34], Bloom, Ésik [6], Bloom, Ésik, Kuich [9]) *Suppose that S is a partial Conway semiring with $D(S) = I$. Then the following equations hold in $S^{k \times k}$:*

1. *The matrix star equation (2) for all possible decompositions of a square matrix over I into four blocks as above such that a and d are square matrices.*

2. *The* permutation equation

$$(\pi A \pi^T)^* = \pi A^* \pi^T,$$

for all A in $I^{k \times k}$ and any $k \times k$ permutation matrix π, where π^T denotes the transpose of π.

We note the following variant of the matrix star equation. Let S be a partial Conway semiring, $I = D(S)$. Then if $A = \begin{pmatrix} a & b \\ c & d \end{pmatrix}$ is a matrix with entries in I, partitioned as above, then

$$A^* = \begin{pmatrix} (a + bd^*c)^* & a^*b(d + ca^*b)^* \\ d^*c(a + bd^*c)^* & (d + ca^*b)^* \end{pmatrix}$$

4 Iteration Semirings

Many important (partial) Conway semirings satisfy the group equations associated with the finite groups, introduced by Conway [13]. When a (partial) Conway semiring satisfies the group equations, it will be called a (partial) iteration semiring. Below we will consider groups of order n defined on the set $\{1, \cdots, n\}$ of positive integers with multiplication $(i, j) \mapsto ij$ and inverse $i \mapsto i^{-1}$.

Definition 4. *We say that the* group equation associated with a finite group G *of order n holds in a partial Conway semiring S if*

$$e_1 M_G^* u_n = (a_1 + \cdots + a_n)^* \tag{3}$$

holds, where $a_1, \cdots, a_n$ are arbitrary elements of $D(S)$, and where M_G is the $n \times n$ matrix whose (i, j)th entry is $a_{i^{-1}j}$, for all $1 \le i, j \le n$, and e_1 is the $1 \times n$ functional matrix whose first entry is 1 and whose other entries are 0, finally u_n is the $n \times 1$ matrix all of whose entries are 1.

Equation (3) asserts that the sum of the entries of the first row of M_G^* is $(a_1 + \cdots + a_n)^*$. For example, the group equation associated with the group of order 2 is

$$(1\, 0) \begin{pmatrix} a_1 & a_2 \\ a_2 & a_1 \end{pmatrix}^* \begin{pmatrix} 1 \\ 1 \end{pmatrix} = (a_1 + a_2)^*$$

which by the matrix star equation can be written as

$$(a_1 + a_2 a_1^* a_2)^* (1 + a_2 a_1^*) = (a_1 + a_2)^*.$$

(It is known that in Conway semirings, this equation is further equivalent to the *power identity* $(a^2)^*(1 + a) = a^*$.)

Definition 5. *We say that a Conway semiring S is an* iteration semiring *if it satisfies all group equations. We say that a partial Conway semiring S is a* partial iteration semiring *if it satisfies all group equations (3) where $a_1, \cdots, a_n$ range over $D(S)$. A* morphism *of (partial) iteration semirings is a (partial) Conway semiring morphism.*

Proposition 1. *Suppose that the partial Conway semiring S satisfies the group equation (3) for all $a_1, \cdots, a_n \in D(S)$. Then S also satisfies*

$$u_n^T M_G^* e_1^T = (a_1 + \cdots + a_n)^*, \tag{4}$$

for all $a_1, \cdots, a_n \in D(S)$, where e_1, M_G and u_n are defined as above. Thus, if S is an iteration semiring, then (4) holds for all finite groups G.

Remark 1. In Conway semirings, the group equation (3) is equivalent to (4).

Remark 2. Let $\mathcal{G}$ denote a class of finite groups. It is known, cf. [13,34] that the defining equations of Conway semirings, in conjunction with the group equations associated with the groups in $\mathcal{G}$ are complete for iteration semirings iff every finite simple group is a quotient of a subgroup of a group in $\mathcal{G}$.

The group equations seem to be extremely difficult to verify in practice. However, they are implied by the simpler functorial star conditions defined below.

Definition 6. *Suppose that S is a partial Conway semiring so that each matrix semiring $S^{n \times n}$ is also a Conway semiring. Let $I = D(S)$, and let C be a class of rectangular matrices over S. We say that S has a functorial star with respect to C if for all matrices $A \in I^{m \times m}$ and $B \in I^{n \times n}$, and for all $m \times n$ matrices C in C, if $AC = CB$ then $A^* C = C B^*$. Finally, we say that S has a functorial star if it has a functorial star with respect to the class of all rectangular matrices.*

Proposition 2. (Bloom, Ésik [6], Bloom, Ésik, Kuich [9]) *Suppose that S is a (partial) Conway semiring.*

1. *S has a functorial star with respect to the class of all injective functional matrices and their transposes.*
2. *If S has a functorial star with respect to the class of functional matrices $m \to 1$, $m \geq 2$, then S has a functorial star with respect to the class of all functional matrices.*
3. *If S has a functorial star with respect to the class of transposes of functional matrices $m \to 1$, $m \geq 2$, then S has a functorial star with respect to the class of transposes of all functional matrices.*
4. *If S has a functorial star with respect to the class of all functional matrices $m \to 1$, $m \geq 2$, or with respect to the class of transposes of functional matrices $m \to 1$, $m \geq 2$, then S is a (partial) iteration semiring.*

Theorem 4. (Conway [13], Krob [34], Ésik [19], Bloom, Ésik, Kuich [9]) *If S is a (partial) iteration semiring, then so is $S^{k \times k}$ for each $k \geq 0$.*

Later we will see that the class of (partial) iteration semirings is also closed under taking power series semirings. We now give two classes of (partial) iteration semirings.

4.1 Partial Iterative Semirings

In this section we exhibit a class of partial iteration semirings. The definition of these semirings is motivated by Salomaa's axiomatization [42] of regular languages.

Definition 7. *A partial iterative semiring is a partial *-semiring S such that for every $a \in D(S)$ and $b \in S$, a^*b is the unique solution of the equation $x = ax + b$. A morphism of partial iterative semirings is a *-semiring morphism.*

We note that any semiring S with a distinguished ideal I such that for all $a \in I$ and $b \in S$, the equation $x = ax + b$ has a unique solution can be turned into a partial iterative semiring, where star is defined on I. Indeed, when $a \in I$, define a^* as the unique solution of the equation $x = ax + 1$. It follows that $aa^*b + b = a^*b$ for all b, so that a^*b is the unique solution of $x = ax + b$. We also note that when S, S' are partial iterative semirings, then any semiring morphism $h : S \to S'$ with $D(S)h \subseteq D(S')$ automatically preserves star.

Theorem 5. (Bloom, Ésik, Kuich [9]) *Every partial iterative semiring is a partial iteration semiring with a functorial star.*

Thus, when S is a partial iterative semiring with $D(S) = I$, then by Definition 3, each matrix semiring $S^{n \times n}$ is a partial iteration semiring with $D(S) = I^{n \times n}$.

Theorem 6. (Bloom, Ésik, Kuich [9]) *Suppose that S is a partial iterative semiring with $D(S) = I$. Then for any $A \in I^{n \times n}$ and $B \in S^{n \times p}$, A^*B is the unique solution of the matrix equation $X = AX + B$. In particular, $S^{n \times n}$ is a partial iterative semiring where the star operation is defined on $I^{n \times n}$.*

Theorem 7. (Bloom, Ésik, Kuich [9]) *Suppose that S is a partial iterative semiring with $D(S) = I$ and $A \in S^{n \times n}$ and $B \in S^{n \times p}$ such that $A^k \in I^{n \times n}$ for some $k \geq 1$. Then the equation $X = AX + B$ in the variable X ranging over $S^{n \times p}$ has $(A^k)^*(A^{k-1}B + \cdots + B)$ as its unique solution.*

We give an example of a partial iterative semiring. Let S be a semiring and M a locally finite partial monoid, and consider the semiring $S\langle\!\langle M \rangle\!\rangle$. We call a series $s \in S\langle\!\langle M \rangle\!\rangle$ proper if $(s, 1) = 0$. Clearly, the proper series form an ideal. For any series s, r, if s is proper, then the equation $x = sx + r$ has a unique solution. (For the case when M is a locally finite monoid, see [16].) Moreover, this unique solution is s^*r, where s^* is the unique solution of the equation $y = sy + 1$.

Proposition 3. *For any semiring S and locally finite partial monoid M, $S\langle\!\langle M \rangle\!\rangle$, equipped with the above star operation defined on the proper series, is a partial iterative semiring and thus a partial iteration semiring.*

Theorem 8. (Bloom, Ésik [6]) *Suppose that S is partial iterative semiring with star operation defined on $D(S) = I$, and suppose that S_0 is a subsemiring of S which is equipped with a unary operation $^\otimes$. Moreover, suppose that S is the direct sum of S_0 and I, so that each $s \in S$ has a unique representation as a sum*

$x + a$ with $x \in S_0$ and $a \in I$. If S_0, equipped with the operation $\otimes$, is a Conway semiring, then there is a unique way to turn S into a Conway semiring whose star operation extends $\otimes$. This operation also extends the star operation defined on I. Moreover, when S_0 is an iteration semiring, then S is also an iteration semiring. In particular, if S is a Conway or an iteration semiring, then so is $S\langle\!\langle A^* \rangle\!\rangle$, for any set A.

The last sentence of the previous result can be generalized.

Theorem 9. *For every locally finite partial monoid M, if S is a Conway or iteration semiring, then so is $S\langle\!\langle M \rangle\!\rangle$ in an essentially unique way.*

4.2 ω-Complete *-Semirings

Power series semirings only have a partial star operation defined on proper series. In order to make star a totally defined operation, Eilenberg introduced complete semirings in [16]. A variant of this notion is defined below, see also Krob [32], Hebisch [27], Karner [29] and Bloom, Ésik [6].

Definition 8. *We call a semiring S ω-complete if it is equipped with a summation operation $\sum_{i \in I} s_i$, defined on countable families s_i, $i \in I$ over S such that $\sum_{i \in \emptyset} s_i = 0$, $\sum_{i \in \{1,2\}} s_i = s_1 + s_2$, moreover,*

$$a\Big(\sum_{i \in I} b_i\Big) = \sum_{i \in I} ab_i \quad \Big(\sum_{i \in I} b_i\Big)a = \sum_{i \in I} b_i a \quad \sum_{j \in J}\sum_{i \in I_j} a_i = \sum_{i \in I} a_i$$

where in the last equation the countable set I is the disjoint union of the sets I_j, $j \in J$. A morphism of ω-complete semirings also preserves summation.

We note that ω-complete semirings are sometimes called *countably complete semirings*. An example of an ω-complete semiring is the semiring $\mathbb{N}_\infty$ obtained by adding a point of infinity ∞ to the semiring $\mathbb{N}$, where a sum $\sum_{i \in I} s_i$ is defined to be ∞ if there is some i with $s_i = \infty$ or the number of i with $s_i \neq 0$ is infinite. In all other cases each s_i is in $\mathbb{N}$ and the number of i with $s_i \neq 0$ is finite, so that we define $\sum_{i \in I} s_i$ as the usual sum of those s_i with $s_i \neq 0$. Suppose that S is an ω-complete semiring. Then we define a star operation by $a^* = \sum_{n \geq 0} a^n$, for all $a \in S$.

Definition 9. *An ω-complete *-semiring is a *-semiring S which is an ω-complete semiring whose star operation is derived from the ω-complete structure as above. A morphism of ω-complete *-semirings is a semiring morphism which preserves all countable sums and thus the star operation.*

Theorem 10. *(Bloom, Ésik [6]) Any ω-complete *-semiring is an iteration semiring with a functorial star.*

The fact that any ω-complete *-semiring is a Conway semiring was shown in [27].

Proposition 4. *If S is an ω-complete semiring, then equipped with the point-wise summation, so is each matrix semiring $S^{n \times n}$ as is each power series semi-ring $S\langle\!\langle M \rangle\!\rangle$, where M is any locally finite partial monoid.*

(Actually for the last fact it suffices that each element of M has an at most countable number of nontrivial decompositions into a product.) Thus, when S is an ω-complete $*$-semiring, where the star operation is derived from an ω-complete structure, then we obtain two star operations on each matrix semiring $S^{n \times n}$: the star operation defined by the matrix star equation, and the star operation derived from the ω-complete structure on $S^{n \times n}$. However, these two star operations are the same. A similar fact holds for each power series semiring $S\langle\!\langle M \rangle\!\rangle$, where M is a locally finite partial monoid.

Theorem 11. (Bloom, Ésik [6]) *Let S be an ω-complete $*$-semiring, so that S is an iteration semiring.*

1. *For each $n \geq 1$, the star operation determined on $S^{n \times n}$ by Definition 3 is the same as the star operation derived from the ω-complete structure on S.*
2. *For each locally finite partial monoid M, the star operation determined on $S\langle\!\langle M \rangle\!\rangle$ by Theorem 8 is the same as that derived from the ω-complete structure on S.*

Actually, the last fact is stated in [6] for free monoids, but the extension is clear. It is easy to show that all free ω-complete $*$-semirings exist.

Theorem 12. *For each set A, $\mathbb{N}_\infty \langle\!\langle A^* \rangle\!\rangle$ is freely generated by A in the class of all ω-complete $*$-semirings. Moreover, for each $k \geq 2$, the $*$-semiring $\mathbf{k}\langle\!\langle A^* \rangle\!\rangle$ is freely generated by A in the class of all ω-complete $*$-semirings satisfying $1^* = k - 1$.*

Note that if an ω-complete semiring S satisfies $1^* = k - 1$, then any countable sum of at least $k - 1$ copies of an element a with itself is $(k - 1)a$, the sum of exactly $k - 1$ copies.

Complete semirings [16] have a summation operation defined on *all* families of elements over the semiring. There is a weaker notion of *rationally additive* semirings, see [21]. These semirings also give rise to iteration semirings with a functorial star.

5   Kleene Theorem

The classical Kleene theorem equates languages recognizable by finite automata with the regular languages, and its generalization by Schützenberger equates power series recognizable by weighted finite automata with rational power series. Since Kleene's theorem can be formulated in equational logic, by the complete-ness results presented in Section 6, it can be proved by equational reasoning using the equational axioms. Actually the Conway semiring equations suffice for that purpose. In this section we show a Kleene theorem for partial Conway semirings. To this end, we define a general notion of (finite) automaton in partial Conway semirings. Our presentation follows [9].

Definition 10. *Suppose that S is a partial Conway semiring, S_0 is a subsemiring of S and A is a subset of $D(S)$. An automaton in S over (S_0, A) is a triplet $\mathbf{A} = (\alpha, M, \beta)$ consisting of an initial vector $\alpha \in S_0^{1 \times n}$, a transition matrix $M \in (S_0 A)^{n \times n}$, where $S_0 A$ is the set of all linear combinations over A with coefficients in S_0, and a final vector $\beta \in S_0^{n \times 1}$. The behavior of $\mathbf{A}$ is $|\mathbf{A}| = \alpha M^* \beta$.*

(Since $M \in D(S)^{n \times n}$, M^* exists.)

Definition 11. *We say that $s \in S$ is recognizable over (S_0, A) if s is the behavior of some automaton over (S_0, A). We let $\mathbf{Rec}_S(S_0, A)$ denote the set of all elements of S which are recognizable over (S_0, A).*

Next we define rational elements.

Definition 12. *Let S, S_0 and A be as above. We say that $s \in S$ is rational over (S_0, A) if $s = x + a$ for some $x \in S_0$ and some $a \in S$ which is contained in the least set $\mathbf{Rat}'_S(S_0, A)$ containing $A \cup \{0\}$ and closed under the rational operations $+, \cdot, {}^+$ and left and right multiplication with elements of S_0. We let $\mathbf{Rat}_S(S_0, A)$ denote the set of rational elements over (S_0, A).*

Note that $\mathbf{Rat}'_S(S_0, A) \subseteq D(S)$.

Proposition 5. *Suppose that S is a partial Conway semiring, S_0 is a subsemiring of S and A is a subset of $D(S)$. Then $\mathbf{Rat}_S(S_0, A)$ contains S_0 and is closed under sum and product. Moreover, it is closed under star iff it is closed under the plus operation.*

Proposition 6. *Suppose that S is a partial Conway semiring, S_0 is a subsemiring of S and A is a subset of $D(S)$. Then $\mathbf{Rat}_S(S_0, A)$ is contained in the least subsemiring of S containing S_0 and A which is closed under star.*

We give two sufficient conditions under which $\mathbf{Rat}_S(S_0, A)$ is closed under star.

Proposition 7. *Let S, S_0 and A be as above. Assume that either $S_0 \subseteq D(S)$ and S_0 is closed under star, or the following condition holds:*

$$\forall x \in S_0 \forall a \in D(S) \quad (x + a \in D(S) \Rightarrow x = 0). \tag{5}$$

Then $\mathbf{Rat}_S(S_0, A)$ is closed under star. Moreover, in either case, $\mathbf{Rat}_S(S_0, A)$ is the least subsemiring of S containing S_0 and A which is closed under star.

Remark 3. Note that the second condition in the above proposition holds whenever each $s \in S$ has at most one representation $s = x + a$ with $x \in S_0$ and $a \in D(S)$. This happens when S is the *direct sum* of S_0 and $D(S)$.

Theorem 13. (Bloom, Ésik, Kuich [9]) *Suppose that S is a partial Conway semiring, S_0 is a subsemiring of S, $A \subseteq D(S)$. Then $\mathbf{Rec}_S(S_0, A) = \mathbf{Rat}_S(S_0, A)$.*

Corollary 1. *Suppose that S is a Conway semiring, S_0 is a Conway subsemiring of S and $A \subseteq S$. Then $\mathbf{Rec}_S(S_0, A) = \mathbf{Rat}_S(S_0, A)$ is the least Conway subsemiring of S which contains $S_0 \cup A$.*

Corollary 2. *Suppose that S is a partial Conway semiring, S_0 is a subsemiring of S and $A \subseteq D(S)$. Suppose that condition (5) holds. Then $\mathbf{Rec}_S(S_0, A) = \mathbf{Rat}_S(S_0, A)$ is the least partial Conway subsemiring of S which contains $S_0 \cup A$.*

The case when the partial Conway semiring is a power series semiring over a free monoid deserves special attention. Let A be set and S be a semiring, and consider the partial iteration semiring $S\langle\!\langle A^* \rangle\!\rangle$ whose star operation is defined on proper series. Alternatively, let S be a Conway semiring, and consider the Conway semiring $S\langle\!\langle A^* \rangle\!\rangle$ (cf. Theorem 8). We denote $\mathbf{Rat}_{S\langle\!\langle A^* \rangle\!\rangle}(S, A)$ by $S^{\mathrm{rat}}\langle\!\langle A^* \rangle\!\rangle$ and $\mathbf{Rec}_{S\langle\!\langle A^* \rangle\!\rangle}(S, A)$ by $S^{\mathrm{rec}}\langle\!\langle A^* \rangle\!\rangle$.

Remark 4. Note that when S is a Conway semiring, then it is also a semiring. So we may view $S\langle\!\langle A^* \rangle\!\rangle$ in two different ways: as a Conway semiring (see Theorem 8), or as a partial Conway semiring where star is defined on the proper series. Thus, we obtain two different notion of rationality: two different definitions of $S^{\mathrm{rat}}\langle\!\langle A^* \rangle\!\rangle$, one being a Conway semiring (in fact an iteration semiring) and the other being a partial Conway semiring. However, by Proposition 7 the elements of these two semirings are the same, and the star operation agrees on proper series in the two semirings.

Corollary 3. *Suppose that S is a semiring and A is a set. Then $S^{\mathrm{rat}}\langle\!\langle A^* \rangle\!\rangle$ is the least partial iteration subsemiring of $S\langle\!\langle A^* \rangle\!\rangle$ containing $S \cup A$. Moreover, $S^{\mathrm{rat}}\langle\!\langle A^* \rangle\!\rangle = S^{\mathrm{rec}}\langle\!\langle A^* \rangle\!\rangle$.*

Corollary 4. *Suppose that S is a Conway semiring. Then $S^{\mathrm{rat}}\langle\!\langle A^* \rangle\!\rangle$ is the least Conway subsemiring of $S\langle\!\langle A^* \rangle\!\rangle$ containing $S \cup A$. Moreover, $S^{\mathrm{rat}}\langle\!\langle A^* \rangle\!\rangle = S^{\mathrm{rec}}\langle\!\langle A^* \rangle\!\rangle$.*

The previous facts can be generalized. Suppose that M is a locally finite partial monoid and let S be a semiring, or a Conway semiring. In the first case, $S\langle\!\langle M \rangle\!\rangle$ is a partial iteration semiring whose star operation is defined on the proper series, and in the second case, $S\langle\!\langle M \rangle\!\rangle$ is a Conway semiring. Moreover, in the first case, each series associated with an irreducible of M is proper. Now let A denote the set of all irreducibles. We let $S^{\mathrm{rat}}\langle\!\langle M \rangle\!\rangle = \mathbf{Rat}_{S\langle\!\langle M \rangle\!\rangle}(S, A)$ and $S^{\mathrm{rec}}\langle\!\langle M \rangle\!\rangle = \mathbf{Rec}_{S\langle\!\langle M \rangle\!\rangle}(S, A)$.

Corollary 5. *Suppose that S is a semiring and M is a locally finite monoid. Then $S^{\mathrm{rat}}\langle\!\langle M \rangle\!\rangle$ is the least partial iteration subsemiring of $S\langle\!\langle M \rangle\!\rangle$ containing $S \cup A$. Moreover, $S^{\mathrm{rat}}\langle\!\langle M \rangle\!\rangle = S^{\mathrm{rec}}\langle\!\langle M \rangle\!\rangle$.*

Corollary 6. *Suppose that S is a Conway semiring and M is locally finite partial monoid. Then $S^{\mathrm{rat}}\langle\!\langle M \rangle\!\rangle$ is the least Conway subsemiring of $S\langle\!\langle M \rangle\!\rangle$ containing $S \cup A$. Moreover, $S^{\mathrm{rat}}\langle\!\langle M \rangle\!\rangle = S^{\mathrm{rec}}\langle\!\langle M \rangle\!\rangle$.*

The above corollaries essentially cover the Kleene theorems for timed automata in [12,15]. To obtain these results, one only needs to specialize M to certain data word monoids.

6 Completeness

Theorem 14. (Bloom, Ésik [8]) *For each set A, the semiring $\mathbb{N}^{\mathrm{rat}}\langle\!\langle A^*\rangle\!\rangle$ is freely generated by A in the class of all partial iteration semirings. In more detail, given any partial iteration semiring S and function $h : A \to S$ with $Ah \subseteq D(S)$, there is a unique morphism of partial iteration semirings $h^{\sharp} : \mathbb{N}^{\mathrm{rat}}\langle\!\langle A^*\rangle\!\rangle \to S$ extending h.*

The proof of Theorem 14 uses the results of Section 5 and some recent results from Béal, Lombardy, Sakarovitch [2,3].

Corollary 7. *For each set A, the semiring $\mathbb{N}^{\mathrm{rat}}\langle\!\langle A^*\rangle\!\rangle$ is freely generated by the set A in the class of all partial iterative semirings.*

The notion of a *(*-semiring) term* (or *rational expression*) over a set A is defined as usual: Each letter $a \in A$ is a term as are the symbols 0 and 1, and if t and t' are terms then so are $t + t'$, tt' and t^*. When S is a *-semiring and t is term, then for any valuation $A \to S$, t evaluates to an element of S. We say that an equation $t = t'$ between terms t and t' over A *holds* in S, or *is satisfied by S*, if t and t' evaluate to the same element for each valuation $A \to S$. We may assume that terms are over a fixed countably infinite set of letters.

The above definition does not make sense for partial *-semirings. Therefore we define *ideal terms* (over A) as follows. Each letter $a \in A$ is an ideal term as is the symbol 0. When t, t' are ideal terms and s is any term over A, then $t + t'$, ts, st are ideal terms. A *guarded term* is a term t such that whenever t has a "subterm" of the form s^*, then s is an ideal term. When S is a partial *-semiring and t is a guarded term over A, then t evaluates to an element of S under each *ideal valuation* $A \to D(S)$. Let t and t' be guarded. We say that $t = t'$ holds in S if t and t' evaluate to the same element of S under each guarded evaluation. The following result follows from Theorem 14.

Theorem 15. *The following conditions are equivalent for guarded terms t and t'.*

1. *The equation $t = t'$ holds in all partial iteration semirings.*
2. *The equation $t = t'$ holds in all partial iterative semirings.*
3. *The equation $t = t'$ holds in all semirings $\mathbb{N}^{\mathrm{rat}}\langle\!\langle A^*\rangle\!\rangle$.*
4. *The equation $t = t'$ holds in $\mathbb{N}^{\mathrm{rat}}\langle\!\langle \{a,b\}^*\rangle\!\rangle$.*

We now turn to the semirings $\mathbb{N}_{\infty}^{\mathrm{rat}}\langle\!\langle A^*\rangle\!\rangle$.

Theorem 16. (Bloom, Ésik [8]) *For any set A, the *-semiring $\mathbb{N}_{\infty}^{\mathrm{rat}}\langle\!\langle A^*\rangle\!\rangle$ is freely generated by A in the class of all iteration semirings satisfying (6), (7) and (8):*

$$1^*1^* = 1^* \tag{6}$$

$$1^*a = a1^* \tag{7}$$

$$(1^*a)^*1^* = 1^*a^*. \tag{8}$$

Moreover, for any $k \geq 2$, the *-semiring $k^{\mathrm{rat}}\langle\!\langle A^*\rangle\!\rangle$ is freely generated by A in the class of all iteration semirings satisfying $1^* = k - 1$.

Using this fact, we obtain:

Theorem 17. (Bloom, Ésik [8]) *The following conditions are equivalent for an equation.*

1. *The equation holds in all ω-complete $*$-semirings.*
2. *The equation holds in all iteration semirings satisfying (6), (7), (8).*
3. *The equation holds in all $*$-semirings $\mathbb{N}_\infty^{\mathrm{rat}}\langle\!\langle A^*\rangle\!\rangle$.*
4. *The equation holds in $\mathbb{N}_\infty^{\mathrm{rat}}\langle\!\langle \{a,b\}^*\rangle\!\rangle$.*

We note that in conjunction with the iteration semiring equations, (6) is equivalent to $1^* = 1^{**}$ and (8) is equivalent to $(1+a)^* = 1^*a^*$ or $a^{**} = 1^*a^*$.

Theorem 18. (Bloom, Ésik [8]) *For any $k \geq 2$, following conditions are equivalent for an equation.*

1. *The equation holds in all ω-complete $*$-semirings satisfying $1^* = k - 1$.*
2. *The equation holds in all iteration semirings satisfying $1^* = k - 1$.*
3. *The equation holds in all $*$-semirings $\mathbf{k}^{\mathrm{rat}}\langle\!\langle A^*\rangle\!\rangle$.*
4. *The equation holds in $\mathbf{k}^{\mathrm{rat}}\langle\!\langle \{a,b\}^*\rangle\!\rangle$.*

In Theorems 16 and 18, the case when $k = 2$ is due to Krob [34].

7   Ordered Iteration Semirings

A semiring S is called *ordered* if it is equipped with a partial order which is preserved by the sum and product operations. An ordered semiring S is *positive* if 0 is the least element of S. Morphisms of (positive) ordered semirings are also monotone.

Note that when S is a positive ordered semiring, then $a \leq a+b$ for all $a, b \in S$. Moreover, the relation $\preceq$ defined by $a \preceq b$ iff there is some c with $a + c = b$ is also a partial order on S which is preserved by the operations. Since $0 \preceq a$ for all $a \in S$, it follows that equipped with the relation $\preceq$, S is a positive ordered semiring, called a *sum ordered semiring*. Every idempotent semiring S is sum ordered, moreover, the sum order agrees with the *semilattice order*: $a \preceq b$ iff $a + b = b$, for all $a, b \in S$. For later use observe that if S and S' are positive ordered semirings such that S is sum ordered, then any semiring morphism $S \to S'$ is monotone and thus a morphism of ordered semirings.

Definition 13. *An ordered iteration semiring is an iteration semiring which is a positive ordered semiring. A morphism of ordered iteration semirings is an iteration semiring morphism which is also an ordered semiring morphism.*

It is clear that if S is an ordered positive semiring, then so is any matrix semiring $S^{n \times n}$ and any power series semiring $S\langle\!\langle A^*\rangle\!\rangle$, or more generally, any semiring $S\langle\!\langle M\rangle\!\rangle$ where M is a locally finite partial monoid. It follows easily that if S is an ordered iteration semiring, then $S^{n \times n}$ and $S\langle\!\langle M\rangle\!\rangle$ are also ordered iteration semirings.

Below we give two classes of ordered iteration semirings.

7.1 ω-Continuous *-Semirings

There are several definitions of continuous, or ω-continuous semirings in the literature. Our definition is taken from [6]. See also [29], where the same semirings are termed *finitary*.[2] Recall that a poset is an ω-cpo if it has a least element and suprema of ω-chains. Moreover, a function between ω-cpo's is ω-*continuous* if it preserves suprema of ω-chains. It is clear that if P is an ω-cpo, then so is each P^n equipped with the pointwise order.

Definition 14. *An ω-continuous semiring is a positive ordered semiring S such that S is an ω-cpo and the sum and product operations are ω-continuous. An idempotent ω-continuous semiring is an ω-continuous semiring which is idempotent. A morphism of (idempotent) ω-continuous semirings is a semiring morphism which is a continuous function.*

Note that each finite positive ordered semiring is ω-continuous. Thus, for each integer $k \geq 1$, the semiring **k**, equipped with the usual order is ω-continuous. Moreover, $\mathbb{N}_\infty$, ordered as usual, is also ω-continuous.

Theorem 19. (Karner [29]) *Every ω-continuous semiring can be turned into an ω-complete semiring by defining*

$$\sum_{i \in I} a_i = \sup_{F \subseteq I} \sum_{i \in F} a_i$$

where F ranges over the finite *subsets of F and a_i, $i \in I$ is a countable family of elements of S. Morphisms of ω-continuous semirings preserve all countable sums.*

Since each ω-continuous semiring is ω-complete and esch ω-complete semiring is a *-semiring, we deduce that each ω-continuous semiring S is a *-semiring where $a^* = \sum_n a^n = \sup_n \sum_{i=0}^n a^i$ for all $a \in S$.

Definition 15. *An ω-continuous *-semiring is an ω-continuous semiring which is a *-semiring where star is determined by the ordered structure as above. A morphism of ω-continuous *-semirings is a morphism of ω-continuous semirings.*

Note that any morphism automatically preserves star. As an example, consider the 3-element semiring $S_0 = \{0, 1, \infty\}$ obtained by adding the element ∞ to the boolean semiring $\mathbb{B}$ such that $a + \infty = \infty + a = \infty$ for all $a \in \{0, 1, \infty\}$ and $b\infty = \infty b = \infty$ if $b \neq 0$. Equipped with the order $0 < 1 < \infty$, it is an

---

[2] In [33,29], a continuous semiring is a semiring equipped with the *sum order* which is a continuous semiring as defined here. Our terminology stems from the commonly accepted definition of an ω-*continuous algebra* [23,26] as an algebra equipped with a partial order which is an ω-cpo such that the operations are ω-continuous in each argument.

ω-continuous semiring and thus an ω-complete semiring. Let us denote by S this ω-complete semiring. Note that in S, a countably infinite sum of 1 with itself is 1 and thus $1^* = 1$ holds. However, there is another way of turning S_0 into an ω-complete semiring. We define infinite summation so that an infinite sum is ∞ iff either one of the summands is ∞, or the number of nonzero summands is infinite. In the resulting ω-complete semiring S', it holds that a countably infinite sum of 1 with itself is ∞ and thus $1^* = \infty$.

Theorem 20. *The class of ω-continuous semirings is closed under taking matrix semirings and power series semirings over all locally finite partial monoids.*

This fact is well-known (the second claim at least for free monoids). In each case, the order is the pointwise order. Since each ω-complete *-semiring is an iteration semiring, the same holds for ω-continuous *-semirings.

Corollary 8. *Every ω-continuous *-semiring is an ordered iteration semiring.*

On the semirings $\mathbf{k}$ and $\mathbb{N}_\infty$, the order is the sum order. It follows that the pointwise order is the sum order on each semiring $\mathbf{k}\langle\!\langle M \rangle\!\rangle$ or $\mathbb{N}_\infty\langle\!\langle M \rangle\!\rangle$. The following fact is easy to establish.

Theorem 21. *For each set A, the ω-continuous *-semiring $\mathbb{N}_\infty\langle\!\langle A^* \rangle\!\rangle$, equipped with the pointwise order, is freely generated by A in the class of all ω-continuous *-semirings. Moreover, for each set A, and for each $k \geq 2$, the ω-continuous *-semiring $\mathbf{k}\langle\!\langle A^* \rangle\!\rangle$, equipped with the pointwise order, is freely generated by A in the class of all ω-continuous semirings satisfying $k - 1 = k$.*

7.2 Inductive *-Semirings

Inductive *-semirings, introduced in [22], are a generalization of ω-continuous *-semirings. As opposed to ω-continuous *-semirings, inductive *-semirings are defined by first-order conditions.

Definition 16. *Suppose that S is a *-semiring which is an ordered semiring. We call S an* inductive *-semiring *if the following hold for all $a, b, x \in S$:*

$$aa^* + 1 \leq a^*$$
$$ax + b \leq x \quad \Rightarrow \quad a^*b \leq x.$$

*A morphism of inductive *-semirings is a *-semiring morphism which is an ordered semiring morphism.*

It then follows that for any a, b in an inductive *-semiring S, a^*b is the *least prefixed point* of the map $x \mapsto ax + b$, which is actually a fixed point. In particular, every inductive *-semiring is a positive ordered semiring. Moreover, it follows that the star operation is also monotone.

Definition 17. *A* symmetric inductive *-semiring *is an inductive* *-semiring *S*
which also satisfies

$$xa + b \leq x \Rightarrow ba^* \leq x$$

for all $a, b, x \in S$. *A morphism of symmetric inductive* *-semirings is an inductive* *-semiring morphism.*

In [31], Kozen defines a *Kleene algebra* as an *idempotent* symmetric inductive *-semiring. A morphism of Kleene algebras is a *-semiring morphism (which is necessarily an ordered semiring morphism).

Theorem 22. (Ésik, Kuich [22]) *Every inductive* *-semiring is an ordered iteration semiring.*

Thus if S is an inductive *-semiring, then, equipped with the star operation of Definition 3, each matrix semiring $S^{n \times n}$ is an iteration semiring. Moreover, for any locally finite partial monoid M, $S\langle\!\langle M \rangle\!\rangle$ is a also an iteration semiring. Actually we have:

Theorem 23. (Ésik, Kuich [22]) *If* S *is a (symmetric) inductive* *-semiring, then each matrix semiring over* S *is a (symmetric) inductive* *-semiring. Moreover, for any locally finite partial monoid* M, $S\langle\!\langle M \rangle\!\rangle$ *is a (symmetric) inductive* *-semiring.*

(The second part of this theorem is proved in [22] only for free monoids.)

8   Completeness, again

Although the sum order and the pointwise order coincide on $\mathbb{N}_\infty \langle\!\langle A^* \rangle\!\rangle$ (and in fact on any semiring $S\langle\!\langle A^* \rangle\!\rangle$, where S is sum ordered), these two orders are in general different on the semiring $\mathbb{N}_\infty^{\mathrm{rat}} \langle\!\langle A^* \rangle\!\rangle$. In this section, we will consider the sum order on the semiring $\mathbb{N}_\infty^{\mathrm{rat}} \langle\!\langle A^* \rangle\!\rangle$. On the other hand, the sum order and the pointwise order coincide on the semirings $\mathbf{k}^{\mathrm{rat}} \langle\!\langle A^* \rangle\!\rangle$. The following fact can be derived from Theorem 16.

Theorem 24. *The semiring* $\mathbb{N}_\infty^{\mathrm{rat}} \langle\!\langle A^* \rangle\!\rangle$, *equipped with the sum order, is freely generated by the set* A *in the class of all ordered iteration semirings satisfying the equations (6),(7) and (8). Moreover, for each* $k \geq 2$ *and any set* A, $\mathbf{k}\langle\!\langle A^* \rangle\!\rangle$, *equipped with the sum order is freely generated by* A *in the class of all ordered iteration semirings satisfying* $1^* = k - 1$.

Theorem 25. (Bloom, Ésik [8]) *For each set* A, $\mathbb{N}_\infty^{\mathrm{rat}} \langle\!\langle A^* \rangle\!\rangle$, *equipped with the sum order, is freely generated by* A *both in the class of all inductive* *-semirings satisfying (6) and in the class of all symmetric inductive* *-semirings. Similarly, for each set* A *and integer* $k > 1$, $\mathbf{k}\langle\!\langle A^* \rangle\!\rangle$ *equipped with the sum order is freely generated by* A *in the class of all inductive (symmetric inductive)* *-semirings satisfying* $k - 1 = k$.

The case when $k = 2$ is due to Krob [34] and Kozen [30,31].

References

1. Arhangelsky, K.B., Gorshkov, P.V.: Implicational axioms for the algebra of regular languages (in Russian). Doklady Akad. Nauk, USSR, ser A. 10, 67–69 (1987)
2. Béal, M.-P., Lombardy, S., Sakarovitch, J.: On the equivalence of ℤ-automata. In: Caires, L., Italiano, G.F., Monteiro, L., Palamidessi, C., Yung, M. (eds.) ICALP 2005. LNCS, vol. 3580, pp. 397–409. Springer, Heidelberg (2005)
3. Béal, M.-P., Lombardy, S., Sakarovitch, J.: Conjugacy and equivalence of weighted automata and functional transducers. In: Grigoriev, D., Harrison, J., Hirsch, E.A. (eds.) CSR 2006. LNCS, vol. 3967, pp. 58–69. Springer, Heidelberg (2006)
4. Berstel, J., Reutenauer, C.: Rational Series and Their Languages. Springer, Heidelberg (1988), October 19, 2007, http://www-igm.univ-mlv.fr/berstel/
5. Bloom, S.L., Ésik, Z.: Equational axioms for regular sets. Mathematical Structures in Computer Science 3, 1–24 (1993)
6. Bloom, S.L., Ésik, Z.: Iteration Theories: The Equational Logic of Iterative Processes. EATCS Monographs on Theoretical Computer Science. Springer, Heidelberg (1993)
7. Bloom, S.L., Ésik, Z.: Two axiomatizations of a star semiring quasi-variety. EATCS Bulletin 59, 150–152 (1996)
8. Bloom, S.L., Ésik, Z.: Axiomatizing rational power series (to appear)
9. Bloom, S.L., Ésik, Z., Kuich, W.: Partial Conway and iteration theories, Fundamenta Informaticae (to appear)
10. Boffa, M.: A remark on complete systems of rational identities (French). RAIRO Inform. Theor. Appl. 24, 419–423 (1990)
11. Boffa, M.: A condition implying all rational identities (French). RAIRO Inform. Theor. Appl. 29, 515–518 (1995)
12. Bouyer, P., Petit, A.: A Kleene/Büchi-like theorem for clock languages. J. Automata, Languages and Combinatorics 7, 167–181 (2001)
13. Conway, J.C.: Regular Algebra and Finite Machines. Chapman and Hall, Boca Raton (1971)
14. Corradini, F., De Nicola, R., Labella, A.: An equational axiomatization of bisimulation over regular expressions. J. Logic Comput. 12, 301–320 (2002)
15. Droste, M., Quaas, K.: A Kleene-Schützenberger theorem for weighted timed automata. In: Amadio, R. (ed.) FOSSACS 2008. LNCS, vol. 4962, pp. 127–141. Springer, Heidelberg (2008)
16. Eilenberg, S.: Automata, Languages, and Machines, vol. A. Academic Press, London (1974)
17. Elgot, C.C.: Monadic computation and iterative algebraic theories. In: Logic Colloquium 1973, Studies in Logic, vol. 80, pp. 175–230. North Holland, Amsterdam (1975)
18. Elgot, C., Bloom, S.L., Tindell, R.: On the algebraic structure of rooted trees. J. Comput. System Sci. 16, 362–399 (1978)
19. Ésik, Z.: Group axioms for iteration. Information and Computation 148, 131–180 (1999)
20. Ésik, Z., Kuich, W.: A generalization of Kozen's axiomatization of the equational theory of the regular sets. In: Words, Semigroups, and Transductions, pp. 99–114. World Scientific, Singapore (2001)
21. Ésik, Z., Kuich, W.: Rationally additive semirings. J. UCS 8, 173–183 (2002)
22. Ésik, Z., Kuich, W.: Inductive *-semirings. Theoret. Comput. Sci. 324, 3–33 (2004)

23. Goguen, J., Thatcher, J., Wagner, E., Wright, J.: Initial algebra semantics and continuous algebras. J. ACM (24), 68–95 (1977)

24. Golan, J.S.: Semirings and their Applications. Kluwer Academic Publishers, Dordrecht (1999)

25. Grätzer, G.: Universal Algebra. Springer, Heidelberg (1979)

26. Guessarian, I.: Algebraic Semantics. LNCS, vol. 99. Springer, Heidelberg (1981)

27. Hebisch, U.: A Kleene theorem in countably complete semirings. Bayreuth. Math. Schr. 31, 55–66 (1990)

28. Ito, T., Ando, S.: A complete axiom system of super-regular expressions. In: Information processing 74 (Proc. IFIP Congress, Stockholm, 1974), pp. 661–665. North-Holland, Amsterdam (1974)

29. Karner, G.: On limits in complete semirings. Semigroup Forum 45, 148–165 (1992)

30. Kozen, D.: A completeness theorem for Kleene algebras and the algebra of regular events, Technical report, Cornell University, Department of Computer Science (1990)

31. Kozen, D.: A completeness theorem for Kleene algebras and the algebra of regular events. Inform. and Comput. 110, 366–390 (1994)

32. Krob, D.: Complete semirings and monoids (French). Semigroup Forum 3, 323–329 (1987)

33. Krob, D.: Continuous semirings and monoids (French). Semigroup Forum 37, 59–78 (1988)

34. Krob, D.: Complete systems of B-rational identities. Theoretical Computer Science 89, 207–343 (1991)

35. Krob, D.: Matrix versions of aperiodic K-rational identities. Theoretical Informatics and Applications 25, 423–444 (1991)

36. Milner, R.: A complete inference system for a class of regular behaviours. J. Comput. System Sci. 28, 439–466 (1984)

37. Milner, R.: A complete axiomatisation for observational congruence of finite-state behaviours. Inform. and Comput. 81, 227–247 (1989)

38. Morisaki, M., Sakai, K.: A complete axiom system for rational sets with multiplicity. Theoretical Computer Science 11, 79–92 (1980)

39. Rabinovitch, A.: A complete axiomatisation for trace congruence of finite state behaviors. In: Mathematical Foundations of Programming Semantics 1993. LNCS, vol. 802, pp. 530–543. Springer, Heidelberg (1994)

40. Redko, V.N.: On the determining totality of relations of an algebra of regular events (in Russian). Ukrainian Math. Ž. 16, 120–126 (1964)

41. Redko, V.N.: On algebra of commutative events (in Russian). Ukrainian Math. Ž. 16, 185–195 (1964)

42. Salomaa, A.: Two complete axiom systems for the algebra of regular events. Journal of the Association for Computing Machinery 13, 158–169 (1966)

Various Aspects of Finite Quantum Automata

(Extended Abstract)

Mika Hirvensalo[1,2,3]

[1] Department of Mathematics, University of Turku, FI-20014, Turku, Finland
[2] TUCS – Turku Centre for Computer Science
[3] Supported by Turku University Foundation
mikhirve@utu.fi

1 Introduction

Determining the birth date of computer science is a very complicated task and certainly reliant to the standpoint chosen. Some may point out the work of Kurt Gödel [18], Alan Turing [31], and Alonso Church [11], thus locating the appearance of computer science to 1930's. Some want to mention Charles Babbage's engines, some Gottfried Leibniz' Calculus Ratiocinator, and some refer back to the Euclidean algorithm.

If the first occurrence of computer science is hard to locate in a satisfactory way, the quest for the most significant open problem in computer science is presumably much more complicated task. Fortunately there is always the easy way: to listen to the money talking. Clay Mathematics Institute has offered an award of million dollars to whom can resolve the deep **P** vs. **NP** question [12]. Even without the prize, this problem is generally accepted very important in theoretical computer science, and some are willing to date the birth of computer science as late as 1970's, to the days when Stephen Cook introduced ingredients to formally state the problem which we now know as **P** vs. **NP** problem [13].

The question **P** $\neq$ **NP**? is a very natural one, but the seek for the solution is a hopelessly shaggy pathway. The problem has resisted all solution attempts, and there is even theoretical reasoning given by Alexander Razborov and Steven Rudich, explaining why most known resolution attempts are doomed to remain powerless [28].

Problems analogous to **P** vs. **NP** also occur in novel computing models. The idea of quantum computing was first introduced by Richard Feynman [16] (1982), and the formal model was further developed by David Deutsch [14]. More sophisticated constructions were subsequently developed by Ethan Bernstein and Umesh Vazirani [7], and problems analogous to **P** vs. **NP** have also emerged: what is the relation between polynomial-time quantum computing and polynomial time deterministic and nondeterministic computing? More precisely, one may ask whether **P** $\neq$ **BQP** or whether **NP** $\subseteq$ **BQP** (**BQP** being the complexity class of problems solvable with a quantum computer in polynomial time, allowing a bounded error probability). Solutions to such questions are unknown so far, and those problems seem to be as tricky as the old classic **P** $\neq$ **NP**?

M. Ito and M. Toyama (Eds.): DLT 2008, LNCS 5257, pp. 21–33, 2008.

As our present knowledge is too frail to allow answers to above questions, we can always restrict the model of computation. Turing machine (TM) as a computational device is the most powerful of those that allow finite description, and restricting to a weaker model may allow us to provide solutions to analogous questions. This is one of the motivations to study *finite quantum automata*.

Finite automata (FA) are very simple computational devices. Shortly we will remind the reader of the formal definitions of FA and their variants, but a brief comparison of TM's and FA's shows that both models are *uniform*, meaning that the same device can handle inputs of all lengths, but that the computation of Turing machine is *non-oblivious* in the sense that the trajectory of the read-write head depends on the input word and can be different on input words of the same length. Finite automata, on their part, always handle the input words in an oblivious manner: one input letter at time. When studying the various types of automata, one can eventually realize that in the presence of non-obliviousness, complexity estimates are essentially harder than for oblivious computation.

2   Variants of Finite Automata

In this section, we represent the definitions needed in later in this article, but for broader presentation and details on finite automata we refer to [15] and [32]. A *finite automaton* is a quadruple $\mathcal{F} = (Q, \Sigma, I, \delta, F)$, where Q is a finite set of *states*, Σ is the *alphabet*, $I \subseteq Q$ is the set of *initial states*, $\delta \subseteq Q \times \Sigma \times Q$ is the *transition relation*, and F is the set of final (accepting) states. If $|I| = 1$ and $|\{(p, \sigma, q) \in \delta \mid q \in Q\}| \leq 1$ for each choice of p and σ. we say that $\mathcal{F}$ is *deterministic*. If $|\{(p, \sigma, q) \in \delta \mid q \in Q\}| = 1$ for each choice of p and σ, we then say that the deterministic automaton is *complete*. It is well-known that for any deterministic automaton one can create an equivalent complete deterministic automaton with only one more state [15]; equivalent in the sense that the automata accept exactly the same language.

2.1   Probabilistic Automata

For a treatment on probabilistic automata, we refer to [26], but the basic definition is as follows: A probabilistic automaton is a quadruple $\mathcal{P} = (Q, \Sigma, I, \delta, F)$, where Q, Σ, and F are as for deterministic automata, $I : Q \to [0, 1]$ is the *initial distribution*, and $\delta : Q \times \Sigma \times Q \to [0, 1]$ is the transition probability function which satisfies

$$\sum_{q \in Q} \delta(p, \sigma, q) = 1 \tag{1}$$

for any pair $p, \sigma \in Q \times \Sigma$. The value $\delta(p, \sigma, q)$ stands for the probability that being in state p and reading letter σ, the automaton enters state q. Thus (1) is a very natural constraint.

Another, equivalent description of probabilistic automata is obtained by enumerating $Q = \{q_1, \ldots, q_n\}$ and defining an $n \times n$ matrix M_a for each $a \in \Sigma$ by $(M_a)_{ij} = \delta(q_j, a, q_i)$. The columns of these matrices are probability distributions by condition (1) which we express in other words by saying that the matrices are *stochastic* or *Markov* matrices. We also define the *initial vector* $x \in \mathbb{R}^n$ by $x_i = I(q_i)$ and the *final vector* y by $y_i = 1$, if $q_i \in F$ and $y_i = 0$ otherwise. This other description of the automaton is thus a triplet $\mathcal{P} = (y, \{M_a \mid a \in \Sigma\}, x)$, and this description allows an easy definition of the function computed by automaton: For any probabilistic automaton $\mathcal{P}$ we define a function $f_\mathcal{P} : \Sigma^* \to [0,1]$ by

$$f_\mathcal{P}(w) = y M_{w^R} x^T,$$

where $M_w = M_{a_1} \ldots M_{a_l}$ for any word $w = a_1 \ldots a_l$, and $w^R = a_l \ldots a_1$ stands for the *mirror image* of word w ($M_1 = I$ for the empty word 1).

The primary function of a probabilistic automaton is hence to compute a function $\sigma^* \to [0,1]$, which is to say that a probabilistic automaton defines a *fuzzy subset* of Σ^*. Ordinary subsets of Σ^* (languages) recognized by these automata can then be defined in various ways; most traditional way is to introduce a *cut point* $\lambda \in [0,1]$ and to define

$$L_{\geq \lambda} = \{w \in \Sigma^* \mid f_\mathcal{P}(w) \geq \lambda\}$$

as a *cut point language* determined by λ. A variant is obtained by defining

$$L_{> \lambda} = \{w \in \Sigma^* \mid f_\mathcal{P}(w) > \lambda\},$$

which we call a *strict cut point language* determined by λ.

An interesting class of cut point languages are those with *isolated* cut point: We say that the cut point is *isolated* if there is an $\epsilon > 0$ so that $f_\mathcal{P}(w) \notin [\lambda - \epsilon, \lambda + \epsilon]$ for all $w \in \Sigma^*$. From the *practical* point of view, the isolation of a cut point λ is necessary to discover whether $f_P(w) \geq \lambda$ or not (without knowing about isolation, it is not possible to know how many experiments (computations) are needed to determine that $f_P(w) \geq \lambda$). On the other hand, the problem of *deciding*, given a probabilistic automaton, whether the cut-point is isolated, is undecidable [8].

In general, cut point languages can also be non-regular, but a cornerstone result on probabilistic automata says that cut-point languages with an isolated cut point are always regular [26], meaning that for any cut point language with isolated cut point there is a deterministic automaton recognizing the same language as well. On the other hand, probabilistic automata with isolated cut point and deterministic automata may be exponentially separated [17]. By this we mean that there is an infinite sequence of languages $L_1, L_2, L_3, \ldots$ so that L_i is an isolated cut-point language to a probabilistic automaton with $O(i)$ states, whereas any deterministic automaton recognizing L_i has $\Omega(c^i)$ states for a constant $c > 1$.

2.2   Quantum Automata

From the above description of probabilistic automata it is easy to see that complete deterministic automata are special cases of them. When going over to quantum automata, this is not so straightforward. Later on, we will introduce notions on quantum mechanics, but at this moment we present the traditional quantum automata.

Definition 1. *A quantum automaton with n states is a triplet $Q = (P, \{M_a \mid a \in \Sigma\}, \boldsymbol{x})$, where $\boldsymbol{x} \in \mathbb{C}^n$ is the initial superposition, each M_a is a unitary matrix, and P is the final measurement projection.*

In the above definition and hereafter, a fixed orthonormal basis of $\mathbb{C}^n$ is identified with the state set of the quantum automaton. Hence it is also possible to define the final measurement projection by a set of final states.

As in the case of probabilistic automata, the primary task quantum automata is also to compute a function $\Sigma^* \to [0, 1]$. However, there are several ways how to define this function. One, introduced by Crutchfield and Moore [24] is the most obvious variant of those functions associated to probabilistic automata:

$$f_Q(w) = \|PQ_{w^R}\boldsymbol{x}\|^2 , \tag{2}$$

where $\|\boldsymbol{x}\| = \sqrt{|x_1|^2 + \ldots + |x_n|^2}$ stands for the L_2-norm in $\mathbb{C}^n$. Quantum automata equipped with function of type (2) is referred to as *measure-once quantum automata or MO-QFA*.

Another way to define the function was introduced by Kondacs and Watrous [22], a definition which we will discuss later, but notions such as cut point languages can be straightforwardly defined to quantum automata, no matter how the function $\Sigma^* \to [0, 1]$ is actually defined.

3   Quantum Mechanics: Formalism

3.1   Pure States

For a more detailed exposition on the mathematical formalism of quantum mechanics, [19] is referred to; in this section, we merely give the guidelines.

The formalism of quantum automata has traditionally been written by using *pure state* quantum mechanics, which is explained as follows: An *n-level system* is a physical system capable of being in n perfectly distinguishable states, which can be denoted by $|1\rangle, |2\rangle, \ldots, |n\rangle$. In the mathematical formalism, the distinguishable states $|1\rangle, |2\rangle, \ldots, |n\rangle$ form an orthonormal basis of an n-dimensional vector space $H_n \simeq \mathbb{C}^n$ over complex numbers (*Hilbert space*). The basis states $|1\rangle, |2\rangle, \ldots, |n\rangle$ are not the only states the system can be in: Any unit-length vector

$$\alpha_1 |1\rangle + \alpha_2 |2\rangle + \ldots + \alpha_n |n\rangle \tag{3}$$

in H_n represents a state of the system, although there may be several vectors of type (3) representing the *same* state. As (3) has unit length, $|\alpha_1|^2 + |\alpha_2|^2 + \ldots + |\alpha_n|^2 = 1$, and the coefficients $\alpha_1, \ldots, \alpha_n$ have the following interpretation: When observing state (3) with respect to basis $\{|1\rangle, \ldots, |n\rangle\}$, one has a probability $|\alpha_i|^2$ of seeing the system in state $|i\rangle$.

Quantum mechanics is fundamentally probabilistic theory, meaning that (3) carries the maximal information about the system state. In particular, this means that (3) cannot be explained by an underlying deterministic theory with *ignorance interpretation* telling that the use of probabilities $|\alpha_1|^2, \ldots, |\alpha_n|^2$ emerges from our incomplete knowledge about the system state.

The *time evolution* of a quantum system can be mathematically described by setting several conditions (see [19]) involving that unit-length vectors are mapped back to unit-length vectors and the requirement of the continuous nature of the evolution. From quite a natural set of requirements, one can derive the *Schrödinger equation* of motion, or equivalently, that the evolution of a quantum system must be *unitary* (mapping is unitary if $U^*U = UU^* = I$, where U^* stands for the adjoint mapping of U. In matrix formalism, the adjoint mapping is obtained by taking the complex conjugate of the transpose). In the discrete-time systems this is expressed by relating systems states x and y via $y = Ux$, where $U : H_n \rightarrow H_n$ is a unitary mapping depending on the physical system.

The *observation* of a quantum system is mathematically described by giving a decomposition $H_n = E_1 \oplus \ldots \oplus E_k$ into orthogonal subspaces. The state $x \in H_n$ can always be written as

$$x = \alpha_1 x_1 + \ldots + \alpha_k x_k, \tag{4}$$

where $x_i \in E_i$ has unit norm. Now each E_i corresponds to a property that the systems can have (equivalently, a value of an observable) and $|\alpha_i|^2$ is interpreted as the probability that the system is seen to have property E_i. This interpretation of (3) is referred as to the *minimal interpretation* of quantum mechanics. if E_i was observed, the post-observation state of the system is x_i. It should be mentioned that this *projection postulate* is of ad hoc -nature and not consistent with the unitary time evolution. The quest for a consistent description of the measurement process faces severe problems, collectively referred as to the *measurement paradox* of quantum physics.

3.2   Mixed States

Mathematical descriptions of quantum computing devices such as Turing machines, quantum circuits, and quantum finite automata are based, with only some exceptions, on a quantum system with pure states. The pure state formalism has an advantage of being mathematically simpler than the general one, but it is evidently insufficient in many cases. For instance, in general it is impossible to assign a pure state to a subsystem even if the whole system is in a pure state. For a more detailed treatment of the topics in this section, see [19], for instance.

In a *general formalism*, the states of quantum system are elements of $L(H_n)$ (linear mappings $H_n \rightarrow H_n$) instead of those in H_n. As there was a restriction to pure states (unit norm), also the elements that depict the states of quantum systems are those of $L(H_n)$ that are 1) self-adjoint, 2) of unit trace 3) positive (positive semidefinite in mathematical terminology). Elements of H_n are frequently identified with matrices that represent them in some fixed basis, and in the matrix language the conditions 1 – 3 read as: 1) The transpose of the complex conjugate of the matrix is the same as the original matrix, 2) the sum of the diagonal elements equals to 1, and 3) the eigenvalues of the matrix are nonnegative (from condition 1 it follows that the eigenvalues are real).

As the matrices are obviously more complicated objects than the vectors, the opportunity of having some general structure is welcome. One of the basic pillars of quantum mechanics is given by the *spectral theorem*: For any $T \in L(H_n)$ satisfying 1 – 3 there is an orthonormal basis $x_1, \ldots, x_n$ of H_n so that

$$T = \lambda_1 |x_1\rangle\langle x_1| + \ldots + \lambda_n |x_n\rangle\langle x_n|. \tag{5}$$

In the *spectral representation* (5), $\lambda_1, \ldots, \lambda_n$ are indeed the eigenvalues of A, and x_i is an eigenvector belonging to λ_i. Notation $|x\rangle\langle x|$ is due to P. Dirac and stands for the projection onto one-dimensional subspace spanned by x. In a matrix form, $|x\rangle\langle y|$ can be more generally interpreted as a tensor product (Kronecker product) of a row vector x and a column vector y. In the matrix language (5) simply says that any matrix satisfying conditions 1 – 3 has a diagonal form in some orthonormal basis.

A state is said to be *pure* if it is a projection onto one-dimensional subspace, meaning that its spectral representation is of form $T = |x\rangle\langle x|$ for some unit-length $x \in H_n$. This gives also rise to the pure state formalism described in the earlier section: Instead of projection to one-dimensional subspace, we choose a unit-length vector (the choice is not unique) to present that subspace. States that are not pure, are *mixed*. As any state T satisfies conditions 1 – 3 described earlier, it follows that $\lambda_i \geq 0$ and $\lambda_1 + \ldots + \lambda_n = 1$, meaning that any state T can be represented as a *convex combination* of pure states. Unfortunately, the spectral representation (5) is unique (if and) only if the eigenvalues of T are distinct, so it is not generally possible to assign an ignorance interpretation to a quantum state T. This means that in the degenerate case (multiple eigenvalues), one cannot interpret (5) as a probability distribution of pure states $|x_1\rangle\langle x_1|, \ldots, |x_n\rangle\langle x_n|$.

The minimal interpretation turns into the following form: For a state T and decomposition $H_n = E_1 \oplus \ldots \oplus E_k$, $\mathrm{Tr}(TE_i)$ is the probability that the system is observed to have property E_i ($\mathrm{Tr}(A)$ means the trace of mapping A). The minimal interpretation gives also the operational means to define *subsystem states*. Two subsystem states $T_1 \in L(H_n)$ and $T_2 \in L(H_m)$ can always be joined into a compound *decomposable* state $T_1 \otimes T_2$, but in a compound system state space $H_n \otimes H_m$ there may be also states that cannot be represented as tensor products of subsystem states. For a general state $T \in L(H_n \otimes H_m)$, the subsystem state $T_1 \in L(H_n)$ is defined by requiring that $\mathrm{Tr}(T_1 P) = \mathrm{Tr}(T(P \otimes I))$ for each projection $P \in L(H_n)$ ($I \in L(H_m)$ stands for the identity mapping). It can be shown that T_1 is uniquely determined, and sometimes notation $T_1 = \mathrm{Tr}_{H_m}(T)$ is used for it [19] (we say that T_1 is obtained by *tracing over H_m*).

From the viewpoint of this story, it is important to give thoughts to the time evolution in the general formalism, if only in the discrete-time case. The requirements are as follows: If states T and T' of the system at consecutive times are related via $T' = V(T)$, then 1) V must be linear and 2) if $T \otimes S$ is self-adjoint, unit-trace, positive mapping, then also $V(T) \otimes S$ is for any state $S \in L(H_m)$. Conditions 1 – 2 can be expressed by saying that V is a *completely positive* trace-preserving, self-adjointness preserving linear mapping $L(H_n) \mapsto L(H_n)$. The reason why we also require $V(T) \otimes S$ to be a state is evident: we want that if the original system is included as a subsystem in any compound system, the time evolution still keeps that as a "legal" state. For such a general time evolution, the following representations are known:

$$V(T) = \sum_{i=1}^{n^2} V_i A V_i^*, \tag{6}$$

where mappings $V_i \in L(H_n)$ satisfy $\sum_{i=1}^{n^2} V_i^* V_i = I$ (*Kraus representation*), and

$$V(T) = \mathrm{Tr}_{H_m}(U(T \otimes S)U^*), \tag{7}$$

where $S \in L(H_m)$ is a fixed state (*Stinespring representation*). From both representations one can see that the unitary evolution described earlier is a special case of completely positive mappings: indeed, it is easy to see that $|Ux\rangle\langle Ux| = U|x\rangle\langle x|U^*$. We say that the time evolution of form $T \mapsto UTU^*$ is that of a *closed* system (opposite: *open*).

It is noteworthy that the latter representation (7) shows that an arbitrary (open) time evolution (completely positive mapping) is obtained via embedding the quantum system in a larger system, which then evolves as a closed system.

4   Quantum Automata (Continued)

4.1   Measure-Once Automata

Quantum automata by Moore and Crutchfield were introduced already in section 2.2. This model is the most natural one for quantum computing, but the model is indeed a variant, not a generalization of deterministic finite automata. In [1] if was noted that the cut point languages accepted by QFA with isolated cut point are all regular.

The closure properties of languages (here $f_Q : \Sigma^* \to [0, 1]$ is regarded as a fuzzy language) defined by MO-QFA are elegant: These languages are closed under convex combination, product, complement ($f \mapsto 1 - f$), and inverse homomorphism. Moreover, they satisfy a version of the pumping lemma [24].

However, MO-QFA cannot recognize all regular languages for most natural acceptance modes. Moore and Crutchfield [24] noticed that if the characteristic function of a regular language L equals to f_Q for some quantum automaton Q, then the language L is a *group language*, meaning that its syntactic monoid is a group. This result can be extended to all languages recognized by MO-automata with an isolated cut point [10].

So the language recognition power of MO-QFA (with isolated cut point) is quite weak. But as a compensation, when they do recognize a language, they can be exponentially more succinct than the corresponding deterministic or probabilistic automata [4]. This means that one can construct an infinite sequence of languages L_n so that MO-QFA with $O(n)$ states can recognize L_n, but any probabilistic automaton recognizing L_n has $2^{\Omega(n)}$ states.

It is also interesting to notice that if the cut point is not isolated, then there is an opportunity of recognizing also non-regular languages. It is an easy exercise to find a two-state quantum automaton Q such that $f_Q(w) = 0$, if $|w|_a = |w|_b$ and $f_Q(w) > 0$ otherwise.

4.2   Measure-Many Quantum Automata

MM-QFA defined by Kondacs and Watrous [22] have the same ingredients as MO-QFA, except that the state set is divided into *accepting*, *rejecting*, and *neutral* states. The computation of an MM-QFA differs from that of MO-QFA essentially: after reading an input symbol, the unitary mapping is applied as in the MO-model, but also the state of the automaton is observed. If the automaton is seen in an accepting (resp. rejecting) state, the input word is accepted (resp. rejected) immediately, otherwise the computation proceeds to the next input symbol; from the state where the automaton was left after the observation procedure.

It is also assumed that the input word is surrounded with special endmarkers not occurring in the input alphabet, and this computational procedure evidently determines an acceptance probability $f_Q(w)$ for each input word w. For a given MO-QFA it is not difficult to construct an MM-QFA recognizing essentially the same language. Hence MM-QFA can be seen as generalizations of MO-QFA.

As we did for MO-QFA, we assume here that the language recognition is with an isolated cut point. The first observation is that also the class of languages recognized by MM-QFA is a proper subclass of regular languages: all languages recognized by MM-QFA are regular, but even the very simple language $\{a, b\}^*a$ cannot be recognized by an MM-QFA (with isolated cut point).

Unlike in the case of MO-QFA, the class of languages recognized by MM-QFA has not very elegant properties. MO-QFA recognize exactly the group languages, but no similar algebraic characterization is known for languages recognized by MM-QFA. These languages are trivially closed under complement and inverse morphisms, but they are not closed under union or intersection [5].

Another peculiar feature of MM-QFA is that the magnitude of the cut point isolation is very fragile. On of the first results in this direction was given in [4], where it was shown, that if an MM-automaton gives the correct answer with a probability greater than $\frac{7}{9}$, then the language is a group language, but this result is not true for probabilities smaller than $0.68\ldots$. Later on, this result was improved by discovering the smallest probability $0.7726\ldots$ for which the result holds [6].

In [3] the authors construct a sequence of languages L_n recognizable by MM-automata, but no MM-QFA can recognize L_n with a correctness probability greater than $\frac{1}{2} + \frac{3}{\sqrt{n-1}}$, thus showing that the cut point isolation, even the best possible one, can decrease arbitrarily small.

It is a well-known fact that two-way deterministic automata cannot recognize more than the one-way automata do (two-way automata may be exponentially more succinct, though) [15]. Therefore it may come as a surprise that the two-way variant of MM-QFA recognize more than the one-way model. Kondacs and Watrous [22] showed that *two-way MM-QFA can recognize all regular languages* (with isolated cut point), and that language $\{a^n b^n \mid n \geq 1\}$ can be recognized by a two-way MM-QFA.

4.3 Latvian Quantum Automata

The notion of Latvian QFA [2] generalizes that one of MM-QFA. Instead of observing the state of quantum automaton, an arbitrary observable on the state space is allowed. Formally, $M = (Q, \Sigma, \{U_a\}, \{M_a\}, q_0, F)$, where each U_a is a unitary matrix, and M_a a measurement defined as $M_a \colon H_n = E_1^a \oplus \ldots \oplus E_k^a$ (orthogonal subspaces). q_0 is the initial state and F is the set of final states. It is also required that for the right endmarker \$, $P_\$$ is a measurement with respect to decomposition $H_n = \langle F \rangle \oplus \langle F \rangle^\perp$, where $\langle F \rangle$ stands for the subspace generated by the final states.

This generalization has much more elegant properties than the MM-QFA do: In [2] it is shown that the languages recognized by Latvian QFA are closed under Boolean operations and inverse homomorphisms, but yet they cannot recognize all regular languages. Also an algebraic characterization is known: The Latvian automata recognize exactly those languages whose syntactic semigroup is of form $J * G$ (wreath product), where J is a $\mathcal{J}$-trivial monoid and G a group (see [2] for definitions).

5 Some Decidability Properties

The topic in this section is somewhat remote from the rest of the tale, but here we have a very delicate issue certainly worth mentioning. We will consider both probabilistic automata and MO-QFA, but in this section, the cut point is *not isolated*. The basic problem under study is as follows: given an automaton A and cut point λ, is it possible to determine whether the cut point language $L_{\geq\lambda}(A)$ (or the strict cut-point language $L_{>\lambda}(A)$) is empty or not?

For probabilistic automata, problems $L_{\geq\lambda}(A) = \emptyset$? and $L_{>\lambda}(A) = \emptyset$? turn out to be both undecidable [8], [21]. The method of proving this begins with embedding the *Post Correspondence Problem* [27] into integer matrices (a method introduced by Paterson [25]), and continues by a procedure introduced by P. Turakainen [30] to convert any set of matrices into doubly stochastic ones, still preserving some essential properties of the original matrix set. It is possible to show that the emptiness problem for cut point and strict cut point languages is in fact undecidable for 25-state probabilistic automata over a binary alphabet.

By using a different method, one can show that for MO-automata with 21 states over binary alphabet, the emptiness problem $(L_{\geq\lambda}(A) = \emptyset?)$ of cut point languages is undecidable as well [21]. As a big surprise, it turns out that the emptiness problem for *strict* cut point languages $(L_{>\lambda}(A) = \emptyset?)$ is *decidable*

[9]. The decision procedure is eventually based on two pillars: One being the fact that the closure of the semigroup generated by unitary matrices is a group, and as such, an algebraic set. The second pillar is the famous Tarski decision procedure for the theory of real numbers (see [9] for details).

6   Quantum Automata with Open Time Evolution

In this final section we will (re)introduce another model of quantum automata. As seen before, measure-once, measure-many, and the Latvian quantum automata are *variants but not generalizations* of deterministic finite automata in the sense that all of these automata cannot recognize all regular languages (even though they can win the traditional automata when thinking about the number of states).

The model was already introduced in [20], but we will reintroduce it here, hopefully with improved arguments talking for the reasonableness of the definition.

Definition 2. *A quantum automaton with open time evolution is a quintuple $\mathcal{Q} = (Q, \Sigma, \delta, q_0, F)$, where Q is the state set, Σ is the alphabet, q_0 is the initial state, and $F \subseteq Q$ the set of final states. Function δ is a transition function from Σ to set of completely positive, trace-preserving mappings on the state space.*

As the other quantum automata, open evolution automata compute a function $f_{\mathcal{Q}} : \Sigma^* \to [0, 1]$. To describe this function, we let $n = |Q|$ and $B = \{\boldsymbol{q} \mid q \in Q\}$, and orthonormal basis of H_n. If there is no danger of confusion, we identify state $q \in Q$ and projection $|\boldsymbol{q}\rangle\langle\boldsymbol{q}|$ (a pure state). The transition function δ is extended to Σ^* by defining $\delta(1) = I$ (the identity mapping) and $\delta(aw) = \delta(w)\delta(a)$ for any $w \in \Sigma^*$ and $a \in \Sigma$.

The final states of the automaton determines the *final projection* by $P = \sum_{q \in Q} |\boldsymbol{q}\rangle\langle\boldsymbol{q}|$, and then the function computed by the automaton is defined as

$$f_{\mathcal{Q}}(w) = \mathrm{Tr}(P\delta(w)|\boldsymbol{q_0}\rangle\langle\boldsymbol{q_0}|).$$

It is worth noticing that it is also possible to modify the definition by replacing $|\boldsymbol{q_0}\rangle\langle\boldsymbol{q_0}|$ by an arbitrary initial state $Q_0 \in L(H_n)$, and the final projection determined by the final states by an arbitrary projection in $L(H_n)$.

The reasons to introduce this definition are quite natural: The first reason is that completely positive mappings provide a very general description of the dynamics of open quantum systems, exactly in the same way as unitary mappings give that for closed systems. The second one is that Definition 2 is indeed a generalization of DFA, and also a generalization of MO, MM, and Latvian quantum automata. In fact, to see that Definition 2 generalizes probabilistic automata (and hence also DFA), it is enough to define, for each Markov matrix M, a mapping $V_{ij} = \sqrt{M_{ij}}|\boldsymbol{q_i}\rangle\langle\boldsymbol{q_j}| \in L(H_n)$, and a trace-preserving completely positive mapping by

$$V_M(A) = \sum_{i,j=1}^{n} V_{ij} A V_{ij}^*.$$

As the action of V_M on a pure state $|\boldsymbol{q}_j\rangle\langle\boldsymbol{q}_j|$ can be easily recovered:

$$V_M(|\boldsymbol{q}_j\rangle\langle\boldsymbol{q}_j|) = \sum_{i=1}^{n} M_{ij}|\boldsymbol{q}_i\rangle\langle\boldsymbol{q}_i|,$$

it is easy to see that the function computed by a probabilistic automaton with Markov matrices $\{M_a \mid a \in \Sigma\}$ and an open-evolution quantum automaton with completely positive, trace-preserving mappings V_M coincide.

To see that the Definition 2 extends also that of all other quantum automata discussed here, it is enough to consider Latvian QFA. To perform unitary evolutions is an easy task: For any unitary $U_a \in L(H_n)$, mapping $\overline{U}_a : A \mapsto U_a A U_a^*$ is trace-preserving and completely positive, and it faithfully reproduces the effect of unitary mapping $\boldsymbol{x} \mapsto U_a\boldsymbol{x}$. Then, for any measurement decomposition $E_a : H_n = E_1 \oplus \ldots \oplus E_k$, we define

$$M_a(A) = P_1 A P_1^* + \ldots + P_k A P_k^*,$$

where P_i is a projection onto the subspace E_i. Then it is not difficult to see that also M_a is a trace-preserving completely positive mapping. If we further define

$$V_a = M_a \overline{U}_a,$$

it is easy to see a quantum automaton with open time evolution defined by mappings V_a determines the same function $\Sigma^* \to [0,1]$ as a Latvian automaton with unitary mappings U_a and measurements E_a (details are left to the reader).

It is possible to claim that Definition 2 is not genuine in the sense that open time evolution for quantum systems requires an auxiliary system (see equation (7)) to be interpreted as a closed system. And this is true also for each computational step, there should be a "fresh" auxiliary system to perform each computational step as a closed system. Hence the Hilbert space consumed by open time quantum automata is much larger than that of the definition. In fact, the actual space dimension depends on the input word, as for each computational step there is need for an auxiliary system in a known state.

The obvious reply to the criticism above is that why should anyone require the evolution of quantum automata to be *closed* for finite automata? The time evolution of classical computations models (Turing machines, finite automata) is definitely not closed – on the contrary, information loss is perfectly acceptable for them, why shouldn't that be acceptable for quantum models, too? In fact, the information loss is present in all real-world computers: the processors are irreversible, and consequently, in a course of the computational processes, they generate heat and are cooled by external equipment designed for that purpose. Therefore, it should be perfectly acceptable also for quantum models to be consider as open system with non-unitary time evolution.

References

1. Ablayev, F., Gainutdinova, A.: On the Lower Bounds for One-Way Quantum Automata. In: Nielsen, M., Rovan, B. (eds.) MFCS 2000. LNCS, vol. 1893, pp. 132–140. Springer, Heidelberg (2000)
2. Ambainis, A., Beaudry, M., Golovkins, M., Ķikusts, A., Mercer, M., Thérien, D.: Algebraic Results on Quantum Automata. Theory of Computing Systems 39, 165–188 (2006)
3. Ambainis, A., Bonner Rūsiņš, R.F., Ķikusts, A.: Probabilities to Accept Languages by Quantum Finite Automata. In: Asano, T., Imai, H., Lee, D.T., Nakano, S.-i., Tokuyama, T. (eds.) COCOON 1999. LNCS, vol. 1627, pp. 174–185. Springer, Heidelberg (1999)
4. Ambainis, A., Freivalds, R.: 1-way quantum finite automata: strengths, weaknesses and generalizations. In: Proceedings of the 39th FOCS, pp. 376–383 (1998)
5. Ambainis, A., Ķikusts, A., Valdats, M.: On the class of languages recognizable by 1-way quantum finite automata. In: Ferreira, A., Reichel, H. (eds.) STACS 2001. LNCS, vol. 2010, pp. 75–86. Springer, Heidelberg (2001)
6. Ambainis, A., Ķikusts, A.: Exact results for accepting probabilities of quantum automata. Theoretical Computer Science 295(1), 3–25 (2003)
7. Bernstein, E., Vazirani, U.: Quantum complexity theory. SIAM Journal on Computing 26(5), 1411–1473 (1997)
8. Blondel, V.D., Canterini, V.: Undecidable problems for probabilistic automata of fixed dimension. Theory of Computing systems 36, 231–245 (2003)
9. Blondel, V.D., Jeandel, E., Koiran, P., Portier, N.: Decidable and undecidable problems about quantum automata. SIAM Journal on Computing 34(6), 1464–1473 (2005)
10. Brodsky, A., Pippenger, N.: Characterizations of 1-Way Quantum Finite Automata. SIAM Journal on Computing 31(5), 1456–1478 (2002)
11. Church, A.: An unsolvable problem in elementary number theory. American Journal of Mathematics 58, 345–363 (1936)
12. Clay Mathematics Institute, http://www.claymath.org/
13. Cook, S.A.: The complexity of theorem proving procedures. In: Proceedings of the Third Annual ACM Symposium on the Theory of Computing, pp. 151–158. ACM, New York (1971)
14. Deutsch, D.: Quantum theory, the Church-Turing principle and the universal quantum computer. Proceedings of the Royal Society of London A 400, 97–117 (1985)
15. Eilenberg, S.: Automata, languages, and machines, vol. A. Academic Press, London (1974)
16. Feynman, R.P.: Simulating physics with computers. International Journal of Theoretical Physics 21(6/7), 467–488 (1982)
17. Freivalds, R.: Non-constructive Methods for Finite Probabilistic Automata. In: Harju, T., Karhumäki, J., Lepistö, A. (eds.) DLT 2007. LNCS, vol. 4588, pp. 169–180. Springer, Heidelberg (2007)
18. Gödel, K.: Über formal unentscheidbare Sätze der Principia Mathematica und verwandter Systeme I. Monatshefte für Mathematik und Physik 38, 173–198 (1931)
19. Hirvensalo, M.: Quantum Computing, 2nd edn. Springer, Heidelberg (2004)
20. Hirvensalo, M.: Some Open Problems Related to Quantum Computing. In: Paun, G., Rozenberg, G., Salomaa, A. (eds.) Current Trends in Theoretical Computer Science – The Challenge of the New Century, vol. 1. World Scientific, Singapore (2004)

21. Hirvensalo, M.: Improved Undecidability Results on the Emptiness Problem of Probabilistic and Quantum Cut-Point Languages. In: van Leeuwen, J., Italiano, G.F., van der Hoek, W., Meinel, C., Sack, H., Plášil, F. (eds.) SOFSEM 2007. LNCS, vol. 4362, pp. 309–319. Springer, Heidelberg (2007)

22. Kondacs, A., Watrous, J.: On the power of quantum finite state automata. In: Proceedings of the 38th IEEE Symposium on Foundations of Computer Science, pp. 66–75 (1997)

23. Matiyasevich, Y., Sénizergues, G.: Decision problems for semi-Thue systems with a few rules. Theoretical Computer Science 330(1), 145–169 (2005)

24. Moore, C., Crutchfield, J.P.: Quantum automata and quantum grammars. Theoretical Computer Science 237(1-2), 275–306 (2000)

25. Paterson, M.S.: Unsolvability in 3 X 3 matrices. Studies in Applied Mathematics 49, 105–107 (1970)

26. Paz, A.: Introduction to Probabilistic Automata. Academic Press, London (1971)

27. Post, E.L.: A variant of a recursively unsolvable problem. Bulletin of American Mathematical Society 52, 264–268 (1946)

28. Razborov, A.A., Rudich, S.: Natural Proofs. Journal of Computer and System Sciences 55, 24–25 (1997)

29. Tarski, A.: A decision method for elementary algebra and geometry. University of California Press (1951)

30. Turakainen, P.: Generalized automata and stochastic languages. Proceedings of American Mathematical Society 21, 303–309 (1969)

31. Turing, A.M.: On Computable Numbers, With an Application to the Entscheidungsproblem. Proc. London Math. Soc. 2(42), 230–265 (1936)

32. Rozenberg, G., Salomaa, A.: Regular Languages. In: Rozenberg, G., Salomaa, A. (eds.) Handbook of Formal Languages. Word, Language, Grammar, vol. 1, Springer, Heidelberg (1997)

On the Hardness of Determining Small NFA's and of Proving Lower Bounds on Their Sizes*

Juraj Hromkovič[1] and Georg Schnitger[2]

[1] Department of Computer Science, ETH Zurich, ETH Zentrum, CH-8022 Zurich, Switzerland
[2] Department of Computer Science, Johann-Wolfgang-Goethe Universität, Robert Mayer-Strasse 11-15, D-6054 Frankfurt a. M., Germany

Abstract. In contrast to the minimization of deterministic finite automata (DFA's), the task of constructing a minimal nondeterministic finite automaton (NFA) for a given NFA is PSPACE-complete. This fact motivates the following computational problems:

(i) Find a minimal NFA for a regular language L, if L is given by another suitable formal description, resp. come up with a small NFA.

(ii) Estimate the size of minimal NFA's or find at least a good approximation of their sizes.

Here, we survey the known results striving to solve the problems formulated above and show that also for restricted versions of minimization of NFA's there are no efficient algorithms.

Since one is unable to efficiently estimate the size of a minimal NFA in an algorithmic way, one can ask at least for developing mathematical proof methods that help in proving good lower bounds on the size of a minimal NFA for a given regular language. We show here that even the best known methods for this purpose fail for some concrete regular languages.

Finally, we give an overview of the results about the influence of the degree of ambiguity on the size of NFA's and discuss the relation between the descriptional complexity of NFA's and NFA's with ε-transitions.

1 Introduction

The minimization of nondeterministic finite automata is a hard computational problem. The same is true even if one strives only to approximately estimate the size of a minimal NFA. Hence, searching for small NFA's cannot be automated in an efficient way and so one has to consider estimating the size of minimal NFA's as a research problem for each particular regular language. Thus the question arises whether there exists a robust mathematical method that could be used to prove at least some tight lower bounds on the sizes of minimal NFA's, if its potential is explored in the right way. The goal of this paper is to give an overview

* The work on this paper was supported by SNF-grant 200023-007327/1, DFG-grant SCHN 503/4-1 and was done during the stay of the second author at ETH Zurich.

on the hardness of searching for minimal or small NFA's and on available methods for proving lower bounds for descriptional complexity measures of minimal NFA's. Moreover, we show that the most robust method known for proving lower bounds on the sizes of minimal NFA's is not universal in the sense that it fails for some specific regular languages.

The paper is organized as follows. In Section 2 we give a survey on the computational hardness of determining minimal or small NFA's and present some related open problems. Section 3 is devoted to methods for proving lower bounds on the size of minimal NFA's for given regular languages. The communication complexity approach is known as the best one in the sense that it seems to subsume all other methods used. But even this general approach is not powerful enough in order to reach tight lower bounds for all regular languages. Here we discuss some generalizations of this approach and prove that even they fail for some specific languages. The last section surveys results and open problems with regard to the sizes of minimal NFA's with restricted ambiguity where ambiguity is one of the fundamental measures of the degree of nondeterminism. We close this paper by listing the results related to the comparison of sizes of minimal NFA's and minimal ε-free NFA's. Here size is measured by the number of transitions and the main goal is to show proof techniques that enable to derive nontrivial lower bounds on the number of transitions of ε-free NFA's recognizing concrete regular languages.

2   The Complexity of Determining Small NFA's

In contrast to the problem of minimizing DFA's, which is efficiently possible, the task of minimizing a given NFA or regular expression is extremely hard in the worst case, namely PSPACE-complete. Moreover, the minimization problem for NFA's or regular expressions remains PSPACE-complete even if the regular language is specified by a DFA [15].

However in many cases a small NFA and not necessarily a minimal NFA is required and hence the task of approximating a minimal NFA is of foremost importance. The approximation complexity depends on the input representation and here we want to consider two scenarios. In the first scenario the language is specified by either an NFA or a regular expression. Not surprisingly the approximation problem is considerably harder than in the second scenario where the language is specified by a DFA. In both scenarios our goal is to determine NFA's with either few states or few transitions, resp. to determine regular expressions of small length, where the length of a regular expression R is the number of symbols from the alphabet Σ appearing in R.

2.1   Small NFA's from NFA's

The approximation complexity for determining NFA's or regular expressions in the first scenario can be characterized rather precisely.

Theorem 1. *[5] Unless* P = PSPACE, *it is impossible to efficiently approximate the size of a minimal NFA or regular expression describing $L(A)$ within an approximation factor of $o(n)$ when given an NFA or regular expression A with n states, transitions or symbols, respectively.*

Thus the news is disastrous: no meaningful approximation can be determined efficiently even if a given NFA of size m may have extremely small equivalent NFA.

We sketch the argument which is based on the proof of the PSPACE-completeness of "Regular Expression Non-Universality" [19]: given a regular expression R, is $L(R) \neq \Sigma^*$? First observe that there is a PSPACE-complete language L which is recognizable by a deterministic in-place Turing machine M with (initial state q_0 and) a unique accepting state q_f; moreover M runs for at least 2^n steps on inputs $w \in L$ of length n. We show how to reduce the word problem for $L(M)$ to the minimization problem for regular expressions.

The crucial step is to construct a regular expression R_w for a given input w of M in time polynomial in the length of w. The expression R_w describes all words which are *not* concatenations of consecutive legal configurations starting from configuration $q_0 w$ and ending with the unique accepting configuration q_f. If the efficient construction of R_w as well as its length bound $|R_w| = O(|w|)$ is taken for granted, then the rest of the argument is immediate.

Case 1: M rejects w. Then $L(R_w)$ coincides with Σ^* and R_w is equivalent with a regular expression of size $|\Sigma| = O(1)$.

Case 2: M accepts w. Now $L(R_w)$ coincides with $\Sigma^* \setminus \{y\}$, where y describes the unique accepting computation on input $w \in L$. But the accepting computation, and hence y, has length at least $2^{|w|}$ due to the requirement on M. Certainly DFA's for $\Sigma^* \setminus \{y\}$ have to have size at least $2^{|w|}$ and equivalent regular expressions or NFA's have to have size at least $|w|$. Thus R_w cannot be compressed significantly in this case. As a consequence, if an equivalent regular expression of length $o(|R_w|)$ is determined efficiently, then this is equivalent with M rejecting w and the word problem is solved efficiently.

We have argued that the approximation problem for regular expressions is hard; the analogous statement for NFA's is now an immediate consequence. The approximation problem for unary regular languages remains hard as well, but now we have to use the stronger P $\neq$ NP assumption.

Theorem 2. *[7] Unless* P = NP, *it is impossible to efficiently approximate the size of a minimal equivalent NFA within an approximation factor of $o(n)$, when given an NFA with n states over the unary alphabet.*

2.2   Small NFA's from DFA's

Now we assume that the regular language L is given by a DFA. The rather redundant formalism of DFA's increases the input size in comparison to representing L by regular expressions or NFA's and hence approximation algorithms determining small regular expressions or small NFA's have considerable more allowed running time, if polynomial time is granted. Thus it should not come as a surprise that negative approximation results require stronger hypotheses than the P $\neq$ PSPACE assumption.

Here we work with the cryptographic assumption that strong pseudo-random functions exist in nonuniform NC^1, i.e., are computable by circuits of logarithmic depth and polynomial size. The concept of strong pseudo-random functions is introduced by Razborov and Rudich [24]. Naor and Reingold [20] show that strong pseudo-random functions can even be computed by threshold circuits in constant depth and polynomial size, provided factoring Blum integers requires time 2^{n^ε} for some $\varepsilon > 0$. Actually, a slightly weaker notion of strong pseudo-random functions suffices.

Definition 1. *Let* $f_N = (f_N^s)_{s \in S}$ *be a function ensemble with functions* $f_N^s :$ $\{0,1\}^N \to \{0,1\}$ *for a seed* $s \in S$ *and let* $(r_N^i)_{i \in \{1,\ldots,2^{2^n}\}}$ *be the ensemble of all N-bit boolean functions. Then* f_N *is a strong pseudo-random ensemble with parameter* ε *iff for any deterministic algorithm* A

$$|\mathrm{prob}[A(f_N) = 1] - \mathrm{prob}[A(r_N) = 1]| < \frac{1}{3},$$

provided A runs in time $O(2^{O(N^\varepsilon)})$ *and has access to* f_N^s, *respectively* r_N^i, *via a membership oracle. The probability is defined by the uniform sampling of s from S, resp. of i from* $\{1, \ldots, 2^{2^N}\}$.

The assumption that strong pseudo-random function f_N exist in nonuniform NC^1 implies that the functions f_N^s have boolean formulae[1] of length polynomial in N. Still regular expressions are too weak to express formulae of small size. However, here a trick of Pitt and Warmuth [22] helps: the computing power of regular expressions increases considerably if inputs are repeated. Since verifying that inputs are repeated correctly requires too large regular expressions, we may utilize the tremendous power of nondeterministic description schemes which have no problem to concisely express the complement, namely that inputs are *not* repeated correctly. In particular, for an N-bit boolean function g and $p(c) = N^c$ define

$$L_{p(c)}(g) = \text{the complement of } \{x^{p(c)} : x \in \{0,1\}^N \wedge g(x) = 1\}.$$

The power of regular expressions is shown by the following technical result.

Lemma 1. *[5] Assume that $b \in \mathbb{N}$. If an N-bit boolean function g has a formula of length at most N^b, then there is a constant c such that $L_{p(c)}(g)$ has regular expressions of length polynomial in N.*

Omitting some technical details, the inapproximability of minimal regular expressions or NFA's can now be shown as follows. First observe that the language $L_{p(c)}(g)$ has DFA's of size at most $O(p(c) \cdot 2^N)$: The DFA consists of a complete

---

[1] A formula is a binary tree with $\wedge$ and $\vee$-gates as interior nodes. Leaves are marked by labels from $\{x_1, \neg x_1, \ldots, x_N, \neg x_N\}$. The formula length is its number of leaves.

binary tree of depth N rooted at the initial state. A leaf that corresponds to a word $x \in \{0,1\}^N$ with $g(x) = 0$ gets a self loop, a leaf that corresponds to a word x with $g(x) = 1$ is starting point of a path of length $N \cdot (p(c) - 1)$ that can only be followed by inputs with $p(c) - 1$ repetitions of x. Whereas each such path leads to a rejecting state, any deviation from the path leads to an accepting trap state. Each state is accepting, except for those already described as rejecting.

The DFA is huge, however observe that $O(p(c) \cdot 2^N)$ states suffice for input size $p(c) \cdot N$. A truly random function r_N^i requires regular expressions of length at least $\Omega(2^N)$ to describe $L_{p(c)}(r_N^i)$, however, applying Lemma 1, $L_{p(c)}(f_N^s)$ can be described by regular expressions of length polynomial in N, if f_N is a strong pseudo-random ensemble represented by formulae of polynomial length.

We have opened up a huge gap between truly random and strong pseudo-random functions. As a consequence, according to Definition 1, no efficient approximation algorithm A can provide a meaningful approximation of the minimal length of an equivalent regular expressions. Since the argument can be applied as well to the stronger description mechanism of NFA's we obtain:

Theorem 3. *[5] Suppose that strong pseudo-random functions exist in nonuniform NC^1 for some parameter $\varepsilon > 0$. Let A be any approximation algorithm that approximately determines the length of a shortest equivalent regular expression, the number of transitions (resp. the number of states) of a minimum equivalent NFA, when given a DFA with n states.*

If A runs in time polynomial in n, then A cannot reach the approximation factor $\frac{n}{\log^c n}$ (resp. $\frac{\sqrt{n}}{\log^c n}$ when minimizing the number of states), where c is a sufficiently large constant.

Thus for any efficient approximation algorithm A there are DFA's of size n with equivalent regular expressions of poly-logarithmic size, but A will find equivalent regular expressions only of size $\frac{n}{poly(\log n)}$.

In the case of minimizing NFA's Gruber and Holzer [6] were able to replace the cryptographic assumption by the weaker $P \neq NP$ assumption. The excluded approximation factors however decrease correspondingly.

Theorem 4. *[6] Suppose that $P \neq NP$. Let A be any approximation algorithm that approximately determines the number of transitions (resp. the number of states) of a minimum equivalent NFA, when given a DFA with n states. If A runs in time polynomial in n, then A cannot reach the approximation factor $n^{1/5-\varepsilon}$ (for all $\varepsilon > 0$). This holds when counting states or transitions.*

The inapproximability results can be improved to $n^{1/3-\varepsilon}$, if alphabets of size $O(n)$ are considered and if states are counted.

The approximation complexity for unary languages decreases considerably. (Observe that a DFA for a unary language consists of a path followed by a cycle. Call a DFA cyclic iff its path is empty.)

Theorem 5. *Assume that $|\Sigma| = 1$ for the alphabet Σ.*
(a) [4] If a cyclic DFA with n states is given, then the minimal number of states of an equivalent NFA can be approximated efficiently within a factor of $O(\log_2 n)$.

(b) [7] If a DFA with n states is given, then the minimal number of states of an equivalent NFA can be approximated efficiently within a factor of $O(\sqrt{n})$.

Although the above results show the significant progress made in understanding the approximation complexity of determining small regular expressions or small NFA's, some important questions remain open.

(1) What is the approximation complexity of determining small unambiguous NFA?

(An unambiguous NFA is an NFA with at most one accepting computation per input. Observe that the equivalence and containment problem is efficiently solvable for unambiguous NFA [27].)

(2) When minimizing unary NFA's only the case of a given DFA remains open. Can the positive results of Theorem 5 be improved?

(3) When minimizing the number of states of an NFA equivalent to a given DFA with n states, the approximation factor $\frac{\sqrt{n}}{\log^c n}$ can be excluded. Is it possible to exclude even the approximation factor $\frac{n}{\log^c n}$?

The argument in [5] cannot be improved in straightforward manner, since the language $L_{p(c)}(g)$, for any N-bit boolean function g, can be accepted by NFA's with $O(2^{N/2})$ states. Thus only a gap between poly(N), for boolean functions with polynomial formula length, and $O(2^{N/2})$ in the worst case is opened up.

3   Communication Complexity and Proving Lower Bounds on the Size of NFA's

Since there does not exist any efficient algorithm for estimating the minimal number of states of nondeterministic finite automata accepting a given regular language, one should ask at least for a method to investigate the size of minimal NFA's "by hand". In 1986 communication complexity was proposed for this aim in [14]. Two-party protocols and their communication complexity were introduced by Yao [28] in order to measure the amount of information exchange between different parts of distributed computing systems. The original two-party protocol is a non-uniform computing model for computing Boolean functions from $\{0,1\}^{2n}$ to $\{0,1\}$. It consists of two computers C_1 and C_2 [Fig. 1]. At the beginning C_1 gets the first half of the input bits and C_2 gets the second one. The computers are required to cooperate in order to compute the value of the function corresponding to their common input. They are allowed to communicate by exchanging binary messages and the complexity of their work is measured by the number of bits exchanged. The communication complexity of two-party protocols became one of the most powerful instruments for proving lower bounds on the complexity of various computing models computing concrete tasks (see some surveys in [11, 13, 16]) as well as for investigating the relative power of determinism, nondeterminism and randomness [13, 16, 21].

A special version of protocols, called one-way two-party protocols is related to proving lower bounds on the number of states of finite automata. One-way

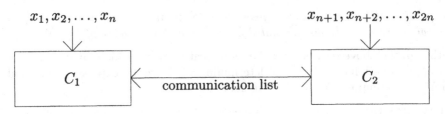

$$x_1, x_2, \ldots, x_n \qquad\qquad x_{n+1}, x_{n+2}, \ldots, x_{2n}$$

$$C_1 \quad \longleftrightarrow \quad \text{communication list} \quad \longrightarrow \quad C_2$$

Fig. 1.

protocols are restricted in that sense, that C_1 is allowed to send only one binary message to C_2 and after that C_2 is required to compute the correct answer. Formally, the work of C_1 can be described by a function $f_1 : \{0,1\}^n \to \{0,1\}^*$, where $f_1(\alpha)$ is the binary message sent to C_2. The work of C_2 can be described by a function $f_2 : \{0,1\}^n \times \{0,1\}^* \to \{0,1\}$. The arguments of f_2 are its n input bits and the message received and the output is the value of the Boolean function computed. The **one-way communication complexity** of a Boolean function f, $cc_1(f)$, is the **communication complexity** of the best one-way protocol computing f.

One-way protocols can be considered for any computation mode (deterministic, nondeterministic or different kinds of randomization). Since each language can be viewed as an infinite sequence of finite functions, one can easily construct a sequence of one-way protocols that simulate the work of a given finite automaton as follows. For each input length m we have one separate one-way protocol $P_m = (C_{1,m}, C_{2,m})$. $C_{1,m}$ with input α of length $\lfloor \frac{m}{2} \rfloor$ sends the binary code of the state q reached by the finite automaton after reading α from its initial sate. Then, $C_{2,m}$ with its input β simulates the work of the finite automaton from the state q on the word β. If and only if the simulation finishes in an accepting state, $C_{2,m}$ outputs the value 1. One can easily observe that this way of simulating finite automata by one-way protocols works for any mode of computation. Since the communication complexity of all protocols simulating an automaton with a state set Q is at most $\lceil \log_2 |Q| \rceil$, one-way communication complexity provides a lower bound on the number of states of finite automata.

In other words, measuring the complexity of one-way protocols as the number $mc_1(n)$ of different messages used in all computations on all inputs of the length n, the message complexity $mc_1(n)$ is a lower bound on the number of states of finite automata. There is one essential drawback of this lower bound technique. The two-party protocol model is non-uniform, while automata are a uniform computing model. Due to this difference in modelling, the gap between the message complexity of one-way protocols and the number of states of finite automata can be arbitrarily large for some languages. For instance, the message complexity of regular languages over one-letter alphabets is only 1. Another example is the finite language $L_n = \{0^{2n}xx | x \in \{0,1\}^n\}$, whose deterministic message complexity is 2, but the size of the minimal NFA's is at least 2^n.

In order to overcome this drawback we have introduced uniform one-way protocols in [10]. A uniform one-way protocol (C_1, C_2) consists again of two computers C_1 and C_2, but inputs $x = x_1, x_2, \ldots, x_n \in \Sigma^*$ for an alphabet Σ are arbitrarily

	λ	0	1	00	01	10	11	...	α	...
λ	a_λ	a_0	a_1	a_{00}	a_{01}					
0	a_0	a_{00}	a_{01}	a_{000}						
1	a_1	a_{10}	a_{11}	a_{100}						
00	a_{00}	a_{000}	a_{001}							
01	a_{01}									
11										
$\vdots$										
β									$a_{\beta\alpha}$	
$\vdots$										

Fig. 2.

divided into a prefix $x_1, \ldots, x_k$ as the input of C_1 and a suffix $x_{k+1}, \ldots, x_n$ as the input of C_2. A uniform protocol is required to provide the correct answer for each of the possible $n + 1$ partitions ($k \in \{0, 1, \ldots, n\}$) of the word $x \in \Sigma^n$. Interestingly, for each regular language L, the message complexity of deterministic uniform one-way protocols accepting L is equal to the size of the minimal deterministic finite automaton for L. To see this fact one has to represent the task of accepting L as the infinite communication matrix M_L for L (Fig. 2)

The rows as well as the columns of M_L are labelled by the words form Σ^* in the canonical order starting with λ. M_L is a 0/1-matrix. The element $\alpha_i\beta_k$ of M_L in the intersection of row R_i and column C_k is 1 iff the word $\alpha_i\beta_k$ belongs to L. Observe, that, for each word $x \in \Sigma^n$, there are exactly $n + 1$ elements in M_L corresponding to x. Hence, we cannot assign a language to any such 0/1-matrix.

Now, we are ready to argue that deterministic message complexity is equal to the size of the minimal deterministic FA for any regular language L. The computer C_1 is required to send a message to C_2. We claim that the number of messages needed is the number of different rows of M_L. Let $R_i = (r_{i1}, r_{i2}, \ldots)$ and $R_j = (r_{j1}, r_{j2}, \ldots)$ be different rows of M_L. For sure, there is a column k labelled by β_k in which they differ, i.e., $r_{ik} \neq r_{jk}$. If C_1 sends the same message m to C_2 for its inputs α_i and α_j corresponding to the rows R_i and R_j, then C_2 either has to accept both $\alpha_i\beta_k$ and $\alpha_j\beta_k$ or to reject both words $\alpha_i\beta_k$ and $\alpha_j\beta_k$ (The arguments of C_2 are the message m and the word β_k and these arguments are the same for inputs $\alpha_i\beta_k$ and $\alpha_j\beta_k$). Since the rows R_i and R_j differ in column k, the number of messages used is at least the number of different rows of M_L. Now, one can easily observe that the number of different rows of M_L is nothing else than the number of the equivalence classes of the Nerode relation for L and we are done (for a detailed argumentation see [10]).

Unfortunately, the situation for the nondeterministic mode of computation is completely different. The nondeterministic message complexity nmc(L) can essentially differ from the size ns(L) of the minimal NFA's for L. Additionally, it is not so easy to estimate nmc(L) for a given regular language L. From communication complexity theory [13, 16] we know that nmc(L) is the minimal number of 1-monochromatic submatrices that cover all 1's in M_L.

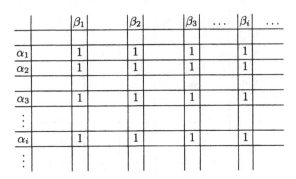

Fig. 3.

One can visualize the argument as follows. The computations of a (nondeterministic) one-way protocol can be represented as $m\#a$ for $a \in \{accept, reject\}$, where m is the message sent from C_1 to C_2. Consider a concrete accepting computation

$$m\#accept.$$

Let

$$(\alpha_1, \beta_1), (\alpha_2, \beta_2), (\alpha_3, \beta_3), \ldots, (\alpha_i, \beta_i), \ldots$$

be inputs with corresponding partitions (C_1 has α_i and C_2 has β_i) for which $m\#accept$ is an accepting computation. Then this accepting computation accepts all words of L corresponding to words on the intersections of

the rows $\alpha_1, \alpha_2, \ldots, \alpha_i, \ldots$ and the columns $\beta_1, \beta_2, \ldots, \beta_i, \ldots$.

In Fig. 3 one immediately sees that this intersection of rows and columns determines unambiguously a 1-monochromatic submatrix of M_L.

To solve the combinatorial problem of covering all 1's of a matrix by the minimal number of potentially overlapping 1-monochromatic submatrices is not easy. One possibility to use this fact is to restrict M_L to a finite submatrix M_L' and then to estimate the largest 1-monochromatic submatrix S of M_L'. As a consequence, the number of 1's in M_L' divided by the number of 1's in S is a lower bound on the message complexity of L.

Another lower bound technique is based on the so-called 1-fooling sets and this technique covers the approach proposed independently by Glaister and Shallit [35], who directly strived to prove lower bounds on the size of NFA's for concrete languages without using the concept of communication complexity. A one-way 1-fooling set for a language L is a finite subset A_L consisting of pairs

$$\{(\alpha, \beta) | \alpha\beta \in L\}$$

[13, 21], such that

if $(\alpha, \beta) \in A_L$ and $(\gamma, \delta) \in A_L$ (i.e., $\alpha\beta \in L$ and $\gamma\delta \in L$),

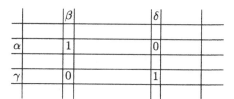

Fig. 4.

then

$$\alpha\delta \notin L \text{ or } \gamma\beta \notin L.$$

If A_L has this property, then for any two elements (α, β) and (γ, δ) from A_L, each protocol accepting L has to send another message $C_1(\alpha)$ from C_1 to C_2 for α than for γ. One can argue this fact as follows. Let, for any $w, v \in \Sigma^*$, $C_1(w, v)$ be the set of all messages submitted from C_1 to C_2 in all possible accepting computations on wv, where the input part of C_1 is w. If $C_1(\alpha, \beta) \cap C_1(\gamma, \delta) \neq \emptyset$ and $\alpha\beta$ is accepted, then unavoidably $\gamma\beta$ is accepted by the protocol as well. Analogously, if $C_1(\alpha, \beta) \cap C_1(\gamma, \delta) \neq \emptyset$ and $\gamma\delta$ is accepted, then $\alpha\delta$ is accepted too. If one wants to argue directly for finite automata, one can say that for any two elements (α, β) and (γ, δ) of A_L, and any NFA accepting L, each state q reached after reading α in an accepting computation on $\alpha\beta$ differs from any state p reached after reading δ in an accepting computation on $\delta\gamma$. Hence, the number of different messages (states) of a protocol (a NFA) accepting L must be at least $|A_L|$. For more details and a survey about methods for proving lower bounds on communication complexity we recommend [11, 13, 16, 29, 30, 31, 32].

To visualize the argument above one can consider Fig. 4 below. We see that the intersection of the rows α and γ and the columns β and δ does not build a 1-monochromatic matrix (because both (α, β) and (δ, γ) are elements of a one-way 1-fooling set). Hence, the 1's corresponding to (α, δ) and to (β, γ) cannot be covered by any 1-monochromatic submatrix because at least one of the elements $a_{\alpha\delta}$ and $a_{\gamma\beta}$ is zero. Therefore, the number of all monochromatic submatrices covering all 1's corresponding to the elements of the one-way 1-fooling sets must be at least the cardinality of the one-way 1-fooling set.

For most regular languages it is not easy to investigate their nondeterministic communication complexity in order to prove a lower bound on the size of minimal NFA's. The situation is still worse. A good lower bound on $\text{nmc}(L)$ need not to be a good lower bound for $\text{ns}(L)$. Consider the language

$$L_{(3,n)} = \{xyz \mid x, y, z \in \{0, 1\}^n, x = y \lor x \neq z\}.$$

In [8] it is shown that

$$\text{nmc}(L_{(3,n)}) \in O(n^2) \text{ and } \text{ns}(L_{(3,n)}) \in 2^{\Omega(n)};$$

i.e., that the message complexity can be logarithmic in the size of minimal NFA's. Let us show that $nmc(L_{(3,n)})$ is small. If C_1 gets at least xy (i.e., the cut is inside of z), then C_1 can check whether $x = y$. If so, C_1 knows that the input word is in $L_{(3,n)}$. If $x \neq y$, C_1 guesses the position in which x and z differ and verifies it by sending the order (index) and the value of this position to C_2.

If C_2 obtains at least the whole suffix yz (i.e., the cut is inside of x), then C_2 checks whether $y = z$. The main point is that if $y = z$, C_2 knows that the input xyz is in the language and accepts independently of the message received from C_1. If $y \neq z$, then the words outside the language $L_{(3,n)}$ must have the form

$$zyz$$

for $y \neq z$. To check the opposite for xyz it is again sufficient to check whether $x \neq y$, which is easy for nondeterministic computation models.

If the cut of xyz is somewhere inside y, one can check the property $x = y$ or $x \neq z$ in a similar way as described above for the other two cases. Observe, that one has to accept all words except xyx for $x \neq y$. To check $x \neq z$ in xyz with a cutpoint in y in a nondeterministic way is easy. To get xxx accepted the protocol accepts if C_1 sees that its part of y is a prefix of x and C_2 sees that its part of y is a suffix of z for an input xyz with a cutpoint in y. If $x = z$ then consequently the input is $xyz = xxx$. If $x \neq z$, we are allowed to accept, because in that case the input cannot be of the forbidden form xyx for $y \neq x$.

The main point is, that independently of the cutpoint, it is sufficient to verify the inequality of two strings and this is an existence task and all existence tasks are easy for nondeterminism.

Now, let us argue that $ns(L_{(3,n)})$ is large. Observe, that $L_{(3,n)}$ contains all words $xxx \in \{0,1\}^{3n}$. For each xxx fix an accepting computation $Com(xxx)$. Let $T\text{-}Com(x,x) = (p,q)$ be the trace of $Com(xxx)$ consisting of the state p reached after reading x and the state q read after reaching xx. Assume, for $x \neq y$,

$$T\text{-}Com(xxx) = T\text{-}Com(yyy) = (p,q)$$

Then one can easily observe that there exists an accepting computation $Com(xyx)$ on xyx with

$$T\text{-}Com(xyx) = T\text{-}Com(xxx) = T\text{-}Com(yyy),$$

because q can be reached from p by reading x as well as y. Hence, the number of different traces must be at least 2^n, which is the cardinality of $\{xxx | x \in \{0,1\}^n\}$. Hence, if Q is the state set of a NFA accepting L, then $|Q|^2 \geq 2^n$, i.e., $ns(L_{(3,n)}) \geq 2^{\frac{n}{2}}$.

The proof above shows that cutting the words into 3 parts may be helpful. Motivated by this observation we proposed to introduce the following generalization of two-party one-way communication protocols.

The uniform k-party communication protocol consists of k agents $A_1, \ldots, A_k$. Inputs $x = x_1 \cdots x_k \in \Sigma^*$ are arbitrarily divided into (possibly empty) substrings such that agent A_i receives substring x_i. The one-way communication starts with agent A_1 and ends with agent A_k who has to decide whether to

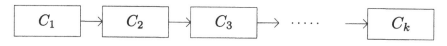

Fig. 5.

accept or reject. Agent A_i, upon receiving a message m from its left neighbor A_{i-1}, determines a message m' based on m and its substring x_i and sends m' to its right neighbor A_{i+1}.

The message complexity of a protocol is the maximum, over all agents A_i, of the number of different messages sent in all computations. The different communication modes such as deterministic, probabilistic or nondeterministic communication are defined in the canonical way.

Let, for any regular language L, $\mathrm{nmc}_k(L)$ denote the message complexity of the best nondeterministic uniform k-party protocol and remember that $\mathrm{ns}(L)$ is the minimal number of states of an NFA recognizing L.

In [33] Adorna established an exponential gap between the message complexities $\mathrm{nmc}_2(L_{(3,n)})$ and $\mathrm{nmc}_3(L_{(3,n)})$, and in his PhD thesis [34] he showed an exponential gap between nmc_k and nmc_{k+1} for any $k \geq 2$. To do so he considered the language

$$L_{(k,n)} = \{x_1, x_2, \ldots, x_k | x_i \in \{0,1\}^n \text{ for } i = 1, 2, \ldots, n$$
$$\text{and } \exists i \in \{1, 2, \ldots, k-2\} \text{ such that}$$
$$x_i = x_{i+1} \vee x_i \neq x_{i+2}\}$$

for any $k \geq 3$. To prove that

$$\mathrm{nmc}_k(L_{(k,n)}) \geq 2^{\frac{n}{k-1}} \text{ (i.e., } \mathrm{ns}(L_{(k,n)}) \geq 2^{\frac{n}{k-1}})$$

one can argue similarly as for $L_{(3,n)}$. The set

$$\{x^k | x \in \{0,1\}^n\}$$

is a subset of $L_{(3,n)}$. If one partitions x^k into k pieces x and each agent A_i gets an x, then one can fix an accepting computation

$$Com(x) = m_1 \# m_2 \# \ldots \# m_{k-1} \# accept$$

of a nondeterministic uniform k-party protocol accepting $L_{(k,n)}$ on the input x^k, where m_i is the message sent by the agent A_i. If, for two different inputs $x^k \neq y^k$,

$$Com(x) = Com(y)$$

then $Com(x)$ is also an accepting computation on the word

$$xyxy \ldots x \text{ (if k is odd) or } xyxy \ldots xy \text{ (if k is even)}.$$

But none of these two words belong to $L_{(k,n)}$. Hence, the number of different accepting computations must be at least 2^n and consequently at least $(2^n)^{\frac{1}{k-1}} = 2^{\frac{n}{k-1}}$ different messages are needed.

To understand that $L_{(k,n)}$ is easy for nmc_{k-1} one has to observe that if a word does not belong to $L_{(k,n)}$ then it has the form

$$vuvuvu\ldots$$

for $v, u \in \{0, 1,\}^n$, and $v \neq u$. Since a nondeterministic protocol can verify a difference of x_i and x_{i+2} for any i easily, one needs only to think how to accept words x^k. If one agent gets two consecutive xx, then it immediately knows that the input is in $L_{(k,n)}$ and we are done. If x^k is partitioned into $k-1$ parts and none contains xx, then each agent getting wh or wxh for a suffix w of x and a prefix h of x, compares all positions i occuring in consecutive fragments of x. One can prove by induction that all positions $i = 1, 2, \ldots, n$ will be checked at least once by the agents during the whole computation. The protocol accepts x^k in computations in which all attempts to prove $x_i \neq x_{i+2}$ failed and all internal comparisons of bits in x_i and x_{i+1} succeeded. This approach works because no such computation can accept words of the form $uvuvuv\ldots$ for $u \neq v$ and these words are the only ones not in $L_{(k,n)}$.

To manage all these comparisons the message of an agent A_j must on one side provide the exact information of the length of the prefix processed up till now and the order of the position to be compared in the test $x_i = x_{i+2}$ for some i. Hence, $O(n^2)$ messages suffice.

The result of Adorna shows, that there is no fixed k such that $\text{nmc}_k(L)$ is polynomially related to $\text{ns}(L)$ for each regular language. Here we prove that the situation is still worse. We strengthen the result by showing that there is a sequence of languages $\{L_n\}_{n \geq 1}$ such that even extremely large values of k (i.e., values of k exponential in $\text{ns}(L_k)$) are insufficient to predict $\text{ns}(L_k)$ with the help of $\text{nmc}_k(L_n)$.

To do so, for every positive integer n, we consider the unary language

$$L_n = \{1^l : l \neq n\}.$$

We work with k parties $A_1, \ldots, A_k$. Agent A_1 nondeterministically selects a prime number $p \leq P_k$ for a parameter P_k to be determined later.

If agent A_i receives the substring 1^{m_i} and if $m_i > n$, it sends a "too long" message to its right neighbor A_{i+1}, resp. passes a received "too long" message on to A_{i+1}. Otherwise, assuming that A_i has received a message $(p, m_1 + \cdots + m_{i-1} \mod p)$ from its left neighbor A_{i-1}, it sends the message $(p, m_1 + \cdots + m_i \mod p)$ to its right neighbor. A_k accepts the joint input if its suffix is too long or if it has received a "too long" message. In this way all words with length at least $k \cdot n + 1$ are accepted, because for each partition of such long inputs at least one of the agents gets an input part of length n. Additionally A_k accepts, if

$$m_1 + \cdots + m_k \not\equiv n \mod p.$$

The protocol accepts only strings from L_n, since it requires a proof that the joint input is different from 1^n. When are all inputs 1^m with $1^m \neq 1^n$ accepted? If

$$m = m_1 + \cdots + m_k \text{ for } m_1, \ldots, m_k \leq n,$$

then $m \leq k \cdot n$ and consequently

$$|m - n| \leq (k - 1) \cdot n.$$

Thus if we require

$$k \cdot n \leq \Pi_{p \leq P_k} p, \qquad (1)$$

then $m \equiv n \mod \Pi_{p \leq P_k} p$ implies that $\Pi_{p \leq P_k} p$ divides $m - n$ which is only possible if $m = n$. Thus the protocol is correct, since all strings in L_n are indeed accepted.

In summary, L_n can be accepted by the uniform k-party communication model even if each agent sends only $O(P_k^2)$ messages. How large is P_k and how many states are required for NFA's recognizing L_n?

Theorem 6. *Let $L_n = \{1^l : l \neq n\}$. Then*

$$\Omega(\sqrt{n}) = \mathrm{nfa}(L_n) \leq n + 1,$$

and

L_n has k-party protocols with message complexity $O(\log_2^2(k \cdot n))$.

In particular, even for $k = 2^{c \cdot n^{1/4}}$ agents, message complexity is smaller than state complexity, provided c is sufficiently small.

Proof. We first give a lower bound for $\mathrm{ns}(L_n)$. Let N_n be some unary NFA recognizing L_n with s states. We apply the Chrobak normal form for unary NFA's [2] and can assume that there is an equivalent NFA N'_n which consists of an initial path of length at most s^2 and subsequent cycles with at most s states altogether. But, if $s = o(\sqrt{n})$, then inputs 1^n and $1^{n+n!}$ are treated alike on each cycle, since $n \equiv n + n! \mod r$ holds for any cycle length $r \leq n$. But N_n has to treat 1^n and $1^{n+n!}$ differently and $\mathrm{ns}(L_n) = \Omega(\sqrt{n})$ follows.

Our next goal is to show that $P_k = O(\log_2(k \cdot n))$ holds. If true, then, as claimed, our protocol achieves message complexity $O(\log_2^2(k \cdot n))$ for k parties. Now consider $\vartheta(x) = \sum_{p \leq x} \ln p$, where we sum all primes $p \leq x$. Then $\vartheta(x) \sim x$ [1] and hence $\ln(\Pi_{p \leq x} p) = \vartheta(x) \geq x/2$. In particular, $\Pi_{p \leq x} p \geq e^{x/2}$ and the requirement (1) is indeed satisfied for $P_k = O(\log_2(k \cdot n))$.

Since our protocol has message complexity $O(P_k^2)$, for $k = 2^{c \cdot n^{1/4}}$ parties its message complexity is bounded by $O(c^2 \cdot \sqrt{n})$ and hence smaller than the state complexity, if c is sufficiently small.   □

We call attention to the fact, that Holger Petersen [36] proved the upper bound $\mathrm{nfa}(L_n) \in O(\sqrt{n})$ and hence our lower bound is asymptotically optimal. For further reading related to the minimization of NFAs and to proving lower bounds on the size of minimal nondeterministic automata we recommend [37, 38, 39, 40, 41, 42].

4 NFAs and Related Nondeterministic Formalisms

4.1 Ambiguity

Let N be an NFA with alphabet Σ and let $T_N(x)$ be the computation tree of N on input x. There are several ways to measure the degree of nondeterminism employed by N as a function of the input size. In the first measure define $\mathrm{advice}_N(x)$ to be the maximal number of nondeterministic guesses in any computation for x; in other words, $\mathrm{advice}_N(x)$ is the maximum, over all paths in $T_N(x)$, of the number of nodes with at least two children. Finally

$$\mathrm{advice}_N(n) = \max\{\mathrm{advice}_N(x) : x \in \Sigma^n\}$$

is the advice-complexity of N. The leaves-complexity of N determines the maximal number of computations for inputs of length n. Thus, if $\mathrm{leaf}_N(x)$ is the number of leaves of $T_N(x)$, then

$$\mathrm{leaf}_N(n) = \max\{\mathrm{leaf}_N(x) : x \in \Sigma^n\}.$$

The ambiguity $\mathrm{ambig}_N(x)$ of N on input $x \in L(N)$ is the number of accepting computations of N on x and

$$\mathrm{ambig}_N(n) = \max\{\mathrm{ambig}_N(x) : x \in \Sigma^n\}$$

is the ambiguity of N. The three measures can be related as follows.

Theorem 7. *[8] For any minimal NFA N*

$$\mathrm{advice}_N(n), \mathrm{ambig}_N(n) \le \mathrm{leaf}_N(n) = O(\mathrm{advice}_N(n) \cdot \mathrm{ambig}_N(n)).$$

As a first consequence, since the advice complexity is at most linear in the input length, NFA's with bounded ambiguity have at most $O(\mathrm{advice}_N(n)) \subseteq O(n)$ different computations for inputs of size n.

The leaf complexity turns out to be either bounded by a constant or at least linear but polynomially bounded, or otherwise exponential in the input length.

Theorem 8. *[8] Let N be an NFA with size_N states. Then*

$$either\ \mathrm{leaf}_N(n) \le \mathrm{size_N}^{\mathrm{size}_N}\ or\ \mathrm{leaf}_N(n) \ge n/\mathrm{size}_N(n) - 1.$$

Moreover,

$$either\ \mathrm{leaf}_N(n) \le (n \cdot \mathrm{size_N})^{\mathrm{size}_N}\ or\ \mathrm{leaf}_N(n) = 2^{\Theta(n)}.$$

To study the degree of ambiguity four classes of NFA's, namely UNA (unambiguous NFA), FNA (finitely ambiguous NFA), PNA (polynomially ambiguous NFA) and ENA (exponentially ambiguous NFA) are introduced in [23]. Of particular interest are FNA's, since important algorithmic problems such as equivalence and containment turn out to be tractable for NFA's with finite ambiguity [27].

As a consequence of Theorems 7 and 8, if an NFA has at least linear ambiguity, then it is either a PNA or an ENA. Leung [17] shows that ENA's can be far more succinct than PNA's; subsequently a more general argument was given in [8].

Theorem 9. *[17] For every $n \in \mathbb{N}$ there is an ENA N_n with n states such that any equivalent PNA has at least $2^n - 1$ states.*

The arguments in [8] implicitly use the following decomposition result for PNA's, which shows structural limitations of PNA's.

Theorem 10. *Let N be a PNA. Then there is a finite set I and finitely many UNA's $U_{i,1}, \ldots, U_{i,k_i}$ for each $i \in I$ such that*

$$L(N) = \bigcup_{i \in I} L(U_{i,1}) \cdots L(U_{i,k_i}).$$

Any $U_{i,j}$ has exactly one final state. Moreover, for any i, the combined number of states or transitions of $U_{i,1}, \ldots, U_{i,k_i}$ is bounded by the number of states or transitions of N.

If we place the UNA's $U_{i,1}, \ldots, U_{i,k_i}$ in sequence, then the resulting NFA has polynomial ambiguity. Thus, for any given PNA N we obtain an equivalent PNA N' by first nondeterministically selecting a "simple" NFA from a finite set of simple NFA's. A simple NFA itself is obtained after "concatenating" relatively few, small UNA's.

Proof of Theorem 10: Interpret the transition diagram of N as a directed graph G_N by selecting the states of N as the nodes of G_N and by inserting an edge (p, q) into G_N iff there is a transition from state p to state q in N.

Let C be an arbitrary strongly connected component of G_N and let $p, q \in C$ be two arbitrary nodes of C. If there are two distinct paths in G_N from p to q, then these two paths cannot be computation paths in N for a common input w: otherwise use the strong connectivity in C to extend both paths to two cycles by appending a path from q back to p. We obtain two distinct cycles which share the state p and are traversed in N by a common input w'. As a consequence the ambiguity of N has to be exponential.

Thus, if we select an arbitrary state p of C as the initial state and an arbitrary state q as the unique final state we obtain a UNA $U_{p,q}$. Since the graph G_N^* of strongly connected components is acyclic, there are only finitely many paths in G_N^*. We choose the index set I in the claim of Theorem 10 as the set of all paths in G_N^*, additionally noting the states of arrival and departure for any edge between strongly connected components. If a path $i \in I$ traverses the edges $(e_0, \ldots, e_m)$ and if e_j leaves component C_j in state r_j and enters component C_{j+1} in state q_{j+1} with the transition $r_j \xrightarrow{a_j} q_{j+1}$, then i contributes the concatenation

$$L(U_{q_0, r_0}) \cdot a_0 \cdot L(U_{q_1, r_1}) \cdot a_1 \cdots L(U_{q_j, r_j}) \cdot a_j \cdots .$$

Obviously the combined number of states or transitions required for i is bounded by the number of states or transitions of N. $\qquad\square$

Can FNA's and PNA's be separated? In other words, is there a family P_n with PNA's P_n of size n such that no equivalent family of FNA's F_n has size polynomial in n? Surprisingly the answer is negative for unary languages.

Theorem 11. *For any unary NFA with s states there is an equivalent FNA with $O(s^2)$ states.*

The statement is an immediate consequence of the Chrobak normal form for NFA's [2]. The separation problem has remained open for NFA's over general alphabets, however even among FNA's succinctness increases dramatically when increasing ambiguity.

Theorem 12. *[8] There is a family $(L_n : n \in \mathbb{N})$ of languages which are recognizable by FNA's with ambiguity n and size $O(n^2)$, but any FNA with ambiguity k has to have at least $\Omega(2^{(n/k)})$ states.*

Certainly a separation of FNA's and PNA's is one of the most important open problems for ambiguity. Moreover, is there an analogue of Theorem 10 for FNA's? For instance is it possible to represent any language $L(F)$ as a finite union $L(U_i)$ of unambiguous NFA's U_i of size at most polynomial in the size of F?

4.2 ε-Transitions and Succinctness

We study the role of ε-transitions in increasing succinctness by first investigating the size of ε-free NFA's recognizing the language expressed by a given regular expression. Then we compare the size of NFA's with ε-transitions and the size of NFA's without ε-transitions. In both cases we measure size by the number of transitions.

From Regular Expressions to ε-Free NFA's. A typical application in lexicographical analysis starts with a regular expression that has to be converted into an ε-free nondeterministic finite automaton. Thus the task of converting regular expressions into small ε-free NFA's is of practical importance.

All classical conversions produce ε-free NFAs with worst-case size quadratic in the length of the given regular expression and for some time this was assumed to be optimal [26] until Hromkovic, Seibert and Wilke [12] constructed ε-free NFAs with surprisingly only $O(n(\log_2 n)^2)$ transitions for regular expressions of length n. Subsequently Geffert [3] noticed the impact of the size of the alphabet, he showed that even ε-free NFAs with $O(n \cdot k \cdot \log_2 n)$ transitions suffice for alphabets of size k. Actually it turned out that regular expressions over very small alphabets are recognizable by ε-free NFA's of almost linear size:

Theorem 13. *[25]*

(a) *Every regular expression R of length n over an alphabet of size k can be recognized by an ε-free NFA with at most*

$$O(n \cdot \min\{\log_2 n \cdot \log_2 2k, k^{1+\log^* n}\})$$

transitions.

(b) *There are regular expressions of length n over an alphabet of size k such that any equivalent ε-free NFA has at least $\Omega(n \cdot \log_2^2 2k)$ transitions.*

As a first consequence of part (a) there are ε-free NFA's of size $O(n \cdot \log_2 n \cdot \log_2 2k)$ for regular expressions of length n over an alphabet of size k. For small alphabets, for instance if $k = O(\log_2 \log_2 n)$, the upper bound $O(n \cdot k^{1+\log^* n})$ is better. In particular, $O(n \cdot 2^{\log_2^* n})$ transitions and hence almost linear size suffice for the binary alphabet.

A first lower bound was also given in [12], where it is shown that the regular expression

$$E_n = (1 + \varepsilon) \cdot (2 + \varepsilon) \cdots (n + \varepsilon)$$

over the alphabet $\{1, \ldots, n\}$ requires NFAs of size at least $\Omega(n \cdot \log_2 n)$. Lifshits [18] improves this bound to $\Omega(n(\log_2 n)^2 / \log_2 \log_2 n)$. As a consequence of the improved lower bound in part (b), the construction of [12] is optimal for large alphabets, i.e., if $k = n^{\Omega(1)}$. Since Theorem 13 is almost optimal for alphabets of fixed size, only improvements for alphabets of intermediate size, i.e., $\omega(1) = k = n^{o(1)}$, are still required.

We conclude this section by showing that E_n can indeed by recognized by ε-free NFA's with $\Theta(n \log_2^2 n)$ transitions. This example shows the main ingredients of the $O(n \cdot \log_2^2 n)$ upper bound of [12].

Example 1. Assume that n is a power of two. We recursively construct NFAs A_n to recognize E_n. $\{0, 1, \ldots, n-1, n\}$ is the set of states of A_n; state 0 is the initial and state n is the unique final state of A_n. To obtain A_n place two copies of $A_{n/2}$ in sequence: $\{0, 1, \ldots, n/2 - 1, n/2\}$ and $\{n/2, n/2+1, \ldots n-1, n\}$ are the sets of states of the first and second copy respectively, where the final state $n/2$ of the first copy is also the initial state of the second copy.

Observe that E_n is the set of all strictly increasing sequences with elements from the set $\{1, \ldots, n\}$. Now, if $(a_1, \ldots, a_r, a_{r+1}, \ldots a_s)$ is any increasing sequence with $a_r \leq n/2 < a_{r+1}$, then the sequence has an accepting path which starts in 0, reaches state $n/2$ when reading a_r and ends in state n when reading a_s. But increasing sequences ending in a letter $a \leq n/2$, resp. starting in a letter $a > n/2$ have to be accepted as well. Therefore direct all transitions, ending in the final state $n/2$ of the first copy, also into the final state n. Analogously, direct all transitions, starting from the initial state $n/2$ of the second copy, also out of initial state 0.

Now unroll the recursion and visualize A_n on the complete binary tree T_n with $n - 1$ nodes (thus we disregard the initial state 0 and the final state n). The root of T_n plays the role of state $n/2$. In particular, for any node v of height h there are 2^h transitions between v and the root. Thus the root is the target of $\sum_h \frac{n}{2^h} \cdot 2^h = n \cdot \log_2 n$ transitions and more generally, nodes u of height h have $2^h \cdot h$ transitions connecting u with a descendant in T_n. All in all, A_n has $O(n \cdot \log_2^2 n)$ transitions if transitions incident with states 0 or n are disregarded.

From NFA's with ε Transitions to NFA's without ε-Transitions. The standard conversion of an ε-free NFA N' from a given NFA N inserts a transition $p \xrightarrow{a} q$ into N' whenever there is an ε-path in N from p to some state r followed by the transition $r \xrightarrow{a} q$. Hence the bound

$$\mathrm{nt}(L) \leq |\Sigma| \cdot \mathrm{nt}_\varepsilon(L)^2 \tag{2}$$

is obvious, where $nt_\varepsilon(L)$, $nt(L)$ are the minimal number of transitions of an NFA for L with, resp. without ε-transitions. Since ε-free NFA turn out to be surprisingly powerful when simulating regular expressions, it is natural to ask whether the bound (2) can be improved.

Theorem 14. *[9] There are regular languages $\{L_n\}_{n=1}^\infty$ and $\{K_n\}_{n=1}^\infty$ such that*

(a) $nt_\varepsilon(L_n) = O(n \cdot 2^n)$, *but* $nt(L_n) = \Omega(2^{2n})$. L_n *is defined over the alphabet* $\{0,1\}^n$.

(b) $nt_\varepsilon(K_n) = O(n \cdot 2^n)$, *but* $nt(K_n) = \Omega(2^{n+c\sqrt{n}})$ *for every constant* $c < 1/2$. K_n *is defined over the alphabet* $\{0,1\}$.

Hence, for $m = 2^n$, L_n and K_n can be accepted by NFAs with $O(m \log_2 m)$ transitions, but every ε-free NFA for L_n has at least $\Omega(m^2)$ transitions, whereas K_n requires ε-free NFAs of size $O(m \cdot 2^{c \cdot \sqrt{\log_2 m}})$ for every $c < 1/2$. Thus ε-free NFAs for L_n require almost quadratically more transitions than unrestricted NFAs for L_n and the obvious transformation from NFAs to ε-free NFAs cannot be drastically improved. Observe however that L_n is defined over an alphabet of size m and the lower bound for L_n is far from establishing that bound (2) for the standard conversion is tight. The provable gap for binary alphabets is smaller, but still a far larger than poly-logarithmic gap is shown to be necessary.

We describe the weaker lower bound $nt(L_n') = \Omega(2^{3n/2})$ for the language

$$L_n' = \{uv : u, v \in \{0,1\}^n, \langle u, v \rangle \equiv 0 \bmod 2 \}$$

of length-2 words over the alphabet $\Sigma_n = \{0,1\}^n$; thus $uv \in L_n'$ iff the inner product of u and v is zero modulo 2. We do not show the upper bound $nt_\varepsilon(L_n') = O(n \cdot 2^n)$, but remark that there are "switching circuits for L_n'" with $O(n \cdot 2^n)$ edges: such a switching circuit contains a path from a source $u \in \Sigma_n$ to a sink $v \in \Sigma_n$ if and only if $uv \in L_n'$. To transform a switching circuit into an NFA, first translate its edges into ε-transitions and subsequently add branches from an initial state to the sources and from the sinks to the unique final state.

The $2^n \times 2^n$ matrix H_n with $H_n[u,v] = \langle u, v \rangle \equiv 1 \bmod 2$ is the nth Hadamard matrix if we replace a matrix entry 0 by -1. H_n has the remarkable property that submatrices consisting only of zeroes have to be small. In particular assume that a submatrix $X \times Y$ of H_n consists only of 0-entries. We may close X and Y under addition modulo two to obtain vector spaces $\overline{X}$ and $\overline{Y}$ (with $X \subseteq \overline{X}$ and $Y \subseteq \overline{Y}$) such that the submatrix $\overline{X} \times \overline{Y}$ still consists of zeroes only. Thus $\overline{X}$ and $\overline{Y}$ are orthogonal vector spaces and $\dim(\overline{X}) + \dim(\overline{Y}) \leq n$ follows. Hence

$$|X| \cdot |Y| \leq 2^n. \tag{3}$$

Now consider an arbitrary NFA N without ε-transitions for L_n'. We may assume that N consists of an initial state q_0, a final state q_f and "inner" states q; moreover all accepting paths are of the form $q_0 \xrightarrow{u} q \xrightarrow{v} q_f$. For an inner state q let $fanin(q)$ (resp. $fanout(q)$) be the set of labels of incoming (resp. leaving) edges. Then any inner state q has the "submatrix"-property: $\text{fanin}(q) \times \text{fanout}(q)$ is a submatrix of H_n covering only zeroes. We apply (3) and obtain

$$|\text{fanin}(q)| \cdot |\text{fanout}(q)| \leq 2^n. \tag{4}$$

But $t(N) = \sum_q |\text{fanin}(q)| + \sum_q |\text{fanout}(q)|$ is the number of transitions of N. Thus we obtain a lower bound for the size of N, if we minimize $t(N)$ subject to the inequalities (4) and to the requirement that all zeroes of H_n are "covered" by at least one inner state, i.e., that $\sum_q |\text{fanin}(q)| \cdot |\text{fanout}(q)| \geq 2^{2n-1}$. This optimization problem is solved optimally if $|\text{fanin}(q)| = |\text{fanout}(q)| = 2^{n/2}$ for 2^{n-1} inner states, implying $t(N) = \Omega(2^{3n/2})$.

Observe that the submatrix–property makes this an a communication complexity argument: the zeroes of the communication matrix H_n are to be covered by 0-chromatic submatrices such that the sum of row and column sizes is as small as possible.

References

1. Bach, E., Shallit, J.: Algorithmic Number Theory 1. MIT Press, Cambridge (1996)
2. Chrobak, M.: Finite automata and unary languages. Theor. Comput. Sci. 47(3), 149–158 (1986)
3. Geffert, V.: Translation of binary regular expressions into nondeterministic ε-free automata with $O(n \log n)$ transitions. J. Comput. Syst. Sci. 66, 451–472 (2003)
4. Gramlich, G.: Probabilistic and nondeterministic unary automata. In: Proc. of 28th MFCS, pp. 460–469 (2003)
5. Gramlich, G., Schnitger, G.: Minimizing nfa's and regular expressions. J. Comput. Syst. Sci. 73, 909–923 (2007)
6. Gruber, H., Holzer, M.: Inapproximability of nondeterministic state and transition complexity assuming P $\neq$ NP. In: Harju, T., Karhumäki, J., Lepistö, A. (eds.) DLT 2007. LNCS, vol. 4588, pp. 205–216. Springer, Heidelberg (2007)
7. Gruber, H., Holzer, M.: Computational complexity of NFA minimization for finite and unary languages. In: Proc. 1st LATA, pp. 261–272 (2007)
8. Hromkovič, J., Karhumäki, J., Klauck, H., Seibert, S., Schnitger, G.: Communication Complexity method for measuring nondeterminism in finite automata. Inf. Comput. 172(2), 202–217 (2002)
9. Hromkovič, J., Schnitger, G.: Comparing the size of NFAs with and without ε-transitions. Theor. Comput. Sci. 380(1-2), 100–114 (2007)
10. Hromkovič, J., Schnitger, G.: On the power of Las Vegas for one-way communication complexity, OBDD's, and finite automata. Information and Computation 169, 284–296 (2001)
11. Hromkovič, J., Schnitger, G.: Communication Complexity and Sequential Computation. In: Pr* ara, I., Ružička, P. (eds.) MFCS 1997. LNCS, vol. 1295, pp. 71–84. Springer, Heidelberg (1997)
12. Hromkovič, J., Seibert, S., Wilke, T.: Translating regular expression into small ε-free nondeterministic automata. J. Comput. Syst. Sci. 62(4), 565–588 (2001)
13. Hromkovič, J.: Communication Complexity and Parallel Computating. Springer, Heidelberg (1997)
14. Hromkovič, J.: Relation Between Chomsky Hierarchy and Communication Complexity Hierarchy. Acta Math. Univ. Com 48-49, 311–317 (1986)

15. Jiang, T., Ravikumar, B.: Minimal NFA problems are hard. SIAM J. Comput. 22(6), 1117–1141 (1993)
16. Kushilevitz, E., Nisan, N.: Communication Complexity. Cambridge University Press, Cambridge (1997)
17. Leung, H.: Separating exponential ambiguous finite automata from polynomially ambiguous finite automata. SIAM J. Comput. 27(4), 1073–1082 (1998)
18. Lifshits, Y.: A lower bound on the size of ε-free NFA corresponding to a regular expression. Inf. Process. Lett. 85(6), 293–299 (2003)
19. Meyer, A.R., Stockmeyer, L.J.: The equivalence problem for regular expressions with squaring requires exponential space. In: Proc. 13th Ann. IEEE Symp. on Switching and Automate Theory, pp. 125–129 (1972)
20. Naor, M., Reingold, O.: Number-theoretic constructions of efficient pseudo-random functions. J. ACM 51(2), 231–262 (2004)
21. Papadimitriou, C., Sipser, M.: Communication Complexity. In: Proc. 14th ACM STOC, pp. 196–200 (1982)
22. Pitt, L., Warmuth, M.K.: Prediction-preserving reducibility. J. Comput. Syst. Sci. 41(3), 430–467 (1990)
23. Ravikumar, B., Ibarra, O.H.: Relating the type of ambiguity of finite automata to the succinctness of their presentation. SIAM J. Comput. 18(6), 1263–1282 (1989)
24. Razborov, A.A., Rudich, S.: Natural proofs. J. Comput. Syst. Sci. 55(1), 24–35 (1997)
25. Schnitger, G.: Regular expressions and NFAs without ε transitions. In: Durand, B., Thomas, W. (eds.) STACS 2006. LNCS, vol. 3884, pp. 432–443. Springer, Heidelberg (2006),
    www.thi.informatik.uni-frankfurt.de
26. Sippu, S., Soisalon-Soininen, E.: Parsing Theory. Languages and Parsing, vol. I. Springer, Heidelberg (1988)
27. Stearns, R.E., Hunt III, H.B.: On the equivalence and containment problems for unambiguous regular expressions, regular grammars and finite automata. SIAM J. Comput. 14(3), 598–611 (1985)
28. Yao, A.C.: Some Complexity Questions Related to Distributed Computing. In: Proc. 11th ACM STOC, pp. 209–213 (1979)
29. Dietzfelbinger, M., Hromkovič, J., Schnitger, G.: A comparison of two lower bound methods for communication complexity. Theoretical Computer Science 168, 39–51 (1996)
30. Lovász, L.: Communication Complexity. A survey. In: Korte, L., Promel, S. (eds.) Paths, Flows, and VLSI Layout. Springer, Berlin (1990)
31. Hromkovič, J.: Randomized communication protocols (A survey). In: Steinhöfel, K. (ed.) SAGA 2001. LNCS, vol. 2264, pp. 1–32. Springer, Heidelberg (2001)
32. Hromkovič, J.: Communicatoin protocols - an exemplary study of the power of randomness. In: Rajasekharan, S., Pardalos, P.M., Reif, J.H., Rolim, J. (eds.) Handbook of Randomized Computing, vol. II, pp. 533–596
33. Adorna, H.N.: 3-party message complexity is better than 2-party ones for proving lower bounds on the size of minimal nondeterministic finite state automata. In: Proc. 3rd Int. Workshop on Descriptional Complexity of Automata, Grammars and Related Structures, pp. 23–34. Univ. Magdeburg (2001), Preprint No. 16; See also Journal of Automata, Languages and Combinatorics 7 (4), 419–432 (2002)
34. Adorna, H.N.: On the separation between k-party and (k+1)-party nondeterministic message complexity. In: Ito, M., Toyama, M. (eds.) DLT 2002. LNCS, vol. 2450, pp. 152–161. Springer, Heidelberg (2003)

35. Glaister, I., Shallit, J.: A lower bound technique for the size of nondeterministic finite automata. Information Processing Letters 59, 75–77 (1996)

36. Petersen, H.: personal communication

37. Arnold, A., Dicky, A., Nivat, M.: A note about minimal non-deterministic automata. Bulletin of the EATCS 47, 166–169 (1992)

38. Carrez, C.: On the minimalization of non-deterministic automaton, Laboratoire de Calcul de la Faculté des Sciences de l'Université de Lille (1970)

39. Birget, J.-C.: Partial orders on words, minimal elements of regular languages and state complexity. Theoret. Comput. Sci. 119, 267–291 (1993)

40. Courcelle, B., Niwinski, D., Podelski, A.: A Geometrical View of the Determinization and Minimization of Finite-State Automata. Mathematical Systems Theory 24(2), 117–146 (1991)

41. Gruber, H., Holzer, M.: Finding Lower Bounds for Nondeterministic State Complexity is Hard. Developments in Language Theory 2006, 363–374 (2006)

42. Salomaa, K.: Descriptional Complexity of Nondeterministic Finite Automata. Developments in Lanugage Theory 2007, 31–35 (2007)

Selected Ideas Used for Decidability and Undecidability of Bisimilarity

Petr Jančar*

Center for Applied Cybernetics,
Dept. of Computer Science, Technical University of Ostrava,
17. listopadu 15, 708 33 Ostrava-Poruba, Czech Republic
petr.jancar@vsb.cz

Abstract. The paper tries to highlight some crucial ideas appearing in the decidability and undecidability proofs for the bisimilarity problem on models originating in language theory, like context-free grammars and pushdown automata. In particular, it focuses on the method of finite bases of bisimulations in the case of decidability and the method of "Defender's forcing" in the case of undecidability. An intent was to write an easy-to-read article in a slightly informal way, which should nevertheless convey the basic ideas with sufficient precision.

Keywords: bisimulation equivalence, decidability.

1 Introduction

In concurrency theory, process theory, theory of reactive systems etc., bisimilarity has been established as a fundamental behavioural equivalence (see, e.g., [1]).

Bisimilarity, also called bisimulation equivalence, is a finer relation than classical language equivalence, and serves as a basis for expressing when two systems (e.g., a specification and an implementation) exhibit the same behaviour. One natural research topic is thus the decidability and complexity questions for deciding bisimilarity on various models, including those well established in the theory of languages and automata. A survey of results and techniques in this area appears in [2]; some later results are summarized in [3] and [4] (the latter being regularly updated on the web).

This text is not meant as another survey. The author's aim has been just to highlight some (subjectively) selected ideas and techniques used in the area, mainly concentrating on the models originating in language theory. An intent was to write an easy-to-read article in a slightly informal way, which should nevertheless convey some basic ideas with sufficient precision.

We will start by defining the bisimulation equivalence via its characterization in terms of two-player games, played between Attacker and Defender. We note that at finite state systems (coinciding with nondeterministic finite automata, in

---

* Supported by Grant No. 1M0567 of the Ministry of Education of the Czech Republic.

M. Ito and M. Toyama (Eds.): DLT 2008, LNCS 5257, pp. 56–71, 2008.

fact), the bisimilarity problem is polynomial (though PSPACE-complete for language equivalence). Then we consider processes generated by context-free grammars, with Greibach normal form type of rules $X \xrightarrow{a} \alpha$ (α being a sequence of variables). These grammars yield the class BPA (Basic Process Algebra) – in the case when concatenation is viewed as sequential composition; the class BPP (Basic Parallel Processes) arises when concatenation is viewed as parallel (commutative) operation. In both cases (for BPA and BPP) we show the decidability of bisimilarity, viewing the proofs as instances of a general decidability scheme.

Pushdown automata can be viewed as finite collections of rules $\alpha \xrightarrow{a} \beta$, yielding the class of processes denoted PDA. When the same collections of rules are used in the parallel (i.e., commutative) setting, we get the class PN of processes generated by Petri nets. Here we just recall the decidability of bisimilarity for PDA (and the connection to the famous problem of language equivalence for deterministic PDA), and the undecidability for PN. We then demonstrate the so called Defender's forcing method on a recent undecidability result, for (Type -1) systems which constitute a generalization of PDA, having the rules (or rule schemes) of the type $R \xrightarrow{a} \alpha$ where R is a regular language (such a rule represents the rules $\beta \xrightarrow{a} \alpha$ for all $\beta \in R$).

We finish by mentioning some open problems.

2 Bisimulation Equivalence

Our central notion, bisimulation equivalence, also called bisimilarity, is defined on (the semantic model called) labelled transition systems; these can be viewed as nondeterministic automata, possibly with infinitely many states.

A *labelled transition system* (LTS) is a triple $(S, \mathcal{A}, \longrightarrow)$, where S is a set of *states*, $\mathcal{A}$ is a set of *actions*, and $\longrightarrow \subseteq S \times \mathcal{A} \times S$ is a *transition relation*. We write $s \xrightarrow{a} s'$ instead of $(s, a, s') \in \longrightarrow$ and we extend this notation to sequences $w \in \mathcal{A}^*$ in the natural way. We write $s \longrightarrow s'$ if there is $a \in \mathcal{A}$ such that $s \xrightarrow{a} s'$ and $s \longrightarrow^* s'$ if $s \xrightarrow{w} s'$ for some $w \in \mathcal{A}^*$ (i.e., s' is reachable from s).

We can thus see that in the finite case, when the sets $S, \mathcal{A}$ are finite, an LTS is just an NFA, a nondeterministic finite automaton, without specified initial and accepting states.

Given an LTS $(S, \mathcal{A}, \longrightarrow)$, we now aim at defining when a binary relation $\mathcal{R} \subseteq S \times S$ is a bisimulation (relation). We do this by help of a certain functional $\mathcal{F} : 2^{S \times S} \to 2^{S \times S}$, defined in terms of a simple ("one-round") game between Attacker and Defender; to make later discussions easier, we consistently view Attacker as "him" and Defender as "her".

Given a relation $\mathcal{R} \subseteq S \times S$, $\mathcal{F}(\mathcal{R})$ is defined as the set of pairs $(s, t) \in S \times S$ for which Defender surely wins in the following game: Attacker "performs" a move $s \xrightarrow{a} s'$, or $t \xrightarrow{a} t'$ (for some action a); if there is no such move (there are no outgoing arcs from s or t), Attacker loses (and Defender wins). After Attacker's move (when it was possible) there is Defender's turn and she has to "perform a matching move" by choosing a transition from the other state ($t \xrightarrow{a} t'$ when Attacker performed $s \xrightarrow{a} s'$, and $s \xrightarrow{a} s'$ when Attacker performed $t \xrightarrow{a} t'$), under *the same label a*, so that the resulting pair (s', t') is in $\mathcal{R}$; if she can do this, she wins, if she has no such matching move, she loses.

We now define that a relation $\mathcal{R} \subseteq S \times S$ is a *bisimulation* if it is a postfixpoint of $\mathcal{F}$, i.e., if $\mathcal{R} \subseteq \mathcal{F}(\mathcal{R})$.

We note that when $\mathcal{R}$ is a bisimulation then Defender has a winning strategy from every $(s, t) \in \mathcal{R}$ in the (possibly infinite) *iteration of the above simple game*: if Defender has won the first round, the second round starts from the resulting pair (s', t') (where Attacker can freely choose any move from s' or t'), etc. Since $\mathcal{R} \subseteq \mathcal{F}(\mathcal{R})$, Defender can maintain that all the pairs (s'', t'') reached after her moves (in the iterated game) belong to $\mathcal{R}$. Any play of this iterated game (from $(s, t) \in \mathcal{R}$) thus either finishes with a lose of Attacker in some m-th round, or goes forever, in which case Defender is the winner.

Mapping $\mathcal{F}$ is clearly monotonic ($\mathcal{R} \subseteq \mathcal{R}'$ implies $\mathcal{F}(\mathcal{R}) \subseteq \mathcal{F}(\mathcal{R}')$), and there is thus the greatest (post)fixpoint of $\mathcal{F}$, the union of all postfixpoints (i.e., of all bisimulations). This greatest fixpoint, denoted $\sim$, is the *bisimulation equivalence*, or *bisimilarity* (on S in the LTS $(S, \mathcal{A}, \longrightarrow)$). Thus

$$\sim = \bigcup \{\mathcal{R} \mid \mathcal{R} \subseteq \mathcal{F}(\mathcal{R})\} \,.$$

(Reflexivity, symmetricity, and transitivity of $\sim$ can be easily checked.)

Thus two states s and t are *bisimulation equivalent (bisimilar)*, written $s \sim t$, iff they are related by some bisimulation. We note that we can also relate states of different LTSs, viewing them as states in the disjoint union of those LTSs.

What does it mean when $s \not\sim t$? It is not difficult to realize that then Attacker has a winning strategy, WS for short, in the above iterated game, even when there is no constraint on the resulting pairs: Defender is allowed to match each move so that the result (s', t') is in $S \times S$; she can only lose when she has no outgoing arc with the relevant label from the relevant state (in some m-th round). When Attacker applies his WS, each play is finite and finishes with his win (where Defender is unable to match his move).

A strategy of Attacker (analogously for Defender) from a pair (s_0, t_0) can be naturally viewed as a tree with the root labelled with (s_0, t_0): any branch (a sequence of nodes labelled with pairs (s', t')) corresponds to a play; each node in which it is Attacker's turn has just one successor, and each node in which it is Defender's turn has successors corresponding to all (legal) moves of Defender. If the underlying LTS is *image finite*, i.e., for every state s and action a there are only finitely many states s' such that $s \xrightarrow{a} s'$, then a WS of Attacker corresponds to a finitely branching tree where every branch is finite – so the whole tree is finite then (recalling König's lemma). In such a case there is even a bound $m \in \mathbb{N}$ such that Attacker can guarantee his win within first m rounds.

In other words, at image finite systems we have $\sim = \bigcap_{m \in \mathbb{N}} \mathcal{F}^{(m)}(S \times S)$, or

$$\sim = \bigcap_{m \in \mathbb{N}} \sim_m$$

where $s \sim_m t$ means that Defender can "survive" at least m rounds in the unconstrained iterated game. ($\mathcal{F}^{(m)}(\mathcal{R})$ is a shorthand for $\mathcal{F}(\mathcal{F}(\ldots \mathcal{F}(\mathcal{R})))$ where

$\mathcal{F}$ is applied m times.) We can easily note that each $\sim_m$ is also an equivalence relation.

Remark. Our unconstrained iterated game is usually called the *bisimulation game*; such game-theoretic characterizations of behavioural equivalences are standard in this area.

3 Rewrite Systems as LTSs Generators

In theory of processes, it is usual to define the syntax of a language for describing systems and/or their specifications as terms (over some variables and operations) of an appropriate algebra, and to provide semantics by attaching labelled transition systems to these terms; the transitions are often defined by structural induction, by structural operational semantics (SOS) rules. Among usual ingredients of such algebras are the operations of *sequential composition* and *parallel composition*. There are usually other operations, allowing communication between parallel components etc., but we only concentrate on these basic operations here.

Our main (syntactic) systems which generate possibly infinite-state LTSs are the following rewrite systems (which can be viewed as simple process algebras).

By a *rewrite system* we understand a structure $(\mathcal{V}, \mathcal{A}, \Delta)$ where $\mathcal{V}$ is a finite set of *variables*, $\mathcal{A}$ a finite set of *actions* (corresponding to terminals in the usual grammar setting), and Δ a finite set of *(basic rewrite) rules* $\alpha \xrightarrow{a} \beta$ where $\alpha \in \mathcal{V}^+$, $\beta \in \mathcal{V}^*$, and $a \in \mathcal{A}$. We usually identify Δ with the whole rewrite system (meaning that $\mathcal{V}, \mathcal{A}$ are determined by Δ).

We note that if the left-hand side of each rule $\alpha \xrightarrow{a} \beta$ in Δ is a variable ($\alpha = X \in \mathcal{V}$), we get an analogue of a context-free grammar in Greibach normal form.

System Δ generates the LTS $\mathcal{L}_\Delta$ where the set od states is $\mathcal{V}^*$ and the transitions are induced by the following (deduction, or SOS) rule.

$$\frac{(\alpha \xrightarrow{a} \beta) \in \Delta,\ \gamma \in \mathcal{V}^*}{\alpha\gamma \xrightarrow{a} \beta\gamma} \tag{1}$$

This rule can be aptly called the prefix-rewriting rule. In the case of a context-free grammar (in Greibach normal form), possible evolutions of any $\delta \in \mathcal{V}^*$ correspond to the left derivations; in other words, the operation of concatenation is here viewed as *sequential* composition.

The introduced (sequential) rewrite systems are also called *PDA-systems* (or pushdown systems), constituting the class denoted PDA, and generating PDA-LTSs (or PDA-graphs). The subclass generated by context-free grammars (in Greibach normal form), is called BPA (Basic Process Algebra).

In fact, one would expect that $\mathcal{V}$ in a PDA-system is the disjoint union $Q \cup \Gamma$ for some set Q of control states and a set Γ of stack symbols, that the (pushdown) rules are of the form $pX \xrightarrow{a} q\alpha$, where $p, q \in Q$, $X \in \Gamma$ and $\alpha \in \Gamma^*$, and that the transitions are captured by the following deduction rule.

$$\frac{(pX \xrightarrow{a} q\alpha) \in \Delta, \ \beta \in \Gamma^*}{pX\beta \xrightarrow{a} q\alpha\beta}$$

But it is not hard to show that the LTSs generated by systems with these special rules are isomorphic with the LTSs generated by the general rules $\alpha \xrightarrow{a} \beta$ (see [5]).

Remark. We do not consider ε-moves (of usual pushdown automata) at the moment.

In the case when concatenation is taken as *parallel* composition, and is thus viewed as a *commutative* operation, a sequence α can be conveniently identified with its Parikh image, the vector $(k_1, k_2, \ldots, k_m) \in \mathbb{N}^m$ where $\mathcal{V} = \{X_1, X_2, \ldots, X_m\}$ and k_i is the number of occurrences of X_i in α (for $i = 1, 2, \ldots, m$).

In this parallel setting, we can still use (an analogue of) the SOS-rule (1) where the conclusion can be more conveniently expressed as $\alpha + \gamma \xrightarrow{a} \beta + \gamma$. Variables can now be seen as places in a Petri net, and (the Parikh image of) sequence α can be seen as the appropriate marking (multiset of variables). The basic rewrite rules naturally correspond to labelled Petri net transitions. We thus call these (parallel rewrite) systems as PN-systems, constituting the class PN. In the case of context-free grammars (in Greibach normal form), the corresponding subclass of PN is called BPP (Basic Parallel Processes); possible evolutions of $\delta \in \mathcal{V}^*$ now correspond to *all* derivations (not only to the left ones).

We finish this definition section with the notion of *norm*. For $\alpha \in \mathcal{V}^*$, the norm of α, denoted $\|\alpha\|$, is the length of a shortest word $u \in \mathcal{A}^*$ such that $\alpha \xrightarrow{u} \varepsilon$ (where ε denotes the empty sequence); $\|\alpha\|$ is infinite, denoted $\|\alpha\| = \omega$, when there is no such word, in which case we also say that α is *unnormed*.

4   Some Ideas for Decidability

4.1   Finite LTSs

We have already noted that finite LTSs are, in fact, nondeterministic finite automata, NFA. They can be viewed as generated by rewrite systems with rules of the type $X \xrightarrow{a} Y$, $X \xrightarrow{a} \varepsilon$.

For NFA, (the problem of deciding) language equivalence is well known to be PSPACE-complete but we now easily observe that bisimilarity is polynomial.

Indeed, given a finite-state LTS A, with the state set S, we can perform the standard greatest fixpoint computation: starting with $\mathcal{R}_0 = S \times S$, we can compute $\mathcal{R}_0 \supseteq \mathcal{F}(\mathcal{R}_0) \supseteq \mathcal{F}(\mathcal{F}(\mathcal{R}_0))\ldots$ until a fixpoint, in fact $\sim$, is reached. Polynomiality is clear but we would have to go into more details when looking for more efficient implementations. Such algorithms were devised by Kanellakis and Smolka [6] and Paige and Tarjan [7]; they are also explained in [8].

We now turn our attention to infinite state LTSs (generated by rewrite systems).

4.2 A General Decidability Scheme for Infinite State Systems

We first note that the LTSs $\mathcal{L}_\Delta$ generated by our rewrite systems $(\mathcal{V}, \mathcal{A}, \Delta)$ are *image finite*; in fact, they are even *finitely branching* (the outdegree of every node of the corresponding graph is finite) since $\mathcal{A}$ is finite.

So we have $\sim = \bigcap_m \sim_m$, and a winning strategy of Attacker (if it exists) can be presented as a finite tree; thus *nonbisimilarity* is surely *semidecidable* for our rewrite systems: given Δ and a pair (α_0, β_0), we can generate all Attacker's finite strategy-trees from (α_0, β_0), and verify for each of them if it represents a WS for him). An idea of semidecidability (and thus decidability) of bisimilarity is to present a finite witness of a winning strategy of Defender, i.e., of a (possibly infinite) bisimulation. We now sketch a general scheme which can be used to this aim.

Suppose we have Δ (a finite set of rules), and a pair (α_0, β_0) of processes (i.e., states in $\mathcal{L}_\Delta$) for which we should decide if $\alpha_0 \sim \beta_0$. We know that this is the case iff there is a bisimulation $\mathcal{R}$ containing (α_0, β_0); unfortunately, all such bisimulations might be infinite.

By a *finite base* we mean a finite set B of pairs (α, β), which somehow generates a bigger (maybe infinite) set $G(B) \supseteq B$ so that the following is guaranteed:

- $\forall m : B \subseteq \sim_m \Rightarrow G(B) \subseteq \sim_m$,
- $B \subseteq \mathcal{F}(G(B))$, i.e., every move by Attacker from $(\alpha, \beta) \in B$ can be matched by Defender so that the resulting pair is in $G(B)$.

Proposition 1. *If B is a finite base then $B \subseteq \sim$.*

Proof. Suppose there is a pair $(\alpha, \beta) \in B$ such that $\alpha \sim_m \beta$ and $\alpha \not\sim_{m+1} \beta$, for the least m.

But then $\alpha' \sim_m \beta'$ for all pairs $(\alpha', \beta') \in G(B)$ and thus $\alpha'' \sim_{m+1} \beta''$ for all pairs $(\alpha'', \beta'') \in \mathcal{F}(G(B)) \supseteq B$ – a contradiction. $\square$

We are intentionally not specific regarding the (generating) mapping G but we can easily observe validity of the following *equivalence rule*: we always assume that $G(B)$ contains the (least) equivalence generated by B. (Recall that $\sim_m$ is an equivalence relation for each m.)

Let us now imagine an agent CLEVER who aims at constructing a finite base B containing (α_0, β_0) when $\alpha_0 \sim \beta_0$ (i.e., when $\alpha_0 \sim_m \beta_0$ for all m). He performs the following (nondeterministic) procedure in which he uses his remarkable abilities: given (α, β), he immediately recognizes if $\alpha \sim \beta$, given a set C, he recognizes if $(\alpha, \beta) \in G(C)$, etc. The procedure also uses a (partial) order on the set of pairs (derived from their structure) to which we refer when saying that (α, β) is *smaller than* (α', β'); this order is supposed to be well-founded (there are no infinite decreasing chains).

CLEVER initializes B (i.e., "a program variable containing a set of pairs") by (the value) $\{(\alpha_0, \beta_0)\}$. He successively includes new elements in B, declaring them as "unprocessed" ((α_0, β_0) is also initially unprocessed), and he either finishes with a finite set (a finite base, in fact) when all elements in B are processed, or works forever. More concretely, he repeatedly performs the following step:

- Take an "unprocessed" element (α, β) in B, guess a minimal (finite) "extension" set $E \subseteq \sim$, $E \cap B = \emptyset$, such that $(\alpha, \beta) \in \mathcal{F}(B \cup E)$ (E is rich enough to enable Defender to match all Attacker's moves from (α, β)), and declare (α, β) as processed.
- Consider "adding" each $(\alpha', \beta') \in E$ to B. This can mean a real including in B but there is also another ("clever nondeterministic") option (which includes (α', β') in $G(B)$):
  - if you find a finite set M of bisimilar pairs $(\alpha_1, \beta_1), (\alpha_2, \beta_2), \ldots, (\alpha_n, \beta_n)$ which are smaller than (α', β') and generate (α', β'), i.e. $(\alpha', \beta') \in G(M \cup B)$, then consider adding the pairs from M instead of (α', β') (and proceed recursively).

It is obvious that any completed finite run of the above procedure (finishing with all pairs processed) has, in fact, constructed a finite base. If CLEVER has no such run at his disposal then there is an infinite sequence of bisimilar pairs in which none can be generated from smaller pairs (by our unspecified G).

In some concrete cases it can be shown that such an infinite sequence cannot exist, which means that there must be a finite base. This then implies (semi)decidability of bisimilarity, under the condition that it is (semi)decidable whether $B \subseteq \mathcal{F}(G(B))$ for a given B; in this case we speak about an *effective finite base*. When we are even able to bound the size of such (smallest) base B (if it exists), we can get a complexity upper bound.

4.3 BPA (Sequential Context-Free Processes)

Language equivalence is well-known to be undecidable for context-free grammars, also for those having no redundant nonterminals, i.e., for those grammars which generate only BPA processes with finite norms. In the case of bisimilarity, Baeten, Bergstra, and Klop [9] showed the decidability for this class. This first proof was lengthy and technically nontrivial; later, simpler proof were developed which also work for the whole class BPA. We now sketch the idea behind the proof of Christensen, Hüttel, and Stirling [10], viewing it as an instance of our previously described general scheme. This means that we show the following theorem.

Theorem 2. *For any BPA system Δ and a pair (α_0, β_0) such that $\alpha_0 \sim \beta_0$ (in $\mathcal{L}_\Delta$), there is an effective finite base B containing (α_0, β_0).*

It is convenient (and harmless) to assume that every variable in Δ appears on the left-hand side of at least one rewrite rule; then surely $\alpha \not\sim \varepsilon$ when $\alpha \neq \varepsilon$.

Besides the previously discussed equivalence rule, it is sufficient to enhance "generator" G by the *congruence rule* based on the following proposition.

Proposition 3. *(In BPA,) if $\alpha \sim_m \alpha'$ and $\beta \sim_m \beta'$ then $\alpha\beta \sim_m \alpha'\beta'$.*

Proof. Starting from $(\alpha\beta, \alpha'\beta')$, Defender uses the strategy guaranteeing her to survive m rounds from (α, α'), until possibly β or β' is exposed; in fact, if this happens before the m-th round, both β and β' must be exposed simultaneously (the pair (β, β') is reached after Defender's response), and Defender can continue by using the appropriate strategy for (β, β').   □

Let us observe that if $\alpha = \beta X \gamma$ and $\|X\| = \omega$ (X is unnormed) then γ is irrelevant and can be safely omitted. So any α is tacitly assumed to have either finite norm or to be of the form βX where β is (the maximal) prefix with finite norm and X is an unnormed variable; $\|\beta\|$ is then viewed as the *prefix-norm* of α.

We can primarily order processes α according to their prefix-norms, and secondarily (those with the same prefix-norm) lexicographically, say. We thus get a linear well-ordering which can be extended to pairs: e.g., $(\alpha, \beta) \leq (\alpha', \beta')$ iff $max\{\alpha, \beta\} < max\{\alpha', \beta'\}$ or $max\{\alpha, \beta\} = max\{\alpha', \beta'\}$ and $min\{\alpha, \beta\} \leq min\{\alpha', \beta'\}$.

The following proposition shows the theorem (CLEVER can always succeed).

Proposition 4. *There is no infinite sequence SEQ of bisimilar pairs such that none of them can be generated from smaller bisimilar pairs by a sequence of applications of the (equivalence and) congruence rule.*

Proof. We assume such SEQ exists. For some fixed "head-variables" X, Y, we can assume that the pairs in (an infinite subsequence of) SEQ are:

$$(X\alpha_1, Y\beta_1), (X\alpha_2, Y\beta_2), (X\alpha_3, Y\beta_3), \ldots \text{ (where } X\alpha_i \sim Y\beta_i \text{ for all } i).$$

We note that if α_i is smaller than α_j then $\alpha_i \nsim \alpha_j$: otherwise the smaller pairs (α_i, α_j) and $(X\alpha_i, Y\beta_j)$ generate $(X\alpha_j, Y\beta_j)$ (using $(X\alpha_i, X\alpha_j)$). This holds analogously for β_i, β_j, so we can even assume that SEQ also satisfies the following: either all α_i are the same or they create an increasing sequence and are pairwise nonbisimilar; the same holds for β_i.

Suppose first that α_i are all the same, equal to α_1; β_i are necessarily pairwise nonbisimilar then, and Y has finite norm. Consider now all pairs $(X\alpha_1, Y\beta_i)$ and suppose that in each of them Attacker starts with a sequence of moves $Y \overset{w}{\longrightarrow} \varepsilon$, thus exposing β_i (on the right-hand side). Defender necessarily matches this infinitely many times by installing the same γ ($X\alpha_1 \overset{w}{\longrightarrow} \gamma$) on the left-hand side; thus $\gamma \sim \beta_i$ for infinitely many i, which is a contradiction (β_i are pairwise nonbisimilar). Analogously, β_i cannot be the same either.

Thus both α_i and β_i are increasing, and both X, Y have finite norm; wlog we assume $\|X\| \geq \|Y\|$. We let Attacker start with a shortest sequence of moves $Y \overset{w}{\longrightarrow} \varepsilon$ in all $(X\alpha_i, Y\beta_i)$ and we thus discover that $\gamma\alpha_i \sim \beta_i$ for a fixed γ and infinitely many i (we can also have $\gamma = \varepsilon$).

So we can assume that we have chosen SEQ so that for all i: $X\alpha_i \sim Y\gamma\alpha_i$.

We cannot have $X \sim Y\gamma$ (where neccessarily $\|X\| = \|Y\gamma\|$, which implies $\|\gamma\| < \|X\|$): $(X\alpha_{i+1}, Y\beta_{i+1})$ is generated by the smaller pairs $(X, Y\gamma)$ and $(\beta_{i+1}, \gamma\alpha_{i+1})$ (using generated $(X\alpha_{i+1}, Y\gamma\alpha_{i+1})$).

So $X \nsim Y\gamma$. Now assume Attacker starts playing his WS for $(X, Y\gamma)$ in $(X\alpha_i, Y\gamma\alpha_i)$ while Defender plays her WS for $(X\alpha_i, Y\gamma\alpha_i)$. During the play they necessarily expose α_i on one side but not on the other (when finishing a round); this happens within m rounds for some fixed m. This implies that $\alpha_i \sim \delta\alpha_i$ for a fixed *nonempty* δ and infinitely many i. For these i, $\alpha_i \sim \delta\alpha_i \sim \delta\delta\alpha_i \sim \delta\delta\delta\alpha_i \sim \cdots \sim \delta^\omega$. (Extending the LTS-semantics to infinite sequences is obvious.) This is a contradiction since α_i are pairwise nonbisimilar.   □

We have thus demonstrated the decidability of bisimilarity on the class BPA. It is not difficult to get some intuition that the size of the smallest possible base (of the discussed type), if there is any, can be bounded (by a function of the size of $\Delta, \alpha_0, \beta_0$) if we submerged into careful technical calculations (and developed further ideas).

Burkart, Caucal, and Steffen [11], in fact, did this; it seems that a close analysis of their algorithm would reveal a doubly exponential complexity bound (according to [2]).

In the normed case, where all variables have finite norm, the complexity results look much better. Based on so called "prime decomposition", Hirshfeld, Jerrum, and Moller [12] showed a polynomial algorithm with complexity in $O(n^{13})$. This was recently improved to $O(n^8\,polylog\,n)$ by Lasota and Rytter [13].

4.4   BPP (Parallel Context-Free Processes)

The decidability of bisimilarity on BPP (generated by context-free grammars where concatenation is commutative) was shown by Christensen, Hirshfeld, and Moller [14]; it can be presented as another instance of our general scheme.

Theorem 5. *For any BPP system Δ and a pair (α_0, β_0) such that $\alpha_0 \sim \beta_0$ (in $\mathcal{L}_\Delta$), there is an effective finite base B containing (α_0, β_0).*

We recall that each α is viewed as a (Parikh) vector $(k_1, k_2, \ldots, k_n) \in \mathbb{N}^n$ where $\mathcal{V} = \{X_1, X_2, \ldots, X_n\}$ and k_i is the number of occurrences of X_i in α. Validity of the *congruence rule* is again obvious.

Proposition 6. *(In BPP,) if $\alpha \sim_m \alpha'$ and $\beta \sim_m \beta'$ then $\alpha + \beta \sim_m \alpha' + \beta'$.*

(Defender matches Attacker's move from α (β) by the relevant move from α' (β') and vice versa.)

We now say that a pair (α_1, β_1) is smaller than (α_2, β_2), denoted $(\alpha_1, \beta_1) <_{lex} (\alpha_2, \beta_2)$, if it is lexicographically smaller as a vector in $\mathbb{N}^{2n}$ (the leftmost component in which the vectors differ decides the order); the well-foundedness can be easily checked.

It is also useful to recall the following simple fact (which can be easily established by induction on m).

Proposition 7. *(Dickson's lemma) In every infinite sequence $v_1, v_2, \ldots$ of vectors from $\mathbb{N}^m$ there is $i < j$ such that $v_i \leq v_j$ (where $\leq$ is taken componentwise).*

Now we are ready to show the theorem (CLEVER can always succeed).

Proposition 8. *There is no infinite sequence SEQ of bisimilar pairs such that none of them can be generated from smaller bisimilar pairs by a sequence of applications of the (equivalence and) congruence rule.*

Proof. Suppose such SEQ (α_1, β_1), (α_2, β_2), ..., where we can assume (i.e., switch the components so) that $\alpha_i <_{lex} \beta_i$ for all i.

Dickson's lemma implies $(\alpha_i, \beta_i) \leq (\alpha_j, \beta_j)$ for some $i < j$; since $(\alpha_i, \beta_i) \neq (\alpha_j, \beta_j)$ we thus have $(\alpha_i, \beta_i) <_{lex} (\alpha_j, \beta_j)$.

But this is impossible since the smaller pairs (α_i, β_i) and $(\alpha_j, \beta_j - \beta_i + \alpha_i)$ generate (α_j, β_j).

(We note that (α_i, β_i) generates $(\alpha_i + (\beta_j - \beta_i), \beta_i + (\beta_j - \beta_i))$ and thus $(\beta_j, \beta_j - \beta_i + \alpha_i)$.) □

Remark. In principle, we have thus also shown a short proof that every congruence on a finitely generated commutative semigroup is finitely generated (see [2] for further references).

Dickson's lemma guarantees finiteness but provides no upper bound directly.

An upper (and optimal) bound was provided in [15], by developing another method. A method of so called dd-functions enables to provide a succint semilinear description of the bisimulation equivalence for a given BPP system, showing that the problem is PSPACE-complete; PSPACE-hardness for both BPP and BPA was shown by Srba [16].

In the normed case, the bisimilarity problem for BPP is again polynomial; prime decomposition was used to establish this in [17] but dd-functions seem again more effective, enabling to derive $O(n^3)$ bound [18].

4.5 PDA

Sénizergues [19] showed by an involved construction that bisimilarity is decidable on the whole class PDA (in fact, on a slightly more general class). This was achieved by elaboration on his techniques developed for solving the famous question of decidability for DPDA (deterministic PDA) language equivalence [20]. Stirling found a shorter exposition of the result for DPDA [21] on which he based a shorter proof for bisimilarity on PDA as well [22].

Remark. The solution of the DPDA problem still remains a bit "mysterious". Sénizergues published a simplified version in [23], while Stirling published another simplified version in [24] (which also provides a complexity upper bound). In principle, the proofs can be seen as technically more demanding instances of the general scheme. (I will not try to sketch them in this text but I plan to give an overview in the conference talk.)

5   Some Ideas for Undecidability

In this section we sketch a method, called Defender's forcing, which is useful for showing undecidability (or complexity lower bounds); for illustration we show undecidability of bisimilarity for so called Type -1 systems (by which an open question of Sénizergues and Stirling has been answered). The presentation is based on [25] where further results and references can be found.

5.1   A Variant of Post Correspondence Problem

We start by defining a variant of Post's Correspondence Problem (PCP) which will serve us for the illustrative reduction.

A *PCP-instance* INST is here defined as a nonempty sequence $(u_1, v_1), (u_2, v_2), \ldots, (u_n, v_n)$ of pairs of nonempty words over the alphabet $\{A, B\}$ where $|u_i| \leq |v_i|$ for all $i \in \{1, 2, \ldots, n\}$ ($|u|$ denoting the length of u).

An *infinite initial solution* of a given PCP-instance is an infinite sequence of indices $i_1, i_2, i_3, \ldots$ from the set $\{1, 2, \ldots, n\}$ such that $i_1=1$ and the infinite words $u_{i_1} u_{i_2} u_{i_3} \cdots$ and $v_{i_1} v_{i_2} v_{i_3} \cdots$ are equal.

By *inf-PCP* we denote the problem to decide whether a given PCP-instance has an infinite initial solution.

Proposition 9. *Problem inf-PCP is Π_1^0-complete (i.e., its complement is equivalent to the halting problem).*

Remark. Our requirement $|u_i| \leq |v_i|$ is non-standard but it can be easily checked to be harmless for the validity of Proposition 9 (as follows directly from an inspection of the standard textbook reduction of the halting problem to PCP); we use this for technical convenience.

We note the following useful fact. By a *partial solution* of a PCP-instance $(u_1, v_1), (u_2, v_2), \ldots, (u_n, v_n)$ we mean a finite sequence $i_1, i_2, i_3, \ldots, i_\ell$ such that $u_{i_1} u_{i_2} \ldots u_{i_\ell}$ is a prefix of $v_{i_1} v_{i_2} \ldots v_{i_\ell}$. For a PCP-instance, a sequence $i_1, i_2, i_3, \ldots$ of indices where $i_1 = 1$ is an infinite initial solution iff for each ℓ the sequence $i_1, i_2, i_3, \ldots, i_\ell$ is a partial solution.

5.2   Type -1 Systems

Type -1 systems (in terminology used by Stirling) generalize the sequential rewrite systems by allowing the rewrite rules of the type $R \xrightarrow{a} \alpha$ where $a \in \mathcal{A}$, $\alpha \in \mathcal{V}^*$, and R is a regular language over $\mathcal{V}$ such that $\varepsilon \notin R$. (For concreteness, we can assume that R is given by a regular expression.)

A Type -1 system Δ defines the LTS $\mathcal{L}_\Delta$ where $\mathcal{V}^*$ is the set of states, and the deduction rule showing transitions in $\mathcal{L}_\Delta$ is the following.

$$\frac{(R \xrightarrow{a} \alpha) \in \Delta, \quad \beta \in R, \quad \gamma \in \mathcal{V}^*}{\beta\gamma \xrightarrow{a} \alpha\gamma}$$

Thus any rule $(R \xrightarrow{a} \alpha) \in \Delta$ represents possibly infinitely many (basic) rewrite rules $\beta \xrightarrow{a} \alpha$ where $\beta \in R$. We note that the out-degree of each node in $\mathcal{L}_\Delta$ is still finite, though the in-degree can be infinite.

5.3   A Reduction of inf-PCP to Bisimilarity on Type -1 Systems

Let us consider a fixed instance INST of inf-PCP, i.e., a sequence of pairs $(u_1, v_1), (u_2, v_2), \ldots, (u_n, v_n)$ over the alphabet $\{A, B\}$; symbols $I_1, I_2, \ldots, I_n$ will represent the indices $1, 2, \ldots, n$. We can imagine the following game: Starting with the one-element sequence I_1, Attacker repeatedly asks Defender to prolong the current sequence $I_{i_1} I_{i_2} \ldots I_{i_\ell}$ (where $i_1 = 1$) by one I_i (of her choice); this can be viewed as a *generating phase* of the game, and if Attacker lets go this phase forever, he loses. But Attacker has always the possibility to *switch* to a *verification phase* in which it is checked whether the current sequence $I_{i_1} I_{i_2} \ldots I_{i_\ell}$ represents a partial solution (i.e., whether $u_{i_1} u_{i_2} \ldots u_{i_\ell}$ is a prefix of $v_{i_1} v_{i_2} \ldots v_{i_\ell}$); the negative case is a win for Attacker, the positive case is a win for Defender. It is obvious that INST has an (infinite initial) solution iff Defender has a WS in the game.

We now want to implement the described game as the bisimulation game starting with the pair $(q_0 I_1 \bot, q'_0 I_1 \bot)$ of processes (states) of the below described Type -1 system.

Since we use (sequential) prefix rewriting, it is convenient to represent the current sequences of indices in the reversed order, as $I_{i_\ell} I_{i_{\ell-1}} \ldots I_{i_1} \bot$, and prolong them to the left. The special symbol $\bot$ is used as an endmarker (the "bottom-of-the-stack symbol") which here just helps to guarantee normedness (as discussed later).

In (a play of) the bisimulation game, starting from $(q_0 I_1 \bot, q'_0 I_1 \bot)$, the players will be creating two copies of a sequence of indices; one copy is intended to be interpreted over u_i's, the other over v_i's (in the verification phase).

The main problem is that we have to arrange that it is Defender who decides by which symbol I_i will both sequences be prolonged (during the generating phase).

To this aim, we use a general idea, the essence of which is sketched on Figure 1.

Imagine we start from a pair (α, β) and there are two "legal" next-step pairs, (α_1, β_1) and (α_2, β_2). (For example, (α, β) can represent two slightly differing copies of a configuration C during a (nondeterministic) computation, where C

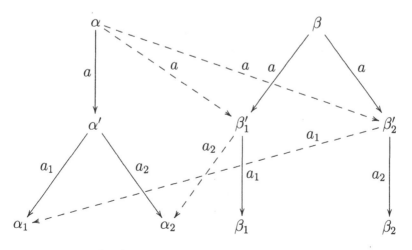

Fig. 1. From (α, β) Defender chooses and forces (α_1, β_1) or (α_2, β_2)

has two possible one-step successors C_1, C_2 – these are represented by (α_1, β_1) and (α_2, β_2), respectively.) The drawn (fragment of an) LTS guarantees that it is Defender who can freely choose one of those pairs:

Attacker is, in fact, forced to play $\alpha \xrightarrow{a} \alpha'$ since otherwise Defender can install a pair with equal components – an obvious win for her. So Defender's turn is in the (intermediate) pair (α', β), and she freely chooses $i \in \{1, 2\}$ and plays $\beta \xrightarrow{a} \beta'_i$. In the resulting pair (α', β'_i), Attacker is forced to use the action a_i (otherwise Defender installs an equal-component pair), and the only answer of Defender then installs the pair (α_i, β_i).

Schematically, we can express this by the following "rules" (where i, j range over $\{1, 2, \ldots, n\}$ for some n).

$$\alpha \xrightarrow{a} \alpha'$$
$$\boxed{\alpha \xrightarrow{a} \beta'_i} \qquad \beta \xrightarrow{a} \beta'_i$$

$$\alpha' \xrightarrow{a_i} \alpha_i \qquad \beta'_i \xrightarrow{a_i} \beta_i$$
$$\boxed{\beta'_i \xrightarrow{a_j} \alpha_j} \text{ for } i \neq j$$

(The rule $\alpha \xrightarrow{a} \beta'_i$ thus stands for the n rules $\alpha \xrightarrow{a} \beta'_1$, $\alpha \xrightarrow{a} \beta'_2$, ..., $\alpha \xrightarrow{a} \beta'_n$, the rule $\beta'_i \xrightarrow{a_j} \alpha_j$, $i \neq j$, stands for $n(n-1)$ rules like $\beta'_1 \xrightarrow{a_2} \alpha_2$, $\beta'_8 \xrightarrow{a_5} \alpha_5$, etc.)

By using frames we have highlighted the use of Defender's forcing (DF); Attacker must make sure that the framed rules (corresponding to the dashed arrows in Fig. 1) are never used (neither by him or her) since otherwise Defender can install a pair with equal components — an obvious win for her.

We now give all the rules of our particular Type -1 system, and then explain them in detail; the frames highlight the use of Defender's forcing.

Notation. We let I^* stand for the regular expression $(I_1 + I_2 + \cdots + I_n)^*$. By u^R we denote the reverse image of u. By $head(w)$ we denote the first symbol of w; $tail(w)$ is the rest of w. By $h(w)$ (head-action) we mean a if $head(w) = A$, and b if $head(w) = B$. Subscripts i, j range over $\{1, 2, \ldots, n\}$; thus the rule $q_0 \xrightarrow{g} p_i$ stands for the n rules $q_0 \xrightarrow{g} p_1$, $q_0 \xrightarrow{g} p_2$, ..., $q_0 \xrightarrow{g} p_n$, the rule $p_i \xrightarrow{a_j} q_0 I_j$, $i \neq j$, stands for $n(n-1)$ rules like $p_1 \xrightarrow{a_2} q_0 I_2$, $p_8 \xrightarrow{a_5} q_0 I_5$, etc.

(G1) rules: $q_0 \xrightarrow{g} t$
$$\boxed{q_0 \xrightarrow{g} p_i} \qquad q'_0 \xrightarrow{g} p_i$$

$$t \xrightarrow{a_i} q_0 I_i \qquad p_i \xrightarrow{a_i} q'_0 I_i$$
$$\boxed{p_i \xrightarrow{a_j} q_0 I_j} \text{ where } i \neq j$$

(S1) rules: $q_0 \xrightarrow{s} q_u$
$$\boxed{q_0(I^*)I_i \xrightarrow{s} q_v w} \qquad q'_0(I^*)I_i \xrightarrow{s} q_v w \quad \text{for all suffixes } w \text{ of } v_i^R$$

(V1) rules: $\quad q_u I_i \xrightarrow{h(u_i^R)} q_u \ tail(u_i^R) \qquad q_v I_i \xrightarrow{h(v_i^R)} q_v \ tail(v_i^R)$
$$q_u A \xrightarrow{a} q_u \qquad\qquad\qquad\quad q_v A \xrightarrow{a} q_v$$
$$q_u B \xrightarrow{b} q_u \qquad\qquad\qquad\quad q_v B \xrightarrow{b} q_v$$
$$q_u \bot \xrightarrow{e} \varepsilon \qquad\qquad\qquad\quad\ q_v \bot \xrightarrow{e} \varepsilon$$

To show that $q_0 I_1 \bot \sim q_0' I_1 \bot$ iff INST has an (infinite initial) solution, we first assume that there is such a (fixed) solution $i_1, i_2, i_3, \ldots$ and describe a WS for Defender. The play starts with the pair $(q_0 I_1 \bot, q_0' I_1 \bot)$, and as long as Attacker uses (G1)-rules (the generating phase), Defender forces that the play goes through longer and longer pairs

$$(q_0 I_{i_\ell} I_{i_{\ell-1}} \ldots I_{i_1} \bot, \ q_0' I_{i_\ell} I_{i_{\ell-1}} \ldots I_{i_1} \bot) \quad (*)$$

where $i_1 = 1$ and $I_{i_1}, I_{i_2}, \ldots, I_{i_\ell}$ represents a prefix of the assumed (fixed) solution $i_1, i_2, i_3, \ldots$. We observe that Defender can guarantee this since it is her who chooses $I_{i_2}, I_{i_3}, \ldots$.

Hence if Attacker wants to win, he has to *switch* (from generating to verification), i.e., to use (S1)-rules in some pair $(*)$; he is then forced to use $q_0 \xrightarrow{s} q_u$. Defender answers by shortening the "right-hand side" sequence so that the resulting pair

$$(q_u I_{i_\ell} I_{i_{\ell-1}} \ldots I_{i_1} \bot, \ q_v w \, I_{i_m} I_{i_{m-1}} \ldots I_{i_1} \bot) \quad (**)$$

satisfies

$$(u_{i_\ell})^R (u_{i_{\ell-1}})^R \ldots (u_{i_1})^R = w \, (v_{i_m})^R (v_{i_{m-1}})^R \ldots (v_{i_1})^R . \qquad (2)$$

Finally the (deterministic) (V1)-rules clearly show that Defender wins.

If INST has no solution then there is an obvious WS for Attacker. He repeatedly uses (G1) until a pair $(*)$ which does not correspond to a partial solution appears. This will eventually happen. Then Attacker switches, using $q_0 \xrightarrow{s} q_u$, and after Defender's response we must get a pair $(**)$ where the condition (2) does not hold. Thus the following verification phase is clearly winning for Attacker.

The processes $q_0 I_1 \bot, q_0' I_1 \bot$ (of the Type -1 system (G1), (S1), (V1)) are obviously normed, i.e., each process (state) reachable from them has finite norm (can be rewritten to ε). We have thus shown the following theorem.

Theorem 10. *The bisimilarity problem for (normed) Type -1 systems is Π_1^0-complete.*

5.4 Petri Nets

For completeness, we can mention that the bisimilarity problem on PN (Petri net processes) is also Π_1^0-complete. This can be also easily shown by using Defender's forcing; to this aim, the nonhalting problem for Minsky counter machines is more convenient than the (sequential) inf-PCP. (We refer the reader to [25] and the references there.)

6 Some Open Problems

One open decidability question, closely related to the discussed problems, concerns so called PA (Process Algebra) processes; the rewrite rules are $X \xrightarrow{a} \alpha$ where α is a term created from variables by (a mixture) of sequential and parallel composition. Hirshfeld and Jerrum [26] showed a procedure working in doubly-exponential nondeterministic time for the normed PA but the question is open for the general class PA.

Remark. A simple subcase of this problem is the "BPA vs. BPP" problem, i.e., deciding if a given BPA process is bisimilar with a given BPP process. In the normed case, an exponential algorithm is shown in [27]; this has been recently improved by showing a polynomial algorithm in [28]. In the general BPA vs. BPP problem, just decidability is known so far [29].

The situation regarding decidability of *weak bisimilarity*, which abstracts away from silent (internal) actions, is much worse. The problem is then highly undecidable for PDA, PA, PN (we again refer to [25]). The decidability questions are open for both BPA and BPP.

References

1. Milner, R.: Communication and Concurrency. Prentice-Hall, Englewood Cliffs (1989)
2. Burkart, O., Caucal, D., Moller, F., Steffen, B.: Verification on infinite structures. In: Bergstra, J., Ponse, A., Smolka, S. (eds.) Handbook of Process Algebra, pp. 545–623. Elsevier Science, Amsterdam (2001)
3. Kučera, A., Jančar, P.: Equivalence-checking on infinite-state systems: Techniques and results. Theory and Practice of Logic Programming 6(3), 227–264 (2006)
4. Srba, J.: Roadmap of infinite results. In: Current Trends In Theoretical Computer Science, The Challenge of the New Century. Formal Models and Semantics, vol. 2, pp. 337–350. World Scientific Publishing Co., Singapore (2004), http://www.brics.dk/~srba/roadmap/
5. Caucal, D.: On the regular structure of prefix rewriting. Theoretical Computer Science 106(1), 61–86 (1992)
6. Kanellakis, P., Smolka, S.: CCS expressions, finite state processes, and three problems of equivalence. Information and Computation 86(1), 43–68 (1990)
7. Paige, R., Tarjan, R.E.: Three partition refinement algorithms. SIAM Journal on Computing 16(6), 973–989 (1987)
8. Cleaveland, R., Sokolsky, O.: Equivalence and preorder checking for finite-state systems. In: Bergstra, J., Ponse, A., Smolka, S. (eds.) Handbook of Process Algebra, pp. 391–424. Elsevier Science, Amsterdam (2001)
9. Baeten, J., Bergstra, J., Klop, J.: Decidability of bisimulation equivalence for processes generating context-free languages. Journal of the ACM 40(3), 653–682 (1993)
10. Christensen, S., Hüttel, H., Stirling, C.: Bisimulation equivalence is decidable for all context-free processes. Information and Computation 121, 143–148 (1995)
11. Burkart, O., Caucal, D., Steffen, B.: An elementary decision procedure for arbitrary context-free processes. In: Hájek, P., Wiedermann, J. (eds.) MFCS 1995. LNCS, vol. 969, pp. 423–433. Springer, Heidelberg (1995)

12. Hirshfeld, Y., Jerrum, M., Moller, F.: A polynomial algorithm for deciding bisimilarity of normed context-free processes. Theoretical Computer Science 158, 143–159 (1996)
13. Lasota, S., Rytter, W.: Faster algorithm for bisimulation equivalence of normed context-free processes. In: Královič, R., Urzyczyn, P. (eds.) MFCS 2006. LNCS, vol. 4162, pp. 646–657. Springer, Heidelberg (2006)
14. Christensen, S., Hirshfeld, Y., Moller, F.: Bisimulation is decidable for all basic parallel processes. In: Best, E. (ed.) CONCUR 1993. LNCS, vol. 715, pp. 143–157. Springer, Heidelberg (1993)
15. Jančar, P.: Strong bisimilarity on Basic Parallel Processes is PSPACE-complete. In: Proc. 18th LiCS, pp. 218–227. IEEE Computer Society, Los Alamitos (2003)
16. Srba, J.: Strong bisimilarity of simple process algebras: Complexity lower bounds. Acta Informatica 39, 469–499 (2003)
17. Hirshfeld, Y., Jerrum, M., Moller, F.: A polynomial-time algorithm for deciding bisimulation equivalence of normed Basic Parallel Processes. Mathematical Structures in Computer Science 6, 251–259 (1996)
18. Jančar, P., Kot, M.: Bisimilarity on normed Basic Parallel Processes can be decided in time O(n3). In: Bharadwaj, R. (ed.) Proceedings of the Third International Workshop on Automated Verification of Infinite-State Systems - AVIS 2004 (2004)
19. Sénizergues, G.: The bisimulation problem for equational graphs of finite outdegree. SIAM Journal on Computing 34(5), 1025–1106 (2005); (a preliminary version appeared at FOCS 1998)
20. Sénizergues, G.: L(A)=L(B)? Decidability results from complete formal systems. Theoretical Computer Science 251(1-2), 1–166 (2001); (a preliminary version appeared at ICALP 1997)
21. Stirling, C.: Decidability of DPDA equivalence. Theoretical Computer Science 255(1-2), 1–31 (2001)
22. Stirling, C.: Decidability of bisimulation equivalence for pushdown processes. Research Report EDI-INF-RR-0005, School of Informatics, Edinburgh University, The latest version is downloadable from the author's home-page (January 2000)
23. Sénizergues, G.: L(A)=L(B)? a simplified decidability proof. Theoretical Computer Science 281(1-2), 555–608 (2002)
24. Stirling, C.: Deciding DPDA equivalence is primitive recursive. In: Widmayer, P., Triguero, F., Morales, R., Hennessy, M., Eidenbenz, S., Conejo, R. (eds.) ICALP 2002. LNCS, vol. 2380, pp. 821–832. Springer, Heidelberg (2002)
25. Jančar, P., Srba, J.: Undecidability of bisimilarity by defender's forcing. Journal of the ACM 55(1), 1–26 (2008)
26. Hirshfeld, Y., Jerrum, M.: Bisimulation equivalence is decidable for normed process algebra. In: Wiedermann, J., Van Emde Boas, P., Nielsen, M. (eds.) ICALP 1999. LNCS, vol. 1644, pp. 412–421. Springer, Heidelberg (1999)
27. Černá, I., Křetínský, M., Kučera, A.: Comparing expressibility of normed BPA and normed BPP processes. Acta Informatica 36, 233–256 (1999)
28. Jančar, P., Kot, M., Sawa, Z.: Normed BPA vs. normed BPP revisited. In: Proceedings of CONCUR 2008. LNCS. Springer, Heidelberg (to appear, 2008)
29. Jančar, P., Kučera, A., Moller, F.: Deciding bisimilarity between BPA and BPP processes. In: Amadio, R., Lugiez, D. (eds.) CONCUR 2003. LNCS, vol. 2761, pp. 159–173. Springer, Heidelberg (2003)

The Frobenius Problem and Its Generalizations

Jeffrey Shallit*

School of Computer Science, University of Waterloo
Waterloo, ON N2L 3G1, Canada
shallit@cs.uwaterloo.ca
http://www.cs.uwaterloo.ca/~shallit

1 Introduction

Let $x_1, x_2, \ldots, x_n$ be positive integers. It is well-known that every sufficiently large integer can be represented as a non-negative integer linear combination of the x_i if and only if $\gcd(x_1, x_2, \ldots, x_n) = 1$. The *Frobenius problem* is the following: given positive integers $x_1, x_2, \ldots, x_n$ with $\gcd(x_1, x_2, \ldots, x_n) = 1$, compute the largest integer *not* representable as a non-negative integer linear combination of the x_i. This largest integer is sometimes denoted $g(x_1, \ldots, x_n)$.

As an example, consider the following problem that appears frequently in books of puzzles (e.g., [24]):

The Chicken McNuggets Problem:

> At McDonald's, Chicken McNuggets are available in packs of either 6, 9, or 20 nuggets. What is the largest number of McNuggets that one cannot purchase?

The answer is $g(6, 9, 20) = 43$. To see that 43 is not representable, observe that we can choose either 0, 1, or 2 packs of 20. If we choose 0 or 1 or 2 packs, then we have to represent 43 or 23 or 3 as a linear combination of 6 and 9, which is impossible.

To see that every larger number is representable, note that

$$44 = 1 \cdot 20 + 0 \cdot 9 + 4 \cdot 6$$
$$45 = 0 \cdot 20 + 3 \cdot 9 + 3 \cdot 6$$
$$46 = 2 \cdot 20 + 0 \cdot 9 + 1 \cdot 6$$
$$47 = 1 \cdot 20 + 3 \cdot 9 + 0 \cdot 6$$
$$48 = 0 \cdot 20 + 0 \cdot 9 + 8 \cdot 6$$
$$49 = 2 \cdot 20 + 1 \cdot 9 + 0 \cdot 6$$

and every larger number can be written as a multiple of 6 plus one of these numbers.

In this survey, I will briefly discuss what is known about the Frobenius problem and then turn to a recent generalization of the problem to words.

---

* Research of this author supported in part by NSERC.

M. Ito and M. Toyama (Eds.): DLT 2008, LNCS 5257, pp. 72–83, 2008.

2 Brief History of the Frobenius Problem

The problem was discussed by Frobenius (1849–1917) in his lectures in the late 1800's — but apparently Frobenius never published anything. A related problem was discussed by Sylvester [23, p. 134] in 1882: he gave a formula for $h(x_1, x_2, \ldots, x_n)$, the total number of non-negative integers not representable as a linear combination of the x_i, in the case $n = 2$. The modern study of the Frobenius problem began with the 1942 paper of Brauer [3]. Applications of the Frobenius problem occur in number theory, automata theory, sorting algorithms, and many other areas. For a good survey of the Frobenius problem, see Ramírez-Alfonsín [15].

3 Research on the Frobenius Problem

Previous research on the Frobenius problem can be divided into four different areas:

1. Explicit formulas or algorithms for computing g when the dimension is bounded;
2. Upper and lower bounds for g;
3. Formulas for g in special cases;
4. Computational complexity of g.

3.1 Explicit Formulas for g

In the case where $n = 2$, we have

Theorem 1. $g(x, y) = xy - x - y$.

Proof. Suppose $xy - x - y$ is representable as $ax + by$. Then, taking the result modulo x, we have $-y \equiv by \pmod{x}$, so $b \equiv -1 \pmod{x}$. Similarly, modulo y, we get $-x \equiv ax$, so $a \equiv -1 \pmod{y}$. But then $ax + by \geq (y - 1)x + (x - 1)y = 2xy - x - y$, a contradiction. So $xy - x - y$ is not representable.

To prove every integer larger than $xy-x-y$ is representable, let $c = x^{-1} \bmod y$ and $d = y^{-1} \bmod x$. Then a simple calculation shows that $(c - 1)y + (d - 1)x = xy-x-y+1$, so this gives a representation for $g(x, y)+1$. To get a representation for larger numbers, we use the extended Euclidean algorithm to find integers e, f such that $ex - fy = 1$. We just add the appropriate multiple of this equation, reducing, if necessary, by $(-y)x + xy$ or $yx + (-x)y$ if a coefficient becomes negative. ∎

For example, for $[x, y] = [13, 19]$, we find $[2, 10] \cdot [x, y] = 216 = g(13, 19) + 1$. Also $[3, -2] \cdot [x, y] = 1$. To get a representation for 217, we just add these two vectors to get $[5, 8]$.

For 3 numbers, more complicated (but still polynomial-time) algorithms have been given by Greenberg [8] and Davison [5].

Kannan [10,11] has given a polynomial-time algorithm for any fixed dimension, but the time depends at least exponentially on the dimension and the algorithm is very complicated.

3.2 Upper and Lower Bounds for g

A simple upper bound can be obtained by dynamic programming [26].

Theorem 2. *If $a_1 < a_2 < \cdots a_n$, then $g(a_1, a_2, \ldots, a_n) < a_n^2$.*

Proof. Consider testing each number $0, 1, 2, \ldots$ in turn to see if it is representable as a non-negative integer linear combination.

Then r is representable if and only if at least one of $r - a_1, r - a_2, \ldots, r - a_n$ is representable. Now group the numbers in blocks of size a_n, and write a 1 if the number is representable, 0 otherwise. Clearly if j is representable, so is $j + a_n$, so each consecutive block has 1's in the same positions as the previous, plus maybe some new 1's. In fact, new 1's must appear in each consecutive block, until it is full of 1's, for otherwise the Frobenius number would be infinite. So we need to examine at most a_n blocks. Once a block is full, every subsequent number is representable. Thus we have shown $g(a_1, a_2, \ldots, a_n) < a_n^2$. ∎

Davison [5] found lower bounds for $g(x_1, x_2, x_3)$. For more general lower bounds, see [15, §3.6].

3.3 Formulas for g in Special Cases

Brauer [3] found an explicit formula for the Frobenius number in the case where the x_i are consecutive integers. This was generalized by Roberts [16] to the case where the x_i are in arithmetical progression.

3.4 Computational Complexity of g

Ramírez-Alfonsín [14] has proven that computing g is NP-hard under Turing-reductions, by reducing from the integer knapsack problem. (No NP-hardness result under Cook-reductions is currently known.)

The integer knapsack problem is

> Given $x_1, x_2, \ldots, x_n$, and a target t, do there exist non-negative integers a_i such that $\sum_{1 \leq i \leq n} a_i x_i = t$?

His reduction requires 3 calls to a subroutine for the Frobenius number g.

Despite this result, it is currently computationally feasible to solve the Frobenius problem for dimensions up to 13, even with very large inputs [2,6,17].

4   Applications of the Frobenius Number

The Frobenius number has applications to sorting; specifically, to the analysis of Shell sort, a sorting algorithm devised by D. Shell in 1959 [22]. The basic idea is to arrange the list to be sorted in an array with j columns for some j, then insertion sort each column, then decrease j and repeat.

For example, suppose we start with 10 5 12 13 4 6 9 11 8 1 7. We arrange the list in 5 columns:

$$10\ 5\ 12\ 13\ 4$$
$$6\ \ 9\ 11\ \ 8\ 1$$
$$7$$

Then insertion sort each column:

$$6\ \ 5\ 11\ \ 8\ 1$$
$$7\ \ 9\ 12\ 13\ 4$$
$$10$$

Next, we arrange the list in 3 columns:

$$6\ \ 5\ 11$$
$$8\ \ 1\ \ 7$$
$$9\ 12\ 13$$
$$4\ 10$$

Then we insertion sort each column:

$$4\ \ 1\ \ 7$$
$$6\ \ 5\ 11$$
$$8\ 10\ 13$$
$$9\ 12$$

Finally, we use insertion sort to sort the remaining elements:
1 4 5 6 7 8 9 10 11 12 13.

The running time of Shell sort depends on the increments used. In its original version, the increments were powers of 2, but this gives a quadratic running time. The running time decreases to $O(n^{3/2})$ if the increments $1, 3, 7, 15, 31, \ldots$ are used. (These numbers are the powers of 2, minus 1.) The running time decreases further to $O(n^{4/3})$ if the increments $1, 8, 23, 77, \ldots$ are used. (These numbers are those of the form $4^{j+1} + 3 \cdot 2^j + 1$). Finally, the running time is $O(n(\log n)^2)$ if the increments $1, 2, 3, 4, 6, 9, 8, 12, 18, 27, 16, 24, \ldots$ are used. (These numbers are those of the form $2^i 3^j$).

The following result links the Frobenius problem with Shell sort.

Theorem 3. *The number of steps required to r-sort a file $a[1..N]$ that is already $r_1, r_2, \ldots, r_t$-sorted is $\leq \frac{N}{r} g(r_1, r_2, \ldots, r_t)$.*

Proof. The number of steps to insert $a[i]$ is the number of elements in $a[i - r], a[i - 2r], \ldots$ that are greater than $a[i]$. But if x is a linear combination of $r_1, r_2, \ldots, r_t$, then $a[i - x] < a[i]$, since the file is $r_1, r_2, \ldots, r_t$-sorted. Thus the number of steps to insert $a[i]$ is $\leq$ the number of multiples of r that are not linear combinations of $r_1, r_2, \ldots, r_t$. This number is $\leq g(r_1, r_2, \ldots, r_t)/r$. ∎

For more details about the connection between the Frobenius number and Shell sort, see [9,18,25,19].

4.1   The Frobenius Problem and NFA to DFA Conversion

As is well-known, when converting an NFA of n states to an equivalent DFA via the subset construction, 2^n states are sufficient. What may be less well-known is that this construction is optimal in the case of a binary or larger input alphabet, in that there exist languages L that can be accepted by an NFA with n states, but no DFA with $< 2^n$ states accepts L. However, for unary languages, the 2^n bound is not attainable. It can be proved that approximately $e^{\sqrt{n \log n}}$ states are necessary and sufficient in the worst case to go from a unary n-state NFA to a DFA.

Chrobak [4] showed that any unary n-state NFA can be put into a certain normal form, where there is a "tail" of $< n^2$ states, followed by a single nondeterministic state which has branches into different cycles, where the total number of states in all the cycles is $\leq n$. The bound of n^2 for the number of states in the tail comes from the bound we have already seen on the Frobenius problem.

5   Related Problems

As we already have seen, Sylvester published a paper in 1882 where he defined $h(x_1, x_2, \ldots, x_n)$ to be the total number of integers not representable as an integer linear combination of the x_i. He also gave the formula $h(x_1, x_2) = \frac{1}{2}(x_1 - 1)(x_2 - 1)$.

There is a very simple proof of this formula: consider all the numbers between 0 and $(x_1 - 1)(x_2 - 1)$. Then it is not hard to see that every representable number in this range is paired with a non-representable number via the map $c \to c'$, where $c' = (x_1 - 1)(x_2 - 1) - c - 1$, and vice-versa.

However, the complexity of computing h is still open.

5.1   The Local Postage Stamp Problem

In this problem, we are given a set of denominations $1 = x_1, x_2, \ldots, x_k$ of stamps, and an envelope that can contain at most t stamps. We want to determine the *smallest* amount of postage we *cannot* provide. Call it $N_t(x_1, x_2, \ldots, x_k)$. For example, $N_3(1, 4, 7, 8) = 25$.

Many papers have been written about this problem, especially in Germany and Norway, and algorithms have been given for many special cases. Alter and Barnett asked [1] if $N_t(x_1, x_2, \ldots, x_k)$ can be "expressed by a simple formula". The answer is, probably not, because it is known that computing $N_t(x_1, x_2, \ldots, x_k)$ is NP-hard [20].

5.2   The Global Postage-Stamp Problem

The global postage-stamp problem is yet another variant: now we are given a limit t on the number of stamps to be used, and an integer k, and the goal is to find a set of k denominations $x_1, x_2, \ldots, x_k$ that maximizes $N_t(x_1, x_2, \ldots, x_k)$.

The computational complexity of this problem is unknown.

5.3   The Optimal Coin Change Problem

Yet another variant is the optimal change problem: here we are given a bound on the number of distinct coin denominations we can use (but allowing arbitrarily many of each denomination), and we want to find a set that minimizes the average number of coins needed to make each amount in some range.

For example, in the US we currently use 4 denominations less than $1 for change: 1¢, 5¢, 10¢, and 25¢. These can make change for every amount between 0¢ and 99¢, with an average cost of 4.7 coins per amount. It turns out that the system of 4 denominations $(1, 5, 18, 25)$ is optimal, with an average cost of only 3.89 coins per amount [21]. For Canada, where 1-dollar and 2-dollar coins are also in general circulation, the best 6-coin systems are $(1, 6, 14, 62, 99, 140)$ and $(1, 8, 13, 69, 110, 160)$, each of which give an expected 4.67 coins per transaction.

5.4   Improving the Current Coin System

One could also ask, what single denomination could we add to the current US system to improve its efficiency in making change? The answer is, add a 32¢ piece. For Canada, where 1-dollar and 2-dollar coins are also in general circulation, the best coin to add is an 83¢ piece [21].

Both Europe and China use a system of denominations based on the recurring pattern

$$1, 2, 5, \quad 10, 20, 50, \quad 100, 200, 500, \ldots$$

This may seem natural, but a small change to

$$1, 3, 4, \quad 10, 30, 40, \quad 100, 300, 400, \ldots$$

would decrease the average number of coins per transaction. This new system has the following advantages:

- change can still be made on a digit-by-digit basis. For example, to make change for 348, first do the hundreds digit (getting 300), then the tens (getting 40), and then the ones (getting 4+4).
- the greedy algorithm can be used in all cases but one. The exception is that $6 = 3+3$ and not $4+1+1$. (Similarly, $60 = 30+30$, etc.)
- assuming the uniform distribution of change denominations, on all scales (10, 100, 1000, etc.) the new system is about 6% better.
- if one assumes change denominations are distributed by Benford's law, the new system is about 7% better up to 10, about 6% better up to 100, and about 6% better up to 1000.

Japan currently uses coins based on the system

$$1, 5, \quad 10, 50, \quad 100, 500, \ldots$$

This could be improved by changing to

$$1, x, \quad 10, 10x, \quad 100, 100x, \ldots$$

where x is either 3 or 4.

6 Generalizing the Frobenius Problem to Words

Above we defined $g(x_1, x_2, \ldots, x_k)$ to be the largest integer not representable as a non-negative integer linear combination of the x_i. We can now replace the integers x_i with words (strings of symbols over a finite alphabet Σ), and ask, what is the right generalization of the Frobenius problem?

There are several possible answers.

- Instead of non-negative integer linear combinations of the x_i, we could consider the regular expression $x_1^* x_2^* \cdots x_k^*$;
- Or we could consider $\{x_1, x_2, \ldots, x_k\}^*$.

Instead of the condition that $\gcd(x_1, x_2, \ldots, x_k) = 1$, which was used to ensure that the number of unrepresentable integers is finite, we could demand that

$$\Sigma^* - x_1^* x_2^* \cdots x_k^*$$

or

$$\Sigma^* - \{x_1, x_2, \ldots, x_k\}^*$$

be finite, or in other words, that

$$x_1^* x_2^* \cdots x_k^*$$

or

$$\{x_1, x_2, \ldots, x_k\}^*$$

be *co-finite*.

And instead of looking for the largest non-representable integer, we could ask for the *length of the longest word* not in $x_1^* x_2^* \cdots x_k^*$ or $\{x_1, x_2, \ldots, x_k\}^*$. This gives us a natural generalization of the Frobenius problem to the noncommutative setting of a free monoid.

However, the first choice, $x_1^* x_2^* \cdots x_k^*$, is not very fruitful, as the following result shows.

Theorem 4. *Let $x_1, x_2, \ldots, x_k \in \Sigma^+$. Then $x_1^* x_2^* \cdots x_k^*$ is co-finite if and only if $|\Sigma| = 1$ and $\gcd(|x_1|, \ldots, |x_k|) = 1$.*

Proof. Let $Q = x_1^* x_2^* \cdots x_k^*$.

If $|\Sigma| = 1$ and $\gcd(|x_1|, \ldots, |x_k|) = 1$, then every sufficiently long unary word can be obtained by concatenations of the x_i, so Q is co-finite.

For the other direction, suppose Q is co-finite. If $|\Sigma| = 1$, let $\gcd(|x_1|, \ldots, |x_k|) = d$. If $d > 1$, Q contains only words of length divisible by d, and so is not co-finite. So $d = 1$.

Hence assume $|\Sigma| \geq 2$, and let a, b be distinct letters in Σ. Let $\ell = \max_{1 \leq i \leq k} |x_i|$, the length of the longest word among the x_i. Let $Q' = ((a^{2\ell} b^{2\ell})^k)^+$. Then we claim that $Q' \cap Q = \emptyset$. For if none of the x_i consists of powers of a single letter, then the longest block of consecutive identical letters in any word in Q is $< 2\ell$, so no word in Q' can be in Q.

Otherwise, say some of the x_i consist of powers of a single letter. Take any word w in Q, and count the number $n(w)$ of maximal blocks of 2ℓ or more consecutive identical letters in w. (Here "maximal" means such a block is delimited on both sides by either the beginning or end of the word, or a different letter.) Clearly $n(w) \leq k$. But $n(w') \geq 2k$ for any word w' in Q'. Thus Q is not co-finite, as it omits all the words in Q'.   ∎

Now let's turn back to $\{x_1, x_2, \ldots, x_k\}^*$; this case is much more fruitful.

Example. Suppose $S = \Sigma^m \cup \Sigma^n$, where $\gcd(m, n) = 1$. Then S^* is co-finite and the length of the longest word not in S^* is $g(m, n)$. In this case, the length of the longest omitted word is quadratic in the length of the longest word in $S_{\dot{\iota}}$.

Example. Define $U_1 = \{1\}$, $U_2 = U_1 \cup \{00, 01\}$, and $U_k = U_{k-1} \cup \{01^{k-2}0, 001^{k-3}0\}$ for $k \geq 3$. Then it is not hard to see that U_n^* omits $1^{i-1}0$ for $1 \leq i \leq n$, and these are all the strings of length $\leq n$ omitted. If we define $V_n = U_n \cup \{1^{n-2}0\}$, then V_n^* is co-finite and the longest word omitted is $1^{n-3}0$. In this case the length of the longest word omitted is only linear in the length of the longest word in V_n. This example is due to Jui-Yi Kao.

Now suppose $\max_{1 \leq i \leq k} |x_i| = n$. We can obtain an exponential upper bound on length of the longest omitted word, as follows:

Given $x_1, x_2, \ldots, x_k$, create a DFA accepting $\Sigma^* - \{x_1, x_2, \ldots, x_k\}^*$. This DFA keeps track of the last $n - 1$ symbols seen, together with markers indicating all positions within those $n - 1$ symbols where a partial factorization of the input into the x_i could end.

Since this DFA accepts a finite language, the longest word it accepts is bounded by the number of states. Thus we have [12]:

Theorem 5. *Let $S = \{x_1, x_2, \ldots, x_k\}$ be a finite set with $\max_{1 \leq i \leq k} |x_i| = n$, that is, the longest word is of length n. Then if S^* is co-finite, the length of the longest word not in S^* is bounded above by $\frac{2}{2|\Sigma|-1}(2^n|\Sigma|^n - 1)$.*

But is an exponential upper bound attainable? Surprisingly, the answer is yes. My Ph. D. student Zhi Xu has recently produced a class of examples $\{x_1, x_2, \ldots, x_k\}$ in which the length of the longest word is n, but the longest word in $\Sigma^* - \{x_1, x_2, \ldots, x_k\}^*$ is exponential in n. Here are his examples:

Let $r(n, k, l)$ denote the word of length l representing n in base k, possibly with leading zeros. For example, $r(3, 2, 3) = 011$. Let

$$T(m, n) = \{r(i, |\Sigma|, n - m)0^{2m-n}r(i + 1, |\Sigma|, n - m) \ : \ 0 \leq i \leq |\Sigma|^{n-m} - 2\}.$$

Then we have [12]:

Theorem 6. *Let m, n be integers with $0 < m < n < 2m$ and $\gcd(m, n) = 1$, and let $S = \Sigma^m + \Sigma^n - T(m, n)$. Then S^* is co-finite and the longest words not in S^* are of length $g(m, l)$, where $l = m|\Sigma|^{n-m} + n - m$.*

Example. Let $m = 3, n = 5, \Sigma = \{0, 1\}$. In this case, $l = 3 \cdot 2^2 + 2 = 14$, $S = \Sigma^3 + \Sigma^5 - \{00001, 01010, 10011\}$. Then a longest word not in S^* is

$$00001010011 \; 000 \; 00001010011$$

of length $25 = g(3, 14)$.

Exponential length for the longest omitted word comes from choosing, for example, $n = 2m - 1$.

Interestingly enough, the language $S(m, n) = \Sigma^m + \Sigma^n - T(m, n)$ can be accepted by an NFA with a relatively small number of states. With this observation, we get a class of NFA's with a polynomial number of states, accepting a finite language L such that L^* is co-finite such that the longest word not in L^* is of exponential length.

Theorem 7. *The language $S(m, n)$ can be accepted by an NFA with $O(n^2)$ states and specified by a regular expression with $O(n^2 \log n)$ symbols.*

Proof. We can construct an NFA with $m + 1$ states for $\{0, 1\}^m$, so it suffices to construct an NFA for the words of length n. The NFA has seven parts that accept seven different sublanguages, as follows. (Here # is short for 0^{2m-n}.)

1. $\{0, 1\}^{n-m} (\{0, 1\}^{2m-n} - \{0^{2m-n}\}) \{0, 1\}^{n-m}$;
2. $\bigcup_{2 \le a < n-m} \{0, 1\}^{n-m-a-1} \, 0 \, L_a \, \# \, \{0, 1\}^{n-m-a-1} \, 1 \, \{0, 1\}^a$;
3. $\bigcup_{2 \le a < n-m} \{0, 1\}^{n-m-a-1} \, 1 \, L_a \, \# \, \{0, 1\}^{n-m-a-1} \, 0 \, \{0, 1\}^a$;
4. $\bigcup_{1 \le a < n-m} \{0, 1\}^{n-m-a-1} \, 0 \, 1^a \, \# \, \{0, 1\}^{n-m-a-1}(\{0, 1\}^{a+1} - \{10^a\})$;
5. $\bigcup_{1 \le a < n-m} \{0, 1\}^{n-m-a-1} \, 1 \, 0^a \, \# \, \{0, 1\}^{n-m-a-1}(\{0, 1\}^{a+1} - \{10^{a-1}1\})$;
6. $0^{n-m} \#(\{0, 1\}^{n-m} - \{0^{n-m-1}1\})$;
7. $1^{n-m} \# \{0, 1\}^{n-m}$.

Here L_a is the language of all words of length a that include at least one 1 and one 0; this language can be accepted by an NFA with $O(a)$ states. It can also be specified by a regular expression with $O(n \log n)$ symbols, using divide and conquer [7], via the following identities:

$$L_{2a} = L_a\{0, 1\}^a \cup \{0, 1\}^a L_a \cup \{0^a 1^a, 1^a 0^a\};$$
$$L_{a+1} = L_a\{0, 1\} \cup \{0^a 1, 1^a 0\}.$$

The idea behind the construction is that we are trying to accept all words not of the form $r(i, 2, n - m)\#r(i + 1, 2, n - m)$. So if the first half differs from the second half of the word, this can occur either at the bits at the end containing at least one 0 and one 1, where carries could occur, or in more significant digits. Expressions 2 and 3 handle the more significant digits. Expressions 4 and 5 handle numbers whose base-2 expansions end in $011 \cdots 1$ or $10 \cdots 0$, and the error occurs there. Expression 6 handles the case $i = 0$ and expression 7 handles the case $i = 2^{n-m} - 1$. Finally, expression 1 handles the case where the word in the middle is not $\# = 0^{2n-m}$. ∎

Zhi Xu has also generated some examples where the number of omitted words is doubly exponential in n, the length of the longest word. Let

$$T'(m,n) = \{r(i,|\Sigma|,n-m)0^{2m-n}r(j,|\Sigma|,n-m) \ : \ 0 \le i < j \le |\Sigma|^{n-m} - 1\}.$$

Theorem 8. *Let m,n be integers with $0 < m < n < 2m$ and $\gcd(m,n) = 1$, and let $S = \Sigma^m + \Sigma^n - T'(m,n)$. Then S^* is co-finite and S^* omits at least $2^{|\Sigma|^{n-m}} - |\Sigma|^{n-m} - 1$ words.*

Example. Let $m = 3, n = 5, \Sigma = \{0,1\}$. Then

$$S = \Sigma^3 + \Sigma^5 - \{00001, 00010, 00011, 01010, 01011, 10011\}$$

and S^* omits $1712 > 11 = 2^{2^2} - 2^2 - 1$ words.

7   Other Possible Generalizations

Instead of considering the longest word omitted by $x_1^* x_2^* \cdots x_k^*$ or $\{x_1, x_2, \ldots, x_k\}^*$, we might consider their state complexity.

The *state complexity* of a regular language L is the smallest number of states in any DFA that accepts L [13,28]. It is written $\mathrm{sc}(L)$.

It turns out that the state complexity of $\{x_1, x_2, \ldots, x_k\}^*$ can be exponential in both the length of the longest word and the number of words.

Theorem 9. *Let t be an integer ≥ 2, and define words as follows:*

$$y := 01^{t-1}0$$

and

$$x_i := 1^{t-i-1}01^{i+1}$$

for $0 \le i \le t-2$. Let $S_t := \{0, x_0, x_1, \ldots, x_{t-2}, y\}$. Then S_t^* has state complexity $3t2^{t-2} + 2^{t-1}$.

Example. For $t = 6$ the words in S_t are 0 and

$$y = 0111110$$
$$x_0 = 1111101$$
$$x_1 = 1111011$$
$$x_2 = 1110111$$
$$x_3 = 1101111$$
$$x_4 = 1011111$$

Using similar ideas, we can also create an example achieving subexponential state complexity for $x_1^* x_2^* \cdots x_k^*$.

Theorem 10. *Let y and x_i be as defined above. Let $L = (0^* x_1^* x_2^* \cdots x_{n-1}^* y^*)^e$ where $e = (t+1)(t-2)/2 + 2t$. Then $\mathrm{sc}(L) \ge 2^{t-2}$.*

This example is due to Jui-Yi Kao [12].

8 Computational Complexity

If S is represented as an NFA or regular expression, then we can show that the problem of determining if S^* is co-finite is NP-hard and is in PSPACE [27]. However, if S is represented as merely a list of words, we do not currently know the computational complexity of the problem.

In a recent e-mail conversation, Oscar Ibarra suggested looking at the complexity of determining if S^* is co-finite in the case where S is over a unary alphabet. In that case, even if S is infinite and represented by an NFA, the decision problem is in P.

Theorem 11. *Let M be an n-state unary NFA. Then $L(M)^*$ is co-finite if and only if the gcd of the lengths of all words accepted by M of length $\leq n^2 + 2n$ is equal to 1.*

Proof. Evidently $L(M)^*$ is co-finite if and only if the gcd of the lengths of all words in $L(M)$ is 1. To see that it suffices to consider words of length $\leq n^2 + 2n$, put M in Chrobak normal form. As mentioned previously, this is a normal form where there is a "tail" of at most n^2 states followed by a single nondeterministic state that goes to at most n different cycles, each of which has cycle length $\leq n$. Consider computing the gcd of all word lengths of M iteratively starting with the shortest word; at some finite length ℓ we reach the final gcd of the lengths of all strings. Assume $\ell > n^2 + 2n$. This corresponds to following the "tail" and then going around a cycle twice and then a bit more. Therefore ℓ is a linear combination of a shorter word and the cycle length. But the cycle length can be written as the difference of a word accepted by going around the cycle 1 time and 2 times. Therefore the gcd is unchanged if we omit ℓ, a contradiction.

9 Open Problems

We conclude by reprising some of the open problems mentioned in this paper.

1. What is the complexity of computing $h(a_1, a_2, \ldots, a_n)$, the total number of non-negative integers not representable as an integer linear combination of the a_i?
2. Is there a Cook reduction for the NP-hardness of the Frobenius problem?
3. What is the complexity of the following problem? Given a finite list of words $S = \{x_1, x_2, \ldots, x_k\}$, determine if S^* is co-finite.
4. What is the computational complexity of the global postage stamp problem?

References

1. Alter, R., Barnett, J.A.: A postage stamp problem. Amer. Math. Monthly 87, 206–210 (1980)
2. Beihoffer, D., Hendry, J., Nijenhuis, A., Wagon, S.: Faster algorithms for Frobenius numbers. Elect. J. Combinatorics 12(1) (2005), Paper R27,
   http://www.combinatorics.org/Volume_12/Abstracts/v12i1r27.html

3. Brauer, A.: On a problem of partitions. Amer. J. Math. 64, 299–312 (1942)
4. Chrobak, M.: Finite automata and unary languages. Theoret. Comput. Sci. 47, 149–158 (1986); Errata 302, 497–498 (2003)
5. Davison, J.L.: On the linear diophantine problem of Frobenius. J. Number Theory 48, 353–363 (1994)
6. Einstein, D., Lichtblau, D., Strzebonski, A., Wagon, S.: Frobenius numbers by lattice point enumeration. Integers 7, A15 (2007) (electronic)
7. Ellul, K., Krawetz, B., Shallit, J., Wang, M.-w.: Regular expressions: new results and open problems. J. Autom. Lang. Combin. 10, 407–437 (2005)
8. Greenberg, H.: Solution to a linear Diophantine equation for nonnegative integers. J. Algorithms 9, 343–353 (1988)
9. Incerpi, J., Sedgewick, R.: Improved upper bounds on shellsort. J. Comput. System Sci. 31, 210–224 (1985)
10. Kannan, R.: Solution of the Frobenius problem. In: Veni Madhavan, C.E. (ed.) Proc. 9th Conf. Found. Software Tech. Theor. Comput. Sci. LNCS, vol. 405, pp. 242–251. Springer, Heidelberg (1989)
11. Kannan, R.: Lattice translates of a polytope and the Frobenius problem. Combinatorica 12, 161–177 (1992)
12. Kao, J.-Y., Shallit, J., Xu, Z.: The Frobenius problem in a free monoid. In: Albers, S., Weil, P. (eds.) STACS 2008, 25th Annual Symposium on Theoretical Aspects of Computer Science, Dagstuhl Seminar Proceedings, Germany, pp. 421–432 (2008)
13. Maslov, A.N.: Estimates of the number of states of finite automata. Dokl. Akad. Nauk. SSSR 194, 1266–1268 (1970); In Russian. English translation in Soviet Math. Dokl. 11, 1373–1375 (1970)
14. Ramírez-Alfonsín, J.L.: Complexity of the Frobenius problem. Combinatorica 16, 143–147 (1996)
15. Ramírez-Alfonsín, J.L.: The Diophantine Frobenius Problem. Oxford University Press, Oxford (2005)
16. Roberts, J.B.: Note on linear forms. Proc. Amer. Math. Soc. 7, 465–469 (1956)
17. Roune, B.H.: Solving thousand-digit Frobenius problems using Gröbner bases. J. Symbolic Comput. 43, 1–7 (2008)
18. Sedgewick, R.: A new upper bound for shellsort. J. Algorithms 7, 159–173 (1986)
19. Selmer, E.S.: On shellsort and the Frobenius problem. BIT 29, 37–40 (1989)
20. Shallit, J.: The computational complexity of the local postage stamp problem. SIGACT News 33(1), 90–94 (2002)
21. Shallit, J.: What this country needs is an 18-cent piece. Math. Intelligencer 25(2), 20–23 (2003)
22. Shell, D.L.: A high-speed sorting procedure. Commun. ACM 27, 30–32 (1959)
23. Sylvester, J.J.: On subinvariants, i.e. semi-invariants to binary quantics of an unlimited order. Amer. J. Math. 5, 119–136 (1882)
24. Vardi, I.: Computational Recreations in Mathematica. Addison-Wesley, Reading (1991)
25. Weiss, M.A., Sedgewick, R., Hentschel, E., Pelin, A.: Shellsort and the Frobenius problem. Congr. Numer. 65, 253–260 (1988)
26. Wilf, H.S.: A circle-of-lights algorithm for the money-changing problem. Amer. Math. Monthly 85, 562–565 (1978)
27. Xu, Z., Shallit, J.: An NP-hardness result on the monoid Frobenius problem (preprint, 2008), http://arxiv.org/abs/0805.4049
28. Yu, S., Zhuang, Q., Salomaa, K.: The state complexities of some basic operations on regular languages. Theoret. Comput. Sci. 125, 315–328 (1994)

Well Quasi-orders in Formal Language Theory[*]

Flavio D'Alessandro[1] and Stefano Varricchio[2]

[1] Dipartimento di Matematica, Università di Roma "La Sapienza"
Piazzale Aldo Moro 2, 00185 Roma, Italy
`dalessan@mat.uniroma1.it`
[2] Dipartimento di Matematica, Università di Roma "Tor Vergata"
via della Ricerca Scientifica, 00133 Roma, Italy
`varricch@mat.uniroma2.it`

Abstract. The concept of well quasi-order is a generalization of the classical notion of well order and plays a role in the studying of several problems of Mathematics and Theoretical Computer Science. This paper concerns some applications of well quasi-orders to Formal Language Theory. In particular, we present a survey of classical and recent results, based upon such structures, concerning context-free and regular languages. We also focus our attention to some application of well quasi-orders in the studying of languages obtained by using the operators of shuffle and iterated shuffle of finite languages.

Keywords: Well quasi-orders, finite automata, context-free languages, shuffle, iterated shuffle.

1 Introduction

The concept of well quasi-order is a generalization of the classical notion of well order. A *quasi-order* on a set S is called a *well quasi-order* (*wqo*) if every non-empty subset X of S has at least one minimal element in X but no more than a finite number of (non-equivalent) minimal elements. There exist various characterizations of this concept which was often rediscovered by different authors (see [20]). The concept of well quasi-order plays a role in the studying of many problems of Mathematics and Theoretical Computer Science. For this reason, well quasi-orders have been widely investigated in the past and there exists a large literature on this subject. Recently, in the theory of language equations, remarkable results based on wqo's have been obtained by M. Kunc [22]. These results have been culminating in the negative solution of the famous conjecture by Conway claiming the regularity of the maximal solutions of the *commutative language equation* $XL = LX$ where L is a finite language of words [21]. On the other hand, using wqo's, in [22] it is proved that the maximal solution of the inequality $XK \subseteq LX$ is a regular language whenever L is so. In this paper, we offer a survey of some classical and recent results about the applications of well

[*] The first author acknowledges the partial support of ``fundings ``Facoltà di Scienze MM. FF. NN. 2006'' of the University of Rome ``La Sapienza''.

quasi-orders in Formal Language Theory. The first part of the paper presents two basic theorems that give a deep insight into combinatorics on words and languages. The first is due to Higman [15] and it gives a very general theorem on division orders in abstract algebras that in the case of semigroups becomes: *Let S be a semigroup quasi-ordered by a division order $\leq$. If there exists a generating set of S well quasi-ordered by $\leq$, then S will also be so.* The second is a remarkable generalization of the famous Myhill-Nerode theorem on regular languages. In [11] Ehrenfeucht et al. proved that a language is regular if and only if it is upwards closed with respect to a monotone well quasi-order. From this result many regularity conditions have been derived (see for instance [1,8,9,10]). Monotone quasi-orders can be associated naturally with the derivation relations of suitable semi-Thue systems, so that one can prove the regularity of a language generated by a semi-Thue system by showing the wqo property of the corresponding derivation relation. In [11] a class of semi-Thue systems called *unitary* is studied. In particular unitary systems whose derivation relation is a wqo are characterized. By applying this result and the generalized Myhill-Nerode theorem, one can obtain a remarkable condition that assures that a language generated by a unitary system is regular. Another important application is the regularity of the languages on a binary alphabet generated by copying systems [1]. We also present a new generalization [2,3] of Higman's theorem to context-free languages while in Section 5 an improvement [4,5] of the above mentioned result for unitary systems is described. In the last section, we consider some applications of well quasi-orders in the studying of languages obtained by using the operators of shuffle and iterated shuffle of finite languages.

2   Preliminaries

The main notions and results concerning quasi-orders and languages are shortly recalled in this section. Let A be a finite *alphabet* and let A^* be the free monoid generated by A. The elements of A are usually called *letters* and those of A^* *words*. The identity of A^* is denoted ϵ and called the *empty word*. A non-empty word $w \in A^*$ can be written uniquely as a sequence of letters as $w = a_1 a_2 \cdots a_n$, with $a_i \in A$, $1 \leq i \leq n$, $n > 0$. The integer n is called the *length* of w and denoted $|w|$. For all $a \in A$, $|w|_a$ denotes the number of occurrences of the letter a in w. If w is the empty word, then we set $|w| = 0$ and, for any $a \in A$, $|w|_a = 0$. Let $w \in A^*$. The word $u \in A^*$ is a *factor* of w if there exist $p, q \in A^*$ such that $w = puq$. If $w = uq$, for some $q \in A^*$ (resp. $w = pu$, for some $p \in A^*$), then u is called a *prefix* (resp. a *suffix*) of w.

The set of all prefixes (resp. suffixes, factors) of w is denoted Pref(w) (resp. Suff(w), Fact(w)). A word u is a *subsequence* of a word v if $u = a_1 a_2 \cdots a_n$, $v = v_1 a_1 v_2 a_2 \cdots v_n a_n v_{n+1}$ with $a_i \in A$, $v_i \in A^*$. A subset L of A^* is called a *language*. If L is a language of A^*, then Alph(L) is the smallest subset B of A such that $L \subseteq B^*$. Moreover, Pref(L) denotes the set of the prefixes of all words of L. A subset X of a semigroup S is called *recognizable* if there exists a finite index congruence of S that saturates X, that is X is a union of cosets of the congruence. The family of recognizable languages of S is denoted $Rec(S)$. Let P be a subset of a semigroup S. Then the sets $P^{-1}S$ and SP^{-1} are defined as:

$$P^{-1}S = \{t \in S \mid \exists\, p \in P,\, s \in S \mid s = pt\},$$

and

$$SP^{-1} = \{t \in S \mid \exists\, p \in P,\, s \in S \mid s = tp\}.$$

A binary relation $\leq$ on a set S is a *quasi-order* (qo) if $\leq$ is reflexive and transitive. Moreover, if $\leq$ is symmetric, then $\leq$ is an equivalence relation. The meet $\leq \cap \leq^{-1}$ is an equivalence relation $\sim$ and the quotient of S by $\sim$ is a *poset* (partially ordered set). A quasi-order $\leq$ in a semigroup S is *monotone on the right (resp. on the left)* if for all $x_1, x_2, y \in S$

$$x_1 \leq x_2 \text{ implies } x_1 y \leq x_2 y \text{ (resp. } y x_1 \leq y x_2).$$

A quasi-order is *monotone* if it is monotone on the right and on the left.

An element $s \in X \subseteq S$ is *minimal* in X with respect to $\leq$ if, for every $x \in X$, $x \leq s$ implies $x \sim s$. For $s, t \in S$ if $s \leq t$ and s is not equivalent to t mod $\sim$, then we set $s < t$.

A quasi-order in S is called a *well quasi-order* (wqo) if every non-empty subset X of S has at least one minimal element but no more than a finite number of (non-equivalent) minimal elements. We say that a set S is *well quasi-ordered* (wqo) by $\leq$, if $\leq$ is a well quasi-order on S.

There exist several conditions which characterize the concept of well quasi-order and that can be assumed as equivalent definitions (cf. [10]).

Theorem 1. *Let S be a set quasi-ordered by $\leq$. The following conditions are equivalent:*

   i. $\leq$ is a well quasi-order;
   ii. every infinite sequence of elements of S has an infinite ascending subsequence;
   iii. if $s_1, s_2, \ldots, s_n, \ldots$ is an infinite sequence of elements of S, then there exist integers i, j such that $i < j$ and $s_i \leq s_j$;
   iv. there exists neither an infinite strictly descending sequence in S (i.e., $\leq$ is well founded), nor an infinity of mutually incomparable elements of S.

A partial order satisfying the wqo property is also called a *well partial order*. The quasi-orders considered in this paper are actually partial orders. However, according to the current terminology, we refer to them as quasi-orders. Let $\sigma = \{s_i\}_{i \geq 1}$ be an infinite sequence of elements of S. Then σ is called *good* if it satisfies condition *iii.* of Theorem 1 and it is called *bad* otherwise, that is, for all integers i, j such that $i < j$, $s_i \not\leq s_j$. It is worth noting that, by condition *iii.* above, a useful technique to prove that $\leq$ is a wqo on S is to prove that no bad sequence exists in S.

Let $\leq$ be a quasi-order on a set S and let X be a subset of S. We say that X is *upwards closed*, or simply *closed*, with respect to $\leq$, if $x \leq y$ and $x \in S$ implies $y \in S$.

Following [10], we recall that a *rewriting system*, or *semi-Thue system*, on an alphabet A is a pair (A, π) where π is a binary relation on A^*. Any pair of words $(p, q) \in \pi$ is called a *production* and denoted by $p \to q$. Let us denote by $\Rightarrow_\pi$ the derivation relation of π, that is, for $u, v \in A^*$, $u \Rightarrow_\pi v$ if

$$\exists\, (p, q) \in \pi \text{ and } \exists\, h,\, k \in A^* \text{ such that } u = hpk, \quad v = hqk.$$

The *derivation relation* $\Rightarrow_\pi^*$ is the transitive and reflexive closure of $\Rightarrow_\pi$. One easily verifies that $\Rightarrow_\pi^*$ is a monotone quasi-order on A^*.

3 Generalized Myhill-Nerode Theorem and Highman Theorem

According to the classical Myhill-Nerode theorem, one can obtain a characterization of recognizable subsets of a semigroup in terms of finite index congruence of the semigroup. A remarkable extension of this theorem was obtained in [11] in terms of wqo.

Theorem 2. *A subset X of a semigroup S is recognizable if and only if X is closed with respect to a monotone well quasi-order in S.*

In Sections 4 and 5 we will consider some applications of the previous theorem to formal languages. It is useful to recall that recognizable sets of a semigroup can be described also by using equivalence relations, monotone on the right or on the left. More precisely, a classical theorem by Nerode states that a subset X is recognizable if and only if there exists a finite index equivalence, monotone on the right (resp. on the left) that saturates X. In this context, a result connected with Theorem 2 was proposed in [9,10]. If X is a subset of a semigroup S, then we associate with X a quasi-order $\leq_X^r$ defined as: for any $s, t \in S$,

$$s \leq_X^r t \iff s^{-1}X \subseteq t^{-1}X.$$

The relation $\leq_X^r$ is monotone on the right. Similarly, one can associate a quasi-order, monotone on the left $\leq_X^l$ defined as: $s \leq_X^l t \iff Xs^{-1} \subseteq Xt^{-1}$. In analogy with the theorem by Nerode, one can ask whether the wqo property of $\leq_X^r$ implies the regularity of X. The answer is negative. Indeed, for instance, one can check that the language $L = \{a^n b^m \mid n \geq m \geq 0\}$ is not regular while, on the other hand, $\leq_X^l$ is a wqo. However, a partial generalization of Nerode's theorem and of Theorem 2 as well is the following.

Theorem 3. *A subset X of a semigroup S is recognizable if and only if the quasi-orders $\leq_X^r$ and $\leq_X^l$ are wqo.*

As a consequence one has:

Corollary 1. *A subset X of a semigroup S is recognizable if and only if X is closed with respect to a left and to a right monotone well quasi-order in S.*

Another important result proved in the wqo theory is the Higman theorem. We recall that a quasi-order $\leq$ in a semigroup S is said to be a *division order* or a *divisibility order* if it is monotone and, moreover, for all $s \in S$ and $x, y \in S^1$, $s \leq xsy$. The ordering by divisibility in abstract algebras was studied by Higman who proved in [15] a very general theorem that, in the case of semigroups, has the following statement.

Theorem 4. *Let S be a semigroup quasi-ordered by a divisibility order $\leq$. If there exists a generating set of S well quasi-ordered by $\leq$, then S will be also so.*

It is worth recalling that in [20] Kruskal extends Higman's result, proving that certain embeddings on finite trees are well quasi-orders. Moreover, in [17] some extensions of Higman and Kruskal's theorem to regular languages and rational trees have been given. In particular, we recall the following generalization of Kruskal's theorem:

Theorem 5. *Let A be a wqo alphabet and let T be the family of the rational k-ary trees, with nodes labeled by A. Then the natural embedding relation induced on T is a wqo.*

A remarkable consequence of Theorem 4 is the following. Let $S = A^*$ be the free monoid generated by an alphabet A quasi-ordered by a relation $\leq$. The relation $\leq$ can be extended to A^* as follows. Let $u, v \in A^*$. We set $u \leq v$ if

$$u = a_1 \cdots a_n, a_i \in A, \ i = 1, ..., n,$$

$$v \in A^* b_1 A^* b_2 A^* \cdots A^* b_n A^*, b_i \in A, \ i = 1, ..., n,$$

where

$$a_i \leq b_i, i = 1, ..., n.$$

Trivially, the relation defined above is a division order, called *subsequence ordering*, and if $\leq$ is a wqo on A, then, by Higman theorem, its extension is a wqo on A^*. In the sequel, we refer to this result as the Higman theorem in the free monoid. It can be proved that the subsequence ordering is the smallest division order in A^*.

In [3] a new generalization of Higman theorem has been given. This result is based upon the notion of *division order* on a language: given a language L over the alphabet A, a quasi order $\leq$ on A^* is called a *division order* on L if it is monotone and for any $u, v \in L$ if u is factor of v then $u \leq v$. When L is the whole free monoid A^* this notion is equivalent to the classical one, but, in general, a quasi-order on A^* could be a division order on a set L and not on A^*. Let $G = (V, A, P)$ be a context-free grammar, where $V = \{A_1, \ldots, A_k\}$ is the alphabet of the variables, A is the alphabet of the terminal symbols and P is the set of the productions. For any i, $1 \leq i \leq k$, denote L_i the language generated by G assuming the variable A_i as start symbol. The following theorem holds [3].

Theorem 6. *Let $G = (V, A, P)$ be a context-free grammar and, according to the previous notation, let $L = \bigcup_{i=1}^{n} L_i$ be the union of all languages generated by the variables of G. If $\leq$ is a division order on L, then $\leq$ is a well quasi-order on L.*

As an immediate corollary of the previous theorem, we have that if L is a context-free language generated by a grammar with only one variable, then any division order on L is a wqo on L. This generalizes Higman theorem on finitely generated free monoids since, for any finite alphabet A, the set A^* can be generated by a context-free grammar having only one variable. It is possible to give a slight generalization of the notion of division order on languages as follows.

Definition 1. *Let $L \subseteq A^*$ be a language and let $\leq$ be a monotone quasi-order. Then $\leq$ is a weak division order on L if for any $u, x, y \in A^*$ such that $u, xuy, xy \in L$, one has $u \leq xuy$.*

We observe that any division order on L is a weak division order on L but the converse is false. Moreover, any weak division order on A^* is a division order. By using some combinatorial arguments akin to that used to prove Theorem 6, one can prove the following theorem.

Theorem 7. *Let L be a context-free language containing the empty word and generated by a context-free grammar with only one variable. Then any weak division order on L is a wqo on L.*

4   Copying Systems

In this section we describe how Theorem 2 has been used to prove the regularity of some relevant formal languages. We consider the case of copying systems and languages generated by them, introduced in [14]. In that paper it is proved that, when the alphabet has cardinality at least three, such languages are not, in general, regular (see Theorem 10). In the case of a binary alphabet, the languages generated by copying systems are actually all regular [1].

Let $A = \{a, b\}$ and let (A, π) be the rewriting system with $\pi = \{(x, xx) \mid x \in A^*\}$. The derivation relation $\Rightarrow_{\pi}^{*}$ is called *copying relation*. We can also consider a restricted copying relation denoted as $\Rightarrow_{\pi'}^{*}$ where

$$\pi' = \{(a, aa), (b, bb), (ab, abab), (ba, baba)\}.$$

Trivially $\Rightarrow_{\pi'}^{*} \subseteq \Rightarrow_{\pi}^{*}$.

Theorem 8. *The derivation relation $\Rightarrow_{\pi'}^{*}$ is a well quasi-order on A^*.*

Let us remark that one can easily prove (*cf* [1]) that the rewriting system π' is, in fact, equivalent to π. Moreover, π' is the smallest set of rules among those which are equivalent to π. Therefore, the following result easily follows.

Theorem 9. *The derivation relation $\Rightarrow_{\pi}^{*}$ is a well quasi-order on A^*.*

Corollary 2. *Let $L \subseteq A^*$ be a language which is closed with respect to $\Rightarrow_\pi^*$. Then L is a regular language.*

Proof. The statement is a consequence of Theorem 9 and Theorem 2.

Let us now consider a free monoid B^* and the copying relation $\Rightarrow_\pi^*$ in B^*. For any $w \in B^*$ we consider the set $L_{w,\pi}$ defined as

$$L_{w,\pi} = \{u \in B^* \mid w \Rightarrow_\pi^* u\}.$$

If a word w contains at least three distinct letters, then the language $L_{w,\pi}$ is not regular [14].

Theorem 10. *Let $w \in B^*$ be a word such that $\mathrm{Card}(\mathrm{Alph}(w)) \geq 3$. Then $L_{w,\pi}$ is not regular.*

Proposition 1. *Let B be a finite alphabet and $w \in B^*$. Then $L_{w,\pi}$ is regular if and only if w contains at most two distinct letters.*

Proof. By Theorem 10, if w is a word containing at least three distinct letters, then $L_{w,\pi}$ is not a regular language. Hence, if $L_{w,\pi}$ is regular, then $\mathrm{Card}(\mathrm{Alph}(w)) \leq 2$. Conversely, suppose that $d = \mathrm{Card}(\mathrm{Alph}(w)) \leq 2$. If $d = 0$, then $w = \epsilon$ and $L_{w,\pi} = \{\epsilon\}$ is regular. If $d = 1$, then $w \in a^*$ with $a \in B$ and $L_{w,\pi} = a^{|w|}a^*$ is regular. If $d = 2$, since $L_{w,\pi}$ is closed with respect to $\Rightarrow_\pi^*$, then from Corollary 2 the result follows.

5   Well Quasi-orders and Unitary Grammars

Other applications of great interest of wqo to Formal Language Theory are based upon the notion of *unitary grammar* introduced in [11]. Let us present these results. A semi-Thue system is called *unitary* if π is a finite set of productions of the kind

$$\epsilon \to u, \ u \in I, \ I \subseteq A^+.$$

Such a system, also called *unitary grammar*, is then determined by the finite set $I \subseteq A^+$. Its derivation relation is denoted by $\Rightarrow_I^*$ (or, simply, $\Rightarrow^*$). We set $L_I^\epsilon = \{u \in A^* \mid \epsilon \Rightarrow^* u\}$. A language L is called *unitary* if there exists a finite set of words I such that $L = L_I^\epsilon$. Unitary grammars have been introduced in order to study the relationships between the classes of context-free and regular languages. Let us consider this aspect with more attention. Unitary languages are context-free since, given a language L_I^ϵ, a context-free grammar generating L_I^ϵ can be constructed from the set I in the obvious way.

Example 1. Let $A = \{a, b\}$ and let $I = \{ab\}$. One can verify that the language L_I^ϵ is the language of the so called *semi Dyck words over* A. We recall that a word u over the alphabet A is said to be a semi-Dyck word if $|u|_a = |u|_b$ and, moreover, for every prefix p of u, $|p|_a \geq |p|_b$. This language is context-free non regular. Similarly, if $I = \{ab, ba\}$, then L_I^ϵ is the language of *Dyck words over* A, that is, of all words u such that $|u|_a = |u|_b$. The very same result holds for every alphabet $A = \{a_1, ..., a_k, b_1, ..., b_k\}$.

By the well-known Chomsky-Schützenberger theorem, every context-free language is the homomorphic image of the intersection of a Dyck language with a regular one. Since the class of regular languages is closed under homomorphism and intersection and since Dyck languages are unitary, these facts indicate that, at least, some unitary languages capture the non regular aspect of a context-free language. This argument eventually lead to investigate the conditions assuring the regularity of a unitary language. In this theoretical setting, an important theorem proven in [11] is based upon the notion of unavoidable set. This notion is classical and well-known in the field of Combinatorics on Words (see [23], Ch. 1). A set I of words is said to be *unavoidable* (on the set $A = \mathrm{Alph}(I)$) if every sufficiently long word over A has a factor that belongs to I. A set is said to be *avoidable* if it is not unavoidable. The next two examples prefigurate an important characterization, given with Theorem 11 below, of unavoidable sets of words in terms of the wqo property of the unitary grammars.

Example 2. Let A^* be the free monoid generated by the alphabet $A = \{a, b\}$. Set $I = \{a, bb\}$. Then the set I is clearly unavoidable since, any word of length at least 2 contains a factor in I. On the other hand, one can check that the derivation relation $\Rightarrow_I^*$ is a wqo on A^*. The same result holds in the case $I = \{aa, ab, ba, bb\}$

Example 3. Let A^* be the free monoid generated by the alphabet $A = \{a, b\}$. Set $I = \{ab\}$. Then the set I is clearly avoidable since, for instance, every power of the letter a avoids I. On the other hand the derivation relation $\Rightarrow_I^*$ is not a wqo on A^*. Indeed, one can verify that the sequence $\{a^n\}_{n \geq 0}$ is bad with respect to the relation $\Rightarrow_I^*$.

Theorem 11. *Let I be a finite set of A^+ and assume that $A = \mathrm{Alph}(I)$. Then the derivation relation $\Rightarrow_I^*$ is a wqo on A^* if and only if the set I is unavoidable.*

The following remark concerns a noteworth application of Theorem 11.

Example 4. Let A^* be the free monoid genereted by the alphabet A. Obviously, A is unavoidable in A^*. Set $I = A$. Then the derivation relation $\Rightarrow_I^*$ is the subsequence ordering on A^*. According to Theorem 11, the derivation relation $\Rightarrow_I^*$ is a wqo on A^*. Thus we obtain Higman Theorem in the free monoid.

A straighforward corollary of Theorem 11 gives a regularity condition for languages generated by unitary grammars.

Corollary 3. *Let I be a finite set of A^+ and assume that $A = \mathrm{Alph}(I)$. The following conditions are equivalent:*

   i. the derivation relation $\Rightarrow_I^*$ is a wqo on A^*;
  ii. the set I is unavoidable;
 iii. the language L_I^ϵ is regular.

Example 5. Let us consider again Example 1. Since $I = \{ab\}$, the language L_I^ϵ is the language of *semi Dyck words*. This language is context-free non regular and I is avoidable.

Example 6. Let us consider again Example 2. One can easily check that L_I^ϵ is the shuffle of a^* and $\{bb\}^*$, and thus it is regular.

A short comment on the proof of the corollary above. Since the language L_I^ϵ is closed with respect to the derivation relation $\Rightarrow_I^*$, the implication *ii.* $\Rightarrow$ *iii.* is immediately obtained by applying Theorem 2 to L_I^ϵ. On the other hand, one can prove the implication *iii.* $\Rightarrow$ *i.*, by showing that, if I is an avoidable set, then the language L_I^ϵ is not regular. This last task can be done by using a suitable *anti-pumping argument.*

One can ask if, in Corollary 3, the condition *i.* can be replaced by the weaker condition that the relation $\Rightarrow_I^*$ is a wqo on L_I^ϵ. The positive answer to this question was given in [4,5], by proving the following Theorem 12.

Theorem 12. *The derivation relation $\Rightarrow_I^*$ is a wqo on A^* if and only if $\Rightarrow_I^*$ is a wqo on L_I^ϵ.*

We mention that another important contribution to the field of formal languages whose proof is based upon Corollary 3 was given by Senizergues in [26]. Here it is proved that every rational subset of a free group is either recognizable or *disjunctive*, that is the syntactic congruence associated with the set is the identical relation. This result can be viewed as an extension of the classical Kleene theorem to rational sets of free groups and gives a positive answer to an open problem raised by Sakarovitch. The reader is referred to [25] for a complete survey on this problem.

6 On Other Well Quasi-orders

One can consider a possible extension of the results presented in the previous section, Theorem 12, Corollary 3 and Theorem 11, with respect to other significant quasi orders. If I is a finite set of words, let us associate with I a binary relation $\vdash_I^*$ defined as the transitive and reflexive closure of $\vdash_I$ where $v \vdash_I w$ if

$$v = v_1 v_2 \cdots v_{n+1},$$

$$w = v_1 a_1 v_2 a_2 \cdots v_n a_n v_{n+1},$$

where the a_i's are letters, and $a_1 a_2 \cdots a_n \in I$. We set $L_{\vdash_I}^\epsilon = \{w \in A^* \mid \epsilon \vdash_I^* w\}$. In [13], the following theorem has been proved.

Theorem 13. *Let $I \subseteq A^+$ and assume that $A = \mathrm{Alph}(I)$. The following conditions are equivalent:*

i. the derivation relation $\vdash_I^*$ is a wqo on A^*;
ii. the set I is subsequence unavoidable in A^*, that is there exists a positive integer k such that any word $u \in A^*$, with $|u| \geq k$, contains as a subsequence a word of I;
iii. the language $L_{\vdash_I}^\epsilon$ is regular.

In [13] it is also proved that I is subsequence unavoidable if and only if, for every $a \in A$, $I \cap \{a\}^+ \neq \emptyset$.

Example 7. Let A^* be the free monoid generated by the alphabet $A = \{a, b\}$. Set $I = \{ab\}$. Then the set I is clearly subsequence avoidable since, for instance, every power of the letter a avoids I. Moreover, it is easily seen that the derivation relation $\vdash_I^*$ is not a wqo on A^*. Indeed, one can verify that the sequence $\{a^n\}_{n \geq 0}$ is bad with respect to the relation $\vdash_I^*$. On the other hand, one can verify that $L_{\vdash_I}^\epsilon = L_I^\epsilon$, so that this language is equal to the language of the semi Dyck words.

Another interesting property of the relation $\vdash_I^*$ is the following consequence of Theorem 7 proven in [3].

Proposition 2. *Let $I \subseteq A^+$. Then $\vdash_I^*$ is a well quasi order on L_I^ϵ.*

Proof. The language L_I^ϵ is generated by a context-free grammar with only one variable and $\epsilon \in L_I^\epsilon$. Moreover, the relation $\vdash_I^*$ is a weak division order over L_I^ϵ. The statement, then, follows from Theorem 7.

By the previous proposition, it is natural to ask whether $\vdash_I^*$ is a wqo on $L_{\vdash_I}^\epsilon$ or not. The answer is negative. In fact, we can exhibit a set I such that the quasi-order $\vdash_I^*$ is not a wqo on $L_{\vdash_I}^\epsilon$. For this purpose, let $A = \{a, b, c, d\}$ be a four-letter alphabet and let $\bar{A} = \{\bar{a}, \bar{b}, \bar{c}, \bar{d}\}$ be a disjoint copy of A. Let $\tilde{A} = A \cup \bar{A}$ and let $I = \{a\bar{a}, b\bar{b}, c\bar{c}, d\bar{d}\}$. Now consider the sequence $\{S_n\}_{n \geq 1}$ of words of $\tilde{A}^*$ defined as: for every $n \geq 1$,

$$S_n = ad\bar{b}c\bar{c}\bar{a}(a\bar{d}dc\bar{c}c\bar{c}\bar{a})^n a\bar{d}b\bar{b}\bar{a}.$$

The following result holds.

Proposition 3. *The sequence $\{S_n\}_{n \geq 1}$ is bad with respect to $\vdash_I^*$. Moreover, the elements of $\{S_n\}_{n \geq 1}$ belong to $L_{\vdash_I}^\epsilon$ and so $\vdash_I^*$ is not a wqo on $L_{\vdash_I}^\epsilon$.*

We can summarize the relationships between, on one hand, the quasi-orders $\vdash_I^*$ and $\Rightarrow_I^*$, and, on the other hand, the languages $L_{\vdash_I}^\epsilon$ and L_I^ϵ by the following list:

- There exists a finite set I such that $\Rightarrow_I^*$ is not a wqo on L_I^ϵ;
- There exists a finite set I such that $\vdash_I^*$ is not a wqo on $L_{\vdash_I}^\epsilon$;
- For any finite set I the relation $\vdash_I^*$ is a wqo on L_I^ϵ.

The theoretical setting we have described, suggests to ask whether Theorem 13 may be extended by replacing condition (i) with the weaker condition that the derivation relation $\vdash_I^*$ is a wqo on $L_{\vdash_I}^\epsilon$. Unfortunately this is not true. Indeed, by the previous Example 7, if $I = \{ab\}$, $L_{\vdash_I}^\epsilon = L_I^\epsilon$ is the language of all *semi-Dyck words* over the alphabet $\{a, b\}$. By Proposition 2, $\vdash_I^*$ is a well quasi order on $L_{\vdash_I}^\epsilon = L_I^\epsilon$ while this language is not regular. This example lead us to further investigate the relation between $L_{\vdash_I}^\epsilon$ and $\vdash_I^*$. The results of this investigation will be presented in the next section.

7 Well Quasi-orders and Shuffle Closure of Finite Languages

Given a set I of word, the set $L_{\vdash_I}^\epsilon$ is, actually, the set of all the words obtained by the shuffle of (copies of) words of I. Moreover, the relation $\vdash_I^*$ is a natural partial order over $L_{\vdash_I}^\epsilon$. Observe also that for any u, v in $L_{\vdash_I}^\epsilon$, $u \vdash_I^* v$ if and only if v is the shuffle of u and another word of $L_{\vdash_I}^\epsilon$. In [5], the authors have opened the problem of the characterization of the finite sets I such that $\vdash_I^*$ is a well quasi-order on $L_{\vdash_I}^\epsilon$. In this section we present the results of [6,7], where a complete answer is given in the case when I consists of a single word w.

In this context, it is worth noticing that in [5] is proved that $\vdash_{\{w\}}^*$ is not a wqo on $L_{\vdash_{\{w\}}}^\epsilon$ if $w = abc$. A simple argument allows one to extend the result above in the case that $w = a^i b^j c^h$, $i, j, h \geq 1$. By using a simple technical argument, this implies that if a word w contains three distinct letters at least, then $\vdash_{\{w\}}^*$ is not a wqo on $L_{\vdash_{\{w\}}}^\epsilon$. Therefore, in order to characterize the word w such that $\vdash_{\{w\}}^*$ is a wqo on $L_{\vdash_{\{w\}}}^\epsilon$, one can consider only the case when w is a word on the binary alphabet $\{a, b\}$. Let E be the exchange morphism $(E(a) = b, E(b) = a)$, and let $\tilde{w}$ be the mirror image of w.

Definition 2. *A word w is called* bad *if one of the words w, $\tilde{w}$, $E(w)$ and $E(\tilde{w})$ has a factor of one of the two following forms*

$$a^k b^h \quad \text{with } k, h \geq 2 \tag{1}$$

$$a^k b a^l b^m \text{ with } k > l \geq 1, m \geq 1 \tag{2}$$

A word w is called good *if it is not bad.*

One can prove that a word is good if and only if it is a factor of $(ba^n)^\omega$ or $(ab^n)^\omega$ for some $n \geq 0$. The following result characterizes the set of good words in terms of the wqo property [6,7].

Theorem 14. *Let w be a word over the alphabet $\{a, b\}$. The derivation relation $\vdash_{\{w\}}^*$ is a wqo on $L_{\vdash_{\{w\}}}^\epsilon$ if and only if w is good.*

Corollary 4. *Let w be a word over the alphabet $\{a, b\}$. The derivation relation $\vdash_{\{w\}}^*$ is a wqo on $L_{\vdash_{\{w\}}}^\epsilon$ if and only if w is a factor of $(ba^n)^\omega$ or $(ab^n)^\omega$ for some $n \geq 0$.*

References

1. Bovet, D.P., Varricchio, S.: On the regularity of languages on a binary alphabet generated by copying systems. Information Processing Letters 44, 119–123 (1992)
2. D'Alessandro, F., Varricchio, S.: On well quasi-orders on languages. In: Ésik, Z., Fülöp, Z. (eds.) DLT 2003. LNCS, vol. 2710, pp. 230–241. Springer, Heidelberg (2003)
3. D'Alessandro, F., Varricchio, S.: Well quasi-orders and context-free grammars. Theoretical Computer Science 327(3), 255–268 (2004)

4. D'Alessandro, F., Varricchio, S.: Avoidable sets and well quasi orders. In: Calude, C.S., Calude, E., Dinneen, M.J. (eds.) DLT 2004. LNCS, vol. 3340, pp. 139–150. Springer, Heidelberg (2004)

5. D'Alessandro, F., Varricchio, S.: Well quasi-orders, unavoidable sets, and derivation systems. RAIRO Theoretical Informatics and Applications 40, 407–426 (2006)

6. D'Alessandro, F., Richomme, G., Varricchio, S.: Well quasi orders and the shuffle closure of finite sets. In: H. Ibarra, O., Dang, Z. (eds.) DLT 2006. LNCS, vol. 4036, pp. 260–269. Springer, Heidelberg (2006)

7. D'Alessandro, F., Richomme, G., Varricchio, S.: Well quasi-orders and context-free grammars. Theoretical Computer Science 377(1-3), 73–92 (2007)

8. de Luca, A., Varricchio, S.: Some regularity conditions based on well quasi-orders. In: Simon, I. (ed.) LATIN 1992. LNCS, vol. 583, pp. 356–371. Springer, Heidelberg (1992)

9. de Luca, A., Varricchio, S.: Well quasi-orders and regular languages. Acta Informatica 31, 539–557 (1994)

10. de Luca, A., Varricchio, S.: Finiteness and regularity in semigroups and formal languages. EATCS Monographs on Theoretical Computer Science. Springer, Berlin (1999)

11. Ehrenfeucht, A., Haussler, D., Rozenberg, G.: On regularity of context-free languages. Theoretical Computer Science 27, 311–332 (1983)

12. Harju, T., Ilie, L.: On well quasi orders of words and the confluence property. Theoretical Computer Science 200, 205–224 (1998)

13. Haussler, D.: Another generalization of Higman's well quasi-order result on Σ^*. Discrete Mathematics 57, 237–243 (1985)

14. Ehrenfeucht, A., Rozenberg, G.: On regularity of languages generated by copying systems. Discrete Applied Mathematics 8, 313–317 (1984)

15. Higman, G.H.: Ordering by divisibility in abstract algebras. Proc. London Math. Soc. 3, 326–336 (1952)

16. Ilie, L., Salomaa, A.: On well quasi orders of free monoids. Theoretical Computer Science 204, 131–152 (1998)

17. Intrigila, B., Varricchio, S.: On the generalization of Higman and Kruskal's theorems to regular languages and rational trees. Acta Informatica 36, 817–835 (2000)

18. Ito, M., Kari, L., Thierrin, G.: Shuffle and scattered deletion closure of languages. Theoretical Computer Science 245(1), 115–133 (2000)

19. Jantzen, M.: Extending regular expressions with iterated shuffle. Theoretical Computer Science 38, 223–247 (1985)

20. Kruskal, J.: The theory of well quasi-ordering: a frequently discovered concept. J. Combin. Theory, Ser. A 13, 297–305 (1972)

21. Kunc, M.: The power of commuting with finite sets of words. In: Diekert, V., Durand, B. (eds.) STACS 2005. LNCS, vol. 3404, pp. 569–580. Springer, Heidelberg (2005)

22. Kunc, M.: Regular solutions of language inequalities and well quasi-orders. Theoretical Computer Science 348(2-3), 277–293 (2005)

23. Lothaire: Algebraic combinatorics on words. In: Encyclopedia of Mathematics and its applications. Cambridge University Press, Cambridge (2002)

24. Puel, L.: Using unavoidable sets of trees to generalize Kruskal's theorem. J. Symbolic Comput. 8(4), 335–382 (1989)

25. Sakarovitch, J.: Éléments de théorie des automates, Vuibert, Paris (2003)

26. Senizergues, G.: On the rational subsets of the free group. Acta Informatica 33(3), 281–296 (1996)

On the Non-deterministic Communication Complexity of Regular Languages

Anil Ada*

School of Computer Science, McGill University
aada@cs.mcgill.ca

Abstract. In this paper we study the non-deterministic communication complexity of regular languages. We show that a regular language has either constant or at least logarithmic non-deterministic communication complexity. We prove several linear lower bounds which we know cover a wide range of regular languages with linear complexity. Furthermore we find evidence that previous techniques (Tesson and Thérien 2005) for proving linear lower bounds, for instance in deterministic and probabilistic models, do not work in the non-deterministic setting.

1 Introduction

The notion of communication complexity was introduced by Yao [16] in light of its applications to parallel computers. Following this seminal work, it has been shown to have many more applications where the need for communication is not explicit and thus has become the "Swiss Army knife" of complexity theory. These applications include time/space lower bounds for VLSI chips [9], time/space tradeoffs for Turing Machines [3], data structures [9], boolean circuit lower bounds [6,8], pseudorandomness [3], separation of proof systems [4] and lower bounds on the size of polytopes representing NP-complete problems [15].

It is an intriguing task to better understand the landscape of communication complexity and thus other areas of complexity theory. A natural starting point is to comprehend the complexity of regular languages, which in some sense are the simplest languages with respect to the usual time/space complexity framework. Perhaps surprisingly, regular languages form a non-trivial case study with respect to communication complexity. There are hard regular languages even in very powerful models of communication complexity. Furthermore, some of the very well-known and studied functions in this area such as Disjointness and Inner Product are equivalent to regular languages from a communication complexity perspective.

In [13], it was established that the class of regular languages having $O(f)$ deterministic communication complexity forms a language variety and so the question of the communication complexity of regular languages has an algebraic answer. In a follow up work [14], a complete algebraic characterization of the

* Supported by the research grants of Prof. Denis Thérien.

communication complexity of regular languages was established in the deterministic, simultaneous, probabilistic, simultaneous probabilistic and Mod_p-counting models. These results unmasked an interesting complexity gap: In all of the above models, the complexity of a regular language falls into one of four classes $O(1), \Theta(\log \log n), \Theta(\log n)$ or $\Theta(n)$. In contrast, we note that for any function f with $1 \le f \le n$, it is possible to construct a non-regular language with complexity $\Theta(f)$ for any of these models.

In this paper we are interested in the non-deterministic communication complexity of regular languages. To get a similar characterization for the non-deterministic model, one needs the notions of *positive language varieties* and *ordered monoids*. This is because the syntactic monoid of a regular language does not distinguish between a language and its complement. Differing from the models mentioned earlier, non-deterministic complexity of a function and its complement may not be equal. So regular languages having $O(f)$ non-deterministic communication complexity do not form a variety but a positive variety.

Adopting this refined approach, we take the first steps towards a complete classification for the non-deterministic communication complexity of regular languages. We identify the regular languages having constant non-deterministic complexity. We show that if a regular language does not have constant complexity than it has $\Omega(\log n)$ complexity, revealing a complexity gap. We also obtain several linear lower bound results which we know cover a wide range of regular languages having linear complexity. These bounds point out sufficient conditions for not being in the positive variety $Pol(\mathcal{C}om)$, providing us with some nice combinatorial intuition about this variety. Finally we find evidence that previous techniques used in [14] for proving linear lower bounds, for instance in deterministic and probabilistic models, do not work in the non-deterministic setting.

Organization. In Sect. 2 and Sect. 3, we give the necessary background on algebraic automata theory and communication complexity respectively. In Sect. 4, we define the communication complexity of a regular language and a monoid. Furthermore, we show that the non-deterministic communication complexity of regular languages admits an algebraic characterization. Section 5 is devoted to the bounds we have on the non-deterministic communication complexity of regular languages and ordered monoids.

2   Algebraic Automata Theory

We refer the reader to [11] for further background on algebraic automata theory with an emphasis on the more general theory of ordered monoids.

A *monoid* $(M, \cdot)$ is a set M together with an associative binary operation $\cdot$ and an identity $1_M \in M$ which satisfies $1_M \cdot m = m \cdot 1_M = m$ for any $m \in M$. An *order* relation on a set S is a relation that is reflexive, anti-symmetric and transitive and it is denoted by $\le$. We say that $\le$ is a *stable order relation* on a monoid M if for all $x, y, z \in M$, $x \le y$ implies $zx \le zy$ and $xz \le yz$. An *ordered monoid* $(M, \le_M)$ is a monoid M together with a stable order relation $\le_M$ that is defined on M. A *morphism of ordered monoids* $\Phi : (M, \leq_M) \rightarrow (N, \leq_N)$ is a morphism between M and N that also preserves the order relation, i.e. for all $m, m' \in M$, $m \leq_M m'$ implies $\Phi(m) \leq_N \Phi(m')$.

A subset $I \subseteq M$ is called an *order ideal* if for any $y \in I$, $x \leq_M y$ implies $x \in I$. Every order ideal I in a finite monoid M has a generating set $x_1, ..., x_k$ such that $I = \langle x_1, ..., x_k \rangle := \{y \in M : \exists x_i \text{ with } y \leq_M x_i\}$. We say that a language $L \subseteq \Sigma^*$ is *recognized* by an ordered monoid $(M, \leq_M)$ if there exists a morphism of ordered monoids $\Phi : (\Sigma^*, =) \rightarrow (M, \leq_M)$ and an order ideal $I \subseteq M$ such that $L = \Phi^{-1}(I)$.

Define the *syntactic congruence* as follows: $x \equiv_L y$ if for all $u, v \in \Sigma^*$ we have $uxv \in L$ iff $uyv \in L$. The *syntactic monoid* is the quotient monoid $M(L) = \Sigma^* / \equiv_L$. Let $x \preceq_L y$ if for all $u, v \in \Sigma^*$, $uyv \in L \implies uxv \in L$. So $x \equiv_L y$ if and only if $x \preceq_L y$ and $y \preceq_L x$. Now $\preceq_L$ induces a well-defined stable order $\leq_L$ on $M(L)$ given by $[x] \leq_L [y]$ if and only if $x \preceq_L y$. The ordered monoid $(M(L), \leq_L)$ is the *syntactic ordered monoid* of L.

We say that an ordered monoid $(N, \leq_N)$ *divides* an ordered monoid $(M, \leq_M)$ if there exists a surjective morphism of ordered monoids from a submonoid of $(M, \leq_M)$ onto $(N, \leq_N)$. We know that $(M(L), \leq_L)$ recognizes L and divides any other ordered monoid that also recognizes L.

We say that a family of ordered monoids $\mathbf{V}$ is a *variety of ordered monoids* if it is closed under division of ordered monoids and finite direct product[1]. A class of languages $\mathcal{V}$ is called a *positive variety of languages* if it is closed under finite intersection, finite union, inverse morphisms, left and right quotients.

Given a variety of finite ordered monoids $\mathbf{V}$, let $\mathcal{V}$ be the set of languages whose syntactic ordered monoid belongs to $\mathbf{V}$. The Variety Theorem originally due to Eilenberg [5] and adapted to the ordered case by Pin [10] states that $\mathcal{V}$ is a positive variety of languages and the mapping $\mathbf{V} \mapsto \mathcal{V}$ is one to one.

The *polynomial closure* of a set of languages $\mathcal{L}$ in Σ^* is a family of languages such that each of them is a finite union of $L_0 a_1 L_1 ... a_k L_k$, where $k \geq 0$, $a_i \in \Sigma$ and $L_i \in \mathcal{L}$. If $\mathcal{V}$ is a variety of languages, then we denote by $Pol(\mathcal{V})$ the class of languages that is the polynomial closure of $\mathcal{V}$. We know that $Pol(\mathcal{V})$ is a positive variety [12].

We say that the concatenation $L_0 a_1 L_1 ... a_k L_k$ is *unambiguous* if all words $x \in L_0 a_1 L_1 ... a_k L_k$ has a unique factorization $x = w_0 a_1 w_1 ... a_k w_k$ with $w_i \in L_i$. We denote by $UPol(\mathcal{V})$ the variety of languages consisting of *disjoint* unions of unambiguous concatenations $L_0 a_1 L_1 ... a_k L_k$ with $L_i \in \mathcal{V}$ (in some sense, there is only one witness for x in $L \in UPol(\mathcal{V})$). Similarly we denote by $Mod_p Pol(\mathcal{V})$ the language variety generated by the languages for which membership depends on the number of factorizations mod p.

An element $e \in M$ is called *idempotent* if $e^2 = e$. For any finite M, there is a number $k > 0$ such that for every element $m \in M$, m^k is an idempotent. We call k an *exponent* of M.

---

[1] The order in a finite direct product $M_1 \times ... \times M_n$ is given by $(m_1, ..., m_n) \leq (m'_1, ..., m'_n)$ iff $m_i \leq m'_i \quad \forall i \in [n]$.

3   Communication Complexity

We present here a quick introduction to communication complexity but refer the reader to the great book of Kushilevitz and Nisan [9] for further details.

In the deterministic model, two players, Alice and Bob, wish to compute a function $f : S^{n_A} \times S^{n_B} \to T$ where S and T are finite sets. Alice is given $x \in S^{n_A}$ and Bob $y \in S^{n_B}$ and they collaborate in order to obtain $f(x,y)$ by exchanging bits using a common blackboard according to some predetermined *communication protocol* $\mathcal{P}$. This protocol determines whose turn it is to write, furthermore what a player writes is a function of that player's input and the information exchanged thus far. When the protocol ends, its output $\mathcal{P}(x,y) \in T$ is a function of the blackboard's content. We say that $\mathcal{P}$ computes f is $\mathcal{P}(x,y) = f(x,y)$ for all x, y and define the *cost* of $\mathcal{P}$ as the maximum number of bits exchanged for any input. The *deterministic communication complexity* of f, denoted $D(f)$ is the cost of the cheapest protocol computing f. We will be interested in the complexity of functions $f : S^* \times S^* \to T$ and will thus consider $D(f)$ as a function from $\mathbb{N} \times \mathbb{N}$ to $\mathbb{N}$ and study its asymptotic behaviour.

In a *non-deterministic communication protocol* $\mathcal{P}$ another player, say God, having access to *both* x and y first sends to Alice and Bob a proof π. Alice and Bob then follow an ordinary deterministic protocol $\mathcal{P}'$ with output in $\{0, 1\}$. The protocol $\mathcal{P}$ accepts the input (x, y) if and only if there is some proof π such that the output of the ensuing deterministic protocol $\mathcal{P}'$ outputs 1. The cost of a non-deterministic protocol is the maximum number of bits exchanged in the protocol (*including* the bits of π) for any input (x, y). We denote the non-deterministic communication complexity of a language L as $N^1(L)$. The co-non-deterministic communication complexity of L, denoted $N^0(L)$ is the non-deterministic communication complexity of L's complement.

Let $PDISJ$ be the following promise problem. Alice gets a set $x \subseteq [n]$ and Bob a set $y \subseteq [n]$ with the guarantee that $|x \cap y| \leq 1$ and $PDISJ(x, y) = 1$ if and only if $x \cap y = \emptyset$. One can show $N^1(PDISJ) = \Omega(n)$ (see [1]). Define two more problems: $LT(x, y) = 1$ iff $x \leq y$ when x and y are viewed as n-bit integers; $IP_q(x, y) = 1$ iff $\sum_{i=1}^n x_i y_i \equiv 0 \mod q$. It is well known that both functions have $\Omega(n)$ non-deterministic communication complexity.

Communication complexity classes were introduced in [2] in which an "efficient" protocol was defined to have cost no more than poly-logarithmic, i.e. $O(\log^c n)$ for a constant c. Thus one obtains communication complexity classes analogous to P and NP in the following way: $P^{cc} := \{f | D(f) = polylog(n)\}$, $NP^{cc} := \{f | N^1(f) = polylog(n)\}$.

4   Algebraic Approach to Communication Complexity

In general, we want to study the communication complexity of functions which do not explicitly have two inputs. In the case of regular languages and ordered monoids we use a form of *worst-case partition* definition. Formally, the communication complexity of a pair (M, I) where M is a finite ordered monoid and I is an order ideal in M is the communication complexity of the monoid evaluation problem corresponding to M and I: Alice is given $m_1, m_3, ..., m_{2n-1}$ and Bob is given $m_2, m_4, ..., m_{2n}$ such that each $m_i \in M$. They want to decide if the product $m_1 m_2 ... m_{2n}$ is in I. The communication complexity of M is the maximum complexity of (M, I) where I ranges over all order ideals in M.

Similarly, the *communication complexity of a regular language* $L \subseteq A^*$ is the communication complexity of the following problem: Alice and Bob respectively receive $a_1, a_3, ... a_{2n-1}$ and $a_2, a_4, ..., a_{2n}$ where each a_i is either in A or is the neutral letter ϵ and they want to determine whether $a_1 a_2 ... a_{2n}$ belongs to L.

The following two lemmas establish the soundness of an algebraic approach to the communication complexity of regular languages.

Lemma 1. *Let $L \subseteq A^*$ be regular and $M = M(L)$. Then $N^1(M) = \Theta(N^1(L))$.*

Proof. It is straightforward to show $N^1(L) = O(N^1(M))$. To show $N^1(M) = O(N^1(L))$, we present a protocol for (M, I) where $I = \langle i_1, ..., i_k \rangle$ is some order ideal in M.

Let Φ be the accepting morphism. For each monoid element m, fix a word that is in the preimage of m under Φ, and denote it by w_m. Let $Y_a := \{(u, v) : uav \in L\}$. Recall that $a \preceq_L b$ if for all $u, v \in \Sigma^*$, $ubv \in L \implies uav \in L$. So $\Phi(a) \leq_L \Phi(b)$ iff $a \preceq_L b$ iff $Y_b \subseteq Y_a$. For each Y_a and Y_b with $Y_b \not\subseteq Y_a$, pick (u, v) such that $(u, v) \in Y_b$ but $(u, v) \notin Y_a$. Let K be the set of all these (u, v). One can think of K as containing a witness for $Y_b \not\subseteq Y_a$ for each such pair. Note that K is finite. Now pad each w_m and each word appearing in a pair in K with the neutral letter ϵ so that each of these words have the same constant length.

Now the protocol is as follows. Suppose Alice is given $m_1^a, m_2^a, ..., m_n^a$ and Bob is given $m_1^b, m_2^b, ..., m_n^b$. For each i_j they want to determine if $m_1^a m_1^b ... m_n^a m_n^b \leq_L i_j$. This is equivalent to determining if $w_{m_1^a m_1^b ... m_n^a m_n^b} \preceq_L w_{i_j}$, which is equivalent to $w_{m_1^a} w_{m_1^b} ... w_{m_n^a} w_{m_n^b} \preceq_L w_{i_j}$. If this is not the case, $Y_{w_{i_j}} \not\subseteq Y_{w_{m_1^a} w_{m_1^b} ... w_{m_n^a} w_{m_n^b}}$ and so there will be a witness of this in K, i.e. there exists (u, v) such that $u w_{i_j} v \in L$ but $u w_{m_1^a} w_{m_1^b} ... w_{m_n^a} w_{m_n^b} v \notin L$. If indeed $w_{m_1^a} w_{m_1^b} ... w_{m_n^a} w_{m_n^b} \preceq_L w_{i_j}$ then for each $(u, v) \in K$ with $u w_{i_j} v \in L$, we will have $u w_{m_1^a} w_{m_1^b} ... w_{m_n^a} w_{m_n^b} v \in L$. Using the protocol for L, Alice and Bob check which of the two cases is true. $\square$

In particular the non-deterministic complexity of an ordered monoid M is, up to a constant, the maximal communication complexity of any regular language that it can recognize.

Lemma 2. *For any increasing $f : \mathbb{N} \to \mathbb{N}$ the class of monoids such that $N^1(M)$ is $O(f)$ forms a variety of ordered monoids.*

Proof. The closure of this class under direct product is obvious. Suppose $N \prec M$, so there is a surjective morphism ϕ from a submonoid M' of M onto N. Denote by $\phi^{-1}(n)$ a fixed element from the preimage of n. Let I be an order ideal in N. A protocol for (N, I) is as follows. Alice is given $n_1^a, n_2^a, ..., n_t^a$ and Bob is given $n_1^b, n_2^b, ..., n_t^b$. They want to decide if $n_1^a n_1^b ... n_t^a n_t^b \in I$. This is equivalent to deciding if $\phi^{-1}(n_1^a) \phi^{-1}(n_1^b) ... \phi^{-1}(n_t^a) \phi^{-1}(n_t^b) \in \phi^{-1}(I)$. It is easy to see $\phi^{-1}(I)$

is an order ideal in M' so Alice and Bob can use the protocol for M' to decide if the above is true. Therefore we have $N^1(N) \leq N^1(M')$. It is straightforward to check that $N^1(M') \leq N^1(M)$ and so $N^1(N) \leq N^1(M)$ as required.   □

To compare the communication complexity of two languages K, L in different models, Babai et al. [2] defined *rectangular reductions* from K to L which are, intuitively, reductions which can be computed privately by Alice and Bob without any communication cost. We give here a form of this definition which specifically suits our needs. Let $u = u_1 u_2 \dots u_k$ be a word over M, i.e. $u \in M^*$. We denote by $eval(u)$ the corresponding monoid element, i.e. $eval(u) = u_1 \cdot \dots \cdot u_k$.

Definition 3. *Let $f : \{0,1\}^n \times \{0,1\}^n \to \{0,1\}$, M a finite ordered monoid and I an order ideal in M. A* rectangular reduction *of length t from f to (M, I) is a sequence of $2t$ functions $a_1, b_2, a_3, \dots, a_{2t-1}, b_{2t}$ with $a_i : \{0,1\}^n \to M$ and $b_i : \{0,1\}^n \to M$ and such that for every $x, y \in \{0,1\}^n$ we have $f(x,y) = 1$ if and only if $eval(a_1(x)b_2(y)\dots b_{2t}(y))$ is in I.*

Such a reduction transforms an input (x, y) of the function f into a sequence of $2t$ monoid elements $m_1, m_2, \dots, m_{2t}$ where the odd-indexed m_i are obtained as a function of x only and the even-indexed m_i are a function of y.

We write $f \leq_r^t (M, I)$ to indicate that f has a rectangular reduction of length t to (M, I). When $t = O(n)$ we omit the superscript t. It should be clear that if $f \leq_r^t (M, I)$ and f has communication complexity $\Omega(g(n))$, then (M, I) has communication complexity $\Omega(g(t^{-1}(n)))$.

We will be interested in a special kind of rectangular reduction which we call a *local rectangular reduction*. In a local rectangular reduction, Alice converts each bit x_i to a sequence of s monoid elements $m_{i,1}^a, m_{i,2}^a, \dots, m_{i,s}^a$ by applying a fixed function $a : \{0,1\} \to M^s$. Similarly Bob converts each bit y_i to a sequence of s monoid elements $m_{i,1}^b, m_{i,2}^b, \dots, m_{i,s}^b$ by applying a fixed function $b : \{0,1\} \to M^s$. $f(x,y) = 1$ iff $eval(m_{1,1}^a m_{1,1}^b \dots m_{1,s}^a m_{1,s}^b \dots \dots m_{n,1}^a m_{n,1}^b \dots m_{n,s}^a m_{n,s}^b) \in I$. The reduction transforms an input (x, y) into a sequence of $2sn$ monoid elements. Let $a(z)_k$ denote the kth coordinate of the tuple $a(z)$. We specify this kind of local transformation with a $2 \times 2s$ matrix:

$a(0)_1$	$b(0)_1$	$a(0)_2$	$b(0)_2$	...	...	$a(0)_s$	$b(0)_s$
$a(1)_1$	$b(1)_1$	$a(1)_2$	$b(1)_2$	...	...	$a(1)_s$	$b(1)_s$

It is convenient to see which words the transformation produces for all possible values of x_i and y_i. For simplicity let us assume s is even.

x_i y_i	corresponding word over M
0 0	$a(0)_1 b(0)_1 \dots a(0)_s b(0)_s$
0 1	$a(0)_1 b(1)_1 a(0)_2 b(1)_2 \dots a(0)_s b(1)_s$
1 0	$a(1)_1 b(0)_1 a(1)_2 b(0)_2 \dots a(1)_s b(1)_s$
1 1	$a(1)_1 b(1)_1 \dots a(1)_s b(1)_s$

5   Bounds for Regular Languages and Monoids

Lemma 4 ([14]). *If M is commutative then $D(M)=O(1)$ and thus $N^1(M)=O(1)$.*

Lemma 5 (Adapted from [14]). *If M is not commutative then for any order on M we have $N^1(M) = \Omega(\log n)$.*

Proof. Since M is not commutative, there must be $a, b \in M$ such that $ab \neq ba$. Therefore either $ab \not\leq_M ba$ or $ba \not\leq_M ab$. W.l.o.g. assume $ba \not\leq_M ab$. Let $I = \langle ab \rangle$. We show that $LT \leq_r^{2^n} (M, I)$. Alice gets x and constructs a sequence of 2^n monoid elements in which a is in position x and 1_M is in everywhere else. Bob gets y and constructs a sequence of 2^n monoid elements in which b is in position y and 1_M is in everywhere else. If $x \leq y$ then the product of the monoid elements is ab which is in I. If $x > y$ then the product is ba which is not in I.   □

Denote by $\mathcal{C}om$ the positive language variety corresponding to the variety of commutative monoids **Com**. The above two results show that regular languages that have constant non-deterministic communication complexity are exactly those languages in $\mathcal{C}om$.

Lemma 6. *If $L \subseteq A^*$ is a language of $Pol(\mathcal{C}om)$ then $N^1(L) = O(\log n)$.*

Proof. Suppose L is a union of t languages of the form $L_0 a_1 L_1 ... a_k L_k$. Alice and Bob know beforehand the value of t and the structure of each of these t languages. So a protocol for L is as follows. Assume Alice is given $x_1^a, ..., x_n^a$ and Bob is given $x_1^b, ..., x_n^b$. God communicates to Alice and Bob which of the t languages the word $x_1^a x_1^b ... x_n^a x_n^b$ resides in. This requires a constant number of bits to be communicated since t is a constant. Then God communicates the positions of each a_i. This requires $k \log n$ bits of communication where k is a constant. The validity of the information communicated by God can be checked by Alice and Bob by checking if the words in between the a_i's belong to the right languages. Since these languages are in $\mathcal{C}om$, this can be done in constant communication. Therefore in total we require only $O(\log n)$ communication.   □

From the above proof, we see that we can actually afford to communicate $O(\log n)$ bits to check that the words between the a_i's belong to the corresponding language. In other words, we could have $L_i \in Pol(\mathcal{C}om)$. Note that this does not mean that this protocol works for a strictly bigger class since $Pol(Pol(\mathcal{C}om)) = Pol(\mathcal{C}om)$.

Denote by UP the subclass of NP in which the languages are accepted by a non-deterministic Turing Machine having *exactly* one accepting path (or one witness) for each string in the language. It is known that $UP^{cc} = P^{cc}$ ([15]). From [14] we know that regular languages having $O(\log n)$ deterministic communication complexity are exactly those languages in $UPol(\mathcal{C}om)$ and regular languages having $O(\log n)$ Mod_p counting communication complexity are exactly those languages in $Mod_p Pol(\mathcal{C}om)$. Furthermore, it was shown that any regular language outside of $UPol(\mathcal{C}om)$ has linear deterministic complexity and any regular language outside of $Mod_p Pol(\mathcal{C}om)$ has linear Mod_p counting complexity. So with respect to regular languages, $UP^{cc} = P^{cc} = UPol(Com)$ and $Mod_pP^{cc} = Mod_pPol(Com)$. Similarly we conjecture that with respect to regular languages $NP^{cc} = Pol(Com)$ and that other regular languages have linear non-deterministic complexity.

Conjecture 7. *If $L \subseteq \Sigma^*$ is a regular language that is not in $Pol(Com)$, then $N^1(L) = \Omega(n)$. Thus we have*

$$N^1(L) = \begin{cases} O(1) & \text{if and only if } L \in Com; \\ \Theta(\log n) & \text{if and only if } L \in Pol(Com) \text{ but not in } Com; \\ \Theta(n) & \text{otherwise.} \end{cases}$$

In general, the gap between deterministic and non-deterministic communication complexity of a function can be exponentially large. However, it has been shown that the deterministic communication complexity of a function f is bounded above by the product $cN^0(f)N^1(f)$ for a constant c and that this bound is optimal [7]. The above conjecture, together with the result of [14] imply the following much tighter relation for regular languages.

Conjecture 8 (Corollary to Conjecture 7). *If L is a regular language then $D(L) = \max\{N^1(L), N^0(L)\}$.*

For any variety $\mathcal{V}$, we have that $Pol(\mathcal{V}) \cap co\text{-}Pol(\mathcal{V}) = UPol(\mathcal{V})$ [11]. This implies that $N^1(L) = O(\log n)$ and $N^0(L) = O(\log n)$ iff $D(L) = O(\log n)$, proving a special case of the above corollary.

Conjecture 7 suggests that when faced with a non-deterministic communication problem for regular languages, the players have three options. They can either follow a trivial protocol that does not exploit the power of non-determinism or apply non-determinism in the most natural way as for the complement of the functions Disjointness and Equality. Otherwise the best protocol up to a constant factor is for one of the players to send all of his/her bits to the other player, a protocol that works for any function in any model. So with respect to regular languages, there is no "tricky" way to apply non-determinism to obtain cleverly efficient protocols.

To prove a linear lower bound for the regular languages outside of $Pol(Com)$, we need a convenient algebraic description for the syntactic monoids of these languages since in most cases lower bound arguments rely on these algebraic properties. So an important question that arises in this context is: What does it mean to be outside of $Pol(Com)$? An algebraic description exists based on a result of [12] that describes the ordered monoid variety corresponding to $Pol(Com)$.

Lemma 9. *Suppose L is not in $Pol(Com)$ and M is the syntactic ordered monoid of L with exponent ω. Then there exists $u, v \in M^*$ such that*

(i) *for any monoid $M' \in \mathbf{Com}$ and any morphism $\phi : M \to M'$, we have $\phi(eval(u)) = \phi(eval(v))$ and $\phi(eval(u)) = \phi(eval(u^2))$,*

(ii) *$eval(u^\omega vu^\omega) \not\leq eval(u^\omega)$.*

We now present the linear lower bound results. The proofs of the next two lemmas can be adapted from [14] to the non-deterministic case using the following simple fact.

Proposition 10. *Any stable order defined on a group G must be the trivial order (equality).*

Proof. Assume the claim is false, so there exists $a, b \in G$ such that $a \neq b$ and $a \leq b$. This means $1 \leq a^{-1}b =: g$. Since $1 \leq g$, we have $1 \leq g \leq g^2 \leq \ldots \leq g^k = 1$ for some k. This implies $1 = g$, i.e. $a = b$. □

Lemma 11. *If M is a non-commutative group then $N^1(M) = \Omega(n)$.*

We say that M is a T_q monoid if there exists idempotents $e, f \in M$ such that $(ef)^q e = e$ but $(ef)^r e \neq e$ when q does not divide r.

Lemma 12. *If M is a T_q monoid for $q > 1$ then $N^1(M) = \Omega(n)$.*

The next lemma captures regular languages that come close to the description of Lemma 9. A word w is a *shuffle* of n words $w_1, \ldots, w_n$ if

$$w = w_{1,1}w_{2,1}\ldots w_{n,1}w_{1,2}w_{2,2}\ldots w_{n,2}\cdots\cdots w_{1,k}w_{2,k}\ldots w_{n,k}$$

with $k \geq 0$ and $w_{i,1}w_{i,2}\ldots w_{i,k} = w_i$ is a partition of w_i into subwords for $1 \leq i \leq n$.

Lemma 13. *If M and $u, v \in M^*$ are such that (i) $u = w_1w_2$ for $w_1, w_2 \in M^*$, (ii) v is a shuffle of w_1 and w_2, (iii) $eval(u)$ is an idempotent, and (iv) $eval(uvu) \not\leq eval(u)$, then $N^1(M) = \Omega(n)$.*

Proof. We show that $PDISJ \leq_r (M, I)$ where $I = \langle eval(u)\rangle$. Since v is a shuffle of w_1 and w_2, there exists $k \geq 0$ such that $v = w_{1,1}w_{2,1}w_{1,2}w_{2,2}\ldots w_{1,k}w_{2,k}$. The reduction is essentially local and is given by the following matrix when $k = 3$. The transformation easily generalizes to any k.

w_1	ϵ	ϵ	ϵ	ϵ	$w_{2,1}$	ϵ	$w_{2,2}$	ϵ	$w_{2,3}$
$w_{1,1}$	$w_{2,1}$	$w_{1,2}$	$w_{2,2}$	$w_{1,3}$	$w_{2,3}$	ϵ	ϵ	ϵ	ϵ

$x_i\ y_i$	corresponding word
0 0	$w_1w_{2,1}w_{2,2}w_{2,3} = u$
0 1	$w_1w_{2,1}w_{2,2}w_{2,3} = u$
1 0	$w_{1,1}w_{1,2}w_{1,3}w_{2,1}w_{2,2}w_{2,3} = u$
1 1	$w_{1,1}w_{2,1}w_{1,2}w_{2,2}w_{1,3}w_{2,3} = v$

After x and y have been transformed into words, Alice prepends her word with u and appends it with $|u|$ many ϵ's, where $|u|$ denotes the length of the word u. Bob prepends his word with $|u|$ many ϵ's and appends it with u. Let $a(x)$ be the word Alice has and let $b(y)$ be the word Bob has after these transformations. If $PDISJ(x, y) = 0$, there exists i such that $x_i = y_i = 1$. By the transformation, this means $a(x)_1 b(x)_1 a(x)_2 b(x)_2 ... a(x)_s b(x)_s$ is of the form $u...uvu...u$ and since $eval(u)$ is idempotent, $eval(a(x)_1 b(x)_1 a(x)_2 b(x)_2 ... a(x)_s b(x)_s) = eval(uvu) \not\leq eval(u)$. If $PDISJ(x,y) = 1$, then by the transformation, $a(x)_1 b(x)_1 ... a(x)_s b(x)_s$ is of the form $u...u$ and so $eval(a(x)_1 b(x)_1 a(x)_2 b(x)_2 ... a(x)_s b(x)_s) = eval(u)$. Thus $PDISJ \leq_r (M, \langle eval(u) \rangle)$.   □

The conditions of this lemma imply the conditions of Lemma 9: since $eval(u)$ is idempotent, for any monoid $M' \in \mathbf{Com}$ and any morphism $\phi : M \to M'$, we have $\phi(eval(u)) = \phi(eval(u^2))$ and since v is a shuffle of w_1 and w_2 we have $\phi(eval(u)) = \phi(eval(v))$. Also, since $eval(u)$ is idempotent, $eval(u^\omega) = eval(u)$, and in this case $eval(uvu) \not\leq eval(u)$ is equivalent to $eval(u^\omega v u^\omega) \not\leq eval(u^\omega)$.

Lemma 13 gives us a corollary about the monoid BA_2^+ which is defined to be the syntactic ordered monoid of $(ab)^* \cup a(ba)^*$. The syntactic ordered monoid of the complement of this language is BA_2^-. The unordered syntactic monoid is denoted by BA_2 and is known as the Brandt monoid.

Corollary 14. $N^1(BA_2^+) = \Omega(n)$.

Proof. It is easy to verify that BA_2^+ is the monoid $\{a,b\}^*$ with the relations $aa = bb, aab = aa, baa = aa, aaa = a, aba = a, bab = b$. All we need to know about the order relation is that $eval(aa)$ is greater than any other element. This can be derived from the definition of the syntactic ordered monoid since for any w_1 and w_2, $w_1 aa w_2$ is not in L. So $w_1 aa w_2 \in L \implies w_1 x w_2 \in L$ trivially holds for any word x. Let $u = ab$ and $v = ba$. These u and v satisfy the four conditions of Lemma 13. The last condition is satisfied because $eval(uvu) = eval(abbaab) = eval(aa)$ and $eval(ab) \neq eval(aa)$. Thus $N^1(BA_2^+) = \Omega(n)$.   □

Denote by U^- the syntactic ordered monoid of the regular language $(a \cup b)^* aa(a \cup b)^*$. The syntactic ordered monoid of the complement of this language is U^+. The unordered syntactic monoid is denoted by U. Observe that $N^1(U^-) = O(\log n)$ since all we need to do is check if there are two consecutive a's. One also easily sees that $N^1(BA_2^-) = O(\log n)$. By an argument similar to the one for Corollary 14, one can show that $N^1(U^+) = \Omega(n)$.

Combining the linear lower bound results we can conclude the following.

Theorem 15. *If M is a T_q monoid for $q > 1$ or is divided by one of BA_2^+, U^+ or a non-commutative group, then $N^1(M) = \Omega(n)$.*

We underline the relevance of the above result by stating a theorem which we borrow from [14].

Theorem 16 (implicit in [14]). *If M is such that $D(M) \neq O(\log n)$ then M is either a T_q monoid for some $q > 1$ or is divided by one of BA_2, U or a non-commutative group.*

As a consequence, we know that if an ordered monoid M is such that $N^1(M) \neq O(\log n)$ then M is either a T_q monoid or is divided by one of BA_2^+, BA_2^-, U^+, U^- or a non-commutative group.

As a corollary to Theorem 15 and Lemma 6 we have:

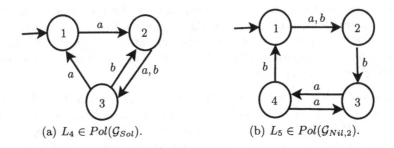

(a) $L_4 \in Pol(\mathcal{G}_{Sol})$.   (b) $L_5 \in Pol(\mathcal{G}_{Nil,2})$.

Fig. 1. Two examples. The missing arrows go to an accepting sink state.

Corollary 17. *If $M(L)$ is a T_q monoid or is divided by one of BA_2^+, U^+ or a non-commutative group, then L is not in $Pol(Com)$.*

Consider the syntactic ordered monoid of the regular language recognized by the automaton in Fig. 1(a). One can show that it does not contain a non-commutative group, is not a T_q monoid and is not divided by BA_2^+ nor U^+. On the other hand, using Lemma 13 with $u = abbaa$ and $v = aabab$ we can show that it requires linear non-deterministic communication. Thus this lower bound is not achievable by previously known methods and highlights the importance of Lemma 13.

The regular language L_5 accepted by the automaton in Fig. 1(b) is a concrete example of a language not in $Pol(Com)$ and where all our techniques fail. In particular, it shows that the conditions in Lemma 13 do not cover every regular language outside of $Pol(Com)$. We know that L_5 lies in $Pol(\mathcal{G}_{Nil,2})$ where $\mathcal{G}_{Nil,2}$ denotes the variety of languages whose syntactic monoid is a nilpotent group of class 2. These groups are "almost" commutative so L_5 in some sense comes close to being in $Pol(Com)$.

For the deterministic and probabilistic models where $PDISJ$ is a hard function, one can observe that all the linear lower bounds obtained in [14] go through a local rectangular reduction from $PDISJ$ since $PDISJ$ reduces both to Disjointness and Inner Product. One might hope to obtain all the non-deterministic lower bounds in this manner as well. Given Lemma 9, one would want a local reduction of the form

x_i y_i	corresponding word
0 0	u^ω
0 1	u^ω
1 0	u^ω
1 1	v

where u and v satisfy the conditions of Lemma 9.

Lemma 18. *There is no local reduction from $PDISJ$ to L_5 as described above.*

Acknowledgements. The author gratefully acknowledges Pascal Tesson and Denis Thérien for introducing him to the problem and for very insightful discussions. We also thank Jean-Eric Pin whose valuable input has been acquired through Pascal Tesson.

References

1. Ada, A.: Non-deterministic communication complexity of regular languages. Master's thesis (2008)
2. Babai, L., Frankl, P., Simon, J.: Complexity classes in communication complexity theory (preliminary version). In: FOCS 1986: Proceedings of the 27th Annual IEEE Symposium on Foundations of Computer Science, pp. 337–347 (1986)
3. Babai, L., Nisan, N., Szegedy, M.: Multiparty protocols, pseudorandom generators for logspace, and time-space trade-offs. J. Comput. Syst. Sci. 45(2), 204–232 (1992)
4. Beame, P., Pitassi, T., Segerlind, N.: Lower bounds for Lovasz-Schrijver systems and beyond follow from multiparty communication complexity. SIAM Journal on Computing 37(3), 845–869 (2007)
5. Eilenberg, S.: Automata, Languages, and Machines. Academic Press, Inc., Orlando (1974)
6. Grolmusz, V.: Separating the communication complexities of MOD m and MOD p circuits. In: IEEE Symposium on Foundations of Computer Science, pp. 278–287 (1992)
7. Halstenberg, B., Reischuk, R.: On different modes of communication. In: STOC 1988: Proceedings of the twentieth annual ACM symposium on Theory of computing, pp. 162–172. ACM, New York (1988)
8. Håstad, J., Goldmann, M.: On the power of small-depth threshold circuits. Computational Complexity 1, 113–129 (1991)
9. Kushilevitz, E., Nisan, N.: Communication Complexity. Cambridge University Press, Cambridge (1997)
10. Pin, J.-E.: A variety theorem without complementation. Russian Mathematics (Izvestija vuzov.Matematika) 39, 80–90 (1995)
11. Pin, J.-E.: Syntactic semigroups. In: Rozenberg, G., Salomaa, A. (eds.) Handbook of formal languages, ch. 10, vol. 1, pp. 679–746. Springer, Heidelberg (1997)
12. Pin, J.-E., Weil, P.: Polynomial closure and unambiguous product. In: Fülöp, Z., Gecseg, F. (eds.) ICALP 1995. LNCS, vol. 944, pp. 348–359. Springer, Heidelberg (1995)
13. Raymond, J.-F., Tesson, P., Thérien, D.: An algebraic approach to communication complexity. In: Larsen, K.G., Skyum, S., Winskel, G. (eds.) ICALP 1998. LNCS, vol. 1443, pp. 29–40. Springer, Heidelberg (1998)
14. Tesson, P., Thérien, D.: Complete classifications for the communication complexity of regular languages. Theory Comput. Syst. 38(2), 135–159 (2005)
15. Yannakakis, M.: Expressing combinatorial optimization problems by linear programs. Journal of Computer and System Sciences 43(3), 441–466 (1991)
16. Yao, A.C.-C.: Some complexity questions related to distributive computing (preliminary report). In: STOC 1979: Proceedings of the eleventh annual ACM symposium on Theory of computing, pp. 209–213. ACM Press, New York (1979)

General Algorithms for Testing
the Ambiguity of Finite Automata

Cyril Allauzen[1,*], Mehryar Mohri[1,2], and Ashish Rastogi[1,*]

[1] Google Research,
76 Ninth Avenue, New York, NY 10011
[2] Courant Institute of Mathematical Sciences,
251 Mercer Street, New York, NY 10012

Abstract. This paper presents efficient algorithms for testing the finite, polynomial, and exponential ambiguity of finite automata with ϵ-transitions. It gives an algorithm for testing the exponential ambiguity of an automaton A in time $O(|A|_E^2)$, and finite or polynomial ambiguity in time $O(|A|_E^3)$, where $|A|_E$ denotes the number of transitions of A. These complexities significantly improve over the previous best complexities given for the same problem. Furthermore, the algorithms presented are simple and based on a general algorithm for the composition or intersection of automata. We also give an algorithm to determine in time $O(|A|_E^3)$ the degree of polynomial ambiguity of a polynomially ambiguous automaton A. Finally, we present an application of our algorithms to an approximate computation of the entropy of a probabilistic automaton.

1 Introduction

The question of the ambiguity of finite automata arises in a variety of contexts. In some cases, the application of an algorithm requires an input automaton to be finitely ambiguous, in others, the convergence of a bound or guarantee relies on finite ambiguity, or the asymptotic rate of increase of ambiguity as a function of the string length. Thus, in all these cases, an algorithm is needed to test the ambiguity, either to determine if it is finite, or to estimate its asymptotic rate of increase.

The problem of testing ambiguity has been extensively analyzed in the past [3,6,7,9,12,13,14,15,16]. The problem of determining the degree of ambiguity of an automaton with finite ambiguity was shown by Chan and Ibarra to be PSPACE-complete [3]. However, testing finite ambiguity can be achieved in polynomial time using a characterization of exponential and polynomial ambiguity given by Ibarra and Ravikumar [6] and Weber and Seidel [15]. The most efficient algorithms for testing polynomial and exponential ambiguity, thereby testing finite ambiguity, were given by Weber and Seidel [14,16]. The algorithms they presented in [16] assume the input automaton to be ϵ-free, but they are extended by Weber to the case where the automaton has ϵ-transitions in [14]. In the presence of ϵ-transitions, the complexity of the algorithms given by Weber[14] is

---

* Research done at the Courant Institute, partially supported by the New York State Office of Science Technology and Academic Research (NYSTAR).

M. Ito and M. Toyama (Eds.): DLT 2008, LNCS 5257, pp. 108–120, 2008.
© Springer-Verlag Berlin Heidelberg 2008

$O((|A|_E + |A|_Q^2)^2)$ for testing the exponential ambiguity of an automaton A and $O((|A|_E + |A|_Q^2)^3)$ for testing polynomial ambiguity, where $|A|_E$ stands for the number of transitions and $|A|_Q$ the number of states of A.

This paper presents significantly more efficient algorithms for testing finite, polynomial, and exponential ambiguity for the general case of automata with ϵ-transitions. It gives an algorithm for testing the exponential ambiguity of an automaton A in time $O(|A|_E^2)$, and finite or polynomial ambiguity in time $O(|A|_E^3)$. The main idea behind our algorithms is to make use of the composition or intersection of finite automata with ϵ-transitions [11,10]. The ϵ-filter used in these algorithms crucially helps in the analysis and test of the ambiguity. The algorithms presented in this paper would not be valid and would lead to incorrect results without the use of the ϵ-filter. We also give an algorithm to determine in time $O(|A|_E^3)$ the degree of polynomial ambiguity of a polynomially ambiguous automaton A. Finally, we present an application of our algorithms to an approximate computation of the entropy of a probabilistic automaton.

The remainder of the paper is organized as follows. Section 2 presents general automata and ambiguity definitions. In Section 3, we give a brief description of existing characterizations for the ambiguity of automata and extend them to the case of automata with ϵ-transitions. In Section 4, we present our algorithms for testing finite, polynomial, and exponential ambiguity, and the proof of their correctness. Section 5 shows the relevance of the computation of the polynomial ambiguity to the approximation of the entropy of probabilistic automata.

2   Preliminaries

Definition 1. *A finite automaton A is a 5-tuple (Σ, Q, E, I, F) where Σ is a finite alphabet; Q is a finite set of states; $I \subseteq Q$ the set of initial states; $F \subseteq Q$ the set of final states; and $E \subseteq Q \times (\Sigma \cup \{\epsilon\}) \times Q$ a finite set of transitions, where ϵ denotes the empty string.*

We denote by $|A|_Q$ the number of states, by $|A|_E$ the number of transitions, and by $|A| = |A|_E + |A|_Q$ the size of an automaton A. Given a state $q \in Q$, $E[q]$ denotes the set of transitions leaving q. For two subsets $R \subseteq Q$ and $R' \subseteq Q$, we denote by $P(R, x, R')$ the set of all paths from a state $q \in R$ to a state $q' \in R'$ labeled with $x \in \Sigma^*$. We also denote by $p[\pi]$ the origin state, by $n[\pi]$ the destination state, and by $i[\pi] \in \Sigma^*$ the label of a path π.

A string $x \in \Sigma^*$ is accepted by A if it labels an accepting path, that is a path from an initial state to a final state. A finite automaton A is said to be *trim* if all its states lie on some accepting path. It is said to be *unambiguous* if no string $x \in \Sigma^*$ labels two distinct accepting paths; otherwise, it is said to be *ambiguous*. The *degree of ambiguity* of a string x in A is denoted by $\mathrm{da}(A, x)$ and defined as the number of accepting paths in A labeled by x. Note that if A contains an ϵ-cycle, there exists $x \in \Sigma^*$ such that $\mathrm{da}(A, x) = \infty$. Using a depth-first search of A restricted to ϵ-transitions, it can be decided in linear time if A contains ϵ-cycles. Thus, in the following, we will assume, without loss of generality, that A is ϵ-cycle free.

The *degree of ambiguity* of A is defined as $\mathrm{da}(A) = \sup_{x \in \Sigma^*} \mathrm{da}(A, x)$. A is said to be *finitely ambiguous* if $\mathrm{da}(A) < \infty$ and *infinitely ambiguous* if $\mathrm{da}(A) = \infty$. It is said to be *polynomially ambiguous* if there exists a polynomial h in $\mathbb{N}[X]$ such that $\mathrm{da}(A, x) \leq h(|x|)$ for all $x \in \Sigma^*$. The minimal degree of such a polynomial is called the *degree of polynomial ambiguity* of A and is denoted by $\mathrm{dpa}(A)$. By definition, $\mathrm{dpa}(A) = 0$ iff A is finitely ambiguous. When A is infinitely ambiguous but not polynomially ambiguous, it is said to be *exponentially ambiguous* and $\mathrm{dpa}(A) = \infty$.

3   Characterization of Infinite Ambiguity

The characterization and test of finite, polynomial, and exponential ambiguity of finite automata without ϵ-transitions are based on the following three fundamental properties [6,15,14,16].

Definition 2. *The properties (EDA), (IDA), and (EDA) for A are defined as follows.*

(a) *(EDA): there exists a state q with at least two distinct cycles labeled by some $v \in \Sigma^*$ (see Figure 1(a)) [6].*

(b) *(IDA): there exist two distinct states p and q with paths labeled with v from p to p, p to q, and q to q, for some $v \in \Sigma^*$ (see Figure 1(b)) [15,14,16].*

(c) *(IDA$_d$): there exist $2d$ states $p_1, \ldots p_d, q_1, \ldots, q_d$ in A and $2d - 1$ strings $v_1, \ldots, v_d$ and $u_2, \ldots u_d$ in Σ^* such that for all $1 \leq i \leq d$, $p_i \neq q_i$ and $P(p_i, v_i, p_i)$, $P(p_i, v_i, q_i)$, and $P(q_i, v_i, q_i)$ are non-empty, and, for all $2 \leq i \leq d$, $P(q_{i-1}, u_i, p_i)$ is non-empty (see Figure 1(c)) [15,14,16].*

Observe that (EDA) implies (IDA). Assuming (EDA), let e and e' be the first transitions that differ in the two cycles at state p, then, since Definition 1 disallows multiple transitions between the same two states with the same label, we must have $n[e] \neq n[e']$. Thus, (IDA) holds for the pair $(n[e], n[e'])$.

In the ϵ-free case, it was shown that a trim automaton A satisfies (IDA) iff A is infinitely ambiguous [15,16], that A satisfies (EDA) iff A is exponentially ambiguous [6], and that A satisfies (IDA$_d$) iff $\mathrm{dpa}(A) \geq d$ [14,16]. In the following proposition, these characterizations are straightforwardly extended to the case of automata with ϵ-transitions.

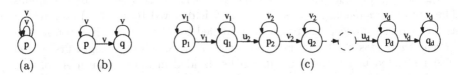

(a)   (b)   (c)

Fig. 1. Illustration of the properties: (a) (EDA); (b) (IDA); and (c) (IDA$_d$)

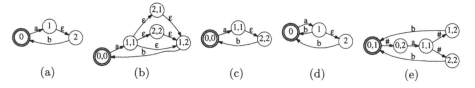

Fig. 2. *ε-filter and ambiguity*: (a) Finite automaton A; (b) $A \cap A$ without using ε-filter, which incorrectly makes A appear as exponentially ambiguous; (c) $A \cap A$ using an ε-filter. *Weber's processing of ε-transitions*: (d) Finite automaton B; (e) ε-free automaton B' such that $\mathrm{dpa}(B) = \mathrm{dpa}(B')$.

Proposition 1. *Let A be a trim ε-cycle free finite automaton.*

(i) A is infinitely ambiguous iff A satisfies (IDA).
(ii) A is exponentially ambiguous iff A satisfies (EDA).
(iii) $\mathrm{dpa}(A) \geq d$ iff A satisfies (IDA_d).

Proof. The proof is by induction on the number of ε-transitions in A. If A does not have any ε-transition, then the proposition holds as shown in [15,16] for (i), [6] for (ii) and [16] for (iii).

Assume now that A has $n+1$ ε-transitions, $n \geq 0$, and that the statement of the proposition holds for all automata with n ε-transitions. Select an ε-transition e_0 in A, and let A' be the finite automaton obtained after application of ε-removal to A limited to transition e_0. A' is obtained by deleting e_0 from A and by adding a transition $(p[e_0], l[e], n[e])$ for every transition $e \in E[n[e_0]]$. It is clear that A and A' are equivalent and that there is a label-preserving bijection between the paths in A and A'. Thus, (a) A satisfies (IDA) (resp. (EDA), (IDA_d)) iff A' satisfies (IDA) (resp. (EDA), (IDA_d)) and (b) for all $x \in \Sigma^*$, $\mathrm{da}(A, x) = \mathrm{da}(A', x)$. By induction, Proposition 1 holds for A' and thus, it follows from (a) and (b) that Proposition 1 also holds for A. □

These characterizations have been used in [14,16] to design algorithms for testing infinite, polynomial, and exponential ambiguity, and for computing the degree of polynomial ambiguity in the ε-free case.

Theorem 1 ([14,16]). *Let A be a trim ε-free finite automaton.*

1. *It is decidable in time $O(|A|_E^3)$ whether A is infinitely ambiguous.*
2. *It is decidable in time $O(|A|_E^2)$ whether A is exponentially ambiguous.*
3. *The degree of polynomial ambiguity of A, $\mathrm{dpa}(A)$, can be computed in $O(|A|_E^3)$.*

The first result of Theorem 1 has also been generalized by [14] to the case of automata with ε-transitions but with a significantly worse complexity.

Theorem 2 ([14]). *Let A be a trim ε-cycle free finite automaton. It is decidable in time $O((|A|_E + |A|_Q^2)^3)$ whether A is infinitely ambiguous.*

The algorithms designed for the ε-free case cannot be readily used for finite automata with ε-transitions since they would lead to incorrect results (see Figure 2(a)-(c)). Instead, [14] proposed a reduction to the ε-free case. First, [14]

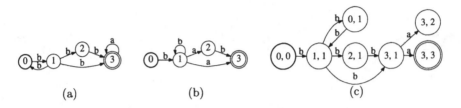

Fig. 3. Example of finite automaton intersection. (a) Finite automata A_1 and (b) A_2. (c) Result of the intersection of A_1 and A_2.

gave an algorithm to test if there exist two states p and q in A with two distinct ϵ-paths from p to q. If that is the case, then A is exponentially ambiguous (complexity $O(|A|_Q^4 + |A|_E)$). Otherwise, [14] defined from A an ϵ-free automaton A' over the alphabet $\Sigma \cup \{\#\}$ such that A is infinitely ambiguous iff A' is infinitely ambiguous, see Figure 2(d)-(e).[1] However, the number of transitions of A' is $|A|_E + |A|_Q^2$. This explains why the complexity in the ϵ-transition case is significantly worse than in the ϵ-free case. The same approach can be used to test the exponential ambiguity of A in time $O((|A|_E + |A|_Q^2)^2)$ and to compute dpa(A) when A is polynomially ambiguous in $O((|A|_E + |A|_Q^2)^3)$. Note that we give tighter estimates of the complexity of the algorithms of [14,16] where the authors gave complexities using the loose inequality: $|A|_E \leq |\Sigma| |A|_Q^2$.

4 Algorithms

Our algorithms for testing ambiguity are based on a general algorithm for the composition or intersection of automata, which we briefly describe in the following section.

4.1 Intersection of Finite Automata

The intersection of finite automata is a special case of the general composition algorithm for weighted transducers [11,10]. States in the intersection $A_1 \cap A_2$ of two finite automata A_1 and A_2 are identified with pairs of a state of A_1 and a state of A_2. The following rule specifies how to compute a transition of $A_1 \cap A_2$ in the absence of ϵ-transition from appropriate transitions of A_1 and A_2: (q_1, a, q_1') and $(q_2, a, q_2') \implies ((q_1, q_2), a, (q_1', q_2'))$. Figure 3 illustrates the algorithm. A state (q_1, q_2) is initial (resp. final) when q_1 and q_2 are initial (resp.

---

[1] Observe that A' is not the result of applying the classical ϵ-removal algorithm to A, since ϵ-removal does not preserve infinite ambiguity and would lead be an even larger automaton. Instead [14] used a more complex algorithm where ϵ-transitions are replaced by regular transitions labeled with a special symbol while preserving infinite ambiguity, dpa(A) = dpa(A'), even though A' is not equivalent to A. States in A' are pairs (q, i) with q a state in A and $i \in \{1, 2\}$. There is a transition from $(p, 1)$ to $(q, 2)$ labeled by $\#$ if q belongs to the ϵ-closure of p and from $(p, 2)$ to $(q, 1)$ labeled by $\sigma \in \Sigma$ if there was such a transition from p to q in A.

final). In the worst case, all transitions of A_1 leaving a state q_1 match all those of A_2 leaving state q_2, thus the space and time complexity of composition is quadratic: $O(|A_1||A_2|)$, or $O(|A_1|_E|A_2|_E)$ when A_1 and A_2 are trim.

4.2   Epsilon-Filtering

A straightforward generalization of the ϵ-free case would generate redundant ϵ-paths. This is a crucial issue in the more general case of the intersection of weighted automata over a non-idempotent semiring, since it would lead to an incorrect result. The weight of two matching ϵ-paths of the original automata would then be counted as many times as the number of redundant ϵ-paths generated in the result, instead of once. It is also a crucial problem in the unweighted case since redundant ϵ-paths can affect the test of infinite ambiguity, as we shall see in the next section. A critical component of the composition algorithm of [11,10] consists however of precisely coping with this problem using an *epsilon-filtering* mechanism.

Figure 4(c) illustrates the problem just mentioned. To match ϵ-paths leaving q_1 and those leaving q_2, a generalization of the ϵ-free intersection can make the following moves: (1) first move forward on an ϵ-transition of q_1, or even a ϵ-path, and remain at the same state q_2 in A_2, with the hope of later finding a transition whose label is some label $a \neq \epsilon$ matching a transition of q_2 with the same label; (2) proceed similarly by following an ϵ-transition or ϵ-path leaving q_2 while remaining at the same state q_1 in A_1; or, (3) match an ϵ-transition of q_1 with an ϵ-transition of q_2.

Let us rename existing ϵ-labels of A_1 as ϵ_2, and existing ϵ-labels of A_2 ϵ_1, and let us augment A_1 with a self-loop labeled with ϵ_1 at all states and similarly, augment A_2 with a self-loop labeled with ϵ_2 at all states, as illustrated by Figures 4(a) and (b). These self-loops correspond to remaining at the same state in that machine while consuming an ϵ-label of the other transition. The three moves just described now correspond to the matches (1) $(\epsilon_2 : \epsilon_2)$, (2) $(\epsilon_1 : \epsilon_1)$, and (3) $(\epsilon_2 : \epsilon_1)$. The grid of Figure 4(c) shows all the possible ϵ-paths between intersection states. We will denote by $\tilde{A}_1$ and $\tilde{A}_2$ the automata obtained after application of these changes.

For the result of intersection not to be redundant, between any two of these states, all but one path must be disallowed. There are many possible ways of selecting that path. One natural way is to select the shortest path with the diagonal transitions (ϵ-matching transitions) taken first. Figure 4(c) illustrates in boldface the path just described from state $(0,0)$ to state $(1,2)$. Remarkably, this filtering mechanism itself can be encoded as a finite-state transducer such as the transducer M of Figure 4(d). We denote by $(p,q) \preceq (r,s)$ to indicate that (r,s) can be reached from (p,q) in the grid.

Proposition 2. *Let M be the transducer of Figure 4(d). M allows a unique path between any two states (p,q) and (r,s), with $(p,q) \preceq (r,s)$.*

Proof. The full proof of this proposition is given in [2].   □

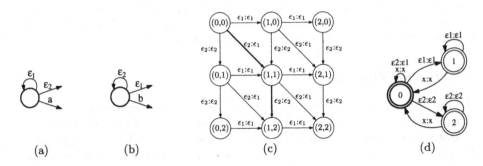

Fig. 4. Marking of automata, redundant paths and filter. (a) $\tilde{A}_1$: self-loop labeled with ϵ_1 added at all states of A_1, regular ϵs renamed to ϵ_2. (b) $\tilde{A}_2$: self-loop labeled with ϵ_2 added at all states of A_2, regular ϵs renamed to ϵ_1. (c) Redundant ϵ-paths: a straightforward generalization of the ϵ-free case could generate all the paths from $(0,0)$ to $(2,2)$ for example, even when composing just two simple transducers. (d) Filter transducer M allowing a unique ϵ-path.

Thus, to intersect two finite automata A_1 and A_2 with ϵ-transitions, it suffices to compute $\tilde{A}_1 \circ M \circ \tilde{A}_2$, using the ϵ-free rules of composition. States in the intersection are now identified with triplets made of a state of A_1, a state of M, and a state of A_2. A transition (q_1, a_1, q_1') in $\tilde{A}_1$, a transition (f, a_1, a_2, f') in M, and a transition (q_2, a_2, q_2') in $\tilde{A}_2$ are combined to form the following transition in the intersection: $((q_1, f, q_2), a, (q_1', f', q_2'))$, with $a = \epsilon$ if $\{a_1, a_2\} \subseteq \{\epsilon_1, \epsilon_2\}$ and $a = a_1 = a_2$ otherwise. In the rest of the paper, we will assume that the result of intersection is trimmed after its computation, which can be done in linear time.

Theorem 3. *Let A_1 and A_2 be two finite automata with ϵ-transitions. To each pair (π_1, π_2) of accepting paths in A_1 and A_2 sharing the same input label $x \in \Sigma^*$ corresponds a unique accepting path π in $A_1 \cap A_2$ labeled with x.*

Proof. This follows straightforwardly from Proposition 2. □

4.3 Ambiguity Tests

We start with a test of the exponential ambiguity of A. The key is that the (EDA) property translates into a very simple property for $A^2 = A \cap A$.

Lemma 1. *Let A be a trim ϵ-cycle free finite automaton. A satisfies (EDA) iff there exists a strongly connected component of $A^2 = A \cap A$ that contains two states of the form (p, p) and (q, q'), where p, q and q' are states of A with $q \neq q'$.*

Proof. Assume that A satisfies (EDA). There exist a state p and a string v such that there are two distinct cycles c_1 and c_2 labeled by v at p. Let e_1 and e_2 be the first edges that differ in c_1 and c_2. We can then write $c_1 = \pi e_1 \pi_1$ and $c_2 = \pi e_2 \pi_2$. If e_1 and e_2 share the same label, let $\pi_1' = \pi e_1$, $\pi_2' = \pi e_2$, $\pi_1'' = \pi_1$ and $\pi_2'' = \pi_2$. If e_1 and e_2 do not share the same label, exactly one of them must be an ϵ-transition. By symmetry, we can assume without loss of generality that e_1 is the ϵ-transition. Let $\pi_1' = \pi e_1$, $\pi_2' = \pi$, $\pi_1'' = \pi_1$ and $\pi_2'' = \epsilon_2 \pi_2$. In both cases, let $q = n[\pi_1'] = p[\pi_1'']$ and $q' = n[\pi_2'] = p[\pi_2'']$. Observe that $q \neq q'$. Since $i[\pi_1'] = i[\pi_2']$, π_1' and π_2' are matched by intersection resulting in a path in A^2 from (p, p) to (q, q'). Similarly, since $i[\pi_1''] = i[\pi_2'']$, π_1'' and π_2'' are matched by intersection resulting in a path from (q, q') to (p, p). Thus, (p, p) and (q, q') are in the same strongly connected component of A^2.

Conversely, assume that there exist states p, q and q' in A such that $q \neq q'$ and that (p, p) and (q, q') are in the same strongly connected component of A^2. Let c be a cycle in (p, p) going through (q, q'), it has been obtained by matching two cycles c_1 and c_2. If c_1 were equal to c_2, intersection would match these two paths creating a path c' along which all the states would be of the form (r, r), and since A is trim this would contradict Theorem 3. Thus, c_1 and c_2 are distinct and (EDA) holds.   □

Observe that the use of the ϵ-filter in composition is crucial for Lemma 1 to hold (see Figure 2). The lemma leads to a straightforward algorithm for testing exponential ambiguity.

Theorem 4. *Let A be a trim ϵ-cycle free finite automaton. It is decidable in time $O(|A|_E^2)$ whether A is exponentially ambiguous.*

Proof. The algorithm proceeds as follows. We compute A^2 and, using a depth-first search of A^2, trim it and compute its strongly connected components. It follows from Lemma 1 that A is exponentially ambiguous iff there is a strongly connected component that contains two states of the form (p, p) and (q, q') with $q \neq q'$. Finding such a strongly connected component can be done in time linear in the size of A^2, i.e. in $O(|A|_E^2)$ since A and A^2 are trim. Thus, the complexity of the algorithm is in $O(|A_E|^2)$.   □

Testing the (IDA) property requires finding three paths sharing the same label in A. As shown below, this can be done in a natural way using the automaton $A^3 = (A \cap A) \cap A$, obtained by applying twice the intersection algorithm.

Lemma 2. *Let A be a trim ϵ-cycle free finite automaton. A satisfies (IDA) iff there exist two distinct states p and q in A with a non-ϵ path in $A^3 = A \cap A \cap A$ from state (p, p, q) to state (p, q, q).*

Proof. Assume that A satisfies (IDA). Then, there exists a string $v \in \Sigma^*$ with three paths $\pi_1 \in P(p, v, p)$, $\pi_2 \in P(p, v, q)$ and $\pi_3 \in P(q, v, p)$. Since these three paths share the same label v, they are matched by intersection resulting in a path π in A^3 labeled with v from $(p[\pi_1], p[\pi_2], p[\pi_3]) = (p, p, q)$ to $(n[\pi_1], n[\pi_2], n[\pi_3]) = (p, q, q)$.

Conversely, if there is a non-ϵ path π form (p, p, q) to (p, q, q) in A^3, it has been obtained by matching three paths π_1, π_2 and π_3 in A with the same input $v = i[\pi] \neq \epsilon$. Thus, (IDA) holds.   □

This lemma appears already as Lemma 5.10 in [8]. Finally, Theorem 4 and Lemma 2 can be combined to yield the following result.

Theorem 5. *Let A be a trim ϵ-cycle free finite automaton. It is decidable in time $O(|A|_E^3)$ whether A is finitely, polynomially, or exponentially ambiguous.*

Proof. First, Theorem 4 can be used to test whether A is exponentially ambiguous by computing A^2. The complexity of this step is $O(|A|_E^2)$.

If A is not exponentially ambiguous, we proceed by computing and trimming A^3 and then testing whether A^3 verifies the property described in Lemma 2. This is done by considering the automaton B on the alphabet $\Sigma' = \Sigma \cup \{\#\}$ obtained from A^3 by adding a transition labeled by $\#$ from state (p, q, q) to state (p, p, q) for every pair (p, q) of states in A such that $p \neq q$. It follows that A^3 verifies the condition in Lemma 2 iff there is a cycle in B containing both a transition labeled by $\#$ and a transition labeled by a symbol in Σ. This property can be checked straightforwardly using a depth-first search of B to compute its strongly connected components. If a strongly connected component of B is found that contains both a transition labeled with $\#$ and a transition labeled by a symbol in Σ, A verifies (IDA) but not (EDA) and thus A is polynomially ambiguous. Otherwise, A is finitely ambiguous. The complexity of this step is linear in the size of B: $O(|B|_E) = O(|A_E|^3 + |A_Q|^2) = O(|A_E|^3)$ since A and B are trim.

The total complexity of the algorithm is $O(|A|_E^2 + |A|_E^3) = O(|A|_E^3)$.

When A is polynomially ambiguous, we can derive from the algorithm just described one that computes dpa(A).

Theorem 6. *Let A be a trim ϵ-cycle free finite automaton. If A is polynomially ambiguous, dpa(A) can be computed in time $O(|A|_E^3)$.*

Proof. We first compute A^3 and use the algorithm of Theorem 5 to test whether A is polynomially ambiguous and to compute all the pairs (p, q) that verify the condition of Lemma 2. This step has complexity $O(|A|_E^3)$.

We then compute the component graph G of A, and for each pair (p, q) found in the previous step, we add a transition labeled with $\#$ from the strongly connected component of p to the one of q. If there is a path in that graph containing d edges labeled by $\#$, then A verifies (IDA$_d$). Thus, dpa(A) is the maximum number of edges marked by $\#$ that can be found along a path in G. Since G is acyclic, this number can be computed in linear time in the size of G, i.e. in $O(|A|_Q^2)$. Thus, the overall complexity of the algorithm is $O(|A|_E^3)$. □

5   Application to Entropy Approximation

In this section, we describe an application in which determining the degree of ambiguity of a *probabilistic* automaton helps estimate the quality of an approximation of its entropy. Weighted automata are automata in which each transition carries some weight in addition to the usual alphabet symbol. The weights are elements of a semiring, that is a ring that may lack negation. The following is a more formal definition.

Definition 3. *A weighted automaton A over a semiring $(\mathbb{K}, \oplus, \otimes, \bar{0}, \bar{1})$ is a 7-tuple $(\Sigma, Q, I, F, E, \lambda, \rho)$ where Σ is a finite alphabet, Q a finite set of states, $I \subseteq Q$ the set of initial states, $F \subseteq Q$ the set of final states, $E \subseteq Q \times \Sigma \cup \{\epsilon\} \times \mathbb{K} \times Q$ a finite set of transitions, $\lambda : I \to \mathbb{K}$ the initial weight function mapping I to $\mathbb{K}$, and $\rho : F \to \mathbb{K}$ the final weight function mapping F to $\mathbb{K}$.*

Given a transition $e \in E$, we denote by $w[e]$ its weight. We extend the weight function w to paths by defining the weight of a path as the $\otimes$-product of the weights of its constituent transitions: $w[\pi] = w[e_1] \otimes \cdots \otimes w[e_k]$. The weight associated by a weighted automaton A to an input string $x \in \Sigma^*$ is defined by $[\![A]\!](x) = \bigoplus_{\pi \in P(I,x,F)} \lambda[p[\pi]] \otimes w[\pi] \otimes \rho[n[\pi]]$. The entropy $H(A)$ of a probabilistic automaton A is defined as:

$$H(A) = - \sum_{x \in \Sigma^*} [\![A]\!](x) \log([\![A]\!](x)). \tag{1}$$

The system $(\mathbb{K}, \oplus, \otimes, (0,0), (1,0))$ with $\mathbb{K} = (\mathbb{R} \cup \{+\infty, -\infty\}) \times (\mathbb{R} \cup \{+\infty, -\infty\})$ and $\oplus$ and $\otimes$ defined as follows defines a commutative semiring called the *entropy semiring* [4]: for any two pairs (x_1, y_1) and (x_2, y_2) in $\mathbb{K}$, $(x_1, y_1) \oplus (x_2, y_2) = (x_1 + x_2, y_1 + y_2)$ and $(x_1, y_1) \otimes (x_2, y_2) = (x_1 x_2, x_1 y_2 + x_2 y_1)$. In [4], the authors showed that a generalized shortest-distance algorithm over this semiring correctly computes the entropy of an unambiguous probabilistic automaton A. The algorithm starts by mapping the weight of each transition to a pair where the first element is the probability and the second the entropy: $w[e] \mapsto (w[e], -w[e] \log w[e])$. The algorithm then proceeds by computing the generalized shortest-distance defined over the entropy semiring, which computes the $\oplus$-sum of the weights of all accepting paths in A.

Here, we show that the same shortest-distance algorithm yields an approximation of the entropy of an ambiguous probabilistic automaton A, where the approximation quality is a function of the degree of polynomial ambiguity, dpa(A). Our proofs make use of the standard log-sum inequality [5], a special case of Jensen's inequality, which holds for any positive reals $a_1, \ldots, a_k$, and $b_1, \ldots, b_k$:

$$\sum_{i=1}^{k} a_i \log \frac{a_i}{b_i} \geq \left(\sum_{i=1}^{k} a_i \right) \log \frac{\sum_{i=1}^{k} a_i}{\sum_{i=1}^{k} b_i}. \tag{2}$$

Lemma 3. *Let A be a probabilistic automaton and let $x \in \Sigma^+$ be a string accepted by A on k paths $\pi_1, \ldots, \pi_k$. Let $w[\pi_i]$ be the probability of path π_i. Clearly, $[\![A]\!](x) = \sum_{i=1}^{k} w[\pi_i]$. Then, $\sum_{i=1}^{k} w[\pi_i] \log w[\pi_i] \geq [\![A]\!](x)(\log[\![A]\!](x) - \log k)$.*

Proof. The result follows straightforwardly from the log-sum inequality, with $a_i = w[\pi_i]$ and $b_i = 1$:

$$\sum_{i=1}^{k} w[\pi_i] \log w[\pi_i] \geq \left(\sum_{i=1}^{k} w[\pi_i] \right) \log \frac{\sum_{i=1}^{k} w[\pi_i]}{k} = [\![A]\!](x)(\log[\![A]\!](x) - \log k). \tag{3}$$

$\square$

Let $S(A)$ be the quantity computed by the generalized shortest-distance algorithm over the entropy semiring or a probabilistic automaton A. When A is unambiguous, it is shown by [4] that $S(A) = H(A)$.

Theorem 7. *Let A be a probabilistic automaton and let L denote the expected length of the strings accepted by A (i.e. $L = \sum_{x \in \Sigma^*} |x| [\![A]\!](x)$). Then,*

1. *if A is finitely ambiguous with $\mathrm{da}(A) = k$ for some $k \in \mathbb{N}$, then $H(A) \leq S(A) \leq H(A) + \log k$;*
2. *if A is polynomially ambiguous with $\mathrm{dpa}(A) = k$ for some $k \in \mathbb{N}$, then $H(A) \leq S(A) \leq H(A) + k \log L$.*

Proof. The lower bound $S(A) \geq H(A)$ follows from the observation that for a string x that is accepted in A by k paths $\pi_1, \ldots, \pi_k$,

$$\sum_{i=1}^{k} w[\pi_i] \log(w(\pi_i)) \leq \left(\sum_{i=1}^{k} w[\pi_i]\right) \log\left(\sum_{i=1}^{k} w[\pi_i]\right). \tag{4}$$

Since the quantity $-\sum_{i=1}^{k} w[\pi_i] \log(w[\pi_i])$ is string x's contribution to $S(A)$ and the quantity $-(\sum_{i=1}^{k} w[\pi_i]) \log(\sum_{i=1}^{k} w[\pi_i])$ its contribution to $H(A)$, summing over all accepted strings x, we obtain $H(A) \leq S(A)$.

Assume that A is finitely ambiguous with degree of ambiguity k. Let $x \in \Sigma^*$ be a string that is accepted on $l_x \leq k$ paths $\pi_1, \ldots, \pi_{l_x}$. By Lemma 3, we have $\sum_{i=1}^{l_x} w[\pi_i] \log w[\pi_i] \geq [\![A]\!](x)(\log[\![A]\!](x) - \log l_x) \geq [\![A]\!](x)(\log[\![A]\!](x) - \log k)$. Thus, $S(A) = -\sum_{x \in \Sigma^*} \sum_{i=1}^{l_x} w[\pi_i] \log w[\pi_i] \leq H(A) + \sum_{x \in \Sigma^*} (\log k)[\![A]\!](x) = H(A) + \log k$. This proves the first statement of the theorem.

Next, assume that A is polynomially ambiguous with degree of polynomial ambiguity k. By Lemma 3, we have $\sum_{i=1}^{l_x} w[\pi_i] \log w[\pi_i] \geq [\![A]\!](x)(\log[\![A]\!](x) - \log l_x) \geq [\![A]\!](x)(\log[\![A]\!](x) - \log(|x|^k))$. Thus,

$$S(A) \leq H(A) + \sum_{x \in \Sigma^*} k[\![A]\!](x) \log |x| = H(A) + k\mathbb{E}_A[\log |x|] \tag{5}$$

$$\leq H(A) + k \log \mathbb{E}_A[|x|] = H(A) + k \log L, \qquad \text{(by Jensen's inequality)}$$

which proves the second statement of the theorem. $\square$

The theorem shows in particular that the quality of the approximation of the entropy of a polynomially ambiguous probabilistic automaton can be estimated by computing its degree of polynomial ambiguity, which can be achieved efficiently as described in the previous section. This also requires the computation of the expected length L of an accepted string. L can be computed efficiently for an arbitrary probabilistic automaton using the entropy semiring and the generalized shortest-distance algorithms, using techniques similar to those described in [4]. The only difference is in the initial step, where the weight of each transition in A is mapped to a pair of elements by $w[e] \mapsto (w[e], w[e])$.

6 Conclusion

We presented simple and efficient algorithms for testing the finite, polynomial, or exponential ambiguity of finite automata with ϵ-transitions. We conjecture that the time complexity of our algorithms is optimal. These algorithms have a variety of applications, in particular to test a pre-condition for the applicability of other automata algorithms. Our application to the approximation of the entropy gives another illustration of their usefulness. Our algorithms also demonstrate the prominent role played by the intersection or composition of automata and transducers with ϵ-transitions [11,10] in the design of *testing algorithms*. Composition can be used to devise simple and efficient testing algorithms. We have shown elsewhere how it can be used to test the functionality of a finite-state transducer, or the twins property for weighted automata and transducers [1].

References

1. Allauzen, C., Mohri, M.: Efficient Algorithms for Testing the Twins Property. Journal of Automata, Languages and Combinatorics 8(2), 117–144 (2003)
2. Allauzen, C., Mohri, M.: 3-way composition of weighted finite-state transducers. In: CIAA 2008. LNCS, vol. 5148, pp. 262–273. Springer, Heidelberg (2008)
3. Chan, T., Ibarra, O.H.: On the finite-valuedness problem for sequential machines. Theoretical Computer Science 23, 95–101 (1983)
4. Cortes, C., Mohri, M., Rastogi, A., Riley, M.: Efficient computation of the relative entropy of probabilistic automata. In: Correa, J.R., Hevia, A., Kiwi, M. (eds.) LATIN 2006. LNCS, vol. 3887, pp. 323–336. Springer, Heidelberg (2006)
5. Cover, T.M., Thomas, J.A.: Elements of Information Theory. John Wiley & Sons, Inc., New York (1991)
6. Ibarra, O.H., Ravikumar, B.: On sparseness, ambiguity and other decision problems for acceptors and transducers. In: Monien, B., Vidal-Naquet, G. (eds.) STACS 1986. LNCS, vol. 210, pp. 171–179. Springer, Heidelberg (1985)
7. Jacob, G.: Un algorithme calculant le cardinal, fini ou infini, des demi-groupes de matrices. Theoretical Computer Science 5(2), 183–202 (1977)
8. Kuich, W.: Finite automata and ambiguity. Technical Report 253, Institute für Informationsverarbeitung - Technische Universität Graz und ÖCG (1988)
9. Mandel, A., Simon, I.: On finite semigroups of matrices. Theoretical Computer Science 5(2), 101–111 (1977)
10. Mohri, M., Pereira, F.C.N., Riley, M.: Weighted Automata in Text and Speech Processing. In: Proceedings of ECAI 1996, Workshop on Extended finite state models of language, Budapest, Hungary. John Wiley and Sons, Chichester (1996)
11. Pereira, F., Riley, M.: Speech Recognition by Composition of Weighted Finite Automata. In: Finite State Language Processing. MIT Press, Cambridge (1997)
12. Ravikumar, B., Ibarra, O.H.: Relating the type of ambiguity of finite automata to the succinctness of their representation. SIAM Journal on Computing 18(6), 1263–1282 (1989)

13. Reutenauer, C.: Propriétés arithmétiques et topologiques des séries rationnelles en variable non commutative. Thèse de troisième cycle, Université Paris VI (1977)
14. Weber, A.: Über die Mehrdeutigkeit und Wertigkeit von endlichen, Automaten und Transducern. Dissertation, Goethe-Universität Frankfurt am Main (1987)
15. Weber, A., Seidl, H.: On the degree of ambiguity of finite automata. In: Wiedermann, J., Gruska, J., Rovan, B. (eds.) MFCS 1986. LNCS, vol. 233, pp. 620–629. Springer, Heidelberg (1986)
16. Weber, A., Seidl, H.: On the degree of ambiguity of finite automata. Theoretical Computer Science 88(2), 325–349 (1991)

Emptiness of Multi-pushdown Automata Is 2ETIME-Complete

Mohamed Faouzi Atig[1], Benedikt Bollig[2], and Peter Habermehl[1,2]

[1] LIAFA, CNRS and University Paris Diderot, France
atig+haberm@liafa.jussieu.fr
[2] LSV, ENS Cachan, CNRS, Inria
bollig@lsv.ens-cachan.fr

Abstract. We consider *multi-pushdown automata*, a multi-stack extension of pushdown automata that comes with a constraint on stack operations: a pop can only be performed on the first non-empty stack (which implies that we assume a linear ordering on the collection of stacks). We show that the emptiness problem for multi-pushdown automata is 2ETIME-complete wrt. the number of stacks. Containment in 2ETIME is shown by translating an automaton into a grammar for which we can check if the generated language is empty. The lower bound is established by simulating the behavior of an alternating Turing machine working in exponential space. We also compare multi-pushdown automata with the model of bounded-phase multi-stack (visibly) pushdown automata.

1 Introduction

Various classes of pushdown automata with multiple stacks have been proposed and studied in the literature. The main goals of these efforts are twofold. First, one may aim at extending the expressive power of pushdown automata, going beyond the class of context-free languages. Second, multi-stack systems may model recursive concurrent programs, in which any sequential process is equipped with a finite-state control and, in addition, can access its own stack to connect procedure calls to their corresponding returns. In general, however, multi-stack extensions of pushdown automata are Turing powerful and therefore come along with undecidability of basic decision problems. To retain desirable decidability properties of pushdown automata, such as emptiness, one needs to restrict the model accordingly. In [3], Breveglieri et al. define *multi-pushdown automata* (MPDA), which impose a linear ordering on stacks. Stack operations are henceforth constrained in such a way that a pop operation is reserved to the first non-empty stack. These automata are suitable to model client-server systems of processes with remote procedure calls. Another possibility to regain decidability in the presence of several stacks is to restrict the domain of input words. In [8], La Torre et al. define *bounded-phase multi-stack visibly pushdown automata* (bounded-phase MVPA). Only those runs are taken into consideration that can be split into a given number of phases, where each phase admits pop operations of one particular stack only. In the above-mentioned cases, the respective emptiness problem is decidable. In [9], the results of [8] are used to show decidability results for restricted queue systems.

M. Ito and M. Toyama (Eds.): DLT 2008, LNCS 5257, pp. 121–133, 2008.
© Springer-Verlag Berlin Heidelberg 2008

In this paper, we resume the study of MPDA and, in particular, consider their emptiness problem. The decidability of this problem, which is to decide if an automaton admits some accepting run, is fundamental for verification purposes. We show that the emptiness problem for MPDA is 2ETIME-complete. Recall that 2ETIME is the class of all decision problems solvable by a deterministic Turing machine in time $2^{2^{dn}}$ for some constant d. In proving the upper bound, we correct an error in the decidability proof given in [3].[1] We keep their main idea: MPDA are reduced to equivalent *depth-n-grammars*. Deciding emptiness for these grammars then amounts to checking emptiness of an ordinary context-free grammar. For proving 2ETIME-hardness, we borrow an idea from [10], where a 2ETIME lower bound is shown for bounded-phase pushdown-transducer automata. We also show that $2m$-MPDA are strictly more expressive than m-phase MVPA providing an alternative proof of decidability of the emptiness problem for bounded-phase MVPA.

The paper is structured as follows: In Section 2, we introduce MPDA formally, as well as depth-n-grammars. Sections 3 and 4 then establish the 2ETIME upper and, respectively, lower bound of the emptiness problem for MPDA, which constitutes our main result. In Section 5, we compare MPDA with bounded-phase MVPA. We conclude by identifying some directions for future work. Missing proofs can be found in [1].

2   Multi-pushdown Automata and Depth-n-grammars

In this section we define *multi-pushdown automata* with $n \geq 1$ pushdown stacks and their corresponding grammars. We essentially follow the definitions of [3].

Multi-pushdown Automata. Our automata have one read-only left to right input tape and $n \geq 1$ read-write memory tapes (stacks) with a last-in-first-out rewriting policy. In each move, the following actions are performed:

- read one or zero symbol from the input tape and move past the read symbol
- read the symbol on the top of the first non-empty stack starting from the left
- switch the internal state
- for each $i \in \{1, \ldots, n\}$, write a finite string α_i on the i-th pushdown stack

Definition 1. *For $n \geq 1$, an (n-)multi-pushdown automaton (n-MPDA or MPDA) is a tuple $M = (Q, \Sigma, \Gamma, \delta, q_0, F, Z_0)$ where:*

- *Q is a finite non-empty set of internal states,*
- *Σ (input) and Γ (memory) are finite disjoint alphabets,*
- *$\delta : Q \times (\Sigma \uplus \{\epsilon\}) \times \Gamma \to 2^{Q \times (\Gamma^*)^n}$ is a transition mapping,*
- *q_0 is the initial state,*
- *$F \subseteq Q$ is the set of final states, and*
- *$Z_0 \in \Gamma$ is the initial memory symbol.*

---

[1] A similar correction of the proof has been worked out independently by the authors of [3] themselves [4]. They gave an explicit construction for the case of three stacks that can be generalized to arbitrarily many stacks.

Table 1. A 2-MPDA for $\{\epsilon\} \cup \{a^{i_1} b^{i_1} c^{i_1} a^{i_2} b^{i_2} c^{i_2} \cdots a^{i_k} b^{i_k} c^{i_k} \mid k \geq 1 \text{ and } i_1, \ldots, i_k > 0\}$

$$M = (\{q_0, \ldots, q_3, q_f\}, \{a, b, c\}, \{A, B, Z_0, Z_1\}, \delta, q_0, \{q_f\}, Z_0)$$

$\delta(q_0, \epsilon, Z_0) = \{(q_f, \epsilon, \epsilon)\}$	$\delta(q_2, b, A) = \{(q_2, \epsilon, \epsilon)\}$
$\delta(q_0, a, Z_0) = \{(q_1, AZ_0, BZ_1)\}$	$\delta(q_2, \epsilon, Z_0) = \{(q_3, \epsilon, \epsilon)\}$
$\delta(q_1, \epsilon, A) = \{(q_2, A, \epsilon)\}$	$\delta(q_3, \epsilon, Z_1) = \{(q_0, Z_0, \epsilon)\}$
$\delta(q_1, a, A) = \{(q_1, AA, B)\}$	$\delta(q_3, c, B) = \{(q_3, \epsilon, \epsilon)\}$

A *configuration* of M is an $(n + 2)$-tuple $\langle q, x; \gamma_1, \ldots, \gamma_n \rangle$ with $q \in Q$, $x \in \Sigma^*$, and $\gamma_1, \ldots, \gamma_n \in \Gamma^*$. The *transition relation* $\vdash_M^*$ is the transitive closure of the binary relation $\vdash_M$ over configurations, defined as follows:

$$\langle q, ax; \epsilon, \ldots, \epsilon, A\gamma_i, \ldots, \gamma_n \rangle \vdash_M \langle q', x; \alpha_1, \ldots, \alpha_{i-1}, \alpha_i \gamma_i, \ldots, \alpha_n \gamma_n \rangle$$

if $(q', \alpha_1, \ldots, \alpha_n) \in \delta(q, a, A)$, where $a \in \Sigma \cup \{\epsilon\}$.

The *language of M accepted by final state* is defined as the set of words $x \in \Sigma^*$ such that there are $\gamma_1, \ldots, \gamma_n \in \Gamma^*$ and $q \in F$ with $\langle q_0, x; Z_0, \epsilon, \ldots \epsilon \rangle \vdash_M^* \langle q, \epsilon; \gamma_1, \ldots, \gamma_n \rangle$. The *language of M accepted by empty stacks*, denoted by $L(M)$, is defined as the set of words $x \in \Sigma^*$ such that there is $q \in Q$ with $\langle q_0, x; Z_0, \epsilon, \ldots \epsilon \rangle \vdash_M^* \langle q, \epsilon; \epsilon, \ldots, \epsilon \rangle$.

Lemma 2 ([3]). *The languages accepted by n-MPDA by final state are the same as the languages accepted by n-MPDA by empty stacks.*

Table 1 shows an example of a 2-MPDA. Notice that it accepts the same language by final state and by empty stacks.

We need the following normal form of n-MPDA for the proof of our main theorem. The normal form restricts the operation on stacks 2 to n: pushing one symbol on these stacks is only allowed while popping a symbol from the first stack, and popping a symbol from them pushes a symbol onto the first stack. Furthermore, the number of symbols pushed on the first stack is limited to two and the stack alphabets are distinct.

Definition 3. *A n-MPDA $(Q, \Sigma, \Gamma, \delta, q_0, F, Z_0)$ with $n \geq 2$ is in normal form if*

- *$\Gamma = \bigcup_{i=1}^{n} \Gamma^{(i)}$ where the $\Gamma^{(i)}$'s are pairwise disjoint memory alphabets whose elements are denoted by $A^{(i)}, B^{(i)}$, etc., and $Z_0 \in \Gamma^{(1)}$.*
- *Only the following transitions are allowed:*
  - *For all $A^{(1)} \in \Gamma^{(1)}$ and $a \in \Sigma \cup \{\epsilon\}$, $\delta(q, a, A^{(1)}) \subseteq \{(q', \epsilon, \ldots, \epsilon) \mid q' \in Q\} \cup \Delta_1 \cup \Delta_2$ with*
    - *$\Delta_1 = \{(q', B^{(1)} C^{(1)}, \epsilon, \ldots, \epsilon) \mid q' \in Q \wedge B^{(1)}, C^{(1)} \in \Gamma^{(1)}\}$,*
    - *$\Delta_2 = \{(q', \epsilon, \ldots, \epsilon, A^{(i)}, \epsilon, \ldots, \epsilon) \mid q' \in Q \wedge A^{(i)} \in \Gamma^{(i)} \wedge 2 \leq i \leq n\}$.*
  - *For all i with $2 \leq i \leq n$ and $a \in \Sigma \cup \{\epsilon\}$,*
    $\delta(q, a, A^{(i)}) \subseteq \{(q', B^{(1)}, \epsilon, \ldots, \epsilon) \mid q' \in Q \wedge B^{(1)} \in \Gamma^{(1)}\}$.

Lemma 4. *An n-MPDA M can be transformed into an n-MPDA M' in normal form with linear blowup in its size such that $L(M) = L(M')$.*

Proof. The proof makes use of the ideas from [3], where a proof for a normal form for D^n-grammars (see below) is given. Notice, however, that we do not use the same normal form as the one of [3] for MPDA.   □

Next, we recall some properties of the class of languages recognized by n-MPDA. We start by defining a renaming operation: A *renaming* of Σ to Σ' is a function $f :$ $\Sigma \rightarrow \Sigma'$. It is extended to strings and languages in the natural way: $f(a_1 \ldots a_k) =$ $f(a_1) \cdot \ldots \cdot f(a_k)$ and $f(L) = \bigcup_{x \in L} f(x)$. The following can be shown following [3].

Lemma 5. *(Closure Properties) The class of languages recognized by n-MPDA is closed under union, concatenation, and Kleene-star. Moreover, given an n-MPDA M over the alphabet Σ and a renaming function $f : \Sigma \rightarrow \Sigma'$, it is possible to construct an n-MPDA M' over Σ' such that $L(M') = f(L(M))$.*

Depth-n-grammars. We now define the notion of a depth-n-grammar. Let V_N and V_T be finite disjoint alphabets and let "(" and ")$_i$" for $i \in \{1, \ldots, n\}$ be $n + 1$ characters not in $V_N \cup V_T$. An *n-list* is a finite string of the form $\overline{\alpha} = w(\alpha_1)_1(\alpha_2)_2 \ldots (\alpha_n)_n$ where $w \in V_T^*$ and $\alpha_i \in V_N^*$ for all i with $1 \leq i \leq n$.

Definition 6. *A depth-n-grammar (D^n-grammar) is a tuple $G = (V_N, V_T, P, S)$ where V_N and V_T are the finite disjoint sets of non-terminal and terminal symbols, respectively, $S \in V_N$ is the axiom, and P is a finite set of productions of the form $A \rightarrow \overline{\alpha}$ with $A \in V_N$ and $\overline{\alpha}$ an n-list.*

For clarity, we may drop empty components of n-lists in the productions as follows: $A \rightarrow w(\epsilon)_1 \ldots (\epsilon)_n$ is written as $A \rightarrow w$, $A \rightarrow (\epsilon)_1 \ldots (\epsilon)_n$ is written as $A \rightarrow \epsilon$, and $A \rightarrow w(\epsilon)_1 \ldots (\epsilon)_{i-1}(\alpha_i)_i(\epsilon)_{i+1} \ldots (\epsilon)_n$ is written as $A \rightarrow w(\alpha_i)_i$.

We define the *derivation relation* on n-lists as follows. Let $i \in \{1, \ldots, n\}$ and let $\overline{\beta} = (\epsilon)_1 \ldots (\epsilon)_{i-1}(A\beta_i)_i(\beta_{i+1})_{i+1} \ldots (\beta_n)_n$ be an n-list, where $\beta_j \in V_N^*$ for all $j \in \{i, \ldots, n\}$. Then,

$$x\overline{\beta} \Rightarrow xw(\alpha_1)_1(\alpha_2)_2 \ldots (\alpha_{i-1})_{i-1}(\alpha_i\beta_i)_i(\alpha_{i+1})_{i+1} \ldots (\alpha_n\beta_n)_n$$

if $A \rightarrow w(\alpha_1)_1(\alpha_2)_2 \ldots (\alpha_n)_n$ is a production and $x \in V_T^*$. Notice that only leftmost derivations are defined. As usual we denote by $\Rightarrow^*$ the reflexive and transitive closure of $\Rightarrow$. A terminal string $x \in V_T^*$ is *derivable* from S if $(S)_1(\epsilon)_2 \ldots (\epsilon)_n \Rightarrow^* x(\epsilon)_1 \ldots (\epsilon)_n$. This will be also denoted by $S \Rightarrow^* x$. The language generated by a D^n-grammar G is $L(G) = \{x \in V_T^* \mid S \Rightarrow^* x\}$.

Definition 7. *Let $G = (V_N, V_T, P, S)$ be a D^n-grammar. Then, the underlying context-free grammar is $G_{cf} = (V_N, V_T, P_{cf}, S)$ with $P_{cf} = \{A \rightarrow w\alpha_1 \ldots \alpha_n \mid A \rightarrow w(\alpha_1)_1 \ldots (\alpha_n)_n \in P\}$.*

The following lemma from [3] is obtained by observing that the language generated by a D^n-grammar is empty iff the language generated by its underlying context-free grammar G_{cf} is empty. Furthermore, it is well-known that emptiness of context-free grammars can be decided in time linear in its size.

Lemma 8. *The emptiness problem of D^n-grammars is decidable in linear time.*

3   Emptiness of MPDA is in 2ETIME

In this section, we show that the emptiness problem of n-MPDA is in 2ETIME. We first show that n-MPDA correspond to D^n-grammars with a double exponential number of non-terminal symbols. To do so, we correct a construction given in [3]. Then, emptiness of D^n-grammars is decidable using the underlying context-free grammar (Lemma 8).

Theorem 9. *A language L is accepted by an n-MPDA iff it is generated by a D^n-grammar.*

In the following we give a sketch of the proof. The "if"-direction is obvious, since a grammar is just an automaton with one state. For the "only if"-direction, let L be a language accepted by empty stacks by an n-MPDA $M = (Q, \Sigma, \Gamma, \delta, q_0, F, Z_0)$. By Lemma 4, we assume, without loss of generality, that M is in normal form. We will construct a D^n-grammar $G_M = (V_N, \Sigma, P, S)$ such that $L(G_M) = L$.

Intuitively, we generalize the proof for the case of 2-MPDA [7]. In [3], an incorrect proof was given for the case of n-MPDA. Recently, the authors of [3] independently gave a generalizable proof for 3-MPDA, which is similar to ours [4]. The general proof idea is the same as for the corresponding proof for pushdown automata. To eliminate states, one has to guess the sequence of states through which the automaton goes by adding pairs of state symbols to the non-terminal symbols of the corresponding grammar. We do this for the first stack. However, when the first stack gets empty, the other stacks may be not empty and one has to know the state in which the automaton is in this situation. For this, we have to guess for all the other non-empty stacks and each of their non-terminal symbols the state in which the automaton will be when reading these symbols. [2]

To do this for the n-th stack, a pair of state symbols is enough. For the $(n-1)$-th stack, in addition to guessing the state, we also have to know the current state on top of the n-th stack to be able to push correctly symbols onto the n-th stack. Therefore, a pair of pairs of states (4 in total) is needed. For the $(n-2)$-th stack, we need to remember the current state and the states on top of the $(n-1)$-th stack and on top of the n-th stack (in total 8 states) and so on. Therefore, there will be 2^n state symbols to be guessed in the first stack. Furthermore we have special state symbols (denoted q_i^e) to indicate that the i-th stack is empty. In Fig. 1 we give an intuitive example illustrating the construction.

Now we define the grammar $G_M = (V_N, \Sigma, P, S)$ formally. To define V_N, we first provide symbols of level i denoted by V_i. For i with $2 \leq i \leq n$, let q_i^e be states pairwise different and different from any state of Q (these are the symbols indicating that the corresponding stack is empty). States of level i are denoted by Q_i and defined as follows : $Q_n = Q \cup \{q_n^e\}$ and for all i such that $2 \leq i < n$, $Q_i = (Q \times Q_{i+1} \times \cdots \times Q_n) \cup \{q_i^e\}$, and $Q_1 = Q \times Q_2 \times \cdots \times Q_n$. We denote by q_i states of Q_i. Then, $V_i = Q_i \times \Gamma \times Q_i$ and $V_N = \{S\} \cup \bigcup_{i=1}^n V_i$. Notice that a state in Q_i different from q_i^e has exactly 2^{n-i} components. Therefore $|V_N| \leq (|Q|+1)^{2^{n+1}} |\Gamma|$. The set P contains exactly the following productions, which are partitioned into five types ($a \in \Sigma \cup \{\epsilon\}$):

---

[2] The proof in [3] incorrectly assumes that this state is the same for each stack when the first stack gets empty.

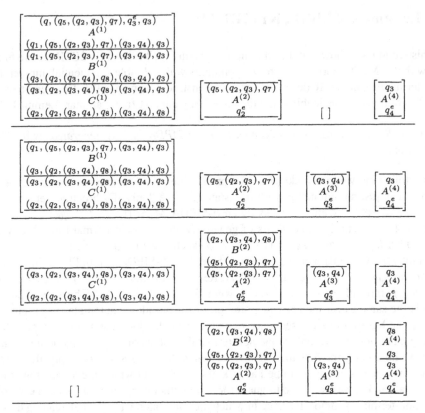

Fig. 1. A sketch of a partial derivation (from top to bottom) of a depth-4-grammar corresponding to a run of a 4-MPDA where three symbols are popped from the first stack while three symbols are pushed onto the other stacks. In each configuration, if the first stack is non-empty, then the state symbols on top of the other stacks can be found on top of the first stack as well. In the last configuration, the top symbols of the other stacks can be found on top of the second stack.

T1 $S \rightarrow ([(q_0, q_2^e, \ldots, q_n^e), Z_0, (q^1, q_2^1, \ldots, q_n^1)])_1$
    if there is k with $2 \leq k \leq n+1$ such that
    - for all i with $2 \leq i < k$ we have $q_i^1 = q_i^e$
    - if $k \leq n$, then $q_k^1 = (q^1, q_{k+1}^1, \ldots, q_n^1)$
T2 $[(q^1, q_2^1, \ldots, q_n^1), A^{(1)}, q_1^2] \rightarrow a([(q^4, q_2^1, \ldots, q_n^1), B^{(1)}, q_1^3][q_1^3, C^{(1)}, q_1^2])_1$
    if $(q^4, B^{(1)}C^{(1)}, \epsilon, \ldots, \epsilon) \in \delta(q^1, a, A^{(1)})$
T3 $[(q^1, q_2^1, \ldots, q_{j-1}^1, q_j^1, q_{j+1}^1, \ldots, q_n^1), A^{(1)}, (q^2, q_2^1, \ldots, q_{j-1}^1, q_j^2, q_{j+1}^1, \ldots, q_n^1)]$
    $\rightarrow a([q_j^2, B^{(j)}, q_j^1])_j$ if $q_j^2 \neq q_j^e$ and $(q^2, \epsilon, \ldots, \epsilon, B^{(j)}, \epsilon, \ldots, \epsilon) \in \delta(q^1, a, A^{(1)})$
T4 $[(q^1, q_{j+1}^1, \ldots, q_n^1), A^{(j)}, q_j^1]$
    $\rightarrow a([(q^4, q_2^e, \ldots, q_{j-1}^1, q_j^1, q_{j+1}^1, \ldots, q_n^1), B^{(1)}, (q^2, q_2^1, \ldots, q_n^2)])_1$
    if $(q^4, B^{(1)}, \epsilon, \ldots, \epsilon) \in \delta(q^1, a, A^{(j)})$, and there is k with $2 \leq k \leq n+1$ such that
    - for all i with $2 \leq i < min(k, j)$ we have $q_i^2 = q_i^e$
    - for all i with $min(k, j) \leq i < k$ we have $q_i^1 = q_i^2 = q_i^e$
    - if $k \leq n$, then $q_k^2 = (q^2, q_{k+1}^2, \ldots, q_n^2)$
T5 $[(q^1, q_2^1, \ldots, q_n^1), A^{(1)}, (q^2, q_2^1, \ldots, q_n^1)] \rightarrow a$ if $(q^2, \epsilon, \ldots, \epsilon) \in \delta(q^1, a, A^{(1)})$

The grammar corresponding to the example in Table 1 can be found in [1]. The following key lemma formalizes the intuition about derivations of the grammar G_M by giving invariants satisfied by them (illustrated in Fig. 1). This lemma is the basic ingredient of the full proof of Theorem 9, which can be found in [1]. Intuitively, condition 1 says that the first element of the first stack contains the state symbols on top of the other stacks. Condition 2 says that the last state symbols in the first stack are of the form allowing condition 3 to be true when the corresponding symbol is popped. Condition 3 says that if the first stack is empty, then the top of the first non-empty stack contains the same state symbols as the top of the other stacks. Conditions 4 and 5 say that the state symbols guessed form a chain through the stacks.

Lemma 10. *Let $w(\gamma_1)(\gamma_2)\ldots(\gamma_n)$ be an n-list different from $(\epsilon)_1 \ldots (\epsilon)_n$ appearing in a derivation of the grammar G_M.*

1. *If $\gamma_1 = [(q^1, q_2^1, \ldots, q_n^1), A^{(1)}, (q^2, q_2^2, \ldots, q_n^2)]\gamma_1'$ with $\gamma_1' \in V_1^*$, then for all i with $2 \le i \le n$, if γ_i is empty, then $q_i^1 = q_i^e$, else $\gamma_i = [q_i^1, B^{(i)}, q_i^3]\gamma_i'$ with $\gamma_i' \in V_i^*$.*
2. *If $\gamma_1 = \gamma_1'[(q^1, q_2^1, \ldots, q_n^1), A^{(1)}, (q^3, q_2^3, \ldots, q_n^3)]$ with $\gamma_1' \in V_1^*$, then there exists k with $2 \le k \le n+1$ such that we have both for all i with $2 \le i < k$, $q_i^3 = q_i^e$ and $k \le n$ implies $q_k^3 = (q^3, q_{k+1}^3, \ldots, q_n^3)$.*
3. *Suppose that $\gamma_1 = \epsilon$. Let i be the smallest k such that γ_k is not empty and let $\gamma_i = [(q^1, q_{i+1}^1, \ldots, q_n^1), A^{(i)}, q_i^2]\gamma_i'$ with $\gamma_i' \in V_i^*$. Then, for all $j > i$, we have: if γ_j is empty, then $q_j^1 = q_j^e$, else $\gamma_j = [q_j^1, A^{(j)}, q_j^3]\gamma_j'$ with $\gamma_j' \in V_j^*$.*
4. *For all i with $2 \le i \le n$, if γ_i is not empty then for some $j \ge 1$,*
   $$\gamma_i = [q_i^1, A_1^{(i)}, q_i^2][q_i^2, A_2^{(i)}, q_i^3]\ldots[q_i^{j-1}, A_{j-1}^{(i)}, q_i^j][q_i^j, A_j^{(i)}, q_i^e] \text{ and for all } l \text{ with}$$
   $1 \le l \le j, q_i^l \ne q_i^e$.
5. *If γ_1 is not empty, then for some $j \ge 1$,*
   $$\gamma_1 = [q_1^1, A_1^{(1)}, q_1^2][q_1^2, A_2^{(1)}, q_1^3]\ldots[q_1^{j-1}, A_{j-1}^{(1)}, q_1^j][q_1^j, A_j^{(1)}, q_1^{j+1}].$$

By observing that the size of the grammar G_M corresponding to an MPDA M in the construction used in the proof of Theorem 9 is double exponential in the number of stacks and using Lemma 8 we obtain the following corollary.

Corollary 11. *The emptiness problem of MPDA is in 2ETIME.*

In the next Section, it is shown that the double exponential upper bound is tight.

4 Emptiness of MPDA Is 2ETIME-Hard

In this section, we prove that the emptiness problem of MPDA is 2ETIME-hard. This is done by adapting a construction in [10], where it is shown that certain bounded-phase pushdown-transducer automata capture precisely the class 2ETIME.

Theorem 12. *The emptiness problem for MPDA is 2ETIME-hard under logspace reductions.*

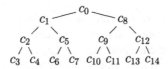

Fig. 2. A run of an alternating Turing machine

Proof. It is well-known that the class of problems solvable by alternating Turing machines in space bounded by 2^{dn} for some d (call it AESPACE) equals 2ETIME [5]. Thus, it is sufficient to show that any problem in AESPACE can be reduced, in logarithmic space, to the emptiness problem for MPDA.

So let T be an alternating Turing machine working in space bounded by 2^{dn}. Let furthermore w be an input for T of length n. We construct (in logarithmic space) from T and w an MPDA M with $2dn + 4$ stacks such that the language of M is non-empty iff w is accepted by T. The simulation of T proceeds in two phases: (1) M guesses a possible accepting run of T on w; (2) M verifies if the guess is indeed a run.

Without loss of generality, we can assume that a transition of T is basically of the form $c \to (c_1 \wedge c_2) \vee (c_3 \wedge c_4) \vee \ldots \vee (c_{h-1} \wedge c_h)$ (where configuration changes are local), i.e., from configuration c, we might switch to both c_1 and c_2 or both c_3 and c_4 and so on. This allows us to represent a run of T as a complete finite binary tree, as shown in Fig. 2, whose nodes are labeled with configurations. Note that each configuration will be encoded as a string, as will be made precise below. The run is accepting if all leaf configurations are accepting. Following the idea of [10], we write the labeled tree as the string (let c^r denote the reverse of c)

$$c_0|c_1|c_2|c_3 \parallel c_3^r \parallel c_4 \parallel c_4^r|c_2^r \parallel c_5|c_6 \parallel c_6^r \parallel c_7 \parallel c_7^r|c_5^r|c_1^r \parallel$$
$$c_8|c_9|c_{10} \parallel c_{10}^r \parallel c_{11} \parallel c_{11}^r|c_9^r \parallel c_{12}|c_{13} \parallel c_{13}^r \parallel c_{14} \parallel c_{14}^r|c_{12}^r|c_8^r|c_0^r$$

It is generated by the (sketched) context-free grammar

$$A \to \alpha_i A \alpha_i + \alpha_i B \alpha_i + \alpha_i \| \alpha_i$$
$$B \to |A \| A|$$

where the α_i are the atomic building blocks of an encoding of a configuration of T. This string allows us to access locally those pairs of configurations that are related by an edge in the tree and thus need to agree with a transition. Finally, the grammar can make sure that all leafs are accepting configurations and that the initial configuration corresponds to the input w. Using two stacks, we can generate such a word encoding of a (possible) run of T and write it onto the second stack, say with c_0 at the top, while leaving the first stack empty behind us (cf. Fig. 3(a)).

The MPDA M now checks if the word written onto stack 2 stems from a run of T. To this aim, we first extract from stack 2 any pair of configurations that needs to be compared wrt. the transition relation of T. For this purpose, some of the configurations need to be duplicated. Corresponding configurations are written side by side as follows: By means of two further stacks, 3 and 4, we transfer the configurations located on stack 2 and separated by the symbol "|" onto the third stack (in reverse order), hereby copying some configuration by writing it onto the fourth stack (cf. Fig. 3(b)).

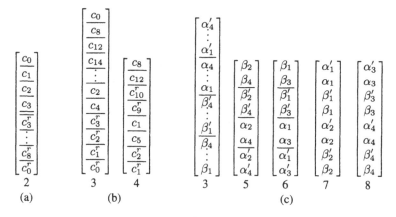

Fig. 3. Guessing and verifying a run of an alternating Turing machine

It still remains to verify that c_0 and c_8 belong to a transition of T, as well as c_{12} and c_{14}, etc. The encoding of one single configuration $a_1 \ldots (q, a_i) \ldots a_{2^{dn}}$ will now allow us to compare two configurations letter by letter. It has the form $(-, a_1, a_2, e)$ $(a_1, a_2, a_3, e) \ldots (a_{i-1}, (q, a_i), a_{i+1}, e) \ldots (a_{2^{dn}-1}, a_{2^{dn}}, -, e)$ where the component e denotes a "transition" $c \to c' \wedge c''$, which has been selected to be executed next and which has been guessed in the above grammar. We would like to compare the k-th letter of one with the k-th letter of another configuration. To access corresponding letters simultaneously, we divide the configurations on stacks 3 and 4 into two, using two further stacks, 5 and 6. We continue this until corresponding letters are arranged one below the other. This procedure, which requires $2dn$ additional stacks, is illustrated in Fig. 3(c) where each α_i and β_i stands for an atomic symbol of the form (a_1, a_2, a_3, e). Note that, in some cases, we encounter pairs of the form (c, c') whereas in some other cases, we face pairs of the form $(c^r, (c')^r)$. Whether we deal with the reverse of a configuration or not can be recognized on the basis of its border symbols (i.e., $(-, a_1, a_2, e)$ or $(a_{2^{dn}-1}, a_{2^{dn}}, -, e)$). Consider, for example, stacks 3 and 4 in Fig. 3(b). We want to compare c_0 and c_8 where c_0 is of the form $(-, a_1, a_2, e) \ldots$, i.e., it is read in the correct order. Suppose e is of the form $c_0 \to c \wedge c'$. Then, locally comparing c_0 and c_8, we can check whether $c' = c_8$. If, at the bottom of stack 3, we compare $c_1^r = (a_{2^{dn}-1}, a_{2^{dn}}, -, e) \ldots$ with c_0^r and e is of the form $c_0 \to c \wedge c'$, then we need to check if $c = c_1$. In other words, the order in which a configuration is read indicates if we follow the right or left successor in the (tree of the) run.   □

From Corollary 11 and Theorem 12, we deduce our main result:

Theorem 13. *The emptiness problem of MPDA is 2ETIME-complete under logspace reductions.*[3]

---

[3] The emptiness problem of MPDA is 2EXPTIME-complete, too. Hereby, 2EXPTIME denotes the class of all decision problems solvable by a deterministic Turing machine in time $exp(exp(n^d))$ for some constant d ($exp(x)$ denoting 2^x). Note that 2EXPTIME is a robust complexity class. On the other hand, 2ETIME is not robust, as it is not closed under logspace reductions.

5   Comparison to Bounded-Phase Multi-stack Pushdown Automata

In this section, we recall m-phase *multi-stack (visibly) pushdown automata* ($m \geq 1$) defined in [8] and show that they are strictly less expressive than $2m$-MPDA.

Multi-stack Visibly Pushdown Automata. For $n \geq 1$, an *n-stack call-return alphabet* is a tuple $\widetilde{\Sigma}_n = \langle \{(\Sigma_c^i, \Sigma_r^i)\}_{i \in \{1,\dots,n\}}, \Sigma_{int} \rangle$ of pairwise disjoint finite alphabets. For $i \in \{1, \dots, n\}$, Σ_c^i is the set of *calls of the stack i*, Σ_r^i is the set of *returns of the stack i*, and Σ_{int} is the set of *internal actions*. For any such $\widetilde{\Sigma}_n$, let $\Sigma_c = \bigcup_{i=1}^n \Sigma_c^i$, $\Sigma_r = \bigcup_{i=1}^n \Sigma_r^i$, $\Sigma^i = \Sigma_c \cup \Sigma_r^i \cup \Sigma_{int}$, for every $i \in \{1, \dots, n\}$, and $\Sigma = \Sigma_c \cup \Sigma_r \cup \Sigma_{int}$.

Definition 14. *A* multi-stack visibly pushdown automaton (MVPA) *over the n-stack call-return alphabet $\widetilde{\Sigma}_n = \langle \{(\Sigma_c^i, \Sigma_r^i)\}_{i \in \{1,\dots,n\}}, \Sigma_{int} \rangle$ is a tuple $N = (Q, \Gamma, \Delta, q_0, F)$ where Q is a finite set of* states, *Γ is a finite* stack alphabet *containing a distinguished stack symbol $\bot$, $\Delta \subseteq (Q \times \Sigma_c \times Q \times (\Gamma \setminus \{\bot\})) \cup (Q \times \Sigma_r \times \Gamma \times Q) \cup (Q \times \Sigma_{int} \times Q)$ is the* transition relation, *$q_0 \in Q$ is the* initial state, *and $F \subseteq Q$ is the set of* final states.

A configuration of N is an $(n+2)$-tuple $\langle q, x; \gamma_1, \dots, \gamma_n \rangle$ where $q \in Q$, $x \in \Sigma^*$, and for all $i \in \{1, \dots, n\}$, $\gamma_i \in \Gamma^*$ is the content of stack i. The *transition relation* $\vdash_N^*$ is the transitive closure of the binary relation $\vdash_N$ over configurations, defined as follows: $\langle q, ax; \gamma_1, \dots, \gamma_n \rangle \vdash_N \langle q', x; \gamma_1', \dots, \gamma_n' \rangle$ if one of the following cases holds:

1. **Internal move:** $a \in \Sigma_{int}$, $(q, a, q') \in \Delta$, and $\gamma_i = \gamma_i'$ for every $i \in \{1, \dots, n\}$.
2. **Push onto stack i:** $a \in \Sigma_c^i$, $\gamma_j' = \gamma_j$ for every $j \neq i$, and there is $A \in \Gamma \setminus \{\bot\}$ such that $(q, a, q', A) \in \Delta$ and $\gamma_i' = A\gamma_i$.
3. **Pop from stack i:** $a \in \Sigma_r^i$, $\gamma_j' = \gamma_j$ for every $j \neq i$, and there is $A \in \Gamma$ such that $(q, a, A, q') \in \Delta$ and either $A \neq \bot$ and $\gamma_i = A\gamma_i'$, or $A = \bot$ and $\gamma_i = \gamma_i' = \bot$.

A string $x \in \Sigma^*$ is *accepted* by N if there are $\gamma_1, \dots, \gamma_n \in \Gamma^*$ and $q \in F$ such that $\langle q_0, x; \bot, \dots, \bot \rangle \vdash_N^* \langle q, \epsilon; \gamma_1, \dots, \gamma_n \rangle$. The language of N, denoted $L(N)$, is the set of all strings accepted by N.

Definition 15. *For $m \geq 1$, an* m-phase multi-stack visibly pushdown automaton *(m-MVPA) over the n-stack call-return alphabet $\widetilde{\Sigma}_n$ is a tuple $K = (m, Q, \Gamma, \Delta, q_0, F)$ where $N = (Q, \Gamma, \Delta, q_0, F)$ is an MVPA over $\widetilde{\Sigma}_n$. The language accepted by K is*
$$L(K) = \bigcup_{i_1,\dots,i_m \in \{1,\dots,n\}} \left(L(N) \cap \left((\Sigma^{i_1})^* \cdots (\Sigma^{i_m})^*\right)\right).$$

Finally, we recall that the class of languages accepted by m-MVPA is closed under union, intersection, renaming, and complementation [8]. However, one easily shows:

Lemma 16. *The class of languages of m-MVPA is not closed under Kleene-star.*

$2m$-MPDA are Strictly More Expressive than m-MVPA. We now show that, for any $m \geq 1$, $2m$-MPDA are strictly more expressive than m-MVPA. Let us fix an m-MVPA $K = (m, Q, \Gamma, \Delta, q_0, F)$ over $\widetilde{\Sigma}_n = \langle \{(\Sigma_c^i, \Sigma_r^i)\}_{i \in \{1,\dots,n\}}, \Sigma_{int} \rangle$, with $N = (Q, \Gamma, \Delta, q_0, F)$ an MVPA.

Proposition 17. *For every sequence* $i_1, \ldots, i_m \in \{1, \ldots, n\}$, *it is possible to construct a* $2m$-*MPDA* M *such that* $L(M) = L(N) \cap \left((\Sigma^{i_1})^* \cdots (\Sigma^{i_m})^* \right)$.

In the following, we sketch the proof. Intuitively, any computation of N accepting a string $x \in L(N) \cap \left((\Sigma^{i_1})^* \cdots (\Sigma^{i_m})^* \right)$ can be decomposed into m phases, where in each phase (say j), N can only pop from the stack i_j (but it can push onto all stacks).

Let $j \in \{1, \ldots, m\}$ be the current phase of N and for every $l \in \{1, \ldots, n\}$, let $k_l^j = min\left(\{k \mid j \le k \le m \wedge i_k = l\} \cup \{m+1\} \right)$ denote the closest phase in $\{j, \ldots, m\}$ such that N can pop from the l-th stack if the phase is k_l^j (note that $k_{i_j}^j = j$), if such phase does not exist, then $k_l^j = m + 1$.

We construct a $2m$-MPDA M over Σ such that the following invariant is preserved during the simulation of N when its current phase is j: the content of the l-th stack of N is stored in the $(2k_l^j - 1)$-th stack of M if $k_l^j \ne m + 1$. Then, an internal move (labeled by $a \in \Sigma_{int}$) of N is simulated by an internal move (labeled by a) of M; a pop rule (labeled by $a \in \Sigma_r^{i_j}$) of N from the i_j-th stack corresponds to a pop rule (labeled by a) of M from the $(2j - 1)$-th stack; and a push rule (labeled by $a \in \Sigma_c^l$) onto the l-th stack of N is simulated by a push rule (labeled by a) of M onto the $(2k_l^j - 1)$-th stack if $k_l^j \ne (m + 1)$, else by an internal move (labeled by a) of M.

On switching phase from j to $(j + 1)$ if $k_{i_j}^{j+1} \ne m + 1$, when N is able once again to pop from the (i_j)-th stack, M moves the content of the $(2j - 1)$-th stack onto the $(2k_{i_j}^{j+1} - 1)$-th stack using the $(2j)$-th stack as an intermediary one, else it removes the content of the $(2j-1)$-th stack. Observe that all the above described behaviors maintain the stated invariant since $k_l^{j+1} = k_l^j$ for every $l \ne i_j$.

We are now ready to present the main result of this section.

Theorem 18. *$2m$-MPDA are strictly more expressive than m-MVPA.*

Proof. For every m-MVPA K over the stack alphabet $\widetilde{\Sigma}_n$ one can construct a $2m$-MPDA M over Σ such that $L(M) = L(K)$ by considering all possible orderings of phases (fixing for each phase the stack which can be popped) and using Proposition 17. To prove *strict* inclusion, we notice that the class of languages recognized by $2m$-MPDA is closed under Kleene-star (Lemma 5) but the class of languages of m-MVPA is not (Lemma 16). □

$2m$-MPDA are Strictly More Expressive than m-MPA. In the following, we extend the previous result to m-phase multi-stack pushdown automata over non-visible alphabets (defined in [8]). A multi-stack pushdown automaton (called MPA) over (non-visible) alphabet Σ is simply an n-stack automaton with ϵ-moves, that can push and pop from any stack when reading any letter. Also, we define m-phase version of these (called m-MPA). An m-MPA is an MPA using at most m-phases, where in each phase one can pop from one distinguished stack, and push on any other stack.

Theorem 19. *$2m$-MPDA are strictly more expressive than m-MPA.*

The idea behind proving inclusion is that for any m-MPA K over Σ, it is possible to construct an m-MVPA K' over $\widetilde{\Sigma}'_n = \langle \{ (\Sigma'^i_c, \Sigma'^i_r) \}_{i \in \{1, \ldots, n\}}, \Sigma'_{int} \rangle$, with $\Sigma'^i_c =$

$(\Sigma \cup \{\epsilon\}) \times \{c\} \times \{i\}$, $\Sigma'^{i}_{r} = (\Sigma \cup \{\epsilon\}) \times \{r\} \times \{i\}$, and $\Sigma'_{int} = (\Sigma \cup \{\epsilon\}) \times \{int\}$, such that every transition on $a \in \Sigma \cup \{\epsilon\}$ that pushes onto the stack i is transformed to a transition on (a, c, i), transitions on a that pop the stack i are changed to transitions on (a, r, i), and the remaining a-transitions are changed to transitions over (a, int). Let f be a renaming function that maps each symbol (a, c, i), (a, r, i), and (a, int) to a. Then, $w \in L(K)$ iff there is some $w' \in L(K')$ such that $w = f(w')$. It follows that $L(K) = f(L(K'))$. Consider now the $2m$-MPDA M' over Σ' constructed from K' such that $L(M') = L(K')$, thanks to Theorem 18. Then, it is possible to construct from M' a $2m$-MPDA M over Σ such that $L(M) = f(L(M'))$ (Lemma 5) which implies that $L(M) = L(K)$. To prove the *strict* inclusion we use the easy to see fact that m-MPA are not closed under Kleene-star whereas $2m$-MPDA are (Lemma 5).

6   Conclusion

We have shown that the emptiness problem for multi-pushdown automata (MPDA) is 2ETIME-complete. The study of the emptiness problem is the first step of a comprehensive study of verification problems for MPDA. For standard pushdown automata, a lot of work has been done recently (see for example [2]) concerning various model-checking problems. It will be interesting to see how these results carry over to MPDA and at which cost. A basic ingredient of model-checking algorithms is typically to characterize the set of successors or predecessors of sets of configurations. For MPDA, this problem remains to be studied. Another class of extended pushdown automata has recently been studied extensively: the class of higher-order pushdown automata (HPDA, see for example [6]). It is quite easy to see that HPDA of order n can simulate MPDA with n stacks (which allows us to use all verification results for HPDA also for MPDA). However, the converse is wrong, since emptiness of pushdown automata of order n is $(n - 1)$-EXPTIME-complete [6]. Therefore, it is interesting to study dedicated algorithms for the verification of MPDA.

References

1. Atig, M.F., Bollig, B., Habermehl, P.: Emptiness of multi-pushdown automata is 2ETIME-complete. Research Report LSV-08-16, LSV, ENS Cachan (May 2008),
   http://www.lsv.ens-cachan.fr/Publis/RAPPORTS_LSV/
   PDF/rr-lsv-2008-16.pdf
2. Bouajjani, A., Esparza, J., Maler, O.: Reachability analysis of pushdown automata: Application to model-checking. In: Mazurkiewicz, A., Winkowski, J. (eds.) CONCUR 1997. LNCS, vol. 1243, pp. 135–150. Springer, Heidelberg (1997)
3. Breveglieri, L., Cherubini, A., Citrini, C., Crespi Reghizzi, S.: Multi-push-down languages and grammars. International Journal of Foundations of Computer Science 7(3), 253–292 (1996)
4. Breveglieri, L., Cherubini, A., Crespo Reghizzi, S.: Personal communication
5. Chandra, A.K., Kozen, D.C., Stockmeyer, L.J.: Alternation. J. ACM 28(1), 114–133 (1981)
6. Engelfriet, J.: Iterated stack automata and complexity classes. Information and Computation 95(1), 21–75 (1991)

7. San Pietro, P.: Two-stack automata. Technical Report 92-073, Dipartimento di elettronica e informazione, Politechnico di Milano (1992)
8. La Torre, S., Madhusudan, P., Parlato, G.: A robust class of context-sensitive languages. In: Proceedings of LICS, pp. 161–170. IEEE, Los Alamitos (2007)
9. La Torre, S., Madhusudan, P., Parlato, G.: Context-bounded analysis of concurrent queue systems. In: Ramakrishnan, C.R., Rehof, J. (eds.) TACAS 2008. LNCS, vol. 4963, pp. 299–314. Springer, Heidelberg (2008)
10. La Torre, S., Madhusudan, P., Parlato, G.: An infinite automaton characterization of double exponential time. In: Proceedings of CSL 2008. LNCS. Springer, Heidelberg (to appear, 2008)

The Average State Complexity of the Star of a Finite Set of Words Is Linear

Frédérique Bassino[1], Laura Giambruno[2], and Cyril Nicaud[3]

[1] LIPN UMR CNRS 7030, Université Paris-Nord, 93430 Villetaneuse, France
[2] Dipartimento di Matematica e Applicazioni, Università di Palermo, 90100, Italy
[3] IGM, UMR CNRS 8049, Université Paris-Est, 77454 Marne-la-Vallée, France
`bassino@lipn.univ-paris13.fr, lgiambr@math.unipa.it, nicaud@univ-mlv.fr`

Abstract. We prove that, for the uniform distribution over all sets X of m (that is a fixed integer) non-empty words whose sum of lengths is n, $\mathcal{D}_X$, one of the usual deterministic automata recognizing X^*, has on average $\mathcal{O}(n)$ states and that the average state complexity of X^* is $\Theta(n)$. We also show that the average time complexity of the computation of the automaton $\mathcal{D}_X$ is $\mathcal{O}(n \log n)$, when the alphabet is of size at least three.

1 Introduction

This paper addresses the following issue: given a finite set of words X on an alphabet A and a word $u \in A^*$, how to determine efficiently whether $u \in X^*$ or not?

With a non-deterministic automaton, one can determine whether a word u is in X^* or not in time proportional to the product of the lengths of u and X, where the length of X is the sum of the lengths of its elements.

With a deterministic automaton recognizing X^*, one can check whether a word u is in X^* or not in time proportional to the size of u, once the automaton is computed. But in [5], Ellul, Krawetz, Shallit and Wand found an example where the state complexity of X^*, *i.e.* the number of states of the minimal automaton of X^*, is exponential. More precisely, for every integer $h \geq 3$, they gave a language X_h of length $\Theta(h^2)$, containing $\Theta(h)$ words, whose state complexity is $\Theta(h2^h)$. Using another measure on finite sets of words, Campeanu, Culik, Salomaa and Yu proved in [2,3] that if the set X is a finite language of state complexity $n \geq 4$, the state complexity of X^* is $2^{n-3} + 2^{n-4}$ in the worst case, for an alphabet with at least three letters. Note that the state complexity of X^* is $2^{n-1} + 2^{n-2}$ in the worst case when X is not necessarily finite [14,15].

An efficient alternative using algorithms related to Aho-Corasick automaton was proposed in [4] by Clément, Duval, Guaiana, Perrin and Rindone. In their paper, an algorithm to compute all the decompositions of a word as a concatenation of elements in a finite set of non-empty words is also given.

This paper is a contribution to this general problem, called the noncommutative Frobenius problem by Shallit [10], from the name of the classical problem [8,9] of which it is a generalization. Our study is made from an average point

M. Ito and M. Toyama (Eds.): DLT 2008, LNCS 5257, pp. 134–145, 2008.

of view. We analyse the average state complexity of X^*, for the uniform distribution of sets of m non-empty words, whose sum of lengths is n, and as n tends towards infinity. We use the general framework of analytic combinatorics [6] applied to sets of words and classical automata constructions. Our main result is that, on average, the state complexity of the star of a set X of m non-empty words is linear with respect to the length of X. For an alphabet with at least three letters, we also provide an algorithm to build a deterministic automaton recognizing X^* in average time $\mathcal{O}(n \log n)$, where n is the length of X.

The paper is organized as follows. In Section 2 we recall some definitions, usual automata constructions and combinatorial properties about words. In Section 3 we sketch the proof of the linearity of the average number of states of a deterministic automaton $\mathcal{D}_X$ recognizing X^*. As a consequence of our construction, in Section 4, we prove that the average time complexity for the construction of the automaton $\mathcal{D}_X$ is in $\mathcal{O}(n \log n)$ when the size of the alphabet is at least three. In Section 5, we establish that the average state complexity of the star of a finite set with m non-empty words whose sum of lengths is n is proportional to n. In the case of sets of two words, we prove a stronger result: the average size of the minimal automaton of X^* is equivalent to n. Finally, in Section 6 we give an algorithm to randomly and equiprobably generate sets X of m non-empty words whose sum of lengths is n, and use it to obtain some experimental results about the average number of states of $\mathcal{D}_X$.

2 Preliminary

2.1 Definitions and Constructions

A *finite automaton* $\mathcal{A}$ over a finite alphabet A is a quintuple $\mathcal{A} = (A, Q, T, I, F)$ where Q is a finite set of *states*, $T \subset Q \times A \times Q$ is the set of *transitions*, $I \subset Q$ is the set of *initial states* and $F \subset Q$ is the set of final states. The automaton $\mathcal{A}$ is *deterministic* if it has only one initial state and for any $(p, a) \in Q \times A$ there exists at most one $q \in Q$ such that $(p, a, q) \in T$. It is *complete* if for each $(p, a) \in Q \times A$, there exists at least one $q \in Q$ such that $(p, a, q) \in T$. A deterministic finite automaton $\mathcal{A}$ is *accessible* when for each state q of $\mathcal{A}$, there exists a path from the initial state to q. The *size* $\#\mathcal{A}$ of an automaton $\mathcal{A}$ is its number of states. The *minimal automaton* of a regular language is the unique smallest accessible and deterministic automaton recognizing this language. The *state complexity* of a regular language is the size of its minimal automaton. We refer the readers to [7,13,1] for elements of theory of finite automata.

Any finite automaton $\mathcal{A} = (A, Q, T, I, F)$ can be transformed into a deterministic automaton $\mathcal{B} = (A, \mathcal{P}(Q), T', \{I\}, F')$ recognizing the same language and in which $F' = \{P \in \mathcal{P}(Q) \mid P \cap F \neq \emptyset\}$ and $T' = \{(P, a, R)$ with $P \in \mathcal{P}(Q), a \in A$ and $R = \{q \mid \exists p \in P, (p, a, q) \in T\}\}$. To be more precise only the accessible part of the automaton $\mathcal{B}$ is really built in this *subset construction*.

Let $X \subset A^*$ be a finite set of words. Denote by $\mathrm{Pr}(X)$ the set of all prefixes of elements of X. The automaton $(A, \mathrm{Pr}(X), T_X, \{\varepsilon\}, X)$, where $T_X = \{(u, a, ua) \mid u \in \mathrm{Pr}(X), a \in A, ua \in \mathrm{Pr}(X)\}$, recognizes the set X and the automaton

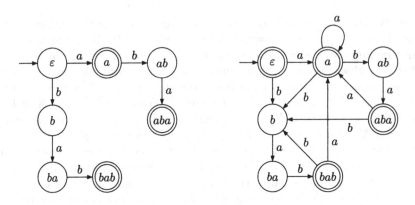

Fig. 1. The automata $(\{a,b\}, \mathrm{Pr}(X), T_X, \{\varepsilon\}, X)$ and $\mathcal{A}_X$, for $X = \{a, aba, bab\}$

$\mathcal{A}_X = (A, \mathrm{Pr}(X), T_X \cup T, \{\varepsilon\}, X \cup \{\varepsilon\})$, where $T = \{(u, a, a) \mid u \in X,\ a \in A \cap \mathrm{Pr}(X)\}$ recognizes X^* (see Fig.1). We denote by $\mathcal{A}_S$ the automaton defined for the set of elements of any sequence S by the above construction. In such an automaton only the states labelled by a letter have more than one incoming transition.

For any finite set of words $X \subset A^*$ (resp. any sequence S), we denote by $\mathcal{D}_X$ (resp. $\mathcal{D}_S$) the accessible deterministic automaton obtained from the automaton $\mathcal{A}_X$ (resp. $\mathcal{A}_S$) making use of the subset construction and by $\mathcal{M}_X$ the minimal automaton of X^*.

Lemma 1. *For any finite set of words $X \subset A^*$, the states of the deterministic automaton $\mathcal{D}_X$ recognizing X^* are non-empty subsets $\{u_1, \cdots, u_\ell\}$ of $\mathrm{Pr}(X)$ such that for all $i, j \in \{1, \cdots, \ell\}$, either u_i is a suffix of u_j or u_j is a suffix of u_i.*

2.2   Enumeration

Let $X \subset A^*$ be a finite set of words. We denote by $|X|$ the cardinality of X and by $\|X\|$ the *length* of X defined as the sum of the lengths of its elements: $\|X\| = \sum_{u \in X} |u|$. Let $Set_{n,m}$ be the set of sets of m non-empty words whose sum of lengths is n:

$$Set_{n,m} = \{X = \{u_1, \cdots, u_m\} \mid \|X\| = n,\ \forall i \in \{1, \cdots, m\}\ u_i \in A^+\}$$

and $\mathcal{S}_{n,m}$ be the set of sequences of m non-empty words whose sum of lengths is n:

$$\mathcal{S}_{n,m} = \{S = (u_1, \cdots, u_m) \mid \|S\| = n,\ \forall i \in \{1, \cdots, m\}\ u_i \in A^+\}$$

We denote by $\mathcal{S}_{n,m}^{\neq} \subset \mathcal{S}_{n,m}$ the set of sequences of pairwise distinct words. Recall that $f(n) = \mathcal{O}(g(n))$ if there exists a positive real number c such that for all n big enough $|f(n)| \le c|g(n)|$.

Proposition 1. *For any fixed integer $m \geq 2$,*

$$|S_{n,m}| = \binom{n-1}{m-1}|A|^n \quad and \quad |Set_{n,m}| = \frac{1}{m!}|S_{n,m}|\left(1 + \mathcal{O}\left(\frac{1}{n^2}\right)\right).$$

Proof. (sketch) Any sequence S of $S_{n,m}$ can be uniquely defined by a word v of length n, which is the concatenation of the elements of S, and a composition of n into m parts, that indicates how to cut the word of length n into m parts. Therefore $|S_{n,m}| = \binom{n-1}{m-1}|A|^n$. Using methods from analytic combinatorics [6], one can prove that

$$|S_{n,m}^{\neq}| = |S_{n,m}|\left(1 + \mathcal{O}\left(\frac{1}{n^2}\right)\right).$$

Furthermore since an element of $Set_{n,m}$ is mapped to exactly $m!$ sequences of $S_{n,m}^{\neq}$, we obtain $|S_{n,m}^{\neq}| = m!|Set_{n,m}|$, concluding the proof. □

We say that the word v is a *proper prefix* (resp. suffix) of a word u if v is a prefix (resp. suffix) of u such that $v \neq \varepsilon$ and $v \neq u$. The word v is called a *border* of u if v is both proper prefix and proper suffix of u. We denote by $\text{Pref}(u)$ (resp. $\text{Suff}(u)$) the set of proper prefixes (resp. suffixes) of u and by $\text{Bord}(u)$ the set of borders of u. A word is *primitive* when it is not the power of another one.

Let u, v and w be three non-empty words such that v is a proper suffix of u and w is a proper suffix of v. We define the three following sets:

$$Q_u = \{\{u\} \cup P \mid P \subset \text{Suff}(u)\}$$
$$Q_{u,v} = \{\{u\} \cup P \mid P \in Q_v\}$$
$$Q_{u,v,w} = \{\{u\} \cup P \mid P \in Q_{v,w}\}.$$

Note that the cardinalities of Q_u, $Q_{u,v}$ and $Q_{u,v,w}$ are respectively equal to $2^{|u|-1}$, $2^{|v|-1}$ and $2^{|w|-1}$.

In the proof of the main result (Theorem 1) of this paper, we count the number of states of automata according to their labels. This enumeration is based on the following combinatorial properties of words whose proofs derived from classical results of combinatorics on words (see [11,12]) are omitted.

Lemma 2. *Let u be a non-empty word of length ℓ. The number of sequences $S \in S_{n,m}$ such that u is a prefix of a word of S is smaller or equal to $m\binom{n-\ell}{m-1}|A|^{n-\ell}$.*

Lemma 3. *Let $u, v \in A^+$ such that v is not a prefix of u, $|u| = \ell$ and $|v| = i$. The number of sequences $S \in S_{n,m}$ such that both u and v are prefixes of words of S is smaller or equal to $m(m-1)|A|^{n-\ell-i}\binom{n-\ell-i+1}{m-1}$.*

Lemma 4 ([12] p. 270). *For $1 \leq i < \ell$, there are at most $|A|^{\ell-i}$ pairs of non-empty words (u,v) such that $|u| = \ell$, $|v| = i$ and v is a border of u.*

Lemma 5. *For $1 \leq j < i < \ell$ such that either $i \leq \frac{2}{3}\ell$ or $j \leq \frac{i}{2}$, there are at most $|A|^{\ell-\frac{i}{2}-j}$ triples of non-empty words (u,v,w) with $|u| = \ell$, $|v| = i$, $|w| = j$ such that v is a border of u and w is a border of v.*

Proposition 2. *For* $1 \leq j < i < \ell$ *such that* $i > \frac{2}{3}\ell$ *and* $j > \frac{i}{2}$ *and for any triple of non-empty words* (u, v, w) *with* $|u| = \ell$, $|v| = i$, $|w| = j$ *such that* v *is a border of* u *and* w *is a border of* v, *there exist a primitive word* x, *with* $1 \leq |x| \leq \ell - i$, *a prefix* x_0 *of* x *and nonnegative integers* $p > q > s > 0$ *such that* $u = x^p x_0$, $v = x^q x_0$ *and* $w = x^s x_0$.

3   Main Result

In this section we give the proof of the following theorem.

Theorem 1. *For the uniform distribution over the sets* X *of* m *(a fixed integer) non-empty words whose sum of lengths is* n, *the average number of states of the accessible and deterministic automata* $\mathcal{D}_X$ *recognizing* X^* *is linear in the length* n *of* X.

First, note that to prove this result on sets it is sufficient to prove it on sequences:

$$\frac{1}{|\mathcal{S}et_{n,m}|} \sum_{X \in \mathcal{S}et_{n,m}} \#\mathcal{D}_X = \frac{1}{m! \, |\mathcal{S}et_{n,m}|} \sum_{S \in \mathcal{S}_{n,m}^{\neq}} \#\mathcal{D}_S \leq \frac{1}{m! \, |\mathcal{S}et_{n,m}|} \sum_{S \in \mathcal{S}_{n,m}} \#\mathcal{D}_S$$

and we conclude using Proposition 1.

Let $Y \subset A^*$ and $S \in \mathcal{S}_{n,m}$, we denote by $\mathfrak{D}et(S, Y)$ the property: Y is the label of a state of $\mathcal{D}_S$. Let P be a property, the operator $[\![\,]\!]$ is defined by $[\![P]\!] = 1$ if P is true and 0 otherwise.

To find an upper bound for the average number of states of the deterministic automaton $\mathcal{D}_S$ when the sequence S ranges the set $\mathcal{S}_{n,m}$, we count the states of all automata according to their labels. More precisely we want to estimate the sum

$$\sum_{S \in \mathcal{S}_{n,m}} \#\mathcal{D}_S = \sum_{S \in \mathcal{S}_{n,m}} \sum_{Y \subset A^*} [\![\mathfrak{D}et(S, Y)]\!],$$

Taking into account the cardinality of the labels of the states:

$$\sum_{S \in \mathcal{S}_{n,m}} \#\mathcal{D}_S = \sum_{S \in \mathcal{S}_{n,m}} \sum_{|Y|=1} [\![\mathfrak{D}et(S, Y)]\!] + \sum_{S \in \mathcal{S}_{n,m}} \sum_{|Y|\geq 2} [\![\mathfrak{D}et(S, Y)]\!].$$

The first sum deals with states labelled by a single word. Since, for each $S \in \mathcal{S}_{n,m}$, the words that appear in the labels of states of $\mathcal{D}_S$ are prefixes of words of S, we have

$$\sum_{S \in \mathcal{S}_{n,m}} \sum_{|Y|=1} [\![\mathfrak{D}et(S, Y)]\!] = \sum_{S \in \mathcal{S}_{n,m}} \sum_{\substack{u \text{ prefix of} \\ \text{a word of } S}} [\![\mathfrak{D}et(S, \{u\})]\!] \leq (n+1)|\mathcal{S}_{n,m}|.$$

It remains to study the sum

$$\Delta = \sum_{S \in \mathcal{S}_{n,m}} \sum_{|Y|\geq 2} [\![\mathfrak{D}et(S, Y)]\!].$$

Let $Y \subset A^*$ be a non-empty set which is not a singleton. By Lemma 1, if Y is the label of a state of an automaton $\mathcal{D}_S$, then Y belongs to a set $Q_{u,v}$, for some non-empty word u and some proper suffix v of u. Therefore

$$\Delta = \sum_{S \in \mathcal{S}_{n,m}} \sum_{u \in A^+} \sum_{v \in \text{Suff}(u)} \sum_{Y \in Q_{u,v}} [\![\mathcal{D}\text{et}(S, Y)]\!].$$

Changing the order of the sums we obtain

$$\Delta = \sum_{u \in A^+} \sum_{v \in \text{Suff}(u)} \sum_{Y \in Q_{u,v}} \sum_{S \in \mathcal{S}_{n,m}} [\![\mathcal{D}\text{et}(S, Y)]\!].$$

We then partition the sum Δ into $\Delta_1 + \Delta_2$ depending on whether the word v is prefix of u or not:

$$\Delta_1 = \sum_{u \in A^+} \sum_{v \in \text{Bord}(u)} \sum_{Y \in Q_{u,v}} \sum_{S \in \mathcal{S}_{n,m}} [\![\mathcal{D}\text{et}(S, Y)]\!]$$

$$\Delta_2 = \sum_{u \in A^+} \sum_{v \in \text{Suff}(u) \backslash \text{Pref}(u)} \sum_{Y \in Q_{u,v}} \sum_{S \in \mathcal{S}_{n,m}} [\![\mathcal{D}\text{et}(S, Y)]\!]$$

To prove Theorem 1, we establish in the following that Δ_1 and Δ_2 are both $\mathcal{O}(n \, |\mathcal{S}_{n,m}|)$.

3.1   Proof for an Alphabet of Size at Least 3

Let $k \geq 3$ be the cardinality of the alphabet A. Using Lemma 3 we have that

$$\Delta_2 \leq \sum_{u \in A^+} \sum_{v \in \text{Suff}(u) \backslash \text{Pref}(u)} \sum_{Y \in Q_{u,v}} m(m-1)k^{n-|u|-|v|} \binom{n-|u|-|v|+1}{m-1}.$$

As $|Q_{u,v}| = 2^{|v|-1}$, with $\ell = |u|$ and $i = |v|$,

$$\Delta_2 \leq \sum_{\ell=2}^{n-m+1} k^{\ell} \sum_{i=1}^{\ell-1} 2^{i-1} m(m-1)k^{n-\ell-i} \binom{n-\ell-i+1}{m-1}.$$

Moreover, since $2^i k^{-i} \leq 1$ and since $\sum_{\ell=2}^{n-m+1} \sum_{i=1}^{\ell-1} \binom{n-\ell-i+1}{m-1} = \binom{n-1}{m}$,

$$\Delta_2 \leq \frac{m(m-1)}{2} k^n \binom{n-1}{m}$$

and thus, by Proposition 1, $\Delta_2 = \mathcal{O}(n \, |\mathcal{S}_{n,m}|)$.
  Now by Lemma 2, we have

$$\Delta_1 \leq \sum_{u \in A^+} \sum_{v \in \text{Bord}(u)} \sum_{Y \in Q_{u,v}} m \binom{n-|u|}{m-1} k^{n-|u|}.$$

Since $|Q_{u,v}| = 2^{|v|-1}$ we get by Lemma 4

$$\Delta_1 \leq \sum_{\ell=2}^{n-m+1} \sum_{i=1}^{\ell-1} m \binom{n-\ell}{m-1} k^{n-\ell} k^{\ell-i} 2^{i-1}.$$

Since $\sum_{i=1}^{\ell-1} \left(\frac{2}{k}\right)^i \leq \frac{2}{k-2}$, when $k \geq 3$, and $\sum_{\ell=2}^{n-m+1} \binom{n-\ell}{m-1} = \binom{n-1}{m}$,

$$\Delta_1 \leq \frac{m}{(k-2)} k^n \binom{n-1}{m}.$$

We use Proposition 1 to conclude that $\Delta_1 = \mathcal{O}(n|\mathcal{S}_{n,m}|)$.

3.2 Proof for an Alphabet of Size 2

The study of Δ_2 is the same as in the previous section. Now we partition the sum Δ_1 into two sums $\Delta_{1,1}$ and $\Delta_{1,2}$ depending on whether the set Y contains exactly two elements or not (and therefore belongs to some set $Q_{u,v,w}$). More precisely,

$$\Delta_{1,1} = \sum_{u \in A^+} \sum_{v \in \text{Bord}(u)} \sum_{S \in \mathcal{S}_{n,m}} [\![\mathfrak{Det}(S, \{u, v\})]\!]$$

and

$$\Delta_{1,2} = \sum_{u \in A^+} \sum_{v \in \text{Bord}(u)} \sum_{w \in \text{Suff}(v)} \sum_{Y \in Q_{u,v,w}} \sum_{S \in \mathcal{S}_{n,m}} [\![\mathfrak{Det}(S, Y)]\!].$$

Using Lemmas 2 and 4, and since $\sum_{i=1}^{\ell-1} 2^{-i} \leq 1$ and $\sum_{\ell=2}^{n-m+1} \binom{n-\ell}{m-1} = \binom{n-1}{m}$, we obtain

$$\Delta_{1,1} \leq \sum_{\ell=2}^{n-m+1} \sum_{i=1}^{\ell-1} m \binom{n-\ell}{m-1} 2^{n-\ell} 2^{\ell-i} \leq m \, 2^n \binom{n-1}{m}.$$

Consequently, by Proposition 1, $\Delta_{1,1} = \mathcal{O}(n|\mathcal{S}_{n,m}|)$.

Next we decompose the sum $\Delta_{1,2}$ into the sums $B_{1,2} + N_{1,2}$ depending on whether w is a prefix (and therefore a border) of v or not.

When w is not a prefix of v, the number of sequences $S \in \mathcal{S}_{n,m}$ such that u and w are prefixes of two distinct words of S is at most $m(m-1)2^{n-\ell-j}\binom{n-\ell-j+1}{m-1}$ from Lemma 3.

Since, from Lemma 4, there are less than $2^{\ell-i}$ pairs (u, v) such that v is a border of u and since $|Q_{u,v,w}| = 2^{|w|-1}$, we get:

$$N_{1,2} = \sum_{u \in A^+} \sum_{v \in \text{Bord}(u)} \sum_{w \in \text{Suff}(v)\backslash\text{Pref}(v)} \sum_{Y \in Q_{u,v,w}} \sum_{S \in \mathcal{S}_{n,m}} [\![\mathfrak{Det}(S, Y)]\!]$$

$$\leq \sum_{\ell=3}^{n-m+1} \sum_{i=2}^{\ell-1} \sum_{j=1}^{i-1} 2^{\ell-i} 2^{j-1} m(m-1) 2^{n-\ell-j} \binom{n-\ell-j+1}{m-1}$$

$$\leq m(m-1) 2^{n-1} \sum_{\ell=3}^{n-m+1} \sum_{i=2}^{\ell-1} 2^{-i} \sum_{j=1}^{i-1} \binom{n-\ell-j+1}{m-1}$$

As $\binom{n-\ell-j+1}{m-1} \leq \binom{n-\ell}{m-1}$, we obtain

$$N_{1,2} \leq m(m-1)2^{n-1} \sum_{\ell=3}^{n-m+1} \binom{n-\ell}{m-1} \sum_{i=2}^{\ell-1}(i-1)2^{-i}$$

Because of the convergence of the series, $\sum_{i=2}^{\ell-1}(i-1)2^{-i}$ is bounded. Therefore, as $\sum_{\ell=3}^{n-m+1} \binom{n-\ell}{m-1} = \binom{n-2}{m}$ and $|S_{n,m}| = \binom{n-1}{m-1}2^n$, we have $N_{1,2} = \mathcal{O}(n|S_{n,m}|)$.

When w is prefix of v, the associated sum $B_{1,2}$ is partitioned into the following sums:

$$B_{1,2} = \sum_{u \in A^+} \sum_{v \in \text{Bord}(u)} \sum_{w \in \text{Bord}(v)} \sum_{Y \in Q_{u,v,w}} \sum_{S \in S_{n,m}} [\![\mathfrak{D}\mathfrak{et}(S,Y)]\!] = B'_{1,2} + B''_{1,2}$$

with

$$B'_{1,2} = \sum_{u \in A^+} \sum_{\substack{v \in \text{Bord}(u) \\ |v| > \frac{2}{3}|u|}} \sum_{\substack{w \in \text{Bord}(v) \\ |w| > \frac{|v|}{2}}} \sum_{Y \in Q_{u,v,w}} \sum_{S \in S_{n,m}} [\![\mathfrak{D}\mathfrak{et}(S,Y)]\!]$$

and $B''_{1,2} = B_{1,2} \setminus B'_{1,2}$. Using Lemma 5, the fact that $|Q_{u,v,w}| = 2^{|w|-1}$ and relaxing the constraints on the lengths of the words v and w, we get

$$B''_{1,2} \leq \sum_{\ell=3}^{n-m+1} \sum_{i=2}^{\ell-1} \sum_{j=1}^{i-1} m\binom{n-\ell}{m-1} 2^{n-\ell} 2^{\ell - \frac{i}{2} - j} 2^{j-1}.$$

Since $\sum_{i=2}^{\ell-1}(i-1)2^{-\frac{i}{2}}$ is bounded by a constant M,

$$B''_{1,2} \leq mM2^{n-1} \sum_{\ell=3}^{n-m+1} \binom{n-\ell}{m-1}.$$

Finally as $\sum_{\ell=3}^{n-m+1} \binom{n-\ell}{m-1} = \binom{n-2}{m}$ and $|S_{n,m}| = \binom{n-1}{m-1}2^n$, $B''_{1,2} = \mathcal{O}(n|S_{n,m}|)$.

Now from Lemma 2 and since $|Q_{u,v,w}| = 2^{|w|-1}$, we get:

$$B'_{1,2} \leq \sum_{u \in A^+} \sum_{\substack{v \in \text{Bord}(u) \\ |v| > \frac{2}{3}|u|}} \sum_{\substack{w \in \text{Bord}(v) \\ |w| > \frac{|v|}{2}}} 2^{|w|-1} m\binom{n-|u|}{m-1} 2^{n-|u|}.$$

Moreover, from Proposition 2, the words u, v and w of length respectively ℓ, i and j are powers of a same primitive word x: $u = x^p x_0$, $v = x^q x_0$ and $w = x^s x_0$, with $p > q > s > 0$ and $x_0 \in \text{Pr}(x)$. Let r be the length of x, then there are less than 2^r such words x and since $1 \leq r \leq \ell - i$ and $i > \frac{2}{3}\ell$, $r < \frac{\ell}{3}$. Finally the lengths of v and w can be written $i = \ell - hr$ where $1 \leq h < \ell/3r$ and $j = \ell - h'r$ where $h < h' < \frac{1}{2}(\frac{\ell}{r} + h)$, since $j > i/2$. Therefore

$$B'_{1,2} \leq \sum_{\ell=3}^{n-m+1} \sum_{r=1}^{\frac{\ell}{3}-1} \sum_{h=1}^{\frac{\ell}{3r}} \sum_{h'=h+1}^{\frac{1}{2}(\frac{\ell}{r}+h)} m\binom{n-\ell}{m-1} 2^{n-\ell} 2^r 2^{\ell-h'r-1}$$

$$\leq m \, 2^{n-1} \sum_{\ell=3}^{n-m+1} \binom{n-\ell}{m-1} \sum_{r=1}^{\frac{\ell}{3}-1} 2^r \sum_{h=1}^{\frac{\ell}{3r}} \sum_{h'=h+1}^{\frac{1}{2}(\frac{\ell}{r}+h)} (2^{-r})^{h'}.$$

As $\sum_{h=1}^{\frac{\ell}{3r}} \sum_{h'=h+1}^{\frac{1}{2}(\frac{\ell}{r}+h)} (2^{-r})^{h'} \leq 4/2^{2r}$ when $r \geq 1$, we obtain

$$B'_{1,2} \leq m2^{n+1} \sum_{\ell=3}^{n-m+1} \binom{n-\ell}{m-1} \sum_{r=1}^{\frac{\ell}{3}-1} 2^{-r} \leq m2^{n+1} \sum_{\ell=3}^{n-m+1} \binom{n-\ell}{m-1}$$

Finally, since $\sum_{\ell=3}^{n-m+1} \binom{n-\ell}{m-1} = \binom{n-2}{m}$ and $|\mathcal{S}_{n,m}| = \binom{n-1}{m-1}2^n$, we obtain that $B'_{1,2} = \mathcal{O}(n|\mathcal{S}_{n,m}|)$, concluding the proof.

4   Average Time Complexity of the Determinization

The state complexity of a language recognized by a non-deterministic automaton with n states is, in the worst case, equal to 2^n. Therefore the lower bound of the worst-case time complexity of the determinization is $\Omega(2^n)$. In such cases, it is interesting to measure the time complexity according to the size of the output of the algorithm and to try to design algorithms whose efficiency is a function of the size of the result instead of the one of the input. In particular they should be fast when the output is small, even if it is not possible to prevent the output from being of exponential size in the worst case.

The complexity of the subset construction basically depends upon the encoding and the storage of the set of states. At each step, for a given set of states P and a letter $a \in A$, the algorithm computes the set $P \cdot a$ of states of the initial automaton that can be reached from a state of P by a transition labelled by a. Then it tests whether this set has already been computed before or not.

For general non-deterministic automata, the choice of an appropriate data structure for the determinization is not easy. The use of a hashtable may not be an efficient strategy: it is hard to choose the size of the table and the time complexity grows when the table has to be resized and new hashvalues have to be computed for every subset.

Here the automata $\mathcal{A}_X$ to be determinized are specific: for any state u and any letter a, the set $u \cdot a$ can only be $\emptyset$, $\{a\}$, $\{ua\}$ or $\{a, ua\}$. The sets of states of $\mathcal{A}_X$ can be encoded with lists ordered according to the suffix order, i.e. $v \leq_{suff} u$ if and only if $v \in \mathrm{Suff}(u) \cup \{\varepsilon\}$. By Lemma 1, it is a total order over the set of states of $\mathcal{D}_X$. Hence for any state P of $\mathcal{D}_X$, which is also a set of states of $\mathcal{A}_X$, and any letter $a \in A$, the set $P \cdot a$ can be computed in $\mathcal{O}(|P|)$ operations using theses data structures. Moreover as the lists are sorted, the comparison of two sets of states P and P' can be done, in the worst case, with $\mathcal{O}(\min\{|P|, |P'|\})$ operations. To store the sets of states of $\mathcal{A}_X$ we use $n+1$ balanced trees $\mathcal{T}_0, \cdots, \mathcal{T}_n$ where each tree $\mathcal{T}_i$ contains only subsets of size i. When a new set of states P is computed, it is inserted in the tree $\mathcal{T}_{|P|}$. To check whether the set of states P has already been created it is sufficient to search P in the tree $\mathcal{T}_{|P|}$. These operations can be done with $\mathcal{O}(\log |\mathcal{T}_{|P|}|)$ set comparisons, therefore their time complexity is $\mathcal{O}(|P| \log |\mathcal{T}_{|P|}|)$. As there are at most $\binom{n}{|P|} \leq n^{|P|}$ elements in $\mathcal{T}_{|P|}$, the insertion or the search of a set of states P can be done in $\mathcal{O}(|P|^2 \log n)$ arithmetic operations.

Using this data representation, we can prove the following result whose proof, similar to the proof of Theorem 1, is omitted.

Theorem 2. *For an alphabet of size at least 3, the average time complexity, for the uniform distribution over the sets X of $Set_{n,m}$, of the construction of the accessible and deterministic automaton $\mathcal{D}_X$ is $\mathcal{O}(n \log n)$.*

The estimation of the time complexity of the determinization of $\mathcal{A}_X$ remains an open problem in the case of a two-letters alphabet.

5   Minimal Automata

In Section 3 we have proved that the average number of states of $\mathcal{D}_X$, for X in $Set_{n,m}$, is linear in the length of X. The same result holds for the average state complexity of X^* since, for each X in $Set_{n,m}$, the size of the minimal automaton $\mathcal{M}_X$ of X^* is smaller or equal to the size of $\mathcal{D}_X$. Moreover, we prove that the average state complexity of X is $\Omega(n)$.

Theorem 3. *For the uniform distribution over the sets X of $Set_{n,m}$ the average state complexity of X^* is $\Theta(n)$.*

Proof. (sketch) Let $\mathcal{S}_{log} \subset \mathcal{S}_{n,m}$ be the subset of sequences $S = (u_1, \ldots, u_m)$ such that for $i \in \{1, \ldots, m\}$, $|u_i| > 2 \lfloor \log n \rfloor$ and the prefixes (resp. suffixes) of length $\lfloor \log n \rfloor$ of words in S are pairwise distinct.

For any $S = (u_1, \ldots, u_m) \in \mathcal{S}_{log}$, the set $\{u_1, \cdots, u_m\}$ is a prefix code. Therefore, making use of a usual construction of the minimal automaton $\mathcal{M}_S$ from the literal automaton of $\{u_1, \cdots, u_m\}$ [1, Prop. 2.4], we prove that $\mathcal{M}_S$ has at least $n - 2m \log n$ states.

Next, using asymptotic estimations, we show that the cardinalities of $\mathcal{S}_{log}$ and $\mathcal{S}_{n,m}$ are asymptotically close: $|\mathcal{S}_{n,m}| = |\mathcal{S}_{log}|(1 + o(1))$. Moreover, as $\mathcal{S}_{log} \subset \mathcal{S}^{\neq}_{n,m}$, we have:

$$\frac{1}{|Set_{n,m}|} \sum_{X \in Set_{n,m}} \#\mathcal{M}_X \geq \frac{1}{m!|Set_{n,m}|} \sum_{S \in \mathcal{S}_{log}} \#\mathcal{M}_S \geq \frac{|\mathcal{S}_{log}|(n - 2m \log n)}{m!|Set_{n,m}|}$$

Finally we conclude the proof using Proposition 1.   □

Corollary 1. *For the uniform distribution over the sets X of $Set_{n,m}$, the average number of states of $\mathcal{D}_X$ is $\Theta(n)$.*

Now we study the case $m = 2$ of sets of two non-empty words:

Theorem 4. *For the uniform distribution over the sets X of $Set_{n,2}$, the average state complexity of X^* is asymptotically equivalent to n.*

Proof. First the proof of Theorem 3 leads to a lower bound asymptotically equivalent to n. Second Kao, Shallit and Xu recently proved [10] that

$$\begin{cases} \#\mathcal{M}_{\{u,v\}} \leq |u| + |v| & \text{if } u, v \in A^+ \text{ are not powers of the same word} \\ \#\mathcal{M}_{\{u,v\}} \leq (|u| + |v|)^2 \text{ otherwise.} \end{cases}$$

Let P_n be the subset of $\mathcal{S}_{n,2}$ containing all sequences (u,v) such that u and v are powers of a same word. For any non-empty word u of size $|u| \leq \frac{n}{2}$ there is at most one word v in A^+ such that $(u,v) \in P_n$. Therefore

$$\sum_{(u,v) \in P_n} \#\mathcal{M}_{\{u,v\}} \leq 2 \sum_{u \in A^+, |u| \leq \frac{n}{2}} n^2 \leq 2n^2 \sum_{i=1}^{\lfloor n/2 \rfloor} |A|^i = \mathcal{O}\left(n^2 |A|^{n/2}\right).$$

Consequently, as $|\mathcal{S}_{n,2}| \sim n|A|^n$ when n tends towards infinity, the contribution of P_n to the average is negligible. And since, for $(u,v) \in \mathcal{S}_{n,2} \setminus P_n$, the size of $M_{\{u,v\}}$ is lower or equal to n, the average state complexity of X^* is equivalent to n.   □

6   Random Generation and Experimental Results

In the following we explain how to build a random generator for the uniform distribution over the set $Set_{n,m}$. Recall that each element of $Set_{n,m}$ corresponds to exactly $m!$ elements of $\mathcal{S}_{n,m}^{\neq}$. Therefore a uniform random generator for $\mathcal{S}_{n,m}^{\neq}$ provides a uniform generator for $Set_{n,m}$.

We use a rejection algorithm to generate elements of $\mathcal{S}_{n,m}^{\neq}$: we repeatedly generate a random element of $\mathcal{S}_{n,m}$, reject it if it is not in $\mathcal{S}_{n,m}^{\neq}$, stop if it is in $\mathcal{S}_{n,m}^{\neq}$. One can show that the average number of elements to be generated is equal to $\frac{1}{p}$, where p is the probability for an element of $\mathcal{S}_{n,m}$ to be in $\mathcal{S}_{n,m}^{\neq}$, which is $\mathcal{O}(1)$ from Proposition 1.

To draw uniformly at random an element $(u_1, \cdots, u_m)$ of $\mathcal{S}_{n,m}^{\neq}$, we first generate the lengths of the u_i. More precisely a random composition of n into m parts is generated making use of the bijection (see Proposition 1) with the subsets of $\{1, \cdots, n-1\}$ of size $m-1$, themself seen as the $m-1$ first values of a random

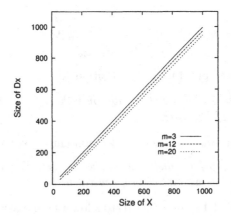

Fig. 2. The average number of states of $\mathcal{D}_X$ for random sets of words $X \in Set_{n,m}$ on a 3-letters alphabet. For each value of m, 20 points have been computed using 1000 random draws each time.

permutation of $\{1, \cdots, n-1\}$. When the lengths of the words are known, each letter is drawn uniformly at random from the alphabet A.

Because of the rejection algorithm, this method may never end, but its average complexity is $\mathcal{O}(n)$. Indeed all algorithms are linear, testing whether the sequence is in $\mathcal{S}_{n,m}^{\neq}$ is also linear, and the average number of rejects is $\mathcal{O}(1)$. This algorithm has been used to obtain the results shown in Figure 2.

From these experimental results, the average number of states of the deterministic automaton $\mathcal{D}_X$ recognizing X^* seems asymptotically of the form $n - c_m + o(1)$, where c_m is a positive number depending on m.

Acknowledgement. The first and third authors were supported by the ANR (project BLAN07-2_195422).

References

1. Berstel, J., Perrin, D.: Theory of Codes. Academic Press, London (1985)
2. Campeanu, C., Culik, K., Salomaa, K., Yu, S.: State complexity of basic operations on finite languages. In: Boldt, O., Jürgensen, H. (eds.) WIA 1999. LNCS, vol. 2214, pp. 60–70. Springer, Heidelberg (2001)
3. Campeanu, C., Salomaa, K., Yu, S.: State complexity of regular languages: finite versus infinite. In: Calude, C.S., Paun, G. (eds.) Finite Versus Infinite: Contributions to an Eternal Dilemma, pp. 53–73. Springer, Heidelberg (2000)
4. Clément, J., Duval, J.-P., Guaiana, G., Perrin, D., Rindone, G.: Parsing with a finite dictionary. Theoretical Computer Science 340, 432–442 (2005)
5. Ellul, K., Krawetz, B., Shallit, J., Wang, M.-W.: Regular expressions: new results and open problems. J. Autom. Lang. Combin. 10, 407–437 (2005)
6. Flajolet, P., Sedgewick, R.: Analytic combinatorics (in preparation, 2008), Version of January 2, 2008, http://www.algo.inria.fr/flajolet/publist.html
7. Hopcroft, J.E., Ullman, J.D.: Introduction to Automata Theory, Languages and Computation. Addison-Wesley Publishing Company, Reading (1979)
8. Ramiréz-Alfonsín, J.L.: Complexity of the Frobenius problem. Combinatorica 16, 143–147 (1996)
9. Ramiréz-Alfonsín, J.L.: The Diophantine Frobenius Problem. Oxford University Press, Oxford (2005)
10. Kao, J.-Y., Shallit, J., Xu, Z.: The Frobenius problem in a free monoid. In: Symposium on Theoretical Aspects of Computer Science 2008, Bordeaux, pp. 421–432 (2008), www.stacs-cong.org
11. Lothaire, M.: Combinatorics on words. Encyclopedia of mathematics and its applications, vol. 17. Addison-Wesley, Reading (1983)
12. Lothaire, M.: Algebraic combinatorics on words. Encyclopedia of mathematics and its applications, vol. 90. Cambridge University Press, Cambridge (2002)
13. Lothaire, M.: Applied combinatorics on words. Encyclopedia of mathematics and its applications, vol. 104. Cambridge University Press, Cambridge (2005)
14. Maslov, A.N.: Estimates of the number of states of finite automata. Dokl. Akad. Nauk. SSRR 194, 1266–1268 (1970) (in Russian); English translation in. Soviet. Math. Dokl. 11, 1373–1375 (1970)
15. Yu, S., Zhuang, Q., Salomaa, K.: The state complexities of some basic operations on regular languages. Theoretical Computer Science 125, 315–328 (1994)

On the Computational Capacity of Parallel Communicating Finite Automata

Henning Bordihn[1,*], Martin Kutrib[2], and Andreas Malcher[2]

[1] Institut für Informatik, Universität Potsdam,
August-Bebel-Straße 89, 14482 Potsdam, Germany
henning@cs.uni-potsdam.de
[2] Institut für Informatik, Universität Giessen
Arndtstraße 2, 35392 Giessen, Germany
{kutrib,malcher}@informatik.uni-giessen.de

Abstract. Systems of parallel finite automata communicating by states are investigated. We consider deterministic and nondeterministic devices and distinguish four working modes. It is known that systems in the most general mode are as powerful as one-way multihead finite automata. Here we solve some open problems on the computational capacity of systems working in the remaining modes. In particular, it is shown that deterministic returning and non-returning devices are equivalent, and that there are languages which are accepted by deterministic returning and centralized systems but cannot be accepted by deterministic non-returning centralized systems. Furthermore, we show that nondeterministic centralized systems are strictly more powerful than their deterministic variants. Finally, incomparability with the class of (deterministic) (linear) context-free languages as well as the Church-Rosser languages is derived.

1 Introduction

The need for a fundamental understanding of parallel processes and cooperating systems is increasing more and more in today's complex world. In the classical theory of formal languages and automata mainly sequential machine models like, for example, finite automata, pushdown automata, or Turing machines are studied. It turned out that this theory is very helpful to describe, analyze, and understand sequential processes. To obtain such a theory also for cooperating systems, it is an obvious generalization to proceed from one sequential automaton to systems of sequential automata. Some questions immediately arising are, for example, whether the input is processed in a parallel or sequential way and how the input is accepted. One may ask how the cooperation between different automata is organized and whether they work in a synchronous or an asynchronous way. One has to define in which way communication between different automata takes place and how appropriate restrictions on the amount of information communicated can be formulated. In the literature, systems of cooperating sequential

---

* Most of the work was done while the author was at Institut für Informatik, Universität Giessen, Germany.

M. Ito and M. Toyama (Eds.): DLT 2008, LNCS 5257, pp. 146–157, 2008.
© Springer-Verlag Berlin Heidelberg 2008

automata appear in many facets. Multi-head finite automata [13] are in some sense the simplest model of cooperating automata, since a finite automaton is provided with a fixed number of reading heads. So, we have some model with one finite state control and the cooperation between the finite state control and the single components is the reading of the input and positioning the heads. This model is generalized to multi-head two-way finite automata [7] and multi-head pushdown automata [6]. Multi-processor automata [2] are in a way restricted multi-head finite automata, and the relation between both classes is investigated in [5]. Systems of different finite automata communicating by appropriate protocols are described in [1,10], and systems of cooperating finite automata working in parallel are introduced in [11]. Apart from systems of cooperating automata there is also the broad field of systems of cooperating grammars [4].

Here, we will focus on parallel communicating finite automata systems which were introduced in [11]. In this model, several finite automata read and process the input in parallel in a synchronized way. The communication between automata is defined in such a way that an automaton can request the current state from another automaton. The system can work in *returning* or *non-returning* mode. In the former case each automaton which sends its current state is reset to its initial state after this communication step. In the latter case the state of the sending automaton is not changed. We also distinguish between *centralized* systems where only one designated automaton, called *master*, can request information from other automata, and *non-centralized* systems where every automaton is allowed to communicate with others. Altogether we obtain four different working modes. One fundamental result shown in [11] is that nondeterministic (deterministic) non-centralized systems working in the non-returning mode are equally powerful as one-way multi-head nondeterministic (deterministic) finite automata. Recently, it has been shown in [3] that the returning and non-returning working modes coincide for nondeterministic non-centralized systems. The authors left as an open question whether the same is also true for deterministic systems. Moreover, the question whether or not centralized systems are equally powerful as non-centralized systems remained open. Here, the first question and, for deterministic systems working in the non-returning mode, the second question are answered.

2 Preliminaries and Definitions

We denote the powerset of a set S by 2^S. The empty word is denoted by λ, the reversal of a word w by w^R, and for the length of w we write $|w|$. We use $\subseteq$ for *inclusions* and $\subset$ for *strict inclusions*.

Next we turn to the definition of the devices in question, which have been introduced in [11]. A parallel communicating finite automata system of degree k is a device of k finite automata working in parallel with each other, synchronized according to a universal clock, on a common one-way read-only input tape. The k automata communicate on request by states, that is, when some automaton enters a distinguished query state q_i, it is set to the current state of automaton A_i.

Concerning the next state of the sender A_i, we distinguish two modes. In *non-returning* mode the sender remains in its current state, whereas in *returning* mode the sender is set to its initial state. Moreover, we distinguish whether all automata are allowed to request communications, or whether there is just one master allowed to request communications. The latter types are called *centralized*.

One of the fundamental results obtained in [11] is the characterization of the computational power of (unrestricted) parallel communicating finite automata systems by multi-head finite automata. Due to this relation, we present a formal definition of language acceptance that suits to the definition given in [15] for one-way multi-head finite automata. To this end, we provide tape inscriptions which are input words followed by an endmarker. Whenever the transition function of (at least) one of the single automata is undefined the whole systems halts. Whether the input is accepted or rejected depends on the states of the automata having undefined transitions. The input is accepted if at least one of them is in an accepting state.

Formally, a *nondeterministic parallel communicating finite automata system of degree k* (PCFA(k)) is a construct $\mathcal{A} = \langle \Sigma, A_1, A_2, \ldots, A_k, Q, \triangleleft \rangle$, where Σ is the set of *input symbols*, each $A_i = \langle S_i, \Sigma, \delta_i, s_{0,i}, F_i \rangle$, $1 \leq i \leq k$, is a *nondeterministic finite automaton* with state set S_i, initial state $s_{0,i} \in S_i$, set of accepting states $F_i \subseteq S_i$, and transition function $\delta_i : S_i \times (\Sigma \cup \{\lambda, \triangleleft\}) \to 2^{S_i}$, $Q = \{q_1, q_2, \ldots, q_k\} \subseteq \bigcup_{1 \leq i \leq k} S_i$ is the set of *query states*, and $\triangleleft \notin \Sigma$ is the *end-of-input symbol*.

The automata $A_1, A_2, \ldots, A_k$ are called *components* of the system $\mathcal{A}$. A *configuration* $(s_1, x_1, s_2, x_2, \ldots, s_k, x_k)$ of $\mathcal{A}$ represents the current states s_i as well as the still unread parts x_i of the tape inscription of all components $1 \leq i \leq k$. System $\mathcal{A}$ starts with all of its components scanning the first square of the tape in their initial states. For input word $w \in \Sigma^*$, the initial configuration is $(s_{0,1}, w \triangleleft, s_{0,2}, w \triangleleft, \ldots, s_{0,k}, w \triangleleft)$. Basically, a computation of $\mathcal{A}$ is a sequence of configurations beginning with an initial configuration and ending with a halting configuration. Each step can consist of two phases. In a first phase, all components are in non-query states and perform an ordinary (non-communicating) step independently. The second phase is the communication phase during which components in query states receive the requested states as long as the sender is not in a query state itself. This process is repeated until all requests are resolved, if possible. If the requests are cyclic, no successor configuration exists. As mentioned above, we distinguish *non-returning* communication, that is, the sender remains in its current state, and *returning* communication, that is, the sender is reset to its initial state.

For the first phase, we define the successor configuration relation $\vdash$ by

$$(s_1, a_1 y_1, s_2, a_2 y_2, \ldots, s_k, a_k y_k) \vdash (p_1, z_1, p_2, z_2, \ldots, p_k, z_k),$$

if $Q \cap \{s_1, s_2, \ldots, s_k\} = \emptyset$, $a_i \in \Sigma \cup \{\lambda, \triangleleft\}$, $p_i \in \delta_i(s_i, a_i)$, and $z_i = \triangleleft$ for $a_i = \triangleleft$ and $z_i = y_i$ otherwise, $1 \leq i \leq k$. For non-returning communication in the second phase, we set $(s_1, x_1, s_2, x_2, \ldots, s_k, x_k) \vdash (p_1, x_1, p_2, x_2, \ldots, p_k, x_k)$, if, for all $1 \leq i \leq k$ such that $s_i = q_j$ and $s_j \notin Q$, we have $p_i = s_j$, and $p_r = s_r$

for all the other r, $1 \leq r \leq k$. Alternatively, for returning communication in the second phase, we set $(s_1, x_1, s_2, x_2, \ldots, s_k, x_k) \vdash (p_1, x_1, p_2, x_2, \ldots, p_k, x_k)$, if, for all $1 \leq i \leq k$ such that $s_i = q_j$ and $s_j \notin Q$, we have $p_i = s_j$, $p_j = s_{0,j}$, and $p_r = s_r$ for all the other r, $1 \leq r \leq k$.

A computation *halts* when the successor configuration is not defined for the current situation. In particular, this may happen when cyclic communication requests appear, or when the transition function of one component is not defined. (We regard the transition function as undefined whenever it maps to the empty set.) The language $L(\mathcal{A})$ accepted by a PCFA(k) $\mathcal{A}$ is precisely the set of words w such that there is some computation beginning with $w\triangleleft$ on the input tape and halting with at least one component having an undefined transition function and being in an accepting state. Let $\vdash^*$ denote the reflexive and transitive closure of $\vdash$ and set $L(\mathcal{A}) = \{\, w \in \Sigma^* \mid (s_{0,1}, w\triangleleft, s_{0,2}, w\triangleleft, \ldots, s_{0,k}, w\triangleleft) \vdash^* (p_1, a_1 y_1, p_2, a_2 y_2, \ldots, p_k, a_k y_k)$, such that $p_i \in F_i$ and $\delta_i(p_i, a_i)$ is undefined, for some $1 \leq i \leq k \,\}$.

If all components A_i are deterministic finite automata, that is, for all $s \in S_i$ the transition function $\delta_i(s, a)$ maps to a set of at most one state and is undefined for all $a \in \Sigma$, whenever $\delta_i(s, \lambda)$ is defined, then the whole system is called *deterministic*, and we add the prefix D to denote it. The absence or presence of an R in the type of the system denotes whether it works in *non-returning* or *returning* mode, respectively. Finally, if there is just one component, say A_1, that is allowed to query for states, that is, $S_i \cap Q = \emptyset$, for $2 \leq i \leq k$, then the system is said to be *centralized*. We denote centralized systems by a C. Whenever the degree is missing we mean systems of arbitrary degree. The *family of languages accepted* by devices of type X (with degree k) is denoted by $\mathscr{L}(X)$ ($\mathscr{L}(X(k))$).

In order to clarify our notation we give an example. We consider the language $\{w\$w \mid w \in \{a, b\}^+\}$ and show that it can be accepted by a DRCPCFA as well as by a DCPCFA with two components. Thus, all types of systems of parallel communicating finite automata accept more than regular languages.

We first describe the construction for the centralized and returning mode. The input can be divided into two halves, namely the halves to the left and to the right of the separating symbol $. The principal idea of the construction is that the non-master component reads the first symbol of the left half and then waits until the master component has been moved to the first symbol of the right half. Then, the master component queries the non-master component and gets the information about the current symbol of the non-master component. Subsequently, the non-master component returns to its initial state. In the next time step, the non-master component reads the next input symbol of the left hand half and the master component checks the information about the current left hand symbol against the next input symbol of the right hand half. If both symbols match, the master component queries again the non-master component and otherwise it stops the computation. This behavior is iterated as long as both symbols match. Finally, the master component enters an accepting state if and only if it reads the end-of-input symbol for the first time and has received the information that the current input symbol of the non-master component is $. The precise rules of the master component (δ_1) and the non-master component (δ_2) are as follows.

$$\delta_1(s_{0,1}, a) = s_{0,1} \qquad \delta_1(s_{0,1}, b) = s_{0,1} \qquad \delta_1(s_{0,1}, \$) = q_2$$
$$\delta_1(s_a, a) = q_2 \qquad \delta_1(s_b, b) = q_2$$
$$\delta_1(s_\$, \triangleleft) = accept$$

$$\delta_2(s_{0,2}, a) = s_a \qquad \delta_2(s_{0,2}, b) = s_b \qquad \delta_2(s_{0,2}, \$) = s_\$$$
$$\delta_2(s_a, \lambda) = s_a \qquad \delta_2(s_b, \lambda) = s_b \qquad \delta_2(s_\$, \lambda) = s_\$$$

The construction for the deterministic centralized non-returning mode is quite different. Here, the rough idea is that in every time step the master component queries the non-master component, and the non-master component reads an input symbol. When the non-master component has read the separating symbol $, which is notified to the master with the help of primed states, then the master component starts to compare its input symbol with the information from the non-master component. If all symbols up to $ match, the input is accepted and in all other cases rejected.

$$\delta_1(s_{0,1}, \lambda) = q_2 \qquad \delta_1(s_a, \lambda) = q_2 \qquad \delta_1(s_b, \lambda) = q_2$$
$$\delta_1(s_\$, \lambda) = q_2 \qquad \delta_1(s'_a, a) = q_2 \qquad \delta_1(s'_b, b) = q_2$$
$$\delta_1(s_\triangleleft, \$) = accept$$

$$\delta_2(s_{0,2}, a) = s_a \qquad \delta_2(s_{0,2}, b) = s_b$$
$$\delta_2(s_a, a) = s_a \qquad \delta_2(s_a, b) = s_b \qquad \delta_2(s_a, \$) = s_\$$$
$$\delta_2(s_b, a) = s_a \qquad \delta_2(s_b, b) = s_b \qquad \delta_2(s_b, \$) = s_\$$$
$$\delta_2(s_\$, a) = s'_a \qquad \delta_2(s_\$, b) = s'_b \qquad \delta_2(s_\triangleleft, \triangleleft) = s_\triangleleft$$
$$\delta_2(s'_a, a) = s'_a \qquad \delta_2(s'_a, b) = s'_b \qquad \delta_2(s'_a, \triangleleft) = s_\triangleleft$$
$$\delta_2(s'_b, a) = s'_a \qquad \delta_2(s'_b, b) = s'_b \qquad \delta_2(s'_b, \triangleleft) = s_\triangleleft$$

3   Deterministic Non-returning Versus Returning

For nondeterministic non-centralized devices it is shown in [3] that returning parallel communicating finite automata systems are neither weaker nor stronger than non-returning ones. In the same paper the question is raised whether the same equivalence is true in the deterministic case. This section is devoted to answering the question in the affirmative. To this end, we first introduce a general method to send information tokens through the returning components cyclically. These tokens can be processed by the components.

Cycling-token-method. Basically, an information token is a finite record of data which can be read or written by the components. So, it can be represented by states. The precise structure of the token depends on the application, but in any case there is one field for the identification of the component that has been processing it at last. Now the idea is to set up the components such that the information token is passed through the components cyclically, that is, from component A_1 to A_2, from A_2 to A_3, and so on until it is passed from A_k back to A_1 what completes the round. Next we show how to set up the components, where we assume for a moment that we have $k \geq 3$ components. Moreover, since at this stage a general framework is provided, we may disregard the current input symbols, that is, we assume that all moves are λ-moves.

The first problem to cope with is to break the synchronization at the beginning. Otherwise, when some component A_{i+1} requests the state of A_i and, thus, A_i reenters its initial state, then A_i will request the state of A_{i-1} and so on. But these cascading communication requests would destroy necessary information. Therefore, we set up the first component A_1 to play a special role, and call it the *guide*. In general, we distinguish four types of components, the guide, the successor A_2 of the guide, the predecessor A_k of the guide, and all remaining components $A_3, A_4, \ldots, A_{k-1}$. For $k = 6$ the following interactions can be seen in Figure 1.

In order to break the synchronization, all components except the guide immediately change from their initial states to the query state q_1, that is, they request the state of the guide which, in turn, changes to the state I indicating that the components are in the first cycling round. In general, the first round is different from later rounds. After being set to state I all components start to count. Component A_2 counts up to 1, and component A_i, $3 \leq i \leq k$, counts up to $3(i - 2)$. Immediately after counting, all cells $2 \leq i \leq k$ change to the query state q_{i-1} in order to receive the state of their predecessors, that is, to receive the token. In order to generate the initial token $t_{0,1}$, the guide changes from state I to state $t_{0,1}$. When a component receives a token it processes it during the next step. Subsequently, it stays in that state until the token is requested by its successor, which causes the component to reenter its initial state. After being set to the initial state, a component requests the state of the guide, say T, that now indicates that the component has completed the first round. So, the guide changes from the state representing the initial token to state T. After sending the token, every component reenters its initial state and thus requests the state T of the guide. At the first time step at which the guide is not being asked when in state T, it changes to the query state q_k. So, it receives the token from its predecessor, and a new round starts. At these points in time, component A_k reenters its initial state and then requests the state of the guide. While in this situation the other components receive state T, component A_k necessarily receives the token which has been processed by A_1 one time step before. But component A_k can interpret its new state appropriately since the identification field of the token shows that it has not been received from its predecessor but the guide.

Now it remains to be described how to set up counters for the second and subsequent rounds. Component A_2 counts up to $3(k - 2) + 1$, component A_i, $3 \leq i \leq k - 1$, counts up to $3(k - 2)$, and component A_k counts up to $3(k - 2) - 2$. Clearly, the states of the counters for the first and subsequent rounds must be different. Again, immediately after counting, all cells $2 \leq i \leq k$ change to the query state q_{i-1} in order to receive the token. From now on the whole process is repeated.

The correctness of the construction can be shown by induction. The computation for $k = 2$ is easily derived. For $k \geq 3$, running the system for six time

t	A_1	A_2	A_3	A_4	A_5	A_6
0	$s_{0,1}$	$s_{0,2}$	$s_{0,3}$	$s_{0,4}$	$s_{0,5}$	$s_{0,6}$
1	I	q_1	q_1	q_1	q_1	q_1
	$s_{0,1}$	I	I	I	I	I
2	I	1	1	1	1	1
3	$t_{0,1}$	q_1	2	2	2	2
	$s_{0,1}$	$t_{0,1}$				
4	I	$t_{0,2}$	3	3	3	3
5	$t_{0,1}$	$t_{0,2}$	q_2	4	4	4
		$s_{0,2}$	$t_{0,2}$			
6	T	q_1	$t_{0,3}$	5	5	5
	$s_{0,1}$	T				
7	I	$1'$	$t_{0,3}$	6	6	6
8	$t_{0,1}$	$2'$	$t_{0,3}$	q_3	7	7
			$s_{0,3}$	$t_{0,3}$		
9	T	$3'$	q_1	$t_{0,4}$	8	8
	$s_{0,1}$		T			
10	I	$4'$	$1'$	$t_{0,4}$	9	9
11	$t_{0,1}$	$5'$	$2'$	$t_{0,4}$	q_4	10
				$s_{0,4}$	$t_{0,4}$	
12	T	$6'$	$3'$	q_1	$t_{0,5}$	11
	$s_{0,1}$			T		
13	I	$7'$	$4'$	$1'$	$t_{0,5}$	12
14	$t_{0,1}$	$8'$	$5'$	$2'$	$t_{0,5}$	q_5
					$s_{0,5}$	$t_{0,5}$
15	T	$9'$	$6'$	$3'$	q_1	$t_{0,6}$
	$s_{0,1}$				T	
16	I	$10'$	$7'$	$4'$	$1'$	$t_{0,6}$
17	$t_{0,1}$	$11'$	$8'$	$5'$	$2'$	$t_{0,6}$
18	T	$12'$	$9'$	$6'$	$3'$	$t_{0,6}$
19	q_6	$13'$	$10'$	$7'$	$4'$	$t_{0,6}$
	$t_{0,6}$					$s_{0,6}$

t	A_1	A_2	A_3	A_4	A_5	A_6
20	$t_{1,1}$	q_1	$11'$	$8'$	$5'$	q_1
	$s_{0,1}$	$t_{1,1}$				$t_{1,1}$
21	I	$t_{1,2}$	$12'$	$9'$	$6'$	$1'$
22	$t_{0,1}$	$t_{1,2}$	q_2	$10'$	$7'$	$2'$
		$s_{0,2}$	$t_{1,2}$			
23	T	q_1	$t_{1,3}$	$11'$	$8'$	$3'$
	$s_{0,1}$	T				
24	I	$1'$	$t_{1,3}$	$12'$	$9'$	$4'$
25	$t_{0,1}$	$2'$	$t_{1,3}$	q_3	$10'$	$5'$
			$s_{0,3}$	$t_{1,3}$		
26	T	$3'$	q_1	$t_{1,4}$	$11'$	$6'$
	$s_{0,1}$		T			
27	I	$4'$	$1'$	$t_{1,4}$	$12'$	$7'$
28	$t_{0,1}$	$5'$	$2'$	$t_{1,4}$	q_4	$8'$
				$s_{0,4}$	$t_{1,4}$	
29	T	$6'$	$3'$	q_1	$t_{1,5}$	$9'$
	$s_{0,1}$			T		
30	I	$7'$	$4'$	$1'$	$t_{1,5}$	$10'$
31	$t_{0,1}$	$8'$	$5'$	$2'$	$t_{1,5}$	q_5
					$s_{0,5}$	$t_{1,5}$
32	T	$9'$	$6'$	$3'$	q_1	$t_{1,6}$
	$s_{0,1}$				T	
33	I	$10'$	$7'$	$4'$	$1'$	$t_{1,6}$
34	$t_{0,1}$	$11'$	$8'$	$5'$	$2'$	$t_{1,6}$
35	T	$12'$	$9'$	$6'$	$3'$	$t_{1,6}$
36	q_6	$13'$	$10'$	$7'$	$4'$	$t_{1,6}$
	$t_{1,6}$					$s_{0,6}$
				$\vdots$		

Fig. 1. Cycling-token-method for six components. We denote by $t_{i,j}$ the token that completed i rounds and has been processed by component j at the latest. Two rows for a time step represent the results after the first phase and after the second (communication) phase.

steps drives the guide into state T for the first time. Next, it is easily verified that the system reaches a configuration at time $3k$ such that, subsequently, the global behavior of the system becomes cyclic with cycle length $3k - 1$, where the token changes due to the processing. The technical details are omitted here. $\square$

In order to prove the equivalence we first show the following simulation.

Lemma 1. *For all $k \geq 1$, the family $\mathscr{L}(\mathrm{DRPCFA}(k))$ includes $\mathscr{L}(\mathrm{DPCFA}(k))$.*

Proof. Given some non-returning parallel communicating finite automata system $\mathcal{A}$ of degree k, we construct an equivalent returning parallel communicating finite automata system $\mathcal{A}'$ of degree k by using the cycling-token-method. The idea is to use tokens that store the states of the components of $\mathcal{A}$ in addition to the data field for the identification of the component that processed it at last. Since at the beginning the guide (of $\mathcal{A}'$) knows that all components (of $\mathcal{A}$) are in their initial states, it can generate the initial token $t_{0,1}$. A component of $\mathcal{A}'$ processes the token by reading the state of the corresponding component of $\mathcal{A}$ from the token, simulating the corresponding transition, and storing the new state of the corresponding component of $\mathcal{A}$ in the token. When the token is received by the guide, it processes it in the same way as the other components but, in addition, it simulates the subsequent communication phase and stores the resolved states in the token, which is now prepared for the next round.

So far, we obtained only a partial solution. The problem to overcome may occur at the end of a computation. All components of $\mathcal{A}$ move at the same time, whereas the components of $\mathcal{A}'$ move one after the other. Assume that for two components of $\mathcal{A}$ the transition functions are undefined at the same time, where one is in an accepting state and the other one in a rejecting state. Then $\mathcal{A}$ accepts the input, but whether $\mathcal{A}'$ accepts or rejects depends on which component receives the token and, thus, halts first. So, we extend the construction as follows. Whenever the transition function of a component of the given system $\mathcal{A}$ is undefined, we define one more step. Depending on whether the component is an accepting or non-accepting state, we let it change to a new accepting state s_a or to a new non-accepting state s_r. The transition functions are undefined for the new states s_a and s_r. In this way the accepted language is not changed, but whether a component halts is determined solely by its state, and not by its current input symbol. Now the guide of the simulating system $\mathcal{A}'$ knows in advance at the beginning of a round whether or not the whole system will halt. Moreover, in the halting case it knows in advance whether the input will be accepted or rejected. So, it remains to let the transition function of the guide be undefined in this case, and to define the set of its accepting states suitably. $\square$

In [11] the equivalence between DPCFA(k) and deterministic one-way k-head finite automata (k-DFA) has been shown. The next theorem concludes the proofs of equivalence.

Theorem 2. *For all $k \geq 1$, the three families $\mathscr{L}(\text{DRPCFA}(k))$, $\mathscr{L}(\text{DPCFA}(k))$, and $\mathscr{L}(k\text{-DFA})$ are equal.*

Proof. It remains to be shown that, for all $k \geq 1$, the family $\mathscr{L}(k\text{-DFA})$ includes $\mathscr{L}(\text{DRPCFA}(k))$. Given some DRPCFA($k$) $\mathcal{A}$, basically, the idea of simulating it by a deterministic one-way k-head finite automaton $\mathcal{A}'$ is to track all current states of the components of $\mathcal{A}$ in the current state of $\mathcal{A}'$ (cf. [11]). So, $\mathcal{A}'$ can simulate all components in parallel. $\square$

4   Deterministic Non-centralized Versus Centralized

This section is devoted to comparing deterministic centralized systems with non-centralized systems. We obtain for centralized systems the surprising result that the returning mode is not weaker than the non-returning mode. Let us consider $L_{\text{rc}} = \{\, uc^x v\$uv \mid u, v \in \{a,b\}^*, x \geq 0 \,\}$.

Theorem 3. *The language L_{rc} belongs to the family $\mathscr{L}(\text{DRCPCFA})$ (and thus to $\mathscr{L}(\text{DRPCFA}) = \mathscr{L}(\text{DPCFA})$), but not to $\mathscr{L}(\text{DCPCFA})$.*

Proof. First, we show that L_{rc} is accepted by a deterministic centralized parallel communicating automata system in returning mode having two components. The construction is similar to the construction of the example in Section 2 and omitted here. Next, in contrast to the assertion we assume that L_{rc} is accepted by a DCPCFA $\mathcal{A}$. First, we derive a contradiction for degree two, that is, $\mathcal{A}$ consists of a master component A_1 allowed to query for states of a second component A_2. The state sets are denoted by S_1 and S_2. In particular, we consider accepting computations on inputs from $\{ uc^x v\$uv \mid u, v \in \{a, b\}^*, |u| = |v| = n, x = n^2 \}$, where n is a fixed constant that is sufficiently large. We distinguish between computations in which the master or the non-master component (component, for short) arrives at the \$-symbol first.

Case 1. The component reaches the \$-symbol before the master in more than the half of the accepting computations. The set of these computations is further partitioned into subclasses depending on the state and the tape position of the master, and the state of the component, all at the time at which the component reaches the \$-symbol. There are 2^{2n} different inputs, $|S_1| \cdot |S_2|$ different state pairs, and $n^2 + 2n$ different tape positions. So, at least one of the subclasses contains at least $\frac{2^{2n}}{2 \cdot |S_1| \cdot |S_2| \cdot (n^2 + 2n)} \geq 2^{2n - c_1 \log(n)}$ inputs, for some constant $c_1 \geq 1$.

Subcase 1.a. The position of the master is somewhere at the infix $c^x v$ when the component reaches the \$-symbol. Here we further partition the subclass into subsubclasses depending on the infix v. Since there are 2^n different infixes v, there is at least one subsubclass containing at least $\frac{2^{2n - c_1 \log(n)}}{2^n} \geq 2^{n - c_1 \log(n)}$ inputs. We consider two of them, say $w = uc^x v\$uv$ and $w' = u'c^x v\$u'v$. If the position of the master is at some symbol c, the accepting computations are as follows: $(s_{0,1}, w\triangleleft, s_{0,2}, w\triangleleft) \vdash^* (p_1, c^i v\$uv\triangleleft, p_2, \$uv\triangleleft) \vdash^* (q_1, z_1, q_2, z_2)$, and $(s_{0,1}, w'\triangleleft, s_{0,2}, w'\triangleleft) \vdash^* (p_1, c^i v\$u'v\triangleleft, p_2, \$u'v\triangleleft) \vdash^* (q_1', z_1', q_2', z_2')$, where $p_1, q_1, q_1' \in S_1$, $p_2, q_2, q_2' \in S_2$, and (q_1, z_1, q_2, z_2) and (q_1', z_1', q_2', z_2') are accepting configurations. This implies that $(s_{0,1}, u'c^x v\$uv\triangleleft, s_{0,2}, u'c^x v\$uv\triangleleft) \vdash^* (p_1, c^i v\$uv\triangleleft, p_2, \$uv\triangleleft) \vdash^* (q_1, z_1, q_2, z_2)$ is an accepting computation, which is a contradiction since $u'c^x v\$uv \notin L_{rc}$. If the position of the master is at some symbol of the subword v, a contradiction is derived in the same way. This concludes Subcase 1.a.

Subcase 1.b. The position of the master is somewhere at the prefix u when the component reaches the \$-symbol. Here we further partition the subclass into subsubclasses depending on the prefix u, the time taken by the component to get from the \$-symbol to the endmarker (or to run into a λ-loop), its state at that time, the possible position and state of the master at that time step as well as on the time taken by the master to reach the \$-symbol from the position (or to halt accepting), and its state at the arrival. There are 2^n different prefixes u. The component takes at most $c_3 n$ time steps to get from the \$-symbol to the endmarker $\triangleleft$ (or to run into a λ-loop), for some constant $c_3 \geq 2$. During this time the master can reach at most $c_3 n$ different positions, and it takes at most

$c_4 n^2$ time steps to reach the \$-symbol (or to halt accepting). In total, there is at least one subsubclass containing at least

$$\frac{2^{2n - c_1 \log(n)}}{2^n \cdot c_3 n \cdot |S_2| \cdot c_3 n \cdot |S_1| \cdot c_4 n^2 \cdot |S_1|} \geq 2^{n - c_5 \log(n)}$$

inputs, for some constant c_5. We consider two of them, say $w = uc^x v\$uv$ and $w' = uc^x v'\$uv'$. Since the component takes at most $c_3 n$ time steps to get from the \$-symbol to the endmarker $\lhd$ (or to run into a λ-loop), but the master needs at least n^2 time steps to pass through the c's, the master reads some symbol of u or a c at that time (if it has not accepted before). All cases are treated in the same way. If the position of the master is at some symbol c and the component reaches the endmarker, the accepting computations are as follows:

$$(s_{0,1}, w\lhd, s_{0,2}, w\lhd) \vdash^* (p_1, u_j u_{j+1} \cdots u_n c^x v\$uv\lhd, p_2, \$uv\lhd)$$
$$\vdash^* (q_1, c^i v\$uv\lhd, q_2, \lhd) \vdash^* (r_1, \$uv\lhd, r_2, \lhd) \vdash^* \cdots$$

$$(s_{0,1}, w'\lhd, s_{0,2}, w'\lhd) \vdash^* (p_1, u_j u_{j+1} \cdots u_n c^x v'\$uv'\lhd, p_2, \$uv'\lhd)$$
$$\vdash^* (q_1, c^i v'\$uv'\lhd, q_2, \lhd) \vdash^* (r_1, \$uv'\lhd, r_2, \lhd) \vdash^* \cdots$$

where $u = u_1 u_2 \cdots u_n$, $p_1, q_1, r_1 \in S_1$, $p_2, q_2, r_2 \in S_2$. But this implies that

$$(s_{0,1}, uc^x v'\$uv\lhd, s_{0,2}, uc^x v'\$uv\lhd) \vdash^* (p_1, u_j u_{j+1} \cdots u_n c^x v'\$uv\lhd, p_2, \$uv\lhd)$$
$$\vdash^* (q_1, c^i v'\$uv\lhd, q_2, \lhd)$$
$$\vdash^* (r_1, \$uv\lhd, r_2, \lhd) \vdash^* \cdots$$

is also an accepting computation, which is a contradiction since $uc^x v'\$uv \notin L_{\mathrm{rc}}$. This concludes Subcase 1.b and, thus, Case 1.

Case 2. The master reaches the \$-symbol not later than the component in at least the half of the accepting computations. The initial partitioning for this case is exactly as for Case 1 with master and component interchanged. So, the set of these computations is partitioned into subclasses depending on the state and the tape position of the component and the state of the master, all at the time at which the master reaches the \$-symbol. Again, at least one of the subclasses contains at least $2^{2n - c_1 \log(n)}$ inputs, for some constant $c_1 \geq 1$. The reasoning for the situation where the position of the component is somewhere at the infix $c^x v\$$ when the master reaches the \$-symbol is dual to Subcase 1.a. Furthermore, if the component does not run into a λ-loop, a synonym for Subcase 1.b does not exist, since in order to avoid infinite loops the component has to read one input symbol every d time steps, where $d \leq |S_2|$ is a constant. Therefore, it has read the n symbols of prefix u before the master can read the n^2 symbols c. If, on the other hand, the component runs into a λ-loop somewhere at the prefix u, then we further partition the subclass into subsubclasses depending on the prefix u and the time taken by the master to accept. There are 2^n different prefixes u, and the master takes at most $c_3 n$ time steps to accept. The further reasoning is as for Subcase 1.b. This concludes Case 2 and the proof of the assertion for DCPCFAs of degree two.

In order to generalize the proof to arbitrary degrees, we first argue that the fundamental situation for Case 2 is almost the same. Roughly speaking, all non-master components have to read symbols after a number of time steps that is bounded by their number of states in order to avoid infinite loops. This implies that only the master can wait at the beginning of the input to match both occurrences of u. Similarly to Case 1, when the first component reaches the \$-symbol all other non-master components are at positions that are close to the \$-symbol, that is, at most at a distance of $d \cdot n$, where d is again a constant depending on the number of states. This, in turn, implies that the components reach the endmarker before the master reaches the infix v. In addition, for all the partitions we obtained the number of inputs in the chosen classes by dividing an exponential function by a polynomial. Dealing with more than two components increases the degree of the polynomials and the constants, but still we have the same situation in the order of magnitude. So, the generalization of the proof is a technical challenge following the idea for degree two. □

Corollary 4. $\mathscr{L}(\text{DCPCFA}) \subset \mathscr{L}(\text{DPCFA}) = \mathscr{L}(\text{DRPCFA})$.

5   Determinism Versus Nondeterminism

In order to show that nondeterministic centralized systems are strictly more powerful than their deterministic variants we consider the complement of the mirror language, that is, $L_{mi} = \overline{\{ww^R \mid w \in \{a, b, c\}^+\}}$.

Lemma 5. The language L_{mi} belongs to $\mathscr{L}(\text{CPCFA})$, but does not belong to $\mathscr{L}(\text{DPCFA})$.

Proof. We omit the construction of a CPCFA accepting L_{mi}, and prove that L_{mi} does not belong to the family $\mathscr{L}(\text{DPCFA}(k))$, for any $k \geq 1$. In contrast to the assertion, assume that there exists a $k \geq 1$ such that $L_{mi} \in \mathscr{L}(\text{DPCFA}(k))$. Since $\mathscr{L}(\text{DPCFA}(k)) \subseteq \mathscr{L}(k\text{-DFA})$ and $\mathscr{L}(k\text{-DFA})$ is closed under complementation, we obtain that $\{ww^R \mid w \in \{a, b, c\}^+\} \in \mathscr{L}(k\text{-DFA})$. This is a contradiction, since $\{ww^R \mid w \in \{a, b, c\}^+\} \notin \mathscr{L}(k\text{-NFA})$, for all $k \geq 1$. □

Corollary 6. 1. $\mathscr{L}(\text{DCPCFA}) \subset \mathscr{L}(\text{CPCFA})$.   2. $\mathscr{L}(\text{DPCFA}) \subset \mathscr{L}(\text{PCFA})$.

Finally, we compare the classes under consideration with some well-known language families.

Lemma 7. The family $\mathscr{L}(\text{PCFA})$ is strictly included in the complexity class NL, hence, in the family of deterministic context-sensitive languages.

Proof. It is well-known that nondeterministic two-way multi-head finite automata characterize the complexity class NL (cf. [14]). Since $L = \{ww^R \mid w \in \{a, b, c\}^+\}$ can be accepted by some two-way NFA with two heads, language L belongs to NL. On the other hand, language L does not belong to $\mathscr{L}(k\text{-NFA})$, for all $k \geq 1$. □

Lemma 8. All language classes accepted by parallel communicating finite automata systems are incomparable to the class of (deterministic) (linear) context-free languages.

Proof. The marked mirror language $\{w\$w^R \mid w \in \{a,b\}^+\}$ is deterministic linear context free, but is not accepted by any k-NFA, for all $k \geq 1$. Thus, the language $\{w\$w^R \mid w \in \{a,b\}^+\}$ does not belong to $\mathscr{L}(\text{PCFA})$.

Conversely, the language $\{w\$w \mid w \in \{a,b\}^+\}$ belongs to $\mathscr{L}(\text{DRCPCFA})$ as well as to $\mathscr{L}(\text{DCPCFA})$ (cf. Section 2), but is not context free. $\qquad\square$

Lemma 9. *All language classes accepted by parallel communicating finite automata systems are incomparable with the class of Church-Rosser languages.*

Proof. The unary language $\{a^{2^n} \mid n \geq 1\}$ is a Church-Rosser language [12]. It does not belong to the family $\mathscr{L}(\text{PCFA})$, since all unary languages in $\mathscr{L}(\text{PCFA})$ are semilinear [8,11].

Conversely, the marked copy language $\{w\$w \mid w \in \{a,b\}^+\}$ belongs to the families $\mathscr{L}(\text{DRCPCFA})$ and $\mathscr{L}(\text{DCPCFA})$, but is not a Church-Rosser language [9]. $\qquad\square$

References

1. Brand, D., Zafiropulo, P.: On communicating finite-state machines. J. ACM 30, 323–342 (1983)
2. Buda, A.: Multiprocessor automata. Inform. Process. Lett. 25, 257–261 (1987)
3. Choudhary, A., Krithivasan, K., Mitrana, V.: Returning and non-returning parallel communicating finite automata are equivalent. RAIRO Inform. Théor. 41, 137–145 (2007)
4. Csuhaj-Varjú, E., Dassow, J., Kelemen, J., Păun, G.: Grammar Systems: A Grammatical Approach to Distribution and Cooperation. Gordon and Breach, Yverdon (1994)
5. Ďuriš, P., Jurdziński, T., Kutyłowski, M., Loryś, K.: Power of cooperation and multihead finite systems. In: Larsen, K.G., Skyum, S., Winskel, G. (eds.) ICALP 1998. LNCS, vol. 1443, pp. 896–907. Springer, Heidelberg (1998)
6. Harrison, M.A., Ibarra, O.H.: Multi-tape and multi-head pushdown automata. Inform. Control 13, 433–470 (1968)
7. Ibarra, O.H.: On two-way multihead automata. J. Comput. System Sci. 7, 28–36 (1973)
8. Ibarra, O.H.: A note on semilinear sets and bounded-reversal multihead pushdown automata. Inform. Process. Lett. 3, 25–28 (1974)
9. Jurdziński, T.: The Boolean closure of growing context-sensitive languages. In: Ibarra, O.H., Dang, Z. (eds.) DLT 2006. LNCS, vol. 4036, pp. 248–259. Springer, Heidelberg (2006)
10. Klemm, R.: Systems of communicating finite state machines as a distributed alternative to finite state machines. Phd thesis, Pennsylvania State University (1996)
11. Martín-Vide, C., Mateescu, A., Mitrana, V.: Parallel finite automata systems communicating by states. Int. J. Found. Comput. Sci. 13, 733–749 (2002)
12. McNaughton, R., Narendran, P., Otto, F.: Church-Rosser Thue systems and formal languages. J. ACM 35, 324–344 (1988)
13. Rosenberg, A.L.: On multi-head finite automata. IBM J. Res. Dev. 10, 388–394 (1966)
14. Wagner, K., Wechsung, G.: Computational Complexity. Reidel, Dordrecht (1986)
15. Yao, A.C., Rivest, R.L.: $k+1$ heads are better than k. J. ACM 25, 337–340 (1978)

On a Generalization of Standard Episturmian Morphisms

Michelangelo Bucci, Aldo de Luca, and Alessandro De Luca

Dipartimento di Matematica e Applicazioni "R. Caccioppoli"
Università degli Studi di Napoli Federico II
Via Cintia, Monte S. Angelo, I-80126 Napoli, Italy
{micbucci,aldo.deluca,alessandro.deluca}@unina.it

Abstract. In a recent paper with L. Q. Zamboni the authors introduced the class of ϑ-*episturmian* words, where ϑ is an involutory antimorphism of the free monoid A^*. In this paper, we introduce and study ϑ-*characteristic morphisms*, that is, morphisms which map standard episturmian words into standard ϑ-episturmian words. They are a natural extension of standard episturmian morphisms. The main result of the paper is a characterization of these morphisms when they are injective.

1 Introduction

The study of combinatorial and structural properties of finite and infinite words is a subject of great interest, with many applications in mathematics, physics, computer science, and biology (see for instance [1,2,3]). In this framework, *Sturmian words* play a central role, as they are the aperiodic infinite words of minimal "complexity" (see [2, Chap. 2]). By definition, Sturmian words are on a binary alphabet; some natural extensions to the case of an alphabet with more than two letters have been given in [4,5], introducing the class of the so-called *episturmian words*.

Several extensions of standard episturmian words are possible. For example, in [6] a generalization was obtained by making suitable hypotheses on the lengths of palindromic prefixes of an infinite word; in [7,8,9,10] different extensions were introduced, all based on the replacement of the reversal operator $R : w \in A^* \mapsto \tilde{w} \in A^*$ by an arbitrary *involutory antimorphism* ϑ of the free monoid A^*. In particular, the so called ϑ-*standard* and *standard* ϑ-*episturmian* words were studied.

In this paper we focus on the study of ϑ-*characteristic morphisms*, a natural extension of standard episturmian morphisms, which map all standard episturmian words on an alphabet X to standard ϑ-episturmian words over some alphabet A. Beside being interesting by themselves, such morphisms are a powerful tool for constructing nontrivial examples of standard ϑ-episturmian words and for studying their properties. The main result of this paper is a characterization of injective ϑ-characteristic morphisms (cf. Theorem 3.2). For the sake of brevity, we shall prove here only this theorem; all the other proofs can be found

M. Ito and M. Toyama (Eds.): DLT 2008, LNCS 5257, pp. 158–169, 2008.

in [11]. For notations and definitions not included here, the reader is referred to [1,2,7,12].

1.1   Standard Episturmian Words and Morphisms

We recall (cf. [4,5]) that an infinite word $t \in A^\omega$ is *standard episturmian* if it is *closed under reversal* (that is, if $w \in$ Fact t then $\tilde{w} \in$ Fact t) and each of its left special factors is a prefix of t. We denote by $SEpi(A)$, or by $SEpi$ when there is no ambiguity, the set of all standard episturmian words over the alphabet A.

Given a word $w \in A^*$, we denote by $w^{(+)}$ its *right palindrome closure*, i.e., the shortest palindrome having w as a prefix (cf. [13]). We define the *iterated palindrome closure* operator $\psi : A^* \to A^*$ by setting $\psi(\varepsilon) = \varepsilon$ and $\psi(va) = (\psi(v)a)^{(+)}$ for any $a \in A$ and $v \in A^*$. From the definition, one easily obtains that the map ψ is injective. Furthermore, for any $u, v \in A^*$, one has $\psi(uv) \in \psi(u)A^* \cap A^*\psi(u)$. The operator ψ can then be naturally extended to A^ω. The following fundamental result was proved in [4]:

Theorem 1.1. *An infinite word t is standard episturmian over A if and only if there exists $\Delta \in A^\omega$ such that $t = \psi(\Delta)$.*

For any $t \in SEpi$, there exists a *unique* Δ such that $t = \psi(\Delta)$. This Δ is called the *directive word* of t. If every letter of A occurs infinitely often in Δ, the word t is called a (standard) *Arnoux-Rauzy word*. In the case of a binary alphabet, an Arnoux-Rauzy word is usually called a *standard Sturmian word* (cf. [2, Chap. 2]).

We report here some properties of the operator ψ which will be useful in the sequel. The first one is known (see for instance [13,4]).

Proposition 1.2. *For all $u, v \in A^*$, u is a prefix of v if and only if $\psi(u)$ is a prefix of $\psi(v)$.*

Proposition 1.3. *Let $x \in A \cup \{\varepsilon\}$, $w' \in A^*$, and $w \in w'A^*$. Then $\psi(w'x)$ is a factor of $\psi(wx)$.*

The following proposition was proved in [4, Theorem 6].

Proposition 1.4. *Let $x \in A$, $u \in A^*$, and $\Delta \in A^\omega$. Then $\psi(u)x$ is a factor of $\psi(u\Delta)$ if and only if x occurs in Δ.*

For each $a \in A$, let $\mu_a : A^* \to A^*$ be the morphism defined by $\mu_a(a) = a$ and $\mu_a(b) = ab$ for all $b \in A \setminus \{a\}$. If $a_1, \ldots, a_n \in A$, we set $\mu_w = \mu_{a_1} \circ \cdots \circ \mu_{a_n}$ (in particular, $\mu_\varepsilon = \mathrm{id}_A$). The next formula, proved in [14], shows a connection between these morphisms, called *pure* standard episturmian morphisms (see [5]), and iterated palindrome closure.

Proposition 1.5. *For any $w, v \in A^*$, $\psi(wv) = \mu_w(\psi(v))\psi(w)$.*

By Theorem 1.1, there exists $v \in A^\omega$ such that $t = \psi(v)$, thus, from Proposition 1.5, one easily derives

$$\psi(wv) = \mu_w(\psi(v)) \ . \tag{1}$$

Furthermore, the following holds (cf. [11]):

Corollary 1.6. *For any* $t \in A^\omega$ *and* $w \in A^*$, $\psi(w)$ *is a prefix of* $\mu_w(t)$.

We recall (cf. [4,14,5]) that a *standard episturmian morphism* is an injective endomorphism φ of A^* such that $\varphi(SEpi) \subseteq SEpi$. As proved in [4], a morphism is standard episturmian if and only if can be written as $\mu_w \circ \sigma$, with $w \in A^*$ and $\sigma : A^* \to A^*$ a morphism extending a permutation on the alphabet A. All these morphisms are injective. The set of all standard episturmian morphisms is a monoid under map composition.

1.2   Involutory Antimorphisms and Pseudopalindromes

An *involutory antimorphism* of A^* is any antimorphism $\vartheta : A^* \to A^*$ such that $\vartheta \circ \vartheta = \mathrm{id}$. The simplest example is the reversal operator. Any involutory antimorphism ϑ satisfies $\vartheta = \tau \circ R = R \circ \tau$ for some morphism $\tau : A^* \to A^*$ extending an involution of A. Conversely, if τ is such a morphism, then $\vartheta = \tau \circ R = R \circ \tau$ is an involutory antimorphism of A^*.

Let ϑ be an involutory antimorphism of A^*. We call ϑ-*palindrome* any fixed point of ϑ, i.e., any word w such that $w = \vartheta(w)$, and denote by PAL_ϑ the set of all ϑ-palindromes. We observe that $\varepsilon \in PAL_\vartheta$ by definition, and that R-palindromes are exactly the usual palindromes. If one makes no reference to the antimorphism ϑ, a ϑ-palindrome is often called a *pseudopalindrome*. Some general properties of pseudopalindromes, have been studied in [7].

In the following, we shall fix an involutory antimorphism ϑ of A^*, and use the notation $\bar{w}$ for $\vartheta(w)$. We denote by $\mathcal{P}_\vartheta$ the set of *unbordered* ϑ-palindromes (i.e., ϑ-palindromes without nontrivial ϑ-palindromic prefixes). We remark that $\mathcal{P}_\vartheta$ is a *biprefix code* (cf. [12]) and that $\mathcal{P}_R = A$. The following result was proved in [9]:

Proposition 1.7. $PAL_\vartheta^* = \mathcal{P}_\vartheta^*$.

This can be equivalently stated as follows: every ϑ-palindrome can be uniquely factorized by the elements of $\mathcal{P}_\vartheta$. For instance, if $\bar{a} = b$ and $\bar{c} = c$, the ϑ-palindrome $abacabcbab$ can be factorized as $ab \cdot acabcb \cdot ab$.

For any nonempty word w, we will denote, from now on, by w^f and w^ℓ respectively the first and the last letter of w. Since $\mathcal{P}_\vartheta$ is a code, the map

$$f : \pi \in \mathcal{P}_\vartheta \longmapsto \pi^f \in A \tag{2}$$

can be extended (uniquely) to a morphism $f : \mathcal{P}_\vartheta^* \to A^*$. Moreover, since $\mathcal{P}_\vartheta$ is a prefix code, any word in $\mathcal{P}_\vartheta^\omega$ can be uniquely factorized by the elements of $\mathcal{P}_\vartheta$, so that f can be naturally extended to $\mathcal{P}_\vartheta^\omega$.

Proposition 1.8. *Let* $\varphi : X^* \to A^*$ *be an injective morphism such that* $\varphi(X) \subseteq \mathcal{P}_\vartheta$. *Then, for any* $w \in X^*$:

1. $\varphi(\tilde{w}) = \overline{\varphi(w)}$,
2. $w \in PAL \Longleftrightarrow \varphi(w) \in PAL_\vartheta$,

1.3   Overlap-Free and Normal Codes

We say that a code Z over A is *overlap-free* if no two of its elements overlap properly, i.e., if for all $u, v \in Z$, Suff $u \cap$ Pref $v \subseteq \{\varepsilon, u, v\}$.

For instance, let $Z_1 = \{a, bac, abc\}$ and $Z_2 = \{a, bac, cba\}$. One has that Z_1 is an overlap-free suffix code, and Z_2 is a prefix code which is not overlap-free.

Let Z be a subset of A^*; we denote by $LS\,Z$ (resp. $RS\,Z$) the set of all words $u \in$ Fact Z which are *left special* (resp. *right special*) in Z, i.e., such that there exist two distinct letters a and b for which $au, bu \in$ Fact Z (resp. $ua, ub \in$ Fact Z).

A code $Z \subseteq A^+$ will be called *right normal* if it satisfies the following condition:

$$(\text{Pref } Z \setminus Z) \cap RS\,Z \subseteq \{\varepsilon\} ,\tag{3}$$

i.e., any proper and nonempty prefix u of any word of Z such that $u \notin Z$ is not right special in Z. In a symmetric way, a code Z is called *left normal* if it satisfies the condition

$$(\text{Suff } Z \setminus Z) \cap LS\,Z \subseteq \{\varepsilon\} .\tag{4}$$

A code Z is called *normal* if it is right and left normal.

As an example, the code $Z_1 = \{a, ab, bb\}$ is right normal but not left normal; the code $Z_2 = \{a, aba, aab\}$ is normal.

Proposition 1.9. *Let Z be a biprefix, overlap-free, and right normal (resp. left normal) code. Then:*

1. *if $z \in Z$ is such that $z = \lambda v \rho$, with $\lambda, \rho \in A^*$ and v a nonempty prefix (resp. suffix) of $z' \in Z$, then $\lambda z'$ (resp. $z' \rho$) is a prefix (resp. suffix) of z, proper if $z \neq z'$.*
2. *for $z_1, z_2 \in Z$, if $z_1^f = z_2^f$ (resp. $z_1^\ell = z_2^\ell$), then $z_1 = z_2$.*

From the preceding proposition, a biprefix, overlap-free, and normal code satisfies both properties 1 and 2 and their symmetrical statements; all the statements of the following propositions can also be applied to codes which are biprefix, overlap-free, and normal.

Proposition 1.10. *Let Z be a suffix, left normal, and overlap-free code over A, and let $a, b \in A$, $v \in A^*$, $\lambda \in A^+$ be such that $a \neq b$, $va \notin Z^*$, $va\lambda \in$ Pref Z^*, and $b\lambda \in$ Fact Z^*. Then $a\lambda \in$ Fact Z.*

Proposition 1.11. *Let Z be a biprefix, overlap-free, and right normal code over A. If $\lambda \in$ Pref $Z^* \setminus \{\varepsilon\}$, then there exists a unique word $u = z_1 \cdots z_k$ with $k \geq 1$ and $z_i \in Z$, $i = 1, \ldots, k$, such that*

$$u = z_1 \cdots z_k = \lambda \zeta, \quad z_1 \cdots z_{k-1} \delta = \lambda ,\tag{5}$$

where $\delta \in A^+$ and $\zeta \in A^*$.

In conclusion of this section, we report (cf. [11]) the following simple general lemma on prefix codes, which will be useful in the next sections:

Lemma 1.12. *Let $g : B^* \to A^*$ be an injective morphism such that $g(B) = Z$ is a prefix code. Then for all $p \in B^*$ and $q \in B^\infty$, one has that p is a prefix of q if and only if $g(p)$ is a prefix of $g(q)$.*

1.4   Standard ϑ-Episturmian Words

In [9] *standard ϑ-episturmian* words were naturally defined by substituting, in the definition of standard episturmian words, the closure under reversal with the *closure under ϑ*. Thus an infinite word s is standard ϑ-episturmian if it satisfies the following two conditions:

1. for any $w \in$ Fact s, one has $\bar{w} \in$ Fact s,
2. for any left special factor w of s, one has $w \in$ Pref s.

We denote by $SEpi_\vartheta$ the set of all standard ϑ-episturmian words over A.

The following proposition, proved in one direction in [9] and completely in [11], is a first tool for constructing nontrivial standard ϑ-episturmian words.

Proposition 1.13. *Let $g : X^* \to A^*$ be an injective morphism such that $g(X) \subseteq \mathcal{P}_\vartheta$ for a fixed ϑ. Then $g(SEpi(X)) \subseteq SEpi_\vartheta(A)$ if and only if each letter of* alph $g(X)$ *appears exactly once in $g(X)$.*

Example 1.14. Let $A = \{a, b, c, d, e\}$, $\bar{a} = b$, $\bar{c} = c$, $\bar{d} = e$, $X = \{a, b\}$, and $s = g(t)$, where $t = aabaaabaaabaab \cdots \in SEpi(X)$, $\Delta(t) = (aab)^\omega$, $g(a) = acb$, and $g(b) = de$, so that

$$s = acbacbdeacbacbacbde \cdots . \tag{6}$$

Proposition 1.13 ensures that g maps $SEpi(X)$ into a subset of $SEpi_\vartheta(A)$, thus s is standard ϑ-episturmian.

In the following, for a given standard ϑ-episturmian word s we shall denote by

$$\Pi_s = \{\pi_n \mid n \geq 1\} \tag{7}$$

the set of words of $\mathcal{P}_\vartheta$ appearing in its unique factorization $s = \pi_1 \pi_2 \cdots$ in unbordered ϑ-palindromes.

The details of the proof of the following useful theorem can be found in [11].

Theorem 1.15. *Let $s \in SEpi_\vartheta$. Then Π_s is an overlap-free and normal code.*

Since for $s \in SEpi_\vartheta$, Π_s is a biprefix, overlap-free, and normal code, by Proposition 1.8, Proposition 1.9, and Lemma 1.12 one can derive the following theorem, proved in [9, Theorem 5.5] in a different way, which shows in particular that any standard ϑ-episturmian word is a morphic image, by a suitable injective morphism, of a standard episturmian word.

Theorem 1.16. *Let s be a standard ϑ-episturmian word. Then $f(s)$ is a standard episturmian word, and the restriction of f to Π_s is injective, i.e., if π_i and π_j occur in the factorization of s over $\mathcal{P}_\vartheta$, and $\pi_i^f = \pi_j^f$, then $\pi_i = \pi_j$.*

2   Characteristic Morphisms

Let X be a finite alphabet. A morphism $\varphi : X^* \to A^*$ will be called ϑ-characteristic if $\varphi(SEpi(X)) \subseteq SEpi_\vartheta$, i.e., φ maps any standard episturmian word over the alphabet X in a standard ϑ-episturmian word on the alphabet A. With this terminology, we observe that an injective morphism $\varphi : X^* \to X^*$ is standard episturmian if and only if it is R-characteristic. A trivial example of a non-injective ϑ-characteristic morphism is the constant morphism $\varphi : x \in X \mapsto a \in A$, where a is a fixed ϑ-palindromic letter; furthermore Proposition 1.13 provides an easy way of constructing injective ϑ-characteristic morphisms, like the one used in Example 1.14.

Let $X = \{x, y\}$, $A = \{a, b, c\}$, ϑ defined by $\bar{a} = a$, $\bar{b} = c$, and $\varphi : X^* \to A^*$ be the injective morphism such that $\varphi(x) = a$, $\varphi(y) = bac$. If t is any standard episturmian word beginning in y^2x, then $s = \varphi(t)$ begins with $bacbaca$, so that a is a left special factor of s which is not a prefix of s. Thus s is not ϑ-episturmian and therefore φ is not ϑ-characteristic.

A first result (cf. [11]) on the structure of ϑ-characteristic morphisms is given by the following:

Proposition 2.1. *Let $\varphi : X^* \to A^*$ be a ϑ-characteristic morphism. For each x in X, $\varphi(x) \in PAL_\vartheta^2$.*

Let $\varphi : X^* \to A^*$ be a morphism such that $\varphi(X) \subseteq \mathcal{P}_\vartheta^*$. For any $x \in X$, let $\varphi(x) = \pi_1^{(x)} \cdots \pi_{r_x}^{(x)}$ be the unique factorization of $\varphi(x)$ by the elements of $\mathcal{P}_\vartheta$. Set

$$\Pi(\varphi) = \{\pi \in \mathcal{P}_\vartheta \mid \exists x \in X, \exists i : 1 \le i \le r_x \text{ and } \pi = \pi_i^{(x)}\} \ . \tag{8}$$

If φ is a ϑ-characteristic morphism, then by Propositions 2.1 and 1.7, we have $\varphi(X) \subseteq PAL_\vartheta^2 \subseteq \mathcal{P}_\vartheta^*$, so that $\Pi(\varphi)$ is well defined.

The following important theorem provides a useful decomposition of injective ϑ-characteristic morphisms (see [11] for a proof).

Theorem 2.2. *Let $\varphi : X^* \to A^*$ be an injective ϑ-characteristic morphism. Then $\Pi(\varphi)$ is an overlap-free and normal code. Furthermore φ can be decomposed as*

$$\varphi = g \circ \mu_w \circ \eta \ , \tag{9}$$

where $\eta : X^* \to B^*$ is an injective literal morphism, $B \subseteq A$, $\mu_w : B^* \to B^*$ is a pure standard episturmian morphism (with $w \in B^*$), and $g : B^* \to A^*$ is an injective morphism such that $g(B) = \Pi(\varphi)$. The above decomposition can always be chosen so that $B = \eta(X) \cup \mathrm{alph}\, w \subseteq A$ and $g(b) \in bA^* \cap \mathcal{P}_\vartheta$ for each $b \in B$.

Example 2.3. Let X, A, ϑ, and g be defined as in Example 1.14 and let $\eta(a) = a$, $\eta(b) = b$. The morphism φ defined by $\varphi(a) = acbdeacb$, $\varphi(b) = acbde$ is decomposable as

$$\varphi = g \circ \mu_{ab} \circ \eta \ .$$

Since g is ϑ-characteristic and $\mu_{ab} \circ \eta = \mu_{ab}$ is R-characteristic, it follows that φ is ϑ-characteristic.

Example 2.4. Let $X = \{x, y\}$, $A = \{a, b, c\}$, and ϑ be the antimorphism of A^* such that $\bar{a} = a$ and $\bar{b} = c$. The morphism $\varphi : X^* \to A^*$ defined by $\varphi(x) = a$ and $\varphi(y) = abac$ is ϑ-characteristic (this will be clear after Theorem 3.2, see Example 3.3). It can be decomposed as $\varphi = g \circ \mu_a \circ \eta$, where $\eta : X^* \to B^*$ (with $B = \{a, b\}$) is the morphism such that $\eta(x) = a$ and $\eta(y) = b$, while $g : B^* \to A^*$ is defined by $g(a) = a$ and $g(b) = bac$. We remark that $(\mu_a \circ \eta)(SEpi(X)) \subseteq SEpi(B)$, but from Proposition 1.13 it follows that $g(SEpi(B)) \not\subseteq SEpi_\vartheta$.

Proposition 2.5 (cf. [11]). *Let $\varphi : X^* \to A^*$ be an injective ϑ-characteristic morphism, decomposed as in (9). The word $u = g(\psi(w))$ is a ϑ-palindrome such that for each $x \in X$, $\varphi(x)$ is either a prefix of u or equal to $ug(\eta(x))$.*

3   Main Result

Before proceeding with the main theorem, which gives a characterization of all injective ϑ-characteristic morphisms, we need the following lemma, again proved in [11].

Lemma 3.1. *Let $t \in SEpi(B)$ with alph $t = B$, and let $s = g(t)$ be a standard ϑ-episturmian word over A, with $g : B^* \to A^*$ an injective morphism such that $g(B) \subseteq \mathcal{P}_\vartheta$. Suppose that $b, c \in A \setminus \text{Suff} \, \Pi_s$ and $v \in \Pi_s^*$ are such that $bv\bar{c} \in \text{Fact} \, \Pi_s$. Then there exists $\delta \in B^*$ such that $v = g(\psi(\delta))$.*

Theorem 3.2. *Let $\varphi : X^* \to A^*$ be an injective morphism. Then φ is ϑ-characteristic if and only if it is decomposable as*

$$\varphi = g \circ \mu_w \circ \eta$$

as in (9), with $B = \eta(X) \cup \text{alph} \, w$ and $g(B) = \Pi \subseteq \mathcal{P}_\vartheta$ satisfying the following conditions:

1. *Π is an overlap-free and normal code,*
2. *$LS(\{g(\psi(w))\} \cup \Pi) \subseteq \text{Pref} \, g(\psi(w))$,*
3. *if $b, c \in A \setminus \text{Suff} \, \Pi$ and $v \in \Pi^*$ are such that $bv\bar{c} \in \text{Fact} \, \Pi$, then $v = g(\psi(w'x))$, with $w' \in \text{Pref} \, w$ and $x \in \{\varepsilon\} \cup (B \setminus \eta(X))$.*

Example 3.3. Let $A = \{a, b, c\}$, $X = \{x, y\}$, $B = \{a, b\}$, and let ϑ and $\varphi : X^* \to A^*$ be defined as in Example 2.4, namely $\bar{a} = a$, $\bar{b} = c$, and $\varphi = g \circ \mu_a \circ \eta$, where $\eta(x) = a$, $\eta(y) = b$, and $g : B^* \to A^*$ is defined by $g(a) = a$ and $g(b) = bac$. Then $\Pi = g(B) = \{a, bac\}$ is an overlap-free, normal code which satisfies condition 2 of Theorem 3.2. The only word verifying the hypotheses of condition 3 is bac, and $bac = ba\bar{b} = g(b) \in \Pi$, with $a \in \Pi^*$ and $b \notin \text{Suff} \, \Pi$. Since $a = g(\psi(a))$ and $B \setminus \eta(X) = \emptyset$, also condition 3 of Theorem 3.2 is satisfied. Hence φ is ϑ-characteristic.

Example 3.4. Let $X = \{x, y\}$, $A = \{a, b, c\}$, ϑ be such that $\bar{a} = a$, $\bar{b} = c$, and the morphism $\varphi : X^* \to A^*$ be defined by $\varphi(x) = a$ and $\varphi(y) = abaac$. In this case we have $\varphi = g \circ \mu_a \circ \eta$, where $B = \{a, b\}$, $g(a) = a$, $g(b) = baac$,

$\eta(x) = a$, and $\eta(y) = b$. Then the morphism φ is not ϑ-characteristic. Indeed, if t is any standard episturmian word starting with yxy, then $\varphi(t)$ has the prefix $abaacaabaac$, so that aa is a left special factor of $\varphi(t)$ but not a prefix of it. In fact, condition 3 of Theorem 3.2 is not satisfied in this case, since $baac = baa\bar{b} = g(b)$, $b \notin \operatorname{Suff} \Pi$, $aa \in \Pi^*$, $B \setminus \eta(X) = \emptyset$, and

$$aa \notin \{g(\psi(w')) \mid w' \in \operatorname{Pref} a\} = \{\varepsilon, a\} .$$

If we choose $X' = \{y\}$ with $\eta'(y) = b$, then

$$g(\mu_a(\eta'(y^\omega))) = (abaac)^\omega \in SEpi_\vartheta ,$$

so that $\varphi' = g \circ \mu_a \circ \eta'$ is ϑ-characteristic. In this case $B = \eta'(X') \cup \operatorname{alph} a$, $B \setminus \eta'(X') = \{a\}$, and $aa = g(\psi(aa)) = g(aa)$, so that condition 3 is satisfied.

Example 3.5. Let $X = \{x, y\}$, $A = \{a, b, c, d, e, h\}$, and ϑ be the antimorphism over A defined by $\bar{a} = a$, $\bar{b} = c$, $\bar{d} = e$, $\bar{h} = h$. Let also $w = adb \in A^*$, $B = \{a, b, d\} = \operatorname{alph} w$, and $\eta : X^* \to B^*$ be defined by $\eta(x) = a$ and $\eta(y) = b$. Finally, set $g(a) = a$, $g(d) = dahae$, and $g(b) = badahaeadahaeac$, so that the morphism $\varphi = g \circ \mu_w \circ \eta$ is such that

$$\varphi(y) = adahaeabadahaeadahaeac \quad \text{and} \quad \varphi(x) = \varphi(y)\, adahaea .$$

Then φ is ϑ-characteristic, as the code $\Pi(\varphi) = g(B)$ and the word $u = g(\psi(w)) = g(adabada) = \varphi(x)$ satisfy all three conditions of Theorem 3.2.

Note that Proposition 1.13 can be derived as a corollary of Theorem 3.2 (cf. [11]).

Remark 3.6. Let us observe that Theorem 3.2 gives an effective procedure to decide whether, for a given ϑ, an injective morphism $\varphi : X^* \to A^*$ is ϑ-characteristic. The procedure runs in the following steps:

1. Check whether $\varphi(X) \subseteq \mathcal{P}_\vartheta^*$.
2. If the previous condition is satisfied, then compute $\Pi = \Pi(\varphi)$.
3. Verify that Π is overlap-free and normal.
4. Compute $B = f(\Pi)$ and the morphism $g : B^* \to A^*$ given by $g(B) = \Pi$.
5. Since $\varphi = g \circ \zeta$, verify that ζ is R-characteristic, i.e., there exists $w \in B^*$ such that $\zeta = \mu_w \circ \eta$, where η is a literal morphism from X^* to B^*.
6. Compute $g(\psi(w))$ and verify that conditions 2 and 3 of Theorem 3.2 are satisfied. This can be effectively done.

Proof (Theorem 3.2). **Necessity:** From Theorem 2.2, we obtain the decomposition (9) where $B = \eta(X) \cup \operatorname{alph} w$ and $g(B) = \Pi(\varphi) \subseteq \mathcal{P}_\vartheta$ is an overlap-free and normal code.

Let us set $u = g(\psi(w))$, and prove that condition 2 holds. We first suppose that card $X \geq 2$, and that $a, a' \in \eta(X)$ are distinct letters. Let Δ be an infinite word such that $\operatorname{alph} \Delta = \eta(X)$. Setting $t_a = \psi(wa\Delta)$ and $t_{a'} = \psi(wa'\Delta)$, by (1) we have

$$t_a = \mu_w(\psi(a\Delta)) \quad \text{and} \quad t_{a'} = \mu_w(\psi(a'\Delta)) ,$$

so that, setting $s_y = g(t_y)$ for $y \in \{a, a'\}$, we obtain

$$s_y = g(\mu_w(\psi(y\Delta))) \in SEpi_\vartheta$$

as $\psi(y\Delta) \in \eta(SEpi(X)) \subseteq SEpi(B)$ and $\varphi = g \circ \mu_w \circ \eta$ is ϑ-characteristic. By Corollary 1.6 and (1), one obtains that the longest common prefix of t_a and $t_{a'}$ is $\psi(w)$. As alph $\Delta = \eta(X)$ and $B = \eta(X) \cup$ alph w, we have alph $t_a =$ alph $t_{a'} = B$, so that $\Pi_{s_a} = \Pi_{s_{a'}} = \Pi$. Since g is injective, by Theorem 1.16 we have $g(a)^f \neq g(a')^f$, so that the longest common prefix of s_a and $s_{a'}$ is $u = g(\psi(w))$. Any word of $LS(\{u\} \cup \Pi)$, being a left special factor of both s_a and $s_{a'}$, has to be a common prefix of s_a and $s_{a'}$, and hence a prefix of u.

Now let us suppose $X = \{z\}$ and denote $\eta(z)$ by a. In this case we have

$$\varphi(SEpi(X)) = \{g(\mu_w(a^\omega))\} = \{(g(\mu_w(a)))^\omega\} \; .$$

Let us set $s = (g(\mu_w(a)))^\omega \in SEpi_\vartheta$. By Corollary 1.6, $u = g(\psi(w))$ is a prefix of s. Let $\lambda \in LS(\{u\} \cup \Pi)$. Since $\Pi = \Pi_s$, the word λ is a left special factor of the ϑ-episturmian word s, so that we have $\lambda \in \mathrm{Pref}\, s$.

If $a \in$ alph w, then $B = \{a\} \cup$ alph $w =$ alph $w =$ alph $\psi(w)$, so that $\Pi \subseteq$ Fact u. This implies $|\lambda| \leq |u|$ and then $\lambda \in \mathrm{Pref}\, u$ as desired.

If $a \notin$ alph w, then by Proposition 2.5 we obtain $\varphi(z) = g(\mu_w(a)) = u\, g(a)$, because $\varphi(z) \notin \mathrm{Pref}\, u$ otherwise by Lemma 1.12 we would obtain $\mu_w(a) \in \mathrm{Pref}\, \psi(w)$, that implies $a \in$ alph w. Hence $s = (u\, g(a))^\omega$. Since $\Pi \subseteq (\mathrm{Fact}\, u) \cup \{g(a)\}$, we have $|\lambda| \leq |u\, g(a)|$, so that $\lambda \in \mathrm{Pref}(u\, g(a))$. Again, if λ is a proper prefix of u we are done, so let us suppose that $\lambda = u\lambda'$ for some $\lambda' \in \mathrm{Pref}\, g(a)$, and that λ is a left special factor of $g(a)$. Then the prefix λ' of $g(a)$ is repeated in $g(a)$. The longest repeated prefix p of $g(a)$ is either a right special factor or a border of $g(a)$. Both possibilities imply $p = \varepsilon$, since $g(a)$ is unbordered and Π is a biprefix and normal code. As $\lambda' \in \mathrm{Pref}\, p$, it follows $\lambda' = \varepsilon$. This proves condition 2.

Finally, let us prove condition 3. Let $b, c \in A \setminus \mathrm{Suff}\, \Pi$, $v \in \Pi^*$, and $\pi \in \Pi$ be such that $bv\bar{c} \in \mathrm{Fact}\, \pi$. Let $t' \in SEpi(X)$ with alph $t' = X$, and set $t = \mu_w(\eta(t'))$, $s_1 = g(t)$. Since φ is ϑ-characteristic, $s_1 = \varphi(t')$ is standard ϑ-episturmian. By Lemma 3.1, we have $v = g(\psi(\delta))$ for some $\delta \in B^*$. If $\delta = \varepsilon$ we are done, as condition 3 is trivially satisfied for $w' = x = \varepsilon$; let us then write $\delta = \delta'a$ for some $a \in B$. The words $bg(\psi(\delta'))$ and $g(a\psi(\delta'))$ are both factors of the ϑ-palindrome π; indeed, $\psi(\delta'a)$ begins with $\psi(\delta')a$ and terminates with $a\psi(\delta')$. Hence $g(\psi(\delta'))$ is left special in π as $b \notin \mathrm{Suff}\, \Pi$ is different from $(g(a))^\ell \in \mathrm{Suff}\, \Pi$. Therefore $g(\psi(\delta'))$ is a prefix of $g(\psi(w))$, as we have already proved condition 2. Since g is injective and Π is a biprefix code, by Lemma 1.12 it follows $\psi(\delta') \in \mathrm{Pref}\, \psi(w)$, so that $\delta' \in \mathrm{Pref}\, w$ by Proposition 1.2. Hence, we can write $\delta = w'x$ with $w' \in \mathrm{Pref}\, w$ and x either equal to a (if $\delta'a \notin \mathrm{Pref}\, w$) or to ε. It remains to show that if $w'x \notin \mathrm{Pref}\, w$, then $x \notin \eta(X)$.

Let us first assume that $\eta(X) = \{x\}$. In this case we have $s_1 = g(\mu_w(\eta(t'))) = g(\psi(wx^\omega))$ by (1). Since $bv = bg(\psi(w'x)) \in \mathrm{Fact}\, \pi$, $g(x)$ is a proper factor of π. Then, as $B = \{x\} \cup$ alph w and $g(x) \neq \pi$, we must have $\pi \in g(\mathrm{alph}\, w)$, so that $bv \in \mathrm{Fact}\, g(\psi(w))$ as alph $w =$ alph $\psi(w)$. By Proposition 1.3, $\psi(w'x)$ is a factor of $\psi(wx)$. We can then write $\psi(wx) = \zeta\psi(w'x)\zeta'$ for some $\zeta, \zeta' \in B^*$. If ζ were empty, by Proposition 1.2 we would obtain $w'x \in \mathrm{Pref}(wx)$; since $w'x \notin \mathrm{Pref}\, w$, we would derive $w = w'$, which is a contradiction since we proved that $bv = bg(\psi(w'x)) \in \mathrm{Fact}\, g(\psi(w))$. Therefore $\zeta \neq \varepsilon$, and v is left special in s, being preceded both by $(g(\zeta))^\ell$ and by $b \notin \mathrm{Suff}\, \Pi$. This implies that v is a prefix of s and then of $g(\psi(w))$ as $|v| \leq |g(\psi(w))|$. By Lemma 1.12, it follows $\psi(w'x) \in \mathrm{Pref}\, \psi(w)$ and then $w'x \in \mathrm{Pref}\, w$ by Proposition 1.2, which is a contradiction.

Suppose now that there exists $y \in \eta(X) \setminus \{x\}$, and let $\Delta \in \eta(X)^\omega$ with $\mathrm{alph}\, \Delta = \eta(X)$. The word $s_2 = g(\psi(wyx\Delta))$ is equal to $g(\mu_w(\psi(yx\Delta)))$ by (1), and is then standard ϑ-episturmian since $\varphi = g \circ \mu_w \circ \eta$ is ϑ-characteristic. By applying Proposition 1.3 to w' and $wy \in w'A^*$, we obtain $\psi(w'x) \in \mathrm{Fact}\, \psi(wyx)$. We can write $\psi(wyx) = \zeta\psi(w'x)\zeta'$ for some $\zeta, \zeta' \in B^*$. As $w'x \notin \mathrm{Pref}\, w$ and $x \neq y$, we have by Proposition 1.2 that $\psi(w'x) \notin \mathrm{Pref}\, \psi(wy)$, so that $\zeta \neq \varepsilon$. Hence $v = g(\psi(w'x))$ is left special in s_2, being preceded both by $(g(\zeta))^\ell$ and by $b \notin \mathrm{Suff}\, \Pi$. This implies that v is a prefix of s_2 and then of $g(\psi(wy))$; by Lemma 1.12, this is absurd since $\psi(w'x) \notin \mathrm{Pref}\, \psi(wy)$.

Sufficiency: Let $t' \in SEpi(\eta(X))$ and $t = \mu_w(t') \in SEpi(B)$. Since $g(B) = \Pi \subseteq \mathcal{P}_\vartheta$, by Proposition 1.8 it follows that $g(t)$ has infinitely many ϑ-palindromic prefixes, so that its set of factors is closed under ϑ.

Thus, in order to prove that $g(t) \in SEpi_\vartheta$, it is sufficient to show that any nonempty left special factor λ of $g(t)$ is in $\mathrm{Pref}\, g(t)$. Since λ is left special, there exist $a, a' \in A$, $a \neq a'$, $v, v' \in A^*$, and $r, r' \in A^\omega$, such that

$$g(t) = va\lambda r = v'a'\lambda r' \ . \tag{10}$$

The word $g(t)$ can be uniquely factorized by the elements of Π. Therefore, $va\lambda$ and $v'a'\lambda$ are in $\mathrm{Pref}\, \Pi^*$. We consider three different cases.

Case 1: $va \notin \Pi^*$, $v'a' \notin \Pi^*$.

Since Π is a biprefix (as it is a subset of $\mathcal{P}_\vartheta$), overlap-free and normal code, by Proposition 1.10 we have $a\lambda, a'\lambda \in \mathrm{Fact}\, \Pi$. Therefore, by condition 2 of Theorem 3.2, it follows $\lambda \in LS\, \Pi \subseteq \mathrm{Pref}\, g(\psi(w))$, so that it is a prefix of $g(t)$ since by Corollary 1.6, $\psi(w)$ is a prefix of $t = \mu_w(t')$.

Case 2: $va \in \Pi^*$, $v'a' \in \Pi^*$.

From (10), we have $\lambda \in \mathrm{Pref}\, \Pi^*$. By Proposition 1.11, there exists a unique word $\lambda' \in \Pi^*$ such that $\lambda' = \pi_1 \cdots \pi_k = \lambda\zeta$ and $\pi_1 \cdots \pi_{k-1}\delta = \lambda$, with $k \geq 1$, $\pi_i \in \Pi$ for $i = 1, \ldots, k$, $\delta \in A^+$, and $\zeta \in A^*$.

Since g is injective, there exist and are unique the words $\tau, \gamma, \gamma' \in B^*$ such that $g(\tau) = \lambda', g(\gamma) = va, g(\gamma') = v'a'$. Moreover, we have $g(\gamma\tau) = va\lambda' = va\lambda\zeta \in \mathrm{Pref}\, g(t)$ and $g(\gamma'\tau) = v'a'\lambda' = v'a'\lambda\zeta \in \mathrm{Pref}\, g(t)$. By Lemma 1.12, we derive $\gamma\tau, \gamma'\tau \in \mathrm{Pref}\, t$. Setting $\alpha = \gamma^\ell, \alpha' = \gamma'^\ell$, we obtain $\alpha\tau, \alpha'\tau \in \mathrm{Fact}\, t$, and $\alpha \neq \alpha'$ as $a \neq a'$. Hence τ is a left special factor of t; since $t \in SEpi(B)$, we have $\tau \in \mathrm{Pref}\, t$, so that $g(\tau) = \lambda' \in \mathrm{Pref}\, g(t)$. As λ is a prefix of λ', it follows $\lambda \in \mathrm{Pref}\, g(t)$.

Case 3: $va \notin \Pi^*$, $v'a' \in \Pi^*$ (resp. $va \in \Pi^*$, $v'a' \notin \Pi^*$).

We shall consider only the case when $va \notin \Pi^*$ and $v'a' \in \Pi^*$, as the symmetric case can be similarly dealt with.

Since $v'a' \in \Pi^*$, by (10) we have $\lambda \in \operatorname{Pref} \Pi^*$. By Proposition 1.11, there exists a unique word $\lambda' \in \Pi^*$ such that $\lambda' = \pi_1 \cdots \pi_k = \lambda \zeta$ and $\pi_1 \cdots \pi_{k-1} \delta = \lambda$, with $k \geq 1$, $\pi_i \in \Pi$ for $i = 1, \ldots, k$, $\delta \in A^+$, and $\zeta \in A^*$. By the uniqueness of λ', $v'a'\lambda'$ is a prefix of $g(t)$.

By (10) we have $va\pi_1 \cdots \pi_{k-1}\delta \in \operatorname{Pref} g(t)$. By Proposition 1.10, $a\lambda \in \operatorname{Fact} \Pi$, so that there exist $\xi, \xi' \in A^*$ and $\pi \in \Pi$ such that

$$\xi a \lambda \xi' = \xi a \pi_1 \cdots \pi_{k-1} \delta \xi' = \pi \in \Pi \ .$$

Since δ is a nonempty prefix of π_k, it follows from Proposition 1.9 that $\pi = \xi a \pi_1 \cdots \pi_k \xi'' = \xi a \lambda' \xi''$, with $\xi'' \in A^*$.

Let p (resp. q) be the longest word in $\operatorname{Suff}(\xi a) \cap \Pi^*$ (resp. in $\operatorname{Pref} \xi'' \cap \Pi^*$), and write $\pi = \xi a \lambda' \xi'' = z p \lambda' q z'$, with $z, z' \in A^*$.

Since λ' and zp are nonempty and Π is a biprefix code, one derives that z and z' cannot be empty. Moreover, $b = z^\ell \notin \operatorname{Suff} \Pi$ and $\bar{c} = (z')^f \notin \operatorname{Pref} \Pi$, for otherwise the maximality of p and q could be contradicted using Proposition 1.9.

By condition 3, we have $p\lambda' q = g(\psi(w'x))$ for some $w' \in \operatorname{Pref} w$ and $x \in \{\varepsilon\} \cup (B \setminus \eta(X))$. Since $p, \lambda', q \in \Pi^*$ and g is injective, we derive $\lambda' = g(\tau)$ for some $\tau \in \operatorname{Fact} \psi(w'x)$. We will show that λ' is a prefix of $g(t)$, which proves the assertion as $\lambda \in \operatorname{Pref} \lambda'$.

Suppose first that $p = \varepsilon$, so that $a = b$ and $\tau \in \operatorname{Pref} \psi(w'x)$. If $\tau \in \operatorname{Pref} \psi(w')$, then $\lambda' \in g(\operatorname{Pref} \psi(w')) \subseteq \operatorname{Pref} g(\psi(w')) \subseteq \operatorname{Pref} g(\psi(w))$, and we are done as $g(\psi(w)) \in \operatorname{Pref} g(t)$. Let us then assume $x \neq \varepsilon$, so that $x \in B \setminus \eta(X)$, and $\psi(w')x \in \operatorname{Pref} \tau$. Moreover, we can assume $w'x \notin \operatorname{Pref} w$, for otherwise we would derive $\lambda' \in \operatorname{Pref} g(\psi(w))$ again. Let $\Delta \in \eta(X)^\omega$ be the directive word of t', so that by (1) we have $t = \psi(w\Delta)$. Since $w' \in \operatorname{Pref} w$, we can write $w\Delta = w'\Delta'$ for some $\Delta' \in B^\omega$, so that $t = \psi(w'\Delta')$.

We have already observed that $v'a'\lambda'$ is a prefix of $g(t)$; as $v'a' \in \Pi^*$, by Lemma 1.12 one derives $\tau \in \operatorname{Fact} t$. Since $\psi(w')x \in \operatorname{Pref} \tau$, it follows $\psi(w')x \in \operatorname{Fact} \psi(w'\Delta')$; by Proposition 1.4, we obtain $x \in \operatorname{alph} \Delta'$. This implies, since $x \notin \eta(X)$, that $w \neq w'$, and we can write $w = w'\sigma x \sigma'$ for some $\sigma, \sigma' \in B^*$. By Proposition 1.3, $\psi(w'x)$ is a factor of $\psi(w'\sigma x)$ and hence of $\psi(w)$, so that, since $\tau \in \operatorname{Pref} \psi(w'x)$, we have $\tau \in \operatorname{Fact} \psi(w)$. Hence we have either $\tau \in \operatorname{Pref} \psi(w)$, so that $\lambda' \in \operatorname{Pref} g(\psi(w))$ and we are done, or there exists a letter y such that $y\tau \in \operatorname{Fact} \psi(w)$, so that $d\lambda' \in \operatorname{Fact} g(\psi(w))$ with $d = (g(y))^\ell \in \operatorname{Suff} \Pi$. In the latter case, since $a = b \notin \operatorname{Suff} \Pi$ and $a\lambda' \in \operatorname{Fact} \Pi$, we have by condition 2 that $\lambda' \in \operatorname{Pref} g(\psi(w))$. Since $g(\psi(w))$ is a prefix of $g(t)$, in the case $p = \varepsilon$ the assertion is proved.

If $p \neq \varepsilon$, we have $a \in \operatorname{Suff} \Pi$. Let then $\alpha, \alpha' \in B$ be such that $(g(\alpha))^\ell = a$ and $(g(\alpha'))^\ell = a'$; as $a \neq a'$, we have $\alpha \neq \alpha'$. Since $p\lambda'$ is a prefix of $g(\psi(w'x))$, $p \in \Pi^*$, and $p^\ell = (g(\alpha))^\ell = a$, by Lemma 1.12 one derives that $\alpha\tau$ is a factor of $\psi(w'x)$. Moreover, as $v'a'\lambda' \in \operatorname{Pref} g(t)$ and $v'a' \in \Pi^*$, we derive that $\alpha'\tau$ is a factor of t.

Let then δ' be any prefix of the directive word Δ of t', such that $\alpha'\tau \in$ Fact $\psi(w\delta')$. By Proposition 1.3, $\psi(w\delta'x)$ contains $\psi(w'x)$, and hence $\alpha\tau$, as a factor. Thus τ is a left special factor of $\psi(w\delta'x)$ and then of the standard episturmian word $\psi(w\delta'x^\omega)$; as $|\tau| < |\psi(w\delta')|$, it follows $\tau \in \text{Pref}\,\psi(w\delta')$ and then $\tau \in \text{Pref}\,t$, so that $\lambda' \in \text{Pref}\,g(t)$. The proof is now complete. $\square$

References

1. Lothaire, M.: Combinatorics on Words. Addison-Wesley, Reading (1983)
2. Lothaire, M.: Algebraic Combinatorics on Words. Cambridge University Press, Cambridge (2002)
3. Lothaire, M.: Applied Combinatorics on Words. Cambridge University Press, Cambridge (2005)
4. Droubay, X., Justin, J., Pirillo, G.: Episturmian words and some constructions of de Luca and Rauzy. Theoretical Computer Science 255, 539–553 (2001)
5. Justin, J., Pirillo, G.: Episturmian words and episturmian morphisms. Theoretical Computer Science 276, 281–313 (2002)
6. Fischler, S.: Palindromic prefixes and episturmian words. Journal of Combinatorial Theory, Series A 113, 1281–1304 (2006)
7. de Luca, A., De Luca, A.: Pseudopalindrome closure operators in free monoids. Theoretical Computer Science 362, 282–300 (2006)
8. Bucci, M., de Luca, A., De Luca, A., Zamboni, L.Q.: On some problems related to palindrome closure. Theoretical Informatics and Applications (to appear, 2008), doi:10.1051/ita:2007064
9. Bucci, M., de Luca, A., De Luca, A., Zamboni, L.Q.: On different generalizations of episturmian words. Theoretical Computer Science 393, 23–36 (2008)
10. Bucci, M., de Luca, A., De Luca, A., Zamboni, L.Q.: On θ-episturmian words. European Journal of Combinatorics (to appear, 2008)
11. Bucci, M., de Luca, A., De Luca, A.: Characteristic morphisms of generalized episturmian words. Preprint n. 18, DMA "R. Caccioppoli" (2008)
12. Berstel, J., Perrin, D.: Theory of Codes. Academic Press, New York (1985)
13. de Luca, A.: Sturmian words: structure, combinatorics, and their arithmetics. Theoretical Computer Science 183, 45–82 (1997)
14. Justin, J.: Episturmian morphisms and a Galois theorem on continued fractions. Theoretical Informatics and Applications 39, 207–215 (2005)

Universal Recursively Enumerable Sets of Strings

Cristian S. Calude[1], André Nies[2], Ludwig Staiger[3], and Frank Stephan[4,*]

[1] Department of Computer Science, The University of Auckland, Private Bag 92019,
Auckland, New Zealand
cristian@cs.auckland.ac.nz
[2] Department of Computer Science, The University of Auckland, Private Bag 92019,
Auckland, New Zealand
andre@cs.auckland.ac.nz
[3] Martin-Luther-Universität Halle-Wittenberg, Institut für Informatik,
D - 06099 Halle, Germany
staiger@informatik.uni-halle.de
[4] Department of Mathematics and School of Computing, National University of
Singapore, Singapore 117543
fstephan@comp.nus.edu.sg

Abstract. The present work clarifies the relation between domains of
universal machines and r.e. prefix-free supersets of such sets. One such
characterisation can be obtained in terms of the spectrum function $s_W(n)$
mapping n to the number of all strings of length n in the set W. An r.e.
prefix-free set W is the superset of the domain of a universal machine iff
there are two constants c, d such that $s_W(n) + s_W(n+1) + \ldots + s_W(n+c)$
is between $2^{n-H(n)-d}$ and $2^{n-H(n)+d}$ for all n; W is the domain of a
universal machine iff there is a constant c such that for each n the pair
of n and $s_W(n) + s_W(n+1) + \ldots + s_W(n+c)$ has at least the prefix-free
Description complexity n. There exists a prefix-free r.e. superset W of a
domain of a universal machine which is the not a domain of a universal
machine; still, the halting probability Ω_W of such a set W is Martin-Löf
random. Furthermore, it is investigated to which extend this results can
be transferred to plain universal machines.

1 Introduction

The present paper provides a classification of recursively enumerable prefix codes
using algorithmic information theory [1,4,5,6,9,10]. The paper combines recursion theoretic arguments with (combinatorial) information theory. It is well-
known that recursion theory does not yield a sufficiently fine distinction between
several classes of recursively enumerable prefix codes, as, for example, the prefix
code $S = \{0^n 1 : n \in W\}$ has the same complexity as the subset $W \subseteq N$ and all
these prefix codes are indistinguishable by their entropy.

---

* F. Stephan is supported in part by NUS grant number R252-000-308-112.

M. Ito and M. Toyama (Eds.): DLT 2008, LNCS 5257, pp. 170–182, 2008.

On the other hand one may assume that recursively enumerable prefix codes are in some sense "maximally complex" if they are the domains of universal prefix-free Turing machines. This observation is supported by Corollary 2 of [3] which states that every recursively enumerable prefix code is one-to-one embeddable into the domain of a universal prefix-free Turing machine by a partial recursive mapping increasing the output length at most by a constant. Moreover, this characterisation yields a connection to the information-theoretic density of a universal recursively enumerable prefix code. Calude and Staiger [3] showed that universal recursively enumerable prefix codes have maximal density — in contrast with the code S discussed above.

The present paper provides a more detailed characterisation of the domains of universal prefix-free Turing machines and universal recursively enumerable (r.e.) prefix codes in terms of the spectrum function. More technically, an r.e. prefix code S is an r.e. set of strings over the fixed alphabet $X = \{0,1\}$ such that there are no two strings $p, q \in S$ with p being a proper prefix of q. A machine U is a partial-recursive function from X^* to X^* and one defines for every x the Description complexity $C_U(x)$ based on U as $C_U(x) = \min\{|p| : U(p) \downarrow = x\}$. U is called universal if for every further machine V there is a constant c with $\forall x [C_U(x) \leq C_V(x) + c]$. In the case that the domain of U is prefix-free, one denotes the Description complexity based on U as H_U and calls U universal iff for every further machine V with prefix-free domain there is a constant c with $\forall x [H_U(x) \leq H_V(x) + c]$. A basic result of algorithmic information theory says that such universal machines exist [1,10]. In general, the underlying machine is fixed to some default and the complexities C (plain) and H (prefix-free) are written without any subscript [7]. Now a prefix code is called universal iff it is the superset of the domain of a prefix-free universal domain.

For a prefix-free set V, let Ω_V be (the set representing the binary course-of-values of the real number) $\sum_{p \in V} 2^{-|p|}$. Ω-numbers turned out to be left-r.e. (as immediate by the definition). Chaitin [4] proved that if V is the domain of a universal prefix-free machine, then Ω_V is Martin-Löf random. Here a set A is Martin-Löf random iff $H(A(0)A(1)A(2)\ldots A(n)) \geq n$ for almost all n. Calude, Hertling, Khoussainov and Wang [2] and Kučera and Slaman [11] showed that the converse is also true and every left-r.e. Martin-Löf random set corresponds to the halting probability of some universal machine. Later, Calude and Staiger [3] extended this work by considering the relations between domains of prefix-free machines and their r.e. prefix-free supersets. They established basic results and showed that such supersets cannot be recursive. In the present work, their results are extended as follows:

1. Let $s_W(n)$ denote the number of strings of length n in W and $s_W(n,m) = \sum_{i=n}^{n+m} s_W(i)$. A prefix-free r.e. set W is the superset of the domain of a prefix-free universal machine iff there is a constant c such that $s_W(n,c) \geq 2^{n-H(n)}$ for all n.

2. A prefix-free r.e. set W is the domain of some universal machine iff there exists a constant c such that $H(\langle n, s_W(n,c)\rangle) \geq n$ for all n.

3. There are prefix-free r.e. sets which satisfy the second but not the first condition; an example is any prefix-free r.e. set which has for almost all n that $s_W(n) = 2^{n-H(n)}$.
4. If W is an r.e. prefix-free superset of the domain of a universal machine U, then Ω_U is Solovay reducible to Ω_W, Ω_W is Martin-Löf random and W is wtt-complete.

To some extend, these results transfer to plain universal machines and their supersets as well. Furthermore, the question is investigated when an r.e. but not necessarily prefix-free set is a superset of the domain of a universal prefix-free machine. In particular the following natural question remained open: Is the domain of every plain universal machine the superset of the domain of some prefix-free universal machine? The reader should be reminded of the following additional notions used in this paper.

The ordering $\leq_{qlex}$ is called the quasi-lexicographical, length-lexicographical or military ordering of X^*: $\lambda <_{qlex} 0 <_{qlex} 1 <_{qlex} 00 <_{qlex} 01 <_{qlex} 10 <_{qlex} 11 <_{qlex} 000 <_{qlex} 001$ and so on. Furthermore, the sets of natural numbers $\mathbb{N}$ and strings X^* are identified by saying that $n \in \mathbb{N}$ represents the unique string x with $\#\{y \in X^* : y <_{qlex} x\} = n$. This is in particular useful in order to extend concepts like complexity to natural numbers without defining these concepts twice.

The function $a, b \mapsto \langle a, b \rangle$ is Cantor's pairing function of a and b: $\langle a, b \rangle = (a + b)(a + b + 1)/2 + b$.

A real number q is Solovay reducible to a real number r if there is an infinite approximation $a_0, a_1, a_2, \ldots$ of q from below such that there is some positive real constant $c > 0$ and some recursive approximation $b_0, b_1, b_2, \ldots$ of r from below such that $(a_{a+1} - a_s)c > b_{s+1} - b_s$ for all s. Similarly a set A is Solovay reducible to B if $\sum_{n \in A} 2^{-n}$ is Solovay reducible to $\sum_{n \in B} 2^{-n}$ as real numbers.

Further unexplained notation can be found in the books of Odifreddi [14], Calude [1] and Li and Vitányi [10].

2   Universal r.e. Prefix Codes

Recall that a *prefix-free universal machine* U is a prefix-free machine such that for every further machine V there is a constant c such that for every $p \in \text{dom}(V)$ there is a $q \in \text{dom}(U)$ with $U(q) = V(p) \wedge |q| \leq |p| + c$. Following [3], a *universal r.e. prefix code* $A \subset X^*$ is an r.e. prefix-free set containing the domain of a prefix-free universal machine. The major goal of this section is to clarify the relation between domains of prefix-free universal machines and universal r.e. prefix codes.

For every $V \subset X^*$, let the spectrum function $s_V : X^* \to \mathbb{N}$ be defined as $s_V(n) = \#(V \cap X^n)$ and $s_V(n, m) = \sum_{i=n}^{n+m} s_V(i)$. Furthermore, for machines U, $s_U(n)$ is just $s_{\text{dom}(U)}(n)$.

Theorem 1. *If U is a universal prefix-free machine then there exists a constant c such that $H(\langle n, s_U(n, c) \rangle) \geq n$ for all n.*

Proof. Assume by way of contradiction that this fails. Now choose c to be a multiple of 3 such that:

1. for every $p \in \mathrm{dom}(U)$ there is a $q \in \mathrm{dom}(U)$ with $|q| < |p| + c/3$ and $U(q) >_{qlex} U(p)$;
2. for every $p \in \mathrm{dom}(U)$, $H(U(p)) < H(p) + c/3$;
3. $H(p_n) \le H(\langle n, s_U(n,c) \rangle) + c/3$, where p_n is the quasi-lexicographically smallest string in $\mathrm{dom}(U)$ such that $n \le |p_n| \le n + c$ and $U(p_n) \ge_{qlex} U(q)$ for all $q \in \mathrm{dom}(U)$ with $n \le |q| \le n + c$.

Note that the third condition can be satisfied as there is a three-place partial-recursive function with inputs m, n and c with the following properties: this function simulates U until U has halted on a set R of m strings r with $n \le |r| \le n + c$ and it then outputs the length-lexicographic first $r' \in R$ for which $U(r')$ is length-lexicographically maximal: $U(r') \ge_{qlex} U(r)$ for all $r \in R$. The function terminates and outputs p_n in the case that $m = s_U(n,c)$. Now the complexity of the output of this two-place function is bounded by $H(\langle n, s_U(n,c) \rangle) + 2\log(c) + c'$, for some constant c' and hence for all sufficiently large c the third condition is satisfied.

Note that by the first item it holds that $U(q) <_{qlex} U(p_n)$ for all $q \in \mathrm{dom}(U)$ with $|q| \le n + 2c/3$. Hence $|p_n| \ge n + 2c/3$. By the second item it holds that $H(p_n) \ge n + c/3$. By the third item it then follows that $H(\langle n, s_U(n,c) \rangle) \ge n. \square$

Theorem 2. *There exists a prefix-free machine W and a universal prefix-free machine U such that $\mathrm{dom}(U) \subset \mathrm{dom}(W)$ and W is not universal.*

Proof. Let U be a universal prefix-free machine such that $\Omega_U < 1/2$. Now one can build by the Kraft-Chaitin Theorem a prefix-free set W such that for all n either $s_U(n) = s_W(n) = 0$ or there is a natural number m with $2^m \le s_U(n) < s_W(n) = 2^{m+1}$. As $s_U(n) \le s_W(n)$ for all n, one can make a partial-recursive one-one function f from $\mathrm{dom}(U)$ into W such that $|f(p)| = |p|$ for all $p \in \mathrm{dom}(U)$; this defines a further partial function from $f(U)$ to X^* by mapping $f(p) \mapsto U(p)$ for all $p \in \mathrm{dom}(U)$ which is a universal machine whose domain is a subset of W. It follows that W is a prefix-free superset of the domain of some universal function. Furthermore, for every constant c, the machine

$$n \mapsto H(\langle n, s_W(n,c) \rangle)$$

is logarithmic in n as for each value $s_W(m)$ has only $n+1$ many possible choices: either 0 or 2^m for some $m \in \{0, 1, \ldots, n\}$. Hence, by Theorem 1, the set W cannot be the domain of a prefix-free universal machine. $\square$

Although the complexity of a universal prefix code might not be large up to a given length n, the next result shows that the number

$$\Omega_W = \sum_{x \in W} 2^{-|x|}$$

is Martin-Löf random, a property shared with the domains of prefix-free universal machines. Note that there is no contradiction as for every left-r.e. real number $\rho > 0$ one can find a recursive prefix-free set W such that $\Omega_W = \rho$, see [2].

Theorem 3. *Let W be an r.e. universal prefix code. Then Ω_W is Martin-Löf random.*

Proof. Assume that U is a prefix-free universal machine whose domain is contained in the prefix-free r.e. set W. The basic idea of the proof is to show that Ω_U is Solovay reducible to Ω_W. This is done by approximating the halting probability of U such that $\Omega_{U,0} = 0$ and for every u one can compute a natural number k_u with $\Omega_{U,u+1} - \Omega_{U,u} = 2^{-k_u}$. Next one constructs a sequence $t_0, t_1, \ldots$ of integers such that there is a rational constant $\delta > 0$ with the property:

$$\forall u \left[\delta \cdot 2^{-k_u} \leq \Omega_{W,t_{u+1}} - \Omega_{W,t_u} \right].$$

This property is a reformulation of the fact that there is a Solovay-reduction from Ω_U to Ω_W. As Ω_U is Solovay-reducible to a left-r.e. set iff the latter is Martin-Löf random, the theorem follows once that δ is found [16].

The constant δ and the sequence $t_0, t_1, t_2, \ldots$ will come out of the following inductive construction: Using the Fixed-point Theorem, one can construct a r.e. prefix-free set V using a constant c such that for every $x \in V$ there is a $p \in \mathrm{dom}(U)$ with $U(p) = x \wedge |p| \leq |x| + c$. Now one defines V in stages:

1. An invariance of the construction is $\Omega_{V,u} = \Omega_{U,u}$ for all u.
2. The initialisation is $t_0 = 0$ and $V_0 = \emptyset$ which is consistent with the given invariance.
3. At stage u, assume that t_u, V_u and W_u are defined. Let k_u be the unique integer with
$$2^{-k_u} = \Omega_{U,u+1} - \Omega_{U,u}.$$

   Find a natural number m_u which is so large that $2|W_{t_u}| < 2^{m_u}$. By the Kraft-Chaitin Theorem one can select 2^{m_u} strings of length $k_u + m_u$ which are not yet in V_u and put them as new elements into V_{u+1}. This adds 2^{-k_u} to Ω_V giving

$$\Omega_{V,u+1} = \Omega_{V,u} + 2^{m_u} \cdot 2^{-k_u - m_u} = \Omega_{U,u} + 2^{-k_u} = \Omega_{U,u+1}.$$

   Furthermore, one can select t_{u+1} to be the first stage beyond t_u where for every string $x \in V_{u+1}$ there is an $y \in \mathrm{dom}(U_{t_{u+1}}) \cap W_{t_{u+1}}$ such that $|y| \leq |x| + c$ and $U(y) = x$; as at least half of these strings y had not been in W_{t_u} it follows that
$$\Omega_{W,t_{u+1}} - \Omega_{W,t_u} \geq 2^{-k_u - c - 1}.$$

4. The last equation of the activity at stage u permits to choose $\delta = 2^{-c-1}$.

Hence Ω_U is Solovay reducible to Ω_W and Ω_W is Martin-Löf random [16].   □

Theorem 4. *If W is an r.e. universal prefix code then there exist two constants c, d such that*
$$\forall n \left[2^{n-H(n)-d} \leq s_W(n,c) \leq 2^{n-H(n)+d} \right]. \tag{1}$$

Proof. It is well-known that for each r.e. prefix-free set there is a constant d' such that

$$\forall n \left[s_W(n) \leq 2^{n-H(n)+d'} \right] .$$

Therefore given c one can select d such that $d \geq d' + c + 2$ in order to get the inequality of the right hand side in (1). For the left hand side, take c so large that $\forall n \left[H(\langle n, s_U(n,c) \rangle) \geq n \right]$. The prefix-free machine V codes pairs $\langle n, m \rangle$ of natural numbers in a prefix-free way: $V(p0^e1q) = \langle n, m \rangle$ if $U(p) = n$, m is the binary value of q and $|q| = n - |p| - 2e$. Thus there is a constant c_V depending on the machine V such that $H(\langle n, m \rangle) \leq n - e + 1 + c_V$ for all $m < 2^{n-|p|-2e}$.

Since $\forall n \left[H(\langle n, s_U(n,c) \rangle) \geq n \right]$, it follows that $s_U(n,c) \leq 2^{n-H(n)-2e}$ can hold only for $e < c_V + 1$, that is, there is a maximal value e for which there are values of n with

$$s_U(n,c) \leq 2^{n-H(n)-2e} .$$

Taking now d to be the maximum of $c + d' + 2$ from above and $2e + 2$ from the current choice of e establishes this theorem.   □

If W is an r.e. universal prefix code, then one can use the constants c, d above to compute for every n the value $H(n)$ up to a constant error. It follows that one can find for every number n a number m with $H(m) > n$: one just takes that m below 4^n for which $m - \log(s_W(m,c))$ is maximal and the choice is right in all but finitely many places. Using Merkle's result on complex sets [8] or Arslanov's completeness criterion for weak truth-table reducibility in combination with the fact that W has r.e. dnr Turing degree [14], one obtains that W is wtt-complete.

Corollary 5. *If W is an r.e. universal prefix code then W is weak truth-table complete, that is, $\mathbb{K} \leq_{wtt} W$.*

The next result is the converse of Theorem 1 and had been deferred to this place as it builds on the above results. This permits to give a characterisation of the domains of prefix-free universal machines in terms of the complexity of the function $s_V(n,m)$. The constant c comes in as there are universal machines which use only programs of even length and so on.

Theorem 6. *Assume that W is an r.e. prefix-free set such that there is a constant c with $\forall n \left[H(\langle n, s_W(n,c) \rangle) \geq n \right]$. Then W is the domain of a universal prefix-free machine.*

Proof. Let c as fixed above. First note that there is a constant d such that

$$\forall n \left[H(\langle n, s_W(n,c) \rangle) \leq n + d \right] .$$

The reason is that there is a constant e such that

$$\forall n \left[s_W(n,c) \leq 2^{n-H(n)+e} \right] ,$$

by Theorem 4; hence one can code n with a program p having the length of $H(n)$ bits and then $s_W(n,c)$ given n with $n + e - |p|$ bits. The constant d might be a bit larger than e as one has to translate this coding into the language of the universal machine used.

Let $p_0, p_1, p_2, \ldots$ be a recursive one-one enumeration of the domain of some prefix-free universal machine U. Now one builds, using the Recursion Theorem, a recursive sequence $t_0, t_1, t_2, \ldots$ such that for some constant b the following holds for all s:

1. $H(s_{W_{t_s}}(m, c)) < |p_s| + (m + b - |p_s|)/2$ for all s and $m \geq |p_s|$.
2. For every s there is a string $q_s \in W_{t_{s+1}} - W_{t_s}$ with $|q_s| \leq |p_s| + b + c$.

Note that the first condition together with Theorem 1 implies that there exists a string q_s as desired in $W - W_{t_s}$. The second condition then allows us to choose t_{s+1} so large that the string q_s is actually in $W_{t_{s+1}}$.

Finally, one defines the following machine V defined on the domain of W: For any $q \in W$ find the unique s such that $q \in W_{t_{s+1}} - W_{t_s}$ and let $V(q) = U(p_s)$.

As $|q_s| \leq |p_s| + b + c$ and $q_s \in W_{t_{s+1}} - W_{t_s}$, it follows that $U(p_s)$ has a program at the machine V which is at most $b + c$ bits longer than p_s, hence V is a universal prefix-free machine with domain W. □

3   Plain Versus Prefix-Free Description Complexity

The main result of this section is the following theorem which parallels Theorems 1, 4 and 6 in the previous section for universal plain machines. Note that X^* would be a legitimate superset of the domain of a plain universal machine in the context of this section, as there are no such requirements like prefix-freeness.

Theorem 7. *Given an r.e. set W, the equivalences (1) $\Leftrightarrow$ (2) and (3) $\Leftrightarrow$ (4) hold for the following four conditions.*
(1) *There is a constant c such that $s_W(n, c) \geq 2^n$ for all n.*
(2) *W is the superset of a domain of a plain universal machine.*
(3) *There is a constant c with $C(s_W(n, c)) \geq n$ for all n.*
(4) *W is the domain of a plain universal machine.*

Proof. (1) $\Rightarrow$ (2): One can construct, for every n which is a multiple of $c + 1$ and uniformly recursive in n, a one-one mapping from $A_n = X^n \cup X^{n+1} \cup \ldots \cup X^{n+c}$ into W such that all $p \in A_n$ is mapped into $W \cap A_{n+c+1}$; these mappings just enumerate the first 2^{n+c+1} elements of $W \cap A_{n+c+1}$ and then map those in A_n in a one-one manner into these elements. This mapping has a partial-recursive and one-one inverse f whose domain is a subset of W and whose range is the full set X^*; note that $|f(p)| \geq |p| - 2c - 2$ for all p where $f(p)$ is defined.

If U is a plain universal machine, then the mapping $p \mapsto U(f(p))$ is also a plain universal machine with its domain being a subset of W; this completes the proof for case (1).

(2) $\Rightarrow$ (1): There is a constant c such that every string of length $n + 1$ has at most plain Description complexity $n + c$. At least half of these strings does not have plain Description complexity below n. Thus it follows that for at least half

of the 2^{n+1} strings x of length $n+1$ there is a $p \in W$ with $n \leq |p| \leq n+c$ and $U(p) = x$. Thus $s_W(n, c) \geq 2^n$.

$(3) \Rightarrow (4)$: Fix the number c and follow closely the proof of Theorem 6. First note that there is a constant d such that

$$\forall n \ [C(s_W(n, c)) \leq n + d].$$

Let $p_0, p_1, p_2, \ldots$ be a recursive one-one enumeration of the domain of a plain universal machine U. Now one builds, using the Recursion Theorem, some recursive sequence $t_0, t_1, t_2, \ldots$ such that for some constant b the following holds for all s:

1. $C(s_{W_{t_s}}(m, c)) < |p_s| + (m + b - |p_s|)/2$, for all s and $m \geq |p_s|$.
2. For every s there is a string $q_s \in W_{t_{s+1}} - W_{t_s}$ with $|q_s| \leq |p_s| + b + c$.

Note that the first condition together with Theorem 1 imply that there exists a string q_s as desired in $W - W_{t_s}$; by virtue of the second condition one can choose t_{s+1} so large that the string q_s is actually in $W_{t_{s+1}}$.

Now the following machine V is defined on the domain of W: For any $q \in W$ find the unique s such that $q \in W_{t_{s+1}} - W_{t_s}$ and let $W(q) = U(p_s)$.

As $|q_s| \leq |p_s| + b + c$ and $q_s \in W_{t_{s+1}} - W_{t_s}$, it follows that $U(p_s)$ has a program for the machine V which is at most $b + c$ bits longer than p_s, hence V is a plain universal machine with domain W.

$(4) \Rightarrow (3)$: Let U be the universal machine with domain W. For each n, let x_n be that string in W which is enumerated last into $W \cap (X^0 \cup X^1 \cup X^2 \cup \ldots \cup X^n)$. Note that one can compute from x_n and $(n - |x_n|)/2$ a string y_n of length n which is not in W; taking s to be the first number with $x_n \in W_s$, y_n is just the length lexicographic first string of X^n which is outside the set $\{U(p) : p \in W_s \wedge |p| < n\}$. On the one hand, one has that

$$C(y_n) \leq C(x_n) + (n - |x_n|)/2 + c' \leq |x_n| + (n - |x_n|)/2 + c'',$$

for some constants c', c'' and all n; on the other hand one has that $C(y_n) \geq n$. It follows that $|x_n| \geq n - 2c''$ and $C(x_n) \geq n - c' - c''$ for all n.

Assume now by way of contradiction that for every $c > c' + c''$ there exists an n_c with $C(s_W(n_c, c)) < n_c$. Then it follows that $C(x_{n_c+c}) \leq n_c + c/2 + c'''$ for some constant c''' and all $c > c' + c''$. To see this, note that one can code this $s_W(n_c, c)$ by a string u. Furthermore, one can code x_{n_c+c} by a string of the form $a1^b0^{b'}1u$ where $a \in \{0, 1\}$, $c = 2b + a$ and $b' = |n_c| - |u| > 0$. Now one can compute a, b, b', u from $a1^b0^{b'}1u$ and has that $n_c = |u| + b'$ and $c = 2b + a$. Afterwards one can compute $s_W(n_c, c)$ from u and has that x_{n_c+c} is the string number $s_W(n_c, c)$ among those strings enumerated into W which have at least length n_c and at most length $n_c + c$. Hence, as said above, $C(x_{n_c+c}) \leq n_c + c/2 + c'''$ for some constant c''', the value of c''' depends then on the translation of the description $a1^b0^{b'}1u$ into the universal machine on which C is based. Hence $n_c + c - c' - c'' \leq n_c + c/2 + c'''$ and $c/2 \leq c' + c'' + c'''$, a contradiction to the assumption that c

could take any value greater than $c' + c''$. Thus there is a $c > c' + c''$ for which n_c does not exist and it follows for this c that $\forall n \, [C(s_W(n, c)) \geq n]$. This completes the proof.   □

A consequence of Theorem 7 is that the compressible strings (for the plain Description complexity) form a domain of a universal machine.

Corollary 8. *Let $W = \{p \in X^* : C(p) < |p|\}$. Then there is a universal plain machine with domain W.*

Proof. Let C_s be an approximation of the complexity C from above and let U be the underlying plain universal machine. Now define a machine V on input of the form $0^i 1^j 0p$ as follows:

1. Let $n = |p| + i + 1$.
2. Determine $m = U(p)$.
3. If m is found, search for the first stage s such that there are at least m strings in the set $\{q : n \leq |q| \leq n + 2j \wedge C_s(q) < |q|\}$.
4. If m, s are found, let $V(0^i 1^j 0p) = r$ be the lexicographic first string of length $n + 2j$ with $C_s(r) \geq |r|$.

Note that $V(0^i 1^j 0p)$ is defined iff the second and third step of this algorithm terminate. There is a constant d such that

$$\forall i, j > 0 \; \left[C(V(0^i 1^j 0p)) < i + j + |p| + d \right].$$

Let $c = 2d$ and assume by way of contradiction that there is a number n with $C(s_W(n, c)) < n$. Then there would be a p with $|p| < n$ and $U(p) = s_W(n, c)$. Let $i = n - |p| - 1$ and let $j = d$. By construction, $V(0^i 1^j 0p)$ is a string of length $n + c$ not in W and

$$C(V(0^i 1^j 0p)) \leq i + j + |p| + d = n + c - 1 < n + c.$$

These two facts contradict together the definitions of c, d and W. Hence W is the domain of a universal machine by Theorem 7.   □

It is easy to see that the domain of a plain universal machine cannot be the subset of any prefix-free set. But the converse question is more interesting. The first theorem gives some minimum requirement on the function s_V.

Theorem 9. *Assume that V is the superset of the domain of a prefix-free universal machine. Then either there is a constant c such that $s_V(n, c) \geq 2^n$ for all n or the Turing degree of s_V is that of the halting problem.*

Proof. Let V be an r.e. superset of the domain of the universal machine U and assume that for every constant c there is a natural number n with $s_V(n, c) < 2^n$.

Now one defines a further prefix-free machine W as follows: for every $p \in dom(U)$, let t be the time the computation of $U(p)$ needs to converge and let n be the first number such that $s_{V,t}(n, 4|p|) < 2^n$. Now let $W(q) = q$ for all $q \in \{p\} \cdot X^{n+|p|}$

By definition, there is a constant c such that for every q in the domain of W there is an r in the domain of U with $U(r) = q$ and $|r| \le |q| + c$. It follows that $s_U(n, 4|p|) \ge 2^{|p|+n} - 2^n \ge 2^n$ for all $p \in \mathrm{dom}(U)$ with $|p| > c$. Hence there is a string of length up to $4|p| + n$ in $V - V_s$.

Now $\mathrm{dom}(U) \le_T V$ by the following algorithm: on input p, search the first n such that $s_V(n, 4|p|) < 2^n$. This number exists by assumption on V. Then determine the time t such that $V_t(q) = V(q)$ for all q with $|q| \le n + 4|p|$ — this can be done easily relative to the oracle V. If $U(p)$ is defined within t steps then output "$p \in \mathrm{dom}(U)$" else output "$p \notin \mathrm{dom}(U)$". It can easily be verified that the whole knowledge needed about V is only the values of s_V and $s_{V,t}$, hence one has even that $\mathrm{dom}(U) \le_T s_V$.   □

Note that for each constant c the set $\{0^c p : |p| \text{ is a multiple of } c\}$ is a superset of the domain of some universal prefix-free machine; hence the "either-condition" Theorem 9 cannot be dropped. The next result shows that the "or-condition" is not sufficient to guarantee that some subset is the domain of a prefix-free universal machine.

Theorem 10. *Let V be an r.e. set such that for every c there is an n with $s_V(n, c) < 2^n$. Then there is an r.e. set V' with $s_V = s_{V'}$ such that V' does not contain the domain of any prefix-free universal machine.*

Proof. The central idea is to construct by induction relative to the halting problem a sequence $p_0, p_1, p_2, \ldots$ of strings such that each p_{e+1} extends p_e and $p_e \in W_e$ whenever this can be satisfied without violating the extension-condition. Furthermore, the set V' is constructed such that for each length n one enumerates $s_V(n)$ many strings of length n into V' and chooses each string $w \in X^n$ such that w is different from the strings previously enumerated into V' and one satisfies that w extends the approximations $p_{0,n}, p_{1,n}, \ldots, p_{e,n}$ of $p_0, p_1, \ldots, p_e$ for the largest possible e which can be selected.

For any fixed e it holds for almost all n that $p_{e,n} = p_e$ and that $s_V(n) \le 2^{n-|p_e|}$ implies that all members of $V' \cap X^n$ extend p_e. By assumption there is for each constant $c > |p_e|$ a sufficiently large n such that $s_{V,4c} < 2^n$ and all members of V' of length $n + c, n + c + 1, \ldots, n + 4c$ extend p_e. Assume now that W_e is the domain of a universal machine. Then, for one of these constants c the corresponding n has in addition the property that there is a member of W_e of between length $n + c$ and $n + 2c$. If this member of W_e is not in V' then W_e is not a subset of V'. If this member of W_e is in V' then it is an extension of p_e and by the way p_e is chosen it follows that also $p_e \in W_e$, a contradiction to the assumption that W_e is prefix-free. Hence none of the W_e is a subset of V' and the domain of a prefix-free universal machine.   □

The previous result is contrasted by the following example.

Example 11. *Assume that V is an r.e. set (not prefix-free) such that there is a real constant $c > 0$ with $s_V(n) \cdot 2^{-n} > c$ for all n and assume that f is a recursive function with $\sum_n 2^{-n} f(n) < c$. Then there is a prefix-free recursive subset $W \subseteq V$ with $s_W(n) = f(n)$ for all n.*

The set W can be constructed by simply picking, for $n = 0, 1, 2, 3, \ldots$, exactly $f(n)$ strings of length n out of V which do not extend previously picked shorter strings.

The main question remains which conditions on s_V guarantee that V has a subset which is the domain of a prefix-free universal machine. In the light of Theorem 10 a necessary condition is that $\exists c \forall n \, [s_V(n, c) \geq 2^n]$. One might ask whether this condition is also sufficient. By Theorem 7 this condition characterises the supersets of plain universal machines; hence one can restate the question as follows.

Open Problem 12. *Is the domain of every plain universal machine a superset of the domain of a prefix-free universal machine?*

4   Discussion

The major goal was to investigate, which prefix-free r.e. sets of strings is a universal prefix code [3], that is, a superset of the domain of a universal machine. The result is that these sets V can be characterised using the function of finite sum of the spectrum function s_V: roughly speaking, $s_V(n, c)$ has to be near to $2^{n-H(n)}$. The reason is that there are universal machines having only strings of even length and so forth. Furthermore, universal prefix codes and domains of universal machines share the property that their halting probability is Martin-Löf random. But it could also be shown that not all universal prefix codes are the domain of a universal machine: while there is a universal prefix code for which $s_V(n) = 2^{n-H(n)}$ for all n, no domain of a universal machine has this property. The reason is that for such a domain there is a constant c such that $H(s_V(n, c))$ is near to n.

A further interesting question is to characterise those r.e. sets in general which are a superset of the domain of a prefix-free universal machine. Combining of Theorem 10 with the fact that $s_U(n) \cdot 2^{-n}$ goes to 0 for n to ∞ for any prefix-free machine U, one can deduce that this characterisation cannot depend on s_V alone, but also on the way the strings are placed. It remains an interesting open problem whether every r.e. set V satisfying $\exists c \forall n \, [s_V(n, c) \geq 2^n]$ contains the domain of a universal prefix-free machine. Note that this question is equivalent to asking whether the domain of every plain universal machine is a superset of the domain of some prefix-free universal machine.

Furthermore, there are various definitions of universality and this paper is based on that definition where one says that U is universal if the Description complexity based on U cannot be improved by more than a constant. The most prominent alternative notion says that U *is universal by adjunction* or *prefix-universal* if for every further machine V there is a finite string q such that $U(qp) = V(p)$ for all $p \in \text{dom}(V)$. Universality by adjunction is quite restrictive and one cannot characterise in terms of the spectrum function s_W when a prefix-free set W is the domain of a machine which is universal by adjunction; however, this is done for normal universal machines in Theorems 1 and 6. Nevertheless, due to the more restrictive nature, prefix-free machines which are universal by adjunction have the property

$$\exists c \, \forall n \, [H(s_U(n)) \geq n - H(n) - c].$$

This property is more natural as the one in Theorem 1. Hence, it is easy to obtain machines which are universal but not universal by adjunction. An example would be a machine U obtained from V such that for all $p \in \mathrm{dom}(V)$, $U(p0) = U(p1) = V(p)$ if $|p|$ is odd and $U(p) = V(p)$ if $|p|$ is even; it is easy to see that U inherits prefix-freeness and universality from V. Calude and Staiger [3, Fact 5] provide more information about this topic.

As the topic of the paper are mostly supersets of domains of universal machines, one could ask what can be said about the r.e. subsets of such domains. Indeed, these subsets are easy to characterise: A prefix-free r.e. set $V \subseteq X^*$ is the subset of the domain of a prefix-free universal machine iff there is a string p such that no q comparable to p is in V; an r.e. set $V \subseteq X^*$ is the subset of the domain of a plain universal machine iff there is a constant c such that $s_{X^*-V}(n,c) \geq 2^n$ for all n. Note that a subset of the domain of a prefix-free machine is also the subset of the domain of a plain universal machine, but not vice versa. Indeed, every prefix-free subset of X^* is the subset of the domain of a plain universal machine.

Acknowledgment. The authors would like to thank Wang Wei for discussions on the topic of this paper.

References

1. Calude, C.S.: Information and Randomness: An Algorithmic Perspective, 2nd edn., Revised and Extended. Springer, Berlin (2002)
2. Calude, C.S., Hertling, P.H., Khoussainov, B., Wang, Y.: Recursively enumerable reals and Chaitin Ω numbers. Theoretical Computer Science 255, 125–149 (2001)
3. Calude, C.S., Staiger, L.: On universal computably enumerable prefix codes. Mathematical Structures in Computer Science (accepted)
4. Chaitin, G.J.: A theory of program size formally identical to information theory. Journal of the Association for Computing Machinery 22, 329–340 (1975)
5. Chaitin, G.J.: Information-theoretic characterizations of recursive infinite strings. Theoretical Computer Science 2, 45–48 (1976)
6. Chaitin, G.J.: Algorithmic information theory. IBM Journal of Research and Development 21, 350–359+496 (1977)
7. Downey, R., Hirschfeldt, D., LaForte, G.: Randomness and reducibility. Journal of Computer and System Sciences 68, 96–114 (2004)
8. Kjos-Hanssen, B., Merkle, W., Stephan, F.: Kolmogorov Complexity and the Recursion Theorem. In: Durand, B., Thomas, W. (eds.) STACS 2006. LNCS, vol. 3884, pp. 149–161. Springer, Heidelberg (2006)
9. Kolmogorov, A.N.: Three approaches to the definition of the concept "quantity of information". Problemy Peredachi Informacii 1, 3–11 (1965)
10. Li, M., Vitányi, P.: An Introduction to Kolmogorov Complexity and Its Applications, 2nd edn. Springer, Heidelberg (1997)

11. Kučera, A., Slaman, T.: Randomness and recursive enumerability. SIAM Journal on Computing 31, 199–211 (2001)
12. Martin-Löf, P.: The definition of random sequences. Information and Control 9, 602–619 (1966)
13. Nies, A.: Computability and Randomness. Oxford University Press, Oxford (to appear)
14. Odifreddi, P.: Classical Recursion Theory. North-Holland, Amsterdam (1989)
15. Schnorr, C.P.: Process complexity and effective random tests. Journal of Computer and System Sciences 7, 376–388 (1973)
16. Solovay, R.: Draft of paper on Chaitin's work Unpublished notes, 215 pages (1975)

Algorithmically Independent Sequences

Cristian S. Calude[1,*] and Marius Zimand[2,**]

[1] Department of Computer Science, University of Auckland, New Zealand
www.cs.auckland.ac.nz/ ~cristian
[2] Department of Computer and Information Sciences, Towson University,
Baltimore, MD, USA
http://triton.towson.edu/~mzimand

Abstract. Two objects are independent if they do not affect each other. Independence is well-understood in classical information theory, but less in algorithmic information theory. Working in the framework of algorithmic information theory, the paper proposes two types of independence for arbitrary infinite binary sequences and studies their properties. Our two proposed notions of independence have some of the intuitive properties that one naturally expects. For example, for every sequence x, the set of sequences that are independent with x has measure one. For both notions of independence we investigate to what extent pairs of independent sequences, can be effectively constructed via Turing reductions (from one or more input sequences). In this respect, we prove several impossibility results. For example, it is shown that there is no effective way of producing from an arbitrary sequence with positive constructive Hausdorff dimension two sequences that are independent (even in the weaker type of independence) and have super-logarithmic complexity. Finally, a few conjectures and open questions are discussed.

1 Introduction

Intuitively, two objects are independent if they do not affect each other. The concept is well-understood in classical information theory. There, the objects are random variables, the information in a random variable is its Shannon entropy, and two random variables X and Y are declared to be independent if the information in the join (X, Y) is equal to the sum of the information in X and the information in Y. This is equivalent to saying that the information in X conditioned by Y is equal to the information in X, with the interpretation that, on average, knowing a particular value of Y does not affect the information in X.

The notion of independence has been defined in algorithmic information theory as well for finite strings [Cha82]. Our approach is very similar. This time the information in a string x is the complexity (plain or prefix-free) of x,

---

* Calude was supported in part by UARC Grant 3607894/9343 and CS-PBRF Grant.
** Zimand was supported by NSF grant CCF 0634830. Part of this work was done while visiting the CDMTCS of the University of Auckland, New Zealand.

M. Ito and M. Toyama (Eds.): DLT 2008, LNCS 5257, pp. 183–195, 2008.

and two strings x and y are independent if the information in the join string $\langle x, y \rangle$ is equal to the sum of the information in x and the information in y, up to logarithmic (or, in some cases, constant) precision.

The case of infinite sequences (in short, sequences) has been less studied. An inspection of the literature reveals that for this setting, independence has been considered to be synonymous with pairwise relative randomness, i.e., two sequences x and y are said to be independent if they are (Martin-Löf) random relative to each other (see [vL90, DH]). As a consequence, the notion of independence is confined to the situation where the sequences are random.

The main objective of this paper is to put forward a concept of independence that applies to *all* sequences. One can envision various ways for doing this. One possibility is to use Levin's notion of mutual information for sequences [Lev84] (see also the survey paper [GV04]) and declare two sequences to be independent if their mutual information is small. If one pursues this direction, the main issue is to determine the right definition for "small." We take another approach, which consists in extending in the natural way the notion of independence from finite strings to sequences. This leads us to two concepts: *independence* and *finitary-independence*. We say that (1) two sequences x and y are independent if, for all n, the complexity of $x{\restriction}n$ (the prefix of x of length n) and the complexity of $x{\restriction}n$ relativized with y are within $O(\log n)$ (and the same relation holds if we swap the roles of x and y), and (2) two sequences x and y are finitary-independent if, for all n and m, the complexity of $x{\restriction}n$ and the complexity of $x{\restriction}n$ given $y{\restriction}m$ are within $O(\log n + \log m)$ (and the same relation holds if we swap the roles of x and y). We have settled for the additive logarithmical term of precision (rather than some higher accuracy) since this provides robustness with respect to the type of complexity (plain or prefix-free) and other technical advantages.

We establish a series of basic facts regarding the proposed notions of independence. We show that independence is strictly stronger than finitary-independence. The two notions of independence apply to a larger category of sequences than the family of random sequences, as intended. However, they are too rough for being relevant for computable sequences. It is not hard to see that a computable sequence x is independent with any other sequence y, simply because the information in x can be obtained directly. In fact, this type of trivial independence holds for a larger type of sequences, namely for any H-trivial sequence, and trivial finitary-independence holds for any sequence x whose prefixes have logarithmic complexity. It seems that for this type of sequences (computable or with very low complexity) a more refined definition of independence is needed (perhaps, based on resource-bounded complexity). We show that the two proposed notions of independence have some of the intuitive properties that one naturally expects. For example, for every sequence x, the set of sequences that are independent with x has measure one.

We next investigate to what extent pairs of independent, or finitary-independent sequences, can be effectively constructed via Turing reductions. For example, is there a Turing reduction f that given oracle access to an arbitrary sequence x produces a sequence that is finitary-independent with x? Clearly, if we allow the output of f to be a computable sequence, then the answer is positive by the type of trivial finitary-independence that we have noted above. We show that if we insist that the output of f has super-logarithmic complexity whenever x has positive constructive Hausdorff dimension, then the answer is negative. In the same vein, it is shown that there is no effective way of producing from an arbitrary sequence x with positive constructive Hausdorff dimension two sequences that are finitary-independent and have super-logarithmic complexity.

Similar questions are considered for the situation when we are given two (finitary-) independent sequences. It is shown that there are (finitary-) independent sequences x and y and a Turing reduction g such that x and $g(y)$ are not (finitary-)independent. This appears to be the only counter-intuitive effect of our definitions. Note that the definition of constructive Hausdorff dimension (or of partial randomness) suffers from the same problem. For example, there exist a sequence x with constructive Hausdorff dimension 1 and a computable g such that $g(x)$ has constructive Hausdorff dimension $\leq 1/2$. It seems that if one wants to extend the notion of independence to non random sequences (in particular to sequences that have arbitrary positive constructive Hausdorff dimension) such counter-intuitive effects cannot be avoided. On the other hand, for any independent sequences x and y and for any Turing reduction g, x and $g(y)$ are finitary-independent.

We also raise the question on whether given as input finitely many (finitary-) independent sequences it is possible to effectively build a new sequence that is (finitary-) independent (in a non-trivial way) with each sequence in the input. It is observed that the answer is positive if the sequences in the input are random, but for other types of sequences the question remains open. The same issue can be raised regarding finite strings and for this case a positive answer is obtained. Namely, it is shown that given three independent finite strings x, y and z with linear complexity, one can effectively construct a new string that is independent with each of x, y and z, has high complexity and its length is a constant fraction of the length of x, y and z.

Because of space limitations, this extended abstract contains no proof. All proofs are available in the full version of the paper [CZ07].

1.1 Preliminaries

Let $\mathbb{N}$ denote the set of non-negative integers; the size of a finite set A is denoted $||A||$. Unless stated otherwise, all numbers are in $\mathbb{N}$ and all logs are in base 2. We work with the binary alphabet $\{0,1\}$. A string is an element of $\{0,1\}^*$ and a sequence is an element of $\{0,1\}^\infty$. If x is a string, $|x|$ denotes its length; xy denotes the concatenation of the strings x and y. If x is a string or a sequence, $x(i)$ denotes the i-th bit of x and $x{\restriction}n$ is the substring $x(1)x(2)\cdots x(n)$. For two sequences x and y, $x \oplus y$ denotes the sequence $x(1)y(1)x(2)y(2)x(3)y(3)\cdots$ and x XOR y denotes the sequence $(x(1)$ XOR $y(1))(x(2)$ XOR $y(2))(x(3)$ XOR $y(3))\cdots$, where $(x(i)$ XOR $y(i))$ is the sum modulo 2 of the bits $x(i)$ and $y(i)$. We identify a sequence x with the set $\{n \in \mathbb{N} \mid x(n) = 1\}$. We say that a sequence x is computable (computably enumerable, or c.e.) if the corresponding set is computable (respectively, computably enumerable, or c.e.). If x is c.e., then for every $s \in \mathbb{N}$, x_s is the sequence corresponding to the set of elements enumerated within s steps by some (given) machine M that enumerates x. We also identify a sequence x with the real number in the interval $[0, 1]$ whose binary writing is $0.x(1)x(2)\cdots$. A sequence x is said to be left c.e. if the corresponding real number x is the limit of a computable increasing sequence of rational numbers. The plain and the prefix-free complexities of a string are defined in the standard way; however we need to provide a few details regarding the computational models. The machines that we consider process information given in three forms: (1) the input, (2) the oracle set, (3) the conditional string. Correspondingly, a universal machine has 3 tapes: (i) one tape for the input and work, (ii) one tape for storing the conditional string, (iii) one tape (called the oracle-query tape) for formulating queries to the oracle.

The oracle is a string or a sequence. If the machine enters the query state and the value written in binary on the oracle-query tape is n, then the machine gets the n-th bit in the oracle, or if n is larger than the length of the oracle, the machine enters an infinite loop.

We fix such a universal machine U. The notation $U^w(u \mid v)$ means that the input is u, the conditional string is v and the oracle is w, which is a string or a sequence. The plain complexity of a string x given the oracle w and the conditional string v is $C^w(x \mid v) = \min\{|u| \mid U^w(u \mid v) = x\}$. There exists a constant c such that for every x, v and w $C^w(x \mid v) < |x| + c$.

A machine is prefix-free (self-delimiting) if its domain is a prefix-free set. There exist universal prefix-free machines. We fix such a machine U; the prefix-free complexity of a string x given the oracle w and the conditional string v is $H^w(x \mid v) = \min\{|u| \mid U^w(u \mid v) = x\}$.

In case w or v are the empty strings, we omit them in $C(\cdot)$ and $H(\cdot)$. Throughout this paper we use the $O(\cdot)$ notation to hide constants that depend only on the choice of the universal machine underlying the definitions of the complexities C and H. There are various equivalent definitions for (algorithmic) random sequences as defined by Martin-Löf [ML66] (see [C02]). In what follows we will use the (weak) complexity-theoretic one [Cha75] using the prefix-free complexity: A sequence x is Martin-Löf random (in short, random) if there is a constant c such that for every n, $H(x{\upharpoonright}n) \geq n - c$. The set of random sequences has constructive (Lebesgue) measure one [ML66]. The sequence x is random relative to the sequence y if there is a constant c such that for every n, $H^y(x{\upharpoonright}n) \geq n - c$.

The constructive Hausdorff dimension of a sequence x—which is the direct effectivization of "classical Hausdorff dimension"—defined by $\dim(x) = \liminf_{n\to\infty} C(x{\upharpoonright}n)/n$ $(= \liminf_{n\to\infty} H(x{\upharpoonright}n)/n)$, measures intermediate levels of randomness (see [Rya84, Sta93, Tad02, May02, Lut03, Rei04], [Sta05, CST06, DHNT06]).

A Turing reduction f is an oracle Turing machine; $f(x)$ is the language computed by f with oracle x, assuming that f halts on all inputs when working with oracle x (otherwise we say that $f(x)$ does not exist). In other words, if $n \in f(x)$ then the machine f on input n and oracle x halts and outputs 1 and if

$n \notin f(x)$ then the machine f on input n and oracle x halts and outputs 0. The function use is defined as follows: $use_f^x(n)$ is the index of the rightmost position on the tape of f accessed during the computation of f with oracle x on input n. The Turing reduction f is a *wtt-reduction* if there is a computable function q such that $use_f^x(n) \le q(n)$, for all n. The Turing reduction f is a *truth-table reduction* if f halts on all inputs for every oracle. A truth-table reduction is a wtt-reduction.

2 Defining Independence

Two objects are independent if none of them contains significant information about the other one. Thus, if in some formalisation, $I(x)$ denotes the information in x and $I(x \mid y)$ denotes the information in x given y, x and y are independent if $I(x) - I(x \mid y)$ and $I(y) - I(y \mid x)$ are both small. In this paper we work in the framework of algorithmic information theory. In this setting, in case x is a string, $I(x)$ is the complexity of x (where for the "complexity of x" there are several possibilities, the main ones being the plain complexity or the prefix-free complexity).

The independence of strings was studied in [Cha82]: two strings are independent if $I(xy) \approx I(x) + I(y)$. This approach motivates our Definition 1 and Definition 2.

In case x is an infinite sequence, the information in x is characterised by the sequence $(I(x{\restriction}n))_{n \in \mathbb{N}}$ of information in the initial segments of x. For the information upon which we condition (e.g., the y in $I(x \mid y)$), there are two possibilities: either the entire sequence is available in the form of an oracle, or we consider initial segments of it. Accordingly, we propose two notions of independence.

Definition 1. (The "integral" type of independence) *Two sequences x and y are* independent *if $C^x(y{\restriction}n) \ge C(y{\restriction}n) - O(\log n)$ and $C^y(x{\restriction}n) \ge C(x{\restriction}n) - O(\log n)$.*

Definition 2. (The finitary type of independence) *Two sequences x, y are* finitary-independent *if for all natural numbers n and m,*

$$C(x{\restriction}n \; y{\restriction}m) \ge C(x{\restriction}n) + C(y{\restriction}m) - O(\log(n) + \log(m)).$$

Remark 1. We will show in Proposition 1, that the inequality in Definition 2 is equivalent to saying that for all n and m, $C(x{\restriction}n \mid y{\restriction}m) \ge C(x{\restriction}n) - O(\log n + \log m)$, which is the finite analogue of the property in Definition 1 and is in line with our discussion above.

Remark 2. If x and y are independent, then they are also finitary-independent (Proposition 2). The converse is not true (Corollary 1).

Remark 3. The proposed definitions use the plain complexity $C(\cdot)$, but we could have used the prefix-free complexity as well, because the two types of complexity are within an additive logarithmic term. Also, in Definition 2 (and throughout this paper), we use concatenation to represent the joining of two strings. However, since any reasonable pairing function $\langle x, y \rangle$ satisfies $|\ |\langle x, y \rangle| - |xy|\ | < O(\log |x| + \log |y|)$, it follows that $|C(< x, y >) - C(xy)| < O(\log |x| + \log |y|)$, and thus any reasonable pairing function could be used instead.

Remark 4. A debatable issue is the subtraction of the logarithmic term. Indeed, there are other natural possibilities. We argue that our choice has certain advantages over other possibilities that come to mind.

Let us focus on the definition of finitary-independence. We want $C(x{\restriction}n\ y{\restriction}m) \geq C(x{\restriction}n) + C(y{\restriction}n) - O(f(x) + f(y))$, for all n, m, where f should be some "small" function. We would like the following two properties to hold:

(A) the sequences x and y are finitary-independent iff $C(x{\restriction}n \mid y{\restriction}m) > C(x{\restriction}n) - O(f(x{\restriction}n) + f(y{\restriction}m))$, for all n and m,

(B) if x is "somewhat" random and $y = 00 \cdots 000 \cdots$, then x and y are finitary-independent.

Other natural possibilities for the definition could be:
(i) if $f(x) = C(|x|)$, the definition of finitary independence–(i) would be:

$$C(x{\restriction}n\ y{\restriction}m) \geq C(x{\restriction}n) + C(y{\restriction}m) - O(C(n) + C(m)),$$

or (ii) if $f(x) = \log C(x)$, the definition of finitary-independence–(ii) would be:

$$C(x{\restriction}n\ y{\restriction}m) \geq C(x{\restriction}n) + C(y{\restriction}m) - O(\log C(x{\restriction}n) + \log C(y{\restriction}m)).$$

If sequences x and y satisfy (i), or (ii), then they also satisfy Definition 2.

Variant (i) implies (B), but not(A) (for example, consider sequences x and y with $C(n) << \log C(x{\restriction}n)$ and $C(m) << \log C(y{\restriction}m)$, for infinitely many n and m). Variant (ii) implies (A), but does not imply (B) (for example if for infinitely many n, $C(x{\restriction}n) = O(\log^3 n)$; take such a value n, let p be a shortest description of $x{\restriction}n$, and let m be the integer whose binary representation is $1p$. Then $x{\restriction}n$ and $0^{\omega}{\restriction}m$, do not satisfy (B)). The proposed definition implies both (A) and (B).

Another advantage is the robustness discussed in Remark 3.

Remark 5. If the sequence x is computable, then x is independent with every sequence y. In fact a stronger fact holds. A sequence is called H-trivial if, for all n, $H(x{\restriction}n) \leq H(n) + O(1)$. This is a notion that has been intensively studied recently (see [DHNT06]). Clearly every computable sequence is H-trivial, but the converse does not hold [Zam90, Sol75]. If x is H-trivial, then it is independent with every sequence y. Indeed, $H^y(x{\restriction}n) \geq H(x{\restriction}n) - O(\log n)$, because $H(x{\restriction}n) \leq H(n) + O(1) \leq \log n + O(1)$, and $H^x(y{\restriction}n) \geq H(y{\restriction}n) - O(\log n)$, because, in fact, $H^x(y{\restriction}n)$ and $H(y{\restriction}n)$ are within a constant of each other [Nie05]. The same inequalities hold if we use the $C(\cdot)$ complexity (see Remark 3).

For the case of finitary-independence, a similar phenomenon holds for a (seemingly) even larger class.

Definition 3. *A sequence* x *is called* C*-logarithmic if* $C(x{\restriction}n) = O(\log n)$.

It can be shown (for example using Proposition 1, (a)) that if x is C-logarithmic, then it is finitary-independent with every sequence y.

Note that every sequence x that is the characteristic sequence of a c.e. set is C-logarithmic. This follows from the observation that, for every n, the initial segment $x{\restriction}n$ can be constructed given the number of 1's in $x{\restriction}n$ (an information which can be written with $\log n$ bits) and the finite description of the enumerator of the set represented by x. If a sequence is H-trivial then it is C-logarithmic, but the converse probably does not hold.

In brief, the notions of independence and finitary-independence are relevant for strings having complexity above that of H-trivial sequences, respectively C-logarithmic sequences. The cases of independent (finitary-independent) pairs (x, y), where at least one of x and y is H-trivial (respectively, C-logarithmic) will be referred to as *trivial independence*.

Remark 6. Some desirable properties of the independence relation are:

P1. Symmetry: x is independent with y iff y is independent with x.
P2. Robustness under type of complexity (plain or prefix-free).
P3. If f is a Turing reduction, except for some special cases, x and $f(x)$ are dependent ("independence cannot be created").
P4. For every x, the set of sequences that are dependent with x is small (i.e., it has measure zero).

Clearly both the independence and the finitary-independence relations satisfy P1. They also satisfy P2, as we noted in Remark 3. It is easy to see that the independence relation satisfies P3, whenever we require that the initial segments of x and $f(x)$ have plain complexity $\omega(\log n)$ (because $C^x(f(x){\restriction}n) = O(\log n)$, while $C(f(x){\restriction}n) = \omega(\log n)$). We shall see that the finitary-independence relation satisfies P3 under some stronger assumptions for f and $f(x)$ (see Section 4.1 and in particular Theorem 6). Theorem 3 shows that the (finitary-) independence relation satisfies P4.

2.1 Properties of Independent and Finitary-Independent Sequences

The following simple properties of finitary-independent sequences are technically useful in some of the next proofs.

Proposition 1. (a) *Two sequences* x *and* y *are finitary-independent iff for all* n *and* m, $C(x{\restriction}n \mid y{\restriction}m) \geq C(x{\restriction}n) - O(\log n + \log m)$.
(b) *Two sequences* x *and* y *are finitary-independent iff for all* n, $C(x{\restriction}n\, y{\restriction}n) \geq C(x{\restriction}n) + C(y{\restriction}n) - O(\log(n))$.
(c) *Two sequences* x *and* y *are finitary-independent iff for all* n, $C(x{\restriction}n \mid y{\restriction}n) \geq C(x{\restriction}n) - O(\log(n))$.
(d) *If* x *and* y *are not finitary-independent, then for every constant* c *there are infinitely many* n *such that* $C(x{\restriction}n\, y{\restriction}n) < C(x{\restriction}n) + C(y{\restriction}n) - c\log n$.

(e) *If x and y are not finitary-independent, then for every constant c there are infinitely many n such that $C(x{\upharpoonright}n \mid y{\upharpoonright}n) < C(x{\upharpoonright}n) - c\log n$.*

Proposition 2. *If the sequences x and y are independent, then they are also finitary-independent.*

Proposition 3. *If $\dim(x) = \sigma$ and the sequences (x,y) are finitary-independent, then $\dim(x \text{ XOR } y) \geq \sigma$.*

Proposition 4. *(a) If x is random and the sequences (x,y) are finitary-independent, then $(y, x \text{ XOR } y)$ are finitary-independent.*
(b) If x is random and (x,y) are independent, then $(y, x \text{ XOR } y)$ are independent.

Proposition 5. *There are sequences x, y, and z such that (x,y) are independent, (x,z) are independent, but $(x, y \oplus z)$ are not finitary-independent.*

In Remark 5, we have listed several types of sequences that are independent or finitary-independent with any other sequence. The next result goes in the opposite direction: it exhibits a pair of sequences that can not be finitary-independent (and thus not independent).

Proposition 6. [Ste07] *If x and y are left c.e. sequences, $\dim(x) > 0$, and $\dim(y) > 0$, then x and y are not finitary-independent.*

3   Examples of Independent and Finitary-Independent Sequences

We give examples of pairs of sequences that are independent or finitary-independent (other than the trivial examples from Remark 5).

Theorem 1. *Let x be a random sequence and let y be a sequence that is random relative to x. Then x and y are independent.*

Theorem 2. *Let x be an arbitrary sequence and let y be a sequence that is random relative to x. Then x and y are finitary-independent.*

As expected, most pairs of sequences are independent (and thus also finitary-independent).

Theorem 3. *For every x, the set $\{y \mid y \text{ independent with } x\}$ has (Lebesgue) measure one.*

4   Effective Constructions of Finitary-Independent Sequences

The examples of (finitary-) independent sequences provided so far are existential (i.e., non-constructive). In this section we investigate to what extent it is possible to effectively construct such sequences. We show some impossibility results and therefore we focus on the weaker type of independence, finitary-independence. Informally speaking, we investigate the following questions:

Question (a). Is it possible to effectively construct from a sequence x another sequence y finitary-independent with x, where the finitary-independence is not trivial (recall Remark 5)? This question has two variants depending on whether we seek a uniform procedure (i.e., one procedure that works for all x), or whether we allow the procedure to depend on x.

Question (b). Is it possible to effectively construct from a sequence x two sequences y and z that are finitary-independent, where the finitary-independence is not trivial? Again, there are uniform and non-uniform variants of this question.

We analyse these questions in Section 4.1. Similar questions for the case when the input consists of two sequences x_1 and x_2 are discussed in Section 4.2.

4.1   If We Have One Source

We first consider the uniform variant of Question (a): Is there a Turing reduction f such that for all $x \in \{0,1\}^*$, $(x, f(x))$ are finitary-independent? We even relax the requirement and demand that f should achieve this objective only if x has positive constructive Hausdorff dimension (this only makes the following impossibility results stronger).

As noted above, the question is interesting if we require $f(x)$ to have some "significant" amount of randomness whenever x has some "significant" amount of randomness. The answer should be negative, because, intuitively, one should not be able to produce independence (this is property P3 in Remark 6).

We consider two situations depending on two different meanings of the concept of "significant" amount of randomness.

Case 1: We require that $f(x)$ is not C-logarithmic. We do not solve the question, but we show that every reduction f that potentially does the job must have non-polynomial *use*.

Proposition 7. *Let f be a Turing reduction. For every sequence x, if the function $use_f^x(n)$ is polynomially bounded, then x and $f(x)$ are not finitary-independent, unless one of them is C-logarithmic.*

Case 2: We require that $f(x)$ has complexity just above that of C-logarithmic sequences (in the sense below). We show that in this case, the answer to the uniform variant of Question (a) is negative: there is no such f.

Definition 4. *A sequence x is C-superlogarithmic if for every constant $c > 0$, $C(x{\restriction}n) > c \log n$, for almost every n.*

The next theorems in this section are based on results from [NR06], [BDS07], and [Zim08].

We proceed to the impossibility results related to **Case 2**. To simplify the structure of quantifiers in the statement of the following result, we posit here the following task for a function f mapping sequences to sequences:

TASK A: for every $x \in \{0,1\}^\infty$ with $\dim(x) > 0$, the following should hold: (a) $f(x)$ exists, (b) $f(x)$ is C-superlogarithmic, (c) x and $f(x)$ are finitary-independent.

Theorem 4. *There is no Turing reduction f that satisfies TASK A.*

We next consider the uniform variant of Question (b). We posit the following task for two functions f_1 and f_2 mapping sequences to sequences:

TASK B: for every $x \in \{0,1\}^\infty$ with $\dim(x) > 0$, the following should hold: (a) $f_1(x)$ and $f_2(x)$ exist, (b) $f_1(x)$ and $f_2(x)$ are C-superlogarithmic, (c) $f_1(x)$ and $f_2(x)$ are finitary-independent.

Theorem 5. *There are no Turing reductions f_1 and f_2 satisfying TASK B.*

The non-uniform variants of Questions (a) and (b) remain open. In the particular case when f is a wtt-reduction, we present impossibility results analogous to those in Theorem 4 and Theorem 5.

Theorem 6. *For all rational $\sigma \in (0,1)$, there exists $\dim(x) = \sigma$ such that for every wtt-reduction f, at least one of the following statements (a), (b), (c) holds true: (a) $f(x)$ does not exist, (b) $f(x)$ is not finitary-independent with x, (c) $f(x)$ is not C-superlogarithmic.*

Theorem 7. *For all rational $\sigma \in (0,1)$, there exists x with $\dim(x) = \sigma$ such that for every wtt-reductions f_1 and f_2, at least one of the following statements (a), (b), (c) holds true: (a) $f_1(x)$ does not exist or $f_2(x)$ does not exist, (b) $f_1(x)$ and $f_2(x)$ are not finitary-independent, (c) $f_1(x)$ is not C-superlogarithmic or $f_2(x)$ is not C-superlogarithmic.*

4.2   If We Have Two Sources

We have seen some limits on the possibility of constructing a finitary-independent sequences starting from one sequence. What if we are given two finitary-independent sequences: is it possible to construct from them more finitary-independent sequences?

First we observe that if x and y are two (finitary-) independent sequences and g is an arbitrary Turing reduction, then it does not necessarily follow that x and $g(y)$ are (finitary-) independent (as one may expect). On the other hand, if x and y are independent, it does follow that x and $g(y)$ are finitary-independent.

Proposition 8. *(a) [Ste07] There are two independent sequences x and y and a Turing reduction g such that x and $g(y)$ are not independent.*

(b) There are two finitary-independent sequences x and y and a Turing reduction g such that x and $g(y)$ are not finitary-independent.

Proposition 9. *If x and y are independent, and g is a Turing reduction, then x and $g(y)$ are finitary-independent (provided $g(y)$ exists).*

Corollary 1. *There are sequences that are finitary-independent but not independent.*

By Proposition 8, we see that (finitary-) independence is not preserved by computable functions. However, we note that there exists a simple procedures that, starting with a finitary-independent pair (x, y), produces a new pair of finitary-independent sequences. Namely, the pair (x, y_{odd}) is finitary-independent. Another question is whether given a pair of (finitary-)independent strings (x, y), it is possible to effectively produce another sequence that is (finitary-)independent with both x and y. The answer is positive in the case when x and y are both random. Indeed, if x and y are random and independent (respectively finitary-independent), then x XOR y is independent (respectively, finitary-independent) with both x and y. The similar question for non-random x and y remains open. (See Section 4.3 for some results for finite strings).

4.3   Producing Independence: The Finite Case

In what follows we attack the question on whether it is possible to effectively produce an object which is independent to each of several given independent objects for the simpler case of strings. In this setting we are able to give a positive answer for the situation when we start with three[1] input strings that are independent (and not necessarily random). First we define the analogue of independence for strings.

Definition 5. *Let $c \in \mathbb{R}^+$ and $k \in \mathbb{N}$. We say that strings $x_1, x_2, \ldots, x_k$ in $\{0, 1\}^*$ are c-independent if*

$$C(x_1 x_2 \ldots x_k) \geq C(x_1) + C(x_2) + \ldots + C(x_k) - c(\log |x_1| + \log |x_2| + \ldots + \log |x_k|).$$

The main result of this section is the following theorem, whose proof draws from the techniques of [Zim08].

Theorem 8. *For all constants $\sigma > 0$ and $\sigma_1 \in (0, \sigma)$, there exists a computable function $f : \{0, 1\}^* \times \{0, 1\}^* \times \{0, 1\}^* \to \{0, 1\}^*$ with the following property: For every $c \in \mathbb{R}^+$ there exists $c' \in \mathbb{R}^+$ such that if the input consists of a triplet of c-independent strings having sufficiently large length n and plain complexity at least $\sigma \cdot n$, then the output is c'-independent with each element in the input triplet and has length $\lfloor \sigma_1 n \rfloor$.*
  More precisely, if

(i) *(x, y, z) are c-independent,*
(ii) *$|x| = |y| = |z| = n$, and*
(iii) *$C(x) \geq \sigma \cdot n, C(y) \geq \sigma \cdot n, C(z) \geq \sigma \cdot n$,*

then, provided n is large enough, the following pairs of strings $(f(x, y, z), x)$, $(f(x, y, z), y)$, $(f(x, y, z), z)$ are c'-independent, $|f(x, y, z)| = \lfloor \sigma_1 n \rfloor$, and $C(f(x, y, z)) \geq \lfloor \sigma_1 n \rfloor - O(\log n)$.

---

[1] The case when the input consists of two independent strings remains open.

Acknowledgments

We are grateful to André Nies and Frank Stephan for their insightful comments. In particular, Definition 1 has emerged after several discussions with André, and Proposition 6 and Proposition 8, (a) are due to Frank [Ste07]. We also thank Jan Reimann for his assistance in modifying a result from [NR06], which was needed for Theorem 6 and Theorem 7. We thank Alexander Shen for suggesting Theorem 3 and Proposition 8, (b).

References

[BDS07]   Bienvenu, L., Doty, D., Stephan, F.: Constructive dimension and weak truth-table degrees. In: Cooper, S.B., Löwe, B., Sorbi, A. (eds.) CiE 2007. LNCS, vol. 4497, Springer, Heidelberg (to appear, 2007) Available as Technical Report arXiv:cs/0701089 ar arxiv.org

[C02]   Calude, C.S.: Information and Randomness: An Algorithmic Perspective, Revised and Extended, 2nd edn. Springer, Berlin (2002)

[CST06]   Calude, C., Staiger, L., Terwijn, S.: On partial randomness. Annals of Pure and Applied Logic 138, 20–30 (2006)

[CZ07]   Calude, C.S., Zimand, M.: Algorithmically Independent Sequences. CDMTCS Research Report 317, 25 (2008)

[Cha75]   Chaitin, G.: A theory of program size formally identical to information theory. Journal of the ACM 22, 329–340 (1975)

[Cha82]   Chaitin, G.: Gödel's theorem and information. International Journal of Theoretical Physics 21, 941–954 (1982)

[DH]   Downey, R., Hirschfeldt, D.: Algorithmic Randomness and Complexity. Springer, Heidelberg (to be published)

[DHNT06] Downey, R., Hirschfeldt, D., Nies, A., Terwijn, S.: Calibrating randomness. The Bulletin of Symbolic Logic 12(3), 411–492 (2006)

[GV04]   Grünwald, P., Vitanyi, P.: Shannon information and Kolmogorov complexity, 2004. CORR Technical report arxiv:cs.IT/0410002, revised (May 2006)

[Kau03]   Kautz, S.M.: Independence properties of algorithmically random sequences, CORR Technical Report arXiv:cs/0301013 (2003)

[Lev84]   Levin, L.: Randomness conservation inequalities: information and independence in mathematical theories. Information and Control 61(1) (1984)

[Lut03]   Lutz, J.: The dimensions of individual strings and sequences. Information and Control 187, 49–79 (2003)

[May02]   Mayordomo, E.: A Kolmogorov complexity characterization of constructive Hausdorff dimension. Information Processing Letters 84, 1–3 (2002)

[ML66]   Martin-Löf, P.: The definition of random sequences. Information and Control 9, 602–619 (1966)

[Nie05]   Nies, A.: Lowness properties and randomness. Advances in Mathematics 197, 274–305 (2005)

[NR06]   Nies, A., Reimann, J.: A lower cone in the wtt degrees of non-integral effective dimension. In: Proceedings of IMS workshop on Computational Prospects of Infinity, Singapore (to appear, 2006)

[Rei04]   Reimann, J.: Computability and fractal dimension, Technical report, Universität Heidelberg, Ph.D. thesis (2004)

[Rya84]   Ryabko, B.: Coding of combinatorial sources and Hausdorff dimension. Doklady Akademii Nauk SSR 277, 1066–1070 (1984)

[Sol75]   Solovay, R.: Draft of a paper (or series of papers) on Chaitin's work, unpublished manuscript, IBM Thomas J. Watson Reserach Center, p. 215 (1975)

[Sta93]   Staiger, L.: Kolmogorov complexity and Hausdorff dimension. Inform. and Comput. 103, 159–194 (1993)

[Sta05]   Staiger, L.: Constructive dimension equals Kolmogorov complexity. Information Processing Letters 93, 149–153 (2005)

[Ste07]   Stephan, F.: Email communication (May 2007)

[Tad02]   Tadaki, K.: A generalization of Chaitin's halting probability Ω and halting self-similar sets. Hokkaido Math. J. 31, 219–253 (2002)

[vL90]   van Lambalgen, M.: The axiomatization of randomness. The Journal of Symbolic Logic 55, 1143–1167 (1990)

[Zam90]   Zambella, D.: On sequences with simple initial segments, ILLC Technical Report ML 1990-05, University of Amsterdam (1990)

[Zim08]   Zimand, M.: Two sources are better than one for increasing the Kolmogorov complexity of infinite sequences. Proceedings of CSR 2008, Moscow (June 2008) (Also available as CORR Techical Report. arXiv:0705.4658)

[ZL70]   Zvonkin, A., Levin, L.: The complexity of finite objects and the development of the concepts of information and randomness by means of the theory of algorithms. Russian Mathematical Surveys 25(6), 83–124 (1970)

Relationally Periodic Sequences and Subword Complexity

Julien Cassaigne[1], Tomi Kärki[2], and Luca Q. Zamboni[3,4]

[1] CNRS, Institut de Mathématiques de Luminy,
Case 907, 163 avenue de Luminy, 13288 Marseille Cedex 9, France
[2] Department of Mathematics and Turku Centre for Computer Science,
University of Turku, 20014 Turku, Finland
[3] Université de Lyon, Université Lyon 1, CNRS UMR 5208 Institut Camille Jordan,
Bâtiment du Doyen Jean Braconnier, 43, blvd du 11 novembre 1918,
F-69622 Villeurbanne Cedex, France
[4] Reykjavik University, School of Computer Science,
Kringlan 1, 103 Reykjavik, Iceland

Abstract. By the famous theorem of Morse and Hedlund, a word is ultimately periodic if and only if it has bounded subword complexity, *i.e.*, for sufficiently large n, the number of factors of length n is constant. In this paper we consider relational periods and relationally periodic sequences, where the relation is a similarity relation on words induced by a compatibility relation on letters. We investigate what would be a suitable definition for a relational subword complexity function such that it would imply a Morse and Hedlund-like theorem for relationally periodic words. We consider strong and weak relational periods and two candidates for subword complexity functions.

Keywords: Period, subword complexity, similarity relation, relational period, partial word.

1 Introduction

Let w be an infinite word, *i.e.*, an infinite sequence of letters from a finite alphabet $\mathcal{A}$. One way to describe the structure of w is to consider its complexity function. Define the subword complexity $c_w(n)$ to be the number of different blocks of letters of length n occurring in w. In "very random" sequences, every block of letters of length n appears with a frequency asymptotic to $1/|\mathcal{A}|^n$. In other words, these sequences have maximal complexity $c_w(n) = |\mathcal{A}|^n$. Similarly, ultimately periodic sequences are also characterized by their subword complexity. In fact, in the first half of the 20th century, Morse and Hedlund [5] proved the following theorem:

Theorem 1 (Morse, Hedlund, 1940). *Let w be an infinite word and let k be the number of different letters occurring in w. The following properties are equivalent:*

M. Ito and M. Toyama (Eds.): DLT 2008, LNCS 5257, pp. 196–205, 2008.
© Springer-Verlag Berlin Heidelberg 2008

(i) w *is ultimately periodic,*
(ii) $c_w(n) = c_w(n+1)$ *for some* n,
(iii) $c_w(n) < n + k - 1$ *for some* $n \geq 1$,
(iv) $c_w(n)$ *is bounded.*

In other words, ultimately periodic sequences are precisely those sequences having bounded complexity. Using the notion of complexity we may also define another very well studied set of binary infinite words, namely Sturmian words. These words have remarkable properties and their subword complexity satisfies $c_w(n) = n + 1$; see [7, Chapter 2].

In this paper we investigate sets of infinite words which are ultimately periodic "up to a given similarity relation". A *similarity relation* is a relation on words induced by a compatibility relation on letters. Three types of so called *relational periods*, were introduces in [3] and [4], where their properties with respect to the theorem of Fine and Wilf were considered. In this paper we concentrate on strong and weak relational periods. Our goal is to characterize ultimately relationally periodic words using a suitable definition for relational subword complexity. It is not clear what would be a good definition of such complexity giving a Morse and Hedlund-like theorem. We discuss this problem and show by examples why some natural candidates are not suitable.

2   Similarity Relations

Let $R \subseteq X \times X$ be a relation on a set X. We usually write $x \, R \, y$ instead of $(x, y) \in R$. The identity relation on X is denoted by ι_X and the universal relation is denoted by Ω_X. For a subset Y of X, we define $R_Y = R \cap (Y \times Y)$. The relation R is a *compatibility relation* if it is both reflexive and symmetric, *i.e.*, (i) $\forall x \in X : x \, R \, x$, and (ii) $\forall x, y \in X : x \, R \, y \implies y \, R \, x$. In this presentation we consider special kind of relations on words defined in the following way.

Definition 1. *Let A be an alphabet. A relation on words over A is called a similarity relation if its restriction on letters is a compatibility relation and, for any words $u = u_1 \cdots u_m$ and $v = v_1 \cdots v_n$ $(u_i, v_j \in A)$, the relation R satisfies*

$$u_1 \cdots u_m \, R \, v_1 \cdots v_n \iff m = n \text{ and } u_i \, R \, v_i \text{ for all } i = 1, 2, \ldots, m .$$

The restriction of R on letters, denoted by R_A, is called the generating relation of R. Words u and v satisfying $u \, R \, v$ are said to be R-similar or R-compatible.

Since a similarity relation R is induced by its restriction on letters, it can be presented by listing all pairs $\{a, b\}$ $(a \neq b)$ such that $(a, b) \in R_A$. We use the notation

$$R = \langle r_1, \ldots, r_n \rangle ,$$

where $r_i = (a_i, b_i) \in A \times A$ for $i = 1, 2, \ldots, n$, to denote that R is the similarity relation generated by the symmetric closure of $\iota_A \cup \{r_1, \ldots, r_n\}$. For example, let $A_\diamond = \{a, b, \diamond\}$ and set $R = \langle (a, \diamond), (b, \diamond) \rangle$. Then we have

$$R_{A_\diamond} = \iota_{A_\diamond} \cup \{(a, \diamond), (\diamond, a), (b, \diamond), (\diamond, b)\} .$$

For example, we have $ab\diamond aa\,R\,abb\diamond a$ but, for instance, $ab\diamond$ and $\diamond a\diamond$ are not R-similar. Note that the relation R corresponds to the compatibility relation of binary partial words with a "do not know"-symbol $\diamond$. Partial words were introduced by Berstel and Boasson in [1]. For more on partial words, see [2].

More on properties of similarity relations can be found in [6]. For example, the connection between similarity relations and the compatibility relation of partial words is discussed in detail.

3   Relational Periods

For a compatibility relation on letters (and for the corresponding similarity relation on words) we will now define two *relational periods*.

Definition 2. *Let R be a compatibility relation on an alphabet $\mathcal{A}$ and denote the ith letter of a finite word x over $\mathcal{A}$ by x_i. An integer $p \geq 1$ is*

(i) a strong R-period of x if, for all $i, j \in \{1, 2, \ldots, |w|\}$, we have

$$i \equiv j \pmod{p} \implies x_i\,R\,x_j ,$$

(ii) a weak R-period of x if, for all $i \in \{1, 2, \ldots, |w| - p\}$, we have $x_i\,R\,x_{i+p}$.

This definition can be generalized naturally to infinite words by letting i and j be any positive integers. We want to point out that in the literature strong R-periods are sometimes called *global* and weak R-periods are called *local* as is the case of strong and weak periods of partial words. However, the term local period has also another meaning and is therefore discarded here. Note also that the definition of a (normal) period of a word coincides with the definitions of both the strong and the weak R-period where $R = \iota_{\mathcal{A}}$.

Next let us consider an example which shows that the above defined two types of relational periods are different. For instance, they can have different minimal periods.

Example 1. Let $\mathcal{A} = \{a, b, c, d\}$, $R = \langle (a, b), (b, c), (c, d), (d, a) \rangle$ and denote $x = babbbcbd$. Clearly, the minimal (normal) period is 8. By the definition of R, we see that 2 is a weak R-period of x. Since $(x_7, x_8) = (b, d) \notin R$, 1 is not a weak period and therefore the smallest weak R-period is 2. Since $(b, d) \notin R$, the minimal strong R-period must be at least 6. Indeed, it is exactly 6 because of the relation $a\,R\,d$.

By the definition, it is evident that a strong R-period of a word x is also a weak R-period of x, whereas the previous example shows that the converse statement does not hold in general. Moreover, it is easy to prove that the above definitions of strong and weak relational periods coincide if the compatibility relation R is also transitive, in other words, if R is an equivalence relation. Note that in this case relational periods can be seen as normal periods by replacing the letters with the corresponding equivalence classes.

In the sequel we will examine infinite words which are relationally periodic at least after a finite prefix.

Definition 3. *An infinite word* $x = x_1 x_2 x_3 \cdots$ *is ultimately strongly (resp. weakly) R-periodic if for some integers n and p the suffix* $x_n x_{n+1} x_{n+2} \cdots$ *has a strong (resp. weak) R-period p. The set of all ultimately strongly R-periodic infinite words is denoted by* $\mathcal{S}_R$ *and the set of all ultimately weakly R-periodic infinite words is denoted by* $\mathcal{W}_R$.

By the above considerations it is clear that $\mathcal{S}_R \subseteq \mathcal{W}_R$. Our goal is to characterize these sets using some suitable complexity functions. More precisely, we would like to define a function $c_{R,w} \colon \mathbb{N} \to \mathbb{N}$ such that a word w belongs to $\mathcal{S}_R$ (resp. to $\mathcal{W}_R$) if and only if $c_{R,w}(n)$ is bounded. This kind of result would be a generalization of Theorem 1 for relational periods. In the next section we will define two candidates and show that they do not entirely satisfy our objectives, but give some insight into tackling the problem.

4   Relational Subword Complexity

A word u is a *factor* (or a *subword*) of a word v, if there exist words x and y such that $v = xuy$. The set of factors of length n of a word u, is denoted by $F_w(n)$. A meaningfull relational subword complexity function should somehow describe the number of relationally different factors of the word. It is evident that the usual subword complexity function $c_w(n) = |F_w(n)|$ is not suitable for this purpose. For example, any word with maximal subword complexity $c_w(n) = |\mathcal{A}|^n$ is strongly $\Omega_\mathcal{A}$-periodic, since all factors of same length are compatible with each other.

In this section we consider two more suitable candidates for the relational subword complexity function. It is easiest to defined these functions using graphs of subwords.

Definition 4. *A relational subword graph* $G_{R,w}(n)$ *is a graph, where the set of of vertices is* $F_w(n)$ *and there is an edge between vertices u and v if u and v are R-compatible.*

Our two candidates for the relational subword complexity function are the following:

1. Let $c_{R,w}^T(n)$ be the number of connected components of $G_{R,w}(n)$. Note that the number of connected components is the number of R^T-equivalence classes of $F_w(n)$, where R^T denotes the transitive closure of R.
2. Let $c_{R,w}^I(n)$ be the maximal cardinality of a set of pairwise incompatible elements of $F_w(n)$. Note that a set with maximal number of pairwise incompatible elements of $F_w(n)$ is a maximal independent set of $G_{R,w}(n)$ and this set need not be unique.

Denote the set of infinite words over the alphabet $\mathcal{A}$ by $\mathcal{A}^\omega$. For the above complexity functions, we next define two subsets of $\mathcal{A}^\omega$ with bounded R-relational subwords complexity, where the relation R is a compatibility relation on the alphabet:

$$\mathcal{T}_R = \{w \in \mathcal{A}^\omega \mid \exists B \in \mathbb{N} : c_{R,w}^T(n) < B\}$$

and
$$\mathcal{I}_R = \{w \in \mathcal{A}^\omega \mid \exists B \in \mathbb{N} : c^I_{R,w}(n) < B\} \,.$$

Recall that $\mathcal{S}_R$ is the set of ultimately strongly R-periodic words and $\mathcal{W}_R$ is the set of ultimately weakly R-periodic words. Unfortunately, it turns out that all the sets $\mathcal{S}_R$, $\mathcal{W}_R$, $\mathcal{T}_R$ and $\mathcal{I}_R$ can be different. However, the following theorem reveals how the above defined complexity functions are connected to the ultimately relationally periodic words.

Theorem 2. *For any similarity relation R, we have $\mathcal{S}_R \subseteq \mathcal{I}_R \subseteq \mathcal{W}_R \subseteq \mathcal{T}_R$. Moreover, if R is not transitive, then $\mathcal{S}_R \neq \mathcal{I}_R \neq \mathcal{W}_R \neq \mathcal{T}_R$. Otherwise, all the sets $\mathcal{S}_R, \mathcal{I}_R, \mathcal{W}_R, \mathcal{T}_R$ are equal.*

We divide the proof of the theorem into several small lemmata.

Lemma 1. *If an infinite word w is ultimately weakly R-periodic, then $c^T_{R,w}(n)$ is bounded ($\mathcal{W}_R \subseteq \mathcal{T}_R$).*

Proof. Let w be an ultimately weakly R-periodic infinite word. Suppose that p is a weak R-period of a suffix w' of w. Then there are at most p connected components in $G_{R,w'}(n)$ for all $n \geq 1$. Namely, by the definition of a weak R-period, all letters occurring in positions congruent to m (mod p) are R^T-related. Hence, all factors in $F_{w'}(n)$ starting from congruent positions modulo p belong to the same connected component. Thus, $c^T_{R,w'}(n) \leq p$ and, consequently, $c^T_{R,w}(n)$ is bounded, since $w = uw'$ for some finite word u. $\qquad\square$

Lemma 2. *There exist a relation R and a word w such that w is not ultimately weakly R-periodic but the function $c^T_{R,w}(n)$ is bounded ($\mathcal{W}_R \neq \mathcal{T}_R$).*

Proof. Consider an infinite word $v = (v_j)_{j\geq 1}$ over the infinite alphabet $\mathbb{N} = \{0, 1, 2, \ldots\}$ defined by
$$v_j = \max\{k \mid j \equiv 0 \pmod{2^k}\} \,.$$

Define $u = \varphi(v)$, where $\varphi(k) = ab^k c$ for each $k \in \mathbb{N}$ and b^k denotes the concatenation of k letters b. Set $R = \langle(a, b), (b, c)\rangle$.

Now $G_{R,u}(n)$ is connected, since $b^n \in F_u(n)$ for every n and b^n is R-similar with all factors of length n. On the other hand, u is not ultimately weakly R-periodic. Namely, if we assume that p is a weak R-period of some suffix u' of u, this contradicts with the fact that the word $ab^{p-1}c$ occurs infinitely often in u'. Hence, p cannot be a weak R-period of any suffix u'. $\qquad\square$

Lemma 3. *If the subword complexity function $c^I_{R,w}(n)$ is bounded, then w is ultimately weakly R-periodic ($\mathcal{I}_R \subseteq \mathcal{W}_R$).*

Proof. Let $w = (w_j)_{j\geq 1}$ be an infinite word and assume that $c^I_{R,w}(n) \leq B$ for some natural number B. Consider factors $w^{(i_n)} = w_{i_n} w_{i_n+1} \cdots w_{i_n+n-1}$ for $i_n = 1, 2, \ldots, B + 1$ and for some $n \geq 1$. At least two of them are R-similar by the assumption. Suppose that these words are $w^{(r_n)}$ and $w^{(s_n)}$, where $r_n < s_n$.

Hence, we have $w_{r_n+j} R w_{s_n+j}$ for all $j = 0, 1, 2, \ldots, n-1$ and, consequently, $p_n = s_n - r_n$ is a weak R-period of the finite word $w_{r_n} w_{r_n+1} \cdots w_{s_n+n-1}$.

We may assume that r_n is the smallest position and s_n is the smallest position with respect to r_n satisfying $w^{(r_n)} R w^{(s_n)}$. Thus, the sequence $(p_n)_{n\geq 1}$ is uniquely defined. Since $1 \leq p_n \leq B$ for all n, some number l occurs infinitely often in $(p_n)_{n\geq 1}$. Consider the infinite set of indices I such that $p_i = l$ for all $i \in I$. Since also $1 \leq r_n \leq B$ for all n, some number r occurs infinitely often in $(r_i)_{i\in I}$. This means that there exist arbitrarily long prefixes of $w' = (w_j)_{j\geq r}$ which have l as a weak R-period. Hence, l is a weak R-period of w', since w' is the limit of these prefixes. Thus, w is ultimately weakly R-periodic.   □

Lemma 4. *There exists a relation R and a word w such that w is ultimately weakly R-periodic but the function $c^I_{R,w}(n)$ is not bounded ($\mathcal{I}_R \neq \mathcal{W}_R$).*

Proof. Consider the morphism $\phi\colon \{a, c\}^* \to \{a, c\}^*$, where $\phi(a) = ac$ and $\phi(c) = a$. The word

$$f = (f_j)_{j\geq 1} := \lim_{m\to\infty} \phi^m(a) = acaacacaacaac\cdots$$

obtained by iterating the morphism ϕ is called the *Fibonacci word*. We modify this word by adding a letter b in front of every letter of f. Hence, we obtain the word $u = (u_j)_{j\geq 1}$ such that $u_{2j} = f_j$ and $u_{2j-1} = b$. Clearly, 1 is a weak R-period of u for $R = \langle(a, b), (b, c)\rangle$. However, $c^I_{R,w}(n)$ is not bounded. Consider the factors of u of even length n starting from the even positions. Since Fibonacci word is a Sturmian sequence, there are exactly $n+1$ different words of length n in f. By the definition of u, all the above mentioned factors have b's in even positions and a's or c's in odd positions. Hence, there are exactly $n/2+1$ different factors of even length n starting from the even positions of u and they are all pairwise R-incompatible, since $(a, c) \notin R$. Thus, we have $u \notin \mathcal{I}_R$.   □

Lemma 5. *If an infinite word w is ultimately strongly periodic, then $c^I_{R,w}(n)$ is bounded ($\mathcal{S}_R \subseteq \mathcal{I}_R$).*

Proof. Consider an infinite word $w = uw'$, where u is a finite word and p is a strong R-period of w'. Then any set of pairwise R-incompatible factors of w' can contain at most p words. Namely, consider a pairwise R-incompatible subset of $F_{w'}(n)$ with $p+1$ elements. By the pigeon hole principle, at least two elements u and v in the set are such that $u = w_i w_{i+1} \cdots w_{i+n-1}$ and $v = w_j w_{j+1} \cdots w_{j+n-1}$ for some i and j satisfying $i \equiv j \pmod{p}$. Hence, by the definition of a strong R-period, these words are R-similar, which is a contradiction. Since u is a finite word, we have $c^I_{R,w}(n) < p+1+|u|$, and therefore $w \in \mathcal{I}_R$.   □

Lemma 6. *There exist a relation R and a word w such that w is not ultimately strongly R-periodic but the function $c^I_{R,w}(n)$ is bounded ($\mathcal{S}_R \neq \mathcal{I}_R$).*

Proof. Consider the alphabet $\mathcal{A} = \{a, b, c\}$ and the relation $R = \langle(a, b), (b, c)\rangle$. We construct an infinite word $s \in \mathcal{A}^\omega$ such that it is not ultimately strongly R-periodic and $c^I_{R,w}(n) = 2$ for all $n \geq 1$.

Let s be of the form $ab^{i_1} cb^{i_2} ab^{i_3} cb^{i_4} a \cdots$, where the indices i_j are such that the following two conditions are satisfied:

(i) $i_k \equiv k - 1 \pmod{k}$ for all $k \geq 1$,
(ii) $p_{k+1} - p_k > 2p_k$, where p_k is the position of the kth non-b letter.

It is clear that such a word s exists. Now it follows from (i) that $2j - 1$ is not a strong R-period of s since the distance between the jth a and jth c is a multiple of $2j - 1$ for each $j \geq 1$. Similarly, it follows that $2j$ is not a strong R-periods of s since the distance between the jth c and $(j + 1)$st a is a multiple of $2j$ for each $j \geq 1$. Moreover, by the construction, it follows for each j, that $2j$ and $2j - 1$ are not strong R-periods of any suffix of s. Namely, by the same reason as above, $2m(2j)$ and $2m(2j - 1)$ are not strong R-periods of the suffix for some $m \in \mathbb{N}$ large enough. Hence, condition (i) implies that s is not ultimately strongly R-periodic.

Moreover, condition (ii) ensures that the following holds:

(I) For all $x \in \{a, c\}$, we have $D_{xx} \cap D_{x\bar{x}} = D_{xx} \cap D_{\bar{x}x} = \emptyset$.
(II) For all $x, y, z \in \{a, c\}$, we have $(D_{xz} + D_{\bar{z}y}) \cap D_{\bar{x}\bar{y}} = \emptyset$.

Here $\bar{a} = c$, $\bar{c} = a$ and D_{xy} is the set of all distances between the kth occurrence of x in s and the lth occurrence of y in s for all integers k and l where $k < l$. By replacing x, y and z by the letters a and c in (I) and (II) we get conditions concerning altogether 12 intersections, which we should prove to be empty. These conditions guarantee that there cannot be three pairwise R-incompatible factors of s of the same length. Let us assume, on the contrary, that there are three pairwise incompatible factors w_1, w_2 and w_3 of length n. Denote the ith letter of a word w by $w(i)$. There are two possibilities (see Figure 1):

(a) There exist positions k and l such that $(w_1(k), w_2(k)) \notin R$, $(w_1(k), w_3(k)) \notin R$ and $(w_2(l), w_3(l)) \notin R$.
(b) There exist positions k, l and m such that $(w_1(k), w_2(k)) \notin R$ and $w_3(k) = b$, $(w_1(l), w_3(l)) \notin R$ and $w_2(l) = b$ and $(w_2(m), w_3(m)) \notin R$ and $w_1(m) = b$.

By the conditions (I) and (II), situations (a) and (b) never occur. Hence, the maximal cardinality of a pairwise R-incompatible set of words of $F_w(n)$ is two, i.e., $c^I_{R,w}(n) = 2$ for all $n \geq 1$.

It remains to prove (I) and (II). This can be done by induction on k for the prefixes of s of length p_k. For a prefix of length $p_0 = 0$ the conditions

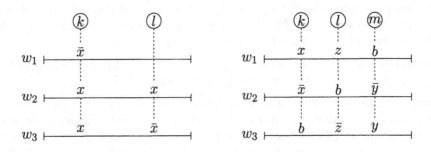

Fig. 1. Illustrations of three pairwise R-incompatible words

hold trivially. Let us now assume that (I) and (II) hold for the prefix of s of length p_k, where $s(p_k) = a$. We prove that (I) and (II) hold also for the prefix of length p_{k+1}. The case where $s(p_k) = c$ is proved similarly. Denote by $D_{xy}(k)$ the above defined set of distances D_{xy} for a prefix of s of length p_k.

For the proof of (I), we note that $D_{aa}(k+1) = D_{aa}(k)$ and $D_{ca}(k+1) = D_{ca}(k)$, since $s(p_{k+1}) = c$ by the construction of s. Moreover, we have

$$D_{ac}(k+1) = D_{ac}(k) \cup N(a, k+1) \text{ and } D_{cc}(k+1) = D_{cc}(k) \cup N(c, k+1) ,$$

where $N(x, k+1) = \{p_{k+1} - p_l \mid l < k+1, s(p_l) = x\}$ corresponds to all distances between x and the new letter $s(p_{k+1})$. Since, by condition (ii), we have $p_{k+1} - p_l \geq p_{k+1} - p_k > 2p_k$ for $l < k+1$, it follows that $d > 2p_k$ for every $d \in N(x, k+1)$. On the other hand, $d' \leq p_k$ for all $d' \in D_{yz}(k)$ and therefore $D_{yz}(k) \cap N(x, k+1) = \emptyset$. Hence, $D_{aa}(k+1) \cap D_{ac}(k+1) = D_{aa}(k) \cap D_{ac}(k) = \emptyset$, where the last equality follows from the induction hypothesis. Similarly, we conclude that $D_{cc}(k+1) \cap D_{ca}(k+1) = D_{cc}(k) \cap D_{ca}(k) = \emptyset$. Moreover, by the induction hypothesis, it holds that $D_{aa}(k+1) \cap D_{ca}(k+1) = \emptyset$. Since $N(x, k+1)$ contains only distances between the letter x and the letter in the fixed position p_{k+1}, we have $N(a, k+1) \cap N(c, k+1) = \emptyset$. Therefore, it also follows that $D_{cc}(k+1) \cap D_{ac}(k+1) = (D_{cc}(k) \cap D_{ac}(k)) \cup (D_{cc}(k) \cap N(a, k+1)) \cup (N(c, k+1) \cap D_{ac}(k)) \cup (N(c, k+1) \cap N(a, k+1)) = \emptyset$. Hence, we have proved (I).

In order to prove (II), we show that $d_1 + d_2 \neq d$ for all distances $d_1 \in D_{xz}(k+1)$, $d_2 \in D_{zy}(k+1)$ and $d \in D_{\bar{x}\bar{y}}(k+1)$. By the induction hypothesis, we have to consider only cases where at least one of the distances d_1, d_2 and d is new, i.e., belongs to the set $N(X, k+1)$ for some $X \in \{x, \bar{z}, \bar{x}\}$. Note that, for $d \in N(X, k+1)$, we have

$$d > 2p_k \tag{1}$$

as above. We consider four different cases.

Assume firstly that $y = z = a$. Then it follows that $D_{xz}(k+1) = D_{xa}(k)$, $D_{\bar{z}y}(k+1) = D_{ca}(k)$ and $D_{\bar{x}\bar{y}}(k+1) = D_{\bar{x}c}(k+1) = D_{\bar{x}c}(k) \cup N(\bar{x}, k+1)$. If $d \in N(\bar{x}, k+1)$, then we have $d > 2p_k$ by (1) but $d_1 + d_2 \leq 2p_k$. Hence, $d_1 + d_2 \neq d$.

Assume secondly that $y = z = c$. This implies that $D_{xz}(k+1) = D_{xc}(k) \cup N(x, k+1)$, $D_{\bar{z}y}(k+1) = D_{ac}(k) \cup N(a, k+1)$ and $D_{\bar{x}\bar{y}}(k+1) = D_{\bar{x}a}(k)$. If either $d_1 \in N(x, k+1)$ or $d_2 \in N(a, k+1)$, then $d_1 + d_2 > 2p_k > p_k \geq d$.

Assume thirdly that $y = c$ and $z = a$. Now we have $D_{xz}(k+1) = D_{xa}(k)$, $D_{\bar{z}y}(k+1) = D_{cc}(k) \cup N(c, k+1)$ and $D_{\bar{x}\bar{y}}(k+1) = D_{\bar{x}a}(k)$. If $d_2 \in N(c, k+1)$, then $d_1 + d_2 > d$ as in the previous case.

Finally, assume that $y = a$, $z = c$. Then we have $D_{xz}(k+1) = D_{xc}(k) \cup N(x, k+1)$, $D_{\bar{z}y}(k+1) = D_{aa}(k)$ and $D_{\bar{x}\bar{y}}(k+1) = D_{\bar{x}c}(k) \cup N(\bar{x}, k+1)$. If there is only one new distance, either d_1 or d, then it follows that $d_1 + d_2 \neq d$ by the same reasoning as above. Hence, consider the case where both distances are new, i.e., $d_1 \in N(x, k+1)$ and $d \in N(\bar{x}, k+1)$. Since d is a distance between $\bar{x}$ and $\bar{y}$, where $\bar{y}$ is in the position p_{k+1}, and d_1 is a distance between x and $z = \bar{y}$, where z is in the position p_{k+1}, we must have $d_1 \neq d$. If $d = d_1 + d_2$, then we conclude that $d > d_1$. Assume that $d_1 = p_{k+1} - p_l$ and $d = p_{k+1} - p_{l'}$ for some l

and l' smaller that $k+1$. By substituting these to the equation $d-d_1 = d_2$, we get $p_l - p_{l'} = d_2$. Since $p_l - p_{l'} \in D_{\bar{x}x}(k)$ and $d_2 \in D_{\bar{z}y}(k) = D_{aa}(k)$, this contradicts with the induction hypothesis $D_{aa}(k) \cap D_{ac}(k) = D_{aa}(k) \cap D_{ca}(k) = \emptyset$.

Hence, we have shown (II) and the proof of the lemma is hereby concluded. $\quad\square$

The proof of Theorem 2 is an easy consequence of the lemmata.

Proof (Proof of Theorem 2). By Lemmata 1, 3 and 5, it follows that $\mathcal{S}_R \subseteq \mathcal{I}_R \subseteq \mathcal{W}_R \subseteq \mathcal{T}_R$. If R is not transitive, then there must exist letters a, b and c such that $a\,R\,b$, $b\,R\,c$ and $(a,c) \notin R$. Hence, the examples of Lemmata 2, 4 and 6 imply that $\mathcal{S}_R \neq \mathcal{I}_R \neq \mathcal{W}_R \neq \mathcal{T}_R$. If R is transitive, then R is an equivalence relation and, as mentioned above, $\mathcal{S}_R = \mathcal{W}_R$. For such R we may replace the word $w = w_1w_2w_3\cdots$ by the word $[w] = [w_1][w_2][w_3]\cdots$, where $[w_i]$ denotes the R-equivalence class of w_i. Since in this case $c_{R,[w]}^T(n) = |F_{[w]}(n)|$, we conclude by Theorem 1 that, for a word $w \in \mathcal{W}_R$, the complexity function $c_{R,[w]}^T(n)$ is bounded and therefore also $c_{R,w}^T(n)$ must be bounded. Hence, we have $\mathcal{S}_R = \mathcal{I}_R = \mathcal{W}_R = \mathcal{T}_R$. $\quad\square$

5 Future Work

It remains an open question whether there exists a complexity function $c_{R,w}(n)$ such that an infinite word w is strongly or weakly R-periodic if and only if $c_{R,w}(n)$ is bounded. The following function could be worth of studying. Let $c_{R,w}^D(n)$ be the minimal size of a dominating set of the graph $G_{R,w}(n)$, *i.e.*, the minimal size of a set $S \in F_w(n)$ such that every factor of $F_w(n)$ is R-compatible with at least one element of S. Another possibility is to allow the dominating set S to be any subset of $\mathcal{A}^n$. This definition of complexity could be related to the ultimately externally R-periodic words, which are infinite words such that for some suffix there exists an external R-period p. These periods were not considered in this paper, but as a final remark we will give the definition, which can also be found in [4]. For a word $x = x_1x_2x_3\cdots$, where $x_i \in \mathcal{A}$, an integer p is an *external R-period* of x if there exists a word $y = y_1\cdots y_p$ such that, for all $i \in \mathbb{N}$ and $j \in \{1, 2, \ldots, p\}$, $i \equiv j \pmod{p}$ implies $x_i\,R\,y_j$. In this case, the word y is called an *external word* of x.

References

1. Berstel, J., Boasson, L.: Partial words and a theorem of Fine and Wilf. Theoret. Comput. Sci. 218, 135–141 (1999)
2. Blanchet-Sadri, F.: Algorithmic Combinatorics on Partial Words. Chapman & Hall/CRC Press, Boca Raton (2007)
3. Halava, V., Harju, T., Kärki, T.: The theorem of Fine and Wilf for relational periods. TUCS Tech. Rep. 786. Turku Centre for Computer Science, Finland (2006)
4. Halava, V., Harju, T., Kärki, T.: Interaction properties of relational periods. Discrete Math. Theor. Comput. Sci. 10, 87–112 (2008)

5. Hedlund, G.A., Morse, M.: Symbolic dynamics II: Sturmian trajectories. Amer. J. Math. 62, 1–42 (1940)
6. Kärki, T.: Similarity Relations on Words: Relational Codes and Periods (PhD thesis). TUCS Dissertations No 98. Turku Centre for Computer Science, Finland (2008)
7. Lothaire, M.: Algebraic combinatorics on words. In: Encyclopedia of Mathematics and its Applications 90. Cambridge University Press, Cambridge (2002)

Bounds on Powers in Strings

Maxime Crochemore[1,2], Szilárd Zsolt Fazekas[3,*],
Costas Iliopoulos[1], and Inuka Jayasekera[1,**]

[1] King's College London, U.K.
[2] Université Paris-Est, France
[3] Rovira i Virgili University, Tarragona, Spain

Abstract. We show a $\Theta(n \log n)$ bound on the maximal number of occurrences of primitively-rooted k-th powers occurring in a string of length n for any integer k, $k \geq 2$. We also show a $\Theta(n^2)$ bound on the maximal number of primitively-rooted powers with fractional exponent e, $1 < e < 2$, occurring in a string of length n. This result holds obviously for their maximal number of occurrences. The first result contrasts with the linear number of occurrences of maximal repetitions of exponent at least 2.

1 Introduction

The subject of this paper is the evaluation of the number of powers in strings. This is one of the most fundamental topics in combinatorics on words not only for its own combinatorial aspects considered since the beginning of last century by the precursor A. Thue [16], but also because it is related to lossless text compression, string representation, and analysis of molecular biological sequences, to quote a few applications. These applications often require fast algorithms to locate repetitions because either the amount of data to be treated is huge or their flow is to be analysed on the fly, but their design and complexity analysis depends of the type of repetitions considered and of their bounds.

A repetition is a string composed of the concatenation of several copies of another string whose length is called a period. The exponent of a string is informally the number of copies and is defined as the ratio between the length of the string and its smallest period. This means that the repeated string, called the root, is primitive (it is not itself a nontrivial integer power). We consider two types of strings: integer powers—those having an integer exponent at least 2, and fractional powers—those having a fractional exponent between 1 and 2. For both of them we consider their maximal number in a given string as well as their maximal number of occurrences.

It is known that all occurrences of integer powers in a string of length n can be computed in time $O(n \log n)$ (see three different methods in [2], [1], and

---

* Supported by grant no. AP2004-6969 from the Spanish Ministry of Science and Education of Spain. Partially supported by grant. no. MTM 63422 from the Ministry of Science and Education of Spain.

** Supported by a DTA Award from EPSRC.

M. Ito and M. Toyama (Eds.): DLT 2008, LNCS 5257, pp. 206–215, 2008.

[12]). Indeed these algorithms are optimal because the number of occurrences of squares (powers of exponent 2) can be of the order of $n \log n$ [2].

The computation of occurrences of fractional powers with exponent at least 2 has been designed initially by Main [11] who restricted the question to the detection of their leftmost maximal occurrences only. Eventually the notion of runs— maximal occurrences of fractional powers with exponent at least 2—introduced by Iliopoulos et al. [7] for Fibonacci words led to a linear-time algorithm for locating all of them on a fixed-sized alphabet. The algorithm, by Kolpakov and Kucherov [8,9], is an extension of Main's algorithm but their fundamental contribution is the linear number of runs in a string. They proved that the number of runs in a string of length n is at most cn, could not provide any value for the constant c, but conjectured that $c = 1$. Rytter [14] proved that $c \leq 5$, then $c \leq 3.44$ in [15], Puglisi et al. [13] that $c \leq 3.48$, Crochemore and Ilie [3] that $c \leq 1.6$, and Giraud [6] that $c \leq 1.5$. The best value computed so far is $c = 1.048$ [4] (see the Web page http://www.csd.uwo.ca/ ilie/runs.html).

Runs capture all the repetitions in a string but without discriminating among them according to their exponent. For example, the number of runs is not easily related to the number of occurrences of squares. This is why we consider an orthogonal approach here. We count and bound the maximal number of repetitions having a fixed exponent, either an integer larger than 1 or a fractional number between 1 and 2. We also bound the number of occurrences of these repetitions.

After introducing the notations and basic definitions in the next section, Section 3 deals with fractional powers with exponent between 1 and 2. It is shown that the maximum number of primitively-rooted powers with a given exponent e, $1 < e < 2$, in a string can be quadratic as well of course as their maximum number of occurrences. In Section 4, we consider primitively-rooted integer powers and show that the maximum number of occurrences of powers of a given exponent k, $k \geq 2$, is $\Theta(n \log n)$. This latter result contrasts with the linear number of such powers. We also present an efficient algorithm for constructing the strings in question.

2   Preliminaries

In this section we introduce the notation and recall some basic results that will be used throughout the paper. All results stated in this section were obtained in [10]. An *alphabet* A is a finite non-empty set. We call the elements of A *letters*. The set of all finite words over A is A^*, which is a monoid with concatenation (juxtaposition), where the unit element is ϵ, the *empty word*, whereas the set of non-empty words is $A^+ = A^* - \epsilon$. The length of a word w is denoted by $|w|$; $|\epsilon| = 0$. Without loss of generality, we can assume that our alphabet is ordered and hence we can have an order on words. The one we will use is called the *lexicographical order* and it is defined by the following relation:

$$au < bv \Leftrightarrow (a < b \text{ or } (a = b \text{ and } u < v)) \text{ or } a < az$$

where $a, b \in A$, $u, v \in A^*$, $z \in A^+$.

For words $u, v, w \in A^*$, with $w = uv$, we say that u is a *prefix* and v is a *suffix* of w. For a word w and an integer $n \geq 0$, the n-th power of w is defined inductively as $w^0 = \epsilon$, $w^n = ww^{n-1}$. Extending this definition we can talk about non-integer powers too. Take $n = \frac{k}{l} > 1$ with $gcd(k, l) = 1$. We say that a word w is an n-power if both of the following conditions apply:

- $|w| = m \cdot k$ for some integer $m > 0$,
- $m \cdot l$ is a period of w.

The prefix of length $m \cdot l$ of w is a *root* of w.

When $w \neq \epsilon$, w^3 is called a *cube*, with *root* w. A word w is called *primitive* if there is no word u and integer $p \geq 2$ such that $w = u^p$. For a word $w = u^p$ with u primitive and $p \geq 1$. We say that w' is a *conjugate* of w if there exist $u, v \in A^*$ such that $w = uv$ and $w' = vu$. A *Lyndon word* is a (primitive) word which is the lexicographically smallest among its conjugates.

Take a primitive word uv, such that vu forms a Lyndon word and v is nonempty. In the cube $(uv)^3$, we call *central Lyndon position* the one at $|uvu|$, $uvu.vuv$. For two non-empty words u and v it is known that $uv = vu$ implies $u, v \in z^+$ for some $z \in A^*$, therefore every word has a unique Lyndon position.

If a word w can be written as $w = uv = vz$, for some words $u, v, z \in A^+$, then we say that w is *bordered* (v is a *border* of w). If a word w is bordered, then there exists $u \in A^+, v \in A^*$ such that $w = uvu$, that is a bordered word w always has a border of length at most half the length of w. Moreover, it is easy to see that a bordered word uvu cannot be a Lyndon word, because then either uuv (if $u < v$) or vuu (if $v < u$) is lexicographically smaller than uvu.

3   A Bound on Repeats with Exponent e, with $1 < e < 2$

In this section, we show that the number of distinct repetitions with exponent e, with $1 < e < 2$ is bound by $\Theta(n^2)$. We do this by looking at the number of such repetitions that can start at a position in words of the form $a^k ba^{\frac{k}{e-1}-1}$, where k is any positive integer such that $c|k$, where $e = \frac{c+d}{d}$ and $gcd(c + d, d) = 1$.

First we consider an example with $e = \frac{3}{2}$ and $k = 9$, ie. $w = a^9 ba^{17}$ (see Fig 2). At the first position in this word, we can have 5 repetitions of exponent $\frac{3}{2}$, namely $a^9 ba^5, a^9 ba^8, a^9 ba^{11}, a^9 ba^{14}$ and $a^9 ba^{17}$. Moving on to the second position, we will have only 4 repetitions of exponent $\frac{3}{2}$, namely $a^8 ba^6, a^8 ba^9, a^8 ba^{12}$ and $a^8 ba^{15}$. In the third position also, we are able to have the repetitions $a^7 ba^7, a^7 ba^{10}$ and $a^7 ba^{13}$. However, now we will have one extra repetition as we can also have $a^7 ba^4$. It is clear that at every other position in the word, as we get closer to the b, we will have an extra repetition. The number of repetitions of exponent $\frac{3}{2}$ at each position are now $5, 4, 4, 3, 3, 2, 2, 1, 1$ (see Fig. 2). The total number of repetitions can now be summed up to $((5 * 6)/2) + (((5 - 1) * 5)/2) = 25$. We will generalise this example in the next theorem.

Theorem 1. *The maximal number of distinct repetitions of exponent e, with $1 < e < 2$, in a word of length n is $\Theta(n^2)$.*

Proof. The upper bound is trivial because no factor of the string can be counted twice as an e-th power for given e, so let us turn to proving the lower bound.

We shall count the number of repetitions starting at each position in a word. For an exponent, e, with $1 < e < 2$, we consider a word, w, formed as shown in Fig. 1. Here, we concatenate a repetition of exponent, e, with root $a^k b$ and $a^{\frac{k}{e-1}-1}$, where k is any positive integer such that $c|k$, where $e = \frac{c+d}{d}$ and $gcd(c+d,d) = 1$. In this case the length of our string will be $k \cdot \frac{e}{e-1}$.

Fig. 1. Structure of word, w

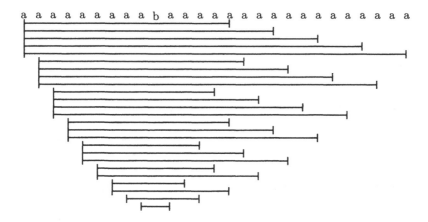

Fig. 2. Repetitions of exponent 1.5 in $a^9 b a^{17}$

For e-powers starting at the first position, the end positions can be $(k+1)(e-1)$, $(k+1)(e-1) + (c+d)$, $(k+1)(e-1) + 2 \cdot (c+d),...$

From here we get that the number of e-th powers starting at the first position is

$$\frac{|w| - (k+1)(e-1)}{c+d} + 1 = \frac{k \cdot \frac{e}{e-1} - (k+1)(e-1)}{c+d} + 1$$

Substituting $\frac{c+d}{d}$ for e in the formula above we get that the number of e-th powers starting at the first position is:

$$k \cdot \frac{d-c}{d \cdot c} - \frac{1}{d} + 1$$

This formula proves useful because by substituting $k - i$ for k and taking the integer part of the result (since we are talking about the number of occurrences) we get the number of e-th powers starting at position $i + 1$. Now let us sum up the number of e-th power occurrences starting at any one of the first k positions:

$$\sum_{i=1}^{k} \lfloor i \cdot \frac{d-c}{d \cdot c} - \frac{1}{d} + 1 \rfloor$$

For any positive n its integer part $\lfloor n \rfloor$ is greater or equal than $n - 1$. As we are trying to give a lower bound to the number of occurrences, it is alright to subtract 1 from the formula instead of taking its integer part:

$$\sum_{i=1}^{k} \left(i \cdot \frac{d-c}{d \cdot c} - \frac{1}{d} \right) = k \cdot (k+1) \cdot \frac{d-c}{2d \cdot c} - \frac{k}{d}$$

This means that the number of e-th powers in our string is quadratic in k. At the same time the length of the string, as we mentioned in the beginning, is $k \cdot \frac{e}{e-1}$, so for a given e, the number of e-th powers in a string of length n is $\Theta(n^2)$.

It is easy to see that every occurrence of an e-th power in this string is unique and this concludes the proof. □

4   A Bound on Primitively Rooted Cubes

After considering powers between 1 and 2, we shall take a look at powers greater than 2. First, we will show that it is possible to construct strings of length n, which have $\Omega(n \log n)$ occurrences of cubes. We can extend the procedure to all integer powers greater than 2, and this, together with the $O(n \log n)$ upper bound implied by the number of squares (see [2]) leads us to the $\Theta(n \log n)$ bound. Finally, we will prove that the sum of all occurrences of powers at least 2 (including non-integer exponents) is quadratic.

Lemma 1. *The maximal number of primitively rooted cubes in a word of length n is $\Omega(n \log n)$.*

Proof. Let us suppose there are two primitively rooted cubes $(uv)^3$ and $(xy)^3$ in w such that their central Lyndon positions $uvu.vuv$ and $xyx.yxy$ are the same. First let us look at the case where the cubes have to be of different length. Without loss of generality we can assume $|uv| < |xy|$. In this case vu is at the same time a prefix and suffix of yx. Hence, yx is bordered and cannot be a Lyndon word contradicting the assumption that $x.y$ is a Lyndon position. This proves that should there be more cubes which have their central Lyndon position identical, they all have to be of the same length. Naturally, the first and last position of a word cannot be central Lyndon to any cube and this gives us the bound $n - 2$ if we disregard cubes of the same length which have their central Lyndon positions at the same place (see Fig. 3). It is easy to see, that because of the periodicity theorem the only string of length n, for which $n - 2$ different positions are central Lyndon ones to some cube, is a^n.

Fig. 3. Cubes of word a^{4k+3}

Now take the word a^{4k+3}. According to our previous argument it has at most $4k + 1$ cubes. However, if we change a's into b's at positions $k + 1, 2k + 2$ and $3k + 3$ we get that the number of primitively rooted cubes in this word is $4k + 1 - 9 + (k + 1) = 5k - 7$. This is because by introducing each b we lose three cubes but in the end we gain another $k + 1$ cubes of the form $(a^j b a^{k-j})^3$ with $0 \leq j \leq k$ (see Fig. 4). Note that these latter cubes all have their central Lyndon position after the first b (assuming $a < b$).

We introduced three b's in the previous step but of course we can repeat the procedure for the four block of a's delimited by these b's and then in turn for the new, smaller blocks of a's that result and so on. In the second step, however, we need to introduce 12 b's - that is, 3 for each of the 4 blocks of a's - not to disrupt the cubes of length $3k + 3$. This way we lose $12 \cdot 3 = 36$ cubes and we gain ($\lfloor (k - 3)/4 \rfloor + 1) \cdot 4$ new ones. Performing the introduction of b's until the number of cubes we lose in a step becomes greater or equal to the ones we gain, gives us a string with the maximal possible number of cubes for its length. If k equals $4j$, $4j + 1$ or $4j + 2$ for some j then according to the formula above the number of cubes we gain is $4j$. Note that if $k = 4j + 3$ than the number of cubes we gain in the second step is $4j + 4 = k + 1$, i.e. the same as in the first step. However, together with the delimiting b's introduced before we would get a big cube which is not primitively rooted anymore, so we need to move the newly introduced b's 1, 2 and 3 positions to the left, respectively. This gives us that in this case too the number of newly formed cubes will be $4j$. The smallest length at which introducing the b's does not induce less cubes is 35 that is with $k = 8$. Summarizing the points above we get that for a string of length n the maximum increase in the number of cubes for the ith $(i > 1)$ consecutive application of our procedure is:

$$\frac{(n - 3)}{4} - 9 \cdot 4^{i-1}$$

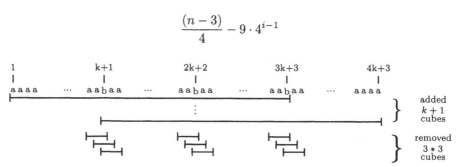

Fig. 4. Cubes of word $a^k b a^k b a^k b a^k$

To be able to sum these increases we have to know the number of steps performed. This is given by solving for i the equation:

$$\frac{n-3}{4} = 9 \cdot 4^{i-1}$$

From here we get that the number of steps performed is $\#steps = \lfloor \log_4 \frac{n-3}{9} \rfloor$, where by $\lfloor x \rfloor$ we mean the integer part of x.

Hence the number of cubes for length $n \geq 39$ is:

$$n - 2 + 1 + \sum_{i=1}^{\#steps} \left(\frac{n-3}{4} - 9 \cdot 4^{i-1} \right)$$

$$= n - 1 + \frac{(n-3)\lfloor \log_4 \frac{(n-3)}{9} \rfloor}{4} - \frac{9(1 - 4^{\lfloor \log_4 \frac{n-3}{9} \rfloor})}{-3}$$

$$= n + 2 + \frac{(n-3)\lfloor \log_4 \frac{(n-3)}{9} \rfloor}{4} - 3 \cdot 4^{\lfloor \log_4 \frac{n-3}{9} \rfloor}$$

The plus one after $n - 2$ comes from the first application of the insertion of b's where we get $(n-3)/4 + 1$ cubes instead of $(n-3)/4$. For strings shorter than 39 therefore the count is one less.   □

Since the first paragraph of the proof is valid for any integer power, we can extend the proof by giving the construction of the strings that prove the lower bound in general for a string of length n and power k (see Fig. 4).

The algorithm above produces strings which have $O(n \log n)$ occurrences of k-th powers. Note, that if we perform the procedure the other way around, we only need $O(\log n)$ cycles and we can eliminate the recursion:

Theorem 2. *Algorithm ConstructStrings2 (see Fig. 4) produces a string of length n that has $\Omega(n \log n)$ occurrences of primitively rooted cubes.*

Proof. Before entering the second **while** loop, the length of *string* and the number of k-th power occurrences in it are both $c = (k+1) \cdot \ell + k$. Now we will show by induction on i that after the i-th iteration of the second **while** loop the length of *string* will be $(k+1)^i \cdot (c+1) - 1$ and the number of occurrences of k-th powers will be $(k+1)^i \cdot c + i \cdot (k+1)^{i-1}(c+1)$.

Note that if the length of *string* was m and the number of k-th power occurrences was p after the previous cycle, then concatenating $k + 1$ copies of *string* delimited by k copies of *delimiter* we get $(k+1) \cdot p + m + 1$ powers in the new *string*, which will have length $(k+1) \cdot m + k$. Therefore, after the first cycle the length of *string* will be

$$(k+1) \cdot c + k = (k+1) \cdot c + (k+1) - 1 = (k+1)^1 \cdot (c+1) - 1$$

At the same time the number of k-th powers will be

$$(k+1) \cdot c + c + 1 = (k+1)^1 \cdot c + 1 \cdot (k+1)^0 \cdot (c+1)$$

Algorithm. *ConstructStrings1 (n, k)*
Input: $n \geq 0$, $k \geq 0$
Output: A string which proves the lower bound of the number of occurrences of integer
      powers.

1. $\ell = n$
2.   string $= a^\ell$
3.   power$(1, \ell)$
4.   Procedure: power(start, end)
5. $\ell =$ end - start
6.   **if** $\ell < k^3 + k^2 + k$
7.   **then** return
8.   **else** string$[$start $+ \lfloor \ell/(k+1) \rfloor] = b$
9.   string$[$start $+ 2 \cdot \lfloor \ell/(k+1) \rfloor] = b$
10.   . . .
11.   string$[$start $+ k \cdot \lfloor ell/(k+1) \rfloor] = b$
12.   **for** $i \leftarrow 0$ **to** k
13.   power$($start $+ i \cdot \ell/(k+1)$, start $+ (i+1) \cdot \ell/(k+1))$

so our statement holds for $i = 1$. Now suppose it is true for some $i \geq 1$. From
here we get that for $i + 1$ the length of *string* will be:

$$(k+1) \cdot ((k+1)^i \cdot (c+1) - 1) + k = (k+1)^{i+1} \cdot (c+1) - 1$$

whereas the number of k-th powers is:

$$(k+1) \cdot ((k+1)^i \cdot c + i \cdot (k+1)^{i-1} \cdot (c+1)) + ((k+1)^i \cdot (c+1) - 1) + 1$$

$$= (k+1)^{i+1} \cdot c + i \cdot (k+1)^i \cdot (c+1) + (k+1)^i \cdot (c+1)$$

$$= (k+1)^{i+1} \cdot c + (i+1) \cdot (k+1)^i (c+1)$$

Now let us look at the running time of the algorithm. In the first **while** loop
we divide the actual length by $k + 1$ and we do it until it becomes smaller than
$k^3 + k^2 + k$ therefore we perform $O(\log n)$ cycles. The second **while** loop has the
same number of cycles, with one string concatenation performed in each cycle,
hence substituting $\log n$ for i in the formula above concludes the proof.   □

Corollary 1. *In a string of length n the maximal number of primitively rooted
k-th powers, for a given integer $k \geq 2$, is $\Theta(n \log n)$.*

Proof. We know from [5] that the maximal number of occurrences of primitively
rooted squares in a word of length n is $O(n \log n)$. This implies that the number
of primitively rooted greater integer powers also have an $O(n \log n)$ upper bound,
while in Theorem 2 we showed the lower bound $\Omega(n \log n)$.   □

Remark 1. The first part of the proof is directly applicable to runs so we have
that in a string of length n the number of runs of length at least $3p - 1$, where
p is the (smallest) period of the run is at most $n - 2$. Unfortunately we cannot
apply the proof directly for runs shorter than that because we need the same
string on both sides of the central Lyndon position.

Algorithm. *ConstructStrings2*(n, k)
Input: $n \geq 0$, $k \geq 0$
Output: A string which proves the lower bound of the number of occurrences of integer
powers.

1. $\ell = n$
2. **while** $\ell \geq k^3 + k^2 + 3k + 2$
3.   **do** $\ell = \frac{\ell - k}{k + 1}$
4. $string = (a^{k^2 + 1} + b)^k + a^{(k+1) \cdot \ell - k^3 - k}$
5. $delimiter = b$
6. **while** $length(string) * (k + 1) + k < n$
7.   **do** $string = (string + delimiter)^k + string$
8.   **if** $delimiter = b$
9.   **then** $delimiter = a$
10.   **else** $delimiter = b$
11.   (* changing the delimiter is needed to stay primitive *)
12. $string = string + a^{n - length(string)}$

We have seen that the number of k-th powers for a given k(≥ 2) in a string of
length n is $\Theta(n \log n)$, but what happens if we sum up the occurrences of k-th
powers for all $k \geq 2$?

Remark 2. The upper bound of the sum of all occurrences of k-th powers with
primitive root, where $k \geq 2$, in a word w with $|w| = n$ is $\frac{n \cdot (n-1)}{2}$. Moreover, the
bound is sharp.

Proof. First consider the word a^n, for some $n > 0$. Clearly, taking any substring
a^k, with $2 \leq k \leq n$, we get a k-th power, so the number of powers greater or
equal to two is given by the number of contiguous substrings of length at least
two, that is $\frac{n \cdot (n-1)}{2}$. Now we will show that this is the upper bound. Let us
suppose that any two positions i and j in the string delimit a k-th power with
$k \geq 2$, just like in the example above. We need to prove that the same string
cannot be considered a k_1-th power and a k_2-th power at the same time, with
$k_1, k_2 \geq 2$ and $k_1 \neq k_2$. Suppose the contrary, that is there are $1 \leq m < \ell \leq \frac{i-i}{2}$
so that both m and ℓ are periods of $w[i, j]$. Since $j - i > m + \ell - gcd(m, \ell)$ the
periodicity lemma tells us that $w[i, j]$ has a period p smaller than m with $p|m$
and $p|\ell$, and this, in turn, means $w[i, i + \ell]$ is not primitive.   □

5   Conclusion

In conclusion, we have proven the following bounds on repetitions in words:

(i) The maximal number of distinct repetitions of exponent, e, with $1 < e < 2$,
    in a word of length n is $\Theta(n^2)$.
(ii) The maximal number of primitively rooted k-th powers in a word of length
    n is $\Omega(n \log n)$.

We have also described an $O(m \log n)$ algorithm which can be used to construct strings to illustrate these bounds. Here $O(m)$ is the time complexity of
concatenating two strings of length n.

References

1. Apostolico, A., Preparata, F.P.: Optimal off-line detection of repetitions in a string. Theoret. Comput. Sci. 22(3), 297–315 (1983)
2. Crochemore, M.: An optimal algorithm for computing the repetitions in a word. Inf. Process. Lett. 12(5), 244–250 (1981)
3. Crochemore, M., Ilie, L.: Maximal repetitions in strings. J. Comput. Syst. Sci. (in press, 2007)
4. Crochemore, M., Ilie, L., Tinta, L.: Towards a solution to the "runs" conjecture. In: Ferragina, P., Landau, G.M. (eds.) Combinatorial Pattern Matching. LNCS. Springer, Berlin (in press, 2008)
5. Crochemore, M., Rytter, W.: Squares, cubes and time-space efficient stringsearching. Algorithmica 13(5), 405–425 (1995)
6. Giraud, M.: Not so many runs in strings. In: Martin-Vide, C. (ed.) 2nd International Conference on Language and Automata Theory and Applications (2008)
7. Iliopoulos, C.S., Moore, D., Smyth, W.F.: A characterization of the squares in a Fibonacci string. Theoret. Comput. Sci. 172(1–2), 281–291 (1997)
8. Kolpakov, R., Kucherov, G.: Finding maximal repetitions in a word in linear time. In: Proceedings of the 40th IEEE Annual Symposium on Foundations of Computer Science, New York, pp. 596–604. IEEE Computer Society Press, Los Alamitos (1999)
9. Kolpakov, R., Kucherov, G.: On maximal repetitions in words. J. Discret. Algorithms 1(1), 159–186 (2000)
10. Lothaire, M.: Applied Combinatorics on Words. Cambridge University Press, Cambridge (2005)
11. Main, M.G.: Detecting leftmost maximal periodicities. Discret. Appl. Math. 25, 145–153 (1989)
12. Main, M.G., Lorentz, R.J.: An O(n log n) algorithm for finding all repetitions in a string. J. Algorithms 5(3), 422–432 (1984)
13. Puglisi, S.J., Simpson, J., Smyth, W.F.: How many runs can a string contain? Personal communication (submitted, 2007)
14. Rytter, W.: The number of runs in a string: Improved analysis of the linear upper bound. In: Durand, B., Thomas, W. (eds.) STACS 2006. LNCS, vol. 3884, pp. 184–195. Springer, Heidelberg (2006)
15. Rytter, W.: The number of runs in a string. Inf. Comput. 205(9), 1459–1469 (2007)
16. Thue, A.: Uber unendliche Zeichenreihen. Norske Vid. Selsk. Skr. I Math-Nat. Kl. 7, 1–22 (1906)

When Is Reachability Intrinsically Decidable?

Barbara F. Csima[1,*] and Bakhadyr Khoussainov[2,**]

[1] Department of Pure Mathematics, University of Waterloo
csima@math.uwaterloo.ca
[2] Department of Computer Science, University of Auckland
bmk@cs.auckland.ac.nz

Abstract. A graph $\mathcal{H}$ is **computable** if there is a graph $\mathcal{G} = (V, E)$ isomorphic to $\mathcal{H}$ where the set V of vertices and the edge relation E are both computable. In this case $\mathcal{G}$ is called a **computable copy** of $\mathcal{H}$. The **reachability problem** for $\mathcal{H}$ in $\mathcal{G}$ is, given $u, w \in V$, to decide whether there is a path from u to w. If the reachability problem for $\mathcal{H}$ is decidable in *all* computable copies of $\mathcal{H}$ then the problem is *intrinsically decidable*. This paper provides syntactic-logical characterizations of certain classes of graphs with intrinsically decidable reachability relations.

1 Introduction

The study of reachability problems has played a central role in many areas of computer science and applications. In the context of finite graphs the problem is reduced to computing the connected components of the graphs. Tarjan's algorithm solves the reachability problem for finite graphs in linear time [18]. In complexity theory reachability for finite graphs has been important in the investigation of subclasses of P (e.g. see [4] and its reference list). The problem plays a valuable role in model checking and verification since model checking tasks are often reduced to reachability problems [2], [3]. There is also a large interest in reachability problems in different types of computational models such as in counter automata, timed automata, pushdown automata, Petri-nets, rewriting systems, hybrid systems, systems with unbounded number of clocks, protocols, communication systems [5], [8], [9]. These all give rise to infinite graphs and many natural tasks for these graphs involve computing the reachability relation. See also [7], [13], [16] and their references. The paper [1] defines a methodology for reachability analysis of infinite-state systems based on the theory of well quasi-orderings.

Suppose that we are given a finite presentation of an infinite graph. For example, the graph can be the space of states of a system with an unbounded number of protocols or the configuration space of a Turing machine. The configuration space of a machine is the graph whose vertices are the configurations of the machine and an edge is put between configurations x and y if there is an

---

* Partially supported by Canadian NSERC Discovery Grant 312501.
** Partially supported by Marsden Fund of Royal New Zealand Society.

M. Ito and M. Toyama (Eds.): DLT 2008, LNCS 5257, pp. 216–227, 2008.
© Springer-Verlag Berlin Heidelberg 2008

instantaneous move of the machine from x to y. In each of these cases the graphs have finite presentations. For example, if T is a Turing machine then T is a finite presentation of its own configuration space. It is clear that the reachability problem for such graphs is computably enumerable (that is, recognizable by Turing machines). Due to the undecidibility of the Halting problem, the reachability problem for graphs that have finite presentations is not always decidable.

The aim of this paper is to study the reachability problem for infinite graphs, and find syntactic conditions under which the reachability problem is always decidable *independent* of the finite presentation. We point out that some finite presentations of graphs automatically imply decidability of the reachability problem. For example, reachability is decidable for the configuration spaces of pushdown automata [7]. Similarly, reachability is decidable for graphs that have certain types of monadic second order interpretations in the binary tree (see for example [10], [11]). In this paper we seek conditions which entail decidability of the reachability problem in a given graph *independent* of its finite presentation. Clearly, these conditions should be intrinsic to the graphs rather than their presentations. We now set up the problem formally.

Let $\mathcal{H}$ be an infinite graph. We always assume that our graphs are undirected. We define finite presentations of graphs via Turing machines as follows:

Definition 1. *The graph $\mathcal{H}$ is* **computable** *if there exists a graph $\mathcal{G} = (V, E)$ isomorphic to $\mathcal{H}$ and there are Turing machines T_V and T_E over an alphabet Σ such that the following two properties hold:* (1) *The machine T_V halts on every input string, and for all $u \in \Sigma^*$, T_V accepts u if and only if $u \in V$.* (2) *The machine T_E halts on every pair of input strings, and for all $u, w \in \Sigma^*$, T_E accepts (u, w) if and only if $(u, w) \in E$. In this case the pair $P = (T_V, T_E)$ is a* **presentation** *of $\mathcal{H}$ and $\mathcal{G}$ is the* **computable copy** *of $\mathcal{H}$ given by P.*

We often abuse notation and identify the presentation P of $\mathcal{H}$ with the computable copy $\mathcal{G}$ given by P.

The reachability relation in $\mathcal{H}$ is the set $\{(x, y) \mid$ there is a path in $\mathcal{H}$ from x to $y\}$. In order to give an algorithmic spin to the reachability relation one needs to employ presentations of the graph $\mathcal{H}$:

Definition 2. *The* **reachability problem** *for $\mathcal{H}$ in presentation P (equivalently in $\mathcal{G}$) is, given $u, w \in V$, to decide whether there exists a path from u to w.*

The reachability problem in a computable graph $\mathcal{G}$ is computably enumerable (that is, recognized by a Turing machine). Indeed, given u and w one systematically searches through all paths starting with u. When a path from u to w is detected the search is terminated. We give several examples.

Example 1. In the graph $\mathcal{G} = (V, E)$ with $V = \{0, 1\}^*$ and $E = \{(u, w) \mid |u| = |v|\}$, where $|x|$ refers to the length of x, the reachability problem is decidable.

Example 2. Let $\mathcal{G}$ be a computable graph such that each connected component of $\mathcal{G}$ embeds exactly one cycle. The reachability problem in $\mathcal{G}$ is decidable. Indeed, on inputs u and w search for cycles embedded into the components of u and w. If the cycles are distinct, then reject the pair (u, w); otherwise, accept.

Example 3. Consider the graph $\mathcal{G} = (Z, E)$, where Z is the set of all integers, and $E = \{(3i, 3i+1) \mid i$ is not negative$\} \cup \{(3i+1, 3i+2) \mid i$ is not negative$\}$. The reachability problem in $\mathcal{G}$ is decidable.

Example 4. This example comes from the verification community. A **parameterized system** S is an infinite sequence M_1, M_2, ... of finite state machines where for each i, we effectively know the number of states in M_i and all the transitions in M_i. Given a specification ϕ (written in a logic such as LTL), the verification of ϕ on S consists of checking whether or not the state space of S satisfies ϕ for all n. The state space of S consists of tuples $(s_1, \ldots, s_n, n)$, where each s_i is a state of M_i and $n \geq 1$. Let $\mathcal{G}$ have an edge between $(s_1, \ldots, s_n, n)$ and $(q_1, \ldots, q_m, m)$ iff $n = m$ and there is a transition from s_i to q_i for $i = 1, \ldots, n$. The reachability problem in $\mathcal{G}$ is decidable.

We stress that decidability of the reachability problem for $\mathcal{H}$ in one presentation does *not* imply reachability is decidable in *all* presentations. In this sense reachability is not absolute in terms of decidability. Here is our central definition:

Definition 3. *If the reachability problem for $\mathcal{H}$ is decidable in all presentations of $\mathcal{H}$ then we say that the problem is **intrinsically decidable**.*

For the graphs in Examples 1 and 2, reachability is intrinsically decidable. For the graphs in Example 3, reachability is not intrinsically decidable (this will follow from Lemma 2 (1); note that any two singleton components arising from negative integers can be disjointly embedded into any 3-vertex component arising from a non-negative integer). In Example 4, intrinsic decidability of reachability depends on the system S.

We note that intrinsic decidability (of the reachability relation) has some similarities with finding lower bounds for regular model checking problems. In this problem the configurations of a transition system are coded by strings in such a way that the edge relation under this coding is recognized by finite automata. Such a coding is called a regular coding (and hence makes the transition system an automatic graph). Under the coding one then checks if a given specification is satisfied by the transition system. Finding lower bounds for this problem requires investigating *all* regular codings.

Our goal is to provide characterizations of certain classes of graphs with intrinsically decidable reachability relations. We work over a language with a single binary relation symbol, E, and equality. Our characterization involves an infinitary logic. Recall that the logic $L_{\infty\omega}$ is an extension of first order logic with infinitary $\bigvee$ and $\bigwedge$ connectives (for a treatment see [15]). We use an effective fragment of this logic, $L_{c\omega}$, defined by Ash and Nerode in [6], as follows.

Definition 4. *We say that a relation $R(\bar{x})$ in graph $\mathcal{H}$ is **existentially definable** if there exists a tuple $\bar{a}$ in $\mathcal{H}$ and a computable sequence of existential first order logic formulas $\psi_0(\bar{x}, \bar{a}), \psi_1(\bar{x}, \bar{a}), \ldots$ such that for all $\bar{v}$ in $\mathcal{H}$ we have*

$$\mathcal{H} \models R(\bar{v}) \iff \psi_0(\bar{v}, \bar{a}) \vee \psi_1(\bar{v}, \bar{a}) \vee \psi_2(\bar{v}, \bar{a}) \vee \ldots.$$

The reachability relation in any graph is existentially definable. The desired formula is the disjunction of formulas $p_i(x, y)$, where $p_i(x, y)$ states that there is a path of length i between x and y. That is, $p_i(x, y) = \exists x_0 ... \exists x_i (x \equiv x_0) \wedge (y \equiv x_i) \wedge (\bigwedge_{m \neq n} \neg x_m \equiv x_n) \wedge (\bigwedge_{0 \leq m < i} E x_m x_{m+1})$. The following is an obvious lemma:

Lemma 1. *If relation R of a graph $\mathcal{H}$ is existentially definable then in all presentations of $\mathcal{H}$ the relation R is computably enumerable.*

Corollary 1. *If the complement of the reachability relation of a graph $\mathcal{H}$ is existentially definable then the reachability problem for $\mathcal{H}$ is intrinsically decidable.*

Proof. Let $\mathcal{G}$ be a computable copy of $\mathcal{H}$. By Lemma 1 the complement of reachability in $\mathcal{G}$ is c.e. We also know that the reachability relation is c.e. So, the reachability problem in $\mathcal{G}$ is decidable. Hence it is intrinsically decidable. □

We investigate the converse of the corollary above. We provide classes of graphs in which the complement of the reachability relation is existentially definable. These results provide logical characterizations of intrinsically decidable reachability relations.

Notations and conventions. Whenever we write "component" we mean "connected component". In the rest of the paper we always assume that our graphs are infinite (save for finite subgraphs of the infinite graphs we are considering). Also, we assume that the set of vertices of computable graphs coincides with the set ω of natural numbers. We note that we may construct an infinite computable graph $\mathcal{G}$ by stages, by at each stage s of an effective construction defining a finite graph $\mathcal{G}_s$ with domain an initial segment of ω, such that $\mathcal{G}_s \subset \mathcal{G}_{s+1}$, and letting $\mathcal{G} = \cup_s \mathcal{G}_s$. The graph $\mathcal{G}$ will have domain ω, and it will be computable since we will know by stage $s = \max\{v, w\}$ whether or not there is an edge between the vertices v and w.

A partial computable function $\Phi : \omega^n \to \omega$ is one that can be computed by a Turing machine. We can systematically list all Turing machines $T_0, T_1, \ldots, T_e, \ldots$, and thus give rise to an effective listing of all partial computable functions $\Phi_0, \Phi_1, \ldots, \Phi_e, \ldots$. Finally, $\Phi_{e,s}$ denotes the following partial function. The domain of $\Phi_{e,s}$ consists of all $i \leq s$ such that $\Phi_e(i)$ is defined, and the value $\Phi_e(i)$ is obtained within s steps of the computation of the Turing machine T_e on input i. Also, $\Phi_{e,s}(i) \downarrow$ means that the value of $\Phi_{e,s}$ on input i is defined.

We let $X^{[2]} = \{\{a, b\} \mid a, b \in X \wedge a \neq b\}$, the set of unordered pairs from X. For other basic notations of computability theory the reader is referred to [17].

Our main proofs involve methods from computability theory (see for example [17]) and computable model theory (see [12]). Our proofs are of two types. The first type of proofs (Lemma 2 and Proposition 1) are finite injury type of construction common in computable model theory and computability. The second type of proofs (Theorems 2 and 3) are based on more complicated constructions known as $\emptyset''$ (the second jump of the computable degree) priority tree constructions (see [17]).

2   Graphs with Computable Size Functions

In this section we consider graphs *all* of whose components are finite. We call these graphs **strongly locally finite**. Our goal is to characterize certain strongly locally finite graphs with intrinsically decidable relations.

Let $\mathcal{H}$ be a strongly locally finite graph. The component of a vertex v of $\mathcal{H}$ is denoted by $C(v)$. The **size function** size$_\mathcal{H}$ of $\mathcal{H}$ gives for each vertex v the cardinality of $C(v)$. Thus, size$_\mathcal{H}(v) = |C(v)|$ for all vertices v of $\mathcal{H}$.

Let $\mathcal{G} = (\omega, E)$ be a computable copy of a strongly locally finite graph $\mathcal{H}$. In this section we consider those computable graphs $\mathcal{G}$ whose size functions size$_\mathcal{G}$ are computable. Clearly, the function size$_\mathcal{G}$ is computable if and only if for each vertex $v \in \omega$ one can effectively compute the number of edges from v. For such a $\mathcal{G}$, we effectively list all connected components of $\mathcal{G}$ as $C_0, C_1, C_2, \ldots$. We fix this listing and use it for the next definitions and results.

Definition 5. *We say that components C_i and C_j **disjointly embed** into C_k if their disjoint union as a graph can be embedded into C_k. We also define the function $h : \omega^{[2]} \to \omega \cup \{\infty\}$ by $h(i,j) = |\{k : C_i \text{ and } C_j \text{ disjointly embed into } C_k\}|$. We call h the **disjoint embedding function** for the presentation $\mathcal{G}$.*

This definition implies that if C_i and C_j disjointly embed into C_k then no edge exists between the images of C_i and C_j in C_k under the embedding.

Lemma 2. *Let $\mathcal{G}$ be a computable copy of $\mathcal{H}$. Under the assumptions above about the graph $\mathcal{G}$ we have the following properties:*

1. *If for all n, there exist $i, j > n$, $i \neq j$, such that $h(i,j) = \infty$, then the reachability problem for $\mathcal{H}$ is not intrinsically decidable.*
2. *If there exists an $n \in \omega$ such that $h(i,j)$ is finite for all $i, j > n$, $i \neq j$, and h is computable, then the reachability problem for $\mathcal{H}$ is intrinsically decidable.*
3. *If there exists an $n \in \omega$ such that $h(i,j)$ is finite for all $i, j > n$, $i \neq j$, and h is not computable, then the reachability for $\mathcal{H}$ is not intrinsically decidable.*

Proof (1). To show that the reachability relation on $\mathcal{G}$ is not intrinsically decidable, we need to exhibit a computable graph $\mathcal{G}' = (\omega, E')$ isomorphic to $\mathcal{G}$ such that the reachability relation is not decidable on $\mathcal{G}'$. For that the graph $\mathcal{G}'$ must satisfy the following requirements:

$$P_e : \quad \Phi_e \text{ does not decide the reachability relation on } \mathcal{G}'$$

where $\Phi_0, \Phi_1, \ldots$ is an effective list of all partial computable functions from ω^2 to $\{0, 1\}$. We consider $\Phi_{e,s+1}(v, w) = 0$ to mean that Φ_e tells us that v and w are not in the same component, and $\Phi_{e,s+1}(v, w) = 1$ to mean that Φ_e tells us that v and w are in the same component.

The requirement P_e has a higher **priority** than P_t if $t > e$. We will construct $\mathcal{G}'$ by stages. At stage s we construct a finite graph $\mathcal{G}'_s$ so that $\mathcal{G}'_s$ is isomorphic to $\mathcal{G}$ restricted to $C_0 \cup \ldots \cup C_{s-1}$, $\mathcal{G}'_s \subset \mathcal{G}'_{s+1}$ for all s, and f_s is the isomorphism constructed at stage s. Our desired graph will be $\mathcal{G}' = \cup_s \mathcal{G}'_s$.

At stage 0, set $\mathcal{G}'_0$ to be the empty graph. Set f_0 to be undefined. Say that all components C_i are free for all requirements P_e.

At stage $s + 1$, consider $\mathcal{G}_s$ obtained by adding C_s to $\mathcal{G}_{s-1}$. Let $C'_0, \ldots, C'_{s-1}$ be all components in $\mathcal{G}'_{s-1}$ where each C'_i is isomorphic to C_i via f_s for $i < s$. Find minimal $e \leq s + 1$ such that for some $\langle i, j \rangle < s$ with $i \neq j$ we have:

1. P_e requires attention and $\Phi_{e,s+1}(v, w) \downarrow$ for some $v \in C'_i$ and $w \in C'_j$
2. C_i and C_j disjointly embed into C_s or $\Phi_{e,s+1}(v, w) \neq 0$.
3. The components C_i and C_j are free for P_e.

If such e does not exist then go to stage $s + 2$. If $\Phi_{e,s+1}(v, w) \neq 0$, declare P_e does not require attention, and declare C_i and C_j not free for all P_t with $t > e$. Otherwise, act as follows: (1) Extend C'_i and C'_j to a single component, denoted by C'_s, such that $C'_s \cong C_s$; (2) Build new copies C'_i and C'_j isomorphic to C_i and C_j; (3) Redefine f_s by mapping C_i to C'_i, C_j to C'_j and C_s to C'_s. Declare C_i, C_j, C_s not free for P_t with $t > e$, declare P_t requires attention for $t > e$, and declare P_e does not require attention. This completes the construction for $\mathcal{G}'_{s+1}$.

The correctness of the construction is now a standard proof. The proof is based on the following two observations. First of all, one inductively shows that each requirement P_e is satisfied. Secondly, one proves that the function $f(v) = \lim_s f_s(v)$ establishes an isomorphism between $\mathcal{G}$ and $\mathcal{G}'$ constructed. □

Proof (2). Suppose $h(i, j)$ is finite for all pairs $i, j > n$ for some fixed n, and that h is computable. We show that the reachability relation for $\mathcal{H}$ is intrinsically decidable. Note that we can view $\mathcal{G}$ as an effective disjoint union of finite graphs $D_0, \ldots, D_n, C_0, C_1, C_2, \ldots$, where $h(i, j)$ is finite for all (i, j) corresponding to C_i, C_j. Let $\mathcal{G}'$ be another computable presentation of $\mathcal{G}$. We want to decide the reachability problem on $\mathcal{G}'$. Since $\mathcal{G}' \cong \mathcal{G}$, we may assume that we are given $D'_0 \cong D_0, \ldots, D'_n \cong D_n$, that is, we can compute membership in the D'_i. Suppose v and w are vertices of $\mathcal{G}'$. Since $\mathcal{G}'$ is computable, we can approximate each component $C(x)$ of vertex x by stages $C_0(x) \subseteq C_1(x) \subseteq \ldots$ so that $C(x) = \cup_s C_s(x)$. We can decide whether w is reachable from v using the following algorithm:

1. If $v \in D'_i$ for some $1 \leq i \leq n$, then $R(v, w) \Longleftrightarrow w \in D'_i$.
2. If at any stage s, a path is found from v to w, then declare v, w connected.
3. Find stage s_1 at which $C_{s_1}(v) = C_i$ and $C_{s_1}(w) = C_j$ for some distinct i, j.
4. Find $C_{k_1}, \ldots, C_{k_{h(i,j)}}$ all the components of $\mathcal{G}$ into which C_i and C_j disjointly embed (These can be found since the size function for $\mathcal{G}$ is computable) .
5. Find the first stage $s_2 \geq s_1$ such that $\mathcal{G}'$ provides components $C'_{k_1}, \ldots, C'_{k_{h(i,j)}}$ isomorphic to $C_{k_1}, \ldots, C_{k_{h(i,j)}}$, and v and w belong to these components.
6. If v and w are in the same component C'_l for some $l \in \{k_1, \ldots, k_{h(i,j)}\}$ then declare v, w connected; otherwise, declare v, w are not connected.

By stage s_2 when the components $C'_{k_1}, \ldots, C'_{k_{h(i,j)}}$ are found the components $C_{s_2}(v)$ and $C_{s_2}(w)$ may now properly extend the old approximations $C_{s_1}(v)$ and $C_{s_1}(w)$. For example, it may be that $C_{s_2}(v) = C_{s_2}(w)$ in which case by item

(2) v and w are declared connected. It is not too hard to see that the algorithm provided decides the reachability problem for the graph $\mathcal{G}'$.   □

Proof (3). We need to build a computable copy $\mathcal{G}' \cong \mathcal{G}$ such that the reachability relation is not decidable on $\mathcal{G}'$. We use the same construction as in the proof of part (1) of this theorem. The only difference is that we modify the list $C_0, C_1, \ldots$ to contain only those components of $\mathcal{G}$ for which $h(i,j)$ is finite.

Suppose P_e is the requirement with the highest priority that is not satisfied. Let s be the stage when all requirements with higher priorities are satisfied. Since Φ_e is the characteristic function of the reachability relation, we can compute the function h as follows. Consider (i,j) such that C_i, C_j are free for P_e. Note that there are only finitely many C_i that are not free for P_e. Let t be the stage $> s$ such that $\Phi_{e,t}(v,w)$ is defined for some $v \in C_i'[t]$ and $w \in C_j'[t]$. Such a stage must exist since Φ_e is total, and since by the construction each C_i' is shifted at most finitely often. We must have $\Phi_{e,t}(v,w) = 0$, as otherwise P_e would have been satisfied at stage t. From this stage on C_i and C_j cannot be disjointly embedded into C_k for all $k > t$. Hence $h(i,j)$ can be computed effectively, a contradiction.   □

Corollary 2. *The reachability relation on $\mathcal{G}$ is intrinsically decidable if and only if $h(i,j)$ is computable and there is an n such that $h(i,j)$ is finite for all $i, j > n$.*

Proof. The conditions on h in parts (1), (2) and (3) cover all possibilities, and only condition (2) gives an intrinsically decidable rechability relation.

Theorem 1. *The reachability problem for $\mathcal{G}$ is intrinsically decidable if and only if both the reachability and its complement are existentially definable.*

Proof. We need only prove the direction that if the reachability problem for $\mathcal{G}$ is intrinsically decidable, then the complement of the reachability relation is existentially definable. Suppose that the reachability problem for $\mathcal{G}$ is intrinsically decidable. Then by Corollary 2, h is computable and has finite values almost everywhere. Let $\mathcal{G}$ be viewed as an effective disjoint union of finite graphs $D_0, \ldots, D_n, C_0, C_1, C_2, \ldots$, where $h(i,j)$ is finite for all (i,j) corresponding to C_i, C_j, as in the proof of Lemma 2 (2). We need to exhibit an existential $\mathcal{L}_{c\omega}$ formula $\psi(v, w, \bar{d})$ such that for any computable presentation $\mathcal{G}'$ of $\mathcal{G}$ there exists some tuple $\bar{d}$ of vertices from $\mathcal{G}'$, such that for any vertices v', w' of $\mathcal{G}'$, we have that $\mathcal{G}' \models \neg R(v', w') \iff \psi(v', w', \bar{d})$. We use the proof of Lemma 2 (2) to provide the formula. For any tuple $\bar{x} = (x_0, \ldots, x_m)$, let $\delta(\bar{x}) := \bigwedge_{p \neq q} \neg x_p = x_q$.

For $1 \leq i \leq n$, let $\varphi_i(v, w, \bar{d}_i) :=$ "$v \in D_i$" $\wedge \neg$"$w \in D_i$". Here "$v \in D_i$" abbreviates the formula $v = d_{i,1} \vee \ldots \vee v = d_{i,n_i}$, where $n_i = |D_i|$.

For all pairs $(i,j) \in \omega^{[2]}$, we can compute $h(i,j)$, and since $\mathcal{G}$ has computable size function, we can compute $C_{k_1}, \ldots C_{k_{h(i,j)}}$, the $h(i,j)$ many distinct components in $\mathcal{G}$ into which C_i and C_j disjointly embed. That is, for each $l \in \{i, j, k_1, \ldots, k_{h(i,j)}\}$, we can compute the vertices $\{c_{l,0}, \ldots, c_{l,n_l}\}$ of C_l. We let $\bar{x}_l = (x_{l,0}, \ldots, x_{l,n_l})$. Now for all pairs $(i,j) \in \omega^{[2]}$, define

$$\psi_{i,j}(v,w) := \exists \overline{x}_i \overline{x}_j \overline{x}_{k_0}...\overline{x}_{k_{h(i,j)}} [\delta(\overline{x}_i, \overline{x}_j, \overline{x}_{k_0}, ..., \overline{x}_{k_{h(i,j)}})$$

$$\wedge \bigvee_{1 \le p \le n_i} v = x_{i,p} \wedge \bigvee_{1 \le p \le n_j} w = x_{j,p} \wedge$$

$$\bigwedge_{\substack{(l,m \in \{i,j,k_1,...,k_{h(i,j)}\} \\ 1 \le p \le n_l, 1 \le q \le n_m)}} [\bigwedge_{\mathcal{G} \models E[c_{l,p},c_{m,q}]} Ex_{l,p}x_{m,q} \wedge \bigwedge_{\mathcal{G} \not\models E[c_{l,p},c_{m,q}]} \neg Ex_{l,p}x_{m,q}]].$$

There are only finitely many formulas of the form φ_i, and the formulas of the form $\psi_{i,j}$ are computable in (i,j). Hence, the following formula is an effective disjunction of existential first order formulas:

$$\psi(v,w,\overline{d}_1,...,\overline{d}_n) := \bigvee_{1 \le i \le n} \varphi_i(v,w,\overline{d}_i) \wedge \bigvee_{(i,j) \in \omega^{[2]}} \psi_{i,j}(v,w).$$

From the proof of Lemma 2 (2) it is not hard to see that the formula above existentially defines the complement of the reachability relation. □

3 Counterexample

A natural question arises whether we can remove the assumption on computability of the size function. The goal of this section is to show the assumption *cannot* be omitted by outlining the proof of the following theorem:

Theorem 2. *There exists a strongly locally finite computable graph $\mathcal{G}$ with intrinsically decidable reachability relation such that the complement of the relation is not existentially definable.*

Proof. We give a stage-wise construction of $\mathcal{G}$ on which the reachability relation is decidable. Our construction must guarantee the following two properties of $\mathcal{G}$.

First, we need to guarantee that the complement of reachability is *not* existentially definable. Let $\psi_0, \psi_1, \psi_e, ...$ be an effective listing of all infinitary effective existential formulas with finitely many parameters from $\mathcal{G}$. For each e the graph $\mathcal{G}$ must provide vertices v and w such that $\mathcal{G} \models R[v,w] \iff \mathcal{G} \models \psi_e[v,w]$. This will guarantee that the complement of reachability is not existentially definable.

The second property is that reachability in $\mathcal{G}$ must be intrinsically decidable. This is done as follows. Let $\mathcal{G}_0, \mathcal{G}_1, \mathcal{G}_2, ...$ be an effective enumeration of all computable graphs. We ensure that if $\mathcal{G}_e \cong \mathcal{G}$ then a computable isomorphism between $\mathcal{G}$ and $\mathcal{G}_e$ exists. This with the fact that reachability will be decidable on $\mathcal{G}$ will guarantee the intrinsic decidability of the reachability relation.

We will construct our graph $\mathcal{G}$ to consist of finite chains, as described below. A **cycle** of length $n > 2$, denoted as $\mathcal{C}_n$, is a graph isomorphic to $\{\{1,...,n\}, E\}$, where $E = \{\{1,2\}, \{2,3\}, ..., \{n-1,n\}, \{n,1\}\}$. We **link** the cycle $\mathcal{C}_n$ to the cycle $\mathcal{C}_m$ (both share no vertex) by adding a single new vertex v that has an edge to n in $\mathcal{C}_n$ and an edge to 1 in $\mathcal{C}_m$. Call the resulting graph the **chain** $\mathcal{C}_n\mathcal{C}_m$. Similarly, chains $\mathcal{C}_{n_1}...\mathcal{C}_{n_k}$ and $\mathcal{C}_{m_1}...\mathcal{C}_{m_l}$ are linked to form the chain $\mathcal{C}_{n_1}...\mathcal{C}_{n_k}\mathcal{C}_{m_1}...\mathcal{C}_{m_l}$.

The strategy to defeat a single $\psi_e = \bigvee_{i \in \omega} \psi_{e,i}$ (in achieving the first property) is as follows. First, construct unique cycles C_{e_1}, C_{e_2}, and C_{e_3} in $\mathcal{G}$ that do not use any parameters mentioned by ψ_e. Choose $v \in C_{e_1}$ and $w \in C_{e_2}$. If at some stage we see that $\mathcal{G}_s \models \psi_{e,i}[v, w]$, then extend the cycle C_{e_3} to a chain $C_{e_1} C_{e_3} C_{e_2}$. Let v' and w' be the images of v and w under the disjoint embedding of C_{e_1} and C_{e_2} into $C_{e_1} C_{e_3} C_{e_2}$. Then since $\psi_{e,i}$ is an existential formula, we must have $\mathcal{G}_{s+1} \models \psi_{e,i}[v', w']$, though also $\mathcal{G}_{s+1} \models R[v', w']$.

The strategy to satisfy the second property for a single $\mathcal{G}_e$ (say $\mathcal{G}_0$) in the presence of one ψ_e is the following. Suppose we knew that $\mathcal{G}_0 \cong \mathcal{G}$, and we wanted to build a computable isomorphism $g_0 : \mathcal{G} \to \mathcal{G}_0$, while still working to defeat ψ_e. In this case we would again begin by constructing cycles C_{e_1}, C_{e_2}, and C_{e_3} in $\mathcal{G}$. However, we would wait until $\mathcal{G}_e$ provided components isomorphic to C_{e_1}, C_{e_2}, and C_{e_3}, and define our computable isomorphism from $\mathcal{G}$ to $\mathcal{G}_0$ accordingly, before choosing $v \in C_{e_1}$ and $w \in C_{e_2}$ and waiting for a stage where $\mathcal{G}_s \models \psi_{e,i}[v, w]$. At that point we would proceed as before, and we would wait for $\mathcal{G}_0$ to extend its component of C_{e_3} to one of the form $C_{e_1} C_{e_3} C_{e_1}$, and we would then define g_0 on this extension. Since we do not know whether $\mathcal{G}_0 \cong \mathcal{G}$, we will need six unique cycles $C_{e0_1}, C_{e0_2}, C_{e0_3}$ and $C_{e1_1}, C_{e1_2}, C_{e1_3}$. The cycles $C_{e0_1}, C_{e0_2}, C_{e0_3}$ will be used to defeat ψ_e under the assumption that $\mathcal{G}_0 \not\cong \mathcal{G}$. They will not wait for $\mathcal{G}_0$ to provide isomorphic copies of them before proceeding against ψ_e. The cycles $C_{e1_1}, C_{e1_2}, C_{e1_3}$ will work under the assumption that $\mathcal{G}_0 \cong \mathcal{G}$. Once $\mathcal{G}_0$ exhibits copies isomorphic to $C_{e1_1}, C_{e1_2}, C_{e1_3}$, then we will use these cycles to work against ψ_e. We will know that we do not need the cycles $C_{e0_1}, C_{e0_2}, C_{e0_3}$, and so we will link new distinct cycles to each of $C_{e0_1}, C_{e0_2}, C_{e0_3}$ and $C_{e0_1} C_{e0_3} C_{e0_2}$ (if present), in order to distinguish them for future definition of the isomorphism g_0.

The general construction is a standard and quite technical $\emptyset''$ priority tree construction, where to defeat ψ_e while still building the possible computable isomorphisms for $\mathcal{G}_i$ with $i < e$, we need to have cycles of the form $C_{\sigma_1}, C_{\sigma_2}, C_{\sigma_3}$, where $\sigma \in \{0, 1\}^e$. That is, the finite binary string σ codes the guess as to which graphs of higher priority are actually isomorphic to $\mathcal{G}$, and waits for those it feels are isomorphic to show isomorphic components before proceeding against ψ_e.

4   Locally Finite Graphs

Here we consider locally finite graphs by allowing infinite components. Recall that a graph is *locally finite* if each vertex belongs to only finitely many edges. A locally finite computable graph $\mathcal{G}$ is *highly computable* if given a vertex v of $\mathcal{G}$ one can compute the number of edges of v. The state spaces of Turing machines are examples of highly computable graphs. Clearly, the graph constructed in Theorem 2 is *not* highly computable.

It is easy to turn the graph built in Theorem 2 into a locally finite graph with intrinsically decidable reachability relation such that (1) *all* components of the graph are infinite and (2) the complement of reachability is not existentially definable. One may assume that the reason for such a phenomenon is that the graph is not highly computable. However, by slightly modifying the construction in Theorem 2, one can prove this:

Theorem 3. *There exists a highly computable graph $\mathcal{G}$ (that necessarily possesses infinite components) with intrinsically decidable reachability relation such that the complement of the relation is not existentially definable.*

To get a positive result we use the Ash-Nerode theorem [6] applied to reachability relations on computable graphs. Let $\mathcal{G} = (V, E)$ be a computable graph and R be its reachability relation. Assume that the following condition, called the *Ash-Nerode condition*, holds: There exists an algorithm that given a FO existential formula $\phi(x, y, \bar{a})$, where $\bar{a} \in V$, decides whether $\mathcal{G} \models \forall x \forall y (\phi(x, y, \bar{a}) \rightarrow \neg R(x, y))$. This implies that R is decidable, and given and FO existential formula $\psi(\bar{a})$, where $\bar{a} \in V$, we can decide whether $\mathcal{G} \models \psi(\bar{a})$.

Theorem 4 (Ash-Nerode [6]). *If the computable graph $\mathcal{G} = (V, E)$ and the reachability relation R satisfies the Ash-Nerode condition then R is intrinsically decidable if and only if R and $V^2 \setminus R$ are both existentially definable.* □

Below we provide a sufficient condition for reachability *not* to be intrinsically decidable. Our goal is to extend Lemma 2 (1) for locally finite computable graphs.
   Assume that $g(i) = \lim_{j \to \infty} f(i, j)$ exists for every i, where $f : \omega \times \omega \to \omega$ is a computable function. Intuitively, the function f approximates the value $g(i)$ by making finitely many guesses about the value of $g(i)$ and eventually makes a correct guess. We say that a set $X \subseteq \omega$ is a Δ_2^0-set if there exists a computable function $f : \omega \times \omega \to \{0, 1\}$ such that $X(i) = \lim_j f(i, j)$ for all $i \in \omega$. The function f is called an approximation to X.

Proposition 1. *Let $\mathcal{G}$ be a locally finite computable graph with decidable reachability relation. For each $s \in \omega$, let $\mathcal{G}_s$ be the restriction of the graph of $\mathcal{G}$ to $\{0, ..., s\}$. Since $\mathcal{G}$ is computable, we can compute $\mathcal{G}_s$. For each $v \in \{0, ..., s\}$, let $C_s(v)$ denote the connected component of v in $\mathcal{G}_s$. Assume that there exists an infinite Δ_2^0-set of vertices X such that (1) Any two distinct elements of X are in distinct components of $\mathcal{G}$; and (2) For all $(x, y) \in X^{[2]}$ and for all $t \geq 1$ there exist infinitely many distinct components of $\mathcal{G}$ into which $C_t(x)$ and $C_t(y)$ disjointly embed. Then the reachability relation on $\mathcal{G}$ is not intrinsically decidable.*

Proof. Since reachability on $\mathcal{G}$ is decidable, we may assume that if $C_{\max(v,w)}(v) \neq C_{\max(v,w)}(w)$, then $C_s(v) \neq C_s(w)$ for all s. We build a computable graph $\mathcal{G}' \cong \mathcal{G}$ by meeting the requirements for $e \in \omega$:

$P_e : \Phi_e$ is not the characteristic function of the reachability relation on $\mathcal{G}'$.

We will construct $\mathcal{G}'$ by stages. At each stage s we will have a function $g_s : \mathcal{G}_s \cong \mathcal{G}'_s$ and we will ensure that $g = \lim_s g_s$ exists. If we declare that $g_s(v) = v'$, then we will define g_s such that $g_s : C_s(v) \cong C_s(v')$. If at a later stage t the component of v in $\mathcal{G}$ grows ($C_s(v) \subsetneq C_t(v)$), and we still have $g_t(v) = g_s(v)$, then we will add a new vertex to $\mathcal{G}'_t$ and define g_t to extend g_s so that $g_t : C_t(v) \cong C_t(v')$. To meet requirement P_e we will find vertices v'_e and w'_e in $\mathcal{G}'$ such that if $\Phi_e(v'_e, w'_e) = 1$ then $\mathcal{G}' \models \neg E v'_e w'_e$ and if $\Phi_e(v'_e, w'_e) = 0$ then $\mathcal{G}' \models E v'_e w'_e$. Let $\{X_s\}_{s \in \omega}$ be a computable approximation to X. Find minimal $e \leq s + 1$ such that P_e requires attention and either: (1) $\Phi_{e,s+1}(v', w') \downarrow \neq 0$ for some v' and w' that are free for P_e and are such that $C_s(v') \neq C_s(w')$; or (2) $\Phi_{e,s+1}(v', w') = 0$ for the least pair $(v, w) \in X_s$ for which v' and w' both free for P_e. (We have used the shorthand $g_s^{-1}(v') = v$.)

If such e does not exist then go on to the next stage. Otherwise, if case (1), declare P_e does not require attention, and declare all current and future members of $C_s(v')$ and $C_s(w')$ not free for all P_t with $t > e$. In case (2), speed up the enumerations of X and $\mathcal{G}$ until either (A) we find a stage $t > s$ where $(v, w) \notin X_t$ or (B) we find a new component $C_t(z)$ on which we have yet to define g and into which $C_s(v)$ and $C_s(w)$ disjointly embed. In case (A), move to stage $t+1$ of the construction. In case (B), extend $C_s(v')$ and $C_s(w')$ to a single component, denoted by $C_t(z')$, such that $C_t(z') \cong C_t(z)$; build new copies $C_t(v')$ and $C_t(w')$ isomorphic to $C_t(v)$ and $C_t(w)$, respectively; redefine g_t by mapping z to z', v to (the new) v', w to (the new) w', and extending the map to an isomorphism $\mathcal{G}_t \cong \mathcal{G}'_t$. Declare all current and future members of $C_t(v')$, $C_t(w')$, and $C_t(z')$ not free for all P_t with $t > e$, declare P_t requires attention for all $t > e$, and declare P_e does not require attention. Continue to stage $t+1$ of the construction. This completes the construction for $\mathcal{G}'_{s+1}$.

One can now show by induction on e that each requirement is satisfied, having only caused finitely many components to be non-free for lower priority requirements. The marking of all relevant components as "not free" for lower priority requirements after an action for P_e ensures that g and g^{-1} are only ever redefined finitely often on any given input. This together with the fact that at each stage s, $g_s : \mathcal{G}_s \cong \mathcal{G}'_s$ shows that g establishes an isomorphism between $\mathcal{G}$ and $\mathcal{G}'$. Thus $\mathcal{G} \cong \mathcal{G}'$, but the reachability relation is not decidable on $\mathcal{G}'$.   □

5   Application

We now apply results obtained to the class of automatic graphs. Recall that a graph is *automatic* if its domain is finite automaton recognizable, and there exists a two tape synchronous finite automaton that recognizes the edge relation. For precise definition and examples see [14]. Given an automatic graph $\mathcal{G}$, a vertex v, and $\Phi(x)$ a first order formula, one can effectively decide if $\Phi(v)$ is true [14]. Hence, if G is automatic then the function that outputs the number of edges for any given vertex v in G is computable. We now apply our results above to the following theorem.

Theorem 5. *Let $\mathcal{G}$ be an automatic graph such that either $\mathcal{G}$ is strongly locally finite or the reachability in $\mathcal{G}$ is finite automata recognizable. Then the reachability problem for $\mathcal{G}$ is intrinsically decidable if and only if the reachability relation and its complement are both existentially definable.*

Proof. If $\mathcal{G}$ is strongly locally finite then Theorem 1 does the job. This follows from the fact that for automatic strongly locally finite graphs the size function is always computable. Now, suppose that $\mathcal{G} = (V, E)$ is automatic and the reachability relation R is finite automata recognizable. Then, (V, E, R) is an automatic structure. Hence, the Ash-Nerod condition holds true for this structure. The rest follows from the Ash-Nerode theorem.

References

1. Abdulla, P.A., Čerāns, K., Jonsson, B., Tsay, Y.: Algorithmic analysis of programs with well quasi-ordered domains. ICom 160, 109–127 (2000)
2. Abdulla, P.A., Collomb-Annichini, A., Bouajjani, A., Jonsson, B.: Using forward reachability analysis for verification of lossy channel systems. Formal Methods in System Design 25(1), 39–65 (2004)
3. Aceto, L., Burgueno, A., Larsen, K.G.: Model checking via reachability testing for timed automata. In: Steffen, B. (ed.) TACAS 1998. LNCS, vol. 1384, pp. 263–280. Springer, Heidelberg (1998)
4. Allender, E.: Reachability problems: An update. In: Cooper, S.B., Löwe, B., Sorbi, A. (eds.) CiE 2007. LNCS, vol. 4497, pp. 25–27. Springer, Heidelberg (2007)
5. Alur, R., Dill, D.: Theory of timed automata. TCS 126, 183–235 (1994)
6. Ash, C.J., Nerode, A.: Intrinsically recursive relations. In: Aspects of effective algebra (Clayton, 1979), pp. 26–41. Upside Down A Book Co., Yarra Glen (1981)
7. Bouajjani, A., Esparza, J., Maler, O.: Reachability analysis of pushdown automata: Application to model checking. In: Mazurkiewicz, A., Winkowski, J. (eds.) CONCUR 1997. LNCS, vol. 1243. Springer, Heidelberg (1997)
8. Bouajjani, A., Esparza, J., Schwoon, S., Strejcek, J.: Reachability analysis of multithreaded software with asynchronous communication. In: Ramanujam, R., Sen, S. (eds.) FSTTCS 2005. LNCS, vol. 3821, pp. 348–359. Springer, Heidelberg (2005)
9. Bouajjani, A., Touili, T.: On computing reachability sets of process rewrite systems. In: Giesl, J. (ed.) RTA 2005. LNCS, vol. 3467, pp. 484–499. Springer, Heidelberg (2005)
10. Caucal, D.: On infinite terms having a decidable monadic theory. In: Diks, K., Rytter, W. (eds.) MFCS 2002. LNCS, vol. 2420, pp. 165–176. Springer, Heidelberg (2002)
11. Colcombet, T., Löding, C.: Transforming structures by set interpretations, Technical report AIB-2006-07 of RWTH Aachen (2006)
12. Ershov, Y.L., Goncharov, S.S., Nerode, A., Remmel, J.B., Marek, V.W.: Handbook of recursive mathematics. Studies in Logic and the Foundations of Mathematics, , vol. 1, 2, vol. 138, 139. North-Holland, Amsterdam (1998)
13. Ibarra, O.H., Bultan, T., Su, J.: On reachability and safety in infinite-state systems. International J. of Foundations of Comp. Sci. 12(6), 821–836 (2001)
14. Khoussainov, B., Nerode, A.: Automatic presentations of structures. In: Leivant, D. (ed.) LCC 1994. LNCS, vol. 960, pp. 367–392. Springer, Heidelberg (1995)
15. Libkin, L.: Elements of finite model theory. Texts in Theoretical Computer Science. An EATCS Series. Springer, Berlin (2004)
16. Rybina, T., Voronkov, A.: A logical reconstruction of reachability. In: Ershov Memorial Conference, pp. 222–237 (2003)
17. Soare, R.I.: Recursively enumerable sets and degrees. Perspectives in Mathematical Logic. Springer, Berlin (1987)
18. Tarjan, R.E.: Depth-first search and linear graph algorithms. SIAM J. Comput. 1(2), 146–160 (1972)

Some New Modes of Competence-Based Derivations in CD Grammar Systems

Erzsébet Csuhaj-Varjú[1,*], Jürgen Dassow[2], and György Vaszil[1,**]

[1] Computer and Automation Research Institute, Hungarian Academy of Sciences
Kende u. 13-17, H-1111 Budapest, Hungary
{csuhaj,vaszil}@sztaki.hu
[2] Otto-von-Guericke-Universität Magdeburg, Fakultät für Informatik PSF 4120,
D-39016 Magdeburg, Germany
dassow@iws.cs.uni-magdeburg.de

Abstract. We introduce some new cooperation protocols for cooperating distributed (CD) grammar systems. They depend on the number of different nonterminals present in the sentential form if a component has finished its work, i.e. on the final competence or efficiency of the grammar on the string (the competence is large if the number of the different nonterminals is small). We prove that if the underlying derivation mode is the t-mode derivation, then some variants of these systems determine the class of random context ET0L languages. If these CD grammar systems use the k step limited derivations (for $k \geq 3$) as underlying derivations, they are able to generate any recursively enumerable language.

1 Introduction

The original motivation for the introduction of cooperating distributed grammar systems (CD grammar system for short) was to model the blackboard type problem solving systems by grammatical means (see [4,6]). In these systems, the grammars generate a common sentential form in turn according to a cooperation protocol (derivation mode). The grammars represent problem solving agents, the sentential form describes the actual state of the problem solving process, and the derivation corresponds to the solving of the problem. Most of the derivation modes of these systems that have been studied so far are based on the so-called competence of the component grammars which is essentially the number of different nonterminals in the sentential form which can be rewritten by the rules of the component (the competence is high, if this number is large). This concept reflects the idea that the nonterminals correspond to unsolved subproblems and their replacement to a step to their solutions. For details, we refer to [3,1,2,5] and the summarizing article [7].

---

* Also works with: Department of Algorithms and Their Applications, Faculty of Informatics, Eötvös Loránd University, H-1117 Budapest, Pázmány Péter sétány 1/c.
** Part of the research was carried out while the author was at the Institute of Computer Science, University of Potsdam.

M. Ito and M. Toyama (Eds.): DLT 2008, LNCS 5257, pp. 228–239, 2008.

The above notion of competence is measured on the current sentential form which is a natural idea, since the current sentential form is given/known, and it is easy to compute the most competent components. On the other hand, the components can create new nonterminals (i.e., new subproblems), such that the sentential form obtained by the work of a competent component can be very difficult with respect to the problem solving. Therefore, another natural concept of competence can be based on the number of different nonterminals (i.e., unsolved subproblems) in the sentential form obtained after the grammar finished the derivation.

In this paper we propose derivation modes based on three variations of this new concept of competence. The first one allows a component to be active if it is able to derive a sentential form with a minimal number of different nonterminals (or unsolved subproblems). Then one of these most competent components performs an arbitrary derivation in the given underlying mode. The second variant differs from the first one only in the requirement that the chosen most competent component has to perform a derivation which leads to a sentential form with a minimal possible number of different nonterminals. In the case of the third variant every component performs a derivation on the sentential form simultaneously and proposes its result; then a proposed string with a minimal number of different nonterminal occurrences is chosen and the derivation continues with this string.

In this paper we study the generative power of the associated CD grammar systems. We prove that if the underlying derivation mode is the t-mode derivation, then some variants of these systems determine the class of random context ET0L languages. If these CD grammar systems use the k step limited derivations (for $k \geq 3$) as underlying derivations, they are able to generate any recursively enumerable language. Due to the lack of space, we present the proofs with only the necessary details; the full proofs can easily be obtained with some simple technical considerations.

2 Definitions

We first present some notations used in the paper. The cardinality of a finite set X is denoted by $card(X)$. The set of all (non-empty) words over an alphabet X is denoted by X^* (and X^+, respectively); the empty word is denoted by λ. For a non-empty word w, $alph(w)$ denotes the minimal alphabet X (with respect to inclusion) such that $w \in X^+$.

Throughout the paper, we assume that the reader is familiar with the basic notions of formal language theory ([10]). For the sake of completeness, we recall the definition of some language families which will be used in the sequel. We first note that we denote by $\mathcal{L}(CF)$ and $\mathcal{L}(RE)$ the class of context-free and recursively enumerable languages, respectively.

An *extended tabled interactionless L system* (an ET0L system) is a construct $G = (V, T, \mathcal{P}, w)$ where V is an alphabet, T is a subset of V, $w \in V^+$ and, $\mathcal{P} = \{P_1, P_2, \ldots, P_r\}$ for some $r \geq 1$ where, for $1 \leq i \leq r$, P_i is a finite subset of

$V \times V^*$ such that, for any $a \in V$, there is at least one element (a, v) in P_i. As usually, we write $a \to v$ instead of (a, v). We say that $x \in V^+$ directly derives $y \in V^*$ in G (written as $x \Longrightarrow y$), if $x = x_1x_2 \ldots x_n$, $y = y_1y_2 \ldots y_n$, $x_i \in V$, $y_i \in V^*$, $1 \le i \le n$, and there is a j, $1 \le j \le r$, such that $x_i \to y_i \in P_j$ for $1 \le i \le n$. The language $L(G)$ generated by the ET0L system G is defined as $L(G) = \{z \mid z \in T^*, \ w \Longrightarrow^* z\}$, where $\Longrightarrow^*$ is the reflexive and transitive closure of $\Longrightarrow$.

A *random context ET0L system* (an RCET0L system, in short) is a construct $G = (V, T, \mathcal{P}, w)$ where V, T and w are specified as in the case of an ET0L system, $\mathcal{P} = \{P_1, P_2, \ldots, P_r\}$ for some $r \ge 1$ such that, for $1 \le i \le r$, $P_i = (P_i', R_i, Q_i)$, where R_i and Q_i are subsets of V, and $G' = (V, T, \{P_1', P_2', \ldots, P_r'\}, w)$ is an ET0L system. We say that $x \in V^+$ directly derives $y \in V^*$ in G (written as $x \Longrightarrow y$), if $x \Longrightarrow y$ holds with respect to P_i' in G' for some i, $1 \le i \le r$, and every letter of R_i occurs in x and no letter of Q_i occurs in x. The language $L(G)$ generated by the RCET0L system G is defined as $L(G) = \{z \mid z \in T^*, \ w \Longrightarrow^* z\}$, where $\Longrightarrow^*$ is the reflexive and transitive closure of $\Longrightarrow$.

By $\mathcal{L}(ET0L)$ and $\mathcal{L}(RCET0L)$ we denote the families of all languages generated by ET0L systems and random context ET0L systems. For detailed information on these language families we refer to [8] and [10].

A *context-free programmed grammar* (with appearance checking) is denoted by $G = (N, T, P, S)$ where N, T with $N \cap T = \emptyset$ are the set of nonterminals and terminals, respectively, $S \in N$ is the start symbol (the axiom), and P is a finite set of rules of the form $(r : A \to z, Succ_r, Fail_r)$, where $r \in lab(P)$ is the label of the rule, $A \in N$, $z \in (N \cup T)^*$, and $Succ_r, Fail_r \subseteq lab(P)$ are the success and failure sets of rule r, respectively. A word $w \in T^*$ is an element of the language $L(G)$ generated by G if and only if there is a derivation $S = y_0 \Longrightarrow_{r_1} y_1 \Longrightarrow_{r_2} y_2 \Longrightarrow_{r_3} \cdots \Longrightarrow_{r_k} y_k = w$ where, for $1 \le i \le k$, $r_i = (r_i : A_i \to z_i, Succ_{r_i}, Fail_{r_i})$ and (1) $y_{i-1} = x_1 A_i x_2$, $y_i = x_1 z_i x_2$ and, if $i < k$, $r_{i+1} \in Succ_{r_i}$, or (2) $|y_{i-1}|_{A_i} = 0$, $y_i = y_{i-1}$ and, if $i < k$, $r_{i+1} \in Fail_{r_i}$. The class of languages generated by context-free programmed grammars is denoted by $\mathcal{L}(PR)$.

We now present the notion of a cooperating distributed grammar system (for details we refer to [6] and [9]).

A *cooperating distributed grammar system* (a CD grammar system, in short) is a construct $G = (N, T, P_1, P_2, \ldots, P_n, S)$, where N is a set of nonterminals and T is a set of terminals, $S \in N$ is the start symbol, and for $1 \le i \le n$, the component P_i is a set of context-free productions.

For $1 \le i \le n$, the set of all nonterminals A such that there is a rule $A \to w$ in P_i is denoted by $dom(P_i)$.

We say that a derivation $D_1 : x \Longrightarrow^* y$ in the CD grammar system G is performed in the terminating mode (for short, in t-mode) if there is an i, $1 \le i \le n$, such that $D_1 : x = x_0 \Longrightarrow_{P_i} x_1 \Longrightarrow_{P_i} x_2 \Longrightarrow_{P_i} \cdots \Longrightarrow_{P_i} x_k = y$, $k \ge 1$, and $dom(P_i) \cap alph(y) = \emptyset$. That is, a component P_i performs a derivation in the t-mode, if it continues rewriting the sentential form as long as it is able to replace a nonterminal. In this case, we write $x \Longrightarrow_{P_i}^t y$.

We say that a derivation $D_2 : x \Longrightarrow^* y$ in G is performed in the k step mode derivation (in short, in the $= k$-mode), where $k \geq 1$, if there is a component P_i, for some i, $1 \leq i \leq n$, such that $D_2 : x = x_0 \Longrightarrow_{P_i} x_1 \Longrightarrow_{P_i} x_2 \Longrightarrow_{P_i} \cdots \Longrightarrow_{P_i} x_k = y$. That is, component P_i rewrites the sentential form according to the $= k$-mode derivation, if it performs exactly k derivation steps on the string. We write then $x \Longrightarrow_{P_i}^{=k} y$.

The language $L(G, \alpha)$, where $\alpha \in \{t\} \cup \{= k \mid k \geq 1\}$, generated by a CD grammar system $G = (N, T, P_1, P_2, \ldots, P_n, S)$, $n \geq 1$, in the α-mode derivation is defined as the set of all words $w \in T^*$ such that there is a derivation $S = w_0 \Longrightarrow_{P_{i_1}}^{\alpha} w_1 \Longrightarrow_{P_{i_2}}^{\alpha} w_2 \Longrightarrow_{P_{i_3}}^{\alpha} \cdots \Longrightarrow_{P_{i_m}}^{\alpha} w_m = w$ in G.

If one considers CD grammar systems as formal language theoretical models of problem solving, then the nonterminals in a sentential form correspond to unsolved subproblems. Therefore, it is natural to say that a component is *more competent* in solving a subproblem than some other one if starting from the same sentential form it derives a string that contains a smaller number of different nonterminals than the string derived by the other grammar. Notice that this concept is strongly related to the notion of *efficiency*, since a component can be considered more efficient than some other one if it is able to reduce the number of different open subproblems more efficiently. We may also call this type of competence *final competence* since it can only be seen after the component finishes the derivation.

In the following, we shall formally define cooperation protocols based on this new concept of competence in CD grammar systems.

Definition 1. *Let $G = (N, T, P_1, P_2, \ldots, P_n, S)$, $n \geq 1$, be a CD grammar system, let $x \in (N \cup T)^+$ and $\alpha \in \{t\} \cup \{= k \mid k \geq 1\}$. We set*

$$fc_\alpha(x) = \min\{\, card(alph(y) \cap N) \mid x \Longrightarrow_{P_i}^{\alpha} y, alph(y) \neq alph(x), 1 \leq i \leq n\}.$$

We say that the component P_i is a component with maximal final competence (or it is a maximally efficient component) on x in the α-mode derivation if there is a derivation $x \Longrightarrow_{P_i}^{\alpha} z$ such that $fc_\alpha(x) = card(alph(z) \cap N)$ holds.

We require $alph(y) \neq alph(x)$ since otherwise the set of occurring nonterminals (or the set of unsolved subproblems) is not changed.

In the following we show that $fc_\alpha(x)$ can be effectively determined. Let $x \in (N \cup T)^*$ and $\alpha \in \{t\} \cup \{= k \mid k \geq 1\}$ be given. Let $Z = alph(x) \cap N = \{A_1, A_2, \ldots A_m\}$, where $A_i \in N$, $1 \leq i \leq m$. Notice that $fc_\alpha(x)$ only depends on the nonterminals occurring in x, since $fc_\alpha(x)$ coincides with $fc_\alpha(A_1 A_2 \ldots A_m)$. Thus, we can define $fc_\alpha(Z) = fc_\alpha(A_1 A_2 \ldots A_m)$. For $Z \subseteq N$ and i, $1 \leq i \leq n$, there is a finite set $M_{Z,i} = \{M_1, M_2, \ldots M_l\}$ of subsets M_j of N, $1 \leq j \leq l$, such that $(alph(v) \cap N) \in M_{Z,i}$ for all words v which can be generated by the component P_i from x in derivation mode α, where $alph(x) \cap N = Z$ and $\alpha \in \{t\} \cup \{= k \mid k \geq 1\}$. Obviously, $M_{Z,i}$ can be effectively determined (one has only to check whether $A_1 A_2 \ldots A_m$ is able to generate by the application of P_i in the α-mode derivation a word over $(T \cup M)$ for subsets M of N). Thus, for a set $Z \subseteq N$, one can effectively compute $fc_\alpha(Z)$ and the set $I_\alpha(Z)$ of components which have maximal final competence on any word x with $alph(x) \cap N = Z$.

Definition 2. *Let $G = (N, T, P_1, P_2, \ldots, P_n, S)$, $n \geq 1$, be a CD grammar system, $\alpha \in \{t\} \cup \{= k \mid k \geq 1\}$, and let $D : S = w_0 \Longrightarrow_{P_{i_1}}^{\alpha} w_1 \Longrightarrow_{P_{i_2}}^{\alpha} \cdots \Longrightarrow_{P_{i_m}}^{\alpha} w_m \in T^*$, $m \geq 1$, be an α-mode derivation in G.*

We say that D is an (fc, α)-mode derivation (a derivation with maximal final competence in the α-mode), if P_{i_j} is a maximally efficient component in the α-mode derivation on w_{j-1} for all $1 \leq j \leq m$.

We say that D is an (sfc, α)-mode derivation (a derivation with strongly maximal final competence in the α-mode), if D is an (fc, α)-mode derivation and for $1 \leq j \leq m$, $card(alph(w_j) \cap N) = fc_\alpha(w_{j-1})$.

If D is an (fc, α)-mode derivation, then the component starting the derivation is one of the maximally efficient grammars on the current sentential form, but the component is allowed to perform an arbitrary α mode derivation. If it is an (sfc, α)-mode derivation, then the component is not only a maximally efficient one, but it performs an α-mode derivation resulting in a sentential form with a minimal number of different nonterminals.

Another natural variant of competence is the case when the *performance of the components is compared*. In this case, all components perform a derivation on a word simultaneously, i.e., every component P_i generates a word x_i, and the generation continues only with those strings x_j where $card(alph(x_j) \cap N) \leq card(alph(x_i) \cap N)$ hold for $1 \leq i \leq n$.

Definition 3. *Let $G = (N, T, P_1, P_2, \ldots, P_n, S)$, $n \geq 1$, be a CD grammar system. For a string $x \in (N \cup T)^+$ and an n-tuple $(y_1, y_2, \ldots, y_n)$ of words such that $x \Longrightarrow_{P_i}^{\alpha} y_i$, $\alpha \in \{t\} \cup \{= k \mid k \geq 1\}$, we define $cc(x, y_1, y_2, \ldots, y_n)$ as the set of words y_j, $1 \leq j \leq n$, such that $card(alph(y_j) \cap N) \leq card(alph(y_i) \cap N)$ for $1 \leq i \leq n$ and $alph(y_j) \neq alph(x)$.*

Let $D : S = w_0 \Longrightarrow_{P_{i_1}}^{\alpha} w_1 \Longrightarrow_{P_{i_2}}^{\alpha} \cdots \Longrightarrow_{P_{i_m}}^{\alpha} w_m \in T^*$, $m \geq 1$, be an α-mode derivation. We say that D is a (cc, α)-mode derivation (a derivation with comparing competence in the α-mode) if there are words $y_{i,j}$, $1 \leq i \leq n$ and $1 \leq j \leq m$, such that

- *$w_{j-1} \Longrightarrow_{P_i}^{\alpha} y_{i,j}$ for $1 \leq i \leq n$ and $1 \leq j \leq m$,*
- *$y_{i_j, j} \in cc(w_{j-1}, y_{1,j}, y_{2,j}, \ldots, y_{n,j})$ for $1 \leq j \leq m$,*
- *$w_j = y_{i_j, j}$.*

The language $L(G, \beta, \alpha)$, for $\beta \in \{fc, sfc, cc\}$ and $\alpha \in \{t\} \cup \{= k \mid k \geq 1\}$ consists of all words $w \in T^*$ that can be derived from S by a (β, α)-mode derivation in G. The family of languages generated by CD grammar systems with (β, α)-mode derivations is denoted by $\mathcal{L}(CD, \beta, \alpha)$.

We illustrate the above introduced notions by an example.

Example 1. Let us consider a CD grammar system $G = (\{S, A, B, C_1, C_2, C_3, C_4, C_5, C_6\}, \{a, b, c\}, P_1, P_2, P_3, P_4, S)$ where $P_1 = \{S \rightarrow AB, A \rightarrow aA', A \rightarrow a, B \rightarrow bB'\}$, $P_2 = \{S \rightarrow AB, A \rightarrow aA', B \rightarrow b, B \rightarrow bB'\}$, $P_3 = \{S \rightarrow AB, A \rightarrow C_1, B \rightarrow C_2, A \rightarrow C_3C_4, B \rightarrow C_5C_6\}$, $P_4 = \{A' \rightarrow A, B' \rightarrow B\} \cup \{C_i \rightarrow c \mid 1 \leq i \leq 6\}$. Let us start with the (fc, t)-derivations. First we have to apply P_1

or P_2 since $fc_t(S) = 1$ and any application of P_3 leads to a word with at least two nonterminals. By by alternating application of P_1 and P_4 or P_2 and P_4 we generate words of the form $a^n Ab^n B$ as long as the chosen component, i.e. P_1 or P_2 does not use its terminating rule, and P_3 cannot be applied to such a word. If we use a terminating rule of the component, then we obtain $w = a^n b^n B'$ or $w' = a^n A' b^n$. After applying P_4, we can only use P_2 in the first case, and only P_1 in the second case (since $fc_t(w) = fc_t(w') = 0$). Thus, $L(G, fc, t) = \{a^n b^m, a^m b^n \mid n > m \geq 1\}$. Using (sfc, t)-derivations, we have to apply P_1 or P_2 to the axiom and we have to derive a word with one nonterminal only, i.e., we obtain abB' or $aA'b$, respectively. After the application of P_4 the derivation has to terminate. Thus, $L(G, sfc, t) = \{aab, abb\}$. In the case of (cc, t)-derivations, we can apply P_3 only if P_1 and P_2 do not use their terminating rules and from P_3 the rules $A \to C_1$ and/or $B \to C_2$ are used. Therefore, we obtain $L(G, cc, t) = \{a^n b^m, a^m b^n \mid n > m \geq 1\} \cup \{a^n cb^m, a^m b^n c \mid n \geq m \geq 1\} \cup \{a^n cb^n c \mid n \geq 0\}$.

3 The Case of Maximal Final Competence

From [4] it is known that $\mathcal{L}(CD, t) = \mathcal{L}(ET0L)$, and $\mathcal{L}(CF) \subset \mathcal{L}(CD, = k) \subseteq \mathcal{L}(MAT)$, for all $k \geq 2$ where $\mathcal{L}(MAT)$ denotes the class of languages generated by matrix grammars without appearance checking. We prove that CD grammar systems working with (fc, t)-mode derivations generate the class of random context ET0L languages and by the $(fc, = k)$-mode derivations, where $k \geq 3$, the class of recursively enumerable languages.

Lemma 1. $\mathcal{L}(RCET0L) \subseteq \mathcal{L}(CD, fc, t)$.

Proof. Let $G = (V, T, \{(P_1, R_1, Q_1), (P_2, R_2, Q_2), \ldots, (P_r, R_r, Q_r)\}, w)$ be a random context ET0L system with $R_i = \{A_{i,1}, A_{i,2}, \ldots, A_{i,m_i}\}$ for $1 \leq i \leq r$. For $1 \leq i \leq r$ and $1 \leq j \leq m_i + 1$, we set $V' = \{x' \mid x \in V\}$, $V'' = \{x'' \mid x \in V\}$, $V^{(i,j)} = \{x^{(i,j)} \mid x \in V\}$, $V^{(i,j)'} = \{x^{(i,j)'} \mid x \in V\}$. Let us define the homomorphisms $h' : V^* \to (V')^*$ by $h'(x) = x'$, $h'' : V^* \to (V'')^*$ by $h''(x) = x''$, $h_{i,j} : V^* \to (V^{(i,j)})^*$ by $h_{i,j}(x) = x^{(i,j)}$, and $h'_{i,j} : V^* \to (V^{(i,j)'})^*$ by $h'_{i,j}(x) = x^{(i,j)'}$. We construct the simulating CD grammar system G' as follows. Let
$$G' = (N, T, P_0, (P'_i)_{1 \leq i \leq r}, ((P_{i,j})_{1 \leq j \leq m_i + 1})_{1 \leq i \leq r}, ((P'_{i,j}, P''_{i,j})_{1 \leq j \leq m_i})_{1 \leq i \leq r},$$
$(P_{fin,i})_{1 \leq i \leq 2}, S)$, where $N = \{X, X^{(i,j)}, X_1^{(i,j)}, X_2^{(i,j)}, X_3, F_1, F_2, F_3 \mid 1 \leq i \leq r, 1 \leq j \leq m_i\} \cup V' \cup V'' \cup \bigcup_{i=1}^{r} \bigcup_{j=1}^{m_i+1} (V^{(i,j)} \cup V^{(i,j)'})$. Now let
$P_0 = \{S \to Xh'(w)\}$, let
$P'_i = \{x' \to x^{(i,1)} \mid x \in V\} \cup \{X \to X^{(i,1)}\}$, and
$P_{i,m_i+1} = \{x^{(i,m_1+1)} \to h'(w) \mid x \in V \setminus Q_i, x \to w \in P_i\} \cup \{x^{(i,m_1+1)} \to F_1 \mid x \in Q_i\} \cup \{X^{(i,m_i+1)} \to X\}$, for all $1 \leq i \leq r$, let
$P_{i,j} = \{X^{(i,j)} \to X_1^{(i,j)} X_2^{(i,j)}\} \cup \{x^{(i,j)} \to x^{(i,j)'} \mid x \in V\}$,
$P'_{i,j} = \{A_{i,j}^{(i,j)} \to A_{i,j}^{(i,j)'} F_1 F_2 F_3, X^{(i,j)} \to F_1\} \cup \{x^{(i,j)} \to x^{(i,j)'} \mid x \in V, x \neq A_{i,j}\}$,
$P''_{i,j} = \{X_1^{(i,j)} \to X^{(i,j+1)}, X_2^{(i,j)} \to \lambda\} \cup \{x^{(i,j)'} \to x^{(i,j+1)} \mid x \in V\}$, for all $1 \leq i \leq r$, $1 \leq j \leq m_i$. Finally, let

$P_{fin,1} = \{x' \to x'' \mid x \in V\} \cup \{X \to X_3\},$
$P_{fin,2} = \{X_3 \to \lambda\} \cup \{x'' \to F_1 \mid x \in V \setminus T\} \cup \{x'' \to x \mid x \in T\}.$

We now discuss the possible derivations in G'. Starting from the startsymbol, S, we obtain $Xh'(w)$, that is, a word of the form $Xh'(z)$. To $Xh'(z)$, we can only apply the components $P_1', P_2', \ldots, P_r'$ and $P_{fin,1}$. These components have maximal final competence since all of them only change the names of the letters, i.e., all the alphabets of the obtained words have the same cardinality.

Let us assume first that we apply $P_{fin,1}$. Then we obtain $X_3 h''(z)$. The only applicable component is $P_{fin,2}$ which yields z if $z \in T^*$ or a word containing at least one occurrence of F_1 if $z \notin T^*$. In the latter case the derivation cannot terminate since there is no rule with left-hand side F_1. Assume that we apply P_i', i.e., we want to simulate the application of (P_i, R_i, Q_i) with $R_i \neq \emptyset$, that is, with $m_i \geq 1$, to z. The application of P_i' gives $X^{(i,1)} h_{i,1}(z)$. Now two components can be used, $P_{i,1}$ and $P_{i,1}'$. Suppose that $A_{i,1}$ occurs in z. Then $P_{i,1}$ yields $X_1^{(i,1)} X_2^{(i,1)} h_{i,1}'(z)$ and the application of $P_{i,1}'$ leads to z' with $alph(z') = alph(h_{i,j}'(z)) \cup \{F_1, F_2, F_3\}$. Therefore, only $P_{i,j}$ has maximal final competence. The only applicable table is $P_{i,1}''$ which results in $X^{(i,2)} h_{i,2}(z)$. If $A_{i,1}$ does not occur in z, then only $P_{i,1}'$ has maximal final competence, but it introduces F_1 and therefore we cannot terminate the derivation. Thus, we can only terminate successfully the derivation if the first letter $A_{i,1}$ of R_i occurs in z. Continuing the procedure, we can check in the same way in succession the presence of $A_{i,2}, A_{i,3}, \ldots, A_{i,m_i}$ and we finally obtain $X^{(i,m_i+1)} h_{i,m_i+1}(z)$ (if $m_i = 0$, we obtain this string already in the first step). Then the only continuation is the application of P_{i,m_i+1} which results in $Xh'(v)$ where $z \Longrightarrow_{P_i} v$ in G (note that we check the absence of the symbols of Q_i in this step, because an occurrence of a letter of Q_i leads to an occurrence of F_1). Therefore, we have simulated one derivation step of G (on primed versions of the words). Now it is very easy to see that $L(G) = L(G, fc, t)$, the further details are left to the reader.

Lemma 2. $\mathcal{L}(CD, fc, t) \subseteq \mathcal{L}(RCET0L)$.

Proof. Let $G = (N, T, P_1, P_2, \ldots, P_n, S)$ be a CD grammar system working with (fc, t)-mode derivations. We construct the simulating RCET0L system $G' = (V, T, \mathcal{P}, XS)$ as follows: Let $V = N \cup T \cup \{X\} \cup \{[i] \mid 1 \leq i \leq n\}$ and $\mathcal{P} = \{R(U) \mid U \subseteq N\} \cup \{R_{fin}\} \cup \bigcup_{i=1}^{n} \{R(i), R(i)'\}$. Let us define $R(U) = (\{X \to [i] \mid i \in I_t(U)\}, U \cup \{X\}, N \setminus U)$ for $U \subseteq N$, $R(i) = (P_i, \{[i]\}, \emptyset)$, for $1 \leq i \leq n$, $R(i)' = (\{[i] \to X\}, \{[i]\}, dom(P_i))$, for $1 \leq i \leq n$, and $R_{fin} = (\{X \to \lambda\}, \{X\}, N)$.

Assume that we have a sentential form Xv with $v \in (N \cup T)^+$. By the permitting and forbidden contexts, we can only apply the table $R(U)$ with $U = alph(v) \cap N$. Its application gives $[i]v$ where $i \in I_t(U)$, i.e., the component P_i applicable with maximal final competence in t-mode to v. Then, the only applicable table is $R(i)$ which has to be applied as long as the sentential form contains a letter of $dom(P_i)$. Thus, by $R(i)$ we simulate the application of P_i in the t-mode. Finally, we get Xv' by an application of $R(i)'$ where $v \Longrightarrow^t v'$. Thus, the derivations in G' have the form $XS \Longrightarrow^* Xv_1 \Longrightarrow^* Xv_2 \Longrightarrow^* Xv_3 \Longrightarrow^* \ldots \Longrightarrow^*$

$Xv_n \Longrightarrow v_n$ (in the last step we apply the table R_{fin} which ensures that $v_n \in T^*$) where $S \Longrightarrow_{P_{i_1}}^t v_1 \Longrightarrow_{P_{i_2}}^t v_2 \Longrightarrow_{P_{i_3}}^t v_3 \Longrightarrow_{P_{i_4}}^t \cdots \Longrightarrow_{P_{i_n}}^t n_n$ is a terminating (fc, t)-mode derivation of G. Now it is easy to show that $L(G, fc, t) = L(G')$, we leave the details to the reader.

Combining Lemma 1 and 2 we obtain

Theorem 1. $\mathcal{L}(CD, fc, t) = \mathcal{L}(RCET0L)$.

Theorem 2. $\mathcal{L}(RE) = \mathcal{L}(CD, fc, = k)$ *for any* $k \geq 3$.

Proof. We show that $\mathcal{L}(PR) \subseteq \mathcal{L}(CD, fc, = k)$, $k \geq 3$. This is sufficient, since $\mathcal{L}(RE) = \mathcal{L}(PR)$ and the inclusion $\mathcal{L}(RE) \supseteq \mathcal{L}(CD, fc, = k)$ can be proved by using standard techniques. First we deal with the case of $k = 3$. Let $G = (N, T, P, S)$ be a programmed grammar. Without loss of generality we may assume that the rules of G are of the form $(r : A \to z, Succ_r, Fail_r)$ where $r \in lab(P)$ and $z = BC$ with $B, C \in N$ or $z \in T$. We construct the simulating CD grammar system

$G' = (N', T, P_0, (P_{r,i})_{r \in lab(P), 1 \leq i \leq 5}, (P_r)_{r \in lab(P)}, (P_{\bar{r},1}, P_{\bar{r},2})_{r \in lab(P)}, (P_A)_{A \in N},$
$P_{t,1}, P_{t,2}, \bar{S})$ as follows. Let
$N' = \{A, A_r, [r], [r]', [r]'', [\bar{r}], [\bar{r}]', [\bar{r}]'' \mid A \in N, r \in lab(P)\} \cup \{[t], [t]', [t]'', X, X',$
$X'', Y, Y', Z, W, F, S', S'', \bar{S}\}$, and let us define the components of the system in the following way:

$P_0 = \{\bar{S} \to S', S' \to S'', S'' \to [r]SXY \mid r \in lab(P)\}$, and for all rules $(r : A \to BC, Succ_r, Fail_r)$ we define

$P_{r,1} = \{A \to B_r C_r W[r]', X \to X', Y \to Y'\},$
$P_{r,2} = \{[r] \to [r]'', [r]'' \to Z, [r]' \to [s], [r]' \to [\bar{s}], [r]' \to [t] \mid s \in Succ_r\},$
$P_{r,3} = \{Z \to \lambda, X' \to X, Y' \to Y\},$
$P_{r,4} = \{B_r \to B, C_r \to C, W \to \lambda\},$
$P_{r,5} = \{[r] \to [r]'', [r]'' \to F, [s]' \to F \mid s \neq r\},$

and let the set $\mathcal{P}_r = \{P_i^{(r)} \mid 1 \leq i \leq 3\}$ contain the following components

$P_1^{(r)} = \{[r]' \to F, X \to F, Y \to F\}, P_2^{(r)} = \{[r]' \to F, X' \to F, Y \to F\},$ and
$P_3^{(r)} = \{[r]' \to F, X \to F, Y' \to F\}.$ Moreover, we define
$P_{\bar{r},1} = \{[\bar{r}] \to [\bar{r}]', [\bar{r}]' \to [\bar{r}]'', [\bar{r}]'' \to [s], [\bar{r}]'' \to [\bar{s}], [\bar{r}]'' \to [t] \mid s \in Fail_r\},$
$P_{\bar{r},2} = \{X \to X', X' \to F, [\bar{r}] \to A\}.$

Now, for each $A \in N$, let $P_A = \{[t] \to A, X \to X', X' \to F\}$, and finally let $P_{t,1} = \{[t] \to [t]', X \to X', X' \to X''\}$, $P_{t,2} = \{[t]' \to \lambda, Y \to \lambda, X'' \to \lambda\}.$

When G' starts working, the only applicable component is P_0, and it produces a sentential form $[r]SXY$ for some rule r. In general, sentential forms w of G correspond to sentential forms $w' = w_1[r]w_2XY$ or $w' = w_1[\bar{r}]w_2XY$ of G' where $w = w_1w_2$, X, Y are additional marker nonterminals, and $[r]$ or $[\bar{r}], r \in lab(P)$, signals that a successful or an unsuccessful application of the rule r, respectively, will follow.

Suppose that the sentential form of G' is of the form $w' = w_1[r_1]w_2XY$ for some $r_1 \in lab(P)$. Then there are one or more applicable components among $P_{r,1}, r \in lab(P)$ where $r = r_1$ not necessarily holds. Let us consider one of these components. If $(r : A \rightarrow BC, Succ_r, Fail_r)$, then the application of $P_{r,1}$ might increase the number of nonterminals in the sentential form by three or four, depending on the number of occurrences of the nonterminal A present in the sentential form.

If the rules of $P_{r,1}$ are applied in such a way that more than one occurrences of the symbol A are rewritten, then at least one of X or Y remains unprimed, thus the components in $\mathcal{P}_r$, component $P_{r,4}$, and $P_{r,5}$ would be able (by the left side of their rules) to rewrite the sentential form. Since if A was rewritten more than once, then there has to be more than one occurrences of the symbol $[r]'$ present, component $P_{r,5}$ cannot decrease the number of nonterminals, therefore some component from $\mathcal{P}_r$ (and possibly component $P_{r,4}$) must be applied (in some order), since one of the elements of $\mathcal{P}_r$ is able to decrease the number of nonterminals by one (and $P_{r,4}$ might also be able to do it, if the sentential form contains additional B or C nonterminals). The elements of $\mathcal{P}_r$ (one of which must be applied in any case) produce a sentential form containing the trap symbol F which cannot be rewritten, thus, a derivation of G' can only be successful, if $P_{r,1}$ is used to rewrite one A to $B_rC_rW[r]'$, and XY to $X'Y'$. Now the component $P_{r,4}$, and in case $r \neq r_1$ also $P_{r,5}$ is able to decrease the number of nonterminals by one (by at least one in case of $P_{r,4}$), thus they have to be applied, and produce a sentential form where W is erased, B_rC_r is changed to BC, and if $r \neq r_1$, then the trap symbol F is present. If we assume that the $r = r_1$, that is, the component $P_{r_1,1}$ corresponding to rule r_1 of G was applied, then the derivation can be continued with component $P_{r_1,2}$ which does not change the number of nonterminals but changes $[r_1]'$ to $[s]$, $[\bar{s}]$, or $[t]$. If now $P_{r_1,3}$ is used which decreases the number of nonterminals by one by erasing Z and changing $X'Y'$ back to XY, then we obtain a sentential form $w_1'[s]w_2'XY$, $w_1'[\bar{s}]w_2'XY$, or $w_1'[t]w_2'XY$ where w_1w_2 can be rewritten to $w_1'w_2'$ by rule r_1 of P, and $s \in Succ_{r_1}$.

If we have a string of the form $w_1[s]w_2XY$, the simulation of the successful application of rule s will follow, as described above. If we have $w_1[\bar{s}]w_2XY$, the failure of the application of rule s will be simulated by the components $P_{\bar{s},i}$, $1 \leq i \leq 2$, both of which are able to take three derivation steps and not increase the number of different nonterminals. If there is no A present, thus, the successful application of rule $(s : A \rightarrow BC, Succ_s, Fail_s)$ is not possible, then none of these components change the number of nonterminals, therefore, the successful derivation has to be continued with $P_{\bar{s},1}$ which introduces a symbol corresponding to a rule in $Fail_s$. If there is an A in the sentential form, that is, when rule s cannot be applied unsuccessfully, then only $P_{\bar{s},2}$ can be used (so the trap symbol is introduced) because in this case it decreases the number of nonterminals.

If we have $w_1[t]w_2XY$, that is, when the nonterminal $[t]$ is introduced, the simulation of G by G' should be finished. This is done by components P_A, and $P_{t,i}$, $1 \leq i \leq 2$ which successfully terminate the derivation if there is no terminal present in the sentential form.

To prove the statement for any $k \geq 3$, we might modify the construction above by adding $k - 3$ different additional marker nonterminals to the sentential form and $k - 3$ additional rules to the components which either prime or un-prime these symbols in $k - 3$ additional steps (or remove them when the derivation is finished). Since priming or unpriming the additional symbols does not change the number of different nonterminals in the sentential form, the arguments above also hold for this modified system which then simulates G with k steps derivations.

4   The Case of Strongly Maximal Final Competence

In this section we first present a lower and an upper bound for the generative power of CD grammar systems using (sfc, t)-mode derivations and then show that these systems with $(sfc, = k)$-mode derivations, for $k \geq 3$, generate all recursively enumerable languages as with $(fc, = k)$-mode derivations.

Lemma 3. $\mathcal{L}(CF) \subset \mathcal{L}(CD, sfc, t)$.

Proof. The inclusion follows from the fact that any context-free grammar can be considered as a CD grammar system with only one component in the t-mode derivation, thus in the (sfc, t)-mode derivation. To prove the strictness of the inclusion, we consider the CD grammar system $G = (\{S, A, B, A', B', A'', B''\}, \{a, b, c\}, P_1, P_2, P_3, S)$ with $P_1 = \{S \rightarrow AB,\ A \rightarrow aA'b,\ B \rightarrow B'c\}$, $P_2 = \{S \rightarrow AB,\ A \rightarrow A'',\ B \rightarrow B''\}$, $P_3 = \{A' \rightarrow A,\ B' \rightarrow B,\ A'' \rightarrow ab,\ B'' \rightarrow c\}$, which generates in the (sfc, t)-mode derivation the non-context-free context-sensitive language $\{a^n b^n c^n \mid n \geq 1\}$.

Lemma 4. $\mathcal{L}(CD, sfc, t) \subseteq \mathcal{L}(RCET0L)$.

Proof. Let $G = (N, T, P_1, P_2, \ldots, P_n, S)$ be a CD grammar system working with (sfc, t)-mode derivations. We now construct the simulating RCET0L system $G' = (V, T, \mathcal{P}, XS)$ with the following components.
    Let $V = N \cup T \cup \{X\} \cup \{[i, U] \mid 1 \leq i \leq n, U \subseteq N\} \cup \{[U] \mid U \subseteq N\}$,
$\mathcal{P} = \{R_{fin}\} \cup \bigcup_{U \subseteq N} \{R(U), R(U)'\} \cup \bigcup_{i=1}^{n} \{R(i), R(i)'\}$, where for $U \subseteq N$
    $R(U) = (\{X \rightarrow [i, U'] \mid i \in Z_U, card(U') = fc(U)\},\ U \cup \{X\},\ N \setminus U)$,
    $R(U)' = (\{[U] \rightarrow X\},\ U \cup \{[U]\},\ N \setminus U)$. For $1 \leq i \leq n$,
    $R(i) = (P_i,\ \emptyset,\ \{X\} \cup \{[j, U'] \mid 1 \leq j \leq n, j \neq i, U' \subseteq N\} \cup \{[U'] \mid U' \subseteq N\})$,
    $R(i)' = (\{[i, U] \rightarrow [U] \mid U \subseteq N\},\ \emptyset,\ dom(P_i)\} \cup \{X\} \cup \{[j, U'] \mid 1 \leq j \leq n, j \neq i, U' \subseteq N\} \cup \{[U'] \mid U' \subseteq N\})$, and $R_{fin} = (\{X \rightarrow \lambda\},\ \{X\},\ N)$.
    As in the proof of Lemma 2, one can prove that $L(G, sfc, t) = L(G')$ taking into consideration that the tables $R(U)'$ are used for checking that the simulation of $w \Longrightarrow_{P_i}^t v$ by tables $R(i)$ and $R(i)'$ ends with a word v where $card(alph(v) \cap N) = fc_t(w)$.

Theorem 3. $\mathcal{L}(RE) = \mathcal{L}(CD, sfc, = k)$, for any $k \geq 3$.

Proof. The statement can be proved by modifying the system constructed in the proof of Theorem 2 by removing the components in the sets $\mathcal{P}_r$ and changing $P_{r,1}$ to $P_{r,1} = \{A \rightarrow B_r C_r W[r]' X' X'',\ X' \rightarrow \lambda,\ X'' \rightarrow \lambda\}$. For $k > 3$, we consider the same modifications as described in the proof of Theorem 2.

5   The Case of Comparing Competence

In this section we show that the generative power of CD grammar systems working with (cc, α)-mode derivations is equal to that of the CD grammar systems with (fc, α)-mode derivations, where $\alpha \in \{t\} \cup \{= k \mid k \geq 3\}$.

Theorem 4. $\mathcal{L}(RCET0L) = \mathcal{L}(CD, cc, t)$.

Proof. To prove the inclusion $\mathcal{L}(RCET0L) \subseteq \mathcal{L}(CD, cc, t)$, we repeat the proof of Lemma 1. In any derivation step of the system we have chosen a component which is in $cc_t(Xh'(z), y_1, y_2, \ldots y_k)$, (for $k = 2r + 2 + \sum_{i=1}^{r} m_i$), too. The reverse inclusion can be proved as follows. Let $G = (N, T, P_1, P_2, \ldots, P_n, S)$ be a CD grammar system working with (cc, t)-mode derivations and let $U = N \cup T$. For $1 \leq i \leq n$, we set $U^{(i)} = \{x^{(i)} \mid x \in U\}$ and define the homomorphism $h_i : U^* \to (U^{(i)})^*$ by $h_i(x) = x^{(i)}$ for $x \in U$. We construct a simulating RCET0L system $G' = (V, T, \mathcal{P}, XS)$ with the following components. Let $V = \{X\} \cup \{[i] \mid 1 \leq i \leq n\} \cup N \cup \bigcup_{i=1}^{n} U^{(i)} \cup \bigcup_{U \subseteq N} \{[U], [U]', [U]''\}$ and let $\mathcal{P} = \{Q_1, Q_2, Q_3, Q_{fin}\} \cup \{R(i) \mid 1 \leq i \leq n\} \cup \{R(\overline{U}) \mid U \subseteq N\} \cup \{R(U, U_1, U_2, \ldots, U_n) \mid U \subseteq N, U_i \subseteq N, 1 \leq i \leq n\}$.

We set $R(U) = (\{X \to [U]\}, U \cup \{X\}, N \setminus U)$. (If a sentential form Xw is given (note that this also holds for the axiom), $R(U)$ is only applicable if $alph(w) = U$; therefore we store $alph(w)$ in the first letter.)

We define $Q_1 = (\{[U] \to [U]' \mid U \subseteq N\} \cup \{A \to A^{(1)}A^{(2)} \ldots A^{(n)} \mid A \in N\}, \emptyset, \{X\} \cup \{[i] \mid 1 \leq i \leq n\} \cup \bigcup_{W \subseteq N} \{[W]', [W]''\})$. (By the application of Q_1, any nonterminal occurring in the current sentential form is replaced by a copy $A^{(i)}$ for $1 \leq i \leq n$).

Let $Q_2 = (\{A^{(i)} \to h_i(w) \mid A \to w \in P_i\}, \emptyset, \{X\} \cup \{[i] \mid 1 \leq i \leq n\} \cup \bigcup_{W \subseteq N} \{[W], [W]''\})$. (We replace in parallel the nonterminals $A^{(i)}$ according to rules of P_i where the upper index i is preserved; this procedure can be repeated which means that we simulate a derivation in the t-mode for any component P_i using the letter indexed by i, $1 \leq i \leq n$).

We set $Q_3 = (\{[U]' \to [U]'' \mid U \subseteq N\}, \emptyset, h_1(dom(P_1)) \cup h_2(dom(P_2)) \cup \ldots \cup h_n(dom(P_n)) \cup \{X\} \cup \{[i] \mid 1 \leq i \leq n\} \cup \bigcup_{W \subseteq N} \{[W], [W]''\})$. (By Q_3 we check whether all components have finished their derivation in the t-mode, thus from Xw we have generated a word $[alph(w)]''v$ where v contains in a partially indexed version (the terminals in w are not indexed) of a word y_i with $w \Longrightarrow_{P_i}^{t} y_i$).

Let $R(U, U_1, U_2, \ldots, U_n) = (\{[U]'' \to [i] \mid h_i(U_i) \neq h_i(U), card(h_i(U_i)) \leq card(h_j(U_j)), h_j(U_j) \neq h_j(U), 1 \leq j \leq n\}, \bigcup_{1 \leq i \leq n} h_i(U_i) \cup \{[U]''\}, \bigcup_{1 \leq i \leq n} (h_i(N) \setminus h_i(U_i)))$. (This table can only be applied if – up to the upper indexes – $alph(y_i) = U_i$; thus from $U, U_1, U_2, \ldots, U_n$ we can compute all i's such that $y_i \in cc_t(w, y_1, y_2, \ldots, y_n)$; one of these values is chosen for the continuation).

Finally, let $R(i) = (\{[i] \to X\} \cup \{x^{(i)} \to x \mid x \in U\} \cup \{x^{(j)} \to \lambda \mid x \in U, 1 \leq j \leq n, i \neq j\}, \{[i]\}, \emptyset)$. (The application of $R(i)$ cancels all letters which are derived by a simulation of a component P_j with $j \neq i$ and all upper indexes are deleted; thus we have $Xw \Longrightarrow^* Xy_i$ in G' where $y_i \in cc_t(w, y_1, y_2, \ldots, y_n)$).

By the explanations added to the tables the reader can easily verify that $L(G') = L(G, cc, t)$.

Theorem 5. $\mathcal{L}(RE) = \mathcal{L}(CD, cc, = k)$, *for* $k \geq 3$.

Proof. It can easily be seen that the CD grammar system G' constructed in the proof of Theorem 2 also simulates the derivations of the programmed grammar G in the $(cc, = k)$-mode derivations.

6 Conclusions

We have shown that most of the language families generated with the t-mode derivation as underlying derivation coincide with the family of random context ETOL languages, while using the $= k$ step derivation modes as underlying derivation mode all recursively enumerable languages can be generated. Since the precise relation of the two language classes is not yet known, our results are contributions to approaching this problem as well.

References

1. ter Beek, M., Csuhaj-Varjú, E., Holzer, M., Vaszil, G.: On competence in CD grammar systems. In: Calude, C.S., Calude, E., Dinneen, M.J. (eds.) DLT 2004. LNCS, vol. 3340, pp. 76–88. Springer, Heidelberg (2004)
2. ter Beek, M., Csuhaj-Varjú, E., Holzer, M., Vaszil, G.: On competence in cooperating distributed grammar systems with parallel rewriting. International Journal of Foundations of Computer Science 18(6), 1425–1439 (2007)
3. Bordihn, H., Csuhaj-Varjú, E.: On competence and completeness in CD grammar systems. Acta Cybernetica 12, 347–361 (1996)
4. Csuhaj-Varjú, E., Dassow, J.: On cooperating/distributed grammar systems. Journal of Information Processing and Cybernetics (EIK) 26, 49–63 (1990)
5. Csuhaj-Varjú, E., Dassow, J., Holzer, M.: CD grammar systems with competence based entry conditions in their cooperation protocols. International Journal of Computer Mathematics 83, 159–169 (2006)
6. Csuhaj-Varjú, E., Dassow, J., Kelemen, J., Păun, Gh.: Grammar Systems - A Grammatical Approach to Distribution and Cooperation. Topics in Computer Mathematics 5. Gordon and Breach Science Publishers, Yverdon (1994)
7. Dassow, J.: On cooperating distributed grammar systems with competence based start and stop conditions. Fundamenta Informaticae 76, 293–304 (2007)
8. Dassow, J., Păun, G.: Regulated Rewriting in Formal Language Theory. EATCS Monograph on Theoretical Computer Science 18. Springer, Heidelberg (1989)
9. Dassow, J., Păun, G., Rozenberg, G.: Grammar systems. In: [10], ch. 4, vol. II, pp. 155–213
10. Rozenberg, G., Salomaa, A. (eds.): Handbook of Formal Languages, vol. I – III. Springer, Heidelberg (1997)

The Synchronization Problem for Strongly Transitive Automata*

Arturo Carpi[1] and Flavio D'Alessandro[2]

[1] Dipartimento di Matematica e Informatica, Università degli Studi di Perugia,
via Vanvitelli 1, 06123 Perugia, Italy
carpi@dipmat.unipg.it
[2] Dipartimento di Matematica, Università di Roma "La Sapienza"
Piazzale Aldo Moro 2, 00185 Roma, Italy
dalessan@mat.uniroma1.it

Abstract. The synchronization problem is investigated for a new class of deterministic automata called strongly transitive. An extension to unambiguous automata is also considered.

Keywords: Černý conjecture, synchronizing automata, rational series.

1 Introduction

The *synchronization problem* for a deterministic n-state automaton consists in the search of an input-sequence, called a *synchronizing word* such that the state attained by the automaton, when this sequence is read, does not depend on the initial state of the automaton itself. If such a sequence exists, the automaton is called *synchronizing*. If a synchronizing automaton is deterministic and complete, a well-known conjecture by Černý claims that it has a synchronizing word of length not larger than $(n-1)^2$ [5]. This conjecture has been shown to be true for several classes of automata (*cf* [1,2,5,7,9,11,14,16]). The interested reader is refered to [12] for a historical survey of the problem. Two of the quoted references deserve a special mention: in [11], Kari proved Černý conjecture for Eulerian automata, that is, for automata whose underlying graph is Eulerian. Dubuc [7] proved the conjecture for circular automata, that is, for automata possessing a letter that acts, as a circular permutation, over the set of states of the automaton. In [1], Béal proposed an unified algebraic approach, based upon rational series, that allows one to obtain quadratic bounds for the minimal length of a synchronizing word of circular or Eulerian automata. In Section 3, by developing the theoretical approach of [1], we study the synchronization problem for a new class of automata called *strongly transitive*. A n-state automaton is said to be strongly transitive if it is equipped by a set of words $\{w_1, ..., w_n\}$, called *independent*, such that, for any pair of states s and t, there exists a word

---

* This work was partially supported by MIUR project ``Linguaggi formali e automi: teoria e applicazioni'' and by fundings ``Facoltà di Scienze MM. FF. NN. 2006'' of the University of Rome ``La Sapienza''.

M. Ito and M. Toyama (Eds.): DLT 2008, LNCS 5257, pp. 240–251, 2008.
© Springer-Verlag Berlin Heidelberg 2008

w_i such that $sw_i = t$. The main result of this section is that any synchronizing strongly transitive n-state automaton has a synchronizing word of length not larger than

$$(n - 2)(n + L - 1) + 1,$$

where L denotes the length of the longest word of an independent set of the automaton. As a straightforward corollary of this result, one can obtain the bound $2(n-2)(n-1)+1$ for the shortest synchronizing word of any n-state synchronizing circular automaton. Together with this result, some basic properties of such automata are investigated. It is shown that circular and synchronizing automata are strongly transitive. In particular it is proved that, if a n-state automaton has a synchronizing word u, then it has an independent set of words of length not larger than $|u| + n - 1$. It is also proved that the previous upper bound is tight. More precisely, we construct an infinite family of synchronizing and thus strongly transitive automata such that any independent set of the automaton contains a word w such that $|w| \geq |u| + n - 1$ where u is the shortest synchronizing word. Moreover we give examples of strongly transitive automata which are neither circular nor synchronizing.

In Section 4, we focus our attention on the class of unambiguous automata. We recall that the synchronization problem is closely related to that of finding short words of minimal rank in an automaton. Here, the rank of a word is the linear rank of the associated transition relation and thus a synchronizing word is a word of rank 1. In general, the length of the shortest word of minimal rank in a nondeterministic automaton is not polynomially upperbounded by the number of states of the automaton [10]. However in the case of unambiguous automata, such a bound exists: in [4], it is shown that for a n-state complete unambiguous automaton, there exists a word of minimal rank r of length less than $\frac{1}{2}rn^3$. Some interesting results on such class of automata have been recently proven in [2].

In this paper, we consider unambiguous automata on an alphabet A satisfying the following combinatorial property: there exist two sets of words V and W such that $A \subseteq V, W$ and, for any state s, one has

$$\sum_{v \in V} \text{Card}(sv) \geq \text{Card}(V) \quad \text{and} \quad \sum_{w \in W} \text{Card}(sw^{-1}) \geq \text{Card}(W).$$

For instance, Eulerian automata satisfy the previous conditions with $V = W = A$. The main result of this section is that a synchronizing unambiguous n-state automaton satisfying the previous conditions has a synchronizing word of length not larger than

$$(n - 2)(n + L - 1) + 1,$$

where L is the maximal length of the words of the set $V \cup W$. In particular, we derive that any synchronizing unambiguous Eulerian n-state automaton has a synchronizing word of length not larger than $(n - 1)^2$.

2 Preliminaries

We assume that the reader is familiar with the theory of automata and rational series. In this section we shortly recall a vocabulary of few terms and we fix the corresponding notation used in the paper.

Let A be a finite alphabet and let A^* be the free monoid of words over the alphabet A. The *empty word* is denoted by ϵ. If n is a positive integer, A^n denotes the set of all words $w \in A^*$ of *length* $|w| = n$. For any $u \in A^*$ and $a \in A$, $|u|_a$ denotes the number of occurrences of the letter a in u. For any finite subset W of A^*, we denote by L_W the length of the longest word in W.

A finite automaton is a triple $\mathcal{A} = (S, A, \delta)$ where S is a finite set of elements called *states* and δ is a map $\delta : S \times A \longrightarrow \Pi(S)$ from $S \times A$ into the family $\Pi(S)$ of all subsets of S. The map δ is called the *transition function* of $\mathcal{A}$. The canonical extension of the map δ to the set $S \times A^*$ is still denoted by δ. For any $u \in A^*$ and $s \in S$, the set of states $\delta(s, u)$ will be also denoted su. If P is a subset of S and u is a word of A^*, we denote by Pu and Pu^{-1} the sets:

$$Pu = \bigcup_{s \in P} su, \quad Pu^{-1} = \{s \in S \mid su \cap P \neq \emptyset\}.$$

If $\mathrm{Card}(sa) \leq 1$ for all $s \in S$, $a \in A$, the automaton $\mathcal{A}$ is *deterministic*; if $Sw \neq \emptyset$ for all $w \in A^*$, $\mathcal{A}$ is *complete*; if $\bigcup_{w \in A^*} sw = S$ for all $s \in S$, $\mathcal{A}$ is *transitive*. If $n = \mathrm{Card}(S)$, we will say that $\mathcal{A}$ is a n-state automaton. Let $\mathcal{A}$ be a deterministic automaton. A *synchronizing* or *reset* word is a word $u \in A^*$ such that $\mathrm{Card}(Su) = 1$. The state q such that $Su = \{q\}$ is called *reset state*. A *synchronizing* deterministic automaton is an automaton that has a reset word. The following conjecture has been raised in [5].

Černý Conjecture. *Each synchronizing complete deterministic n-state automaton has a reset word of length not larger than $(n-1)^2$.*

Let $\mathbb{K}$ be a semiring. We recall that a *formal series* with coefficients in $\mathbb{K}$ and variables in A is a mapping of the free monoid A^* into $\mathbb{K}$. A series $S : A^* \to \mathbb{K}$ is *rational* if there exists a triple (α, μ, β) where $\alpha \in \mathbb{K}^{1 \times n}$, $\beta \in \mathbb{K}^{n \times 1}$, $\mu : A^* \to \mathbb{K}^{n \times n}$ is a morphism of the free monoid A^* in the multiplicative monoid $\mathbb{K}^{n \times n}$ of matrices with coefficients in $\mathbb{K}$, and, for every $u \in A^*$, $S(u) = \alpha\mu(u)\beta$. The triple (α, μ, β) is called *a representation* of S and the integer n is called its *dimension*. With a minor abuse of language, if no ambiguity arises, the number n will be also called the dimension of S. Let $\mathcal{A} = (S, A, \delta)$ be any n-state automaton. One can associate with $\mathcal{A}$ a morphism $\varphi_\mathcal{A} : A^* \to \mathbb{Q}^{S \times S}$ of the free monoid A^* in the multiplicative monoid $\mathbb{Q}^{S \times S}$ of matrices over the set of rational numbers, defined as: for any $a \in A$ and for any $s, t \in S$,

$$\varphi_\mathcal{A}(a)_{st} = \begin{cases} 1 & \text{if } t \in sa \\ 0 & \text{otherwise.} \end{cases}$$

It is worth to recall some well-known properties of the map $\varphi_\mathcal{A}$. For every $u \in A^*$ and for every $s, t \in S$, the coefficient $\varphi_\mathcal{A}(u)_{st}$ is the number of all distinct computations of $\mathcal{A}$ from s to t labelled by u. If every matrix of the monoid $\varphi_{\mathcal{A}}(A^*)$ is such that every row does not contain more than one non-null entry, then $\mathcal{A}$ is deterministic. If $\varphi_{\mathcal{A}}(A^*)$ does not contain the null matrix then $\mathcal{A}$ is complete. The following result is important [3, Corollary 3.6].

Proposition 1. *Let* $S : A^* \to \mathbb{Q}$ *be a rational series of dimension* n *with coefficients in* $\mathbb{Q}$. *If, for every* $u \in A^*$ *such that* $|u| \leq n - 1$, $S(u) = 0$ *the series* S *is null.*

As a corollary we obtain the following well-known result (see [3], [8]).

Theorem 1. *(Moore, Conway) Let* S_1, $S_2 : A^* \to \mathbb{Q}$ *be two rational series with coefficients in* $\mathbb{Q}$ *of dimension* n_1 *and* n_2 *respectively. If, for every* $u \in A^*$ *such that* $|u| \leq n_1 + n_2 - 1$, $S_1(u) = S_2(u)$, *the series* S_1 *and* S_2 *are equal.*

Let P be a subset of S. We associate with P a series S with coefficients in $\mathbb{Q}$ defined as: for every $u \in A^*$, $S(u) = \mathrm{Card}(Pu^{-1}) - \mathrm{Card}(P)$. As proven in [1, Lemma 2], S is a rational series of dimension n. As a straightforward consequence one obtains the following corollary of Proposition 1.

Corollary 1. *Let* $\mathcal{A} = (S, A, \delta)$ *be a deterministic* n-*state automaton and let* P *be a subset of* S. *Suppose that there exists a word* u *such that* $\mathrm{Card}(Pu^{-1}) \neq \mathrm{Card}(P)$. *Then there exists a word satisfying the previous condition whose length is not larger than* $n - 1$.

Remark 1. It is useful to remark that every proper and nonempty subset of a deterministic transitive synchronizing automaton satisfies the hypotheses of Corollary 1. Indeed, if P is such a set, for any state p of P, one can find a reset word w such that $Sw = \{p\}$. This gives $S = Pw^{-1}$ and thus $\mathrm{Card}(Pw^{-1}) > \mathrm{Card}(P)$.

3   Strongly Transitive Automata

In this section, we deal with deterministic complete automata. As pointed out, for instance in [13], the study of Černý conjecture can be always reduced to the case of transitive automata. We now study Černý conjecture for a special class of transitive automata called strongly transitive. Let us first introduce the following definition.

Definition 1. *Let* $\mathcal{A} = (S, A, \delta)$ *be a* n-*state automaton. Then* $\mathcal{A}$ *is called strongly transitive if there exist* n *words* $w_0, \dots, w_{n-1} \in A^*$ *such that*

$$\forall \ s, t \in S, \quad \exists \ i = 0, \dots, n-1, \ sw_i = t. \tag{1}$$

The set $\{w_0, \dots, w_{n-1}\}$ is called *independent*.

We observe that a set $\{w_0, \dots, w_{n-1}\}$ is independent if and only if, for any state s of S, the states sw_i, $i = 0, \dots, n-1$, are pairwise distinct. The following example shows that transitivity does not imply strongly transitivity.

Example 1. Consider the 3-state automaton $\mathcal{A}$ over the alphabet $A = \{a, b\}$ defined by the following graph:

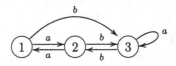

The automaton $\mathcal{A}$ is transitive. Let us prove that it is not strongly transitive. Indeed, by contradiction, let W be an independent set. Then there are words $u, v \in W$ such that $3u = 1, 3v = 2$. One easily derives that $|u|_b$ and $|v|_b$ are odd so that $1u = 1v = 3$. Therefore we have a contradiction because the states $1u$ and $1v$ must be distinct. Hence W cannot be independent.

The following proposition shows that any transitive synchronizing automaton is strongly transitive.

Proposition 2. *Let $\mathcal{A}$ be a transitive synchronizing n-state automaton. If $\mathcal{A}$ has a reset word of length ℓ, then there exists an independent set W for $\mathcal{A}$ such that $L_W < \ell + n$.*

Proof. Let q and u be a reset state and a reset word of $\mathcal{A}$ respectively with $|u| = \ell$. Since $\mathcal{A}$ is transitive, there exist words $u_0, u_1, \ldots, u_{n-1}$ of length $< n$ that label computations from q to all the states of $\mathcal{A}$. The set $W = \{uu_0, uu_1, \ldots, uu_{n-1}\}$ is independent and $L_W < \ell + n$.

By a well known result of [9,13,14], any synchronizing automaton has a reset word of length not larger than $(n^3 - n)/6$. Thus it has an independent set W such that L_W is not larger than $(n^3 - n)/6 + n - 1$. Moreover if Černý conjecture is true, the previous number can be lowered to $n(n-1)$. Now we prove that there exist automata for which the upper bound stated in Proposition 2 is tight. More precisely, we will construct, for any positive integer n, a synchronizing $(2n + 1)$-state automaton $\mathcal{A}$ such that, for any independent set W of $\mathcal{A}$, $L_W \geq \ell + 2n$, where ℓ is the length of the shortest reset word of $\mathcal{A}$.

Example 2. Let n be a positive integer. Consider the automaton $\mathcal{A}_n = (S_n, A, \delta_n)$ where $A = \{a, b, c\}$, $S_n = \{0, 1, 2, \ldots, 2n\}$ and the transition map δ_n is defined as follows:

- $0a = 0c = 0$, $0b = 1$,
- for $i = 1, 2, \ldots, n$, $ia = i - 1$, $ib = i + (-1)^i$, $ic = i - (-1)^i$,
- for $i = n + 1, n + 2, \ldots, 2n - 1$, $ia = i + 1$, $ib = ic = n + 1$,
- $(2n)a = 0$, $(2n)b = (2n)c = n + 1$.

For instance, the graph of the automaton $\mathcal{A}_4$ is drawn below. One can easily check that the word a^4 is a reset word of minimal length while the set $a^4\{\epsilon, b, bc, bcb, bcbc, bcbcb, bcbcba, bcbcba^2, bcbcba^3\}$, is an independent set of words of the automaton $\mathcal{A}_4$.

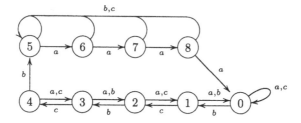

One can easily verify that $S_n a^n = \{0\}$. Thus $\mathcal{A}_n$ is a synchronizing automaton and, by Proposition 2, it is strongly transitive. We shall prove that, for any independent set W, $L_W \geq \mathrm{Card}(S_n) + |a^n| - 1$. We need the following lemma whose proof is omitted.

Lemma 1. *Let W be an independent set of $\mathcal{A}_n$. Then every word of W is a reset word.*

Proposition 3. *For any independent set W of $\mathcal{A}_n$ one has $L_W \geq 3n$.*

Proof. Let $w \in W$ be a word such that $0w = 2n$. The main task amounts to prove that $|w| \geq 3n$. By Lemma 1, we can prove the inequality above in the case that $S_n w = \{2n\}$. For this purpose, one can observe that, for every $i = 0, \ldots, n$ and $\sigma \in A$, $\{0, 1, \ldots, i\}\sigma \supseteq \{0, 1, \ldots, i-1\}$. This implies that, for every $u \in A^n$,

$$0 \in \{0, 1, \ldots, n\}u. \qquad (2)$$

Since the minimal length of a path from 0 to $2n$ in the graph of $\mathcal{A}_n$ is $2n$, one has $|w| \geq 2n$, so that one can factorize $w = uv$ with $u, v \in A^*$ and $|u| = n$. Eq. (2) implies that $0 \in S_n u$ and therefore, $0v \in S_n w = \{2n\}$. By the previous remark, this implies $|v| \geq 2n$, so that $|w| \geq 3n$ and the proof is complete.

The following useful property easily follows from Definition 1.

Lemma 2. *Let $\mathcal{A}$ be a strongly transitive automaton and let W be an independent set of $\mathcal{A}$. Then, for every $u \in A^*$, the set uW is an independent set of $\mathcal{A}$.*

Proposition 4. *Let $\mathcal{A} = (S, A, \delta)$ be a strongly transitive n-state automaton and let W be an independent set of $\mathcal{A}$. Then for every subset P of S:*

$$\sum_{w \in W} \mathrm{Card}(Pw^{-1}) = n\,\mathrm{Card}(P). \qquad (3)$$

Proof. Let $W = \{w_0, \ldots, w_{n-1}\}$ and let $p \in S$. Because of Eq. (1), one has $S = \bigcup_{i=0}^{n-1} \{p\}w_i^{-1}$, and the sets $\{p\}w_i^{-1}$ are pairwise disjoint. This immediately gives:

$$\sum_{i=0}^{n-1} \mathrm{Card}(\{p\}w_i^{-1}) = n. \qquad (4)$$

Let $P = \{p_1, \ldots, p_m\}$ be a set of m states. Since $\mathcal{A}$ is deterministic, for any pair p_i, p_j of distinct states of P and for every $u \in A^*$, one has $\{p_i\}u^{-1} \cap \{p_j\}u^{-1} = \emptyset$, and, along with Eq. (4), this yields:

$$\sum_{i=0}^{n-1} \mathrm{Card}(Pw_i^{-1}) = \sum_{i=0}^{n-1} \sum_{j=1}^{m} \mathrm{Card}(\{p_j\}w_i^{-1}) = mn. \tag{5}$$

Corollary 2. *Let $\mathcal{A} = (S, A, \delta)$ be a synchronizing transitive n-state automaton and let W be an independent set of $\mathcal{A}$. Let P be a proper and non empty subset of S. Then there exists a word $v \in A^*$ such that*

$$|v| \leq n + L_W - 1, \quad \mathrm{Card}(Pv^{-1}) > \mathrm{Card}(P).$$

Proof. Let $W = \{w_0, \ldots, w_{n-1}\}$. We first prove that there exist a word $v \in A^*$ with $|v| \leq n - 1$ and $i = 0, \ldots, n-1$ such that

$$\mathrm{Card}(P(vw_i)^{-1}) \neq \mathrm{Card}(P). \tag{6}$$

If there exists $i = 0, \ldots, n-1$ such that $\mathrm{Card}(Pw_i^{-1}) \neq \mathrm{Card}(P)$, take $v = \epsilon$. Now suppose that the latter condition does not hold so that $\mathrm{Card}(Pw_0^{-1}) = \mathrm{Card}(P)$. Since P is a proper subset of S, by Remark 1 and by applying Corollary 1 to the set Pw_0^{-1}, one has that there exists a word $v \in A^*$ such that $|v| \leq n - 1$ and $\mathrm{Card}(P(vw_0)^{-1}) \neq \mathrm{Card}(P)$.

Thus take words v and w_i that satisfy Eq. (6). If $\mathrm{Card}(P(vw_i)^{-1}) > \mathrm{Card}(P)$, since $|vw_i^{-1}| \leq n-1+L_W$, we are done. Finally suppose that $\mathrm{Card}(P(vw_i)^{-1}) < \mathrm{Card}(P)$. By Lemma 2, the set $vW = \{vw_0, \ldots, vw_{n-1}\}$ is independent for $\mathcal{A}$. Therefore, by Proposition 4,

$$\sum_{i=0}^{n-1} \mathrm{Card}(P(vw_i)^{-1}) = n \, \mathrm{Card}(P),$$

so that Eq. (6) implies the existence of an index j such that $\mathrm{Card}(P(vw_j)^{-1}) > \mathrm{Card}(P)$. Since, as before, $|vw_j^{-1}| \leq n - 1 + L_W$, the claim is proved.

As a consequence of Corollary 2, the following theorem holds.

Theorem 2. *Let $\mathcal{A} = (S, A, \delta)$ be a synchronizing transitive n-state automaton and let W be an independent set of $\mathcal{A}$. Then there exists a reset word for $\mathcal{A}$ of length not larger than $(n-2)(n + L_W - 1) + 1$.*

Proof. Let P be a non-empty subset of S with $\mathrm{Card}(P) < n$. Since $\mathcal{A}$ is synchronizing, there exists some word u such that $\mathrm{Card}(Pu^{-1}) \neq \mathrm{Card}(P)$. By Corollary 2, we can assume that $|u| \leq n + L_W - 1$ and $\mathrm{Card}(Pu^{-1}) > \mathrm{Card}(P)$. Therefore from any subset P of at least 2 states, by applying the previous argument $(n-2)$ times at most, we can construct a word u such that $Su = P$, $|u| \leq (n-2)(n + L_W - 1)$. The claim finally follows from the fact that, in a synchronizing automaton, there always exist a letter $a \in A$ and a set P of two states such that $\mathrm{Card}(Pa) = 1$.

Remark 2. In [7], Dubuc showed that Černý conjecture is true for circular automata. An n-state automaton is called *circular* if its underlying graph has a Hamiltonian cycle labelled by a power of a letter. This is equivalent to say that such a letter, say a, acts, as a circular permutation, on the set of states of the automaton. This implies that the words $\epsilon, a, a^2, \ldots, a^{n-1}$ form an independent set of the automaton. Thus, from Theorem 2, one derives that any circular n-state automaton has a reset word of length not larger than

$$2(n-2)(n-1) + 1.$$

We remark that a similar bound was established in [15] for the larger class of regular automata.

We have seen that circular automata are strongly transitive. However this notion is more general than that of circular automaton as shown in the following example.

Example 3. Consider an automaton $\mathcal{A} = (S, A, \delta)$ where $A = \{a, b, c\}$,

$$S = \{s_{ij} \mid i \in \{0, \ldots, \ell-1\}, \ j \in \{0, \ldots, k-1\}\}, \quad \ell, k \geq 1$$

and the map δ satisfies the following conditions:

1. $\forall\, i = 0, \ldots, \ell-1, \forall\, j = 0, \ldots, k-2,\ s_{ij}a = s_{ij+1},\ \ s_{ik-1}a = s_{i0}$,
2. $\forall\, i = 0, \ldots, \ell-2, \forall\, j = 0, \ldots, k-1,\ s_{ij}b = s_{i+1j'}$ and $s_{ij}c = s_{ij''}$, for some $j', j'' = 0, \ldots, k-1$;
3. $\forall\, j = 0, \ldots, k-1,\ s_{\ell-1j}b = s_{0j'}, \quad$ and $s_{\ell-1j}c = s_{0j''}$, for some $j', j'' = 0, \ldots, k-1$;
4. $\forall\, j = 0, \ldots, k-2,\ s_{0j}c = s_{0j},\ \ s_{0k-1}c = s_{00}$;

It is easily checked that the automaton is not circular and that the words

$$\epsilon, a, \ldots, a^{k-1}, \quad b, ba, \ldots, ba^{k-1}, \quad \ldots \quad, b^{\ell-1}, b^{\ell-1}a, \ldots, b^{\ell-1}a^{k-1}$$

form an independent set of $\mathcal{A}$ and their length is $k + \ell - 2$ at most. One could verify that $\alpha = (b(ca^{k-1})^{k-2}c)^\ell (cb^{\ell-1})^{\ell-2}c(ca^{k-1})^{k-2}c$ is a reset word.

4   Unambiguous Automata

We recall some basic notions and results concerning monoids of $(0,1)$-matrices. Let S be a finite set of indexes and let $\mathbb{Q}^{S \times S}$ be the monoid of $S \times S$ matrices with the usual row-column product. For any $m \in \mathbb{Q}^{S \times S}$ and for any $s \in S$, the symbols m_{s*} and m_{*s} will denote respectively the row and the column of m of index s.

We will denote by $\{0, 1\}^{S \times S}$ the set of the matrices of $\mathbb{Q}^{S \times S}$ whose entries are all 0 and 1. Any submonoid M of $\mathbb{Q}^{S \times S}$ such that $M \subseteq \{0, 1\}^{S \times S}$ will be called a *monoid of* $(0, 1)$-*matrices* (or *monoid of unambiguous relations*)

A monoid of $(0,1)$-matrices is *transitive* if, for any $s, t \in S$, there exists $m \in M$ such that $m_{st} = 1$.

Let M be a monoid of $(0,1)$-matrices. Any row (resp. column) of a matrix of M will be called a *row* (resp. *column*) of M. The sets of the rows and columns of M are ordered in the usual way: $\mathbf{a} \leq \mathbf{b}$ if $a_s \leq b_s$ for all $s \in S$. The *weight* of a row or column $\mathbf{a}$ of M is the integer $\|\mathbf{a}\| = \sum_{s \in S} a_s$. The following two lemmas will be useful in the sequel. Proofs are omitted for the sake of brevity.

Lemma 3. *Let M be a transitive monoid of (0,1)-matrices. If $\mathbf{a} \neq \mathbf{0}$ is a row (resp. column) of M which is not maximal, then one has $\|\mathbf{am}\| > \|\mathbf{a}\|$ (resp. $\|\mathbf{ma}\| > \|\mathbf{a}\|$) for some $m \in M$.*

Lemma 4. *Let M be a transitive monoid of (0,1)-matrices of dimension n. For any row $\mathbf{a}$ and any column $\mathbf{b}$ of M, one has $\|\mathbf{a}\| + \|\mathbf{b}\| \leq n + 1$.*

The minimal ideal of a transitive monoid of (0-1)-matrices has been characterized by Césari [6]. We summarize in the following statement some of the results of [6].

Proposition 5. *Let M be a transitive monoid of (0-1)-matrices which does not contain the null matrix and let D be its minimal ideal and p be the minimal rank of its elements. Then the elements of D are the matrices of M of the form*

$$m = \mathbf{b}_1 \mathbf{a}_1 + \mathbf{b}_2 \mathbf{a}_2 + \cdots + \mathbf{b}_p \mathbf{a}_p + \mu,$$

with $\mathbf{a}_1, \mathbf{a}_2, \ldots, \mathbf{a}_p$ maximal rows of M, $\mathbf{b}_1, \mathbf{b}_2, \ldots, \mathbf{b}_p$ maximal columns of M, and $\mu \in \{0,1\}^{S \times S}$. Moreover, for any such m one has $\mu = 0$.

An automaton $\mathcal{A}$ is said to be *unambiguous* if and only if $M = \varphi_{\mathcal{A}}(A^*)$ is a monoid of $(0,1)$-matrices. This is equivalent to say that, for any pair of states s, t and any word u, there exists at most one computation of $\mathcal{A}$ from s to t labelled by u. A *reset word* of $\mathcal{A}$ is any word w such that $\varphi_{\mathcal{A}}(w)$ has linear rank 1.

In the sequel we shall suppose that $\mathcal{A}$ is a transitive unambiguous n-state automaton. Moreover, we assume that there exists a finite set $V \subseteq A^*$ such that $A \subseteq V$ and for every state p of $\mathcal{A}$

$$\sum_{v \in V} \mathrm{Card}(pv) \geq \mathrm{Card}(V). \tag{7}$$

Notice that if $\mathcal{A}$ is deterministic and complete, then any finite set V satisfies Eq. (7). Under our hypotheses, the following holds.

Lemma 5. *For all $\mathbf{a} \in \mathbb{N}^S$, $\sum_{v \in V} \|\mathbf{a}\varphi_{\mathcal{A}}(v)\| \geq \|\mathbf{a}\| \mathrm{Card}(V)$.*

In view of Proposition 5, in order to find a reset word of $\mathcal{A}$, it would be useful to find a word w of short length such that $\varphi_{\mathcal{A}}(w)$ has a maximal row or column. Next proposition furnishes a tool to produce rows of increasing weight.

Proposition 6. *Let* **a** *be a row of* $\varphi_A(A^*)$ *such that* $\|\mathbf{a}\varphi_A(u)\| \neq \|\mathbf{a}\|$ *for some* $u \in A^*$. *Then there exists a word* w *such that*

$$\|\mathbf{a}\varphi_A(w)\| > \|\mathbf{a}\|, \quad |w| \leq L_V + n - 1.$$

Proof. We notice that the series S defined by $S(u) = \|\mathbf{a}\varphi_A(u)\|$, $u \in A^*$, is a rational series of dimension n. Indeed, S has the linear representation $(\mathbf{a}, \varphi_A, \boldsymbol{\Lambda})$ where $\boldsymbol{\Lambda} = {}^t(1, 1, \ldots, 1)$. On the other side, the series S_0 defined by $S_0(u) = \|\mathbf{a}\|$ for all $u \in A^*$, is a rational series of dimension 1.

Let u be the shortest word such that $\|\mathbf{a}\varphi_A(u)\| \neq \|\mathbf{a}\|$. By Proposition 1, one has $|u| \leq n$. If $\|\mathbf{a}\varphi_A(u)\| > \|\mathbf{a}\|$, then the statement is verified for $w = u$. Thus we assume $\|\mathbf{a}\varphi_A(u)\| < \|\mathbf{a}\|$.

Write $u = u'x$ with $u' \in A^*$ and $x \in A$, and set $\mathbf{b} = \mathbf{a}\varphi_A(u')$. Since $x \in V$, by Lemma 5 one has

$$\sum_{v \in V \setminus \{x\}} \|\mathbf{b}\varphi_A(v)\| \geq \|\mathbf{b}\| \operatorname{Card}(V) - \|\mathbf{b}\varphi_A(x)\|.$$

By the minimality of u, one has $\|\mathbf{b}\| = \|\mathbf{a}\|$ while $\|\mathbf{b}\varphi_A(x)\| = \|\mathbf{a}\varphi_A(u)\| < \|\mathbf{a}\|$. Thus, from the previous equation one obtains

$$\sum_{v \in V \setminus \{x\}} \|\mathbf{b}\varphi_A(v)\| > \|\mathbf{a}\| \operatorname{Card}(V \setminus \{x\}).$$

Consequently, there is $v \in V \setminus \{x\}$ such that $\|\mathbf{b}\varphi_A(v)\| > \|\mathbf{a}\|$. Taking $w = u'v$, one has $\|\mathbf{a}\varphi_A(w)\| = \|\mathbf{b}\varphi_A(v)\| > \|\mathbf{a}\|$. Since, moreover, $|w| = |u| - 1 + |v| \leq |u| + L_V - 1$, the proof is achieved.

Lemma 6. *The automaton* $\mathcal{A}$ *is complete.*

Proof. Let **a** be a row of $\varphi_A(A^*)$ with $\|\mathbf{a}\|$ maximal. By Proposition 6, it follows that $\|\mathbf{a}\varphi_A(u)\| = \|\mathbf{a}\| > 0$ for all $u \in A^*$. Consequently, $\varphi_A(u) \neq 0$ for all $u \in A^*$.

Proposition 7. *Set* $m_1 = \max\{\|\mathbf{a}\| \mid \mathbf{a} \text{ row of } \varphi_A(A^*)\}$. *There exists a word* w *such that* $\varphi_A(w)$ *has a maximal row and*

$$|w| \leq \max\{0, \ 1 + (m_1 - 2)(L_V + n - 1)\}. \tag{8}$$

Proof. If the automaton $\mathcal{A}$ is deterministic, then any row of $\varphi_A(\varepsilon)$ is maximal and the statement is trivially verified. Thus we assume that $\mathcal{A}$ is not deterministic. Hence, there is a letter $x \in A$ and a row $\mathbf{a}_0$ of $\varphi_A(x)$ such that $\|\mathbf{a}_0\| \geq 2$.

In view of Proposition 6 and Lemma 3 one can find words w_i and vectors $\mathbf{a}_i$, $1 \leq i \leq k$ such that

$$\mathbf{a}_i = \mathbf{a}_{i-1}\varphi_A(w_i), \quad \|\mathbf{a}_i\| > \|\mathbf{a}_{i-1}\|, \quad |w_i| \leq L_V + n - 1, \tag{9}$$

and $\mathbf{a}_k$ is a maximal row of $\varphi_A(A^*)$. Set $w = xw_1 w_2 \cdots w_k$. Since $\mathbf{a}_0$ is a row of $\varphi_A(x)$, the vector $\mathbf{a}_k = \mathbf{a}_0\varphi_A(w_1 w_2 \cdots w_k)$ is a row of $\varphi_A(w)$. Moreover, from (9) one has

$$m_1 \geq \|\mathbf{a}_k\| \geq k + \|\mathbf{a}_0\| \geq k + 2, \quad |w| \leq 1 + k(L_V + n - 1).$$

From these inequalities, one easily derives Eq. (8), concluding the proof.

In the sequel we will further suppose that there exists also a finite set $W \subseteq A^*$ such that $A \subseteq W$ and for every state p of $\mathcal{A}$

$$\sum_{w \in W} \mathrm{Card}(pw^{-1}) \geq \mathrm{Card}(W). \tag{10}$$

In such a case, with an argument symmetrical to that used in the proof of Proposition 7 one can prove the following

Proposition 8. Set $m_2 = \max\{\|\mathbf{b}\| \mid \mathbf{b}$ column of $\varphi_A(A^*)\}$. There exists a word v such that $\varphi_A(v)$ has a maximal column and

$$|v| \leq \max\{0,\ 1 + (m_2 - 2)(L_W + n - 1)\}.$$

Now we state the main result of this section.

Proposition 9. Let $\mathcal{A}$ be a synchronizing unambiguous transitive n-state automaton, with $n \geq 2$. Let $V, W \subseteq A^*$ be two finite sets of words satisfying Eq. (7) and Eq. (10), respectively, with $A \subseteq V, W$. Then, $\mathcal{A}$ has a reset word u such that

$$|u| \leq (n - 2)L_{V \cup W} + n^2 - 3n + 3.$$

Proof. First, we consider the case that the parameters m_1, m_2 introduced in Propositions 7 and 8 verify the conditions $m_i \geq 2$, $i = 1, 2$.

By Propositions 7 and 8, there are words w and v and states $p, q \in S$ such that $(\varphi_A(w))_{p*} = \mathbf{a}$ is a maximal row of $\varphi_A(A^*)$, $(\varphi_A(v))_{*q} = \mathbf{b}$ is a maximal column of $\varphi_A(A^*)$, and

$$|w| \leq 1 + (m_1 - 2)(L_V + n - 1), \quad |v| \leq 1 + (m_2 - 2)(L_W + n - 1). \tag{11}$$

Since $\mathcal{A}$ is transitive, there exists a word z such that $p \in qz$ and $|z| \leq n - 1$. One has then $(\varphi_A(z))_{qp} = 1$ and consequently

$$\varphi_A(vzw) = (\varphi_A(v))_{*q}(\varphi_A(z))_{qp}(\varphi_A(w))_{p*} + \mu = \mathbf{ba} + \mu,$$

for some $\mu \in \{0, 1\}^{S \times S}$. By Lemma 6, $\varphi_A(A^*)$ is a transitive monoid of $(0,1)$-matrices without 0, and its minimal rank is 1. By Proposition 5 one derives that $\mu = 0$ and $u = vzw$ is a reset word.

Now we evaluate $|u|$. From (11) one has

$$|u| = |v| + |w| + |z| \leq n + 1 + (m_1 + m_2 - 4)(L_{V \cup W} + n - 1).$$

Since by Lemma 4, $m_1 + m_2 \leq n + 1$, one derives

$$|u| \leq (n - 3)L_{V \cup W} + n^2 - 3n + 4,$$

so that the statement holds true.

Now we consider the case $m_2 = 1$ (the case $m_1 = 1$ is symmetrically dealt with). We can still find a word w and a state $p \in S$ such that $\mathbf{a} = (\varphi_A(w))_{p*}$ is a maximal row of $\varphi_{\mathcal{A}}(A^*)$ and $|w| \leq 1 + (m_1 - 2)(L_V + n - 1)$. Since $m_1 \leq n$ and $L_V \leq L_{V \cup W}$, one obtains $|w| \leq (n - 2)L_{V \cup W} + n^2 - 3n + 3$. Now, to complete the proof, it is sufficient to verify that w is a reset word. Since $m_2 = 1$, the vector $\mathbf{b} = (\varphi_{\mathcal{A}}(\epsilon))_{*p}$ is a maximal column of $\varphi_{\mathcal{A}}(A^*)$. Moreover, $\varphi_{\mathcal{A}}(w) = \mathbf{ba} + \mu$ for some $\mu \in \{0, 1\}^{S \times S}$. By Proposition 5 one derives that w is a reset word. This concludes the proof.

An automaton $\mathcal{A}$ on a k-letter alphabet is *Eulerian* if for any vertex of its graph, there are exactly k in-coming and k out-coming arrows. In [11], Kari showed that Černý conjecture is true for Eulerian deterministic automata. If $\mathcal{A}$ is an Eulerian automaton, then the hypotheses of Proposition 9 are satisfied for $V = W = A$. Thus we may extend Kari's result to unambiguous Eulerian automata.

Corollary 3. *Any transitive, synchronizing, and unambiguous Eulerian n-state automaton has a reset word of length not larger than $(n - 1)^2$.*

References

1. Béal, M.-P.: A note on Černý's Conjecture and rational series, technical report, Institut Gaspard Monge, Université de Marne-la-Vallée (2003)
2. Béal, M.-P., Ceizler, E., Kari, J., Perrin, D.: Unambiguous automata. Math. comput. sci., 14 (2008)
3. Berstel, J., Reutenauer, C.: Rational series and their languages. Springer, Heidelberg (1988)
4. Carpi, A.: On synchronizing unambiguous automata. Theoret. comput. sci. 60, 285–296 (1988)
5. Černý, J., Poznámka, K.: Homogénnym experimenton s konečnými automatmi. Mat. fyz. cas SAV 14, 208–215 (1964)
6. Césari, Y.: Sur l'application du théorème de Suschewitsch à l'étude des codes rationnels complets. In: Loeckx, J. (ed.) Automata, Languages and Programming. LNCS, vol. 14, pp. 342–350. Springer, Berlin (1974)
7. Dubuc, L.: Sur les automates circulaires et la conjecture de Cerny. RAIRO Inform. Théor. Appl. 32, 21–34 (1998)
8. Eilenberg, S.: Automata, Languages and Machines, vol. A. Academic Press, London (1974)
9. Frankl, P.: An extremal problem for two families of sets. Eur. J. Comb. 3, 125–127 (1982)
10. Goralčík, P., Hedrlín, Z., Koubek, V., Ryšlinková, J.: A game of composing binary relations. RAIRO Inform. Théor. 16, 365–369 (1982)
11. Kari, J.: Synchronizing finite automata on Eulerian digraphs. Theoret. comput. sci. 295, 223–232 (2003)
12. Mateescu, A., Salomaa, A.: Many-valued truth functions, Cerny's conjecture and road coloring. EATCS Bull. 68, 134–150 (1999)
13. Pin, J.E.: Le problème de la synchronization et la conjecture de Cerny, Thèse de 3 ème cycle, Université de Paris 6 (1978)
14. Pin, J.E.: Sur un cas particulier de la conjecture de Cerny. In: Ausiello, G., Böhm, C. (eds.) ICALP 1978. LNCS, vol. 62, pp. 345–352. Springer, Heidelberg (1978)
15. Rystov, I.: Almost optimal bound of recurrent word length for regular automata. Cybern. Syst. Anal. 31(5), 669–674 (1995)
16. Trahtman, A.N.: The Cerny conjecture for aperiodic automata. Discrete Math. Theor. Comput. Sci. 9, 3–10 (2007)

On the Decidability of the
Equivalence for k-Valued Transducers
(Extended Abstract)

Rodrigo de Souza*

TELECOM ParisTech, 46, rue Barrault, 75634 Paris Cedex 13
rsouza@enst.fr

Abstract. We give a new proof for the decidability of the equivalence of two k-valued transducers, a result originally established by Culik and Karhümaki and independently by Weber. Our proof relies on two constructions we have recently introduced to decompose a k-valued transducer and to decide whether a transducer is k-valued. As a result, our proof is entirely based on the structure of the transducers under inspection, and the complexity it yields is of single exponential order on the number of states. This improves Weber's result by one exponential.

1 Introduction

This communication is the third part of a complete reworking of the theory of k-valued rational relations and transducers which makes it appear as a natural generalisation of the theory of rational functions and functional transducers, not only at the level of results but *also at the level of proofs*. In [1], we present a construction to decompose a k-valued transducer into a sum of k functional and unambiguous ones of *single exponential size*. The existence of such a decomposition has been settled by Weber in [2]; but this proof yields a bound of double exponential order. And in [3] we generalise a technique of [4] to a new algorithm to decide whether a transducer is k-valued, a result originally due to Gurari and Ibarra [5]. Here, we combine the techniques of [1] and [3] into a procedure to decide the equivalence of k-valued transducers in *single exponential time.*

Equivalence of automata is a most fundamental problem in the field of automata theory and has been studied for several formalisms. For transducers the equivalence reduces to the Post Correspondence Problem and is thus undecidable [6]. This result has been subsequently sharpened by Griffiths [7] for generalised sequential machines with no empty words in the transitions and next by Ibarra [8] for the same device over a unary input (or output) alphabet.

Things change for k-valued transducers. For the functional ones $(k = 1)$, the decidability of the equivalence follows from the *decidability of the functionality*, a particular case of Gurari and Ibarra's theorem which had been established by Schützenberger [9] and independently by Blattner and Head [10]. The general

---

* Research supported by CAPES Foundation (Brazilian government).

M. Ito and M. Toyama (Eds.): DLT 2008, LNCS 5257, pp. 252–263, 2008.
© Springer-Verlag Berlin Heidelberg 2008

case is much more involved. It has been tackled with different methods by Culik and Karhumäki [11] and by Weber, the latter with a bound for the complexity:

Theorem 1 (Weber [12]). *The equivalence of two k-valued transducers is decidable in double exponential time.*

Here, we improve Weber's result by one exponential (which seems to be the optimal for the equivalence of automata is a PSPACE-complete problem [13]):

Theorem 2. *The equivalence of two k-valued transducers is decidable in single exponential time.*

Culik and Karhumäki's proof relies heavily on existential properties of test sets. They show that, for every HDT0L language, there exists a finite test set for the family of k-valued transducers having at most n states (where k and n are fixed parameters). The finiteness of this test set and thus a termination condition for their procedure comes from the existence of a test set for morphisms on every language — the Ehrenfeucht Conjecture, which had just been confirmed. Even thought Culik and Karhumäki's test set can be effectively constructed (which proves indeed the decidability for k-valued transducers on HDT0L languages) no clue for the complexity of the underlying algorithm is given.

Weber's proof starts with a decomposition of the k-valued transducers under inspection, and offers two ways to decide the equivalence. The first one relies on Gurari and Ibarra's procedure to decide the k-valuedness; it yields a counter automaton which accepts some word iff the transducers are not equivalent, and uses a polynomial time procedure due to Ibarra to decide the emptiness of a counter automaton [14]. The second algorithm depends upon a bound for some Ramsey numbers and from it a bound for the length of a witness for the *non-inclusion*. The complexities of both solutions are affected by the size (number of states) of the initial decomposition, which is of double exponential order.

At first sight, our proof may resemble Weber's one: it uses a decomposition of the transducers and next a procedure to test the k-valuedness. But it is at the same time different in both steps, as one can derive from the discussions in [1] and in [3]. In one word, both the decomposition theorem and the Ramsey-type arguments which are developed in [12] are highly combinatorial; the exact overall complexity of Gurari and Ibarra's reduction to decide the k-valuedness is difficult to estimate. Our constructions, which we explain more in detail below, are more oriented towards the structure of the transducers. They yield more accurate complexities, and does not resort to any other object than transducers.

The first ingredient of our proof is the decomposition of a k-valued transducer T into k functional and unambiguous ones we have explained in [1], and which we call here a *lexicographic decomposition* of T. It is based on *covering of automata*. Coverings have been introduced in [15]; roughly speaking, a covering of A is a larger automaton B with a mapping between states and transitions which puts the computations of A and B in bijection. Typically, B allows to distinguish more easily among the computations and contains a subautomata which chooses some special subset of them. We construct two coverings in [1], one for N-automata and other for transducers — respectively, the "multi-skimming covering" and the

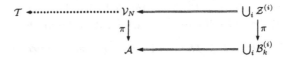

Fig. 1. Lexicographic decomposition of $\mathcal{T}$: $\mathcal{V}_N$ is an equivalent and input-k-ambiguous transducer yielded by the lag separation covering; $\mathcal{A}$ is the underlying input automaton of $\mathcal{V}_N$; $\bigcup_i \mathcal{B}_k^{(i)}$ is a decomposition of $\mathcal{A}$ given by multi-skimming covering; $\bigcup_i \mathcal{Z}^{(i)}$ is the final decomposition, which one obtains by "lifting" the outputs of $\mathcal{T}$ to $\bigcup_i \mathcal{B}_k^{(i)}$

"lag separation covering".[1] Both are based on the same principle of putting a lexicographic ordering on the transitions and from it a lexicographic ordering on the computations. The whole decomposition uses both coverings, as it is pictured in Figure 1: the lag separation covering allows to built a transducer which is equivalent to $\mathcal{T}$ and *input-k-ambiguous*, the multi-skimming covering yields a decomposition of such a transducer. Individually, each covering can provoke an exponential blow up; but the mappings kept between the computations, together with the fact that $\mathcal{T}$ is k-valued, allow to show that the useful states in the lexicographic decomposition are of some very restricted form. One can bound the number of these states by one exponential on the number of states of $\mathcal{T}$.

The algorithm we present in [3] to decide the k-valuedness is completely different from Gurari and Ibarra's one. It starts with a generalisation of the product of $\mathcal{T}^2$ by the Lead or Delay Action (LDA) $\mathcal{G}$, which has been introduced in [4] to characterise the functionality ($k = 1$). In $\mathcal{T}^2 \times \mathcal{G}$, $\mathcal{G}$ measures differences of outputs of pairs of successful computations reading a same word. Then, it can be "seen" in the final states of $\mathcal{T}^2 \times \mathcal{G}$ whether $\mathcal{T}$ is functional. Likewise, the k-valuedness can be read in the product of $\mathcal{T}^{k+1}$ by the *Pairwise Lead or Delay Action* $\mathcal{G}_{k+1}$. But, as we explain in [3], this generalisation is not so straightforward for $\mathcal{T}^{k+1} \times \mathcal{G}_{k+1}$ may be infinite for $k > 1$. To tackle this general case we define a valuation of $\mathcal{T}^{k+1}$ where the value of a state[2] $\mathbf{q}$ is a finite set $\mathsf{m}(\mathbf{q})$ of *partially defined pairwise differences* (PDPD) which "traverse" the (potentially infinite) set of states of $\mathcal{T}^{k+1} \times \mathcal{G}_{k+1}$ projecting on $\mathbf{q}$. Then, it can be read in $\mathsf{m}(\mathbf{q})$ if every computation of $\mathcal{T}^{k+1}$ from some initial state to $\mathbf{q}$ contains a pair of projections with the same output. Clearly, $\mathcal{T}$ is k-valued iff this holds in the final states. We call $\mathsf{m}(\mathbf{q})$ the value of $\mathbf{q}$ and the family of these sets the *Lead or Delay Valuation* of $\mathcal{T}^{k+1}$ (LDV). Our algorithm to test the k-valuedness constructs the LDV in polynomial time with a topological traversal of $\mathcal{T}^{k+1}$.

Our algorithm to decide the inclusion of (the behaviour of) a k-valued transducer $\mathcal{S}$ in (the behaviour of) another one $\mathcal{T}$ builds at first a lexicographic decomposition of $\mathcal{S}$, say $\mathcal{R}^{(0)}, \ldots, \mathcal{R}^{(k-1)}$. This is intended to simplify the problem: as the union of these unambiguous transducers is equivalent to $\mathcal{S}$, it remains to decide whether each of them is included in $\mathcal{T}$ in *polynomial time on their sizes*.

Let $\mathcal{R}$ be any of the $\mathcal{R}^{(i)}$. In order to decide the inclusion of $\mathcal{R}$ in $\mathcal{T}$, one can be tempted to construct a lexicographic decomposition of $\mathcal{T}$, $\mathcal{Z}^{(0)}, \ldots, \mathcal{Z}^{(k-1)}$,

---

[1] The latter is renamed the "lead or delay covering" in the journal version of [1].

[2] We write tuples of states, words or computations with bold letters.

and decide whether $\mathcal{W} = \mathcal{R} \times \mathcal{Z}^{(0)} \times \ldots \times \mathcal{Z}^{(k-1)}$ is k-valued. Such a procedure fails due to the following. If $\mathcal{R}$ is included in $\mathcal{T}$, then $\mathcal{R} \cup \mathcal{T}$ is clearly a k-valued transducer. But the contrary is false in general: it may well exist a word which is read by at most k successful computations in $\mathcal{R} \cup \mathcal{T}$ with the output of the one of $\mathcal{R}$ being different from the outputs of the ones of $\mathcal{T}$. Thus, one has to be more accurate before applying the LDV and be able to know, for every successful computation of $\mathcal{R}$, all the ones in $\mathcal{Z}^{(0)}, \ldots, \mathcal{Z}^{(k-1)}$ reading the same input.

Here comes the trick which allows to combine both constructions. We make a product of $\mathcal{W}$ by an action $\circ_\mu$ of the free monoid A^* of the input words on $\mathbb{N}$-vectors. Roughly speaking, $\circ_\mu$ "counts", for every $u \in A^*$, the ends of the computations in $\mathcal{Z}^{(0)}, \ldots, \mathcal{Z}^{(k-1)}$ which start in some initial state and read u. In $\mathcal{W} \times \circ_\mu$, these $\mathbb{N}$-vectors allow to distinguish the computations of $\mathcal{W}$ satisfying the aforementioned crucial property: the projection on $\mathcal{R}$ is a successful computation; the successful computations of $\mathcal{Z}^{(0)}, \ldots, \mathcal{Z}^{(k-1)}$ reading the same input are entirely included in the other projections (Proposition 6). It follows that we can decide the inclusion of $\mathcal{R}$ in $\mathcal{T}$ by constructing the LDV of $\mathcal{W} \times \circ_\mu$, and analysing the values $\mathsf{m}(\mathbf{q})$ of the ends of these critical computations.

Although the size of the "accessible part" of $\circ_\mu$ seems to be of double exponential order (on the number n of states of $\mathcal{T}$), the properties we have used in [1] to bound the size of the lexicographic decomposition by one exponential (Proposition 4) implies the same for $\circ_\mu$ (Proposition 9). Then we come to our main result: the equivalence can be decided in single exponential time (Theorem 6).

The description of this algorithm is postponed to Section 5. Before, we recall briefly the LDV (Sections 3) and the lexicographic decomposition (Section 4).

Let us stress that in our result the valuedness k is viewed as a fixed parameter and is thus a constant; otherwise, the single exponential complexity does not hold anymore for the expression given in Theorem 6 grows slower than 2^{n^k}. Moreover, it is to be acknowledged that Weber's procedure allows to decide the equivalence for bounded valued transducers whose exact valuedness is not known beforehand. In a sense, both problems are equivalent for, as Weber showed in [16], the valuedness of a bounded valued transducer can be effectively calculated; but it may be of exponential order on the number of states of the transducer. Therefore, in a more honest comparison it has to be added that the complexity of the procedure proposed in [12] remains of double exponential order even under the hypothesis of constant valuedness we are dealing with (this is due, in particular, to the size of the preliminary decomposition used in [12]).

Finally, let us note that Weber also proves in [12] the existence of a witness of double exponential size for the non-inclusion of k-valued transducers. Again, the lexicographic decomposition and the LDV allow to obtain a more concise one. Due to space constraints we postpone to a forthcoming paper the construction of this witness, as well as the details and proofs which are omitted here.

2   Preliminaries

We follow the definitions and notation in [17,18,19].

The set of words over a finite alphabet A (the free monoid over A) is denoted by A^*, and the empty word by 1_{A^*}, or simply 1 in figures.

Let M be a monoid. An automaton $\mathcal{A} = \langle Q, M, E, I, T \rangle$ is a labelled directed graph given by sets Q of states, $I, T \subseteq Q$ of initial and final states, respectively, and $E \subseteq Q \times M \times Q$ of transitions labelled by M. It is finite if Q and E are finite.

A *computation* in $\mathcal{A}$ is a sequence of transitions $c : p_0 \xrightarrow{m_1} p_1 \xrightarrow{m_2} \ldots \xrightarrow{m_l} p_l$, also written $c : p_0 \xrightarrow[\mathcal{A}]{m_1 \ldots m_l} p_l$. The *label* of c is $m_1 \ldots m_l \in M$, and c is a *successful computation* if $p_0 \in I$ and $p_l \in T$. The *behaviour* of $\mathcal{A}$ is the set $|\mathcal{A}| \subseteq M$ of labels of successful computations. The behaviour of finite automata over M coincide with the family $\operatorname{Rat} M$ of the *rational subsets* of M [18].

If M is a free monoid A^* and the labels of transitions are letters, then $\mathcal{A}$ is a (boolean) automaton over A. If M is a product $A^* \times B^*$, then every transition is labelled by an input word $u \in A^*$ and an output one $x \in B^*$ — this is denoted by $u|x$ — and $\mathcal{A}$ is a *transducer* realising a *rational relation* from A^* to B^*.

The *image* of a word $u \in A^*$ by (the behaviour of) a transducer is the set of outputs of successful computations which *read* u. The transducer is called *k-valued*, where k is a positive integer, if, for every input word, the image has at most k words. It is *bounded valued* if there exists such an integer k.

We shall only consider transducers which are *real-time*: every transition is labelled by a pair $a|K$ formed by a letter $a \in A$ and a set $K \in \operatorname{Rat} B^*$, and I, T are functions from Q to $\operatorname{Rat} B^*$. By using classical constructions on automata, every transducer can be transformed into a real-time one. For bounded valued relations, we may suppose that every transition outputs a single word, and that the image of every initial or final state is the empty word. In this case[3], the transducer is rather denoted as $\mathcal{T} = \langle Q, A, B^*, E, I, T \rangle$.

We shall make systematic use of product of (real-time) transducers. This operation is defined in the same way as for boolean automata, with the difference that the outputs have to be taken into account. Formally, the square of a transducer $\mathcal{T} = \langle Q, A, B^*, E, I, T \rangle$ is the transducer $\mathcal{T}^2 = \langle Q^2, A, B^{*2}, F, I^2, T^2 \rangle$ where $(p, q) \xrightarrow{a|(u,v)} (p', q')$ is in F if, and only if, both $p \xrightarrow{a|u} p'$ and $q \xrightarrow{a|v} q'$ are in E (see [4] for details). Likewise, the product $\mathcal{T}_1 \times \ldots \times \mathcal{T}_l$ of several transducers over $A^* \times B^*$ is a transducer labelled by $A \times B^{*l}$. Moreover, all these automata are implicitly assumed to be accessible (as in the convention adopted in [3]).

An $\mathbb{N}$-automaton (where $\mathbb{N}$ is the semiring of the natural numbers) is an automaton labelled by letters with multiplicities in $\mathbb{N}$ attached to transitions and states (for the latter, an initial and a final one). It realises an $\mathbb{N}$-*rational series*: a function $s : A^* \to \mathbb{N}$ mapping every $u \in A^*$ to a multiplicity obtained by summing the weights of the successful computations labelled by u.

A *morphism* from an automaton $\mathcal{B} = \langle R, M, F, J, U \rangle$ (over a monoid M) to $\mathcal{A} = \langle Q, M, E, I, T \rangle$ is a pair of mappings, $R \to Q$ and $F \to E$ (both denoted by φ) which respect adjacency of transitions and where $J\varphi \subseteq I$ and $U\varphi \subseteq T$. The image by φ of every successful computation of $\mathcal{B}$ is a successful one in $\mathcal{A}$ with the same label, hence $|\mathcal{B}| \subseteq |\mathcal{A}|$. The morphism is a *covering* if it is *locally bijective*.

---

[3] A *nondeterministic generalised sequential machine* in some references.

This implies a bijection between the successful computations and thus $|\mathcal{B}| = |\mathcal{A}|$. See [15] for details. Coverings for automata with multiplicity in some semiring $\mathbb{K}$, or $\mathbb{K}$-covering, have been defined in [20]. Rather than put a bijection between the successful computations of the (support of the) automata, a $\mathbb{K}$-covering equals the sums of the multiplicities of computations labelled by the same word.

3   The Lead or Delay Valuation

The Lead or Delay Valuation (LDV) is the second tool we use to decide the equivalence, but we shall discuss it in first place in order to define some notation.

As said in the introduction, we have introduced the LDV of a product T^{k+1} in [3] to decide the k-valuedness of T (our new proof for Gurari and Ibarra's theorem). In order to ease the use of this notion in Section 5, we shall explain here what we call the LDV of a product $W = T_1 \times \ldots \times T_l$ of possibly distinct transducers. There is no novelty with respect to [3], for the valuation of the states of W is contained in the valuation of T^l where $T = T_1 \cup \cdots \cup T_l$.

In order to define the LDV we need an action, the PLDA in Definition 1. Let $F(B)$ be the free group generated by an alphabet B, that is, the quotient of $(B \cup \overline{B})^*$ by the relations $x\overline{x} = \overline{x}x = 1_{B^*}$ ($x \in B$) where $\overline{B}$ a disjoint copy of B. We write $\Delta = B^* \cup \overline{B}^* \cup \{\mathbf{0}\}$, where $\mathbf{0}$ is a new element not in $F(B)$, and define a function $\rho : F(B) \cup \{\mathbf{0}\} \to \Delta$ by $w\rho = w$, if $w \in \Delta$, and $w\rho = \mathbf{0}$ otherwise.[4]

Definition 1. *The LDA, denoted by $\mathcal{G}$, is the action of $B^* \times B^*$ on Δ defined as follows: for every $w \in \Delta$ and $(u, v) \in B^* \times B^*$, $w \cdot (u, v) = (\overline{u}wv)\rho$ (where $\mathbf{0}u = u\mathbf{0} = \mathbf{0}$). For every integer $l > 1$, the Pairwise Lead or Delay Action (PLDA), $\mathcal{G}_l$, is the action of B^{*l} on the set Δ_l of Δ-vectors indexed by the set $\{(i, j) \mid 1 \le i < j \le l\}$ which applies the LDA independently on each coordinate.*

For computations $c : p \xrightarrow{u|x} q$ and $d : r \xrightarrow{u|y} s$ with the same input, let $\mathrm{LD}(c, d)$, their *Lead or Delay*, be the element $1_{B^*} \cdot (x, y)$ of Δ. Intuitively, $\mathrm{LD}(c, d)$ is the "difference" between the outputs of c and d. The PLDA measures the Lead or Delay between every pair of projections of computations in $W = T_1 \times \ldots \times T_l$.

For every state $\mathbf{q}$ of W, let $X(\mathbf{q}) = \{\delta \in \Delta_l \mid (\mathbf{q}, \delta) \text{ in } W \times \mathcal{G}_l\}$. This set may be infinite. The aim of the LDV is to characterise the k-valuedness within a finite object. It attributes to $\mathbf{q}$ a finite set $\mathbf{m}(\mathbf{q})$ of "minimal traverses" for $X(\mathbf{q})$.

In order to define these traverses, let H_l be the union of Δ_l with the vectors (of same dimension) on $\Delta \cup \{\bot\}$, where $\bot$ represents undefined entries; the elements of H_l are called *partially defined pairwise differences*, or PDPD. A traverse for $X \subseteq H_l$ is a PDPD $\gamma \in H_l$ satisfying: no entry of γ is equals to $\mathbf{0}$; for every $\delta \in X$, there exists a coordinate (i, j) such that $\delta_{i,j} \ne \bot$ and $\gamma_{i,j} = \delta_{i,j}$; for every (i, j) such that $\gamma_{i,j} \ne \bot$, there exists at least one δ in X such that $\gamma_{i,j} = \delta_{i,j}$.

The set H_l is naturally ordered by $\beta \sqsubseteq \gamma$ iff γ coincides with β on the defined entries of β. Now we can define the LDV of the states of W:

---

[4] We use a postfix notation for relations: $x\tau$ is the image of x by the relation τ

Definition 2. *For every $X \subseteq H_l$, let* tv (X) *be the set of traverses for X. Denote* $m(X) = \min(\text{tv}(X))$ *(the set of minimal traverses for X). The value of a state* $\mathbf{q}$ *of $\mathcal{W}$ in the LDV is the set* $m(\mathbf{q}) = m(X(\mathbf{q}))$.

The finiteness of the LDV and the useful information on the computations of $\mathcal{W}$ it allows to retrieve are stated in the propositions below. The second one is a direct consequence of the definition of the LDV; the proof of the first one is more involved (see the manuscript version of [3]).

Proposition 1. *For every $l > 1$, for every $X \subseteq \Delta_l$,* $\text{card}(m(X)) \leq 2^{l^4}$.   □

Proposition 2. *For every state $\mathbf{q}$ of $\mathcal{W}$, the following assertions are equivalent: there exists at least one γ in $m(\mathbf{q})$ whose defined entries are all 1_{B^*}; every computation of $\mathcal{W}$ from some initial state to $\mathbf{q}$ has at least one pair of projections with the same output.*   □

As explained in [3], the LDV can be constructed with a traversal of the strongly connected components (SCCs) of $\mathcal{W}$. It builds the values in some topological order and is based on two properties. The first one is a stability property on the valuation within every SCC; it implies that the values of distinct states in a same SCC are interdependent, thus the knowledge of one of them allows to calculate the others. The second property states that every $m(\mathbf{p})$ depends uniquely on the values of the states which precedes and are adjacent to the SCC of $\mathbf{q}$: it is equal to the set of the minimal least upper bounds by the order $\sqsubseteq$ of the previously calculated values (the operation $\oplus$ in Figure 2). Then we have:

Proposition 3. *The LDV of $\mathcal{W}$ can be constructed in time and space complexity $\mathcal{O}(2^{5(k+1)^4} \ell n^{k+1} m^{k+1})$, where n and m are the sum of the number of states and the sum of the number of transitions of the transducers $\mathcal{T}_1, \ldots, \mathcal{T}_l$, respectively, and ℓ is the maximal length of the outputs of the transitions.*   □

4   The Lexicographic Covering

The lexicographic covering is a method to build coverings of automata we have introduced in [1] to construct a "concise" decomposition of a k-valued transducer:

Theorem 3 ([1]). *Every k-valued transducer $\mathcal{T}$ can be effectively decomposed into a sum of k unambiguous transducers with $2^{\mathcal{O}(h\ell k^4 n^{k+4})}$ states, where n and ℓ are respectively the maximal numbers of states and lengths of the outputs of transitions of $\mathcal{T}$ and h is the cardinality of the output alphabet.*

We defined two instances of this method, one for transducers, the *lag separation covering*, and other for $\mathbb{N}$-automata, the *multi-skimming covering*. The idea of both is to order lexicographically computations of automata, inasmuch as it can be made with words on some alphabet. Then, by erasing some parts of the covering, one can find a subautomaton which chooses some special subset of computations — the smallest ones, for example, in the multi-skimming covering.

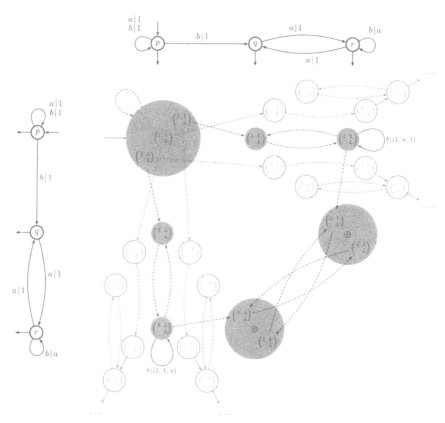

Fig. 2. The LDV of part of the cube of a transducer $\mathcal{T}$. The first coordinate is fixed on p, the second and third ones are the horizontal and vertical projections, respectively. The PDPDs are represented as upper triangular matrices indexed by $\{p,q\} \times \{q,r\}$ (in this order). The values of the states of $\mathcal{T}^3$ are represented inside the gray regions; in two cases, it is the result of the operation $\oplus$ (the least upper bound of PDPDs) on PDPDs coming from other SCCs. Dashed transitions have output equal to $(1_{B^*}, 1_{B^*}, 1_{B^*})$.

The aim of the lag separation covering of $\mathcal{T} = \langle Q, A, B^*, E, I, T \rangle$ is a selection among the computations with the same label and such that the differences of lengths of outputs along them (their "lag") are bounded by an integer N. In this covering, $\mathcal{U}_N$, the states are pairs in $Q \times \mathfrak{P}(B^{\leq N} \cup \overline{B}^{\leq N})^Q$, the vectors in the second component intend to "store" the Lead or Delay between the computations of $\mathcal{T}$ and the smaller ones. Together with a combinatorial property of the LDA, this covering allows to prove that every k-valued rational relation can be realised by a transducer whose underlying input automaton is k-ambiguous:

Theorem 4 ([1]). *For every k-valued transducer $\mathcal{T}$, there exists an index N such that the lag separation covering $\mathcal{U}_N$ contains a subtransducer $\mathcal{V}_N$ which is equivalent to $\mathcal{T}$ and input-k-ambiguous.* □

The aim of the multi-skimming covering is the construction of an automaton of single exponential size which performs "skimmings" on an $\mathbb{N}$-rational series:

Theorem 5 ([1]). *Let $\mathcal{A}$ be an $\mathbb{N}$-automaton with n states. For every $k > 0$, there exists an $\mathbb{N}$-covering $\mathcal{B}_k$ of $\mathcal{A}$ such that: $\mathcal{B}_k$ has at most $n(k+1)^n$ states; $\mathcal{B}_k$ has a subautomaton $\mathcal{D}_k$ which realises $|\mathcal{A}| \dot{-} k$: for every i, $0 \le i < k$, $\mathcal{B}_k$ has an unambiguous subautomaton $\mathcal{B}_k^{(i)}$ which recognises the support of $|\mathcal{A}| \dot{-} i$.*

The multi-skimming covering of $\mathcal{A} = \langle Q, I, E, T \rangle$ is an $\mathbb{N}$-automaton $\mathcal{B}$ of dimension $Q \times \mathbb{N}^Q$; the $\mathbb{N}$-covering $\mathcal{B}_k$ which fits with the statement of Theorem 5 is the finite $\mathbb{N}$-quotient of $\mathcal{B}$ based on the quotient $\mathbb{N}_k$ of $\mathbb{N}$ by the relation $k = k+1$. Roughly speaking, the $\mathbb{N}$-vectors "count", for every successful computation of $\mathcal{A}$, the number of the smaller ones. One can thus chooses the smallest computations to construct the unambiguous subautomaton $\mathcal{B}_k^{(0)}$, and so on.

These coverings allow a decomposition of T in two steps: first, the lag separation covering of T yields an equivalent and input-k-ambiguous transducer $\mathcal{V}_N$; next, the construction of the $\mathbb{N}$-covering $\mathcal{B}_k$ on the underlying input automaton of $\mathcal{V}_N$ followed by a "lifting" of the outputs of the transitions of T to $\mathcal{B}_k$ yields unambiguous transducers $\mathcal{Z}^{(0)}, \ldots, \mathcal{Z}^{(k-1)}$ decomposing T (see Figure 3).

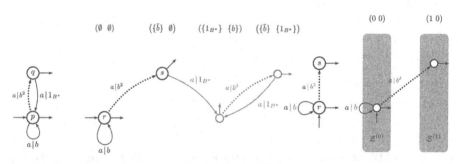

(a) A lag separation covering $\mathcal{U}_N$ with $N = 1$. The $\mathfrak{P}(\Delta_1)^Q$-vectors (vertical projection) are indexed by $\{p, q\}$, in that order.

(b) A 2-skimming of the input automaton of $\mathcal{V}_1$.

Fig. 3. A lexicographic decomposition of a 2-valued transducer into unambiguous ones $\mathcal{Z}^{(0)}$ and $\mathcal{Z}^{(1)}$, which realise the functions $a^n \mapsto b^n$ ($n \ge 0$) and $a^n \mapsto b^{n+1}$ ($n > 0$), respectively. In both coverings, an ordering is put on the transitions leaving the initial state; the solid transition is the smallest one. In 3(a), the input-2-ambiguous subtransducer $\mathcal{V}_1$ is reduced to the states $\{r, s\}$. In 3(b), $\mathcal{Z}^{(0)}$ and $\mathcal{Z}^{(1)}$ are obtained by keeping as final exactly one final state on the indicated column.

Effectiveness and single exponential size are critical properties of the lexicographic decomposition to be used in the equivalence algorithm we are going to present. The latter is not so obvious in view of the use of two constructions of exponential size. But the single exponential size can be achieved if the coverings are restricted to the trim parts of the involved automata: under this hypothesis, we can show (Lemma 4.3 in [1]) that all the built vectors have a linear number of non-null entries. There exists a single exponential number of such vectors:

Proposition 4. *Let T be a k-valued transducer. Let n, ℓ and h be respectively the numbers of states, the maximal length of the outputs of transitions, and the cardinality of the output alphabet of T. In every useful state of a lexicographic decomposition $\mathcal{Z}^{(0)}, \ldots, \mathcal{Z}^{(k-1)}$, the second component is a vector with at most kn entries different from 0, and thus every $\mathcal{Z}^{(i)}$ has $2^{O(h\ell k^4 n^{k+4})}$ useful states.*

5 Putting Everything Together to Decide the Equivalence

Now we show how the LDV applied to the unambiguous transducers given by the lexicographic decomposition yields a procedure to establish the inclusion (and thus the equivalence) of k-valued transducers:

Theorem 6. *Let S and T be two k-valued transducers on $A^* \times B^*$. Let n and ℓ be respectively the maximal numbers of states and lengths of the outputs of transitions of these transducers, and h the cardinality of B. The inclusion of (the behaviour of) S in (the behaviour of) T is decidable in $2^{O(h\ell k^5 n^{k+4})}$.*

Restating the Problem with the Decomposition. Recall that, in order to decide the inclusion of S in T, it suffices to decide the inclusion of every unambiguous transducer of a lexicographic decomposition of S in T. Let $\mathcal{R}$ any of them. In order to decide the inclusion of $\mathcal{R}$ in T, we use now a lexicographic decomposition $\mathcal{Z}^{(0)}, \ldots, \mathcal{Z}^{(k-1)}$ of T. For every computation c in $\mathcal{R} \times \mathcal{Z}^{(0)} \times \ldots \times \mathcal{Z}^{(k-1)}$, let us call its projection on the first component (so, on $\mathcal{R}$) the $\mathcal{R}$-*projection* and every projection on the other ones (so, on some transducer in the decomposition) a T-*projection*. We also say that c is a *full computation* if its $\mathcal{R}$-projection is a successful computation and its T-projections contain all the successful computations in $\mathcal{Z}^{(0)}, \ldots, \mathcal{Z}^{(k-1)}$ reading the input of c. The following is clear:

Proposition 5. *The unambiguous transducer $\mathcal{R}$ is included in T iff every full computation of $\mathcal{R} \times \mathcal{Z}^{(0)} \times \ldots \times \mathcal{Z}^{(k-1)}$ has at least one T-projection which is successful and whose output is equals to the output of its $\mathcal{R}$-projection.* □

We shall show below how to test these conditions on the full computations with the help of the LDV. But before we need to *identify* these full computations. To do this, let $\mathcal{A}$ be the underlying input automaton of the input-k-ambiguous transducer V_N from which the decomposition $\mathcal{Z}^{(0)}, \ldots, \mathcal{Z}^{(k-1)}$ has been extracted. Let (λ, μ, ν) be the matrix representation of $\mathcal{A}$ (see the definition in [19]), and let $\circ_\mu : \mathbb{N}^Q \times A^* \to \mathbb{N}^Q$ be the action defined as follows: for every $\mathbf{v} \in \mathbb{N}^Q$ and $a \in A$, $\mathbf{v} \circ_\mu a = \mathbf{v} \cdot a\mu$. The "initial state" of this action (which can be seen as an automaton whose states are $\mathbb{N}$-vectors) is the vector λ of initial multiplicities of $\mathcal{A}$. We can see the full computations within the product[5] $\mathcal{R} \times \mathcal{Z}^{(0)} \times \ldots \times \mathcal{Z}^{(k-1)} \times \circ_\mu$:

Proposition 6. *Let*

$$C : (\mathbf{i}, \mathbf{i}^{(0)}, \ldots, \mathbf{i}^{(k-1)}, \lambda) \xrightarrow[\mathcal{R} \times \mathcal{Z}^{(0)} \times \ldots \times \mathcal{Z}^{(k-1)} \times \circ_\mu]{u|(x, x^{(0)}, \ldots, x^{(k-1)})} (\mathbf{q}, \mathbf{q}^{(0)}, \ldots, \mathbf{q}^{(k-1)}, \mathbf{v})$$

---

[5] See [4] for a background on actions and product of an automaton by an action.

be a computation starting in an initial state. The projection of C on the product $\mathcal{R} \times \mathcal{Z}^{(0)} \times \ldots \times \mathcal{Z}^{(k-1)}$ is a full computation iff the number of final states in $\mathbf{q}^{(0)}, \ldots, \mathbf{q}^{(k-1)}$ is equal to the sum of the coefficients of the $\mathbb{N}$-vector $\mathbf{v} \cdot \nu$.   □

Accordingly, we say that a state $(\mathbf{q}, \mathbf{q}^{(0)}, \ldots, \mathbf{q}^{(k-1)}, \mathbf{v})$ of $\mathcal{R} \times \mathcal{Z}^{(0)} \times \ldots \times \mathcal{Z}^{(k-1)}$ is *full* if the number of final states in $\mathbf{q}^{(0)}, \ldots, \mathbf{q}^{(k-1)}$ is equal to the sum of the coefficients of the vector $\mathbf{v} \cdot \nu$.

Deciding the Inclusion with the LDV. Let $\mathcal{W}$ be the accessible part of $\mathcal{R} \times \mathcal{Z}^{(0)} \times \ldots \times \mathcal{Z}^{(k-1)} \times \circ_\mu$. It is a direct consequence of the definition of the PLDA $\mathcal{G}_{k+1}$ that Proposition 5 can be restated as follows:

Proposition 7. *The unambiguous transducer $\mathcal{R}$ is included in $\mathcal{T}$ iff for every state $(\mathbf{q}, \delta)$ in $\mathcal{W} \times \mathcal{G}_{k+1}$ such that $\mathbf{q}$ is full in $\mathcal{W}$, there exists a coordinate i, $2 \leq i \leq k+1$, such that $\delta_{1,i} = 1_{B^*}$.*   □

The same can be expressed in the LDV of $\mathcal{W}$. This comes from the definition of traverse of a set of PDPDs (compare with Proposition 2):

Proposition 8. *The unambiguous transducer $\mathcal{R}$ is included in $\mathcal{T}$ iff for every full state $\mathbf{q}$ of $\mathcal{W}$, the value $\mathsf{m}(\mathbf{q})$ contains at least one PDPD γ such that: γ has at least one defined coordinate of form $\gamma_{1,i}$; all the defined coordinates of this form are equal to 1_{B^*}.*   □

Our algorithm to decide the inclusion of $\mathcal{R}$ in $\mathcal{T}$ is a construction of the LDV of $\mathcal{W}$ as described in Section 3 followed by the easy test of the condition stated in Proposition 8 in the full states.

Complexity. As explained in [1], the size of the transducers $\mathcal{R}$ (one of the functional transducers of a lexicographic decomposition of $\mathcal{S}$) and $\mathcal{Z}^{(0)} \times \ldots \times \mathcal{Z}^{(k-1)}$ (a lexicographic decomposition of $\mathcal{T}$), measured as the number of states, is bounded by $2^{\mathcal{O}(h\ell k^4 n^{k+4})}$. In order to evaluate the size of $\mathcal{W}$, it remains to evaluate the number of accessible vectors in the action $\circ_\mu$. This can be made with the same argument we used to show that the number of states in a lexicographic decomposition is bounded by one exponential (Proposition 4):

Proposition 9. *If $\mathcal{T}$ is a trim k-valued transducer, then every accessible vector of the action $\circ_\mu$ has at most nk non-null coordinates.*   □

We conclude that the number of accessible vectors of $\circ_\mu$ and thus the size of $\mathcal{W}$ is bounded by $2^{\mathcal{O}(h\ell k^4 n^{k+4})}$. By putting this upper bound in the expression of the complexity of our algorithm to construct the LDV of $\mathcal{W}$ (Proposition 3), we obtain the complexity claimed in Theorem 6.

Acknowledgements. I am grateful to Jacques Sakarovitch for his support and for the discussions which have led to the result presented in this communication.

References

1. Sakarovitch, J., Souza, R.: On the decomposition of k-valued rational relations. In: Albers, S., Weil, P. (eds.) Proceedings of STACS 2008, pp. 621–632 (2008), http://stacs-conf.org (to appear in Theory of Computing Systems) arXiv:0802.2823v1
2. Weber, A.: Decomposing a k-valued transducer into k unambiguous ones. RAIRO Informatique Théorique et Applications 30(5), 379–413 (1996)
3. Sakarovitch, J., Souza, R.: On the decidability of bounded valuedness for transducers. In: Ochmański, E., Tyszkiewicz, J. (eds.) MFCS 2008. LNCS, vol. 5162, pp. 588–600. Springer, Heidelberg (2008), Preliminary full version with proofs in http://www.infres.enst.fr/~rsouza/DFV.pdf
4. Béal, M.P., Carton, O., Prieur, C., Sakarovitch, J.: Squaring transducers: an efficient procedure for deciding functionality and sequentiality. Theoretical Computer Science 292, 45–63 (2003)
5. Gurari, E., Ibarra, O.: A note on finite-valued and finitely ambiguous transducers. Mathematical Systems Theory 16, 61–66 (1983)
6. Fischer, P.C., Rosenberg, A.L.: Multitape one-way nonwriting automata. Journal of Computer and System Sciences 2(1), 88–101 (1968)
7. Griffiths, T.V.: The unsolvability of the equivalence problem for Λ-free nondeterministic generalized machines. Journal of the ACM 15(3), 409–413 (1968)
8. Ibarra, O.: The unsolvability of the equivalence problem for ϵ-free NGSM's with unary input (output) alphabet and applications. SIAM Journal on Computing 7(4), 524–532 (1978)
9. Schützenberger, M.P.: Sur les relations rationnelles. In: Automata Theory and Formal Languages, 2nd GI Conference. LNCS, vol. 33, pp. 209–213. Springer, Heidelberg (1975)
10. Blattner, M., Head, T.: Single-valued a-transducers. Journal of Computer and System Sciences 15(3), 310–327 (1977)
11. Culik, K., Karhumäki, J.: The equivalence of finite valued transducers (on HDT0L languages) is decidable. Theoretical Computer Science 47(1), 71–84 (1986)
12. Weber, A.: Decomposing finite-valued transducers and deciding their equivalence. SIAM Journal on Computing 22(1), 175–202 (1993)
13. Garey, M., Johnson, D.: Computers and Intractability. Freeman, New York (1979)
14. Gurari, E., Ibarra, O.: The complexity of decision problems for finite-turn multicounter machines. Journal of Computer and System Sciences 22(2), 220–229 (1981)
15. Sakarovitch, J.: A construction on finite automata that has remained hidden. Theoretical Computer Science 204(1–2), 205–231 (1998)
16. Weber, A.: On the valuedness of finite transducers. Acta Informatica 27(8), 749–780 (1989)
17. Berstel, J.: Transductions and Context-Free Languages. B. G. Teubner (1979)
18. Eilenberg, S.: Automata, Languages, and Machines, vol. A. Academic Press, London (1974)
19. Sakarovitch, J.: Éléments de théorie des automates. Vuibert, Paris (2003); English translation: Elements of Automata Theory. Cambridge University Press, Cambridge (to appear)
20. Sakarovitch, J.: The rational skimming theorem. In: Van, D.L., Ito, M. (eds.) Proc. of The Mathematical Foundations of Informatics (1999), pp. 157–172. World Scientific, Singapore (2005)

Decidable Properties of 2D Cellular Automata[*]

Alberto Dennunzio[1] and Enrico Formenti[2,**]

[1] Università degli Studi di Milano–Bicocca
Dipartimento di Informatica, Sistemistica e Comunicazione,
Viale Sarca 336, 20126 Milano Italy
dennunzio@disco.unimib.it
[2] Université de Nice-Sophia Antipolis, Laboratoire I3S,
2000 Route des Colles, 06903 Sophia Antipolis France
enrico.formenti@unice.fr

Abstract. In this paper we study some decidable properties of two-dimensional cellular automata (2D CA). The notion of closingness is generalized to the 2D case and it is linked to permutivity and openness. The major contributions of this work are two deep constructions which have been fundamental in order to prove our new results and we strongly believe it will be a valuable tool for proving other new ones in the near future.

Keywords: cellular automata, decidability, symbolic dynamics.

1 Introduction and Motivations

Cellular automata (CA) are a well-known formal model for complex systems and, at the same time, a paradigmatic model of massive parallel computation. Indeed, a CA is made of an infinite number of identical finite automata arranged on a regular lattice ($\mathbb{Z}^2$ or $\mathbb{Z}$ in this paper). Each automaton assumes a state chosen from a set A, called the *set of states* or the *alphabet*. A *configuration* is a snapshot of all states of the automata. A *local rule* updates the state of each automaton on the basis of its current state and the ones of a finite set of neighboring automata. All the automata are updated synchronously.

Behind the simplicity of the CA definition stands the huge complexity of different dynamical behaviors which captured the attention of researchers all over these last thirty years. We refer the reader to [10, 15, 7, 13, 18] for a review of the main results and for a comprehensive bibliography.

Historically, Von Neumann introduced CA to study formal models for cells self-reproduction. These were two-dimensional models. Paradoxically, the study

---

[*] This work has been supported by the Interlink/MIUR project "Cellular Automata: Topological Properties, Chaos and Associated Formal Languages", by the ANR Blanc "Projet Sycomore" and by the PRIN/MIUR project "Formal Languages and Automata: Mathematical and Applicative Aspects".
[**] Corresponding author.

M. Ito and M. Toyama (Eds.): DLT 2008, LNCS 5257, pp. 264–275, 2008.

of the CA dynamical behavior concentrated essentially on the one-dimensional case except for additive CA and a few others (see, for example [19]).

The reason of this gap is may be twofold: from one hand, there is a common feeling that most of topological results are "automatically" transferred to higher dimensions; from the other hand, researchers mind the complexity gap. Indeed, many CA properties are dimension sensitive i.e. they are decidable in dimension 1 and undecidable in higher dimensions [2, 8, 12, 9, 3].

In this paper, in order to overcome this complexity gap, we use two deep constructions which allow to see a 2D as a 1D CA (see Section 4). In this way, well-known properties of one-dimensional CA can be lifted to the two-dimensional case. The idea is to "cut" the space of configurations of a CA in dimension d into slices of dimension $d - 1$. Hence, the former CA (in dimension d) can be seen as a new CA (in dimension $d - 1$) operating on configurations made of slices. The only inconvenient is that this latter CA has an infinite set of states. However, Theorem 3 and 4 prove that this is not always a problem.

The idea for this construction appeared in the context of additive CA in [17] and it was formalized in [6] . In the present paper, we generalize it to arbitrary 2D CA. Moreover, we further refine it so that slices are translation invariant along some fixed direction. This confers finiteness to the set of states of the sliced CA allowing to lift even more properties.

Finally, we list the new contributions given by this paper besides the afore-mentioned constructions:

- the notion of closingness is generalized to 2D;
- closingness is decidable;
- closing 2D CA have a dense set of periodic orbits (DPO) and are surjective;
- 4-closing 2D CA are open;
- open CA (in any dimension) are surjective;
- permutive 2D CA have DPO and are mixing and surjective;
- bi-permutive 2D CA are strongly transitive.

Remark that permutivity is also a decidable property. We conjecture that open 2D CA are 4-closing and hence openness should also be a decidable property. At present, those three properties seem to be the frontier between decidability and undecidability in CA.

For the sake of simplicity and lack of space, we decided to formulate new results and definition for the directions NE, SE, SW, NW but most of them are generalizable to any arbitrary direction. A complete treatment will appear in an extended forthcoming journal version of the present paper. Moreover, in order to stress on results and new notions, we have chosen to put all the technical parts and the constructions at the end of the paper.

2   Basic Definitions

In this section we briefly recall standard definitions about CA as dynamical systems. For introductory matter and recent results see [14, 1, 16] , for instance.

For all $i, j \in \mathbb{Z}$ with $i \leq j$, let $[i, j] = \{i, i+1, \ldots, j\}$. Let $\mathbb{N}_+$ be the set of positive integers. For a vector $x \in \mathbb{Z}^2$, denote by $|x|$ the infinite norm (in $\mathbb{R}^2$) of x. Let $r \in \mathbb{N}$. Denote by $\mathcal{M}_r$ the set of all the two-dimensional matrices with values in A and entry vectors in the square $[-r, r]^2$. For any matrix $N \in \mathcal{M}_r$, $N(x) \in A$ represents the element of the matrix with entry vector x.

1D CA. Let A be a possibly infinite alphabet. A *1D CA configuration* is a function from $\mathbb{Z}$ to A. The *1D CA configuration set* $A^{\mathbb{Z}}$ is usually equipped with the metric d defined as follows

$$\forall c, c' \in A^{\mathbb{Z}}, \ d(c, c') = 2^{-n}, \text{ where } n = \min\left\{i \geq 0 : c_i \neq c'_i \text{ or } c_{-i} \neq c'_{-i}\right\} .$$

If A is finite, $A^{\mathbb{Z}}$ is a compact, totally disconnected and perfect topological space (i.e., $A^{\mathbb{Z}}$ is a Cantor space). For any pair $i, j \in \mathbb{Z}$, with $i \leq j$, and any configuration $c \in A^{\mathbb{Z}}$ we denote by $c_{[i,j]}$ the word $c_i \cdots c_j \in A^{j-i+1}$, i.e., the portion of c inside the interval $[i, j]$. A *cylinder* of block $u \in A^k$ and position $i \in \mathbb{Z}$ is the set $[u]_i = \{c \in A^{\mathbb{Z}} : c_{[i, i+k-1]} = u\}$. Cylinders are clopen sets w.r.t. the metric d and they form a basis for the topology induced by d.

A *1D CA* is a structure $\langle 1, A, r, f \rangle$, where A is the alphabet, $r \in \mathbb{N}$ is the *radius* and $f : A^{2r+1} \to A$ is the *local rule* of the automaton. The local rule f induces a *global rule* $F : A^{\mathbb{Z}} \to A^{\mathbb{Z}}$ defined as follows,

$$\forall c \in A^{\mathbb{Z}}, \ \forall i \in \mathbb{Z}, \quad F(c)_i = f(x_{i-r}, \ldots, x_{i+r}) .$$

Note that F is a uniformly continuous map w.r.t. the metric d. A 1D CA with global rule F is *right* (resp., *left*) *closing* iff $F(c) \neq F(c')$ for any pair $c, c' \in A^{\mathbb{Z}}$ of distinct left (resp., right) asymptotic configurations, i.e., $c_{(-\infty, n]} = c'_{(-\infty, n]}$ (resp., $c_{[n, \infty)} = c'_{[n, \infty)}$) for some $n \in \mathbb{Z}$, where $a_{(-\infty, n]}$ (resp., $a_{[n, \infty)}$) denotes the portion of a configuration a inside the infinite integer interval $(-\infty, n]$ (resp., $[n, \infty)$). A CA is said to be *closing* if it is either left or right closing. A rule $f : A^{2r+1} \to A$ is *righmost* (resp., *leftmost*) *permutive* iff $\forall u \in A^{2r}, \forall \beta \in A, \exists \alpha \in A$ such that $f(u\alpha) = \beta$ (resp., $f(\alpha u) = \beta$). A 1D CA is said to be *permutive* if its local rule is either rightmost or leftmost permutive.

2D CA. Let A be a finite alphabet. A *2D CA configuration* is a function from $\mathbb{Z}^2$ to A. The *2D CA configuration set* $A^{\mathbb{Z}^2}$ is equipped with the following metric which is denoted for the sake of simplicity by the same symbol of the 1D case:

$$\forall c, c' \in A^{\mathbb{Z}^2}, \quad d(c, c') = 2^{-k} \quad \text{where} \quad k = \min\left\{|x| : x \in \mathbb{Z}^2, c(x) \neq c'(x)\right\} .$$

The 2D configuration set is a Cantor space. A *2D CA* is a structure $\langle 2, A, r, f \rangle$, where A is the alphabet, $r \in \mathbb{N}$ is the *radius* and $f : \mathcal{M}_r \to A$ is the *local rule* of the automaton. The local rule f induces a *global rule* $F : A^{\mathbb{Z}^2} \to A^{\mathbb{Z}^2}$ defined as follows,

$$\forall c \in A^{\mathbb{Z}^2}, \ \forall x \in \mathbb{Z}^2, \quad F(c)(x) = f\left(M_r^x(c)\right) ,$$

where $M_r^x(c) \in \mathcal{M}_r$ is the *finite portion* of c of reference position $x \in \mathbb{Z}^2$ and radius r defined by $\forall k \in [-r, r]^2, M_r^x(c)(k) = c(x + k)$. For any $v \in \mathbb{Z}^2$ the *shift map* $\sigma^v : A^{\mathbb{Z}^2} \to A^{\mathbb{Z}^2}$ is defined by $\forall c \in A^{\mathbb{Z}^2}, \forall \boldsymbol{x} \in \mathbb{Z}^2, \sigma^v(c)(\boldsymbol{x}) = c(\boldsymbol{x} + \boldsymbol{v})$. A function $F : A^{\mathbb{Z}^2} \to A^{\mathbb{Z}^2}$ is said to be *shift-commuting* if $\forall \boldsymbol{k} \in \mathbb{Z}^2, F \circ \sigma^k = \sigma^k \circ F$. Note that 2D CA are exactly the class of all shift-commuting functions which are (uniformly) continuous with respect to the metric d (Hedlund's theorem from [11]). A 2D *subshift* S is a closed subset of the CA configuration space such that for any $\boldsymbol{v} \in \mathbb{Z}^2, \sigma^v(S) \subset S$.

For any fixed vector $\boldsymbol{v}$, we denote by S_v the set of all configurations $c \in A^{\mathbb{Z}^2}$ such that $\sigma^v(c) = c$. Remark that, for any 2D CA global map F and for any $\boldsymbol{v}$, the set S_v is F-invariant, i.e., $F(S_v) \subseteq S_v$.

DTDS. A *discrete time dynamical system (DTDS)* is a pair (X, g) where X is a set equipped with a distance d and $g : X \mapsto X$ is a map which is continuous on X with respect to the metric d. When X is the configuration space of a (either 1D or 2D) CA equipped with the above introduced metric, the pair (X, F) is a DTDS. From now on, for the sake of simplicity, we identify a CA with the dynamical system induced by itself or even with its global rule F.

Given a DTDS (X, g), a point $c \in X$ is *periodic* for g if there exists an integer $p > 0$ such that $g^p(c) = c$. If the set of all periodic points of g is dense in X, we say that the DTDS has the *denseness of periodic orbits (DPO)*. Recall that a DTDS (X, g) is *(topologically) transitive* if for any pair of non-empty open sets $O_1, O_2 \subseteq X$ there exists an integer $n \in \mathbb{N}$ such that $g^n(O_1) \cap O_2 \neq \emptyset$. A DTDS (X, g) is *(topologically) mixing* if for any pair of non-empty open sets $O_1, O_2 \subseteq X$ there exists an integer $n \in \mathbb{N}$ such that for any $t \geq n$ we have $g^t(O_1) \cap O_2 \neq \emptyset$. Trivially, any mixing DTDS is also transitive. A DTDS (X, g) is *(topologically) strongly transitive* if for any non-empty open set $O \subseteq X$, it holds that $\bigcup_{n \in \mathbb{N}} g^n(O) = X$. A DTDS (X, g) is *open* (resp., *surjective*) iff g is open (resp., g is surjective). Remark that any strongly transitive system is surjective. Moreover, in compact spaces, transitive (or mixing) DTDS are surjective.

Recall that two DTDS (X, g) and (X', g') are *isomorphic* (resp., *topologically conjugated*) if there exists a bijection (resp., homeomorphism) $\phi : X \mapsto X'$ such that $g' \circ \phi = \phi \circ g$. (X', g') is a *factor* of (X, g) if there exists a continuous and surjective map $\phi : X \to X'$ such that $g' \circ \phi = \phi \circ g$. Remark that in that case, (X', g') inherits from (X, g) some properties such as surjectivity, transitivity, mixing, and DPO.

3   Main Results

Notation. In the sequel, the symbol U is any direction in $\{NE, SE, SW, NW\}$. Given the direction U, let $\bar{U}$ be the opposite direction, i.e., $\overline{NE} = SW, \overline{SW} = NE, \overline{NW} = SE$ and $\overline{SE} = NW$. Moreover, we associate the NE direction with the vector $\boldsymbol{\lambda} = (1, 1)$ and the SE one with $\boldsymbol{\mu} = (1, -1)$. Then, the vector associated with SW is $-\boldsymbol{\lambda} = (-1, -1)$ and, similarly, $-\boldsymbol{\mu} = (-1, 1)$ is associated with NW. For each direction U, let $\boldsymbol{\nu}$ be the vector associated with U.

A *pattern* P is a function from a finite domain $Dom(P) \subseteq \mathbb{Z}^2$ taking values in A. Given two patterns P and P', denote $P \oplus P'$, the pattern Π such that $Dom(\Pi) = Dom(P) \cup Dom(P')$ and $\forall \boldsymbol{x} \in Dom(\Pi), \Pi(\boldsymbol{x}) = P(x)$ if $\boldsymbol{x} \in Dom(P), P'(\boldsymbol{x})$ otherwise. The notion of cylinder can be conveniently extended to general patterns as follows: for any $P \subseteq A^{\mathbb{Z}^2}$, let $[P]$ be the set $\left\{ c \in A^{\mathbb{Z}^2} \mid \forall \boldsymbol{x} \in Dom(P), c(\boldsymbol{x}) = P(\boldsymbol{x}) \right\}$. As in the 1D case, cylinders form a basis for the open sets.

In the sequel, with a little abuse of notation, for any pattern P, $F(P)$ is the pattern P' such that $dom(P') = \{ \boldsymbol{x} \in dom(P), \mathcal{B}_r(\boldsymbol{x}) \subseteq dom(P) \}$ and forall $\boldsymbol{x} \in dom(P')$, $P'(\boldsymbol{x}) = f(\mathcal{B}_r(\boldsymbol{x}))$, where $\mathcal{B}_r(\boldsymbol{x}) = \left\{ y \in \mathbb{Z}^2, |\boldsymbol{x} - y| \leq r \right\}$.

3.1   Closingness

We now generalize to 2D CA the notion of closingness. This property turns out to be decidable. As a main result, we prove that closing 2D CA have DPO.

Definition 1 (U-asymptotic configurations). *Two configurations $c, c' \in A^{\mathbb{Z}^2}$ are U-asymptotic if there exists $q \in \mathbb{Z}$ such that $\forall \boldsymbol{x} \in \mathbb{Z}^2$ with $\nu \cdot \boldsymbol{x} \geq q$ it holds that $c(\boldsymbol{x}) = c'(\boldsymbol{x})$.*

Definition 2 (Closingness). *A 2D CA F is U-closing is for any pair of $\bar{U}$-asymptotic configurations $c, c' \in A^{\mathbb{Z}^2}$ we have that $c \neq c'$ implies $F(c) \neq F(c')$. A 2D CA is* closing *(resp., 4-closing) if it is U-closing for some U (resp., for all $U \in \{NE, SE, SW, NW\}$).*

We are going to define two families of binary relations on the set of configurations which will greatly help in simplifying the notation. For any $m \in \mathbb{N}$ and for any $c, c' \in A^{\mathbb{Z}^2}$, we write $c \triangle_m^\nu c'$ if and only if $c(\boldsymbol{x}) = c'(\boldsymbol{x})$ for each $\boldsymbol{x} \in \mathbb{Z}^2$ with $|\boldsymbol{x}| \leq m$ and $\nu \cdot \boldsymbol{x} < 0$. For any $m \in \mathbb{N}$ and for any $c, c' \in A^{\mathbb{Z}^2}$, we write $c \square_m c'$ if and only if $\forall \boldsymbol{x} \in \mathbb{Z}^2$ with $|\boldsymbol{x}| \leq m$ it holds that $c(\boldsymbol{x}) = c'(\boldsymbol{x})$. Remark that the definition of these binary relations can be easily extended to work on patterns.

The following proposition gives a combinatorial characterization of closingness for 2D CA. Its proof is very similar to the one-dimensional case (see [14]); we report it here only to show the role played by the relations $\triangle_m^\nu$ and $\square_m$.

Proposition 1. *A 2D CA F is U-closing iff there exists $m \in \mathbb{N}$ such that for all configurations $c, c' \in A^{\mathbb{Z}^2}$, $c \triangle_m^\nu c'$ and $F(c) \square_m F(c')$ implies $c(\boldsymbol{0}) = c'(\boldsymbol{0})$.*

Proof. Assume that F is NE-closing and, by contradiction, that for all $m > 0$ there exist $c, c' \in A^{\mathbb{Z}^2}$ such that $c \triangle_m^\nu c'$ and $F(c) \square_m F(c')$ implies $c(\boldsymbol{0}) \neq c'(\boldsymbol{0})$. For any $m > 0$ define the following sets

$$X_m = \left\{ (c, c') \in A^{\mathbb{Z}^2} \times A^{\mathbb{Z}^2} \mid c \triangle_m^\nu c' \text{ and } F(c) \square_m F(c') \text{ and } c(\boldsymbol{0}) \neq c'(\boldsymbol{0}) \right\} .$$

Remark that for any $m > 0$, $X_m \neq \emptyset$, $X_{m+1} \subseteq X_m$ and X_m is closed. Therefore $X = \cap_{m>0} X_m$ is not empty. In other words, there exist c, c' such that for all $m > 0$, $c \triangle_m^\nu c'$ i.e. c and c' are SW-asymptotic. Moreover, for all $m > 0$, $F(c) \square_m F(c')$ implies $F(c) = F(c')$. Finally, $c(\boldsymbol{0}) \neq c'(\boldsymbol{0})$ (since c, c' belong to some X_m) together with the previous remarks gives the contradiction. For the opposite implication. Assume that there exists $m > 0$ such that for all configurations $c, c' \in A^{\mathbb{Z}^2}$, $c \triangle_m^\nu c'$ and $F(c) \square_m F(c')$ implies $c(\mathbf{0}) = c'(\mathbf{0})$ but F is not NE-closing. Consider a pair of distinct SW-asymptotic configurations c, c' and shift them enough (if necessary) such that $c(\mathbf{0}) \neq c'(\mathbf{0})$. Clearly, $c \triangle_m^\nu c'$ and, since they are SW-asymptotic, we have $F(c) \square_m F(c')$.   □

Proposition 2. *Closingness is a decidable property.*

Proof. Consider a 2D CA F with radius r. Assume that F is NE-closing (the other cases are similar). Consider the following statement $Q(m) =$

$$[\forall\, P, P', C, C', \; (P \triangle_m^\nu P' \text{ and } F(P \oplus C) \square_m F(P' \oplus C')) \Rightarrow C(\mathbf{0}) = C'(\mathbf{0})] \; .$$

where the quantification is made over all patterns P, P' such that $dom(P) = dom(P') = \{x \in \mathbb{Z}^2 \mid |x| \le m + r, \nu \cdot x < 0\}$ and over all pattern C, C' such that $dom(C) = dom(C') = \{x \in \mathbb{Z}^2 \mid |x| \le m + r, \nu \cdot x \ge 0\}$. We claim that $Q(m)$ must be true for some $m \le 2r$. Indeed, assume that this is not the case and let $m > 2r$. Consider the patterns P, P', C, C' satisfying the sufficient part of $Q(2r)$. Let $\bar{P}, \bar{P}', \bar{C}, \bar{C}'$ be the same as P, P', C, C' but such that $\bar{P} \oplus \bar{C}$ and $\bar{P}' \oplus \bar{C}'$ are surrounded by a border of cells in state $a \in A$ of width $m - 2r$. Then, $\bar{P}, \bar{P}', \bar{C}, \bar{C}'$ do not verify $Q(m)$. If $Q(m)$ is never verified for any m, then F is not NE-closing (just extend patterns into configurations by adding some "default" symbol outside the domain of the pattern).   □

Theorem 1. *Any closing 2D CA has DPO.*

Proof. Assume that F is NE-closing. Choose $c \in A^{\mathbb{Z}^2}$ and $\epsilon > 0$. Let $n \in \mathbb{N}$ be such that $\frac{1}{2^n} < \epsilon$. Set $v = n\mu$ (which is perpendicular to λ). Since F is NE-closing, by Lemma 5 and 6, (S_v, F) is topologically conjugated to a 1D CA $(B^{\mathbb{Z}}, F^*)$ where F^* is right closing and B is finite. Since closing 1D CA on finite alphabet have DPO [4], then (S_v, F) has DPO too. Let $c' \in S_v$ be a configuration such that $d(c', c) < \frac{1}{2^n}$. Then there exists a periodic configuration $p \in S_v$ such that $d(c', p) < \frac{1}{2^n}$. This concludes the proof.   □

Corollary 1. *Any closing 2D CA is surjective.*

Proof. Just recall that DPO implies surjectivity, then use Theorem 1.   □

3.2   Openness

In this section we study the relation between openness and closingness. Recall that in 1D CA, a CA is open if and only if it is both left and right closing. Here we prove a weaker result, namely that 4-closing 2D CA are open. We conjecture that also the opposite relation is true.

Notation. For $t, m \ge 1$, $q, q' \in \mathbb{Z}$, we say that a pattern u has *shape* $[m, t, q, q']$ if $dom(u) = \{x \in \mathbb{Z}^2 \mid q \le \lambda x \le q + m - 1 \text{ and } q' \le -\mu x \le q' + t - 1\}$.

Proposition 3. *Consider a 2D NE-closing CA F. Then, for all sufficiently large $m > 0$ and any $t \geq 1, q, q' \in \mathbb{Z}$, if u and v are patterns of shape $[t, m, q, q']$ and $[t, 2m, q, q']$, resp., and $F([u]) \cap [v] \neq \emptyset$, then for each pattern b of shape $[1, t, q + 2m, q']$ there exists a pattern a of shape $[1, t, q + m, q']$ such that*

$$F([u \oplus a]) \cap [v \oplus b] \neq \emptyset \ . \tag{1}$$

Proof. Consider the slicing F^* on $S_{\boldsymbol{k}}$ of F according to a vector $\boldsymbol{k} \perp \lambda$ with $|\boldsymbol{k}|$ bigger than t. Take m like in Proposition 1. By Lemma 6, F^* is right-closing. Let $B = A^{|\boldsymbol{k}|}$. The patterns u, v, b are contained in suitable 1D blocks $u' \in B^m, v' \in B^{2m}$, and a suitable $b' \in B$, respectively. By [14, Thm. 5.44, p. 228], there exists $a' \in B$, such that $W = F^*([u'a']) \cap [v'b'] \neq \emptyset$. To conclude the proof just remark that any $c \in W$ is such that $\Psi^{-1}(c)$ is in the intersection set in (1). □

Similar results as in Proposition 3 hold for all other directions SE, SW, NW.

Theorem 2. *If a 2D CA F is 4-closing, then F is open.*

Proof. We just need to show that the image of a cylinder is an open set. Let m be like in Proposition 3. For any pattern u of shape $[2k+1, 2k+1, -k, -k]$, if v is a pattern of shape $[2k'+1, 2k'+1, -k', -k']$ for $k' = k + m$ and $F([u]) \cap [v] \neq \emptyset$, then, using Proposition 3 and a completeness argument, one can see that any configuration in $[v]$ has a preimage in $[u]$. Therefore,

$$F([u]) = \bigcup \{[v] : F([u]) \cap [v] \neq \emptyset, \text{ and } v \text{ has shape } [2k'+1, 2k'+1, -k', -k']\}$$

is a union of cylinders and hence $F([u])$ is open. □

The following result is well-known for 1D CA, we are not aware of proofs for higher dimensions. Moreover, we underline the fact that the result is obtained using only topological arguments.

Proposition 4. *Any open CA is surjective.*

Proof. For any D-dimensional CA of global rule F, $F(A^{\mathbb{Z}^D})$ is subshift. If F is open, then $F(A^{\mathbb{Z}^D})$ has non-empty interior. It is well-known that $A^{\mathbb{Z}^D}$ is the only subshift with non-empty interior, hence $F(A^{\mathbb{Z}^D}) = A^{\mathbb{Z}^D}$. □

3.3   Permutivity

We now introduce another decidable property for the local rule of a 2D CA.

Definition 3 (Permutivity). *A 2D CA of local rule f and radius r is U-permutive, if for each pair of matrices $N, N' \in \mathcal{M}_r$ with $N(\boldsymbol{x}) = N'(\boldsymbol{x})$ in all vectors $\boldsymbol{x} \neq r\boldsymbol{\nu}$, it holds that $N(r\boldsymbol{\nu}) \neq N'(r\boldsymbol{\nu})$ implies $f(N) \neq f(N')$. A 2D CA is bi-permutive iff it is both U permutive and $\bar{U}$-permutive.*

Proposition 5. *Any U-permutive 2D CA is U-closing.*

Proof. Assume that a 2D CA F is NE-permutive. Take two SW-asymptotic configurations c, c' with $F(c) = F(c')$. Let $q \in \mathbb{Z}$ be the integer such that $c(\boldsymbol{x}) = c'(\boldsymbol{x})$ for any $\boldsymbol{x}$ with $\boldsymbol{\lambda} \cdot \boldsymbol{x} \leq q$. Consider all vectors $\boldsymbol{y}$ with $\boldsymbol{y} \cdot \boldsymbol{\lambda} = q + 1$. By hypothesis, we obtain $c(\boldsymbol{y}) = c(\boldsymbol{y})$ for all these vectors. The repetition of this argument to $q + 2, q + 3, \ldots$ gives $c = c'$.   □

Proposition 6. *Any U-permutive 2D CA has DPO.*

Proof. Use Proposition 5 and Theorem 1.   □

Theorem 3. *Any bi-permutive 2D CA is strongly transitive.*

Proof. Consider a bi-permutive 2D CA F. By Lemma 1, it is a factor of a 1D CA $((A^{\mathbb{Z}})^{\mathbb{Z}}, F^*)$ which, by Lemma 2, is both left and right permutive. Lemma 4 concludes the proof.   □

Theorem 4. *Any U-permutive 2D CA is (topologically) mixing.*

Proof. Consider a U-permutive 2D CA F. By Lemma 1, it is a factor of a 1D CA $((A^{\mathbb{Z}})^{\mathbb{Z}}, F^*)$ which, by Lemma 2, is either left or right permutive. Lemma 3 concludes the proof.   □

4   Constructions

In this section we illustrate the slicing constructions which are fundamental to prove all the results of this paper.

U-Slicing. For each U, define $\boldsymbol{d} = \boldsymbol{\mu}$, if $U \in \{NE, SW\}$, $\boldsymbol{d} = \boldsymbol{\lambda}$, otherwise. We construct the line L_0 generated by the vector $\boldsymbol{d}$. The construction is such that the set $L_0^* = L_0 \cap \mathbb{Z}^2$ contains vectors of the form $\boldsymbol{x} = t\boldsymbol{d}$ where $t \in \mathbb{Z}$. The mapping $\varphi : L_0^* \mapsto \mathbb{Z}$ associating any $\boldsymbol{x} \in L_0^*$ with the integer $\varphi(\boldsymbol{x}) = t$ is a group isomorphism with respect to the standard operations. Consider now the family $\mathcal{L}$ constituted by all the lines parallel to L_0 containing at least a point of integer coordinates (trivially, for any vector $\boldsymbol{x} \in \mathbb{Z}^2$ there exists a line parallel to L_0 which contains this vector). It is clear that $\mathcal{L}$ is in a one-to-one correspondence with $\mathbb{Z}$. We enumerate the lines according to their intersection with the axis l_1 given by the direction $\boldsymbol{e}_1 = (1, 0)$. In other words, for any $i \in \mathbb{Z}$, L_i is the line whose intersection with l_1 is the point $i\boldsymbol{e}_1$. Equivalently, L_i is the line expressed in parametric form by $\boldsymbol{x} = i\boldsymbol{e}_1 + t\boldsymbol{d}$ ($\boldsymbol{x} \in \mathbb{R}^2, t \in \mathbb{R}$). Note that, for any $\boldsymbol{x} \in \mathbb{Z}^2$ there exist $i, t \in \mathbb{Z}$ such that $\boldsymbol{x} = i\boldsymbol{e}_1 + t\boldsymbol{d}$. In particular, one can remark that $\forall i, j \in \mathbb{Z}$, if $\boldsymbol{x} \in L_i$ and $\boldsymbol{y} \in L_j$, then $\boldsymbol{x} + \boldsymbol{y} \in L_{i+j}$.

Let us summarize the construction. We have a countable collection $\mathcal{L} = \{L_i : i \in \mathbb{Z}\}$ of lines parallel to L_0 inducing a partition of $\mathbb{Z}^2$. Define $L_i^* = L_i \cap \mathbb{Z}^2$, then $\mathbb{Z}^2 = \bigcup_{i \in \mathbb{Z}} L_i^*$. Therefore, any configuration $c \in A^{\mathbb{Z}^2}$ can be viewed as a mapping $c : \bigcup_{i \in \mathbb{Z}} L_i^* \mapsto \mathbb{Z}$. For every $i \in \mathbb{Z}$, the *slice* c_i over the lines L_i of the configuration c is the mapping $c_i : L_i^* \to A$, which is the restriction of c to the set $L_i^* \subset \mathbb{Z}^2$.

In this way, a configuration $c \in A^{\mathbb{Z}^2}$ can be expressed as the bi-infinite one-dimensional sequence $\prec c \succ = (\ldots, c_{-2}, c_{-1}, c_0, c_1, c_2, \ldots)$ of its slices $c_i \in A^{L_i^*}$ where the i-th component of the sequence $\prec c \succ$ is $\prec c \succ_i = c_i$. Let us stress that each slice c_i is defined only over the set L_i^*. Moreover, $\forall x \in \mathbb{Z}^2$, $\exists! i \in \mathbb{Z} : x \in L_i^*$ and in this case we identify $\prec c \succ (x) \equiv \prec c \succ_i (x) = c_i(x) = c(x) \in A$. The identification of any configuration $c \in A^{\mathbb{Z}^2}$ with the corresponding bi-infinite sequence of slices $c \equiv \prec c \succ = (\ldots, c_{-2}, c_{-1}, c_0, c_1, c_2, \ldots)$, allows the introduction of a new one-dimensional bi-infinite CA over the alphabet $A^{\mathbb{Z}}$ expressed by a global transition mapping $F^* : (A^{\mathbb{Z}})^{\mathbb{Z}} \mapsto (A^{\mathbb{Z}})^{\mathbb{Z}}$ which associates any configuration $a : \mathbb{Z} \mapsto A^{\mathbb{Z}}$ with a new configuration $F^*(a) : \mathbb{Z} \to A^{\mathbb{Z}}$. The local rule f^* of this new CA we are going to define will take a certain number of configurations of $A^{\mathbb{Z}}$ as input and will produce a new configuration of $A^{\mathbb{Z}}$ as output.

For each $h \in \mathbb{Z}$, define the following bijective mapping $\mathcal{T}_h : A^{L_h^*} \mapsto A^{L_0^*}$ which associates any slice c_h over the line L_h with the slice $\mathcal{T}_h(c_h)$

$$(c_h : L_h^* \to A) \xrightarrow{\mathcal{T}_h} (\mathcal{T}_h(c_h) : L_0^* \to A)$$

defined as $\forall x \in L_0^*, \mathcal{T}_h(c_h)(x) = c_h(x + he_1)$. Remark that the mapping $\mathcal{T}_h^{-1} : A^{L_0^*} \to A^{L_h^*}$ associates any slice c_0 over the line L_0 with the slice $\mathcal{T}_h^{-1}(c_0)$ over the line L_h such that $\forall x \in L_h^*, \mathcal{T}_h^{-1}(c_h)(x) = c_0(x - he_1)$. Denote by $\Phi_0 : A^{L_0^*} \to A^{\mathbb{Z}}$ the bijective mapping putting in correspondence any $c_0 : L_0^* \to A$ with the configuration $\Phi_0(c_0) \in A^{\mathbb{Z}}$,

$$(c_0 : L_0^* \to A) \xrightarrow{\Phi_0} (\Phi_0(c_0) : \mathbb{Z}^2 \to A)$$

defined as follows: $\forall t \in \mathbb{Z}$, $\Phi_0(c_0)(t) := c_0(\varphi^{-1}(t)) \in A$ (equivalently we have that $\forall x \in L_0^*, \Phi_0(c_0)(\varphi(x)) = c_0(x)$). Let us stress that the mapping $\Phi_0^{-1} : A^{\mathbb{Z}} \to A^{L_0^*}$ associates any configuration $a \in A^{\mathbb{Z}}$ with the configuration $\Phi_0^{-1}(a) \in A^{L_0^*}$ in the following way: $\forall x \in L_0^*, \Phi_0^{-1}(a)(x) = a(\varphi(x))$.

Now we have all the necessary formalism to correctly define the radius r^* local rule $f^* : (A^{\mathbb{Z}})^{2r^*+1} \to A^{\mathbb{Z}}$ with $r^* = 2r$ starting from a radius r 2D CA F:

$$\forall (a_{-r^*}, \ldots, a_{r^*}) \in (A^{\mathbb{Z}})^{2r^*+1}, \quad f^*(a_{-r^*}, \ldots, a_{r^*}) = \Phi_0(b)$$

where $b : L_0^* \to A$ is the slice obtained the simultaneous application of the local rule f of the original CA on the slices $c_{-r^*}, \ldots, c_{r^*}$ of any configuration c such that $\forall i \in [-r^*, r^*], c_i = \mathcal{T}_i^{-1}(\Phi_0^{-1}(a_i))$. The global map of this new CA is $F^* : A^{\mathbb{Z}^{\mathbb{Z}}} \to A^{\mathbb{Z}^{\mathbb{Z}}}$ and the link between F^* and f^* is given, as usual, by $(F^*(a))_i = f^*(a_{i-r^*}, \ldots, a_{i+r^*})$ where $a = (\ldots, a_{-1}, a_0, a_1, \ldots) \in (A^{\mathbb{Z}})^{\mathbb{Z}}$ and $i \in \mathbb{Z}$.

Lemma 1. *The DTDS $(A^{\mathbb{Z}^2}, F)$ is isomorphic to the DTDS $((A^{\mathbb{Z}})^{\mathbb{Z}}, F^*)$ by the bijective mapping $\Psi : A^{\mathbb{Z}^2} \to (A^{\mathbb{Z}})^{\mathbb{Z}}$ defined as follows*

$$\forall c \in A^{\mathbb{Z}^2}, \quad \Psi(c) = (\ldots, \Phi_0(\mathcal{T}_{-1}(c_{-1})), \Phi_0(\mathcal{T}_0(c_0)), \Phi_0(\mathcal{T}_1(c_1)), \ldots) .$$

Moreover, the mapping $\Psi^{-1} : (A^{\mathbb{Z}})^{\mathbb{Z}} \mapsto A^{\mathbb{Z}^2}$

$$\forall a \in (A^{\mathbb{Z}})^{\mathbb{Z}}, \quad \Psi^{-1}(a) = (\ldots, \mathcal{T}_{-1}^{-1}(\Phi_0^{-1}(a_{-1})), \mathcal{T}_0^{-1}(\Phi_0^{-1}(a_0)), \mathcal{T}_1^{-1}(\Phi_0^{-1}(a_1)), \ldots)$$

is continuous. Hence, $(A^{\mathbb{Z}^2}, F)$ is a factor of $((A^{\mathbb{Z}})^{\mathbb{Z}}, F^*)$.

Proof. It is clear that Ψ is bijective. We show that $\Psi \circ F = F^* \circ \Psi$, i.e., that $\forall i \in \mathbb{Z}, \forall c \in A^{\mathbb{Z}^2}$, $\Psi(F(c))_i = F^*(\Psi(c))_i$. We have $\Psi(F(c))_i = \Phi_0(\mathcal{T}_i(F(c)_i))$ where the slice $F(c)_i$ is obtained by the simultaneous application of f on the slices $c_{i-r^*}, \ldots, c_{i+r^*}$. On the other hand $F^*(\Psi(c))_i$ is equal to

$$f^*(\Psi(c)_{i-r^*}, \ldots, \Psi(c)_{i+r^*}) = f^*(\Phi_0(\mathcal{T}_{i-r^*}(c_{i-r^*})), \ldots, \Phi_0(\mathcal{T}_{i+r^*}(c_{i+r^*}))) = \Phi_0(b)$$

where, by definition of f^*, b is the slice obtained by the simultaneous application of f on the slices $d_{r^*} = \mathcal{T}_{-r^*}^{-1}(\mathcal{T}_{i-r^*}(c_{i-r^*})), \ldots, d_{r^*} = \mathcal{T}_{r^*}^{-1}(\mathcal{T}_{i+r^*}(c_{i+r^*}))$ which gives $\mathcal{T}_i(F(c)_i)$. We now prove that Ψ^{-1} is continuous mapping from the metric space $(A^{\mathbb{Z}})^{\mathbb{Z}}$ to the metric space $A^{\mathbb{Z}^2}$, both equipped with the suitable metric, which for the sake of simplicity is denoted by the same symbol d. Choose an arbitrary configuration $a = (\ldots, a_{-1}, a_0, a_1, \ldots) \in (A^{\mathbb{Z}})^{\mathbb{Z}}$ and a real number $\epsilon > 0$. Let n be a positive integer such that $\frac{1}{2^n} < \epsilon$. Consider the hyperplanes H_i with $-2n \leq i \leq 2n$. Setting $\delta = \frac{1}{2^{2n}}$, for any configuration $b \in (A^{\mathbb{Z}})^{\mathbb{Z}}$ with $d(b, a) < \delta$, we have that $b_i = a_i$ for each $i \in \mathbb{Z}, -2n \leq i \leq 2n$. This fact implies that $(\Psi^{-1}(b))_i = (\Psi^{-1}(a))_i$, and then $(\Psi^{-1}(b))_i(x) = (\Psi^{-1}(a))_i(x)$, for each i with $-2n \leq i \leq 2n$ and for any $x \in H_i^*$. Equivalently, we have $(\Psi^{-1}(a))(x) = (\Psi^{-1}(b))(x)$, for any $x \in H_i^*$, with $-2n \leq i \leq 2n$ and in particular for any x such that $|x| \leq n$. So we have obtained that $d(\Psi^{-1}(b), \Psi^{-1}(a)) < \epsilon$. Hence, Ψ^{-1} is continuous. $\square$

Lemma 2. *Consider a 2D CA F. If F is U-permutive and $\nu \cdot \lambda + \nu \cdot \mu > 0$ (resp. $\nu \cdot \lambda + \nu \cdot \mu < 0$) then the 1D CA $((A^{\mathbb{Z}})^{\mathbb{Z}}, F^*)$ obtained by the U-slicing construction is rightmost (resp., leftmost) permutive.*

Proof. It immediately follows from the U-slicing construction. $\square$

Lemma 3. *Let F be a 1D CA with local rule f on a possibly infinite alphabet A. If f is either rightmost or leftmost permutive, then F is topologically mixing.*

Proof. The proof is similar to that given in [5] for CA with finite alphabet. $\square$

Lemma 4. *Let F be a 1D CA with local rule f on a possibly infinite alphabet A. If f is both rightmost and leftmost permutive, then F is strongly transitive.*

Proof. Choose $b \in A^{\mathbb{Z}}$, $u \in A^*$, $i \in \mathbb{Z}$, and consider the cylinder $[u]_i$. Let $t \in \mathbb{N}$ be such that $tr > i + |u| - 1$ and $-tr < i$. The value $b_0 \in A$ depends only on the values of any configuration $c \in F^{-t}(b)$ in $[-rt, rt]$. We build a configuration $a \in [u]_i$ such that $F^t(a) = b$. Fix $a_i = c_i$ for each $-rt \leq i \leq rt$, assuring that $a \in [u]_i$ and $F^t(a)_0 = b_0$. Since the local rule $f^{(t)}$ of F^t is both leftmost and rightmost permutive there exist symbols $\alpha_{-1}, \alpha_1 \in A$, such that $F^t(a)_{-1} = b_{-1}$ and $F^t(a)_1 = b_1$, when setting $a_{-rt-1} = \alpha_{-1}$, $a_{rt+1} = \alpha_1$. By repeating the above procedure the thesis is obtained. $\square$

Slicing plus finite alphabet. The following lemmata grant that for any 2D CA F, we can build an associated sliced version F^* with finite alphabet. This is very useful since one can use all the well-known results about 1D CA and try to lift them to F.

Lemma 5. *Let F be a 2D CA. For any vector $v \in \mathbb{Z}^2$ with $v \perp \lambda$ or $v \perp \mu$, (S_v, F) is topologically conjugated to a 1D dimensional CA $(B^{\mathbb{Z}}, F^*)$ on the alphabet $B = A^{|v|}$.*

Proof. Fix a vector $v \perp \lambda$. Consider the slicing construction on the set S_v. According to it, any configuration $c \in S_v$ is identified with the corresponding bi-infinite sequence of slices. Since slices of configurations in S_v are in one-to-one correspondence with symbols of the alphabet B, the slicing construction gives a 1D CA $F^* : B^{\mathbb{Z}} \to B^{\mathbb{Z}}$ such that, by Lemma 1, (S_v, F) is isomorphic to $(B^{\mathbb{Z}}, F^*)$ by the bijective map $\Psi : S_v \to B^{\mathbb{Z}}$. By Lemma 1, Ψ^{-1} is continuous. Since configurations of S_v are periodic with respect to σ^v, Ψ is continuous too.   □

Lemma 6. *Let F be a 2D CA. For any vector $v \in \mathbb{Z}^2$ with $v \perp \lambda$ or $v \perp \mu$, let $(B^{\mathbb{Z}}, F^*)$ be the 1D CA of Lemma 5 which is topologically conjugated to (S_v, F). When $v \perp \lambda$, if F is NE-closing (resp., SW-closing), then F^* is right (resp., left) closing. On the other hand, when $v \perp \mu$, if F is SE-closing (resp., NW-closing), then F^* is right-closing (resp., left-closing).*

Proof. Take $v \perp \lambda$ and assume that F is NE-closing. Since F does not collapse any pair of distinct SW-asymptotic configurations in S_v, the thesis immediately follows by the slicing construction on S_v. The other cases are similar.   □

5   Conclusions

In this paper we studied some decidable properties of 2D CA. In particular, we generalized to the 2D case the notion of closingness and we investigated its relation with permutivity and openness. This has been done by means of a construction which associates any 2D CA with a peculiar 1D CA. We strongly believe that these two constructions are useful to prove other fundamental results about 2D CA dynamics and their view as transformations of picture languages. We are currently investigating these connections.

References

[1] Acerbi, L., Dennunzio, A., Formenti, E.: Shifting and lifting of cellular automata. In: Cooper, S.B., Löwe, B., Sorbi, A. (eds.) CiE 2007. LNCS, vol. 4497, pp. 1–10. Springer, Heidelberg (2007)

[2] Amoroso, S., Patt, Y.N.: Decision procedures for surjectivity and injectivity of parallel maps for tesselation structures. Journal of Computer and System Sciences 6, 448–464 (1972)

[3] Bernardi, V., Durand, B., Formenti, E., Kari, J.: A new dimension sensitive property for cellular automata. Theoretical Computer Science 345, 235–247 (2005)

[4] Boyle, M., Kitchens, B.: Periodic points for cellular automata. Indag. Math. 10, 483–493 (1999)

[5] Cattaneo, G., Dennunzio, A., Margara, L.: Chaotic subshifts and related languages applications to one-dimensional cellular automata. Fundamenta Informaticae 52, 39–80 (2002)

[6] Cattaneo, G., Dennunzio, A., Margara, L.: Solution of some conjectures about topological properties of linear cellular automata. Theoretical Computer Science 325, 249–271 (2004)

[7] Cervelle, J., Dennunzio, A., Formenti, E.: Chaotic behavior of cellular automata. In: Meyers, B. (ed.) Mathematical basis of cellular automata, Encyclopedia of Complexity and System Science. Springer, Heidelberg (2008)

[8] Durand, B.: Global properties of 2d cellular automata: Some complexity results. In: Borzyszkowski, A.M., Sokolowski, S. (eds.) MFCS 1993. LNCS, vol. 711, pp. 433–441. Springer, Heidelberg (1993)

[9] Durand, B.: Global properties of cellular automata. In: Goles, E., Martinez, S. (eds.) Cellular Automata and Complex Systems. Kluwer, Dordrecht (1998)

[10] Formenti, E., Kůrka, P.: Dynamics of cellular automata in non-compact spaces. In: Meyers, B. (ed.) Mathematical basis of cellular automata, Encyclopedia of Complexity and System Science. Springer, Heidelberg (2008)

[11] Hedlund, G.A.: Endomorphisms and automorphisms of the shift dynamical system. Mathematical System Theory 3, 320–375 (1969)

[12] Kari, J.: Reversibility and surjectivity problems of cellular automata. Journal of Computer and System Sciences 48, 149–182 (1994)

[13] Kari, J.: Tiling problem and undecidability in cellular automata. In: Meyers, B. (ed.) Mathematical basis of cellular automata, Encyclopedia of Complexity and System Science. Springer, Heidelberg (2008)

[14] Kůrka, P.: Topological and Symbolic Dynamics. Cours Spécialisés, vol. 11. Société Mathématique de France (2004)

[15] Kůrka, P.: Topological dynamics of one-dimensional cellular automata. In: Meyers, B. (ed.) Mathematical basis of cellular automata, Encyclopedia of Complexity and System Science. Springer, Heidelberg (2008)

[16] Di Lena, P., Margara, L.: Computational complexity of dynamical systems: the case of cellular automata. Information and Computation (to appear, 2008)

[17] Margara, L.: On some topological properties of linear cellular automata. In: Kutyłowski, M., Wierzbicki, T., Pacholski, L. (eds.) MFCS 1999. LNCS, vol. 1672, pp. 209–219. Springer, Heidelberg (1999)

[18] Pivato, M.: The ergodic theory of cellular automata. In: Meyers, B. (ed.) Mathematical basis of cellular automata, Encyclopedia of Complexity and System Science. Springer, Heidelberg (2008)

[19] Theyssier, G., Sablik, M.: Topological dynamics of 2d cellular automata. In: Beckmann, A., Dimitracopoulos, C., Löwe, B. (eds.) CiE 2008. LNCS, vol. 5028, pp. 523–532. Springer, Heidelberg (2008)

Fixed Point and Aperiodic Tilings

Bruno Durand[1], Andrei Romashchenko[2,3], and Alexander Shen[1,3]

[1] LIF, CNRS & Univ. de Provence, Marseille
[2] LIP, ENS de Lyon & CNRS
[3] Institute for Information Transmission Problems, Moscow

Abstract. An aperiodic tile set was first constructed by R. Berger while proving the undecidability of the domino problem. It turned out that aperiodic tile sets appear in many topics ranging from logic (the Entscheidungsproblem) to physics (quasicrystals)

We present a new construction of an aperiodic tile set that is based on Kleene's fixed-point construction instead of geometric arguments. This construction is similar to J. von Neumann self-reproducing automata; similar ideas were also used by P. Gács in the context of error-correcting computations.

The flexibility of this construction allows us to construct a "robust" aperiodic tile set that does not have periodic (or close to periodic) tilings even if we allow some (sparse enough) tiling errors. This property was not known for any of the existing aperiodic tile sets.

1 Introduction

In this paper, *tiles* are unit squares with colored sides. Tiles are considered as prototypes: we may place translated copies of the same tile into different cells of a cell paper (rotations are not allowed). Tiles in the neighbor cells should match (common side should have the same color in both).

Formally speaking, we consider a finite set C of *colors*. A *tile* is a quadruple of colors (left, right, top and bottom ones), i.e., an element of C^4. A *tile set* is a subset $\tau \subset C^4$. A *tiling* of the plane with tiles from τ (τ-*tiling*) is a mapping $U: \mathbb{Z}^2 \to \tau$ that respects the color matching condition. A tiling U is *periodic* if it has a *period*, i.e., a non-zero vector $T \in \mathbb{Z}^2$ such that $U(x + T) = U(x)$ for all $x \in \mathbb{Z}^2$. Otherwise the tiling is *aperiodic*. The following classical result was proved by Berger in a paper [2] where he used this construction as a main tool to prove *Berger's theorem*: the *domino problem* (to find out whether a given tile set has tilings or not) is undecidable.

Theorem 1. *There exists a tile set τ such that τ-tilings exist and all of them are aperiodic.* [2]

The first tile set of Berger was rather complicated. Later many other constructions were suggested. Some of them are simplified versions of the Berger's construction ([16], see also the expositions in [1,5,13]). Some others are based on polygonal tilings (including famous Penrose and Ammann tilings, see [10]). An

M. Ito and M. Toyama (Eds.): DLT 2008, LNCS 5257, pp. 276–288, 2008.
© Springer-Verlag Berlin Heidelberg 2008

ingenious construction suggested in [11] is based on the multiplication in a kind of positional number system and gives a small aperiodic set of 14 tiles (in [3] an improved version with 13 tiles is presented).

In this paper we present yet another construction of aperiodic tile set. It does not provide a small tile set; however, we find it interesting because:

• The existence of an aperiodic tile set becomes a simple application of a classical construction used in Kleene's fixed point (recursion) theorem, in von Neumann's self-reproducing automata [15] and, more recently, in Gács' reliable cellular automata [7,8]; we do not use any geometric tricks. The construction of an aperiodic tile set is not only an interesting result but an important tool (recall that it was invented to prove that domino problem is undecidable); our construction makes this tool easier to use (see Theorem 3 and Section 10 as examples).

• The construction is rather general, so it is flexible enough to achieve some additional properties of the tile set. Our main result is Theorem 6: there exists a "robust" aperiodic tile set that does not have periodic (or close to periodic) tilings even if we allow some (sparse enough) tiling errors. It is not clear whether this can be achieved for previously known aperiodic tile sets; however, the mathematical model for a processes like quasicrystals' growth or DNA-computation should take errors into account. Note that our model (independent choice of place where errors are allowed) has no direct physical meaning; it is just a simple mathematical model that can be used as a playground to develop tools for estimating the consequences of tiling errors.

The paper is organized as follows. In Section 2 we define the notion of a self-similar tile set (a tile set that simulates itself). In Section 3 we explain how a tile set can be simulated by a computation implemented by another tile set. Section 4 shows how to achieve a fixed point (a tile set that simulates itself). Then we provide several applications of this construction: we use it to implement substitution rules (Section 5) and to obtain tile sets that are aperiodic in a strong sense (Section 6) and robust to tiling errors (Sections 7 and 8). Section 9 provides probability estimates that show that tiling errors are correctable with probability 1 (with respect to Bernoulli distribution). Finally, we show some other applications of the fixed point construction that simplify the proof of the undecidability of the domino problem and related results.

2   Macro-Tiles

Fix a tile set τ and an integer $N > 1$. A *macro-tile* is an $N \times N$ square tiled by matching τ-tiles. Every side of a macro-tile carries a sequence of N colors called a *macro-color*.

Let ρ be a set of τ-macro-tiles. We say that τ *simulates* ρ if (a) τ-tilings exist, and (b) for every τ-tiling there exists a unique grid of vertical and horizontal lines that cuts this tiling into $N \times N$ macro-tiles from ρ.

Example 1. Assume that we have only one ('white') color and τ consists of a single tile with 4 white sides. Fix some N. There exists a single macro-tile of size $N \times N$. Let ρ be a singleton that contains this macro-tile. Then every τ-tiling can be cut into macro-tiles from ρ. However, τ does not simulate ρ, since the placement of cutting lines is not unique.

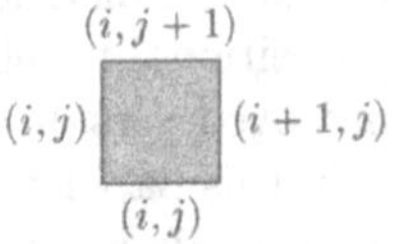

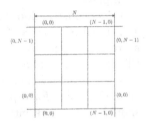

Fig. 1.

Example 2. In this example a set ρ that consists of exactly one macro-tile (that has the same macro-colors on all four sides) is simulated by some tile set τ. The tile set τ consists of N^2 tiles indexed by pairs (i, j) of integers modulo N. A tile from τ has colors on its sides as shown on Fig. 1. The macro-tile in ρ has colors $(0, 0), \ldots, (0, N-1)$ and $(0, 0), \ldots, (N-1, 0)$ on its borders (Fig. 2).

If a tile set τ simulates some set ρ of τ-macro-tiles with zoom factor $N > 1$ and ρ is isomorphic to τ, the set τ is called *self-similar*. Here an *isomorphism* between τ and ρ is a bijection that respects the relations "one tile can be placed on the right of another one" and "one tile can be placed on the top of another one". (An isomorphism induces two bijections between horizontal/vertical colors of τ and horizontal/vertical macro-colors of ρ.)

The idea of self-similarity is used (more or less explicitly) in most constructions of aperiodic tile sets ([11,3] are exceptions); we find the following explicit formulation useful.

Fig. 2.

Theorem 2. *A self-similar tile set τ has only aperiodic tilings.*

Proof. Every τ-tiling U can be uniquely cut into $N \times N$-macro-tiles from ρ. So every period T of U is a multiple of N (since the T-shift of a cut is also a cut). Then T/N is a period of ρ-tiling, which is isomorphic to a τ-tiling, so T/N is again a multiple of N. Iterating this argument, we conclude that T is divisible by N^k for every k, so $T = 0$. □

So to prove the existence of aperiodic tile sets it is enough to construct a self-similar tile set, and we construct it using the fixed-point idea. To achieve this, we first explain how to simulate a given tile set by embedding computations.

3   Simulating a Tile Set

For brevity we say that a tile set τ simulates a tile set ρ when τ simulates some set of macro tiles $\tilde{\rho}$ isomorphic to ρ (e.g., a self-similar tile set simulates itself).

Let us start with some informal discussion. Assume that we have a tile set ρ whose colors are k-bit strings ($C = \mathbb{B}^k$) and the set of tiles $\rho \subset C^4$ is presented as a predicate $R(c_1, c_2, c_3, c_4)$. Assume that we have some Turing machine $\mathcal{R}$ that computes R. Let us show how to simulate ρ using some other tile set τ.

This construction extends Example 2, but simulates a tile set ρ that contains not a single tile but many tiles. We keep the coordinate system modulo N embedded into tiles of τ; these coordinates guarantee that all τ-tilings can be uniquely cut into blocks of size $N \times N$ and every tile "knows" its position in the block (as in Example 2). In addition to the coordinate system, now each tile in τ carries supplementary colors (from a finite set specified below) on its sides. On the border of a macro-tile (i.e., when one of the coordinates is zero) only two supplementary colors (say, 0 and 1) are allowed. So the macro-color encodes a string of N bits (where N is the size of macro-tiles). We assume that $N \geq k$ and let k bits in the middle of macro-tile sides represent colors from C. All other bits on the sides are zeros (this is a restriction on tiles: each tile knows its coordinates so it also knows whether non-zero supplementary colors are allowed).

Now we need additional restrictions on tiles in τ that guarantee that the macro-colors on sides of each macro-tile satisfy the relation R. To achieve this, we ensure that bits from the macro-tile sides are transferred to the central part of the tile where the checking computation of $\mathcal{R}$ is simulated (Fig. 3).

For that we need to fix which tiles in a macro-tile form "wires" (this can be done in any reasonable way; let us assume that wires do not cross each other) and then require that each of these tiles carries equal bits on two sides; again it is easy since each tile knows its coordinates.

Then we check R by a local rule that guarantees that the central part of a macro-tile represents a time-space diagram of $\mathcal{R}$'s computation (the tape is horizontal, time goes up). This is done in a standard way. We require that computation terminates in an accepting state: if not, the tiling cannot be formed.

Fig. 3.

To make this construction work, the size of macro-tile (N) should be large enough: we need enough space for k bits to propagate and enough time and space (=height and width) for all accepting computations of $\mathcal{R}$ to terminate.

In this construction the number of supplementary colors depends on the machine $\mathcal{R}$ (the more states it has, the more colors are needed in the computation zone). To avoid this dependency, we replace $\mathcal{R}$ by a fixed universal Turing machine $\mathcal{U}$ that runs a *program* simulating $\mathcal{R}$. Let us agree that the tape has an additional read-only layer. Each cell carries a bit that is not changed during the computation; these bits are used as a program for the universal machine (Fig. 4). So in the computation zone the columns carry unchanged bits, and the tile set restrictions guarantee that these bits form the program for $\mathcal{U}$, and the central zone represents the protocol of an accepting computation for that program. In this way we get a tile set τ that simulates ρ with zoom factor N using $O(N^2)$ tiles. (Again we need N to be large enough.)

Fig. 4.

4   Simulating Itself

We know how to simulate a given tile set ρ (represented as a program for the universal TM) by another tile set τ with a large enough zoom factor N. Now we want τ to be isomorphic to ρ (then Theorem 2 guarantees aperiodicity). For this we use a construction that follows Kleene's recursion (fixed-point) theorem[1] [12].

Note that most rules of τ do not depend on the program for $\mathcal{R}$, dealing with information transfer along the wires, the vertical propagation of unchanged program bits, and the space-time diagram for the universal TM in the computation zone. Making these rules a part of ρ's definition (we let $k = 2\log N + O(1)$ and encode $O(N^2)$ colors by $2\log N + O(1)$ bits), we get a program that checks that macro-tiles behave like τ-tiles in this respect.

The only remaining part of the rules for τ is the hardwired program. We need to ensure that macro-tiles carry the same program as τ-tiles do. For that our program (for the universal TM) needs to access the bits of its own text. (This self-referential action is in fact quite legal: the program is written on the tape, and the machine can read it.) The program checks that if a macro-tile belongs to the first line of the computation zone, this macro-tile carries the correct bit of the program.

How should we choose N (hardwired in the program)? We need it to be large enough so the computation described (which deals with $O(\log N)$ bits) can fit in the computation zone. The computation is rather simple (polynomial in the input size, i.e., $O(\log N)$), so for large N it easily fits in $\Omega(N)$ available time.

This finishes the construction of a self-similar aperiodic tile set.

5   Substitution System and Tilings

The construction of self-similar tiling is rather flexible and can be easily augmented to get a self-similar tiling with additional properties. Our first illustration is the simulation of substitution rules.

Let A be some finite alphabet and $m > 1$ be an integer. A *substitution rule* is a mapping $s: A \to A^{m \times m}$. By A-configuration we mean an integer lattice filled with letters from A, i.e., a mapping $\mathbb{Z}^2 \to A$ considered modulo translations.

A substitution rule s applied to a configuration X produces another configuration $s(X)$ where each letter $a \in A$ is replaced by an $m \times m$ matrix $s(a)$.

A configuration X is *compatible* with substitution rule s if there exists an infinite sequence $\ldots \xrightarrow{s} X_3 \xrightarrow{s} X_2 \xrightarrow{s} X_1 \xrightarrow{s} X$, where X_i are some configurations.

---

[1] A reminder: Kleene's theorem says that for every transformation π of programs one can find a program p such that p and $\pi(p)$ produce the same output. Proof sketch: since the statement is language-independent (use translations in both directions before and after π), we may assume that the programming language has a function GetText() that returns the text of the program and a function Exec(string s) that replaces the current process by execution of a program s. (Think about an interpreter: surely it has an access to the program text; it can also recursively call itself with another program.) Then the fixed point is Exec(π(GetText())).

Example 3. Let $A = \{0,1\}$, $s(0) = \left(\begin{smallmatrix} 0 & 1 \\ 1 & 0 \end{smallmatrix}\right)$, $s(1) = \left(\begin{smallmatrix} 0 & 1 \\ 1 & 0 \end{smallmatrix}\right)$. It is easy to see that the only configuration compatible with s is the chess-board coloring.

Example 4. Let $A = \{0,1\}$, $s(0) = \left(\begin{smallmatrix} 0 & 1 \\ 1 & 0 \end{smallmatrix}\right)$, $s(1) = \left(\begin{smallmatrix} 1 & 0 \\ 0 & 1 \end{smallmatrix}\right)$. One can check that all configurations that are compatible with this substitution rule (called *Thue – Morse configurations* in the sequel) are aperiodic.

The following theorem goes back to [14]. It says that every substitution rule can be enforced by a tile set.

Theorem 3 (Mozes). *Let A be an alphabet and let s be a substitution rule over A. Then there exists a tile set τ and a mapping $e \colon \tau \to A$ such that*
  (a) *s-image of any τ-tiling is an A-configuration compatible with s;*
  (b) *every A-configuration compatible with s can be obtained in this way.*

Proof. We modify the construction of the tile set τ (with zoom factor N) taking s into account. Let us first consider the very special case when

• the substitution rule maps each A-letter into an $N \times N$-matrix (i.e., $m = N$).
• the substitution rule is easy to compute: given a letter $u \in A$ and (i,j), we can compute the (i,j)-th letter of $s(u)$ in time $\mathrm{poly}(\log|A|) \ll N$.

In this case we proceed as follows. In our basic construction every tile knows its coordinates in the macro-tile and some additional information needed to arrange 'wires' and simulate calculations of the universal TM. Now in addition to this basic structure each tile keeps two letters of A: the first is the label of a tile itself, and the second is the label of the $N \times N$-tile it belongs to. This means that we keep additional $2\log|A|$ bits in each tile, i.e., multiply the number of tiles by $|A|^2$. It remains to explain how the local rules work. We add two requirements:
  (a) the second letter is the same for neighbor tiles (unless they are separated by a border of some $N \times N$ macro-tile);
  (b) the first letter in a tile is determined by the second letter and the coordinates of the tile inside the macro-tile, according to the substitution rule.

Both requirements are easy to integrate in our construction. The requirement (a) is rather trivial; to achieve (b) we need to embed in a macro-tile a calculation of $s(\text{[label on this macro-tile]})$. It is possible when s is easy to compute.

The requirements (a) and (b) ensure that configuration is an s-image of some other configuration. Also (due the self-similarity) we have the same at the level of macro-tiles. But this is not all: we need to guarantee that the first letter on the level of macro-tiles is identical to the second letter on the level of tiles. This is also achievable: the first letter of a macro-tile is encoded by bits on its border, and we can require that these bits match the second letter of the tiles at that place (recall that second letter is the same across the macro-tile). It is easy to see that now τ has the required properties (each tiling projects into a configuration compatible with τ and vice versa).

However, this construction assumes that N (the zoom factor) is equal to the matrix size in the substitution rule, which is usually not the case (m is given, and N we have to choose, and it needs to be large enough). The solution is to let N be equal to m^k for some k, and use the substitution rule s^k, i.e., the k-th iteration of s (a configuration is compatible with s^k if and only if it is compatible with s). Now we do not need s to be easily computed: for large k the computation of s^k will fit into the space available (exponential in k).   □

6   Strong Version of Aperiodicity

Let $\alpha > 0$ be a real number. A configuration $U : \mathbb{Z}^2 \to A$ is *α-aperiodic* if for every nonzero vector $T \in \mathbb{Z}^2$ there exists N such that in every square whose side is at least N the fraction of points x such that $U(x) \neq U(x + T)$ exceeds α.

Remark 1. If U is α-aperiodic, then Besicovitch distance between U and any periodic pattern is at least $\alpha/2$. (The Besicovitch distance is defined as $\limsup_N d_N$ where d_N is the fraction of points where two patterns differ in the $N \times N$ centered square.)

Theorem 4. *There exists a tile set τ such that τ-tilings exist and every τ-tiling is α-aperiodic for every $\alpha < 1/3$.*

Proof. This tile set is obtained by applying Theorem 3 to Thue–Morse substitution rule T (Example 4). Note that any configuration $C = \{c_{ij}\}$ compatible with T is a xor-combination $c_{ij} = a_i \oplus b_j$ of two one-dimensional Thue-Morse sequences a and b, and for a and b a similar result (every shift changes between $1/3$ and $2/3$ of positions in a large block) is well known (see, e.g., [17]).   □

7   Filling Holes

The second application of our flexible fixed-point construction is an aperiodic tile set where isolated defects can be healed.

Let $c_1 < c_2$ be positive integers. We say that a tile set τ is *(c_1, c_2)-robust* if the following holds: For every n and for every τ-tiling U of the $c_2 n$-neighborhood of a square $n \times n$ excluding the square itself there exists a tiling V of the entire $c_2 n$-neighborhood of the square (including the square itself) that coincides with U outside of the $c_1 n$-neighborhood of the square (see Fig. 5).

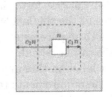

Fig. 5.

Theorem 5. *There exists a self-similar tile set that is (c_1, c_2)-robust for some c_1 and c_2.*

Proof. For every tile set μ it is easy to construct a "robustified" version μ' of μ, i.e., a tile set μ' and a mapping $\delta : \mu' \to \mu$ such that: (a) δ-images of μ'-tilings are exactly μ-tilings; (b) μ' is "5-robust": every μ'-tiling of a 5×5 square minus 3×3 hole can be uniquely extended to the tiling of the entire 5×5 square.

Indeed, it is enough to keep in one μ'-tile the information about, say, 5×5 square in μ-tiling and use the colors on the borders to ensure that this information is consistent in neighbor tiles.

This robustification can be easily combined with the fixed-point construction. In this way we can get a 5-robust self-similar tile set τ if the zoom factors N is large enough. Let us show that this set is also (c_1, c_2)-robust for some c_1 and c_2 (that depend on N, but N is fixed.)

Fig. 6.

Indeed, let us have a tiling of a large enough neighborhood around an $n \times n$ hole. Denote by k the minimal integer such that $N^k \geq n$ (so the k-level macro-tiles are greater than the hole under consideration). Note that the size of k-level macro-tiles is linear in n since $N^k \leq N \cdot n$.

In the tiling around the hole, an $N \times N$ block structure is correct except for the N-neighborhood of the central $n \times n$ hole. For similar reasons $N^2 \times N^2$-structure is correct except for the $N + N^2$-neighborhood, etc. So for the chosen k we get a k-level structure that is correct except for (at most) $9 = 3 \times 3$ squares of level k, and such a hole can be filled (due to 5-robustness) with $N^k \times N^k$ squares, and these squares can be then detalized back.

To implement this procedure (and fill the hole), we need a correct tiling only in the $O(N^k)$-neighborhood of the hole (technically, we need to have a correct tiling in $(3N^k)$-neighborhood of the hole; as $3N^k \leq 3Nn$, we let $c_2 = 3N$). The correction procedure involves changes in another $O(N^k)$-neighborhood of the hole (technically, changes touch $(2N^k)$-neighborhood of the hole; $2N^k \leq 2Nn$, so we let $c_1 = 2N$). □

8 Tilings with Errors

Now we combine our tools to prove that there exists a tile set τ that is aperiodic in rather strong sense: this set does not have periodic tilings or tilings that are close to periodic. Moreover, this remains true if we allow the tiling to have some "sparse enough" set of errors. Tiling with errors is no more a tiling (as defined above): in some places the neighbor colors do not match. Technically it is more convenient to consider tilings with "holes" (where some cells are not tiled) instead of errors but this does not matter: we can convert a tiling error into a hole just by deleting one of two non-matching tiles.

Let τ be a tile set and let $H \subset \mathbb{Z}^2$ be some set (H for "holes"). We consider (τ, H)-*tilings*, i.e., mappings $U: \mathbb{Z}^2 \setminus H \to \tau$ such that every two neighbor tiles from $\mathbb{Z}^2 \setminus H$ match (i.e., have the same color on the common side).

We claim that there exists a tile set τ such that (1) τ-tilings of the entire plane exist and (2) for every "sparse enough" set H every (τ, H)-tiling is far from every periodic mapping $\mathbb{Z}^2 \to \tau$.

To make this claim true, we need a proper definition of a "sparse" set. The following trivial counterexample shows that a requirement of small density is not enough for such a definition: if H is a grid made of vertical and horizontal lines at large distance N, the density of H is small but for any τ there exist (τ, H)-tilings with periods that are multiples of N.

The definition of sparsity we use (see below) is rather technical; however, it guarantees that for small enough ε a random set where every point appears with probability ε independently of other points, is sparse with probability 1. More precisely, for every $\varepsilon \in (0,1)$ consider a Bernoulli probability distribution B_ε on subsets of $\mathbb{Z}^2$ where each point is included in the random subset with probability ε and different points are independent.

Theorem 6. *There exists a tile set τ with the following properties: (1) τ-tilings of $\mathbb{Z}^2$ exist; (2) for all sufficiently small ε for almost every (with respect to B_ε) subset $H \subset \mathbb{Z}^2$ every (τ, H)-tiling is at least $1/10$ Besicovitch-apart from every periodic mapping $\mathbb{Z}^2 \to \tau$.*

Remark 2. Since the tiling contains holes, we need to specify how we treat the holes when defining Besicovitch distance. We do *not* count points in H as points where two mappings differ; this makes our statement stronger.

Remark 3. The constant $1/10$ is not optimal and can be improved by a more accurate estimate.

Proof. Consider a tile set τ such that (a) all τ-tilings are α-aperiodic for every $\alpha < 1/3$; (b) τ is (c_1, c_2)-robust for some c_1 and c_2. Such a tile set can be easily constructed by combining the arguments used for Theorem 5 and Theorem 4.

Then we show (this is the most technical part postponed until Section 9) that for small ε a B_ε-random set H with probability 1 has the following "error-correction" property: every (τ, H)-tiling is Besicovitch-close to some τ-tiling of the entire plane. The latter one is α-aperiodic, therefore (if Besicovitch distance is small compared to α) the initial (τ, H)-tiling is far from any periodic mapping.

For simple tile sets that allow only periodic tilings this error-correction property can be derived from basic results in percolation theory (the complement of H has large connected component etc.) However, for aperiodic tile sets this argument does not work and we need more complicated notion of "sparse" set based on "islands of errors". We employ the technique suggested in [7] (see also applications of "islands of errors" in [9], [6]).

9   Islands of Errors

Let $E \subset \mathbb{Z}^2$ be a set of points; points in E are called *dirty*; other points are *clean*. Let $\beta \geq \alpha > 0$ be integers. A set $X \subset E$ is an (α, β)-*island* in E if:

(1) the diameter of X does not exceed α;
(2) in the β-neighborhood of X there is no other points from E.

(Diameter of a set is a maximal distance between its elements; the distance d is defined as the maximum of distances along both coordinates; β-neighborhood of X is a set of all points y such that $d(y, x) \leq \beta$ for some $x \in X$.)

It is easy to see that two (different) islands are disjoint (and the distance between their points is greater than β).

Let $(\alpha_1, \beta_1), (\alpha_2, \beta_2), \ldots$ be a sequence of pairs of integers and $\alpha_i \leq \beta_i$ for all i. Consider the iterative "cleaning" procedure. At the first step we find all (α_1, β_1)-islands (*rank 1 islands*) and remove all their elements from E (thus getting a smaller set E_1). Then we find all (α_2, β_2)-islands in E_1 (*rank 2 islands*); removing them, we get $E_2 \subset E_1$, etc. Cleaning process is *successful* if every dirty point is removed at some stage.

At the ith step we also keep track of the β_i-neighborhoods of islands deleted during this step. A point $x \in \mathbb{Z}^2$ is *affected* during a step i if x belongs to one of these neighborhoods.

The set E is called *sparse* (for given sequence α_i, β_i) if the cleaning process is successful, and, moreover, every point $x \in \mathbb{Z}^2$ is affected at finitely many steps only (i.e., x is far from islands of large ranks).

The values of α_i and β_i should be chosen in such a way that:

(1) for sufficiently small $\varepsilon > 0$ a B_ε-random set is sparse with probability 1 (Lemma 1 below);

(2) if a tile set τ is (c_1, c_2)-robust and H is sparse, then any (τ, H)-tiling is Besicovitch close to some τ-tiling of the entire plane (Lemmas 2 and 3).

Lemma 1. Assume that $8 \sum_{k<n} \beta_k < \alpha_n \le \beta_n$ for every n and $\sum_i \frac{\log \beta_i}{2^i} < \infty$. Then for all sufficiently small $\varepsilon > 0$ a B_ε-random set is sparse with probability 1.

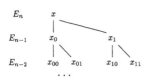

Fig. 7. Explanation tree; vertical lines connect different names for the same points

Proof of Lemma 1. Let us estimate the probability of the event "x is not cleaned after n steps" for a given point x (this probability does not depend on x). If $x \in E_n$, then x belongs to E_{n-1} and is not cleaned during the nth step (when (α_n, β_n)-islands in E_{n-1} are removed). Then $x \in E_{n-1}$ and, moreover, there exists some other point $x_1 \in E_{n-1}$ such that $d(x, x_1)$ is greater than $\alpha_n/2$ but not greater than $\beta_n + \alpha_n/2 < 2\beta_n$. Indeed, if there were no such x_1 in E_{n-1}, then $\alpha_n/2$-neighborhood of x in E_{n-1} is an (α_n, β_n)-island in E_{n-1} and x would be removed.

Each of the points x_1 and x (that we denote also x_0 to make the notation uniform) belongs to E_{n-1} because it belongs to E_{n-2} together with some other point (at the distance greater than $\alpha_{n-1}/2$ but not exceeding $\beta_{n-1} + \alpha_{n-1}/2$). In this way we get a tree (Figure 7) that explains why x belongs to E_n.

The distance between x_0 and x_1 in this tree is at least $\alpha_n/2$ while the diameter of the subtrees starting at x_0 and x_1 does not exceed $\sum_{i<n} 2\beta_i$. Therefore, the Lemma's assumption guarantees that these subtrees cannot intersect and, moreover, that all the leaves of the tree are different. Note that all 2^n leaves of the tree belong to $E = E_0$. As every point appears in E independently from other points, such an "explanation tree" is valid with probability ε^{2^n}. It remains to estimate the number of possible explanation trees for a given point x.

To specify x_1 we need to specify horizontal and vertical distance between x_0 and x_1. Both distances do not exceed $2\beta_n$, therefore we need about $2 \log(4\beta_n)$ bits to specify them (including the sign bits). Then we need to specify the distances between x_{00} and x_{01} as well as distances between x_{10} and x_{11}; this requires at most $4 \log(4\beta_{n-1})$ bits. To specify the entire tree we therefore need

$$2 \log(4\beta_n) + 4 \log(4\beta_{n-1}) + 8 \log(4\beta_{n-2}) + \ldots + 2^n \log(4\beta_1),$$

that is (reversing the sum and taking out the factor 2^n) equal to $2^n(\log(4\beta_1) + \log(4\beta_2)/2 + \ldots)$. Since the series $\sum \log \beta_n/2^n$ converges by assumption, the total number of explanation trees for a given point (and given n) does not exceed $2^{O(2^n)}$, so the probability for a given point x to be in E_n for a B_ε-random E does not exceed $\varepsilon^{2^n} 2^{O(2^n)}$, which tends to 0 (even super-exponentially fast) as $n \to \infty$.

We conclude that the event "x is not cleaned" (for a given point x) has zero probability; the countable additivity guarantees that with probability 1 all points in $\mathbb{Z}^2$ are cleaned.

It remains to show that every point with probability 1 is affected by finitely many steps only. Indeed, if x is affected by step n, then some point in its β_n-neighborhood belongs to E_n, and the probability of this event is at most $O(\beta_n^2)\varepsilon^{2^n} 2^{O(2^n)} = 2^{2\log \beta_n + O(2^n) - \log(1/\varepsilon)2^n}$; the convergence conditions guarantees that $\log \beta_n = o(2^n)$, so the first term is negligible compared to others, the probability series converges and the Borel–Cantelli lemma gives the desired result. $\square$

The following (almost evident) Lemma describes the error correction process.

Lemma 2. Assume that a tile set τ is (c_1, c_2)-robust, $\beta_k > 4c_2\alpha_k$ for every k and a set $H \subset \mathbb{Z}^2$ is sparse (with respect to α_i, β_i). Then every (τ, H)-tiling can be transformed into a τ-tiling of the entire plane by changing it in the union of $2c_1\alpha_k$-neighborhoods of rank k islands (for all islands of all ranks).

Proof of Lemma 2. Note that $\beta_k/2$-neighborhoods of rank k islands are disjoint and large enough to perform the error correction of rank k islands, since $\beta_k > 4c_2\alpha_k$. $\square$

It remains to estimate the Besicovitch size of the part of the plane changed during error correction.

Lemma 3. The Besicovitch distance between the original and corrected tilings (in Lemma 2) does not exceed $O(\sum_k (\alpha_k/\beta_k)^2)$.

(Note that the constant in O-notation depends on c_1.)

Proof of Lemma 3. We need to estimate the fraction of changed points in large centered squares. By assumption, the center is affected only by a finite number of islands. For every larger rank k, the fraction of points affected at the stage k in *any* centered square does not exceed $O((\alpha_k/\beta_k)^2)$: if the square intersects with the changed part, it includes a significant portion of the unchanged part. For smaller ranks the same is true for *all large enough* squares that cover completely the island affecting the center point). $\square$

It remains to chose α_k and β_k. We have to satisfy all the inequalities in Lemmas 1–3 at the same time. To satisfy Lemma 2 and Lemma 3, we may let $\beta_k = ck\alpha_k$ for large enough c. To satisfy Lemma 1, we may let $\alpha_{k+1} = 8(\beta_1 + \ldots + \beta_k) + 1$. Then α_k and β_k grow faster that any geometric sequence (like factorial multiplied by a geometric sequence), but still $\log \beta_i$ is bounded by a polynomial in i and the series in Lemma 1 converges.

With these parameters (taking c large enough) we may guarantee that Besicovitch distance between the original (τ, H)-tiling and the corrected τ-tiling does not exceed, say $1/100$. Since the corrected tiling is $1/5$-aperiodic and $1/10 + 2 \cdot (1/100) < 1/5$, we get the desired result (Theorem 6).   □

10   Other Applications of Fixed Point Self-similar Tilings

The fixed point construction of aperiodic tile set is flexible enough and can be used in other contexts. For example, the "zoom factor" N can depend on the level k (number of grouping steps). This construction can be used to replace the constant $1/10$ in Theorem 6 by any number less that 1, to provide a new proof for the results of [4] (a tileset whose tilings have maximal Kolmogorov complexity) and extend them to tilings with sparse errors; it can be also used in some other applications of tilings. Here is an example. We say that a tile set τ is *m-periodic* if τ-tilings exist and for each of them the set of periods is the set of *all* multiples of m (this is equivalent to the fact that both vectors $(0, m)$ and $(m, 0)$ are periods). Let E [resp. O] be all m-periodic tile sets for all even m [resp. odd m].

Theorem 7. *The sets E and O are inseparable enumerable sets.*

Acknowledgments. The authors thank the participants of the Kolmogorov seminar in Moscow (working on the RFBR project 06-01-00122-a) for many fruitful discussions.

References

1. Allauzen, C., Durand, B.: Appendix A: Tiling Problems. In: Börger, E., Grädel, E., Gurevich, Y. (eds.) The Classical Decision Problems. Springer, Heidelberg (1996)
2. Berger, R.: The Undecidability of the Domino Problem. Mem. Amer. Math. Soc 66 (1966)
3. Culik, K.: An Aperiodic Set of 13 Wang Tiles. Discrete Math. 160, 245–251 (1996)
4. Durand, B., Levin, L., Shen, A.: Complex Tilings. J. Symbolic Logic 73(2), 593–613 (2008)
5. Durand, B., Levin, L., Shen, A.: Local Rules and Global Order, or Aperiodic Tilings. Math. Intelligencer 27(1), 64–68 (2004)
6. Durand, B., Romashchenko, A.: On Stability of Computations by Cellular Automata. In: Proc. European Conf. Compl. Syst., Paris (2005)
7. Gács, P.: Reliable Cellular Automata with Self-Organization. In: Proc. 38th Ann. Symp. Found. Comput. Sci., pp. 90–97 (1997)
8. Gács, P.: Reliable Cellular Automata with Self-Organization. J. Stat. Phys. 103(1/2), 245–267 (2001)
9. Gray, L.: A Reader's Guide to Gács' Positive Rates Paper. J. Stat. Phys. 103(1/2), 1–44 (2001)
10. Grünbaum, B., Shephard, G.C.: Tilings and Patterns. W.H. Freeman and Company, New York (1987)
11. Kari, J.: A Small Aperiodic Set of Wang tiles. Discrete Math. 160, 259–264 (1996)
12. Rogers, H.: The Theory of Recursive Functions and Effective Computability. MIT Press, Cambridge (1987)

13. Levin, L.: Aperiodic Tilings: Breaking Translational Symmetry. Computer J. 48(6), 642–645 (2005), http://www.arxiv.org/cs.DM/0409024
14. Mozes, S.: Tilings, Substitution Systems and Dynamical Systems Generated by Them. J. Analyse Math. 53, 139–186 (1989)
15. von Neumann, J.: Theory of Self-reproducing Automata. Burks, A. (ed.). University of Illinois Press (1966)
16. Robinson, R.: Undecidability and Nonperiodicity for Tilings of the Plane. Inventiones Mathematicae 12, 177–209 (1971)
17. Zaks, M., Pikovsky, A.S., Kurths, J.: On the Correlation Dimension of the Spectral Measure for the Thue–Morse Sequence. J. Stat. Phys. 88(5/6), 1387–1392 (1997)

Extended Multi Bottom-Up Tree Transducers

Joost Engelfriet[1], Eric Lilin[2], and Andreas Maletti[3],[*]

[1] Leiden Institute of Advanced Computer Science
Leiden University, P.O. Box 9512, 2300 RA Leiden, The Netherlands
engelfri@liacs.nl
[2] Université des Sciences et Technologies de Lille
UFR IEEA 59655, Villeneuve d'Ascq, France
eric.lilin@lifl.fr
[3] International Computer Science Institute
1947 Center Street, Suite 600, Berkeley, CA 94704, USA
maletti@icsi.berkeley.edu

Abstract. Extended multi bottom-up tree transducers are defined and investigated. They are an extension of multi bottom-up tree transducers by arbitrary, not just shallow, left-hand sides of rules; this includes rules that do not consume input. It is shown that such transducers can compute any transformation that is computed by a linear extended top-down tree transducer. Moreover, the classical composition results for bottom-up tree transducers are generalized to extended multi bottom-up tree transducers. Finally, a characterization in terms of extended top-down tree transducers is presented.

1 Introduction

In the field of natural language processing, KNIGHT [1,2] proposed the following criteria that any reasonable formal tree-to-tree model of syntax-based machine translation [3] should fulfil:

(a) It should be a genuine generalization of finite-state transducers [4]; this includes the use of epsilon rules, i.e., rules that do not consume any part of the input tree.
(b) It should be efficiently trainable.
(c) It should be able to handle rotations (on the tree level).
(d) Its induced class of transformations should be closed under composition.

GRAEHL and KNIGHT [5] proposed the linear and nondeleting extended (top-down) tree transducer (ln-xtt) [6,7] as a suitable formal model. It fulfils (a)–(c) but fails to fulfil (d). Further models were proposed but, to the authors' knowledge, they all fail at least one criterion. Table 1 shows some important models and their properties.

---

[*] Author was supported by a fellowship within the Postdoc-Programme of the German Academic Exchange Service (DAAD).

M. Ito and M. Toyama (Eds.): DLT 2008, LNCS 5257, pp. 289–300, 2008.

Table 1. Overview of formal models with respect to desired criteria. "x" marks fulfilment; "–" marks failure to fulfil. A question mark shows that this remains open though we conjecture fulfilment.

Model \ Criterion	(a)	(b)	(c)	(d)
Linear and nondeleting top-down tree transducer [8,9]	–	x	–	x
Quasi-alphabetic tree bimorphism [10]	–	?	–	x
Synchronous context-free grammar [11]	x	x	–	x
Synchronous tree substitution grammar [12]	x	x	x	–
Synchronous tree adjoining grammar [13,14,15]	x	x	x	–
Linear and complete tree bimorphism [15]	x	x	x	–
Linear and nondeleting extended top-down tree transducer [5,6,7]	x	x	x	–
Linear multi bottom-up tree transducer [16,17,18]	–	?	x	x
Linear extended multi bottom-up tree transducer [this paper]	x	?	x	x

We propose a formal model that satisfies criteria (a), (c), and (d), and has more expressive power than the ln-xtt. The device is called linear extended multi bottom-up tree transducer, and it is as powerful as the linear model of [16,17,18] enhanced by epsilon rules (as shown in Theorem 5). In this paper we formally define and investigate the extended multi bottom-up tree transducer (xmbutt) and various restrictions (e.g., *linear*, *nondeleting*, and *deterministic*). Note that we consider the xmbutt in general, not just its linear restriction.

We start with normal forms for xmbutts. First, we construct for every xmbutt an equivalent nondeleting xmbutt (see Theorem 3). This can be achieved by guessing the required translations. Though the construction preserves linearity, it obviously destroys determinism. Next, we present a *one-symbol normal form* for xmbutts (see Theorem 5): each rule of the xmbutt either consumes one input symbol (without producing output), or produces one output symbol (without consuming input). This normal form preserves all three restrictions above.

Our main result (Theorem 13) states that the class of transformations computed by xmbutts is closed under pre-composition with transformations computed by linear xmbutts and under post-composition with those computed by deterministic xmbutts. In particular, we also obtain that the classes of transformations computed by linear and/or deterministic xmbutts are closed under composition. These results are analogous to classical results (see [19, Theorems 4.5 and 4.6] and [20, Corollary 7]) for bottom-up tree transducers [21,19] and thus show the "bottom-up" nature of xmbutts. Also, they generalize the composition results of [18, Theorem 11]. As in [18], our proof essentially uses the principle set forth in [20, Theorem 6], but the one-symbol normal form allows us to present a very simple composition construction for xmbutts and verify that it is correct, provided that the first input transducer is linear or the second is deterministic. We observe here that the "extension" of a tree transducer model (or even just the addition of epsilon rules) can, in general, destroy closure under composition, as can be seen from the linear and nondeleting top-down tree transducer. This seems to be due to the non-existence of a one-symbol normal form in the top-down case.

We verify that linear xmbutts have sufficient power for syntax-based machine translation. This is because, as mentioned before (and shown in Theorem 8), they can simulate all ln-xtts. Thus, we have a lower bound to the power of linear xmbutts. In fact, even the composition closure of the class of transformations computed by ln-xtts is strictly contained in the class of transformations computed by linear xmbutts. Finally, we also present exact characterizations (Theorems 7 and 14): xmbutts are as powerful as compositions of an ln-xtt with a deterministic top-down tree transducer. In the linear case the latter transducer has the so-called single-use property [22,23,24,25,26], and similar results hold in the deterministic case. Thus, the composition of two extended top-down tree transducers forms an upper bound to the power of the linear xmbutt. As a side-result we obtain that linear xmbutts admit a semantics based on recognizable rule tree languages. This suggests that linear xmbutts also satisfy criterion (b).

2   Preliminaries

Let A, B, C be sets. A relation from A to B is a subset of $A \times B$. Let $\tau_1 \subseteq A \times B$ and $\tau_2 \subseteq B \times C$. The composition of τ_1 and τ_2 is the relation $\tau_1 \,;\, \tau_2$ given by $\tau_1 \,;\, \tau_2 = \{(a, c) \mid \exists b \in B \colon (a, b) \in \tau_1, (b, c) \in \tau_2\}$. This composition is lifted to classes of relations in the usual manner.

The nonnegative integers are denoted by $\mathbb{N}$ and $\{i \mid 1 \le i \le k\}$ is denoted by $[k]$. A ranked set is a set Σ of symbols with a relation $\mathrm{rk} \subseteq \Sigma \times \mathbb{N}$ such that $\{k \mid (\sigma, k) \in \mathrm{rk}\}$ is finite for every $\sigma \in \Sigma$. Commonly, we denote the ranked set only by Σ and the set of k-ary symbols of Σ by $\Sigma^{(k)} = \{\sigma \in \Sigma \mid (\sigma, k) \in \mathrm{rk}\}$. We also denote that $\sigma \in \Sigma^{(k)}$ by writing $\sigma^{(k)}$. Given two ranked sets Σ and Δ with associated rank relations rk_Σ and rk_Δ, respectively, the set $\Sigma \cup \Delta$ is associated the rank relation $\mathrm{rk}_\Sigma \cup \mathrm{rk}_\Delta$. A ranked set Σ is uniquely-ranked if for every $\sigma \in \Sigma$ there exists exactly one k such that $(\sigma, k) \in \mathrm{rk}$. For uniquely-ranked sets, we denote this k simply by $\mathrm{rk}(\sigma)$. An alphabet is a finite set, and a ranked alphabet is a ranked set Σ such that Σ is an alphabet.

Let Σ be a ranked set. The set of Σ-trees, denoted by T_Σ, is the smallest set T such that $\sigma(t_1, \ldots, t_k) \in T$ for every $k \in \mathbb{N}$, $\sigma \in \Sigma^{(k)}$, and $t_1, \ldots, t_k \in T$. We write α instead of $\alpha()$ if $\alpha \in \Sigma^{(0)}$. Let $\Gamma \subseteq \Sigma$ and $H \subseteq T_\Sigma$. By $\Gamma(H)$ we denote $\{\gamma(t_1, \ldots, t_k) \mid \gamma \in \Gamma^{(k)}, t_1, \ldots, t_k \in H\}$. Now, let Δ be a ranked set. We denote by $T_\Delta(H)$ the smallest set $T \subseteq T_{\Sigma \cup \Delta}$ such that $H \subseteq T$ and $\Delta(T) \subseteq T$.

Let $t \in T_\Sigma$. The set of positions of t, denoted by $\mathrm{pos}(t)$, is defined by $\mathrm{pos}(\sigma(t_1, \ldots, t_k)) = \{\varepsilon\} \cup \{iw \mid i \in [k], w \in \mathrm{pos}(t_i)\}$ for every $\sigma \in \Sigma^{(k)}$ and $t_1, \ldots, t_k \in T_\Sigma$. Note that we denote the empty string by ε and that $\mathrm{pos}(t) \subseteq \mathbb{N}^*$. Let $w \in \mathrm{pos}(t)$ and $u \in T_\Sigma$. The subtree of t that is rooted in w is denoted by $t|_w$, the symbol of t at w is denoted by $t(w)$, and the tree obtained from t by replacing the subtree rooted at w by u is denoted by $t[u]_w$. For every $\Gamma \subseteq \Sigma$ and $\sigma \in \Sigma$, let $\mathrm{pos}_\Gamma(t) = \{w \in \mathrm{pos}(t) \mid t(w) \in \Gamma\}$ and $\mathrm{pos}_\sigma(t) = \mathrm{pos}_{\{\sigma\}}(t)$.

Let $X = \{x_i \mid i \ge 1\}$ be a set of formal variables, each considered to have the unique rank 0. A tree $t \in T_\Sigma(X)$ is linear (respectively, nondeleting) in $V \subseteq X$ if $\mathrm{card}(\mathrm{pos}_v(t)) \le 1$ (respectively, $\mathrm{card}(\mathrm{pos}_v(t)) \ge 1$) for every $v \in V$. The set of variables of t is $\text{var}(t) = \{v \in X \mid \text{pos}_v(t) \neq \emptyset\}$ and the sequence of variables is given by $\text{yield}_X \colon T_\Sigma(X) \to X^*$ with $\text{yield}_X(v) = v$ for every $v \in X$ and $\text{yield}_X(\sigma(t_1, \ldots, t_k)) = \text{yield}_X(t_1) \cdots \text{yield}_X(t_k)$ for every $\sigma \in \Sigma^{(k)}$ and $t_1, \ldots, t_k \in T_\Sigma(X)$. A tree $t \in T_\Sigma(X)$ is normalized if $\text{yield}_X(t) = x_1 \cdots x_m$ for some $m \in \mathbb{N}$. Every mapping $\theta \colon X \to T_\Sigma(X)$ is a substitution. We define the application of θ to a tree in $T_\Sigma(X)$ inductively by $v\theta = \theta(v)$ for every $v \in X$ and $\sigma(t_1, \ldots, t_k)\theta = \sigma(t_1\theta, \ldots, t_k\theta)$ for every $\sigma \in \Sigma^{(k)}$ and $t_1, \ldots, t_k \in T_\Sigma(X)$.

An *extended* (top-down) *tree transducer* (xtt, or *transducteur généralisé descendant*) [6,7] is a tuple $M = (Q, \Sigma, \Delta, I, R)$ where Q is a uniquely-ranked alphabet such that $Q = Q^{(1)}$, Σ and Δ are ranked alphabets that are both disjoint with $Q \cup X$, $I \subseteq Q$, and R is a finite set of rules of the form $l \to r$ with $l \in Q(T_\Sigma(X))$ linear in X, and $r \in T_\Delta(Q(\text{var}(l)))$. The xtt M is linear (respectively, nondeleting) if r is linear (respectively, nondeleting) in $\text{var}(l)$ for every $l \to r \in R$. The semantics of the xtt is given by term rewriting. Let $\xi, \zeta \in T_\Delta(Q(T_\Sigma))$. We write $\xi \Rightarrow_M \zeta$ if there exist a rule $l \to r \in R$, a position $w \in \text{pos}(\xi)$, and a substitution $\theta \colon X \to T_\Sigma$ such that $\xi|_w = l\theta$ and $\zeta = \xi[r\theta]_w$. The tree transformation computed by M is the relation

$$\tau_M = \{(t, u) \in T_\Sigma \times T_\Delta \mid \exists q \in I \colon q(t) \Rightarrow_M^* u\} \ .$$

The class of all tree transformations computed by xtts is denoted by XTOP. We use 'l' and 'n' to restrict to linear and nondeleting devices, respectively. Thus, ln-XTOP denotes the class of all tree transformations computable by linear and nondeleting xtts.

An xtt $M = (Q, \Sigma, \Delta, I, R)$ is a *top-down tree transducer* [8,9] if for every rule $l \to r \in R$ there exist $q \in Q$ and $\sigma \in \Sigma^{(k)}$ such that $l = q(\sigma(x_1, \ldots, x_k))$. The top-down tree transducer M is *deterministic* if (i) $\text{card}(I) = 1$ and (ii) for every $l \in Q(\Sigma(X))$ there exists at most one r such that $l \to r \in R$. Finally, M is *single-use* [22,23,24,25] if for every $q(v) \in Q(X)$, $k \in \mathbb{N}$, and $\sigma \in \Sigma^{(k)}$ there exist at most one $l \to r \in R$ and $w \in \text{pos}(r)$ such that $l(1) = \sigma$ and $r|_w = q(v)$. We use TOP and TOP_{su} to denote the classes of transformations computed by top-down tree transducers and single-use top-down tree transducers, respectively. We also use the prefixes 'l', 'n', and 'd' to restrict to linear, nondeleting, and deterministic devices, respectively.

Finally, we recall top-down tree transducers with regular look-ahead [27]. We use the standard notion of a *recognizable* (or *regular*) tree language [28,29], and we let $\text{Rec}(\Sigma) = \{L \subseteq T_\Sigma \mid L \text{ recognizable}\}$. A *top-down tree transducer with regular look-ahead* is a pair $\langle M, c\rangle$ such that $M = (Q, \Sigma, \Delta, I, R)$ is a top-down tree transducer and $c \colon R \to \text{Rec}(\Sigma)$. We say that such a transducer $\langle M, c\rangle$ is *deterministic* if (i) $\text{card}(I) = 1$ and (ii) for every $l \in Q(\Sigma(X))$ and $t \in T_\Sigma$ there exists at most one r such that $l \to r \in R$, $l(1) = t(\varepsilon)$, and $t \in c(l \to r)$. Similarly, $\langle M, c\rangle$ is *single-use* [26, Definition 5.5] if for every $q(v) \in Q(X)$ and $t \in T_\Sigma$ there exist at most one $l \to r \in R$ and $w \in \text{pos}(r)$ such that $l(1) = t(\varepsilon)$, $t \in c(l \to r)$, and $r|_w = q(v)$. The semantics of $\langle M, c\rangle$ is defined in the same manner as for xtt with the additional restriction that $l\theta|_1 \in c(l \to r)$. We use TOP^{R} and $\text{TOP}^{\text{R}}_{\text{su}}$ to denote the classes of transformations computed by top-down tree transducers with regular look-ahead and single-use top-down tree transducers with regular look-ahead, respectively. We use the prefix 'd' in the usual manner. For further information on tree languages and tree transducers, we refer to [28,29].

3   Extended Multi Bottom-Up Tree Transducers

In this section, we define *S-transducteurs ascendants généralisés* [30], which are a generalization of *S-transducteurs ascendants* (STA) [30,31]. We choose to call them *extended multi bottom-up tree transducers* here in line with [16,17,18], where 'multi' refers to the fact that states may have ranks different from one.

Definition 1. *An* extended multi bottom-up tree transducer *(xmbutt) is a tuple* $(Q, \Sigma, \Delta, F, R)$ *where*

- *Q is a uniquely-ranked alphabet of states, disjoint with $\Sigma \cup \Delta \cup X$;*
- *Σ and Δ are ranked alphabets of input and output symbols, respectively, which are both disjoint with X;*
- *$F \subseteq Q \setminus Q^{(0)}$ is a set of final states; and*
- *R is a finite set of rules of the form $l \to r$ where $l \in T_{\Sigma}(Q(X))$ is linear in X and $r \in Q(T_{\Delta}(\text{var}(l)))$.*

A rule $l \to r \in R$ is an epsilon rule *if $l \in Q(X)$; otherwise it is* input-consuming. *The sets of epsilon and input-consuming rules are denoted by R^{ε} and R^{Σ}, respectively.*

An xmbutt $M = (Q, \Sigma, \Delta, F, R)$ is a *multi bottom-up tree transducer* (mbutt) [respectively, an STA] if $l \in \Sigma(Q(X))$ [respectively, $l \in \Sigma(Q(X)) \cup Q(X)$] for every $l \to r \in R$. Linearity and nondeletion of xmbutts are defined in the natural manner. The xmbutt M is *linear* if r is linear in var(l) for every rule $l \to r \in R$. Moreover, M is *nondeleting* if (i) $F \subseteq Q^{(1)}$ and (ii) r is nondeleting in var(l) for every $l \to r \in R$. Finally, M is *deterministic* if (i) there do not exist two distinct rules $l_1 \to r_1 \in R$ and $l_2 \to r_2 \in R$, a substitution $\theta \colon X \to X$, and $w \in \text{pos}(l_2)$ such that $l_1\theta = l_2|_w$, and (ii) there does not exist an epsilon rule $l \to r \in R$ such that $l(\varepsilon) \in F$. Let us now present a rewrite semantics. In the rest of this section, let $M = (Q, \Sigma, \Delta, F, R)$ be an xmbutt.

Definition 2. *Let Σ' and Δ' be ranked alphabets disjoint with Q. Moreover, let $\xi, \zeta \in T_{\Sigma \cup \Sigma'}(Q(T_{\Delta \cup \Delta'}))$, and $l \to r \in R$. We write $\xi \Rightarrow_M^{l \to r} \zeta$ if there exist $w \in \text{pos}(\xi)$ and $\theta \colon X \to T_{\Delta \cup \Delta'}$ such that $\xi|_w = l\theta$ and $\zeta = \xi[r\theta]_w$, and we write $\xi \Rightarrow_M \zeta$ if there exists $\rho \in R$ such that $\xi \Rightarrow_M^{\rho} \zeta$. The tree transformation computed by M is $\tau_M = \{(t, \xi|_1) \in T_{\Sigma} \times T_{\Delta} \mid \xi \in F(T_{\Delta}), t \Rightarrow_M^* \xi\}$.*

The xmbutt M' is *equivalent* to M if $\tau_{M'} = \tau_M$. We denote by XMBOT the class of tree transformations computed by xmbutts. We use the prefixes 'l', 'n', and 'd' to restrict to linear, nondeleting, and deterministic devices, respectively. For example, l-XMBOT denotes the class of all tree transformations computed by linear xmbutts.

If M is deterministic and $t \in T_{\Sigma}$, then there exists at most one $\xi \in Q(T_{\Delta})$ such that (i) $t \Rightarrow_M^* \xi$ and (ii) there exists no ζ such that $\xi \Rightarrow_M \zeta$. Hence, τ_M is a partial function, if M is deterministic. Moreover, if M is a deterministic mbutt and $t \in T_{\Sigma}$, then there exists at most one $\xi \in Q(T_{\Delta})$ such that $t \Rightarrow_M^* \xi$. A deterministic mbutt M is *total* if for every $t \in T_\Sigma$ there exists $\xi \in Q(T_\Delta)$ such that $t \Rightarrow_M^* \xi$ (an equivalent static definition is easy to formulate).

Our first result shows that every xmbutt is equivalent to a nondeleting one. Unfortunately, the construction does not preserve determinism.

Theorem 3. XMBOT $=$ n-XMBOT *and* l-XMBOT $=$ ln-XMBOT.

Proof. For the xmbutt $M = (Q, \Sigma, \Delta, F, R)$ we construct an equivalent non-deleting xmbutt $M' = (Q', \Sigma, \Delta, F', R')$. The idea is that M' simulates M but guesses at each moment which subtrees of the states will be deleted in the remainder of M's computation. The set Q' of states of M' consists of all pairs $\langle q, J \rangle$ with $q \in Q^{(k)}$ and $J \subseteq [k]$, and the rank of $\langle q, J \rangle$ is $\operatorname{card}(J)$; moreover, $F' = \{\langle q, \{1\}\rangle \mid q \in F\}$. The rules of M' are constructed such that $t \Rightarrow_{M'}^* \langle q, J \rangle (u_{i_1}, \ldots, u_{i_m})$, where $J = \{i_1, \ldots, i_m\}$ and $i_1 < \cdots < i_m$, if and only if there exist $u_i \in T_\Delta$ for every $i \in [k] \setminus J$ such that $t \Rightarrow_M^* q(u_1, \ldots, u_k)$. $\square$

The following normal form will be at the heart of our composition construction in the next section. It says that exactly one input or output symbol occurs in every rule (such rules will be called *one-symbol* rules).

Definition 4. *The xmbutt M is in* one-symbol normal form *if for every rule* $l \to r \in R$ *we have* $\operatorname{card}(\operatorname{pos}_\Sigma(l)) + \operatorname{card}(\operatorname{pos}_\Delta(r)) = 1$.

Theorem 5. *For every xmbutt M there exists an equivalent xmbutt N in one-symbol normal form. Moreover, if M is linear (respectively, nondeleting, deterministic), then so is N.*

Proof. Let us assume, without loss of generality, that all left-hand sides of rules of M are normalized. First we take care of the left-hand sides of rules and decompose rules with more than one input symbol in the left-hand side into several rules, cf. [30, Proposition II.B.5]. Take a uniquely-ranked set P and a bijection $f \colon T_\Sigma(Q(X)) \to P$ such that (i) $Q \subseteq P$, (ii) $f(q(x_1, \ldots, x_n)) = q$ for every $q \in Q^{(n)}$, and (iii) $\operatorname{rk}(f(l)) = \operatorname{card}(\operatorname{var}(l))$ for every $l \in T_\Sigma(Q(X))$. In fact, we will only use $f(l)$ for normalized $l \in T_\Sigma(Q(X))$.

Let $l \to r \in R$ be input-consuming such that $l \notin \Sigma(P(X))$. Suppose that $l = \sigma(l_1, \ldots, l_k)$ for some $\sigma \in \Sigma^{(k)}$ and $l_1, \ldots, l_k \in T_\Sigma(Q(X))$. Moreover, let $\theta_1, \ldots, \theta_k \colon X \to X$ be bijections such that $l_i \theta_i$ is normalized. Finally, for every $i \in [k]$ let $p_i = f(l_i \theta_i)$ and $r_i = p_i(x_1, \ldots, x_m)$ where $m = \operatorname{rk}(p_i)$. We construct the xmbutt $M_1 = (Q \cup \{p_1, \ldots, p_k\}, \Sigma, \Delta, F, (R \setminus \{l \to r\}) \cup R_{1,1} \cup R_{1,2})$ where $R_{1,1} = \{l_i \theta_i \to r_i \mid i \in [k], l_i \notin Q(X)\}$ and $R_{1,2} = \{l' \to r\}$, in which l' is the unique normalized tree of $\{\sigma\}(P(X))$ such that $l'(i) = p_i$ for every $i \in [k]$ (note that if $l_i \in Q(X)$, then $p_i = l_i(\varepsilon)$ and so $l'|_i = l_i$). Repeated application of this construction (keeping P and the mapping f fixed) eventually yields an equivalent xmbutt $M' = (Q', \Sigma, \Delta, F, R')$ such that $l \in \Sigma(P(X))$ for each input-consuming rule $l \to r \in R'$.

Next, we remove all epsilon rules $l \to r \in R'$ such that $r \in Q'(X)$ in the standard way. Finally, we decompose the right-hand sides. Let $M'' = (S, \Sigma, \Delta, F, R'')$ be the xmbutt obtained so far and $l \to r \in R''$ a rule that is not yet a one-symbol rule. Let $r = s(u_1, \ldots, u_{i-1}, \delta(u'_1, \ldots, u'_k), u_{i+1}, \ldots, u_n)$ for some $s \in S^{(n)}$, $i \in [n]$, $\delta \in \Delta^{(k)}$, and $u_1, \ldots, u_{i-1}, u_{i+1}, \ldots, u_n, u'_1, \ldots, u'_k \in T_\Delta(X)$. Also, let $q \notin S$ be a new state of rank $k + n - 1$. We construct the xmbutt $M''_1 = (S \cup \{q\}, \Sigma, \Delta, F, R''_1)$ with $R''_1 = (R'' \setminus \{l \to r\}) \cup R''_{1,1}$ where $R''_{1,1}$ contains the two rules:

- $l \to q(u_1, \ldots, u_{i-1}, u'_1, \ldots, u'_k, u_{i+1}, \ldots, u_n)$ and
- $q(x_1, \ldots, x_{k+n-1}) \to s(x_1, \ldots, x_{i-1}, \delta(x_i, \ldots, x_{i+k-1}), x_{i+k}, \ldots, x_{k+n-1})$.

Repeated application of the procedure yields the desired xmbutt N.   □

Example 6. Let $(Q, \Sigma, \Gamma, Q, R)$ be the xmbutt with $Q = \{q^{(1)}\}$, $\Sigma = \{\sigma^{(1)}, \alpha^{(0)}\}$, $\Gamma = \{\gamma^{(2)}, \alpha^{(0)}\}$, and $R = \{\sigma(\alpha) \to q(\alpha),\ \sigma(q(x_1)) \to q(\gamma(x_1, \alpha))\}$. Clearly, it is linear, nondeleting, and deterministic. Applying the procedure of Theorem 5 we obtain the states $q^{(1)}, q_1^{(0)}, q_2^{(0)}, q_3^{(2)}, q_4^{(1)}$ and the rules $\alpha \to q_1$, $\sigma(q_1) \to q_2$, $q_2 \to q(\alpha)$, $\sigma(q(x_1)) \to q_4(x_1)$, $q_4(x_1) \to q_3(x_1, \alpha)$, $q_3(x_1, x_2) \to q(\gamma(x_1, x_2))$.

In the deterministic case we have an additional normal form: the deterministic mbutt. This allows us to characterize the classes d-XMBOT and ld-XMBOT in terms of top-down tree transducers, using the result of [16].

Theorem 7. *For every deterministic xmbutt M there exists an equivalent total deterministic mbutt N. Moreover, if M is linear, then so is N. Consequently,* d-XMBOT = d-TOP^R *and* ld-XMBOT = d-TOP$_{su}^R$.

Proof. Applying the first construction in the proof of Theorem 5 and then removing all epsilon rules in the usual way, we obtain an equivalent deterministic mbutt N. Obviously, by introducing a dummy state of rank 0, N can be made total. To obtain a deterministic multi bottom-up tree transducer of [16] we also have to add a special root symbol and add rules that, while consuming the special root symbol, project on the first argument of a final state. It is proved in [16] that such deterministic multi bottom-up tree transducers have the same power as deterministic top-down tree transducers with regular look-ahead. The second equality was already suggested in the Conclusion of [17]. We prove it by reconsidering (a minor variation of) the proofs of [16, Lemmata 4.1 and 4.2]. If the mbutt is linear, then the corresponding top-down tree transducer with regular look-ahead will be single-use and vice versa.   □

Finally, we verify that l-XMBOT is suitably powerful for applications in machine translation. We do this by showing that all transformations of l-XTOP are also in l-XMBOT. This shows that xmbutts can handle rotations [2].

Theorem 8. l-XTOP ⊂ l-XMBOT.

Proof. The inclusion can be proved in a similar manner as l-TOP ⊆ l-BOT [19, Theorem 2.8], where l-BOT denotes the class of transformations computed by linear bottom-up tree transducers [21,19]. Every transformation of l-XTOP preserves recognizability [18, Theorem 4], but there is a linear (deterministic) mbutt that computes the transformation $\{(\sigma(t), \delta(t, t)) \mid t \in T_{\Sigma \setminus \{\sigma\}}\}$ where $\sigma \in \Sigma^{(1)}$ and $\delta \in \Delta^{(2)}$. Hence, not every transformation of l-XMBOT preserves recognizability.   □

4   Composition Construction

In this section, we investigate compositions of tree transformations computed by xmbutts. Let us first recall the classical composition results for bottom-up tree transducers [19,20]. Let M and N be bottom-up tree transducers. If M is linear or N deterministic, then the composition of the transformations computed by M and N can be computed by a bottom-up tree transducer. As a special case, the classes of transformations computed by linear, linear and nondeleting, and deterministic bottom-up tree transducers are closed under composition.

In our setting, let M and N be xmbutts. We will prove that if M is linear or N is deterministic, then there is an xmbutt $M ; N$ that computes $\tau_M ; \tau_N$. In particular, we prove that l-XMBOT, d-XMBOT, and ld-XMBOT are closed under composition. The closure of l-XMBOT was first presented in [30, Propositions II.B.5 and II.B.7]. The closure of d-XMBOT is also immediate from Theorem 7 and [27, Theorem 2.11]; in [31, Proposition 2.5] it was shown for a different notion of determinism. The closure of ld-XMBOT is to be expected from Theorem 7 and the fact that the single-use restriction was introduced in [22,23] to guarantee the closure under composition of attribute grammar transformations (see [25, Theorem 3]).

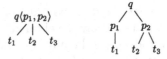

Fig. 1. Tree homomorphism φ where $q \in Q^{(2)}$, $p_1 \in P^{(1)}$, and $p_2 \in P^{(2)}$.

Let us prepare the composition construction. Let $M = (Q, \Sigma, \Gamma, F_M, R_M)$ and $N = (P, \Gamma, \Delta, F_N, R_N)$ be xmbutts such that Q, P, and $\Sigma \cup \Gamma \cup \Delta$ are pairwise disjoint. We define the uniquely-ranked alphabet

$$Q\langle P \rangle = \{q\langle p_1, \ldots, p_n \rangle \mid q \in Q^{(n)}, p_1, \ldots, p_n \in P\}$$

such that $\mathrm{rk}(q\langle p_1, \ldots, p_n \rangle) = \sum_{i=1}^n \mathrm{rk}(p_i)$ for every $q \in Q^{(n)}$ and $p_1, \ldots, p_n \in P$. Let $\Pi = \Sigma \cup \Gamma \cup \Delta \cup X$. We define the mapping $\varphi \colon T_{\Pi \cup Q\langle P \rangle} \to T_{\Pi \cup Q \cup P}$ such that for every $q\langle p_1, \ldots, p_n \rangle \in Q\langle P \rangle^{(k)}$, $\pi \in \Pi^{(k)}$, and $t_1, \ldots, t_k \in T_{\Pi \cup Q\langle P \rangle}$

$$\varphi(q\langle p_1, \ldots, p_n \rangle(t_1, \ldots, t_k)) = q(p_1(\varphi(t_1), \ldots, \varphi(t_l)), \ldots, p_n(\varphi(t_m), \ldots, \varphi(t_k)))$$
$$\varphi(\pi(t_1, \ldots, t_k)) = \pi(\varphi(t_1), \ldots, \varphi(t_k))$$

where $l = \mathrm{rk}(p_1)$ and $m = k - \mathrm{rk}(p_n) + 1$. Thus, we group the subtrees below the corresponding state p_i (see Fig. 1). Note that φ is a linear and nondeleting tree homomorphism, which acts as a bijection from $T_\Sigma(Q\langle P \rangle(T_\Delta(X)))$ to $T_\Sigma(Q(P(T_\Delta(X))))$. In the sequel, we will identify t with $\varphi(t)$ for all trees $t \in T_\Sigma(Q\langle P \rangle(T_\Delta(X)))$.

Definition 9. *Let $M = (Q, \Sigma, \Gamma, F_M, R_M)$ be an xmbutt in one-symbol normal form and $N = (P, \Gamma, \Delta, F_N, R_N)$ an STA. Moreover, let $\mathrm{LHS}(\Sigma)$ and $\mathrm{LHS}(\varepsilon)$ be the sets of normalized trees of $\Sigma(Q\langle P\rangle(X))$ and $Q\langle P\rangle(X)$, respectively. The composition $M \,;\, N = (Q\langle P\rangle, \Sigma, \Delta, F_M\langle F_N\rangle, R)$ of M and N is the STA with $R = R_1 \cup R_2 \cup R_3$ where:*

$$R_1 = \{l \to r \mid l \in \mathrm{LHS}(\Sigma) \text{ and } \exists \rho \in R_M^\Sigma : l \Rightarrow_M^\rho r\},$$
$$R_2 = \{l \to r \mid l \in \mathrm{LHS}(\varepsilon) \text{ and } \exists \rho \in R_N^\varepsilon : l \Rightarrow_N^\rho r\}, \text{ and}$$
$$R_3 = \{l \to r \mid l \in \mathrm{LHS}(\varepsilon) \text{ and } \exists \rho_1 \in R_M^\varepsilon, \rho_2 \in R_N^\Gamma : l \,(\Rightarrow_M^{\rho_1} ; \Rightarrow_N^{\rho_2})\, r\}.$$

To illustrate the implicit use of φ, let us show the "official" definition of R_1:

$$R_1 = \{l \to r \mid l \in \mathrm{LHS}(\Sigma), r \in Q\langle P\rangle(T_\Delta(X)), \text{ and } \exists \rho \in R_M^\Sigma : \varphi(l) \Rightarrow_M^\rho \varphi(r)\} \ .$$

The construction preserves linearity; moreover, it preserves determinism if N is an mbutt. In the rest of this section we investigate when $\tau_{M;N} = \tau_M \,;\, \tau_N$, but we first illustrate the construction on our small running example.

Example 10. Let M be the xmbutt of Example 6 in one-symbol normal form, and let $N = (\{g^{(1)}, h^{(1)}\}, \Gamma, \Delta, \{g\}, R_N)$ be the STA with $\Delta = \Gamma \cup \{\delta^{(1)}\}$ and

$$R_N = \{\alpha \to h(\alpha),\ h(x_1) \to h(\delta(x_1)),\ \gamma(h(x_1), h(x_2)) \to g(\gamma(x_1, x_2))\} \ .$$

Clearly, N computes $\{(\gamma(\alpha, \alpha), \gamma(\delta^i(\alpha), \delta^j(\alpha)) \mid i, j \in \mathbb{N}\}$, and hence $\tau_M \,;\, \tau_N$ is $\{(\sigma(\sigma(\alpha)), \gamma(\delta^i(\alpha), \delta^j(\alpha)) \mid i, j \in \mathbb{N}\}$. The states of $M \,;\, N$ will be

$$\{q\langle g\rangle^{(1)}, q\langle h\rangle^{(1)}, q_1\langle\rangle^{(0)}, q_2\langle\rangle^{(0)}, q_3\langle g, g\rangle^{(2)}, \dots, q_3\langle h, h\rangle^{(2)}, q_4\langle g\rangle^{(1)}, q_4\langle h\rangle^{(1)}\} \ ,$$

of which only $q\langle g\rangle$ is final. We present some relevant rules only [left in official form $l \to r$; right in alternative notation $\varphi(l) \to \varphi(r)$].

$$
\begin{array}{ll}
\alpha \to q_1\langle\rangle & \alpha \to q_1 \\
\sigma(q_1\langle\rangle) \to q_2\langle\rangle & \sigma(q_1) \to q_2 \\
q_2\langle\rangle \to q\langle h\rangle(\alpha) & q_2 \to q(h(\alpha)) \\
q\langle h\rangle(x_1) \to q\langle h\rangle(\delta(x_1)) & q(h(x_1)) \to q(h(\delta(x_1))) \\
\sigma(q\langle h\rangle(x_1)) \to q_4\langle h\rangle(x_1) & \sigma(q(h(x_1))) \to q_4(h(x_1)) \\
q_4\langle h\rangle(x_1) \to q_3\langle h, h\rangle(x_1, \alpha) & q_4(h(x_1)) \to q_3(h(x_1), h(\alpha)) \\
q_3\langle h, h\rangle(x_1, x_2) \to q\langle g\rangle(\gamma(x_1, x_2)) & q_3(h(x_1), h(x_2)) \to q(g(\gamma(x_1, x_2)))
\end{array}
$$

The first, second, and fifth rules are in R_1 (of Definition 9), the fourth rule is in R_2, and the remaining rules in R_3. $\qquad\square$

Next, we will prove that $\tau_M \,;\, \tau_N$ is in XMBOT provided that (i) M is linear or (ii) N is deterministic. We can assume that M is in one-symbol normal form, by Theorem 5, and that it is nondeleting in case (i), by Theorem 3. We can also assume that N is an STA in case (i), by Theorem 5, and a total deterministic mbutt in case (ii), by Theorem 7. Thus, we meet the requirements of Definition 9 and henceforth assume its notation.

We start with a simple lemma. It shows that in a derivation that uses steps of M and N (like the derivations of M ; N) we can always perform all steps of M first and only then perform the derivation steps of N. This already proves one direction needed for the correctness of the composition construction.

Lemma 11. *Let $t \in T_\Sigma$ and $\xi \in Q(P(T_\Delta))$. If $t \Rightarrow^* \xi$ where $\Rightarrow$ is $\Rightarrow_M \cup \Rightarrow_N$, then $t \ (\Rightarrow_M^* \ ; \Rightarrow_N^*) \ \xi$. In particular, $\tau_{M;N} \subseteq \tau_M$; τ_N.*

Proof. It obviously suffices to prove: For every $\xi, \zeta \in T_\Sigma(Q(T_\Gamma(P(T_\Delta))))$, if $\xi \ (\Rightarrow_N \ ; \Rightarrow_M) \ \zeta$, then $\xi \ (\Rightarrow_M \ ; \Rightarrow_N^*) \ \zeta$. Its proof is easy.   □

Next we prove that τ_M ; $\tau_N \subseteq \tau_{M;N}$ under the above assumptions on M and N, by a standard induction over the length of the derivation.

Lemma 12. *Let $t \in T_\Sigma$ and $\xi \in Q(P(T_\Delta))$ be such that $t \ (\Rightarrow_M^* \ ; \Rightarrow_N^*) \ \xi$. If (i) M is linear and nondeleting, or (ii) N is a total deterministic mbutt, then $t \Rightarrow_{M;N}^* \xi$. With $\xi \in F_M(F_N(T_\Delta))$ we obtain τ_M ; $\tau_N \subseteq \tau_{M;N}$.*

Theorem 13. *The three classes l-XMBOT, d-XMBOT, and ld-XMBOT are closed under composition. Moreover,*

l-XMBOT ; XMBOT $\subseteq$ XMBOT   *and*   XMBOT ; d-XMBOT $\subseteq$ XMBOT .

Proof. The inequalities follow directly from Lemmata 11 and 12 using Theorems 3, 5, and 7 to establish the preconditions of Definition 9 and Lemma 12. The closure results follow from the fact that the composition construction preserves linearity and determinism.   □

5   Relation to Top-Down Tree Transducers

Now, let us focus on an upper bound to the power of xmbutts. By [18, Theorem 14] every mbutt computes a transformation of ln-TOP ; d-TOP. Here we prove a similar result for xmbutts.

Theorem 14

l-XMBOT = ln-XTOP ; d-TOP$_{su}$ *and*   XMBOT = ln-XTOP ; d-TOP .

Proof. By Theorems 7, 8, and 13, the inclusions $\supseteq$ are immediate. For the decomposition results, we employ the standard idea of separating the input and output behavior of the given xmbutt M (cf. [19, Theorem 3.15]). For each input tree, the first xtt M_1 outputs "rule trees" that encode which rules could be applied. The second xtt M_2 then deterministically executes these rules and creates the output. Linearity of M implies that M_2 is single-use. More formally, if $t \Rightarrow_M^* q(u_1, \ldots, u_m)$, then there is a "rule tree" $\tilde{t}$ such that $q(t) \Rightarrow_{M_1}^* \tilde{t}$ and $n(\tilde{t}) \Rightarrow_{M_2}^* u_n$ for every $n \in [m]$.   □

We note that the direction $\subseteq$ of all four equalities in Theorems 7 and 14 is proved in essentially the same manner. If we compare the deterministic (Theorem 7) to the nondeterministic case (Theorem 14), then in the former case there exists at most one successful "rule tree", which can be constructed by a deterministic, finite-state bottom-up relabeling [19,27] and has the shape of the input tree. Thus, the deterministic top-down tree transducer can query its look-ahead for the rule that labels its current position in the successful "rule tree".

By Theorem 14, the power of xmbutts is limited by two extended top-down tree transducers. In particular, the first equation shows in a precise way how much stronger l-XMBOT is with respect to ln-XTOP. But Theorem 14 also shows that we can separate nondeterminism and state checking (performed by the linear and nondeleting xtt) from evaluation (performed by the deterministic top-down tree transducer). Since linear xtts preserve recognizability, the *rule trees* mentioned in the proof of Theorem 14 form a recognizable tree language. Formally, the set $\{u \in T_R \mid \exists t \in T_\Sigma \colon (t, u) \in \tau_{M_1}\}$ is recognizable, which shows that the set of rule trees of an xmbutt is recognizable. This is a strong indication toward the existence of efficient training algorithms.

Conclusion and Open Problems. We have shown that xmbutts are suitably powerful to compute any transformation that can be computed by linear extended top-down tree transducers (see Theorem 8). Moreover, we generalized the main composition results of [19,20] for bottom-up tree transducers to xmbutts (see Theorem 13). In particular, we showed that l-XMBOT and d-XMBOT are closed under composition. Finally, we characterized XMBOT as the composition of ln-XTOP and d-TOP (see Theorem 14), which shows that, analogously to bottom-up tree transducers, nondeterminism and evaluation can be separated.

Since linear xmbutts do not necessarily preserve recognizability whereas linear xtts do, it is clear that even the composition closure of l-XTOP is strictly contained in l-XMBOT. This raises two questions: (a) Which xmbutts can be transformed into an xtt? (b) Can we characterize the composition closure of l-XTOP?

References

1. Knight, K.: Criteria for reasonable syntax-based translation models. Personal communication (2007)
2. Knight, K., Graehl, J.: An overview of probabilistic tree transducers for natural language processing. In: Gelbukh, A. (ed.) CICLing 2005. LNCS, vol. 3406, pp. 1–24. Springer, Heidelberg (2005)
3. DeNeefe, S., Knight, K., Wang, W., Marcu, D.: What can syntax-based MT learn from phrase-based MT? In: EMNLP & CoNLL, pp. 755–763 (2007)
4. Hopcroft, J.E., Ullman, J.D.: Introduction to Automata Theory, Languages, and Computation. Addison Wesley, Reading (1979)
5. Graehl, J., Knight, K.: Training tree transducers. In: HLT-NAACL, pp. 105–112 (2004)
6. Arnold, A., Dauchet, M.: Transductions inversibles de forêts. Thèse 3ème cycle M. Dauchet, Université de Lille (1975)
7. Arnold, A., Dauchet, M.: Bi-transductions de forêts. In: ICALP, pp. 74–86. Edinburgh University Press (1976)

8. Rounds, W.C.: Mappings and grammars on trees. Math. Systems Theory 4(3), 257–287 (1970)
9. Thatcher, J.W.: Generalized² sequential machine maps. J. Comput. System Sci. 4(4), 339–367 (1970)
10. Steinby, M., Tîrnăucă, C.I.: Syntax-directed translations and quasi-alphabetic tree bimorphisms. In: Holub, J., Žďárek, J. (eds.) CIAA 2007. LNCS, vol. 4783, pp. 265–276. Springer, Heidelberg (2007)
11. Aho, A.V., Ullman, J.D.: Syntax directed translations and the pushdown assembler. J. Comput. System Sci. 3(1), 37–56 (1969)
12. Shabes, Y.: Mathematical and Computational Aspects of Lexicalized Grammars. PhD thesis, University of Pennsylvania (1990)
13. Shieber, S.M., Shabes, Y.: Synchronous tree-adjoining grammars. In: COLING, pp. 1–6 (1990)
14. Shieber, S.M.: Unifying synchronous tree adjoining grammars and tree transducers via bimorphisms. In: EACL. The Association for Computer Linguistics, pp. 377–384 (2006)
15. Arnold, A., Dauchet, M.: Morphismes et bimorphismes d'arbres. Theoret. Comput. Sci. 20, 33–93 (1982)
16. Fülöp, Z., Kühnemann, A., Vogler, H.: A bottom-up characterization of deterministic top-down tree transducers with regular look-ahead. Inf. Process. Lett. 91(2), 57–67 (2004)
17. Fülöp, Z., Kühnemann, A., Vogler, H.: Linear deterministic multi bottom-up tree transducers. Theoret. Comput. Sci. 347(1–2), 276–287 (2005)
18. Maletti, A.: Compositions of extended top-down tree transducers. Inform. and Comput. (to appear, 2008)
19. Engelfriet, J.: Bottom-up and top-down tree transformations: A comparison. Math. Systems Theory 9(3), 198–231 (1975)
20. Baker, B.S.: Composition of top-down and bottom-up tree transductions. Inform. and Control 41(2), 186–213 (1979)
21. Thatcher, J.W.: Tree automata: An informal survey. In: Currents in the Theory of Computing, pp. 143–172. Prentice Hall, Englewood Cliffs (1973)
22. Ganzinger, H.: Increasing modularity and language-independency in automatically generated compilers. Sci. Comput. Prog. 3(3), 223–278 (1983)
23. Giegerich, R.: Composition and evaluation of attribute coupled grammars. Acta Inform. 25(4), 355–423 (1988)
24. Kühnemann, A.: Berechnungsstärken von Teilklassen primitiv-rekursiver Programmschemata. PhD thesis, Technische Universität Dresden (1997)
25. Kühnemann, A.: Benefits of tree transducers for optimizing functional programs. In: Arvind, V., Ramanujam, R. (eds.) FST TCS 1998. LNCS, vol. 1530, pp. 146–157. Springer, Heidelberg (1998)
26. Engelfriet, J., Maneth, S.: Macro tree transducers, attribute grammars, and MSO definable tree translations. Inform. and Comput. 154(1), 34–91 (1999)
27. Engelfriet, J.: Top-down tree transducers with regular look-ahead. Math. Systems Theory 10(1), 289–303 (1977)
28. Gécseg, F., Steinby, M.: Tree Automata. Akadémiai Kiadó, Budapest (1984)
29. Gécseg, F., Steinby, M.: Tree languages. In: Handbook of Formal Languages, vol. 3, pp. 1–68. Springer, Heidelberg (1997)
30. Lilin, E.: Une généralisation des transducteurs d'états finis d'arbres: les S-transducteurs. Thèse 3ème cycle, Université de Lille (1978)
31. Lilin, E.: Propriétés de clôture d'une extension de transducteurs d'arbres déterministes. In: Astesiano, E., Böhm, C. (eds.) CAAP 1981. LNCS, vol. 112, pp. 280–289. Springer, Heidelberg (1981)

Derivation Tree Analysis for
Accelerated Fixed-Point Computation

Javier Esparza, Stefan Kiefer, and Michael Luttenberger

Institut für Informatik, Technische Universität München, 85748 Garching, Germany
{esparza,kiefer,luttenbe}@model.in.tum.de

Abstract. We show that for several classes of idempotent semirings the least fixed-point of a polynomial system of equations $X = f(X)$ is equal to the least fixed-point of a *linear* system obtained by "linearizing" the polynomials of f in a certain way. Our proofs rely on derivation tree analysis, a proof principle that combines methods from algebra, calculus, and formal language theory, and was first used in [5] to show that Newton's method over commutative and idempotent semirings converges in a linear number of steps. Our results lead to efficient generic algorithms for computing the least fixed-point. We use these algorithms to derive several consequences, including an $O(N^3)$ algorithm for computing the throughput of a context-free grammar (obtained by speeding up the $O(N^4)$ algorithm of [2]), and a generalization of Courcelle's result stating that the downward-closed image of a context-free language is regular [3].

1 Introduction

Systems $X = f(X)$ of fixed-point equations, where f is a system of polynomials, appear naturally in semantics, interprocedural program analysis, language theory, and in the study of probabilistic systems (see e.g. [7,8,10,13]). In all these applications the equations are interpreted over ω-continuous semirings, an algebraic structure that guarantees the existence of a least solution μf. The key algorithmic problem is to compute or at least approximate μf.

In [5,4] we generalized Newton's method—the well-known method of numerical mathematics for approximating a zero of a differentiable function—to arbitrary ω-continuous semirings. Given a polynomial system f, our generalized method computes a sequence of increasingly accurate approximations to μf, called Newton approximants. We showed in [5] that the n-th Newton approximant of a system of n equations over an idempotent (w.r.t. addition) and commutative (w.r.t. multiplication) semiring is already equal to μf. This theorem leads to a generic computing procedure.

Our proof of this result uses a (to the best of our knowledge) novel technique, which we call *derivation tree analysis*. The system f induces a set $\mathcal{T}$ of *derivation trees*, a generalization of the well-known derivation trees of context-free grammars. Each tree can be naturally assigned a semiring element, called the *yield* of the tree. It is easy to show that μf is equal to the sum of the yields of all derivation trees. Derivation tree analysis first identifies a subset T' of derivation trees whose total yield $Y(T')$ is easy to compute in some sense, and then proves that T' satisfies the *embedding property*:

M. Ito and M. Toyama (Eds.): DLT 2008, LNCS 5257, pp. 301–313, 2008.

$Y(t) \sqsubseteq Y(T')$ for every derivation tree t. If the semiring is idempotent, the embedding property implies $Y(\mathcal{T}) = Y(T')$, and so $\mu f = Y(T')$. In [5], the set T' was chosen so that $Y(T')$ is equal to the n-th Newton approximant, and the embedding property was proved using some tree surgery and exploiting the commutativity of the semiring.

The computation of the n-th Newton approximant can still require considerable resources. In this paper we present a further application of derivation tree analysis to idempotent semirings, leading to more efficient computation algorithms. For this, we define the set $\mathcal{B}$ of *bamboos* of a system f. Loosely speaking, bamboos are derivation trees with an arbitrarily long stem but only short branches. We first show that $Y(\mathcal{B})$ is the solution of a linear system of equations whose functions are similar (but not identical) to the straightforward linearisation of f. Then, we prove that the following three classes of semirings satisfy the embedding property:

- *Star-distributive semirings* are idempotent and commutative semirings satisfying the additional axiom $(a+b)^* = a^* + b^*$ (where * is the well-known Kleene iteration operator). The so-called "tropical" $(\min, +)$-semiring over the reals (extended with $+\infty$ and $-\infty$) is star-distributive. Our tree analysis leads to an algorithm for computing μf very similar to the generalized Bellman-Ford algorithm of Gawlitza and Seidl [9]. We use it to derive a new algorithm for computing the throughput of a context-free grammar, a problem introduced and analyzed by Caucal et al. in [2]. Our algorithm runs in $O(N^3)$, a factor N faster than the algorithm presented in [2].
- *Lossy semirings* are idempotent semirings satisfying the additional axiom $a + 1 = a$ where 1 is the neutral element of multiplication. A natural model are downward-closed languages with union and concatenation as operations. Lossy semirings find application in the verification of lossy channel systems, a model of computation thoroughly investigated by Abdulla et al. (see e.g. [1]). Our tree analysis leads to an algebraic proof of Courcelle's theorem stating that the downward closure of a context-free language is effectively regular [3].
- *1-bounded semirings* are idempotent semirings where the equation $a + 1 = 1$ holds. A natural example is the "maximum probability" semiring with the interval $[0, 1]$ as carrier, maximum as addition, and standard multiplication over the reals. Using derivation tree analysis it is very easy to show that the least fixed-point μf of a polynomial system f with n variables is given by $f^n(0)$, the n-fold application of f to $\mathbf{0}$.

The rest of the paper is organized as follows. After the preliminaries in Section 2 we introduce derivation tree analysis in Section 3. Bamboos are defined in Section 4. In the Sections 5, 6 and 7 we apply derivation tree analysis to the semiring classes mentioned above. A technical report [6] includes the missing proofs.

2 Preliminaries

As usual, $\mathbb{N}$ denotes the set of natural numbers including 0.

An *idempotent semiring* $\mathcal{S} = \langle S, +, \cdot, 0, 1 \rangle$ consists of a commutative, idempotent additive monoid $\langle S, +, 0 \rangle$, and a multiplicative monoid $\langle S, \cdot, 1 \rangle$. In the following we often omit the dot $\cdot$ in products. Both algebraic structures are connected by left- and right-distributivity, e.g. $a(b + c) = ab + ac$, and by the requirement that $0 \cdot a = 0$ for all $a \in S$. The *natural-order relation* $\sqsubseteq S \times S$ is defined by $a \sqsubseteq b \Leftrightarrow a + b = b$. The semiring S is *naturally ordered* if $\sqsubseteq$ is a partial order.

An idempotent, naturally ordered semiring S is ω-*continuous*, if countable summation $\sum_{i \in \mathbb{N}} a_i \in S$ is defined (with $a_i \in S$), and satisfies the following requirements: (i) summation is continuous, i.e., $\sup^{\sqsubseteq} \{a_0 + a_1 + \ldots + a_k \mid k \in \mathbb{N}\} = \sum_{i \in \mathbb{N}} a_i$ for all sequences $a : N \to S$; (ii) distributivity extends in the natural way to countable summation; and (iii) $\sum_{j \in J} \sum_{i \in I_j} a_i = \sum_{i \in \mathbb{N}} a_i$ holds for all partitions $(I_j)_{j \in J}$ of $\mathbb{N}$. In every such ω-continuous semiring the Kleene-star operator $^* : S \to S$ is well-defined by $a^* := \sum_{k \in \mathbb{N}} a^k$ for all $a \in S$. In the following we consider only idempotent ω-continuous semirings S. We refer to them as *io-semirings*.

We fix a finite, non-empty set $\mathcal{X}$ of *variables* for the rest of the section, and use n to denote $|\mathcal{X}|$ in the following. A map from $\mathcal{X}$ to S is called a *vector*. The set of all vectors is denoted by V. We write both $v(X)$ and v_X for the value of a vector v at $X \in \mathcal{X}$, also called the X-component of v. Sum of vectors is defined componentwise: given a countable set I and a vector v_i for every $i \in I$, we denote by $\sum_{i \in I} v_i$ the vector given by $\left(\sum_{i \in I} v_i\right)(X) = \sum_{i \in I} v_i(X)$ for every $X \in \mathcal{X}$.

A *monomial of degree* k is a finite expression $a_1 X_1 a_2 \cdots a_k X_k a_{k+1}$ where $k \geq 0$, $a_1, \ldots, a_{k+1} \in S \setminus \{0\}$ and $X_1, \ldots, X_k \in \mathcal{X}$. A *polynomial* is an expression of the form $m_1 + \cdots + m_k$ where $k \geq 0$ and $m_1, \ldots, m_k$ are monomials. Since S is idempotent, we assume w.l.o.g. that all monomials of a polynomial are distinct. The degree of a polynomial is the largest degree of its monomials. We let $S[\mathcal{X}]$ denote the set of all polynomials.

Let $f = \alpha_1 X_1 \alpha_2 \ldots X_k \alpha_{k+1}$ be a monomial and let v be a vector. The *evaluation of f at v*, denoted by $f(v)$, is the product $\alpha_1 v_{X_1} \alpha_2 \cdots \alpha_k v_{X_k} \alpha_{k+1}$. We extend this to any polynomial: if $f = \sum_{i=1}^k m_i$, then $f(v) = \sum_{i=1}^k m_i(v)$.

A *system of polynomials* or polynomial system is a map $f : \mathcal{X} \to S[\mathcal{X}]$. We write f_X for $f(X)$. Every polynomial system induces a map from V to V by componentwise evaluation of the polynomials: $f(v)_X := f_X(v)$ for all $v \in V$, and $X \in \mathcal{X}$. The following proposition, which follows easily from Kleene's theorem and the fact that f is a monotone and continuous mapping, shows that any polynomial system f has a least fixed-point μf, which is by definition the least solution of $X = f(X)$.

Proposition 1. *A polynomial system f has a unique least fixed-point μf, i.e., $\mu f = f(\mu f)$, and $\mu f \sqsubseteq v$ holds for all v with $v = f(v)$. Further, μf is the supremum (w.r.t. $\sqsubseteq$) of the* Kleene *sequence* $(f^i(0))_{i \in \mathbb{N}}$, *where f^i denotes the i-fold application of f.*

3 Derivation Trees

We generalize the notion of derivation tree, as known from formal languages and grammars. We identify a node u of a (ordered) tree t with the subtree of t rooted at u. In particular, we identify a tree with its root.

Let f be a polynomial system over a set $\mathcal{X}$ of variables. A *derivation tree* t of f is an ordered (finite) tree whose nodes are labelled with both a variable X and a monomial m of f_X. We write λ_v, resp. λ_m for the corresponding labelling-functions. Moreover, if the monomial labelling of a node u is $\lambda_m(u) = a_1 X_1 a_2 \ldots X_s a_{s+1}$ for some $s \geq 0$,

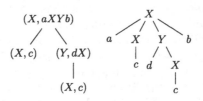

Fig. 1. A derivation tree on the left, and its standard representation on the right

then u has exactly s children $u_1, \ldots, u_s$, ordered from left to right, with $\lambda_v(u_i) = X_i$ for all $i = 1, \ldots, s$. A derivation tree t is an X-*tree* if $\lambda_v(t) = X$. The set of all X-trees of f is denoted by $\mathcal{T}_{f,X}$, or just by $\mathcal{T}_X$ if f is clear from the context.

The left part of Figure 1 shows a derivation tree of the system f over the variables X and Y given by $f_X = aXYb + c$ and $f_Y = dX + Ye$. The derivation trees of f are very similar to the derivation trees of the context-free grammar with productions $X \to aXYb|c$ and $Y \to dX|Ye$. For technical reasons, the nodes of "our" trees are labeled by "productions" (for instance, the label $(X, aXYb)$ corresponds to the production $X \to aXYb$). On the right of Figure 1 we show how the tree would look like according to the standard definition. The height $h(t)$ of a derivation tree t is the length of a longest path from the root to a leaf. The set of X-trees (of f) of height *at most* h is denoted by $\mathcal{T}_X^{(h)}$. The yield $\mathsf{Y}(t)$ of a derivation tree t with $\lambda_m(t) = a_1 X_1 a_2 \cdots X_s a_{s+1}$ is inductively defined to be $\mathsf{Y}(t) = a_1 \mathsf{Y}(t_1) a_2 \cdots \mathsf{Y}(t_s) a_{s+1}$. We extend the definition of Y to sets $T \subseteq \mathcal{T}_X$ by setting $\mathsf{Y}(T) := \sum_{t \in T} \mathsf{Y}(t)$. E.g., the system f defined above has exactly two X-trees of height at most 2: the tree consisting of a single node labeled by (X, c), and the left tree of Figure 1. Their yields are c and $acdcb$, respectively, and so $\mathsf{Y}(\mathcal{T}_X^{(2)}) = c + acdcb$. It follows $\mathsf{Y}(\mathcal{T}_X^{(2)}) = f^3(0)_X$, i.e., the yield of the X-trees of height at most 2 is equal to the "Kleene approximant" $f^3(0)_X$ from Proposition 1. The following proposition, easy to prove [4], shows that this is not a coincidence.

Proposition 2. *For all $h \in \mathbb{N}$ and $X \in \mathcal{X}$, we have $\mathsf{Y}(\mathcal{T}_X^{(h)}) = \left(f^{h+1}(0)\right)_X$.*

Together with Proposition 1 we get:

Corollary 1. $\mu f_X = \mathsf{Y}(\mathcal{T}_X)$.

3.1 Derivation Tree Analysis

We say that a set $\mathcal{T}_X$ of X-trees satisfies the *embedding property* if $\mathsf{Y}(t) \sqsubseteq \mathsf{Y}(\mathcal{T}_X)$ holds for every X-tree t. Loosely speaking, the yield of every X-tree can be "embedded" in the yield of $\mathcal{T}_X$. As addition is idempotent, the embedding property immediately implies that $\mathsf{Y}(\mathcal{T}_X) \sqsubseteq \mathsf{Y}(\mathcal{T}_X)$. Of course, as $\mathcal{T}_X \subseteq \mathcal{T}_X$, we also have the other direction, which leads to the following result.

Proposition 3. *Let f be a system of polynomials over an io-semiring, and let X be a variable of f. If a set $\mathcal{T}_X$ of X-trees of f satisfies the embedding property, then $\mu f = \mathsf{Y}(\mathcal{T}_X)$.*

This proposition suggests a technique for the design of efficient algorithms computing μf: (1) define a set T_X of derivation trees whose yield is "easy to compute" in some io-semiring, and (2) identify "relevant" classes of io-semirings for which T_X satisfies the embedding property. By Proposition 3, μf is "easy to compute" for these classes. We call this technique *derivation tree analysis*.

4   Bamboos and Their Yield

The difficulty of derivation tree analysis lies in finding a set T_X exhibiting a good balance between the contradictory requirements "easy to compute" and "relevant": if $T_X = \emptyset$ then the yield is trivial to compute, but T_X does not satisfy the embedding property in any interesting case. Conversely, $T_X = \mathcal{T}_X$ trivially satisfies the embedding property for every io-semiring, but is not easy to compute. The main contribution of this paper is the identification of a class of derivation trees, *bamboos*, exhibiting this balance. In this section we define bamboos and show that their yield is the least solution of a system of *linear* equations easily derivable from f. The "easy to compute" part is justified by the fact that in most semirings used in practice linear equations are far easier to solve than polynomial equations (e.g. in the real semiring or the language semiring with union and concatenation as operations). The "relevance" of bamboos is justified in the next three sections.

Definition 1. *Let f be a system of polynomials. A tree $t \in \mathcal{T}_{f,X}$ is an X-bamboo if there is a path leading from the root of t to some leaf of t, the* stem, *such that the height of every subtree of t not containing a node of the stem is at most $n - 1$. The set of all X-bamboos of f is denoted by $\mathcal{B}_{f,X}$, or just by $\mathcal{B}_X$ if f is clear from the context.*

In order to define the system of linear equations mentioned above we need the notion of differential of a system of polynomials.

Definition 2. *Let $f \in S[\mathcal{X}]$ be a polynomial and let $v \in V$ be a vector. The differential of f at v w.r.t. a variable X is the map $D_X f|_v \colon V \to S$ inductively defined as follows:*

$$D_X f|_v(a) = \begin{cases} 0 & \text{if } f \in S \text{ or } f \in \mathcal{X} \setminus \{X\} \\ a_X & \text{if } f = X \\ D_X g|_v(a) \cdot h(v) + g(v) \cdot D_X h|_v(a) & \text{if } f = g \cdot h \\ \sum_{i=1}^k D_X m_i|_v(a) & \text{if } f = \sum_{i=1}^k m_i. \end{cases}$$

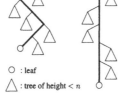

$\bigcirc$: leaf

$\triangle$: tree of height $< n$

Fig. 2. A bamboo with its stem printed bold; on the right it is shown with its stem straightened

Further, we define the differential *of f at v by $Df|_v(a) := \sum_{X \in \mathcal{X}} D_X f|_v(a)$. The differential of a system of polynomials $\boldsymbol{f}$ at v is defined componentwise by $(D\boldsymbol{f}|_v(a))_X :$
$= D(\boldsymbol{f}_X)|_v(a)$ for all $X \in \mathcal{X}$.*

Example 1. *For $f(X,Y) = a \cdot X \cdot X \cdot Y \cdot b$, $v = (v_X, v_Y)$, $c = (c_X, c_Y)$ we have:*

$$D_X f|_v(c) = a \cdot c_X \cdot v_X \cdot v_Y \cdot b + a \cdot v_X \cdot c_X \cdot v_Y \cdot b$$
$$D_Y f|_v(c) = a \cdot v_X \cdot v_X \cdot c_Y \cdot b$$

Using differentials we define a particular linearization of a polynomial system.

Definition 3. *Let $\boldsymbol{f}$ be a system of n polynomials. The* bamboo system $\boldsymbol{f}_B$ *associated to $\boldsymbol{f}$ is the linear system $\boldsymbol{f}_B(\boldsymbol{X}) = D\boldsymbol{f}|_{\boldsymbol{f}^n(0)}(\boldsymbol{X}) + \boldsymbol{f}(0)$. The least solution of the system of equations $\boldsymbol{X} = \boldsymbol{f}_B(\boldsymbol{X})$ is denoted by $\mu\boldsymbol{f}_B$.*

Now we can state the relation between bamboos and bamboo systems.

Theorem 1. *Let $\boldsymbol{f}$ be a system of polynomials over an io-semiring. For every variable X of $\boldsymbol{f}$ we have $\mathsf{Y}(\mathcal{B}_X) = (\mu\boldsymbol{f}_B)_X$, i.e., the yield of the X-bamboos is equal to the X-component of the least solution of the bamboo system.*

Together with Proposition 3 we get the following corollary.

Corollary 2 (derivation tree analysis for bamboos). *Let $\boldsymbol{f}$ be a system of polynomials over an io-semiring. If $\mathcal{B}_X$ satisfies the embedding property for all X, i.e., for all X-trees t it holds $\mathsf{Y}(t) \sqsubseteq \mathsf{Y}(\mathcal{B}_X)$, then $\mu\boldsymbol{f} = \mu\boldsymbol{f}_B$.*

5 Star-Distributive Semirings

Definition 4. *A commutative (w.r.t. multiplication) io-semiring S is* star-distributive *if $(a + b)^* = a^* + b^*$ holds for all $a, b \in S$.*

A commutative io-semiring is star-distributive whenever the natural order $\sqsubseteq$ is total:

Proposition 4. *Any totally ordered commutative io-semiring is star-distributive.*

Proof. Let w.l.o.g. $a \sqsubseteq b$. Then $(a + b)^* = b^* \sqsubseteq a^* + b^* \sqsubseteq (a + b)^*$. $\qquad\square$

In particular, the $(\min, +)$-semiring over the integers or reals is star-distributive.

We have already considered commutative idempotent semirings in [5] where we showed that μf can be computed by solving n *linear* equation systems by means of a Newton-like method, improving the $\mathcal{O}(3^n)$ bound of Hopkins and Kozen [12]. In this section we improve this result even further for star-distributive semirings: One single linear system, the bamboo system $\boldsymbol{f}_B$, needs to be solved. This leads to an efficient algorithm for computing μf in arbitrary star-distributive semirings. In Section 5.1 we instantiate this algorithm for the $(\min, +)$-semiring; in Section 5.2 we use it to improve the algorithm of [2] for computing the throughput of a context-free grammar.

We start by stating two useful properties of star-distributive semirings.

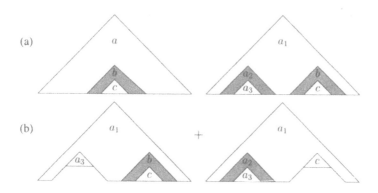

Fig. 3. "Unpumping" trees to make them bamboos

Proposition 5. *In any star-distributive semiring the following equations hold:*

(1) $a^*b^* = a^* + b^*$*, and* *(2)* $(ab^*)^* = a^* + ab^*$.

We can now state and prove our result:

Theorem 2. $\mu f = \mu f_{\mathcal{B}}$ *holds for polynomial systems f over star-distributive semirings.*

Proof Sketch (see [6] for a complete proof). The proof is by derivation tree analysis. So it suffices to discharge the precondition of Corollary 2. More precisely we show for any X-tree t that $Y(t) \sqsubseteq Y(\mathcal{B}_X)$ holds. It suffices to consider the case where t is not an X-bamboo. Then the height of t is at least n, and so t is "pumpable", i.e., one can choose a path p in t from the root to a leaf such that two different nodes on the path share the same variable-label. So t can be decomposed into three (partial) trees with yields a, b, c, respectively, such that $Y(t) = abc$, see the left side of Figure 3(a). Notice that, by commutativity of product, ab^*c is the yield of a set of trees obtained by "pumping" t. We show $ab^*c \sqsubseteq Y(\mathcal{B}_X)$ which implies $Y(t) \sqsubseteq Y(\mathcal{B}_X)$. As t is not an X-bamboo, t has a pumpable subtree disjoint from p. In this sketch we assume that it is a subtree of that part of t whose yield is a, see the right side of Figure 3(a). Now we have $a = a_1a_2a_3$, and so $ab^*c = a_1a_2a_3b^*c \sqsubseteq a_1a_2^*a_3b^*c = a_1a_3b^*c + a_1a_2^*a_3c$, where we used commutativity and Proposition 5(1) in the last step. Both summands in above sum are yields of sets of trees obtained by pumping pumpable trees smaller than t, see Figure 3(b). By an inductive argument those yields are both included in $Y(\mathcal{B}_X)$.   □

5.1   The $(\min, +)$-Semiring

Consider the "tropical" semiring $\mathcal{R} = (\mathbb{R} \cup \{-\infty, \infty\}, \wedge, +_\mathbb{R}, \infty, 0)$. By $\wedge$ resp. $+_\mathbb{R}$ we mean minimum resp. addition over the reals. Observe that the natural order $\sqsubseteq$ is the order $\geq$ on the reals.[1] As $\mathcal{R}$ is totally ordered, Proposition 4 implies that $\mathcal{R}$ is star-distributive. Assume for the rest of this section that f is a polynomial system over $\mathcal{R}$

---

[1] By symmetry, we could equivalently consider maximum instead of minimum.

of degree at most 2. We can apply Theorem 2, i.e., $\mu f = \mu f_B$ holds. This immediately suggests a polynomial algorithm to compute the least fixed-point: Compute $f^n(\infty)$ by performing n Kleene iterations, and solve the linear system $X = Df|_{f^n(\infty)}(X) \wedge f(\infty)$. The latter can be done by means of the Bellman-Ford algorithm.

Example 2. Consider the following equation system.

$$(X, \quad Y, \quad Z) = (-2 \wedge (Y +_\mathbb{R} Z), \quad Z +_\mathbb{R} 1, \quad X \wedge Y) =: f(X)$$

We have $f(\infty) = (-2, \infty, \infty)$, $f^2(\infty) = (-2, \infty, -2)$, $f^3(\infty) = (-2, -1, -2)$. The linear system $X = Df|_{f^n(\infty)}(X) \wedge f(\infty) = f_B(X)$ looks as follows:

$$(X, \quad Y, \quad Z) = (-2 \wedge (-1 +_\mathbb{R} Z) \wedge (Y +_\mathbb{R} -2), \quad Z +_\mathbb{R} 1, \quad X \wedge Y).$$

This equation system corresponds in a straightforward way to the following graph.

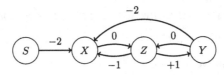

We claim that the V-component of μf_B equals the least weight of any path from S to V where $V \in \{X, Y, Z\}$. To see this, notice that $(f_B^k(\infty))_V$ corresponds to the least weight of any path from S to V of length at most k. The claim follows by Kleene's theorem. So we can compute μf_B with the Bellman-Ford algorithm. In our example, X, Y, Z are all reachable from S via a negative cycle, so $\mu f_B = (-\infty, -\infty, -\infty)$. By Theorem 2, $\mu f = \mu f_B = (-\infty, -\infty, -\infty)$.   □

The Bellman-Ford algorithm can be used here as it handles negative cycles correctly. The overall runtime of our algorithm to compute μf is dominated by the Bellman-Ford algorithm. Its runtime is in $\mathcal{O}(n \cdot m)$, where m is the number of monomials appearing in f. We conclude that our algorithm has the same asymptotic complexity as the "generalized Bellman-Ford" algorithm of [9]. It is by a factor of n faster than the algorithm deducible from [5] because our new algorithm uses the Bellman-Ford algorithm only once instead of n times.

5.2   Throughput of Grammars

In [2], a polynomial algorithm for computing the *throughput* of a context-free grammar was given. Now we show that the algorithm can be both simplified and accelerated by computing least fixed-points according to Theorem 2.

Let us define the problem following [2]. Let Σ be a finite alphabet and $\rho : \Sigma \to \mathbb{N}$ a weight function. We extend ρ to words $a_1 \cdots a_k \in \Sigma^*$ by setting $\rho(a_1 \cdots a_k) := \rho(a_1) + \ldots + \rho(a_k)$.[2] The mean weight of a non-empty word w is defined as $\overline{\rho}(w) := \rho(w)/|w|$. The throughput of a non-empty language $L \subseteq \Sigma^+$ is defined as the infimum of the mean weights of the words in L: $tp(L) := \inf\{\overline{\rho}(w) \mid w \in L\}$. Let

---

[2] We write $+$ for the addition of reals in this section.

$G = (\Sigma, \mathcal{X}, P, S)$ be a context-free grammar and $L = L(G)$ its language. The problem is to compute $tp(L)$. As in [2] we assume that G has at most 2 symbols on the right hand side of every production and that L is non-empty and contains only non-empty words.

Note that we cannot simply construct a polynomial system having $tp(L)$ as its least fixed-point, as the throughput of two non-terminals is not additive. In [2] an ingenious algorithm is proposed to avoid this problem. Assume we already know a routine, the *comparing routine*, that decides for a given $t \in \mathbb{Q}$ whether $tp(L) \geq t$ holds. Assume further that this routine has $\mathcal{O}(N^k)$ time complexity for some k. Using the comparing routine we can approximate $tp(L)$ up to any given accuracy by means of binary search. Let $d = \max_{a \in \Sigma} \rho(a) - \min_{a \in \Sigma} \rho(a)$. A dichotomy result of [2] shows that $\mathcal{O}(N + \log d)$ iterations of binary search suffice to approximate $tp(L)$ up to an ε that allows to compute the exact value of $tp(L)$ in time $\mathcal{O}(N^3)$. This is proved by showing that, once a value t has been determined such that $t - \varepsilon < tp(L) \leq t$, one can:

- transform G in $\mathcal{O}(N^3)$ time into a grammar G' of size $\mathcal{O}(N^3)$ generating a finite language, and having the same throughput as G (this construction does not yet depend on $tp(L)$);
- compute the throughput of G' in linear time in the size of G', i.e., in $\mathcal{O}(N^3)$ time.

The full algorithm for the throughput runs then in $\mathcal{O}(N^k(N + \log d)) + \mathcal{O}(N^3)$ time.

The algorithm of [2] and our new algorithm differ in the comparing routine. In the routine of [2] the transformation of G into the grammar G' is done *before* $tp(L)$ has been determined. Then a linear time algorithm can be applied to G' to decide whether $tp(L) \geq t$ holds. (This algorithm does not work for arbitrary context-free grammars, and that is why one needs to transform G into G'.) Since G' has size $\mathcal{O}(N^3)$, the comparing routine has $k = 3$, and so the full algorithm runs in $\mathcal{O}(N^4 + N^3 \log d)$ time.

We give a more efficient comparing routine with $k = 2$. Given a $t \in \mathbb{Q}$, assign to each word $w \in \Sigma^+$ its *throughput balance* $\sigma_t(w) = \rho(w) - |w| \cdot t$. Notice that $\sigma_t(w) \geq 0$ if and only if $\overline{\rho}(w) \geq t$. Further, for two words w, u we now have $\sigma_t(wu) = \sigma_t(w) + \sigma_t(u)$. So we can set up a polynomial system $\boldsymbol{X} = \boldsymbol{f}(\boldsymbol{X})$ over the tropical semiring $\mathcal{R}$ where $\boldsymbol{f}$ is constructed such that each variable $X \in \mathcal{X}$ in the equation system corresponds to the minimum (infimum) throughput balance of the words derivable from X. More formally, define a map m by setting $m(a) = \rho(a) - t$ for $a \in \Sigma$ and $m(X) = X$ for $X \in \mathcal{X}$. Extend m to words in $(\Sigma \cup \mathcal{X})^*$ by setting $m(\alpha_1 \cdots \alpha_k) = m(\alpha_1) + \cdots + m(\alpha_k)$. Let P_X be the productions of G with X on the left hand side. Then set $\boldsymbol{f}_X(\boldsymbol{X}) := \bigwedge_{(X \to w) \in P_X} m(w)$. For instance, if P_X consists of the rules $X \to aXY$ and $X \to bZ$, we have $\boldsymbol{f}_X(\boldsymbol{X}) = \rho(a) - t + X + Y \wedge \rho(b) - t + Z$.

It is easy to see that the relevant solution of the system $\boldsymbol{X} = \boldsymbol{f}(\boldsymbol{X})$ is the least one w.r.t. $\sqsubseteq$, i.e., $(\mu \boldsymbol{f})_S \geq 0$ if and only if $tp(L) \geq t$. So we can use the algorithm from Section 5.1 as our comparing routine. This takes time $\mathcal{O}(N^2)$ where N is the size of the grammar. With that comparing routine we obtain an algorithm for computing the throughput with $\mathcal{O}(N^3 + N^2 \log d)$ runtime.

6 Lossy Semirings

Definition 5. *An io-semiring S is called* lossy *if $1 \sqsubseteq a$ holds for all $a \neq 0$.*

Note that by definition of natural order the requirement $1 \sqsubseteq a$ is equivalent to $a = a+1$. In the free semiring generated by a finite alphabet Σ, and augmented by the equation $a = a + 1$ ($a \in S \setminus \{0\}$), every language $L \subseteq \Sigma^*$ is "downward closed", i.e. for every word $w = a_1 a_2 \ldots a_l \in L$ all possible subwords $\{a_1' a_2' \ldots a_l' \mid a_i' \in \{\varepsilon, a_i\}\}$ are also included in L. By virtue of Higman's lemma [11] the downward-closure of a context-free language is regular. This has been used in [1] for an efficient analysis of systems with unbounded, lossy FIFO channels. Downward closure was used there to model the loss of messages due to transmission errors.

We say that a system f of polynomials is *clean* if $\mu f_X \neq 0$ for all $X \in \mathcal{X}$. Every system can be *cleaned* in linear time by removing the equations of all variables X such that $\mu f_X = 0$ and setting these variables to 0 in the other equations (the procedure is similar to the one that eliminates non-productive variables in context-free grammars). We consider only clean systems, and introduce a normal form for them.

Definition 6. *Let $f \in S[\mathcal{X}]^{\mathcal{X}}$ be a system of polynomials over a lossy semiring. f is in* quadratic normal form *if every polynomial f_X has the form*

$$c + \sum_{Y, Z \in \mathcal{X}} a_{Y,Z} \cdot Y \cdot Z + \sum_{Y \in \mathcal{X}} b_{l,Y} \cdot Y \cdot b_{r,Y}$$

where (i) $c \in S \setminus \{0\}$, (ii) $a_{Y,Z} \in \{0,1\}$, and (iii) if $\sum_{Z \in \mathcal{X}} a_{Y,Z} \neq 0$, then $b_{l,Y} \neq 0 \neq b_{r,Y}$ for all $Y, Z \in \mathcal{X}$.

Lemma 1. *For every clean $g \in S[\mathcal{X}]^{\mathcal{X}}$ we can construct in linear time a system $f \in S[\mathcal{X}']^{\mathcal{X}'}$ in quadratic normal form, where $\mathcal{X} \subseteq \mathcal{X}'$ and $\mu g_X = \mu f_X$ for all $X \in \mathcal{X}$.*

Proof Sketch. Note that, as g is clean, we have $1 \sqsubseteq \mu g$. Hence, requirement (i) is no restriction. The transformation that normalizes a system is similar to the one that brings a context-free grammar into Chomsky normal-form (CNF). The superset $\mathcal{X}' \supseteq \mathcal{X}$ results from the introduction of new variables by this transformation into CNF. □

Our main result in this section is that for *strongly connected* systems f in quadratic normal form we again have that $\mu f = \mu f_B$. We then show how this result leads to an algorithm for arbitrary systems.

Given two variables $X, Y \in \mathcal{X}$, we say that X depends on Y (w.r.t. f) if Y occurs in a monomial of f_X or there is a variable Z such that X depends on Z and Z depends on Y. The system f is *strongly connected* if X depends on Y for all variables X, Y.

Theorem 3. *$\mu f = \mu f_B$ holds for strongly connected polynomial systems f in quadratic normal form over lossy semirings.*

We again use derivation tree analysis to show that every derivation tree t can be transformed into a bamboo subsuming the yield of t, see [6] for details.

Because of the preceding theorem, given a strongly connected system f, we may use the linear system $f_B(X) = f(0) + Df|_{f^n(0)}(X)$ for calculating μf. As f is strongly connected, f_B is also strongly connected. The least fixed-point of such a strongly connected linear system f_B is easily calculated: all non-constant monomials appearing in f_B have the form $b_l X b_r$ for some $X \in \mathcal{X}$, and $b_l, b_r \in S \setminus \{0\}$. As f_B is strongly connected, every polynomial $(f_B)_Y$ is substituted for Y in $(f_B)_X$ again and again when calculating the Kleene sequence $(f_B^k(0))_{k \in \mathbb{N}}$. So, let l be the sum of all left-handed coefficients b_l (appearing in *any* f_X), and similarly define r. We then have $(\mu f_B)_X = l^* \left(\sum_{Y \in \mathcal{X}} f_Y(0) \right) r^*$ for all $X \in \mathcal{X}$.

If f is not strongly connected, we first decompose f into strongly connected subsystems, and then we solve these systems bottom-up. Note that substituting the solutions from underlying SCCs into a given SCC leads to a new system in normal form. As there are at most $n = |\mathcal{X}|$ many strongly connected components for a given system $f \in S[\mathcal{X}]^{\mathcal{X}}$, we obtain the following theorem which was first stated explicitly for context-free grammars in [3].

Theorem 4. *The least fixed-point μf of a polynomial system f over a lossy semiring is representable by regular expressions over S. If f is in normal form μf can be calculated solving at most n bamboo systems.*

7   1-Bounded Semirings

Definition 7. *An io-semiring S is called 1-bounded if $a \sqsubseteq 1$ holds for all $a \in S$.*

Natural examples are the tropical semiring over the natural numbers $(\mathbb{N} \cup \{\infty\}, \wedge, +, \infty, 0)$ and the "maximum-probability" semiring $([0,1], \vee, \cdot, 0, 1)$, where $\wedge$ and $\vee$ denote minimum and maximum, respectively. Notice that any commutative 1-bounded semiring is star-distributive (as $a^* = 1$ for all a), but not all 1-bounded semirings have commutative multiplication. Consider for example the semiring of those languages L over Σ that are *upward-closed*, i.e., $w \in L$ implies $u \in L$ for all u such that w is a subword of u. This semiring is 1-bounded and has Σ^* as 1-element. Upward-closed languages form a natural dual to downward-closed languages from the previous section. We show that μf can be computed very easily in the case of 1-bounded semirings:

Theorem 5. $\mu f = f^n(0)$ *holds for polynomial systems over 1-bounded semirings.*

Proof Sketch. Recall that, by Proposition 2, we have $(f^n(0))_X = \mathsf{Y}(\mathcal{T}_X^{(n-1)})$, where $\mathcal{T}_X^{(n-1)}$ contains all X-trees of height at most $n - 1$. We proceed by derivation tree analysis, i.e., by discharging the precondition of Proposition 3. So it suffices to show that for any X-tree t there is an X-tree t' of height at most $n - 1$ with $\mathsf{Y}(t) \sqsubseteq \mathsf{Y}(t')$. Such a tree t' can be constructed by pruning t as long as some variable label occurs more than once along any path. $\square$

Theorem 5 appears to be rather easy from our point of view, i.e., from the point of view of derivation trees. However, even this simple result has very concrete applications in the domain of interprocedural program analysis [14]. The main algorithms of [14], the so-called *post** and *pre** algorithms, can be seen as solvers of fixed-point equations over *bounded* semirings, which are semirings that do not have infinite ascending chains. Those solvers are based on Kleene's iteration and the complexity result given there depends on the maximal length of ascending chains in the semiring (cf. [14], page 28). Such a bound may not exist, and does not exist for the tropical semiring over the natural numbers $(\mathbb{N} \cup \{\infty\}, \wedge, +, \infty, 0)$ which is considered as an example in [14], pages 13 and 18. However, Theorem 5 can be applied to this semiring, which shows that the program analysis algorithms of [14] applied to 1-bounded semirings are polynomial-time algorithms, independent of the length of chains in the semiring.

8 Conclusion

We have shown that derivation tree analysis, a proof technique first introduced in [5], is an efficient tool for the design of efficient fixed-point algorithms on io-semirings. We have considered three classes of io-semirings with applications to language theory and verification. We have shown that for star-distributive semirings and lossy semirings the least fixed-point of a polynomial system of equations is equal to the least fixed-point of a *linear system*, the bamboo system. This improves the results of [5]: The generic algorithm given there requires to solve N different systems of linear equations in the star-distributive case (where N is the original number of polynomial equations), and is not applicable to the lossy case.

We have used our results to design an efficient fixed-point algorithm for the $(\min, +)$-semiring. In turn, we have applied this algorithm to provide a cubic algorithm for computing the throughput of a context-free language, improving the $\mathcal{O}(N^4)$ upper bound obtained by Caucal et al. in [2].

For lossy semirings, derivation tree analysis based on bamboos has led to an algebraic generalization of a result of Courcelle stating that the downward-closure of a context-free language is effectively regular. Finally we have used derivation tree analysis to derive a simple proof that $\mu f = f^n(0)$ holds for 1-bounded semirings, with some applications in interprocedural program analysis.

References

1. Abdulla, P.A., Bouajjani, A., Jonsson, B.: On-the-fly analysis of systems with unbounded, lossy FIFO channels. In: Y. Vardi, M. (ed.) CAV 1998. LNCS, vol. 1427, pp. 305–318. Springer, Heidelberg (1998)
2. Caucal, D., Czyzowicz, J., Fraczak, W., Rytter, W.: Efficient computation of throughput values of context-free languages. In: Holub, J., Žďárek, J. (eds.) CIAA 2007. LNCS, vol. 4783, pp. 203–213. Springer, Heidelberg (2007)
3. Courcelle, B.: On constructing obstruction sets of words. EATCS Bulletin 44, 178–185 (1991)
4. Esparza, J., Kiefer, S., Luttenberger, M.: An extension of Newton's method to ω-continuous semirings. In: Harju, T., Karhumäki, J., Lepistö, A. (eds.) DLT 2007. LNCS, vol. 4588, pp. 157–168. Springer, Heidelberg (2007)
5. Esparza, J., Kiefer, S., Luttenberger, M.: On fixed point equations over commutative semirings. In: STACS 2007. LNCS, vol. 4397, pp. 296–307. Springer, Heidelberg (2007)
6. Esparza, J., Kiefer, S., Luttenberger, M.: Derivation tree analysis for accelerated fixed-point computation. Technical report, Technische Universität München (2008)

7. Esparza, J., Kučera, A., Mayr, R.: Model checking probabilistic pushdown automata. Logical Methods in Computer Science (2006)
8. Etessami, K., Yannakakis, M.: Recursive Markov chains, stochastic grammars, and monotone systems of nonlinear equations. In: Diekert, V., Durand, B. (eds.) STACS 2005. LNCS, vol. 3404, pp. 340–352. Springer, Heidelberg (2005)
9. Gawlitza, T., Seidl, H.: Precise fixpoint computation through strategy iteration. In: De Nicola, R. (ed.) ESOP 2007. LNCS, vol. 4421, pp. 300–315. Springer, Heidelberg (2007)
10. Harris, T.E.: The Theory of Branching Processes. Springer, Heidelberg (1963)
11. Higman, G.: Ordering by divisibility in abstract algebras. Proc. London Math. Soc. 2 (1952)
12. Hopkins, M.W., Kozen, D.: Parikh's theorem in commutative Kleene algebra. In: LICS 1999 (1999)
13. Nielson, F., Nielson, H.R., Hankin, C.: Principles of Program Analysis. Springer, Heidelberg (1999)
14. Reps, T., Schwoon, S., Jha, S., Melski, D.: Weighted pushdown systems and their application to interprocedural dataflow analysis. Science of Computer Programming 58(1–2), 206–263 (2005); Special Issue on the Static Analysis Symposium 2003

Tree Automata with Global Constraints

Emmanuel Filiot[1], Jean-Marc Talbot[2], and Sophie Tison[1]

[1] INRIA Lille - Nord Europe, Mostrare Project,
University of Lille 1 (LIFL, UMR 8022 of CNRS)
[2] University of Provence (LIF, UMR 6166 of CNRS), Marseille

Abstract. A tree automaton with global tree equality and disequality constraints, TAGED for short, is an automaton on trees which allows to test (dis)equalities between subtrees which may be arbitrarily faraway. In particular, it is equipped with an (dis)equality relation on states, so that whenever two subtrees t and t' evaluate (in an accepting run) to two states which are in the (dis)equality relation, they must be (dis)equal. We study several properties of TAGEDs, and prove decidability of emptiness of several classes. We give two applications of TAGEDs: decidability of an extension of Monadic Second Order Logic with tree isomorphism tests and of unification with membership constraints. These results significantly improve the results of [10].

1 Introduction

The emergence of XML has strengthened the interest in tree automata, as it is a clean and powerful model for XML tree acceptors [19,20]. In this context, tree automata have been used, for example, to define schemas, and queries, but also to decide tree logics, to type XML transformations, and even to learn queries. However, it is known that sometimes, expressiveness of tree automata is not sufficient. This is the case for instance in the context of non-linear rewriting systems, for which more powerful tree acceptors are needed to decide interesting properties of those rewrite systems. For example, the set of ground instances of $f(x, x)$ is not regular.

Tree automata with constraints have been introduced to overcome this lack of expressiveness [3,9,12,13]. In particular, the transitions of these tree automata are fired as soon as the subtrees of the current tree satisfy some structural (dis)equality. But typically, these constraints are kept local to preserve decidability of emptiness and good closure properties – in particular, tests are performed between siblings or cousins –. In the context of XML, and especially to define tree patterns, one need global constraints. For instance, it might be useful to represent the set of ground instances of the pattern $X(\mathtt{author}(x), \mathtt{author}(x))$, where X is a binary context variable, and x is an author which occurs at least twice (we assume this pattern to be matched against XML trees representing a bibliography). In this example, the two subtrees corresponding to the author might be arbitrarily faraway, making the tree equality tests more global. Patterns might be more complex, by the use of negation (which allow to test tree disequalities), Boolean operations, and regular constraints on variables. In [10], we study the spatial logic TQL, which in particular, allows to define such patterns. We proved decidability

M. Ito and M. Toyama (Eds.): DLT 2008, LNCS 5257, pp. 314–326, 2008.

of a powerful fragment of this logic, by reduction to emptiness test of a new class of tree automata, called *Tree Automata with Global Equalities and Disequalities* (TAGEDs for short). These are tree automata A equipped with an equality $=_A$ and a disequality $\neq_A$ relation on (a subset of) states. A tree t is accepted by A if there is a computation of A on t which leads to an accepting state, and whenever two subtrees of t evaluate to two states q_1, q_2 such that $q_1 =_A q_2$ (resp. $q_1 \neq_A q_2$), then these two subtrees must be structurally equal (resp. structurally different). TAGEDs may be interesting on their own, since they are somehow orthogonal to usual automata with constraints [3]. Indeed, if we view equality tests during computations as an equivalence relation on a subset of nodes (two nodes being equivalent if their rooted subtrees are successfully tested to be equal), in the former, there are a bounded number of equivalence classes of unbounded cardinality, while in the latter, there might be an unbounded number of equivalence classes of bounded cardinality.

The main result of [10] was decidability of emptiness of a subclass of TAGEDs, called bounded TAGEDs, which allow only a bounded number (by some constant independent of the tree) of (dis)equality tests on the run. In this paper, we prove several properties of TAGED-definable languages (closure by union and intersection, non-closure by complement). We prove results on TAGEDs (non-determinization, undecidability of universality). The other main results are decidability of emptiness of several classes of TAGEDs which significantly improves the result of [10], and uses different and simpler techniques. In particular, we prove a pumping lemma for TAGEDs which performs a bounded number of disequality tests along paths (and arbitrarily many equality tests).

We give two applications of TAGEDs. The first is decidability of an extension of MSO with tree isomorphism tests. The second application concerns a first-order disunification problems, with (regular) membership constraints. Dealing with membership constraints has been done in several papers. In [8], the authors prove solvability of first-order formulas whose atoms are either equations between terms or membership constraints $t \in L$ where L is a regular tree language. In [15], the authors propose an algorithm to solve iterated matching of hedges against terms with flexible arity symbols, one-hole context and sequence variables constrained to range over a regular language. In this paper, we extend the logic of [8] with context variables (with arbitrarily many holes, and membership constraints) to allow arbitrary depth matching. Context unification is still an open problem, but motivated by XML tasks, we do not need to do full context unification. We prefer to impose a strong linearity condition on context variables. We prove that, even with this restriction, solvability of first-order formulas over these atoms is undecidable. We introduce an existential fragment for which satisfiability is decidable by reduction to emptiness of a class of TAGEDs.

Related Work. Extensions of tree automata which allow for syntactic equality and disequality tests between subterms have already been defined by adding constraints to the rules of automata. E.g., adding the constraint $1.2 = 2$ to a rule means that one can apply the rule at position π only if the subterm at position $\pi.1.2$ is equal to the subterm at position $\pi.2$. Testing emptiness of the recognized language is undecidable in general [17] but two classes with a decidable emptiness problem have been emphasized. In the first class, automata are deterministic and the number of equality tests along a path is bounded [9] whereas the second restricts tests to sibling subterms [3]. This latter class has recently been extended to unranked trees [13], the former one has been extended to equality modulo equational theories [12]. But, contrarily to TAGEDs, in all these classes, tests are performed locally, typically between sibling or cousin subterms. Automata with local and global equality tests, using one memory, have been considered in [6]. Their emptiness problem is decidable, and they can simulate positive TAGEDs (TAGEDs performing only equality tests) which use at most one state per runs to test equalities. Finally, automata for DAGs are studied in [2,4], they cannot be compared to positive TAGEDs, as they run on DAG representations of trees (with maximal sharing), and in TAGEDs, we cannot impose every equal subtrees to evaluate in the same state in a successful run.

We only sketch the proofs, but all the missing proofs are in the full paper version [23].

2 Trees and TAGED

Binary trees. We start from a ranked alphabet Σ ranged over binary symbols f and constant symbols a. A *binary tree* t is a ground term over Σ. The set of binary trees over Σ is denoted by T_Σ. The set of *nodes* of a tree $t \in T_\Sigma$, denoted by N_t, is defined inductively as a set of words over $\{1, 2\}$ by: $N_a = \{\epsilon\}$ and $N_{f(t_1,t_2)} = \{\epsilon\} \cup \{\alpha.u \mid \alpha \in \{1, 2\}, u \in N_{t_\alpha}\}$ (ϵ denotes the empty word and . the concatenation). For any tree t, and any node $u \in N_t$, we define the subtree at node u, denoted by $t|_u$, inductively by: $t|_\epsilon = t$, $f(t_1, t_2)|_{\alpha.u} = t_\alpha|_u$, $\alpha \in \{1, 2\}$. Note that we have $N_{t|_u} = \{v \mid u.v \in N_t\}$. We also denote by $O_t(u) \in \Sigma$ the label of node u in t. Finally, we denote by $\lhd$ the strict descendant relation between nodes. Hence, for all $u, v \in N_t$, $u \lhd v$ if u is a prefix of v (therefore the root is minimal for $\lhd$).

Tree Automata. We define tree automata on binary trees, but the reader may refer to [7] for more details. A *tree automaton* is a 4-tuple $A = (\Sigma, Q, F, \Delta)$ where Q is a finite set of states, $F \subseteq Q$ is a set of final states, and Δ is a set of rules of the form $a \to q$ and $f(q_1, q_2) \to q$, where f is a binary function symbol, a a constant, and q_1, q_2, q are states from Q. A *run* of A on a tree t is a tree r over Q such that: (i) $N_r = N_t$, (ii) for all leaves $u \in N_r$, we have $O_t(u) \to O_r(u) \in \Delta$, and (iii) for all inner-nodes $u \in N_r$, we have $O_t(u)(O_r(u.1), O_r(u.2)) \to O_r(u) \in \Delta$. A run r is *successful* if $O_r(\epsilon) \in F$. The *language recognized* (or defined) by A, denoted $L(A)$, is the set of trees t for which there exists a successful run of A.

We consider binary and constant symbols only, but the two definitions above can be easily extended to symbols of other arity (in particular, we use unary symbols in several proofs and examples).

Example 1. Let Σ_b be the alphabet consisting of the two binary symbols $\land, \lor$, the unary symbol $\neg$, and the two constant symbols $0, 1$. Trees from T_{Σ_b} represents Boolean formulas. We define an automaton on Σ_b which accepts only Boolean formulas logically equivalent to 1. Its set of states (resp. final states) is defined by $Q_b = \{q_0, q_1\}$ (resp. $F_b = \{q_1\}$), and its set of rules Δ_b by, for all $b, b_1, b_2 \in \{0, 1\}$, and all $\oplus \in \{\land, \lor\}$:

$$b \to q_b \qquad \neg(q_b) \to q_{\neg b} \qquad \oplus(q_{b_1}, q_{b_2}) \to q_{b_1 \oplus b_2} .$$

Definition 1 (TAGED). *A TAGED is a 6-tuple* $A = (\Sigma, Q, F, \Delta, =_A, \neq_A)$ *such that:*
- (Σ, Q, F, Δ) *is a tree automaton;*
- $=_A$ *is a reflexive and symmetric binary relation on a subset of* Q;
- $\neq_A$ *is an irreflexive and symmetric binary relation on* Q.

A TAGED A *is said to be positive (resp. negative) if* $\neq_A$ *(resp.* $=_A$*) is empty.*

The notion of successful run differs from tree automata as we add equality and disequality constraints. A *run* r of the tree automaton (Σ, Q, F, Δ) on a tree t satisfies the equality constraints if $\forall u, v \in N_t$, $O_r(u) =_A O_r(v) \Rightarrow t|_u = t|_v$. Similarly, it satisfies the disequality constraints if $\forall u, v \in N_t$, $O_r(u) \neq_A O_r(v) \Rightarrow t|_u \neq t|_v$.

A run is *successful* (or *accepting*) if it is successful for the tree automaton (Σ, Q, F, Δ) and if it satisfies the constraints. The language accepted by A, denoted $L(A)$, is the set of trees t having a successful run for A. We denote by $\mathrm{dom}(=_A)$ the domain of $=_A$, i.e. $\{q \mid \exists q' \in Q, q =_A q'\}$. The set $\mathrm{dom}(\neq_A)$ is defined similarly. Finally, two TAGEDs are *equivalent* if they accept the same language.

In [10], we introduced the class of bounded TAGED, where in successful runs, the number of occurrences of states from $\mathrm{dom}(=_A) \cup \mathrm{dom}(\neq_A)$ is bounded by some fixed $k \in \mathbb{N}$. We proved emptiness to be decidable for that class. The classes we consider in this paper are either incomparable or strictly more expressive. All the results from Section 3 also hold for bounded TAGED. Note also that TAGED are strictly more expressive than tree automata, as illustrated by the next example.

Example 2. Let $Q = \{q, q_=, q_f\}$, $F = \{q_f\}$, and let Δ be defined as the set of following rules: $a \to q$, $a \to q_=$, $f(q, q) \to q$, $f(q, q) \to q_=$, $f(q_=, q_=) \to q_f$, for all $a, f \in \Sigma$. Let the positive TAGED $A_1 = (\Sigma, Q, F, \Delta, \{q_= =_{A_1} q_=\})$. Then $L(A_1)$ is the set $\{f(t, t) \mid f \in \Sigma, t \in T_\Sigma\}$, which is known to be non regular [7].

Example 3. Let $\mathcal{X}$ be a finite set of variables. We now define a TAGED A_{sat} which accepts tree representations of satisfiable Boolean formulas with free variables $\mathcal{X}$. We let $A_b = (\Sigma_b, Q_b, F_b, \Delta_b)$ be the automaton defined in Example 1. Every variable is viewed as a binary symbol, and we let $\Sigma_\mathcal{X} = \Sigma_b \cup \mathcal{X}$. Every Boolean formula is naturally viewed as a tree, except for variables $x \in \mathcal{X}$ which are encoded as trees $x(0, 1)$ over $\Sigma_\mathcal{X}$. For instance, the formula $(x \wedge y) \vee \neg x$ is encoded as the tree $\vee(t_1, t_2)$, where $t_1 = \wedge(x(0, 1), y(0, 1))$ and $t_2 = \neg(x(0, 1))$.

Now, we let $Q = Q_b \cup \{q_x \mid x \in \mathcal{X}\} \cup \{p_0, p_1\}$, for two fresh states p_0, p_1, and $F = F_b$. The idea is to choose non-deterministically to evaluate the leaf 0 or 1 below x to q_x, but not both, for all $x \in \mathcal{X}$. This means that we affect 0 or 1 to a particular occurrence of x. Then, by imposing that every leaf evaluated to q_x are equal, we can ensure that we have chosen to same Boolean value for all occurrences of x, for all $x \in \mathcal{X}$. This can be done with the set of rules Δ_b extended with the following rules, for all $b \in \{0, 1\}$ and all $x \in \mathcal{X}$:

$$b \to p_b \qquad b \to q_x \qquad x(p_0, q_x) \to q_1 \qquad x(q_x, p_1) \to q_0.$$

Finally, for all $x \in \mathcal{X}$, we let $q_x =_{A_{sat}} q_x$.

The *(uniform) membership problem* is "given a TAGED A, given a tree t, does t belong to $L(A)$?". We can prove the following:

Proposition 1. *Membership is NP-complete for TAGED.*

Proof. Example 3 gives a polynomial reduction of SAT to membership of TAGEDs. To show it is in NP, it suffices to guess a labeling of the nodes of the tree by states, and then to verify that it is a run, and that equality and disequality constraints are satisfied. This can be done in linear time both in the size of the automaton and of the tree.   □

3   Closure Properties of TAGEDs and Decision Problems

In this section, we prove closure properties of TAGED-definable languages.

Proposition 2 (Closure by union and intersection). *TAGED-definable languages are closed by union and intersection.*

Proof. Let $A = (\Lambda, Q, F, \Delta, =_A, \neq_A)$ and $A' = (\Lambda, Q', F', \Delta', =_{A'}, \neq_{A'})$ be two TAGEDs. Wlog, we suppose that $Q \cap Q' = \emptyset$. A TAGED accepting $L(A) \cup L(A')$ is defined by $A \cup A' = (\Lambda, Q \cup Q', F \cup F', \Delta \cup \Delta', =_A \cup =_{A'}, \neq_A \cup \neq_{A'})$.

For the closure by intersection, we use the usual product construction $A \times A'$ [7], whose set of final states if $F \times F'$. State equality $=_{A \times A'}$ is defined by $\{((q, q'), (p, p')) \mid$ *breakq* $=_A$ p or $q =_{A'} p\}$, while $\neq_{A \times A'}$ is defined by $\{((q, q'), (p, p')) \mid q \neq_A$ p or $q \neq_{A'} p\}$.   □

Prop 2 also holds for the class of languages defined by positive or negative TAGEDs. A TAGED is *deterministic* if all rules have different left-hand sides (hence there is at most one run per tree). For a deterministic TAGED A, we can prove that one can compute a non-deterministic TAGED accepting the complement of $L(A)$: we have to check if the tree evaluates in a non-accepting state or in an accepting state but in this case we non-deterministically guess a position where a constraint is not satisfied. However:

Proposition 3. *TAGEDs are not determinizable.*

Proof. Let $\Sigma = \{f, a\}$ an alphabet where f is binary and a a constant. Consider the language $L_0 = \{f(t, t) \mid t \in T_\Sigma\}$ of Example 2. It is obvious that L_0 is definable by a non-deterministic (bounded) TAGED. Suppose that there is a deterministic TAGED $A = (\Sigma, Q, F, \Delta, =_A, \neq_A)$ such that $L(A) = L_0$. Let t be a tree whose height is strictly greater than $|Q|$. Since $f(t, t) \in L_0$, there are a successful run $q_f(r, r)$ of A on $f(t, t)$ for some final state q_f, two nodes u, v and a state $q \in Q$ such that $t|_v$ is a strict subtree of $t|_u$, and $O_r(u) = O_r(v) = q$. Since $f(t|_u, t|_u) \in L_0$, and A is deterministic, there is a final state $q'_f \in F$ and a rule $f(q, q) \to q'_f \in \Delta$. Hence $q'_f(r|_u, r|_v)$ is a run of A on $f(t|_u, t|_v)$. Since $q_f(r, r)$ satisfies the constraints, $q'_f(r|_u, r|_v)$ also satisfies the constraints. Hence $f(t|_u, t|_v) \in L(A)$, which contradicts $t|_u \neq t|_v$.   □

Proposition 3 is not surprising, since:

Proposition 4. *The class of TAGED-definable languages is not closed by complement.*

Proof. (Sketch) We exhibit a tree language whose complement is easily definable by a TAGED, but which is not TAGED-definable. This language is the union of sets T_n, for all $n \in \mathbb{N}$, where $T_n = \{f(g(t,t),t') \mid t \in T_\Sigma, t' \in T_{n-1}\}$, and $T_0 = \{a\}$. To check whether a tree is in T_n, a TAGED would have to perform n equality tests, for each subtree rooted by g. This would require n states. This is only an intuition. The proof is a bit more complicated as the TAGED could also perform inequality tests.   □

We end up this section with an undecidability result:

Proposition 5. *Testing universality of TAGEDs is undecidable.*

Proof. (Sketch) We adapt the proof of [17] for undecidability of emptiness of classical tree automata with equality constraints. We start from an instance of the Post Correspondence Problem (PCP). We encode the set of solutions of PCP as a tree language whose complement is easily definable by a TAGED. Hence, the complement is universal iff PCP has no solution.   □

Even if TAGEDs are not determinizable, we can assume that testing an equality between subtrees can be done using the same state, as stated by the following lemma:

Lemma 1. *Every TAGED A is equivalent to a TAGED A' (whose size might be exponential in the size of A) such that $=_{A'} \subseteq id_{Q_{A'}}$, where $id_{Q_{A'}}$ is the identity relation on $Q_{A'}$. Moreover, A' can be built in exponential time (and may have exponential size).*

Proof. (Sketch) Intuitively, we can view an accepting run r of A on a tree t as a DAG structure. Let $U \subseteq N_t$ such that all subtrees $t|_u$, $u \in U$, have been successfully tested equal by A in the run r (i.e. $\forall u, v \in U$, $O_r(u) =_A O_r(v)$). Let $t_0 = t|_u$, for some $u \in U$. We replace all nodes of U by a single node u_0 which enroots t_0. The parent of any node of U points to u_0. We maximally iterate this construction to get the DAG. Note that this DAG is not maximal sharing[1], since only subtrees which have been successfully tested to be equal are shared. We construct A' such that it simulates a run on this DAG, obtained by overlapping the runs on every equal subtrees for which a test has been done.   □

4   Emptiness of Positive and Negative TAGEDs

In this section we prove decidability of emptiness of positive and negative TAGEDs respectively. For positive TAGEDs, it uses Lemma 1, and the classical reachability method for tree automata. For negative TAGEDs, we reduce the problem to testing satisfiability of set constraints.

Theorem 1. *Testing emptiness of positive TAGEDs is EXPTIME-complete.*

---

[1] There might be two isomorphic subgraphs occurring at different positions.

Proof. **upper bound.** Let A be a positive TAGED such that its equality relation is a subset of the identity relation (otherwise we transform A modulo an exponential blow-up, thanks to Lemma 1). Let A^- be its associated tree automaton (i.e. A without the constraints). We have $L(A) \subseteq L(A^-)$.

Then it suffices to apply a slightly modified version of the classical reachability method used to test emptiness of a tree automaton [7]. In particular, we can make this procedure associate with any state q a unique tree t_q. When a new state is reached, it can possibly activate many rules $f(q_1, q_2) \rightarrow q$ whose rhs are the same state q. The algorithm has to make a choice between this rules in order to associate a unique tree $t_q = f(t_{q_1}, t_{q_2})$ to q. This choice can be done for instance by giving an identifier to each rule and choosing the rule with the least identifier.

If $L(A^-)$ is empty, then $L(A)$ is also empty. If $L(A^-)$ is non-empty, we get a tree t and a run r which obviously satisfies the equality constraints, since a state q such that $q =_A q$ is mapped to unique tree t_q (if q is reachable).

lower bound. We reduce the problem of testing emptiness of the intersection of n tree automata $A_1, \ldots, A_n$ (see [7]), which is known to be EXPTIME-complete. We assume that their sets of states are pairwise disjoint ($Q_i \cap Q_j = \varnothing$ whenever $i \neq j$), and for all $i = 1, \ldots, n$, A_i has exactly one final state q_{f_i}, and q_{f_i} does not occur in lhs of rules of A_i (otherwise we slightly modify A_i, modulo a factor 2 in the size of A_i). We let $L = \{f(t_1, \ldots, t_n) \mid f \in \Sigma, \forall i, t_i \in L(A_i), \forall i, j, t_i = t_j\}$. It is clear that L is empty iff $L(A_1) \cap \ldots L(A_n)$ is empty. It is not difficult to construct a TAGED A (with $|A| = O(\sum_i |A_i|)$), such that $L = L(A)$: it suffices to take the union of $A_1, \ldots, A_n$ and to add the rule $f(q_{f_1}, \ldots, q_{f_n}) \rightarrow q_f$, where q_f is a fresh final state of A. Then we add the following equality constraints: $\forall i, j, q_{f_i} =_A q_{f_j}$. □

If $=_A \subseteq id_Q$, in a successful run we can assume that the subruns rooted at states q such that $q =_A q$ are the same. Hence, we can introduce a pumping technique for positive TAGEDs satisfying this property. The idea is to pump similarly in parallel below all states q such that $q =_A q$, while keeping the equality constraints satisfied. The pumping technique is described in the full version of the paper [23]. Thanks to this, if there is a loop in a successful run, we can construct infinitely many accepted trees. In particular:

Theorem 2. *Let A be a positive TAGED. It is decidable whether $L(A)$ is infinite or not, in $O(|A||Q|^2)$ if $=_A \subseteq id_Q$, and in EXPTIME otherwise.*

We now prove decidability of emptiness of negative TAGEDs ($=_A = \varnothing$), by reduction to positive and negative set constraints (PNSC for short). Set expressions are built from set variables, function symbols, and Boolean operations. Set constraints are either positive, $e_1 \subseteq e_2$, or negative, $e_1 \not\subseteq e_2$, where e_1, e_2 are set expressions. Set expressions are interpreted in the Herbrand structure while set constraints are interpreted by Booleans 0,1. Testing the existence of a solution of a system of set constraints has been proved to be decidable in several papers [5,1,21,11]. In particular, it is known to be NEXPTIME-complete. We do not formally define set constraints and refer the reader to [5,1,21,11]. Consider for instance the constraint $f(X, X) \subseteq X$. It has a unique solution which is the empty set. Consider now $X \subseteq f(X, X) \cup a$, where a is a constant symbol. Every set of terms over $\{f, a\}$ closed by the subterm relation is a solution. More generally, we can encode the emptiness problem of tree automata as a system of set constraints. Let $A = (\Sigma, Q, F, \Delta)$ be a tree automaton. Wlog, we assume all state $q \in Q$ to occur in the rhs of a rule. We associate with A the system S_A defined by:

$$(S_A) \quad \begin{cases} X_q \subseteq \bigcup_{f(q_1,q_2)\to q\in\Delta} f(X_{q_1}, X_{q_2}) \cup \bigcup_{a\to q\in\Delta} a & \text{for all } q \in Q \\ \bigcup_{q\in F} X_q \not\subseteq \varnothing \end{cases}$$

We can prove that $L(A)$ is non-empty iff S_A has a solution. Let $(A, \neq_A)$ be a negative TAGED, and consider the system S'_A consisting in S_A extended with the constraints $X_q \cap X_p = \varnothing$, for all $q, p \in Q$ such that $q \neq_A p$. We can prove that $L(A, \neq_A) \neq \varnothing$ iff S'_A has a solution. Since deciding existence of a solution of a system of PNSC is in NEXPTIME, we get:

Theorem 3. *Emptiness of negative TAGEDs is decidable in NEXPTIME.*

5 Emptiness When Mixing Equality and Disequality Constraints

In this section, we mix equality and disequality constraints. This has already been done in [10] for bounded TAGEDs. Emptiness was proved by decomposition of runs, but here we use a pumping technique that allows to decide emptiness for a class of TAGEDs that significantly extends the class considered in [10]. In particular, we allow an unbounded number of positive tests, but boundedly many negative tests along root-to-leaves paths, *i.e.* branches. While this class subsumes positive TAGEDs, the upper-bound for testing emptiness is bigger than the bound obtained in Section 4.

Formally, a *vertically bounded TAGED* (vbTAGED for short) is a pair (A, k) where A is a TAGED, and $k \in \mathbb{N}$. A run r of (A, k) on a tree $t \in T_\Sigma$ is a run of A on t. It is successful if r is successful for A and the number of states from $\mathrm{dom}(\neq_A)$ occurring along a root-to-leaves path is bounded by k: in other words, for all root-to-leaves path $u_1 \lhd \ldots \lhd u_n$ of t (where each u_i is a node), one has $|\{u_i \mid O_r(u_i) \in \mathrm{dom}(\neq_A)\}| \leq k$.

We now come to the main result of the paper:

Theorem 4. *Emptiness of vbTAGEDs (A, k) is decidable in 2NEXPTIME.*

Proof. Sketch We first transform A so that it satisfies $=_A \subseteq id_Q$, thanks to Lemma 1 (modulo an exponential blow-up). Let $t \in T_\Sigma$, and r a run of A on it which satisfies the equality constraints (but not necesssarily the disequality constraints), and such that its root is labeled by a final state. We introduce sufficient conditions on t and r (which can be verified in polynomial-time, in $|t|$, $|r|$ and $|A|$) to be able to repair the unsatisfied inequality constraints in t in finitely many rewriting steps. These rewritings can be done while keeping the equality constraints satisfied. In particular, since $=_A \subseteq id_Q$, we can assume that for all $u, v \in N_t$ such that $u \sim_{t,r} v$, $r|_u = r|_v$. Hence, we can use a "parallel" pumping technique similar to the pumping technique for positive TAGEDs. The pumping is a bit different however: indeed, if t and r satisfies the sufficient conditions, we increase the size of some contexts of t and r, called *elementary contexts*, in order to repair all the unsatisfied inequality constraints. The repairing process is inductive. In particular, we introduce a notion of frontier below which all inequality constraints have been repaired. The process stops when the frontier reach the top of the tree (and in this case the repaired tree is in the language). From a tree and a run that satisfy the sufficient conditions, and a frontier F, one can create a new tree and a new run satisfying the conditions, and a new frontier which is strictly contained in F. Conversely, if $L(A, k) \neq \varnothing$, then there is a tree t and a run r satisfying the conditions such that the height of t is smaller than $2(k + |Q|)|Q|$ (and by $(k + 2^{|Q|})2^{|Q|+1}$ if $=_A \not\subseteq id_Q$). Hence, it suffices to guess a tree and a run satisfying the conditions to decide emptiness of A.

Since the class of vbTAGEDs subsumes the class of positive TAGEDs, we also get an EXPTIME lower bound for emptiness of vbTAGEDs, by Theorem 1. Moreover, if $=_A \subseteq id_Q$ and $k \leq |Q|$ (or k is unary encoded), emptiness of A is in NEXPTIME.   □

6   Applications

6.1   MSO with Tree Isomorphism Tests

We study an extension of MSO with isomorphism tests between trees. Trees over an alphabet Σ are viewed as structures over the signature consisting of unary predicates O_a, for all $a \in \Sigma$, to test the labels, and the two successor relations S_1 and S_2 which relates the parent to its first child and its second child respectively. The domain of the structure is the set of nodes.

We consider node variables x, y and set variables X, Y. MSO consists of the closure of atomic formulas $O_a(x)$ (for $a \in \Sigma$), $S_1(x, y)$, $S_2(x, y)$, $x \in X$, by conjunction $\wedge$, negation $\neg$, and existential quantifications $\exists x$, $\exists X$. We refer the reader to [16] for the semantics of MSO. It is well-known that MSO sentences and tree automata define the same tree languages [22]. We use similar back and forth translations to prove that an extension of MSO with tree isomorphism tests effectively defines the same language as vertically bounded TAGED. This significantly improves the result of [10].

We consider a predicate $\mathsf{eq}(X)$, which holds in a tree t under assignment $\rho : X \mapsto U$ (denoted by $t, \rho \models \mathsf{eq}(X)$), for some $U \subseteq N_t$, if for all $u, v \in U$, the trees $t|_u$ and $t|_v$ are isomorphic. For all $k \in \mathbb{N}$, we consider the predicate $\mathsf{diff}_k(X, Y)$, which holds in t under assignment ρ if (i) the maximal length of a descendant chain in $\rho(X)$ and $\rho(Y)$ is bounded by k, (ii) for all $u \in \rho(X), v \in \rho(Y)$, the trees $t|_u$ and $t|_v$ are **not** isomorphic. We consider $\mathrm{MSO}_{\eqsim}^{\exists}$ the extension of MSO whose formulas are of the form $\exists X_1 \dots \exists X_n \phi$, where ϕ is an MSO formula extended with atoms $\mathsf{eq}(X_i)$ and $\mathsf{diff}_k(X_i, X_j)$ $(1 \leq i, j \leq n)^2$. $\mathrm{MSO}_{\eqsim}^{\exists}$ is strictly more expressive than MSO as tree isomorphism is not expressible in MSO [7], but as a corollary of Theorem 4, we obtain:

Theorem 5. *$\mathrm{MSO}_{\eqsim}^{\exists}$ and vbTAGEDs effectively define the same tree languages, and satisfiability of $\mathrm{MSO}_{\eqsim}^{\exists}$ formulas is decidable.*

If we allow universal quantification of set variables $X_1, \dots, X_n$, the logic becomes undecidable (even if the X_is denote singletons) [10].

---

2 We assume that $X_1, \dots, X_n$ are not quantified in ϕ

6.2 Unification with Membership Constraints

We show that TAGEDs are particularly suitable to represent sets of ground instances of terms. Then we investigate a particular unification problem with tree and context variables where context variables can occur only in a restricted manner. In particular, we consider first-order logic (FO) over term equations $t \approx t'$, where t, t' are terms with tree and context variables, such that in a formula, every context variable can occur at most once. Tree and context variables might be constrained to range over regular languages (membership constraints). We prove this logic to be undecidable and exhibit a decidable existential fragment. This is particularly relevant for XML queries, as we can express tree patterns with negations. For instance, let L_{dtd} be a regular tree language representing the DTD of a bibliography, d a ground term representing a bibliography, L_{path} the set of unary contexts denoted by the XPath expression $bib/books$ (i.e. contexts whose hole is reachable by the path $bib/books$), and X a unary context variable. The formula $\phi(y, z) = \exists X, d \approx X(book(author(y), title(z))) \wedge d \in L_{dtd} \wedge X \in L_{path}$ checks that d conforms to the DTD and extracts from d all (author,title) pairs reachable from the root by a path $bib/books/book$. The formula $\exists z \exists z', \phi(y, z) \wedge \phi(y, z') \wedge \neg(z \approx z')$ extracts from d all authors y who published at least two books.

The restriction on context variables allows to test (dis)equalities arbitrarily deeply but can not be used to test context (dis)equalities. Even with this restriction, FO is undecidable, while it is known that without context variables, FO on atoms $t \approx t'$ with membership constraints is decidable [8].

Let Σ be a ranked alphabet (assumed to be of binary and constant symbols for the sake of clarity). Let $\mathcal{X}_t$ be a countable set of tree variables x, y, and $\mathcal{X}_c$ a countable set of multi-ary context variables X, Y (we assume the existence of a mapping $ar : \mathcal{X}_c \rightarrow \mathbb{N}$ giving the arity of any context variable). The set of terms over Σ, $\mathcal{X}_t$ and $\mathcal{X}_c$ is denoted by $\mathcal{T}(\Sigma, \mathcal{X}_t, \mathcal{X}_c)$. For instance $X(f(x, X(y), x))$ is a term where $X \in \mathcal{X}_c$ (arity 1), $f \in \Sigma$ (arity 3) and $x, y \in \mathcal{X}_t$. A term is *ground* if it does not contain variables. The set of ground terms over Σ is simply denoted T_Σ. We also denote by $\mathcal{C}_\Sigma$ the set of contexts over Σ, and by $\mathcal{C}_\Sigma^n$ the set of n-ary contexts over Σ, for all $n \in \mathbb{N}$. For all $C \in \mathcal{C}_\Sigma^n$, and terms $t_1, \dots, t_n \in \mathcal{T}(\Sigma, \mathcal{X}_t, \mathcal{X}_c)$, we denote by $C[t_1, \dots, t_n]$ the term obtained by substituting the holes in C by $t_1, \dots, t_n$ respectively (see [7] for a formal definition of contexts). A *ground substitution* σ is a function from $\mathcal{X}_t \cup \mathcal{X}_c$ into $T_\Sigma \cup \mathcal{C}_\Sigma$ such that for all $x \in \mathcal{X}_t$, $\sigma(x) \in T_\Sigma$, and for all $X \in \mathcal{X}_c$, $\sigma(X) \in \mathcal{C}_\Sigma$ and $ar(X) = ar(C)$. The ground term obtained by applying σ on a term t is denoted $t\sigma$. A ground term t' is a *ground instance* of a term t if there is σ such that $t' = t\sigma$. Finally, a term t is *context-linear* if every context variables occurs at most once in t.

Proposition 6. *Let $t \in \mathcal{T}(\Sigma, \mathcal{X}_t, \mathcal{X}_c)$ be context-linear. The set of ground instances of t is definable by a positive TAGED.*

Proof. (Sketch) It suffices to introduce states for each subterm of t, and a special state $q_\forall$ in which every ground term evaluates. Then we add state equalities $q_x =_A q_x$ for all variable x occurring in t. $\square$

We now introduce unification problems. An *equation* e is a pair of terms denoted by $t \approx t'$, where $t, t' \in \mathcal{T}(\Sigma, \mathcal{X}_t, \mathcal{X}_c)$. A ground substitution σ is a solution of e if $t\sigma$

and $t'\sigma$ are ground terms, and $t\sigma = t'\sigma$. Let $n \in \mathbb{N}$. A regular n-ary context language L is a regular language over $\Sigma \cup \{\circ_1, \ldots, \circ_n\}$, where $\circ_1, \ldots, \circ_n$ are fresh symbols denoting the holes, and such that every symbol $\circ_i$ occurs exactly once in terms (this can be ensured by a regular control). A *membership constraint* is an atom of the form $x \in L_x$, or $X \in L_X$, where $x \in \mathcal{X}_t$, $X \in \mathcal{X}_c$, L_x is a regular tree language, and L_X is a regular $\mathrm{ar}(X)$-ary context language.

We consider FO over equations and membership constraint atoms, with the following restriction: for all formulas ϕ, and all context variables $X \in \mathcal{X}_c$, there is at most one equation e, and one term t in e such that X occurs in t. We denote by FO$[\approx, \in]$ this logic. FO$[\approx, \in]$-formulas are interpreted over ground substitutions σ. We define the semantics $\sigma \models \phi$ inductively: $\sigma \models e$ if σ is a solution of e, $\sigma \models x \in L_x$ if $\sigma(x) \in L_x$ (and similarly for $X \in L_X$), $\sigma \models \exists x \phi$ if there is a ground term t such that $\sigma[x \mapsto t] \models \phi$ (and similarly for $\exists X \phi$). Disjunction and negation are interpreted as usual.

We can show the following by reducing PCP:

Proposition 7. *Satisfiability of FO$[\approx, \in]$ is undecidable.*

However, it is known that satisfiability of FO$[\approx, \in]$ in which no context variable occurs is decidable [8]. We consider the existential fragment FO$^\exists[\approx, \in]$ of FO$[\approx, \in]$ formulas where existential quantifiers $\exists x$ or $\exists X$ cannot occur below an odd number of negations.

Theorem 6. *Satisfiability of FO$^\exists[\approx, \in]$ is decidable.*

Proof. (Sketch) Wlog, we consider only closed formulas. We define a normal form which intuitively can be viewed as a set of pairs (E, M), where E is a set of equations e (or negated equations $\neg e$), and M is a set of membership constraints. For each pair (E, M), we construct a vbTAGED $(A_{E,M}, |E|)$ which defines the ground instances of the term t_0 depicted in Fig. 1 satisfying: (i) # is a fresh symbol, (ii) for all terms t, t', there exists $i \in \{1, \ldots, n\}$ s.t. $t = t_i$ and $t' = t'_i$ iff either $(t \approx t') \in E$ or $\neg(t \approx t') \in E$, (iii) if $t_0 \sigma$ is a ground instance of t_0, then the membership constraints are satisfied, and σ is a solution of every equation of E (this can be done for instance by adding state inequalities $q_t \neq_A q_{t'}$, if $\neg(t \approx t') \in E$). The formula is satisfiable iff there is a pair (E, M) such that $L(A_{E,M}, |E|) \neq \varnothing$. $\qquad\square$

Anti-pattern matching. [14] considers terms with negations (called anti-patterns). For instance, the anti-pattern $f(x, \neg x)$ denotes all the ground terms $f(t_1, t_2)$ such that $t_1 \neq t_2$. More generally, negations can occur at any position in the term: $\neg(g(\neg a))$ denotes all ground terms which are not rooted by g or $g(a)$, and $\neg f(x, x)$ denotes ground terms which are not of the form $f(t, t)$. A ground term matches an anti-pattern if it belongs to its denotation. [14] proves it to be decidable. We can easily define a vbTAGED A_p which accepts the denotation of an anti-pattern p

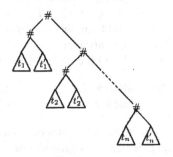

Fig. 1. term t_0

where negations occur at variables only. Thus the anti-pattern matching problem reduces to test membership to $L(A_p)$. When negations occur arbitrarily, the translation is not so clear since the semantics of anti-patterns is universal (a ground term t matches $\neg f(x, x)$ if $\forall x, t \neq f(x, x)$). We let as future work this translation (for instance by pushing down the negations).

7 Future Work

In [10], TAGEDs are based on hedge automata [18], so that they accept unranked trees. We can encode unranked trees over Σ as terms over the signature $\Sigma \cup \{\text{cons}\}$, where cons is a binary symbol denoting concatenation of an unranked tree to an hedge. For instance, $f(a, b, c)$ maps to $f(\text{cons}(a, \text{cons}(b, c)))$. Hedge automata can be translated into tree automata over those encodings. Moreover, the encoding is unique, and any subtree t becomes a subtree rooted by a symbol of Σ in the encoding. Hence testing constraints between subtrees in unranked trees is equivalent to test constraints between subtrees rooted by Σ in binary encodings. Therefore, TAGEDs over unranked trees can be translated into TAGEDs over encodings, and we can prove that all the results presented in this paper carry over to unranked trees. Moreover, in [10], we consider the TQL logic over unranked trees, and prove a fragment of it to be decidable, by reduction to emptiness of bounded TAGEDs (over unranked trees). This work could be used to decide larger fragments of TQL, via a binary encoding. Concerning the unification problem considered here, we would like to use TAGEDs to test whether there are finitely many solutions, and to represent the set of solutions. A question remains: deciding emptiness of full TAGEDs. It is not easy, even for languages of trees of the form $f(t_1, t_2)$, where t_1 and t_2 are unary. Finally, it could be interesting to consider more general tests, like recognizable relations on trees (since tree (dis)equality is a particular recognizable binary relation).

References

1. Aiken, A., Kozen, D., Wimmers, E.L.: Decidability of systems of set constraints with negative constraints. Information and Computation 122(1), 30–44 (1995)
2. Anantharaman, S., Narendran, P., Rusinowitch, M.: Closure properties and decision problems of dag automata. Inf. Process. Lett. 94(5), 231–240 (2005)
3. Bogaert, B., Tison, S.: Equality and disequality constraints on direct subterms in tree automata. In: Finkel, A., Jantzen, M. (eds.) STACS 1992. LNCS, vol. 577, pp. 161–171. Springer, Heidelberg (1992)
4. Charatonik, W.: Automata on dag representations of finite trees. Technical report (1999)
5. Charatonik, W., Pacholski, L.: Set constraints with projections are in NEXPTIME. In: IEEE Symposium on Foundations of Computer Science (1994)
6. Comon, H., Cortier, V.: Tree automata with one memory, set constraints and cryptographic protocols. TCS 331(1), 143–214 (2005)
7. Comon, H., Dauchet, M., Gilleron, R., Löding, C., Jacquemard, F., Lugiez, D., Tison, S., Tommasi, M.: Tree automata techniques and applications (2007), http://www.grappa.univ-lille3.fr/tata
8. Comon, H., Delor, C.: Equational formulae with membership constraints. Information and Computation 112(2), 167–216 (1994)
9. Dauchet, M., Caron, A.-C., Coquidé, J.-L.: Reduction properties and automata with constraints. JSC 20, 215–233 (1995)

10. Filiot, E., Talbot, J.-M., Tison, S.: Satisfiability of a spatial logic with tree variables. In: Duparc, J., Henzinger, T.A. (eds.) CSL 2007. LNCS, vol. 4646, pp. 130–145. Springer, Heidelberg (2007)

11. Gilleron, R., Tison, S., Tommasi, M.: Some new decidability results on positive and negative set constraints. In: Proceedings of the International Conference on Constraints in Computational Logics, pp. 336–351 (1994)

12. Jacquemard, F., Rusinowitch, M., Vigneron, L.: Tree automata with equality constraints modulo equational theories. Research Report, LSV, ENS Cachan (2006)

13. Karianto, W., Löding, C.: Unranked tree automata with sibling equalities and disequalities. Research Report, RWTH Aachen (2006)

14. Kirchner, C., Kopetz, R., Moreau, P.-E.: Anti-pattern matching. In: De Nicola, R. (ed.) ESOP 2007. LNCS, vol. 4421. Springer, Heidelberg (2007)

15. Kutsia, T., Marin, M.: Solving regular constraints for hedges and contexts. In: UNIF 2006, pp. 89–107 (2006)

16. Libkin, L.: Logics over unranked trees: an overview. LMCS 2006 3(2), 1–31 (2006)

17. Mongy, J.: Transformation de noyaux reconnaissables d'arbres. Forêts RATEG. PhD thesis, Université de Lille (1981)

18. Murata, M.: Hedge automata: A formal model for xml schemata. Technical report, Fuji Xerox Information Systems (1999)

19. Neven, F.: Automata, logic, and XML. In: Bradfield, J.C. (ed.) CSL 2002 and EACSL 2002. LNCS, vol. 2471, pp. 2–26. Springer, Heidelberg (2002)

20. Schwentick, T.: Automata for XML – a survey. J. Comput. Syst. Sci. 73(3), 289–315 (2007)

21. Stefansson, K.: Systems of set constraints with negative constraints are nexptime-complete. In: LICS (1994)

22. Thatcher, J.W., Wright, J.B.: Generalized finite automata with an application to a decision problem of second-order logic. Mathematical System Theory 2, 57–82 (1968)

23. Full paper version, http://hal.inria.fr/inria-00292027

Bad News on Decision Problems for Patterns

Dominik D. Freydenberger[1,*] and Daniel Reidenbach[2]

[1] Institut für Informatik, Goethe-Universität, Postfach 111932,
D-60054 Frankfurt am Main, Germany
`freydenberger@em.uni-frankfurt.de`
[2] Department of Computer Science, Loughborough University,
Loughborough, Leicestershire, LE11 3TU, United Kingdom
`D.Reidenbach@lboro.ac.uk`

Abstract. We study the inclusion problem for pattern languages, which is shown to be undecidable by Jiang et al. (J. Comput. System Sci. 50, 1995). More precisely, Jiang et al. demonstrate that there is no effective procedure deciding the inclusion for the class of *all* pattern languages over *all* alphabets. Most applications of pattern languages, however, consider classes over *fixed* alphabets, and therefore it is practically more relevant to ask for the existence of *alphabet-specific* decision procedures. Our first main result states that, for all but very particular cases, this version of the inclusion problem is also undecidable. The second main part of our paper disproves the prevalent conjecture on the inclusion of so-called similar E-pattern languages, and it explains the devastating consequences of this result for the intensive previous research on the most prominent open decision problem for pattern languages, namely the equivalence problem for general E-pattern languages.

1 Introduction

A *pattern* – a finite string that consists of *variables* and of *terminal symbols* (or: *letters*) – is a compact and natural device to define a formal language. It generates a word by a *substitution* of all variables with arbitrary words of terminal symbols (taken from a fixed alphabet Σ) and, hence, its language is the set of all words under such substitutions. More formally, a pattern language thus is the (typically infinite) set of all images of the pattern under *terminal-preserving* morphisms, i. e. morphisms which map each terminal symbol onto itself. For example, if we consider the pattern $\alpha := x_1 \, a \, x_2 \, b \, x_1$ (where the symbols x_1 and x_2 are variables and a and b are terminal symbols) then the language generated by α exactly contains those words which consist of an arbitrary prefix u, followed by the letter a, an arbitrary word v, the letter b and a suffix which equals u again. Consequently, the *pattern language* of α includes, amongst others, the words $w_1 := a \, a \, b \, b \, b \, a$, $w_2 := a \, b \, a \, b \, a \, b \, a \, b$ and $w_3 := a \, a \, a \, b \, a \, a$, and it does not cover the words $w_4 := b \, a$, $w_5 := b \, a \, b \, b \, b \, a$ and $w_6 := a \, b \, b \, a$. It is a well-known fact that pattern languages in general are not context-free.

---

* Corresponding author.

M. Ito and M. Toyama (Eds.): DLT 2008, LNCS 5257, pp. 327–338, 2008.

Basically, two types of pattern languages are considered in literature: *NE*-pattern languages and *E*-pattern languages. The definition of the former was introduced by Angluin [1], and it disallows that variables are substituted with the empty word (hence, "NE" is short for "nonerasing"). The latter kind of pattern languages additionally consider those substitutions which map one ore more variables onto the empty word (so "E" stands for "erasing" or "extended"); this definition goes back to Shinohara [25]. Thus, in our above example, the word w_3 is contained in the E-pattern language of α, but not in its NE-pattern language. Surprisingly, this small difference in the definitions leads to significant differences in the characteristics of the resulting (classes of) languages.

As a consequence of their simple definition, which comprises nothing but finite strings and (a particular type of) morphisms, pattern languages show numerous connections to other fundamental topics in computer science and discrete mathematics, including classical ones such as (un-)avoidable patterns (cf. Jiang et al. [8]), word equations (cf. Mateescu, Salomaa [12], Karhumäki et al. [10]) and equality sets (and, thus, the Post Correspondence Problem, cf. Reidenbach [18]) as well as emerging ones such as extended regular expressions (cf. Câmpeanu et al. [3]) and the ambiguity of morphisms (cf. Freydenberger et al. [6], Reidenbach [18]). In terms of the basic *decision problems*, pattern languages show a wide range of behaviors: trivial (linear time) decidability (e. g., the equivalence of NE-pattern languages), NP-completeness (e. g., the membership in NE- and E-pattern languages) and undecidability (e. g., the inclusion of NE- and E-pattern languages); furthermore, the decidability of quite a number of these problems is still open (e. g., the equivalence problem for E-pattern languages). Surveys on these topics are provided by, e. g., Mateescu and Salomaa [13] and Salomaa [23].

Among the established properties (and even among all results on pattern languages), the proof for the undecidability of the inclusion problem by Jiang, Salomaa, Salomaa and Yu [9] is considered to be one of the most notable achievements, and this is mainly due to the very hard proof, which answers a long-standing open question, and the fact that the result remarkably contrasts with the trivial decidability of the equivalence problem for NE-pattern languages. Furthermore, the inclusion problem is of vital importance for the main field of application of pattern languages, namely *inductive inference*. Inductive inference of pattern languages – which deals with an approach to the important problem of computing a pattern that is common to a given set of strings – is a both classical and active area of research in learning theory; a survey is provided by Ng and Shinohara [16]. It is closely connected to the inclusion problem for pattern languages since, according to the celebrated characterization by Angluin [2], the inferrability of any indexable class of languages largely depends on the inclusion relation between the languages in the class. Consequently, many (both classical and recent) papers on inductive inference of classes of pattern languages nearly exclusively deal with questions related to the inclusion problem for these classes (see, e. g., Mukouchi [15], Reidenbach [19,21], Luo [11]).

Unfortunately, from this rather practical point of view, the inclusion problem for E- and for NE-pattern languages as understood and successfully tackled by Jiang et al. [9] is not very significant, since they prove that there is no *single* procedure which, *for every terminal alphabet* Σ and for every pair of patterns, decides on the inclusion between the languages over Σ generated by these patterns. Hence, slightly more formally, Jiang et al. [9] demonstrate that the inclusion problem is undecidable for (a technical subclass of) the class of all pattern languages over all alphabets, and the requirement for any decision procedure to handle pattern languages over various alphabets is extensively utilized in the proof. Contrary to this, in inductive inference of pattern languages – and virtually every other field of application of pattern languages known to the authors – one always considers a class of pattern languages over a *fixed* alphabet. Consequently, it seems practically more relevant to investigate the problem of whether, for any alphabet Σ, there exists a procedure deciding the inclusion problem for the class of (E/NE-)pattern languages *over this alphabet Σ*.

In the present paper we study and answer this question (or rather: these infinitely many questions). Our considerations reveal that, for every finite alphabet Σ with at least two letters, the inclusion problem is undecidable for the full classes of E-pattern languages over Σ. Furthermore, with regard to the class of NE-pattern languages over any Σ, we prove the equivalent result, but our reasoning does not cover binary and ternary alphabets. Although we thus have the same outcome as Jiang et al. [9] for their variant of the inclusion problem, the proof for our much stronger statement considerably differs from their argumentation; consequently, it suggests that there is no straightforward way from the well-established result to ours. Moreover, we feel that our insights (and our uniform reasoning for all alphabet sizes) are a little surprising, since the inferrability of classes of pattern languages is known to be discontinuous depending on the alphabet size and the question of whether NE- or E-pattern languages are considered (cf. Reidenbach [21]). The second main part of our paper addresses the other major topic in [9]: we discuss the extensibility of a positive decidability result given in [9] on the inclusion problem for the class of *terminal-free* E-pattern languages (generated by those patterns that consist of variables only) to classes of so-called *similar* E-pattern languages. This question is intensively discussed in literature (e. g. by Ohlebusch, Ukkonen [17]) as it is of major importance for the still unresolved equivalence problem for the full class of E-pattern languages. We demonstrate that, in contrast to the prevalent conjecture, the inclusion of similar E-pattern languages does *not* show an analogous behavior to that of terminal-free E-pattern languages, and we explain the fatal impact of this insight on the previous research dealing with the equivalence problem.

2   Preliminaries

Let $\mathbb{N} := \{1, 2, 3, \ldots\}$ and $\mathbb{N}_0 := \mathbb{N} \cup \{0\}$. The symbol ∞ stands for infinity. For an arbitrary alphabet A, a *string* (over A) is a finite sequence of symbols from A, and λ stands for the *empty string*. The symbol A^+ denotes the set of all nonempty strings over A, and $A^* := A^+ \cup \{\lambda\}$. For the *concatenation* of two strings w_1, w_2 we write $w_1 \cdot w_2$ or simply $w_1 w_2$. We say that a string $v \in A^*$ is a *factor* of a string $w \in A^*$ if there are $u_1, u_2 \in A^*$ such that $w = u_1 v u_2$. If $u_1 = \lambda$ (or $u_2 = \lambda$), then v is a *prefix* of w (or a *suffix*, respectively). The notation $|x|$ stands for the size of a set x or the length of a string x. For any $w \in \Sigma^*$ and any $n \in \mathbb{N}_0$, w^n denotes the *n-fold concatenation of* w, with $w^0 := \lambda$. Furthermore, we use $\cdot$ and the regular operations $*$ and $+$ on sets and strings in the usual way.

For any alphabets A, B, a *morphism* is a function $h : A^* \to B^*$ with $h(vw) = h(v)h(w)$ for all $v, w \in A^*$. Given morphisms $f : A^* \to B^*$ and $g : B^* \to C^*$ (for alphabets A, B, C), their *composition* $g \circ f$ is defined as $g \circ f(w) := g(f(w))$ for all $w \in A^*$. A morphism $h : A^* \to B^*$ is *nonerasing* if $h(a) \neq \lambda$ for all $a \in A$.

Let Σ be a (finite or infinite) alphabet of so-called *terminal symbols* (or: *letters*) and X an infinite set of *variables* with $\Sigma \cap X = \emptyset$. Unless specified differently, we assume $X = \{x_i \mid i \in \mathbb{N}\}$, with $x_i \neq x_j$ for all $i \neq j$. A *pattern* is a string over $\Sigma \cup X$, a *terminal-free pattern* is a string over X and a *word* is a string over Σ. The set of all patterns over $\Sigma \cup X$ is denoted by Pat_Σ, the set of terminal-free patterns by $\mathrm{Pat}_{\mathrm{tf}}$. For any pattern α, we refer to the set of variables in α as $\mathrm{var}(\alpha)$ and to the set of terminal symbols as $\mathrm{term}(\alpha)$. Two patterns $\alpha, \beta \in \mathrm{Pat}_\Sigma$ are *similar* if their factors over Σ are identical and occur in the same order in the patterns. More formally, α, β are similar if $\alpha = \alpha_0 u_1 \alpha_1 u_2 \ldots \alpha_{n-1} u_n \alpha_n$ and $\beta = \beta_0 u_1 \beta_1 u_2 \ldots \beta_{n-1} u_n \beta_n$ for some $n \in \mathbb{N}_0$, $\alpha_i, \beta_i \in X^+$ for each $i \in \{1, \ldots, n-1\}$, $\alpha_0, \beta_0, \alpha_n, \beta_n \in X^*$ and $u_j \in \Sigma^+$ for each $j \in \{1, \ldots, n\}$.

A morphism $\sigma : (\Sigma \cup X)^* \to (\Sigma \cup X)^*$ is called *terminal-preserving* if $\sigma(a) = a$ for all $a \in \Sigma$. A terminal-preserving morphism $\sigma : (\Sigma \cup X)^* \to \Sigma^*$ is called a *substitution*. The *E-pattern language* $L_{\mathrm{E},\Sigma}(\alpha)$ of a pattern $\alpha \in \mathrm{Pat}_\Sigma$ is the set of all $w \in \Sigma^*$ such that $\sigma(\alpha) = w$ for some substitution σ; the *NE-pattern language* $L_{\mathrm{NE},\Sigma}(\alpha)$ is defined in the same way, but restricted to nonerasing substitutions. The term *pattern language* refers to any of the definitions introduced above. Two pattern languages are called *similar* if they have generating patterns that are similar. Accordingly, we call a pattern language *terminal-free* if it is generated by a terminal-free pattern. We denote the class of all E-pattern languages over Σ with ePAT_Σ and the class of all NE-pattern languages over Σ with nePAT_Σ.

A *nondeterministic 2-counter automaton without input* (cf. Ibarra [7]) is a 4-tuple $\mathcal{A} = (Q, \delta, q_0, F)$, consisting of a state set Q, a transition relation $\delta :$ $Q \times \{0,1\}^2 \to Q \times \{-1, 0, +1\}^2$, the initial state $q_0 \in Q$ and a set of accepting states $F \subseteq Q$. A *configuration* of $\mathcal{A}$ is a triple $(q, m_1, m_2) \in Q \times \mathbb{N}_0 \times \mathbb{N}_0$, where q indicates the state of $\mathcal{A}$ and m_1 (or m_2) denotes the content of the first (or second, respectively) counter. The relation $\vdash_\mathcal{A}$ on $Q \times \mathbb{N}_0 \times \mathbb{N}_0$ is defined by δ as follows: Let $p, q \in Q$, $m_1, m_2, n_1, n_2 \in \mathbb{N}_0$. Then $(p, m_1, m_2) \vdash_\mathcal{A} (q, n_1, n_2)$ iff there exist $c_1, c_2 \in \{0, 1\}$ and $r_1, r_2 \in \{-1, 0, +1\}$ such that (i) $c_i = 0$ if $m_i = 0$ and $c_i = 1$ if $m_i \geq 1$ for $i \in \{1, 2\}$, (ii) $n_i = m_i + r_i$ for $i \in \{1, 2\}$ and (iii) $(q, r_1, r_2) \in \delta(p, c_1, c_2)$. Furthermore, for $i \in \{1, 2\}$, we assume that $r_i \neq -1$ if $c_i = 0$. Intuitively, in every state $\mathcal{A}$ is only able to check whether the counters equal zero, change each counter by at most one and switch into a new state.

A *computation* is a sequence of configurations, and an *accepting computation* of $\mathcal{A}$ is a sequence $C_1, \ldots, C_n \in Q \times \mathbb{N}_0 \times \mathbb{N}_0$ (for some $n \in \mathbb{N}_0$) with $C_1 = (q_0, 0, 0)$, $C_n \in F \times \mathbb{N}_0 \times \mathbb{N}_0$ and $C_i \vdash_\mathcal{A} C_{i+1}$ for all $i \in \{1, \ldots, n-1\}$. In order to encode configurations of $\mathcal{A}$, we assume that $Q = \{q_0, \ldots, q_s\}$ for some $s \in \mathbb{N}_0$ and define a function cod : $Q \times \mathbb{N}_0 \times \mathbb{N}_0 \to \{0, \#\}^*$ by $\mathrm{cod}(q_i, m_1, m_2) := 0^{i+1} \# 0^{m_1+1} \# 0^{m_2+1}$ and extend this to an encoding of computations by defining $\mathrm{cod}(C_1, C_2, \ldots, C_n) := \#\# \mathrm{cod}(C_1) \#\# \mathrm{cod}(C_2) \#\# \ldots \#\# \mathrm{cod}(C_n) \#\#$ for every $n \geq 1$ and every sequence $C_1, \ldots, C_n \in Q \times \mathbb{N}_0 \times \mathbb{N}_0$. Furthermore, let $\mathrm{VALC}(\mathcal{A}) := \{\mathrm{cod}(C_1, \ldots, C_n) \mid C_1, \ldots, C_n \text{ is an accepting computation of } \mathcal{A}\}$, and $\mathrm{INVALC}(\mathcal{A}) := \{0, \#\}^* \setminus \mathrm{VALC}(\mathcal{A})$. As the emptiness problem for 2-counter automata with input is undecidable (cf. Minsky [14], Ibarra [7]), it is also undecidable whether a nondeterministic 2-counter automaton without input has an accepting computation.

3   The Inclusion of Pattern Languages over Fixed Alphabets

In this section, we discuss the decidability of the inclusion problem for ePAT_Σ and nePAT_Σ. We begin with all non-unary finite alphabets Σ; the special case $|\Sigma| \in \{1, \infty\}$ is studied separately. Jiang, Salomaa, Salomaa and Yu [9] prove the undecidability of the *general inclusion problem for E-pattern languages*:

Theorem 1 (Jiang et al. [9]). *There is no total computable function* χ_E *which, for every alphabet* Σ *and for every pair of patterns* $\alpha, \beta \in \mathrm{Pat}_\Sigma$, *decides on whether or not* $L_{\mathrm{E},\Sigma}(\alpha) \subseteq L_{\mathrm{E},\Sigma}(\beta)$.

Technically, Jiang et al. show that, given a nondeterministic 2-counter automaton without input $\mathcal{A}$, one can effectively construct an alphabet Σ and patterns $\alpha_\mathcal{A}, \beta_\mathcal{A} \in \mathrm{Pat}_\Sigma$ such that $L_{\mathrm{E},\Sigma}(\alpha_\mathcal{A}) \subseteq L_{\mathrm{E},\Sigma}(\beta_\mathcal{A})$ iff $\mathcal{A}$ has an accepting computation. As this problem is known to be undecidable, the general inclusion problem for E-pattern languages must also be undecidable.

In their construction, Σ contains one letter for every state of $\mathcal{A}$, and six further symbols that are used for technical reasons. As limiting the number of states would lead to a finite number of possible automata (and thus trivial and inapplicable decidability), one cannot simply fix the number of states in order to adapt this result to the inclusion problem for ePAT_Σ with some fixed alphabet Σ. Thus, as mentioned by Reidenbach [18] and Salomaa [24], there seems to be no straightforward way from this undecidability result to the undecidability of the inclusion problem for ePAT_Σ, especially when Σ is comparatively small. Nevertheless, our first main theorem states:

Theorem 2. *Let* Σ *be a finite alphabet with* $|\Sigma| \geq 2$. *Then the inclusion problem for* ePAT_Σ *is undecidable.*

The proof of this theorem is rather lengthy and can be found in Section 3.1. It is in principle based on the construction by Jiang et al.[9], with two key differences. First, the problem of an unbounded number of states (and therefore the number of letters necessary to encode these states) is handled by using a unary encoding instead of special letters to designate the states in configurations; second, the special control symbols are encoded over a binary alphabet or removed. These modifications enforce considerable changes to the patterns and the underlying reasoning. But before we go into these details, we first discuss the immediate consequences of Theorem 2. In fact, the proof demonstrates a stronger result:

Corollary 1. *Let Σ be a finite alphabet with $|\Sigma| \geq 2$. Given two patterns $\alpha \in \mathrm{Pat}_{\Sigma}$ and $\beta \in (\{a\} \cup X)^*$ for some terminal $a \in \Sigma$, it is in general undecidable whether $L_{\mathrm{E},\Sigma}(\alpha) \subseteq L_{\mathrm{E},\Sigma}(\beta)$.*

This corollary is the alphabet specific version of Jiang et al.'s Corollary 5.1 in [9] that is used to obtain the following result on the *general inclusion problem for NE-pattern languages*:

Theorem 3 (Jiang et al. [9]). *There is no total computable function χ_{NE} which, for every alphabet Σ and for every pair of patterns $\alpha, \beta \in \mathrm{Pat}_{\Sigma}$, decides on whether or not $L_{\mathrm{NE},\Sigma}(\alpha) \subseteq L_{\mathrm{NE},\Sigma}(\beta)$.*

In the terminology used in the present paper, the proof of Theorem 3 in [9] reduces the inclusion problem for ePAT_{Σ} (for patterns of a restricted form as in Corollary 1) to the inclusion problem for $\mathrm{nePAT}_{\Sigma \cup \{\star, \$\}}$, where $\star$ and $\$$ are two extra letters that are not contained in Σ. Using the same reasoning as Jiang et al. in their proof of Theorem 3, but when substituting their Corollary 5.1 with Corollary 1 above, one immediately achieves the following result:

Theorem 4. *Let Σ be a finite alphabet with $|\Sigma| \geq 4$. Then the inclusion problem for nePAT_{Σ} is undecidable.*

As the construction used in the reduction heavily depends on the two extra letters, the authors do not see a straightforward way to adapt it to binary or ternary alphabets. Therefore, the decidability of the inclusion problem for NE-pattern languages over these alphabets remains open:

Open Problem 1. *Let Σ be an alphabet with $|\Sigma| = 2$ or $|\Sigma| = 3$. Is the inclusion problem for nePAT_{Σ} decidable?*

We now take a brief look at the special cases of unary and infinite alphabets. Here we can state that the inclusion of pattern languages is less complex than in the standard case:

Proposition 1. *Let Σ be an alphabet, $|\Sigma| \in \{1, \infty\}$. Then the inclusion problem is decidable for ePAT_{Σ} and for nePAT_{Σ}.*

The proof for Proposition 1 is omitted due to space constraints.

Obviously, Proposition 1 implies that the *equivalence* problem is decidable, too, for ePAT_{Σ} and nePAT_{Σ} over unary or infinite alphabets Σ. Furthermore, with regard to $2 \leq |\Sigma| < \infty$, it is shown by Angluin [1] that two patterns generate the same *NE*-pattern language iff they are the same (apart from a renaming of variables). Thus, the equivalence problem for nePAT_{Σ} is trivially decidable for every Σ, a result which nicely contrasts with the undecidability of the inclusion problem established above. The equivalence problem for ePAT_{Σ}, however, is still an open problem in case of $2 \leq |\Sigma| < \infty$. In Section 4 we present and discuss a result that has a significant impact on this widely-discussed topic.

3.1   Proof of Theorem 2

Due to space constraints, the proofs of all lemmas in this section are omitted. We begin with the case $|\Sigma| = 2$, so let $\Sigma := \{0, \#\}$. Let $\mathcal{A} := (Q, \delta, q_0, F)$ be a nondeterministic 2-counter automaton; w. l. o. g. let $Q := \{q_0, \ldots, q_s\}$ for some $s \in \mathbb{N}_0$. Our goal is to construct patterns $\alpha_{\mathcal{A}}, \beta_{\mathcal{A}} \in \text{Pat}_\Sigma$ such that $L_{\text{E}, \Sigma}(\alpha_{\mathcal{A}}) \subseteq L_{\text{E}, \Sigma}(\beta_{\mathcal{A}})$ iff $\text{VALC}(\mathcal{A}) = \emptyset$. We define $\alpha_{\mathcal{A}} := vv\#^4 vxvyv\#^4 vuv$, where x, y are distinct variables, $v = 0\#^3 0$ and $u = 0\#\#0$. Furthermore, for a yet unspecified $\mu \in \mathbb{N}$ that shall be defined later, let $\beta_{\mathcal{A}} := (x_1)^2 \ldots (x_\mu)^2 \#^4 \hat{\beta}_1 \ldots \hat{\beta}_\mu \#^4 \ddot{\beta}_1 \ldots \ddot{\beta}_\mu$, with, for all $i \in \{1, \ldots, \mu\}$, $\hat{\beta}_i := x_i \gamma_i x_i \delta_i x_i$ and $\ddot{\beta}_i := x_i \eta_i x_i$, where $x_1, \ldots, x_\mu$ are distinct variables and all $\gamma_i, \delta_i, \eta_i \in X^*$ are terminal-free patterns. The patterns γ_i and δ_i shall be defined later; for now, we only mention:

1. $\eta_i := z_i(\hat{z}_i)^2 z_i$ and $z_i \neq \hat{z}_i$ for all $i \in \{1, \ldots, \mu\}$,
2. $\text{var}(\gamma_i \delta_i \eta_i) \cap \text{var}(\gamma_j \delta_j \eta_j) = \emptyset$ for all $i, j \in \{1, \ldots, \mu\}$ with $i \neq j$,
3. $x_k \notin \text{var}(\gamma_i \delta_i \eta_i)$ for all $i, k \in \{1, \ldots, \mu\}$.

Thus, for every i, the elements of $\text{var}(\gamma_i \delta_i \eta_i)$ appear nowhere but in these three factors. Let H be the set of all substitutions $\sigma : (\Sigma \cup \{x, y\})^* \to \Sigma^*$. We interpret each triple $(\gamma_i, \delta_i, \eta_i)$ as a predicate $\pi_i : H \to \{0, 1\}$ in such a way that $\sigma \in H$ *satisfies* π_i if there exists a morphism $\tau : \text{var}(\gamma_i \delta_i \eta_i)^* \to \Sigma^*$ with $\tau(\gamma_i) = \sigma(x)$, $\tau(\delta_i) = \sigma(y)$ and $\tau(\eta_i) = u$ – in the terminology of word equations (cf. Karhumäki et al. [10]), this means that σ satisfies π_i iff the system consisting of the three equations $\gamma_i = \sigma(x)$, $\delta_i = \sigma(y)$ and $\eta_i = u$ has a solution τ. Later, we shall see that $L_{\text{E}, \Sigma}(\alpha_{\mathcal{A}}) \backslash L_{\text{E}, \Sigma}(\beta_{\mathcal{A}})$ exactly contains those $\sigma(\alpha_{\mathcal{A}})$ for which σ does not satisfy any of π_1 to π_μ, and choose these predicates to describe $\text{INVALC}(\mathcal{A})$. The encoding of $\text{INVALC}(\mathcal{A})$ shall be handled by π_4 to π_μ, as each of these predicates describes a sufficient criterium for membership in $\text{INVALC}(\mathcal{A})$. But at first we need a considerable amount of technical preparations. A substitution σ is of *good form* if $\sigma(x) \in \{0, \#\}^*$, $\sigma(x)$ does not contain $\#^3$ as a factor, and $\sigma(y) \in 0^*$. Otherwise, σ is of *bad form*. The predicates π_1 and π_2 handle all cases where σ is of bad form and are defined through $\gamma_1 := y_{1,1}(\hat{z}_1)^3 y_{1,2}$, $\delta_1 := \hat{y}_1$, $\gamma_2 := y_2$, and $\delta_2 := \hat{y}_{2,1} \hat{z}_2 \hat{y}_{2,2}$, where $y_{1,1}, y_{1,2}, y_2, \hat{y}_1, \hat{y}_{2,1}, \hat{y}_{2,2}, \hat{z}_1$ and $\hat{z}_2$ are pairwise distinct variables. Recall that $\eta_i = z_i(\hat{z}_i)^2 z_i$ for all i. It is not very difficult to see that π_1 and π_2 characterize the morphisms that are of bad form:

Lemma 1. *A substitution $\sigma \in H$ is of bad form iff σ satisfies π_1 or π_2.*

This allows us to make the following observation, which serves as the central part of the construction and is independent from the exact shape of π_3 to π_μ:

Lemma 2. *For every substitution $\sigma \in H$, $\sigma(\alpha_{\mathcal{A}}) \in L_{\text{E}, \Sigma}(\beta_{\mathcal{A}})$ iff σ satisfies one of the predicates π_1 to π_μ.*

Thus, we can select predicates π_1 to π_μ in such a way that $L_{\text{E}, \Sigma}(\alpha_{\mathcal{A}}) \backslash L_{\text{E}, \Sigma}(\beta_{\mathcal{A}}) = \emptyset$ iff $\text{VALC}(\mathcal{A}) = \emptyset$ by describing $\text{INVALC}(\mathcal{A})$ through a disjunction of predicates on H. The proof of Lemma 2 shows that if $\sigma(\alpha_{\mathcal{A}}) = \tau(\beta_{\mathcal{A}})$ for substitutions σ, τ, where σ is of good form, there exists exactly one i ($3 \leq i \leq \mu$) s.t. $\tau(x_i) = 0\#^3 0$. Due to technical reasons, we need a predicate π_3 that, if unsatisfied, sets a lower bound on the length of $\sigma(y)$, defined by $\gamma_3 := y_{3,1}\,\hat{y}_{3,1}\,y_{3,2}\,\hat{y}_{3,2}\,y_{3,3}\,\hat{y}_{3,3}\,y_{3,4}$, and $\delta_3 := \hat{y}_{3,1}\,\hat{y}_{3,2}\,\hat{y}_{3,3}$, where all of $y_{3,1}$ to $y_{3,4}$ and $\hat{y}_{3,1}$ to $\hat{y}_{3,3}$ are pairwise distinct variables. Clearly, if some $\sigma \in H$ satisfies π_3, $\sigma(y)$ is a concatenation of three (possibly empty) factors of $\sigma(x)$. Thus, if σ satisfies none of π_1 to π_3, $\sigma(y)$ must be longer than the three longest non-overlapping sequences of 0s in $\sigma(x)$. This allows us to identify a class of predicates definable by a rather simple kind of expression, which we use to define π_4 to π_μ in a less technical way.

Let $X' := \{\hat{x}_1, \hat{x}_2, \hat{x}_3\} \subset X$, let G denote the set of those substitutions in H that are of good form and let R be the set of all substitutions $\rho : (\Sigma \cup X')^* \to \Sigma^*$ for which $\rho(0) = 0$, $\rho(\#) = \#$ and $\rho(\hat{x}_i) \in 0^*$ for all $i \in \{1,2,3\}$. For patterns $\alpha \in (\Sigma \cup X')^*$, we define $R(\alpha) := \{\rho(\alpha) \mid \rho \in R\}$.

Definition 1. *A predicate* $\pi : G \to \{0,1\}$ *is called a* simple predicate *if there exist a pattern* $\alpha \in (\Sigma \cup X')^*$ *and languages* $L_1, L_2 \in \{\Sigma^*, \{\lambda\}\}$ *such that* σ *satisfies* π *iff* $\sigma(x) \in L_1 R(\alpha) L_2$.

From a slightly different point of view, the elements of X' can be understood as numerical parameters describing (concatenational) powers of 0, with substitutions $\rho \in R$ acting as assignments. For example, if $\sigma \in G$ satisfies a simple predicate π iff $\sigma(x) \in \Sigma^* R(\#\hat{x}_1\#\hat{x}_2 0\#\hat{x}_1)$, we can also write that σ satisfies π iff $\sigma(x)$ has a suffix of the form $\#0^m\#0^n0\#0^m$ (with $m, n \in \mathbb{N}_0$), which could also be written as $\#0^m\#0^*0\#0^m$, as n occurs only once in this expression. Using π_3, our construction is able to express all simple predicates:

Lemma 3. *For every simple predicate* π_S *over* n *variables with* $n \leq 3$, *there exists a predicate* π *defined by terminal-free patterns* γ, δ, η *such that for all substitutions* $\sigma \in G$:

1. *if* σ *satisfies* π_S, *then* σ *also satisfies* π *or* π_3,
2. *if* σ *satisfies* π, *then* σ *also satisfies* π_S.

Roughly speaking, if σ does not satisfy π_3, then $\sigma(y)$ (which is in 0^*, due to $\sigma \in G$) is long enough to provide building blocks for simple predicates using variables from X'.

Our next goal is a set of predicates that (if unsatisfied) forces $\sigma(x)$ into a basic shape common to all elements of $\mathrm{VALC}(\mathcal{A})$. We say that a word $w \in \{0, \#\}^*$ is of *good structure* if $w \in (\#\#0^+\#0^+\#0^+)^+\#\#$. Otherwise, w is of *bad structure*. Recall that due to the definition of cod, all elements of $\mathrm{VALC}(\mathcal{A})$ are of good structure, thus being of bad structure is a sufficient criterion for belonging to $\mathrm{INVALC}(\mathcal{A})$. In order to cover the morphisms σ where $\sigma(x)$ is of bad structure, we define predicates π_4 to π_{13} through simple predicates as follows:

$\pi_4 : \sigma(x) = \lambda$,	$\pi_9 : \sigma(x)$ ends on 0,
$\pi_5 : \sigma(x) = \#$,	$\pi_{10} : \sigma(x)$ ends on 0#,
$\pi_6 : \sigma(x) = \#\#$,	$\pi_{11} : \sigma(x)$ contains a factor $\#\#0^*\#\#$,
$\pi_7 : \sigma(x)$ begins with 0,	$\pi_{12} : \sigma(x)$ contains a factor $\#\#0^*\#0^*\#\#$,
$\pi_8 : \sigma(x)$ begins with #0,	$\pi_{13} : \sigma(x)$ contains a factor $\#\#0^*\#0^*\#0^*\#0$.

Due to Lemma 3, the predicates π_1 to π_{13} do not strictly give rise to a characterization of substitutions with images that are of bad structure, as there are $\sigma \in G$ where $\sigma(x)$ is of good structure, but π_3 is satisfied due to $\sigma(y)$ being too short. But this problem can be avoided by choosing $\sigma(y)$ long enough to leave π_3 unsatisfied, and the following holds:

Lemma 4. *A word $w \in \Sigma^*$ is of good structure iff there exists a substitution $\sigma \in H$ with $\sigma(x) = w$ such that σ satisfies none of the predicates π_1 to π_{13}.*

For every w of good structure, there exist uniquely determined n, i_1, j_1, k_1, ..., i_n, j_n, $k_n \in \mathbb{N}_1$ such that $w = \#\#0^{i_1}\#0^{j_1}\#0^{k_1}\#\#\ldots\#\#0^{i_n}\#0^{j_n}\#0^{k_n}\#\#$. Thus, if $\sigma \in H$ does not satisfy any of π_1 to π_{13}, $\sigma(x)$ can be understood as an encoding of a sequence $T_1, \ldots, T_n$ of triples $T_i \in (\mathbb{N}_1)^3$, and for every sequence of that form, there is a $\sigma \in H$ such that $\sigma(x)$ encodes a sequence of triples of positive integers, and σ does not satisfy any of π_1 to π_{13}.

In the encoding of computations that is defined by cod, $\#\#$ is always a border between the encodings of configurations, whereas single $\#$ separate the elements of configurations. As we encode every state q_i with 0^{i+1}, the predicate π_{14}, which is to be satisfied whenever $\sigma(x)$ contains a factor $\#\#00^{s+1}$, handles all encoded triples (i, j, k) with $i > s + 1$. If σ does not satisfy this simple predicate (in addition to the previous ones), there is a computation $C_1, \ldots, C_n$ of $\mathcal{A}$ with $\mathrm{cod}(C_1, \ldots, C_n) = \sigma(x)$.

All that remains is to choose an appropriate set of predicates that describe all cases where C_1 is not the initial configuration, C_n is not an accepting configuration, or there are configurations C_i, C_{i+1} such that $C_i \vdash_{\mathcal{A}} C_{i+1}$ does not hold (thus, the exact value of μ depends on the number of invalid transitions in $\mathcal{A}$). As this construction is rather lengthy, but similar to the approach of Jiang et al. [9], we abstain from giving a detailed description of the predicates π_{15} to π_μ.

Now, if there is a substitution σ that does not satisfy any of π_1 to π_μ, then $\sigma(x) = \mathrm{cod}(C_1, \ldots, C_n)$ for a computation $C_1, \ldots, C_n$, where C_1 is the initial and C_n a final configuration, and for all $i \in \{1, \ldots, n-1\}$, $C_i \vdash_{\mathcal{A}} C_{i+1}$. Thus, if $\sigma(\alpha_{\mathcal{A}}) \notin L_{E,\Sigma}(\beta_{\mathcal{A}})$, then $\sigma(x) \in \mathrm{VALC}(\mathcal{A})$, which means that $\mathcal{A}$ has an accepting computation.

Conversely, if there is some accepting computation $C_1, \ldots, C_n$ of $\mathcal{A}$, we can define σ through $\sigma(x) := \mathrm{cod}(C_1, \ldots, C_n)$, and choose $\sigma(y)$ to be an appropriately long sequence from 0^*. Then σ does not satisfy any of the predicates π_1 to π_μ defined above, thus $\sigma(\alpha_{\mathcal{A}}) \notin L_{E,\Sigma}(\beta_{\mathcal{A}})$, and $L_{E,\Sigma}(\alpha_{\mathcal{A}}) \nsubseteq L_{E,\Sigma}(\beta_{\mathcal{A}})$.

We conclude that $\mathcal{A}$ has an accepting computation iff $L_{E,\Sigma}(\alpha_{\mathcal{A}})$ is not a subset of $L_{E,\Sigma}(\beta_{\mathcal{A}})$. Therefore, any algorithm deciding the inclusion problem for ePAT_Σ can be used to decide whether a nondeterministic 2-counter automata without input has an accepting computation. As this problem is known to be undecidable, the inclusion problem for ePAT_Σ is also undecidable.

The proof for larger (finite) alphabets requires only little changes to the way the patterns $\alpha_{\mathcal{A}}$ and $\beta_{\mathcal{A}}$ are derived from a given automaton $\mathcal{A}$. Thus, we omit this part of the proof.

This concludes the proof of Theorem 2.

4   The Inclusion of Similar E-Pattern Languages

It can be easily observed that the patterns used for establishing the undecidability of the inclusion problem for E-pattern languages are not *similar* (cf. Section 2). Hence, our reasoning in Section 3.1 does not answer the question of whether the inclusion problem is undecidable for these natural subclasses. In this regard, Jiang et al. [9] demonstrate that for the full class of the *simplest* similar E-pattern languages, namely those generated by *terminal-free* patterns, inclusion is decidable. This insight directly results from the following characterization:

Theorem 5 (Jiang et al. [9]). *Let Σ be an alphabet, $|\Sigma| \geq 2$, and let $\alpha, \beta \in$ Pat$_{tf}$ be terminal-free patterns. Then $L_{E,\Sigma}(\alpha) \subseteq L_{E,\Sigma}(\beta)$ iff there exists a morphism $\phi : X^* \to X^*$ satisfying $\phi(\beta) = \alpha$.*

Note that the decidability of the inclusion problem for terminal-free *NE*-pattern languages is still open.

The problem of the extensibility of Theorem 5 to general similar patterns (replacing $\phi : X^* \to X^*$ by a terminal-preserving morphism $\phi : (\Sigma \cup X)^* \to (\Sigma \cup X)^*$) is not only of intrinsic interest, but it has a major impact on the so far unresolved *equivalence* problem for E-pattern languages (see our explanations below). Therefore it has attracted a lot of attention, and it is largely conjectured in literature (e. g., Dányi, Fülöp [4], Ohlebusch, Ukkonen [17]) that the inclusion of similar E-pattern languages shows the same property as that of terminal-free E-pattern languages. Our main result of the present section, however, demonstrates that, surprisingly, this conjecture is not correct:

Theorem 6. *For every finite alphabet Σ, there exist similar patterns $\alpha, \beta \in$ Pat$_\Sigma$ such that $L_{E,\Sigma}(\alpha) \subset L_{E,\Sigma}(\beta)$ and there is no terminal-preserving morphism $\phi : (\Sigma \cup X)^* \to (\Sigma \cup X)^*$ satisfying $\phi(\beta) = \alpha$.*

Due to space constraints, we do not present a proof for Theorem 6, but we merely give appropriate example patterns for $\Sigma := \{a, b, c, d, e\}$, i. e. $|\Sigma| = 5$:

$$\alpha = x_1 \, a \, x_2 \, a \, x_3 \, b \, x_2 \, b \, x_5 \, c \, x_2 \, c \, x_7 \, d \, x_2 \, d \, x_9 \, e \, x_2 \, e \, x_{11},$$

$$\beta = x_1 \, a \, x_2 x_4 \, a \, x_3 \, b \, x_4 x_6 \, b \, x_5 \, c \, x_6 x_8 \, c \, x_7 \, d \, x_8 x_{10} \, d \, x_9 \, e \, x_{10} x_2 \, e \, x_{11}.$$

The relevance of Theorem 6 for the research on the equivalence problem for E-pattern languages follows from a result by Jiang et al. [8] which says that, for alphabets with at least three letters, two patterns need to be similar if they generate the same E-pattern language:

Theorem 7 (Jiang et al. [8]). *Let Σ be an alphabet, $|\Sigma| \geq 3$, and let $\alpha, \beta \in$ Pat$_\Sigma$. If $L_{E,\Sigma}(\alpha) = L_{E,\Sigma}(\beta)$ then α and β are similar.*

Consequently, in literature the inclusion problem for similar E-pattern languages is mainly understood as a tool for gaining a deeper understanding of the equivalence problem, and the main conjecture by Ohlebusch and Ukkonen [17] expresses the expectation that the relation between inclusion problem for similar E-pattern languages and equivalence problem might be equivalent to the relation between these problems for terminal-free patterns (cf. Theorem 5):

Conjecture 1 (Ohlebusch, Ukkonen [17]). Let Σ be an alphabet, $|\Sigma| \geq 3$, and let $\alpha, \beta \in \mathrm{Pat}_\Sigma$. Then $L_{\mathrm{E},\Sigma}(\alpha) = L_{\mathrm{E},\Sigma}(\beta)$ iff there exist terminal-preserving morphisms $\phi, \psi : (\Sigma \cup X)^* \to (\Sigma \cup X)^*$ satisfying $\phi(\beta) = \alpha$ and $\psi(\alpha) = \beta$.

Note that the existence of ϕ and ψ necessarily implies that α and β are similar.

Ohlebusch and Ukkonen [17] demonstrate that Conjecture 1 holds true for a variety of rich classes of E-pattern languages. In general, however, the conjecture is disproved by Reidenbach [20] using very complex counter example patterns. These patterns are valid for alphabet sizes 3 and 4 only, and their particular construction seems not to be extendable to larger alphabets. Concerning finite alphabets Σ with $|\Sigma| \geq 5$, our result in Theorem 6 does not directly contradict Conjecture 1, since our patterns α, β do not generate identical languages. Thus, there is still a chance that the conjecture is correct for alphabet sizes greater than or equal to 5. Nevertheless, as the considerations by Ohlebusch and Ukkonen [17] are based on a specific expectation concerning the inclusion of similar E-pattern languages which Theorem 6 demonstrates to be incorrect, it seems that the insights given in the present section disprove the very foundations of their approach to the equivalence problem for the full class of E-pattern languages. Therefore we feel that the only remaining evidence that still supports Conjecture 1 for $|\Sigma| \geq 5$ is the lack of known counter-examples.

Furthermore, our result definitely affects the use of the sophisticated proof technique introduced by Filè [5] and Jiang et al. [9] for the proof of Theorem 5. For *terminal-free* patterns α, β and any alphabet Σ with $|\Sigma| \geq 2$, this technique constructs a particular substitution τ_β such that $\tau_\beta(\alpha) \in L_{\mathrm{E},\Sigma}(\beta)$ if *and only if* there is a morphism mapping β onto α. After considerable effort made by Dányi and Fülöp [4], Ohlebusch and Ukkonen [17] and Reidenbach [20] to extend this approach to *general* similar patterns, Theorem 6 demonstrates that such a substitution τ_β does not exist for every pair of such patterns, since, for every finite alphabet Σ, there are similar patterns α, β such that $L_{\mathrm{E},\Sigma}(\beta)$ contains *all* words in $L_{\mathrm{E},\Sigma}(\alpha)$, although there is no terminal-preserving morphism mapping β onto α. Consequently, Theorem 6 shows that the main tool for tackling the inclusion problem for terminal-free E-pattern languages – namely our profound knowledge on the properties of the abovementioned substitution τ_β – necessarily fails if we want to extend it to arbitrary similar patterns, and therefore it seems that the research on the inclusion problem for similar E-pattern languages (and, thus, the equivalence problem for general E-pattern languages) needs to start virtually from scratch again.

References

1. Angluin, D.: Finding patterns common to a set of strings. Journal of Computer and System Sciences 21, 46–62 (1980)
2. Angluin, D.: Inductive inference of formal languages from positive data. Information and Control 45, 117–135 (1980)
3. Câmpeanu, C., Salomaa, K., Yu, S.: A formal study of practical regular expressions. Int. J. Found. Comput. Sci. 14, 1007–1018 (2003)

4. Dányi, G., Fülöp, Z.: A note on the equivalence problem of E-patterns. Information Processing Letters 57, 125–128 (1996)
5. Filè, G.: The relation of two patterns with comparable language. In: Proc. STACS 1988. LNCS, vol. 294, pp. 184–192. Springer, Heidelberg (1988)
6. Freydenberger, D.D., Reidenbach, D., Schneider, J.C.: Unambiguous morphic images of strings. Int. J. Found. Comput. Sci. 17, 601–628 (2006)
7. Ibarra, O.: Reversal-bounded multicounter machines and their decision problems. Journal of the ACM 25, 116–133 (1978)
8. Jiang, T., Kinber, E., Salomaa, A., Salomaa, K., Yu, S.: Pattern languages with and without erasing. Int. J. Comput. Math. 50, 147–163 (1994)
9. Jiang, T., Salomaa, A., Salomaa, K., Yu, S.: Decision problems for patterns. Journal of Computer and System Sciences 50, 53–63 (1995)
10. Karhumäki, J., Mignosi, F., Plandowski, W.: The expressibility of languages and relations by word equations. Journal of the ACM 47, 483–505 (2000)
11. Luo, W.: Compute inclusion depth of a pattern. In: Auer, P., Meir, R. (eds.) COLT 2005. LNCS (LNAI), vol. 3559, pp. 689–690. Springer, Heidelberg (2005)
12. Mateescu, A., Salomaa, A.: Finite degrees of ambiguity in pattern languages. RAIRO Informatique théoretique et Applications 28, 233–253 (1994)
13. Mateescu, A., Salomaa, A.: Patterns. In: [22], pp. 230–242 (1997)
14. Minsky, M.: Recursive unsolvability of Post's problem of "Tag" and other topics in the theory of turing machines. Ann. of Math. 74, 437–455 (1961)
15. Mukouchi, Y.: Characterization of pattern languages. In: Proc. 2nd International Workshop on Algorithmic Learning Theory, ALT 1991, pp. 93–104 (1991)
16. Ng, Y.K., Shinohara, T.: Developments from enquiries into the learnability of the pattern languages from positive data. Theor. Comp. Sci. 397, 150–165 (2008)
17. Ohlebusch, E., Ukkonen, E.: On the equivalence problem for E-pattern languages. Theor. Comput. Sci. 186, 231–248 (1997)
18. Reidenbach, D.: The Ambiguity of Morphisms in Free Monoids and its Impact on Algorithmic Properties of Pattern Languages. Logos Verlag, Berlin (2006)
19. Reidenbach, D.: A non-learnable class of E-pattern languages. Theor. Comput. Sci. 350, 91–102 (2006)
20. Reidenbach, D.: An examination of Ohlebusch and Ukkonen's conjecture on the equivalence problem for E-pattern languages. Journal of Automata, Languages and Combinatorics 12, 407–426 (2007)
21. Reidenbach, D.: Discontinuities in pattern inference. Theor. Comput. Sci. 397, 166–193 (2008)
22. Rozenberg, G., Salomaa, A.: Handbook of Formal Languages, vol. 1. Springer, Berlin (1997)
23. Salomaa, K.: Patterns. In: Martin-Vide, C., Mitrana, V., Păun, G. (eds.) Formal Languages and Applications. Studies in Fuzziness and Soft Computing, vol. 148, pp. 367–379. Springer, Heidelberg (2004)
24. Salomaa, K.: Patterns. Lecture, 5th PhD School in Formal Languages and Applications, URV Tarragona (2006)
25. Shinohara, T.: Polynomial time inference of extended regular pattern languages. In: Proc. RIMS Symposia on Software Sci. Eng. LNCS, vol. 147, pp. 115–127. Springer, Heidelberg (1982)

Finding the Growth Rate of a Regular of Context-Free Language in Polynomial Time

Paweł Gawrychowski[1,*], Dalia Krieger[2], Narad Rampersad[2],
and Jeffrey Shallit[2]

[1] Institute of Computer Science, University of Wrocław
ul. Joliot-Curie 15, PL-50-383 Wrocław, Poland
gawry1@gmail.com
[2] School of Computer Science, University of Waterloo
Waterloo, Ontario, N2L 3G1, Canada
{d2kriege@cs,nrampersad@math,shallit@graceland}.uwaterloo.ca

Abstract. We give an $O(n + t)$ time algorithm to determine whether an NFA with n states and t transitions accepts a language of polynomial or exponential growth. Given a NFA accepting a language of polynomial growth, we can also determine the order of polynomial growth in $O(n+t)$ time. We also give polynomial time algorithms to solve these problems for context-free grammars.

1 Introduction

Let $L \subseteq \Sigma^*$ be a language. If there exists a polynomial $p(x)$ such that $|L \cap \Sigma^m| \leq p(m)$ for all $m \geq 0$, then L has *polynomial growth*. Languages of polynomial growth are also called *sparse* or *poly-slender*. If there exists a real number $r > 1$ such that $|L \cap \Sigma^m| \geq r^m$ for infinitely many $m \geq 0$, then L has *exponential growth*. Languages of exponential growth are also called *dense*. If there exist words $w_1, w_2, \ldots, w_k \in \Sigma^*$ such that $L \subseteq w_1^* w_2^* \cdots w_k^*$, then L is called a *bounded language*.

Ginsburg and Spanier (see [6, Chapter 5], [7]) proved many deep results concerning the structure of bounded context-free languages. One significant result [6, Theorem 5.5.2] is that determining if a context-free grammar generates a bounded language is decidable. However, although it is a relatively straightforward consequence of their work, they did not make the following connection between the bounded context-free languages and those of polynomial growth: a context-free language is bounded if and only if it has polynomial growth. Curiously, this result has been independently discovered at least six times: namely, by Trofimov [23], Latteux and Thierrin [14], Ibarra and Ravikumar [9], Raz [19], Incitti [11], and Bridson and Gilman [2]. A consequence of all of these proofs is that a context-free language has either polynomial or exponential growth; no intermediate growth is possible.

---

* Research supported by MNiSW grant number N N206 1723 33, 2007-2010.

M. Ito and M. Toyama (Eds.): DLT 2008, LNCS 5257, pp. 339–358, 2008.

The particular case of the bounded regular languages was also studied by Ginsburg and Spanier [8], and subsequently by Szilard, Yu, Zhang, and Shallit [22] (see also [10]). Shur [20,21] has also studied the growth rate of regular languages. It follows from the more general decidability result of Ginsburg and Spanier that there is an algorithm to determine whether a regular language has polynomial or exponential growth (see also [22, Theorem 5]). Ibarra and Ravikumar [9] observed that the algorithm of Ginsburg and Spanier runs in polynomial time for NFA's, but they gave no detailed analysis of the runtime. Here we give a linear time algorithm to solve this problem. If the growth rate is polynomial we show how to find the order of polynomial growth in linear time.

For the general case of context-free languages, an analysis of the algorithm of Ginsburg and Spanier shows that it requires exponential time. Assuming that we want to explicitly construct words $w_1, w_2, \ldots, w_k \in \Sigma^*$ such that $L \subseteq w_1^* w_2^* \cdots w_k^*$, this exponential complexity is unavoidable, as there are context-free languages for which an exponentially large value of k is required. Surprisingly, it turns out that using a more complicated algorithm it is possible to check if a given context-free language has polynomial growth in polynomial time. We also give a polynomial time algorithm for finding the exact order of this growth.

Due to space considerations, the proofs of certain lemmas have been removed to the appendix.

2   Regular Languages

2.1   Polynomial vs. Exponential Growth

In this section we give an $O(n+t)$ time algorithm to determine whether an NFA with n states and t transitions accepts a language of polynomial or exponential growth.

Theorem 1. *Given a NFA M, it is possible to test whether $L(M)$ is of polynomial or exponential growth in $O(n + t)$ time, where n and t are the number of states and transitions of M respectively.*

Let $M = (Q, \Sigma, \delta, q_0, F)$ be an NFA. We assume that every state of M is both accessible and co-accessible, i.e., every state of M can be reached from q_0 and can reach a final state. For each state $q \in Q$, we define a new NFA $M_q = (Q, \Sigma, \delta, q, \{q\})$ and write $L_q = L(M_q)$.

Following Ginsburg and Spanier, we say that a language $L \subseteq \Sigma^*$ is *commutative* if there exists $u \in \Sigma^*$ such that $L \subseteq u^*$. The following lemma is essentially a special case of a stronger result for context-free languages (compare [6, Theorem 5.5.1], or in the case of languages specified by DFA's, [22, Lemmas 2 and 3]).

Lemma 1. *The language $L(M)$ has polynomial growth if and only if for every $q \in Q$, L_q is commutative.*

We now are ready to prove Theorem 1.

Proof. Let n denote the number of states of M. The idea is as follows. For every $q \in Q$, if L_q is commutative, then there exists $u \in \Sigma^*$ such that $L_q \subseteq u^*$. For any $w \in L_q$, we thus have $w \in u^*$. If z is the primitive root of w, then z is also the primitive root of u. If $L_q \subseteq z^*$, then L_q is commutative. On the other hand, if $L_q \not\subseteq z^*$, then L_q contains two words with different primitive roots, and is thus not commutative. This argument leads to the following algorithm.

Let $q \in Q$ be a state of M. We wish to check whether L_q is commutative. Any accepting computation of M_q can only visit states in the strongly connected component of M containing q. We therefore assume that M is indeed strongly connected (if it is not, we run the algorithm on each strongly connected component separately; they can be determined in $O(n+t)$ time [3, Section 22.5]).

We first construct the NFA M_q accepting L_q. This takes $O(n+t)$ time. Then we find a word $w \in L(M_q)$, where $|w| < n$. If $L(M_q)$ is non-empty, such a w exists and can be found in $O(n+t)$ time. Next we find the *primitive root* of w, i.e., the shortest word z such that $w = z^k$ for some $k \geq 1$. This can be done in $O(n)$ time using any linear time pattern matching algorithm. To find the primitive root of $w = w_1 \cdots w_\ell$, find the first occurrence of w in $w_2 \cdots w_\ell w_1 \cdots w_{\ell-1}$. If the first occurrence begins at position i, then $z = w_1 \cdots w_i$ is the primitive root of w.

For $i = 0, 1, \ldots, |z| - 1$, let A_i be the set of all $q' \in Q$ such that there is a path from q to q' labeled by a word from $z^* z_1 z_2 \cdots z_i$. Observe that if some q' belongs to A_i and A_j where $i < j$ then we can find two different paths from q to q: $z^a z_1 \cdots z_i s$ and $z^b z_1 \cdots z_j s$, where a and b are non-negative integers and s is the label of some path from q' to q. For L_q to be commutative, both these words must be powers of z, which is impossible: their lengths are different modulo $|z|$. Thus, the A_i's must be disjoint.

We determine the A_i's as follows. To begin, we know $q \in A_0$. For any i, if $q' \in A_i$, then we know $q'' \in A_{(i+1) \bmod |z|}$ for all $q'' \in \delta(q', z_{1+i \bmod |z|})$. Based on this rule, we proceed to iteratively assign states to the appropriate A_i until all states have been assigned. If some q' appears in two distinct A_i's, we terminate and report that L_q is not commutative. Since we never need to examine a state more than once, it follows that the complexity of computing the A_i's is $O(n+t)$.

Next, for each i we check that for each $q' \in A_i$ all outgoing transitions are labeled by $z_{1+i \bmod |z|}$. If not, L_q cannot be commutative (as we could then find a path from q to q that is not a power of z). If this holds for all i, the automaton has a very simple structure: it is an $|z|$-partite graph (the A_i's forming the partition classes) and edges outgoing from one partition class all have the same label. Thus, every path that starts and ends in q is a power of z. Furthermore, every path that starts and ends in some q' is a power of some cyclic shift of z. Thus, $L_{q'}$ is also commutative, so we do not have to repeat the computation for the remaining states of M (i.e., the states in $Q \setminus \{q\}$).

We have therefore determined whether L_q is commutative for all $q \in Q$, and hence whether $L(M)$ has polynomial or exponential growth, in $O(n+t)$ time. $\square$

2.2   Finding the Exact Order of Polynomial Growth

In this section we show that given an NFA accepting a language of polynomial growth, it is possible to efficiently determine the exact order of polynomial growth.

Let M be an NFA accepting a language of polynomial growth. We will need the following definition:

Definition 1. *We call* $x_0 y_1^* x_1 y_2^* x_2 \ldots y_k^* x_k$*, where each* y_i *is non-empty, a star expression of level* k*. We say that it is primitive when each* y_i *is primitive.*

We would like to decompose $L(M)$ into languages described by such expressions. Let us look at the exact order of polynomial growth of the language described by a primitive star expression. It is easy to see that $L(x_0 y_1^* x_1 y_2^* x_2 \ldots y_k^* x_k)$ has $O(m^{k-1})$ growth; getting a lower bound is slightly more involved. Let $T = x_0 y_1^* x_1 y_2^* x_2 \ldots y_k^* x_k$ be a primitive star expression. If there exists $1 \leq i < k$ such that $x_i = y_i^l y_i[1..j]$ and $y_{i+1} = y_i^{(j)}$ (y_i cyclically shifted by j) for some $l \geq 0$ and $j < |y_i|$, then we say that i is a *fake index*.

Lemma 2. *Let* $T = x_0 y_1^* x_1 y_2^* x_2 \ldots y_k^* x_k$ *be a primitive star expression.* $L(T)$ *has* $\Theta(m^{k-1})$ *growth iff* T *has no fake indices.*

Extending the above lemma gives us the following result:

Lemma 3. *Let* $T = x_0 y_1^* x_1 y_2^* x_2 \ldots y_k^* x_k$ *be a primitive star expression.* $L(T)$ *has* $\Theta(m^{k-1-d})$ *growth iff there are exactly* d *fake indices in* T*.*

We thus have an efficient way of calculating the growth order of a language described by a primitive star expression. However, we need a stronger result:

Theorem 2. *Let* $T = x_0 y_1^* x_1 y_2^* x_2 \ldots y_k^* x_k$ *be a star expression.* $L(T)$ *has* $\Theta(m^{k-1-d})$ *growth iff there are exactly* d *indices* $1 \leq i < k$ *such that* $x_i = \mathrm{root}(y_i)^l \mathrm{root}(y_i)[1..j]$ *and* $\mathrm{root}(y_{i+1}) = \mathrm{root}(y_i)^{(j)}$ *for some* $l \geq 0$ *and* $j < |\mathrm{root}(y_i)|$*, where* $\mathrm{root}(w)$ *stands for the primitive root of* w*.*

Proof. Obviously, replacing each y_i by its primitive root cannot remove any word from the described language. Thus $L(T)$ has $O(m^{k-1-d})$ growth. To show that it is $\Omega(m^{k-1-d})$, we use the same method as in the previous two lemmas (see the appendix for the proofs): first we get rid of all d fake indices (here it may happen that we decrease the language in question, as replacing $(y^a)^*(y^b)^*$ by $(y^a)^*$ only might be necessary), then construct the corresponding equation and show that it has at most one solution for each word. □

Corollary 1. *Inserting an additional* y^* *into a star expression* T *and replacing any* y_i *by some power of its primitive root or* y_i^* *by* $y_i^a y_i^* y_i^b$ *do not change the growth order of* $L(T)$*.*

Now we return to the original problem. Given M, we will construct a new automaton M' which will be almost acyclic, the only cycles being self-loops. Recall that as $L(M)$ has polynomial growth, the algorithm of Theorem 1 partitions the vertices of each strongly connected component into sets $A_0, A_1, \ldots, A_{|z|-1}$. Take one such component S and let z be the corresponding primitive word. Now choose any $u, v \in S$ and consider labels of paths from u to v:

Case 1: if $u \in A_i, v \in A_j$ and $i \geq j$, then they are all of the form $z[i + 1..|z|]z^p z[1..j]$. Furthermore, for some $q, r > 0$, there are such paths for every $p \in \{q, q + r, q + 2r, \ldots\}$.

Case 2: if $u \in A_i, v \in A_j$ and $i < j$, then they are all of the form $z[i + 1..j]$ or $z[i + 1..|z|]z^p z[1..j]$. As in the previous case, for some $q, r > 0$, there are paths of the second form for every $p \in \{q, q + r, q + 2r, \ldots\}$.

This suggests the following construction: for each state v create two copies v_{in} and v_{out}. Each edge connecting u and v belonging to two different strongly connected components gets replaced by an edge from u_{out} to v_{in} with the same label. For each non-singleton strongly connected component S we create the following gadget:

$$b_0 \xrightarrow{z[1]} b_1 \xrightarrow{z[2]} b_2 - - \to b_{s-1} \xrightarrow{z[s]} \overset{z}{\underset{}{\circlearrowright}} a_0 \xrightarrow{z[1]} a_1 \xrightarrow{z[2]} a_2 - - \to a_{s-1} \xrightarrow{z[s-1]} a_s$$

For each $u \in S$ we find i such that $u \in A_i$ and add edges $u_{in} \xrightarrow{\epsilon} b_i$, $b_i \xrightarrow{\epsilon} u_{out}$ and $a_i \xrightarrow{\epsilon} u_{out}$. The starting state of M' is simply q_{in}, where q is the start state of M, and v_{out} is a final state of M' whenever v is a final state of M.

$L(M')$ should be viewed as a finite sum of languages described by star expressions corresponding to labels of simple paths from the start state to some final state (whenever there is a self-loop adjacent to some vertex v and labeled with z, we should think that v is labelled with z^*). Let $\mathcal{T}$ denote the set of all such star expressions, so that $L(M') = \sum_{T \in \mathcal{T}} L(T)$.

Theorem 3. *The orders of polynomial growth of $L(M)$ and $L(M')$ are the same.*

Now we can focus on finding the growth order of $L(M')$. Although M' has a relatively simple structure, $\mathcal{T}$ can be of exponential size, so we cannot afford to construct it directly. This turns out to not be necessary due to the characterization of growth orders of primitive star expressions that we have developed. Take an expression $T = x_0 y_1^* x_1 y_2^* x_2 \ldots y_k^* x_k$ and observe that to calculate the growth order of $T y_{k+1}^* x_{k+1}$ we only need to know the growth order of T, the word y_k and if x_k is of the form $y_k^l y_k[1..j]$, the value of j. This suggests a dynamic programming solution: for each vertex v of M' we calculate: (1) the greatest possible growth order of a star expression $T = x_0 y_1^* x_1 y_2^* x_2 \cdots y_k^* x_k$ such that x_k is not a prefix of y_k^∞ and T is a label of a path from the start state to v; (2) for each y_k being a label of some self-loop and $j < |y_k|$, the greatest possible growth order of a star expression $T = x_0 y_1^* x_1 y_2^* x_2 \cdots y_k^* x_k$ such that x_k is of the form $y_k^l y_k[1..j]$ and T is a label of a path from the start state to v.

We process the vertices of M' in topological order (ignoring the self-loops gives us such order). Assuming that we already have the correct information for some vertex, we can iterate through the outgoing transitions and update the information for all its successors. For example, knowing that there is a path ending in u whose label is a star expression $\cdots y_k^* y_k^l y_k[1..j]$ having a growth order of d, a transition $u \xrightarrow{c} v$ and a self-loop $v \xrightarrow{z} v$, we can construct a path ending in v whose label is a star expression $\cdots y_k^* y_k^l y_k[1..j]cz^*$ having a growth order of $d+1$ if $c \neq y_k[j+1]$ or $z \neq y_k^{(j+1)}$, and d otherwise. At the very beginning we know only that for any self-loop $v \xrightarrow{z} v$ there is a path ending in v whose label is a star expression $\cdots z^*$ having a growth order of 0.

We have to calculate $O(|Q|)$ information for each vertex. There are at most $O(|Q||\delta|)$ updates and each of them can be done in $O(1)$ time if we can decide in constant time whether $z' = z^{(j)}$ for any j and labels z, z' of some self-loops in M'. As both z, z' are primitive, there can be at most one such j. We can preprocess it for all pairs of labels in time $O(|Q|^2)$, giving a $O(|Q||\delta|)$ algorithm.

It turns out that we can achieve linear complexity by reducing the amount of information kept for each vertex. First observe that whenever we have two labels z, z' of self-loops such that $z' = z^{(i)}$, any star expression $\cdots z'^* z'^l z'[1..j]$ can be treated as an expression $\cdots z^* z^l z[1..1 + (i + j - 1) \mod |z|]$ having the same growth order. After such reductions for each vertex v we can keep information only about those two expressions of the form $\cdots y_k^* y_k[1..j]$ that have the greatest growth order among all possible y_k and j. Indeed, whenever we have an expression $T_1 T_2$ such that $T_1 = \cdots y_k^* y_k[1..j]$ is a label of a path ending in v and there are two different (with respect to the above reduction) expressions T_1', T_1'' that have greater or equal order of growth and are also labels of paths ending in v, at least one of $T_1' T_2$, $T_1'' T_2$ will have growth order as large as $T_1 T_2$. This decreases the amount of information kept for each vertex to a constant. To get linear total complexity we must improve the runtime of the preprocessing phase. Recall that it is possible to find the lexicographically smallest cyclic shift of a word in linear time, for example by using Duval's algorithm. Such a shift is the same for any two conjugate primitive words, so we calculate the smallest cyclic shifts of all labels of self-loops and then group those labels whose shifts are the same. This can be done by inserting them one-by-one into a trie. After such preprocessing we can find the value of j such that $z' = z^{(j)}$ in constant time for any two labels z, z'.

Theorem 4. *Given an NFA M with n states and t transitions such that $L(M)$ is of polynomial growth, there is an algorithm that finds the exact order of polynomial growth of $L(M)$ in $O(n + t)$ time.*

2.3   An Algebraic Approach for DFA's

We now consider an algebraic approach to determining whether the order of growth is polynomial or exponential, and in the polynomial case, the order of polynomial growth. Shur [21] has recently presented a similar algebraic method for this problem. Let $M = (Q, \Sigma, \delta, q_0, F)$ be a DFA, where $|Q| = n$, and let $A = A(M) = (a_{ij})_{1 \leq i, j \leq n}$ be the *adjacency matrix* of M, that is, a_{ij} denotes the number of paths of length 1 from q_i to q_j. Then $(A^m)_{i,j}$ counts the number of paths of length m from q_i to q_j. Since a final state is reachable from every state q_j, the order of growth of $L(M)$ is the order of growth of A^m as $m \to \infty$. This order of growth can be estimated using nonnegative matrix theory.

Theorem 5 (Perron-Frobenius). *Let A be a nonnegative square matrix, and let r be the spectral radius of A, i.e., $r = \max\{|\lambda| : \lambda$ is an eigenvalue of $A\}$. Then*

1. *r is an eigenvalue of A;*
2. *there exists a positive integer h such that any eigenvalue λ of A with $|\lambda| = r$ satisfies $\lambda^h = r^h$.*

For more details, see [17, Chapters 1, 3].

Definition 2. The number $r = r(A)$ described in the above theorem is called the *Perron-Frobenius eigenvalue of A*. The *dominating Jordan block* of A is the largest block in the Jordan decomposition of A associated with $r(A)$.

Lemma 4. *Let A be a nonnegative $n \times n$ matrix over the integers. Then either $r(A) = 0$ or $r(A) \geq 1$.*

Proof. Let $r(A) = r, \lambda_1, \ldots, \lambda_\ell$ be the distinct eigenvalues of A, and suppose that $r < 1$. Then $\lim_{m \to \infty} r^m = \lim_{m \to \infty} \lambda_i^m = 0$ for all $i = 1, \ldots, \ell$, and so $\lim_{m \to \infty} A^m = 0$ (the zero matrix). But A^m is an integral matrix for all $m \in \mathbb{N}$, and the above limit can hold if and only if A is nilpotent, i.e., $r = \lambda_i = 0$ for all $i = 1, \ldots, \ell$. $\qquad\square$

Lemma 5. *Let A be a nonnegative $n \times n$ matrix over the integers. Let $r(A) = r, \lambda_1, \ldots, \lambda_\ell$ be the distinct eigenvalues of A, and let d be the size of the dominating Jordan block of A. Then $A^m \in \Theta(r^m m^{d-1})$.*

Note: The growth order of A^m supplies an algebraic proof of the fact that regular languages can grow either polynomially or exponentially, but no intermediate growth order is possible. This result can also be derived from a more general matrix-theoretic result of Bell [1].

Lemma 5 implies that to determine the order of growth of $L(M)$, we need to compute the Perron-Frobenius eigenvalue r of $A(M)$: if $r = 0$, then $L(M)$ is finite; if $r = 1$, the order of growth is polynomial; if $r > 1$, the order of growth is exponential. In the polynomial case, if we want to determine the order of polynomial growth, we need to also compute the size of the dominating Jordan block, which is the algebraic multiplicity of r in the minimal polynomial of $A(M)$.

Both computations can be done in polynomial time, though the runtime is more than cubic. The characteristic polynomial, $c_A(x)$, can be computed in $\tilde{O}(n^4 \log \|A\|)$ bit operations (here $\tilde{O}$ stands for soft-O, and $\|A\|$ stands for the L_∞ norm of A). If $c_A(x) = x^n$ then $r = 0$; else, if $c_A(1) \neq 0$, then $r > 1$. In the case of $c_A(1) = 0$, we need to check whether $c_A(x)$ has a real root in the open interval $(1, \infty)$. This can be done using a real root isolation algorithm; it seems the best deterministic one uses $\tilde{O}(n^6 \log^2 \|A\|)$ bit operations [4]. The minimal polynomial, $m_A(x)$, can be computed through the rational canonical form of A in $\tilde{O}(n^5 \log \|A\|)$ bit operations (see references in [5]). All algorithms mentioned above are deterministic; both $c_A(x)$ and $m_A(x)$ can be computed in $\tilde{O}(n^{2.697} \log \|A\|)$ bit operations using a randomized Monte Carlo algorithm [12].

An interesting problem is the following: given a nonnegative integer matrix A, is it possible to decide whether $r(A) > 1$ in time better than $\tilde{O}(n^6 \log^2 \|A\|)$? Using our combinatorial algorithm, we can do it in time $O(n^2 \log \|A\|)$, as follows. We first construct a graph G from A by creating edges (i, j) in G if the ij entry of A is positive. We can do this in $O(n^2 \log \|A\|)$ time. We then find the strongly connected components of G in $O(n^2)$ time. For each edge (i, j), if i and j are in the same strongly connected component and the ij entry of A is > 1, then we may immediately report that $r(A) > 1$. We thus assume henceforth that if i and j are in the same strongly connected component of G, then the ij entry of A is at most 1.

We now consider the strongly connected components of G separately. For each strongly connected component H of G, we turn H into a DFA M (we don't bother specifying an initial state or final states) over an m-letter alphabet, where m is the number of vertices of H, as follows. For each edge (i, j) in H, we turn the edge (i, j) into a single transition labeled by an alphabet symbol that has not already been used for a transition outgoing from i. This is justified by our previous assumption that the ij entries of A within the same strongly connected component are at most 1. Thus, there are at most m outgoing transitions from a given state, and the DFA M has m states and $O(m^2)$ transitions. We now run the algorithm of Theorem 1 to determine if all of the L_q's of M are commutative in $O(m^2)$ time. If so, then H has polynomial growth; otherwise, it has exponential growth.

If all strongly connected components of G have polynomial growth, then $r(A) \leq 1$; otherwise, $r(A) > 1$. The total running time of the algorithm is $O(n^2 \log \|A\|)$.

3   Context-Free Languages

Given a context-free grammar $G = (V, \Sigma, R, S)$, we are interested in checking whether $L(G)$ has polynomial growth. We assume that G is in Chomsky normal form, each nonterminal can be derived from S and languages generated by nonterminals are nonempty. The following result can be found in [6, Theorem 5.5.1].

Lemma 6. *The language generated by a CFG $G = (V, \Sigma, R, S)$ is bounded if and only if for each nonterminal A both $\mathrm{left}(A)$ and $\mathrm{right}(A)$ are commutative, where $\mathrm{left}(A) = \{u : A \overset{*}{\Rightarrow} uAw \text{ for some } w \in \Sigma^*\}$ and $\mathrm{right}(A) = \{u : A \overset{*}{\Rightarrow} wAu \text{ for some } w \in \Sigma^*\}$.*

From now on we focus on testing whether each $\mathrm{left}(A)$ is commutative. To test all $\mathrm{right}(A)$ we reverse all productions and repeat the whole procedure. Note that if $L \subseteq \Sigma^*$ is commutative and $w \in L$, then $L \subseteq \mathrm{root}(w)^*$. So, to check if a nonempty left(A) is commutative we should: (1) take any $u \in$ left(A) and set w to be its primitive root, and (2) verify that each $u \in$ left(A) is a power of w.

Before we proceed further we need a convenient description of left(A). Define a graph $H = (V, E)$ where V is the set of all nonterminals and for each production $A \rightarrow BC$ we put $A \overset{\epsilon}{\rightarrow} B$ and $A \overset{B}{\rightarrow} C$ into E. Each left(A) is a sum of languages generated by labels of paths from A to A, where the label of a path is the concatenation of the labels of all of its edges.

In our algorithm we will make heavy use of results concerning *straight-line programs*. A SLP is a context-free grammar in Chomsky normal form such that each nonterminal occurs on the left side of exactly one production and derives exactly one word. Such grammars should be viewed as a convenient way of describing compressed words. Given a text T and a pattern P, both as SLPs, there are polynomial time algorithms for finding the first occurrence of P in T or detecting that there is no such occurrence (see [13] or [15] for a more efficient version). Given a SLP describing some word w we can easily construct a SLP describing any subword of w.

Constructing a SLP describing some $u \in$ left(A) is quite straightforward. We first define the function length(A) := $\min\{|w| : A \overset{*}{\Rightarrow} w\}$. We need the following:

Lemma 7. *Given a context free grammar, we can calculate for each nonterminal A the value of* length(A) *in polynomial time.*

To construct u, first use the above lemma. For each nonterminal A choose one production: $A \rightarrow a$ if there is such an a, and $A \rightarrow BC$ for which length(A) = length(B)+length(C) otherwise. It is easy to see that after removing all the other productions any nonterminal A still generates some word. Let $X_1 X_2 \ldots X_k$ be a label of one of the simple paths from A to A in H. If $k = 1$ we can take X_1 as the start symbol. Otherwise we add productions $Y_i \rightarrow X_i Y_{i+1}$ for $i = 1, \ldots, k - 2$ and $Y_{k-1} \rightarrow X_{k-1} X_k$ and make Y_1 the start symbol. It is easy to verify that in both cases the resulting grammar is a SLP describing some $u \in$ left(A).

Having a description of some $u \in$ left(A), we construct the description of $u[2..|u|]u[1..|u| - 1]$ and use one of the compressed pattern matching algorithms to find the length of root(u). Having this length, we can easily construct a description of the primitive word w itself.

Now we should verify that $L(X_1 \ldots X_k) \subseteq w^*$ for each label $X = X_1 \ldots X_k$ of a path from A to A in H. If such containment does not hold we say that we found a *contradiction*. Let S be the strongly connected component of H containing A. The verification can be done in two steps:

Lemma 8. *There is a polynomial time algorithm that detects a contradiction or calculates for any $B \in S$ the value* pathlength(B) *such that the label of any path from A to B derives only words of lengths giving the remainder of* pathlength(B) *when divided by $|w|$.*   □

Proof. First we apply Lemma 7. Then we need to verify that for any nonterminal C being a label of some edge connecting two vertices in S, the lengths of all words in $L(C)$ give the same remainder as length(C) when divided by $|w|$. This can be done by checking that $\text{length}(X) \equiv \text{length}(Y) + \text{length}(Z) \mod |w|$ holds for any production $X \to YZ$ that can appear in a derivation of some word in $L(C)$. This condition is clearly necessary: if it does not hold, we can find $u, v \in L(C)$ such that $|u| \not\equiv |v| \mod |w|$ and a path from A to A having a label of the form $X_1 X_2 \ldots X_i C X_{i+2} \ldots X_k$. This label derives two words whose lengths give different remainders when divided by $|w|$ so they cannot both be a power of w and we found a contradiction. To prove that this condition is also sufficient, we use induction to show that in such a case the lengths of all words in any $L(C)$ give the same remainder as $\text{length}(C)$ when divided by $|w|$. Indeed, it holds for all words having a derivation tree of depth 1. Assume that it holds for all words having a derivation tree of depth less than m and take $u \in L(C)$ having a derivation tree of depth $m > 1$. There must be a production $C \to DE$ such that $D \stackrel{*}{\Rightarrow} u_1$, $E \stackrel{*}{\Rightarrow} u_2$ where $u = u_1 u_2$ and the derivation trees of both u_1 and u_2 have depths less than m. Thus from the induction hypothesis $|u_1| \equiv \text{length}(D) \mod |w|$ and $|u_2| \equiv \text{length}(E) \mod |w|$, so $|u| \equiv \text{length}(D) + \text{length}(E) \mod |w|$. We verified that $\text{length}(C) \equiv \text{length}(D) + \text{length}(E) \mod |w|$ so in fact $|u| \equiv \text{length}(C) \mod |w|$ and we are done.

Now we can define the values of $\text{pathlength}(B)$. For each edge in S set its weight to be 0 if its label is ϵ or $\text{length}(B)$ if it is some nonterminal B. Then for each $B \in S$ define $\text{pathlength}(B)$ to be the weight modulo $|w|$ of some path from A to B. Obviously, this value is the only possible candidate for $\text{pathlength}(B)$. We still need to check if it is correct, though. Verify that $\text{pathlength}(B) + \text{length}(s) \equiv \text{pathlength}(C) \mod |w|$ for each edge $B \stackrel{s}{\to} C$ in S where $\text{length}(\epsilon) = 0$. This condition is obviously necessary: otherwise we would have two paths from A to A whose labels derive words having different lengths modulo $|w|$ and a contradiction can be found. To see that it is also sufficient, we use induction to prove that the length of any word that can be derived from a label of any path from A to some B gives the remainder of $\text{pathlength}(B)$ when divided by $|w|$.   □

Having the above lemma, we should check whether for each edge in S outgoing from some B and having a nonempty label of C, each word in $L(C)$ is a prefix of $(w[\text{pathlength}(B) + 1..|w|]w[1..\text{pathlength}(B)])^\infty$. This is clearly both necessary and sufficient, as it ensures that any word that can be derived from a label of any path starting in A and ending in B is of the form $w^* w[1..\text{pathlength}(B)]$.

Lemma 9. *There is a polynomial time algorithm that detects a contradiction or verifies that each word in $L(B)$ is a prefix of $(w[i..|w|]w[1..i-1])^\infty$.*

Proof. We describe the algorithm in terms of constraints. The meaning of a constraint $\langle B, i \rangle$ is that each word from $L(B)$ should be a prefix of $(w[i..|w|]w[1..i-1])^\infty$. We begin with only one such constraint, namely $\langle B, i \rangle$, which is initially marked as unprocessed. While there is an unprocessed constraint $\langle B, i \rangle$ for some B from which it is possible to derive a word of length at least $|w|$, we mark it as processed and add new constraints $\langle C, i \rangle$ and $\langle D, 1 + (i + \text{length}(C) - 1) \mod|w| \rangle$ for each production $B \to CD$. We do not add a new constraint if it has been already processed. To achieve polynomial time we need the following observation: if we have two processed constraints $\langle B, i \rangle$ and $\langle B, j \rangle$ where $i \neq j$ then $L(B)$ contains a word of length at least $|w|$ that should be a prefix of both $(w[i..|w|]w[1..i-1])^\infty$ and $(w[j..|w|]w[1..j-1])^\infty$. But then $w^{(i)} = w^{(j)}$ and w cannot be primitive. Thus we have found a contradiction as soon as the number of processed constraints is greater than the number of nonterminals and we can terminate. Checking if a given $L(B)$ contains a word of length at least $|w|$ can be done by identifying nonterminals that generate infinite languages first. All the others create an *acyclic* part of the grammar, in the sense that we can order them as $A_1, A_2, \ldots A_m$ in such a way that whenever $A_i \rightarrow A_j A_k$ is a production $j, k > i$ holds. Thus we can use a simple dynamic programming to calculate for each A_i the greatest length of a word in $L(A_i)$. This ensures a polynomial running time.

As a result we get a set of unprocessed constraints of the form $\langle A_i, r \rangle$ where all A_i belong to the acyclic part of the grammar. Additionally, we verified earlier that all lengths of words in each $L(A_i)$ give the same remainder when divided by $|w|$. Thus each such constraint can be rewritten as $L(A_i) = \{w'\}$ where w' is described by a SLP of polynomial size. Now we would like to verify this by either using the compressed pattern matching algorithm. Unfortunately, it may happen that the context free grammar describing a given $L(A_i)$ is not a SLP: it may happen that a nonterminal appears on the left side of more than one production. On the other hand, if for each A_i we remove all but one production with A_i on the left side, we get a SLP: each nonterminal generates exactly one word and appears on the left side of exactly one production. If this SLP does not generate w', we found a contradiction. If it does, for each removed production $A_i \rightarrow A_j A_k$ check whether $A_j A_k$ and A_i generate the same word in the constructed SLP. Similarly, for each removed production $A_i \rightarrow a$ check whether A_i generates a in the constructed SLP. This is obviously a necessary condition. To prove that it is also sufficient, we use induction to show that for each $i = m, m - 1, \ldots, 1$ $|L(A_i)| = 1$. It is clear for $i = m$. Assume that it holds for all $j > i$ but A_i contains two different words u, v. We can assume that u is generated by A_i in the SLP. If the uppermost productions in the derivation trees of u and v are the same, we can immediately use the induction hypothesis. Otherwise, the condition we verified with the induction hypothesis give us that $u = v$. □

Combining these two lemmas gives:

Theorem 6. *There is a polynomial time algorithm that checks whether the language generated by a given context-free grammar has polynomial growth.*

Having checked that $L(G)$ is bounded, we would like to calculate the exact order of its polynomial growth as we did in the NFA case. The general idea is almost the same: we decompose $L(G)$ into languages described by star expressions and use dynamic programming (slightly more involved than in the NFA case) to calculate the greatest growth order of those expressions.

First for each nonterminal A we find primitive words $\text{left}_A, \text{right}_A$ using the above method such that $\text{left}(A) \subseteq \text{left}_A^*, \text{right}(A) \subseteq \text{right}_A^*$, and for some $\alpha, \beta \geq 1$, left$_A^\alpha \in$ left(A) and right$_A^\beta \in$ right(A) . In the case that one of the languages in question is empty, we take ϵ as the corresponding word.

We would like to construct a set of star expressions such that the maximum of their growth orders is the same as the growth order of $L(G)$, which was relatively simple in the case of regular languages. When we deal with context-free languages, things get more complicated. Consider a grammar $A \rightarrow uAv|a$ where u and v are different primitive words. Its growth order is clearly 0 while the obvious way of representing the language it generates as a star expression would give u^*av^* with a growth order of 1. On the other hand, adding a production $A \rightarrow u^2Av$ increases the growth order to 1 and in such a case we can represent the language in question as u^*av^*. It turns out that those two extreme situations are in some sense the only possibilities. To formalize this statement, we need

Definition 3. *We call* $x_0y_1^{v_1}x_1y_2^{v_2}x_3 \ldots y_k^{v_k}x_k$, *where each* y_i *is nonempty and all* v_i *are different variables, a generalized star expression. Additionally, we are allowed to add constraints of the form* $(v_i, v_j) \in C \subseteq \mathbb{N}^2$ *as long as each variable is a part of at most one such constraint.*

The set of words described by such an expression contains words that we get by assigning nonnegative values to variables in a way that is consistent with all the constraints. Thus a star expression is just a generalized star expression with no constraints. This notion allows us to represent $L(G)$ in a convenient way. For any nonterminal A define

$$L_1(A) := \{a : A \rightarrow a \text{ is a production}\}$$
$$L_{i+1}(A) := L_i(A) \cup \bigcup_{A \rightarrow BC} \text{left}_A^\alpha L_i(B) L_i(C) \text{right}_A^\beta,$$

where $i = 1, 2, \ldots, n-1$ and $(\alpha, \beta) \in \text{context}(A) := \left\{(\alpha, \beta) : A \overset{*}{\Rightarrow} \text{left}_A^\alpha A \text{ right}_A^\beta\right\}$. Each $L_i(A)$ corresponds in a natural way to a finite sum of languages described by generalized star expressions and it is clear that $L_n(S) = L(G)$. We need to get rid of the constraints without modifying the growth order. This can be done in two phases:

Definition 4. *Given a nonterminal* A *such that* left$_A$, right$_A \neq \epsilon$, *we say that it is independent if* $(\alpha, \beta_1), (\alpha, \beta_2) \in \text{context}(A)$ *for some* $\alpha, \beta_1 \neq \beta_2$.

First we remove constraints concerning independent nonterminals, then we modify constraints concerning dependent nonterminals:

Lemma 10. *Given a generalized star expression, we can remove a constraint* $(v_i, v_j) \in \text{context}(A)$, *where* A *is an independent nonterminal, without changing the growth order of the language in question.*

Lemma 11. *Given a generalized star expression, we can replace each constraint* $(v_i, v_j) \in \text{context}(A)$, *where* A *is a dependent nonterminal, by either* $(v_i, v_j) \in \mathbb{N} \times \{1\}$ *or* $(v_i, v_j) \cup \{1\} \times \mathbb{N}$ *without changing the growth order.*

These two lemmas allow us to modify the definition of each $L_{i+1}(A)$ so as to get a set of primitive star expressions. If A is an independent nonterminal, we set

$$L_{i+1}(A) := L_i(A) \cup \bigcup_{A \to BC} \text{left}_A^* L_i(B) L_i(C) \text{right}_A^*$$

and otherwise we take

$$L_{i+1}(A) := L_i(A) \cup \bigcup_{A \to BC} \text{left}_A L_i(B) L_i(C) \text{right}_A^* \cup \bigcup_{A \to BC} \text{left}_A^* L_i(B) L_i(C) \text{right}_A$$

Before we show how to calculate the greatest growth order of an expression in $L_n(S)$, we need to prove that the above definition can be effectively used:

Lemma 12. *Given a context-free grammar, we can check in polynomial time if a given nonterminal A is independent.*

Now we can focus on finding the greatest growth order of a primitive star expression in $L_n(S)$. As in the NFA case, we observe that calculating the growth order of $\text{left}_A^* T_1 T_2 \text{right}_A^*$ requires only a partial knowledge about the structure of T_1 and T_2. Indeed, if $T_1 = x_0 y_1^* x_1 \cdots y_k^* x_k$ and $T_2 = x_0' y_1'^* x_1' \cdots y_{k'}'^* x_{k'}'$, where $k, k' \geq 1$, we need only to know the words y_k, y_1', the growth orders of T_1, T_2, the length of x_k modulo $|y_k|$ if $x_k \in y_k^* y_k[1..j]$, and the length of x_0' modulo $|y_1'|$ if $x_0' \in y_1'[j'..|y_1'|]y_1'^*$. So, for each $L_i(A)$ we would like to calculate the greatest possible growth order of a star expression $T = x_0 y_1^* x_1 \ldots y_k^* x_k$ from $L_i(A)$ such that $k \geq 1$, y_1 and y_k are some left_B or right_B, and one of the following cases applies:

1. $x_0 \notin y_1[j'..|y_1|]y_1^*$ for all j' and $x_k \notin y_k^* y_k[1..j]$ for all j,
2. $x_0 \notin y_1[j'..|y_1|]y_1^*$ for all j' and $x_k \in y_k^* y_k[1..j]$,
3. $x_0 \in y_1[j'..|y_1|]y_1^*$ and $x_k \notin y_k^* y_k[1..j]$ for all j,
4. $x_0 \in y_1[j'..|y_1|]y_1^*$ and $x_k \in y_k^* y_k[1..j]$,

where $j = 0, 1, \ldots, |y_k| - 1$ and $j' = 1, 2, \ldots, |y_1|$. There is one problem, though: the lengths $|y_1|$ and $|y_k|$ can be exponential and we cannot afford to store information about an exponential number of states. To overcome this, observe that for a fixed y_1, y_k and j', it makes sense to keep information only about two different values of j for which the corresponding growth orders are greatest. The same applies to j'. This allows us to use dynamic programming: now there is only a polynomial number of different states to consider. Assuming that we have calculated the greatest growth orders of expressions in $L_i(A)$ for any nonterminal A, we can calculate growth order of expressions in all $L_{i+1}(A)$. There is one problem, though: in the above reasoning we assumed that both T_1 and T_2 have orders of at least 1 but three other cases are also possible:

Case 1: both T_1 and T_2 have orders of 0. As we are interested in getting an expression of order at least 1, we may assume that $\text{left}_A \neq \epsilon$ or $\text{left}_B \neq \epsilon$. $T_1 T_2$

is just a word that can be derived from $B_n C_n$ using only the following new productions:

$$A_i \to a \text{ for each original production } A \to a$$
$$A_{i+1} \to B_i C_i \text{ for each original production } A \to BC$$

where $i = 1, 2, \ldots, n-1$. W.l.o.g. assume that $\text{left}_A \neq \epsilon$. To update information about growth orders of expressions of the form $\text{left}_A^* T_1 T_2 \text{right}_A^*$, we could calculate all j such that $T_1 T_2$ can be of the form $\text{left}_A^* \text{left}_A[1..j]$ and check if it is possible that $T_1 T_2$ is not of such form. As we mentioned before, in case there are many such j, we need only two of them. So, for each nonterminal A_i, in a bottom-up order, we calculate the set of remainders modulo $|\text{length}_A|$ of lengths of words that can be derived from A_i, which we call $R(A_i)$. If $|R(A_i)| > 1$, we can forget about all but two values so the complexity is polynomial. We can take $R(B_n C_n)$ as the set of j such that $T_1 T_2$ can be of the form $\text{left}_A^* \text{left}_A[1..j]$. Of course even if $j \in R(B_n C_n)$ we do not know if we can find $T_1 T_2$ being the corresponding prefix of left_A^∞, we only know that we can find $T_1 T_2$ having the corresponding length. Fortunately, if $T_1 T_2$ is not a prefix of left_A^∞, the growth order can only increase. What is more, if $|R(B_n C_n)| > 1$ we can forget about the possibility that $T_1 T_2$ is not a prefix of left_A^* because at least one value of j will give us the same growth order. We still need to consider the case of $R(B_n C_n) = \{r\}$, though. In such a case we should check if each word in $L(B_n C_n)$ is of the form $\text{left}_A^* \text{left}_A[1..r]$. We can use Lemma 9 for that; it may not happen that it finds a contradiction as the language in question would not be bounded then.

Case 2: T_1 has an order of 0 but T_2 has a nonzero order. Now we cannot assume that $\text{left}_A \neq \epsilon$ (which was crucial in the previous case), but knowing that T_2 is of the form $x_0 y_1^* \ldots y_k^* x_k$ and has a specified growth order, we can use the same method as above, replacing left_A by y_1.

Case 3: T_1 has a nonzero order but T_2 has an order of 0. Similar as above.

We have proven the following theorem:

Theorem 7. *Given an CFG G such that $L(G)$ is of polynomial growth, there is a polynomial-time algorithm that finds the exact order of polynomial growth of $L(G)$.*

Acknowledgments

We would like to thank Arne Storjohann for his input regarding algorithms for computing the Perron-Frobenius eigenvalue of a nonnegative integer matrix.

References

1. Bell, J.: A gap result for the norms of semigroups of matrices. Linear Algebra Appl. 402, 101–110 (2005)
2. Bridson, M., Gilman, R.: Context-free languages of sub-exponential growth. J. Comput. System Sci. 64, 308–310 (2002)

3. Cormen, T., Leiserson, C., Rivest, R., Stein, C.: Introduction to Algorithms, 2nd edn. MIT Press, Cambridge (2001)
4. Eigenwillig, A., Sharma, V., Yap, C.K.: Almost tight recursion tree bounds for the Descartes method. In: ISSAC 2006, pp. 71–78 (2006)
5. Giesbrecht, M., Storjohann, A.: Computing rational forms of integer matrices. J. Symbolic Comput. 34, 157–172 (2002)
6. Ginsburg, S.: The Mathematical Theory of Context-free Languages. McGraw-Hill, New York (1966)
7. Ginsburg, S., Spanier, E.: Bounded ALGOL-like languages. Trans. Amer. Math. Soc. 113, 333–368 (1964)
8. Ginsburg, S., Spanier, E.: Bounded regular sets. Proc. Amer. Math. Soc. 17, 1043–1049 (1966)
9. Ibarra, O., Ravikumar, B.: On sparseness, ambiguity and other decision problems for acceptors and transducers. In: Monien, B., Vidal-Naquet, G. (eds.) STACS 1986. LNCS, vol. 210, pp. 171–179. Springer, Heidelberg (1985)
10. Ilie, L., Rozenberg, G., Salomaa, A.: A characterization of poly-slender context-free languages. Theoret. Informatics Appl. 34, 77–86 (2000)
11. Incitti, R.: The growth function of context-free languages. Theoret. Comput. Sci. 225, 601–605 (2001)
12. Kaltofen, E., Villard, G.: On the complexity of computing determinants. Comput. Complex. 13, 91–130 (2004)
13. Karpinski, M., Rytter, W., Shinohara, A.: An efficient pattern-matching algorithm for strings with short descriptions. Nordic Journal of Computing 4, 172–186 (1997)
14. Latteux, M., Thierrin, G.: On bounded context-free languages. Elektron. Informationsverarb. Kybernet. 20, 3–8 (1984)
15. Lifshits, Y.: Solving classical string problems on compressed texts. In: CPM 2007, pp. 228–240 (2007)
16. Lyndon, R.C., Schützenberger, M.-P.: The equation $a^M = b^N c^P$ in a free group. Michigan Math. J. 9, 289–298 (1962)
17. Minc, H.: Nonnegative Matrices. Wiley, Chichester (1988)
18. Plandowski, W.: The Complexity of the Morphism Equivalence Problem for Context-Free Languages, PhD thesis
19. Raz, D.: Length considerations in context-free languages. Theoret. Comput. Sci. 183, 21–32 (1997)
20. Shur, A.M.: Combinatorial complexity of rational languages. Discr. Anal. and Oper. Research, Ser. 1 12(2), 78–99 (2005)
21. Shur, A.M.: Combinatorial complexity of regular languages. In: Hirsch, E.A., Razborov, A.A., Semenov, A., Slissenko, A. (eds.) Computer Science – Theory and Applications. LNCS, vol. 5010, pp. 289–301. Springer, Heidelberg (2008)
22. Szilard, A., Yu, S., Zhang, K., Shallit, J.: Characterizing regular languages with polynomial densities. In: Havel, I.M., Koubek, V. (eds.) MFCS 1992. LNCS, vol. 629, pp. 494–503. Springer, Heidelberg (1992)
23. Trofimov, V.I.: Growth functions of some classes of languages. Cybernetics 6, 9–12 (1981)

A   Appendix

In this appendix we give the full proofs of certain lemmas, which were omitted in the main text above due to space considerations.

A.1   Proofs Omitted from Section 2

Lemma 2. *Let $T = x_0 y_1^* x_1 y_2^* x_2 \ldots y_k^* x_k$ be a primitive star expression. $L(T)$ has $\Theta(m^{k-1})$ growth iff T has no fake indices.*

Proof. Assume there is a fake index i. Then $y_i^* x_i y_{i+1}^*$ can be rewritten as $y_i^* y_i^l y_i [1..j] (y_i [j + 1..|y_i|] y_i [1..j])^*$, which is the same as $y_i^* y_i^l y_i^* y_i [1..j]$ $= y_i^* y_i^l y_i [1..j]$. Thus $L(T)$ can be described by a primitive star expression of level $k - 1$, so its growth is $O(m^{k-2})$.

Now assume there is no such i. First we modify T in the following way: whenever some x_i is of the form $y_i z$, replace $y_i^* x_i$ by $y_i y_i^* z$. This does not change $L(T)$. So, we can assume that for each $1 \leq i < k$ either $x_i = y_i[1..j]$ and $y_{i+1} \neq y_i^{(j)}$ or $x_i = y_i[1..j]cz$ where $c \neq y_i[j + 1]$ for some $j < |y_i|$. Then we observe that $L(T') \subseteq L(T)$ where $T' = x_0 y_1'^+ x_1 y_2'^+ x_2 \ldots y_k'^+ x_k$ and each y_i' is a power of y_i chosen so that all lengths $|y_i'|$ are the same.

Now we can show that for each $w \in \Sigma^*$ the equation

$$w = x_0 y_1'^{\alpha_1} x_1 y_2'^{\alpha_2} x_2 \ldots y_k'^{\alpha_k} x_k, \text{ all } \alpha_i \geq 1,$$

has at most one solution. This will prove that $L(T')$ has $\Omega(m^{k-1})$ growth. We apply induction on k. The cases in which $k = 0, 1$ are obvious. Suppose $k > 1$. Assume that there are $a > b \geq 1$ such that both $x_0 y_1'^a x_1 y_2'$ and $x_0 y_1'^b x_1 y_2'$ are prefixes of w. Then $x_1 y_2'$ is a prefix of $y_1'^c x_1 y_2'$ for some $c \geq 1$. Consider two possibilities:

Case 1: $x_1 = y_1'[1..j]$ for some $j < |y_i|$ (y_i' is a power of y_i). This means that y_2' is the same as $y_1'^{(j)}$. Therefore, the primitive roots of these two words should be the same. The length of the primitive root does not change after rotation, so we get $y_2 = y_1^{(j)}$, a contradiction.

Case 2: $x_1 = y_1'[1..j]cz$ for some $j < |y_1|$ and $c \neq y_1[j + 1]$. This means that $y_1'[1..j]c$ is a prefix of y_1', which is impossible.   □

Lemma 3. *Let $T = x_0 y_1^* x_1 y_2^* x_2 \ldots y_k^* x_k$ be a primitive star expression. $L(T)$ has $\Theta(m^{k-1-d})$ growth iff there are exactly d fake indices in T.*

Proof. We have already proved this lemma for $d = 0$. Assume that $d > 0$. We can transform T to get an level $k - d$ primitive star expression containing no fake indices using the following operation: take a maximal contiguous segment of fake indices $i, i + 1, \ldots, i + m - 1$ and observe that the whole expression $y_i^* x_i y_{i+1}^* x_{i+1} \ldots y_{i+m}^*$ can be replaced by $y_i^* y_i^l y_i [1..j]$ for some $l \geq 0$ and $j < |y_i|$ without changing the language in question. Such a replacement decreases the number of fake indices by exactly m: the only possible place where a new fake index could be created is $y_i^* y_i^l y_i [1..j] x_{i+m} y_{i+m+1}^*$, but as we have chosen a maximal segment, there are only two possibilities (y' stands for $y_i^{(j)}$):

Case 1: x_{i+m} is not of the form $y'^* y'[1..j']$. Then $y_i^l y_i[1..j] x_{i+m}$ cannot be written as $y_i^{l'} y_i[1..j'']$;

Case 2: x_{i+m} is of the form $y'^*y'[1..j']$ but $y_{i+m+1} \neq y'^{(j')}$. Then $y_i^l y_i[1..j] x_{i+m} = y_i^{l'} y_i[1..j'']$ but $y_{i+m+1} \neq y_i^{(j'')}$.

Thus, repeating the above operations leaves us with a level $k-d$ primitive star expression having no fake indices and describing the same language (describing only some subset of $L(T)$ would be enough). Thus $L(T)$ has $\Theta(m^{k-d-1})$ growth.

□

Theorem 3. *The orders of polynomial growth of $L(M)$ and $L(M')$ are the same.*

Proof. By the construction of M', it is easy to see that for each $w \in L(M)$ we can find a $T \in \mathcal{T}$ such that $w \in L(T)$.

The other direction is more involved: we must show that for each $T \in \mathcal{T}$ we can find a subset of $L(M)$ having the same order of growth. Of course, there is a natural way of doing that: T corresponds to a sequence of vertices of M:

$$u_1 \xrightarrow{w_1} v_1 \xrightarrow{c_1} u_2 \xrightarrow{w_2} v_2 \xrightarrow{c_2} \ldots \xrightarrow{c_{m-1}} u_m \xrightarrow{w_m} v_m, \qquad T = w_1 c_1 w_2 c_2 \ldots c_{m-1} w_m$$

such that u_k and v_k are in the same strongly connected component of M, $v_k \xrightarrow{c_k} u_{k+1}$ are transitions connecting different strongly connected components in M and w_k is either $z_k[i+1..j]$ or $z_k[i+1..|z_k|] z_k^* z_k[1..j]$, where z_k is the primitive root associated with the strongly connected component containing u_k, v_k. We would like to take the labels of all paths in M going through the above sequence of vertices. Unfortunately, two bad things may happen:

Case 1: $w_k = z_k[i+1..j]$ for some k but there is no path from u_k to v_k labeled with such a w_k in M. However, we know that there are paths labeled with $z_k[i+1..|z_k|] z_k^{q+\alpha r} z_k[1..j]$, so we can modify T, setting $w_k = z_k[i+1..j](t^r)^* t^{1+q}$, where $t = z_k^{(j)}$, which does not decrease the growth order. Then we can observe that setting $w_k = z_k[i+1..|z_k|](z_k^r)^* z_k^q z_k[1..j]$ results in the same language.

Case 2: $w_k = z_k[i+1..|z_k|] z_k^* z_k[1..j]$ but paths from u_k to v_k are of the form $z_k[i+1..|z_k|] z_k^{q+\alpha r} z_k[1..j]$. In this case, we know that replacing z_k^* in T by $(z_k^r)^* z_k^q$ does not change its growth order, so we can set $w_k = z_k[i+1..|z_k|](z_k^r)^* z_k^q z_k[1..j]$.

Thus we can modify T without decreasing its growth order so that $L(T)$ is contained in the set of labels of paths in M going through the above sequence of vertices.

□

Lemma 5. *Let A be a nonnegative $n \times n$ matrix over the integers. Let $r(A) = r, \lambda_1, \ldots, \lambda_\ell$ be the distinct eigenvalues of A, and let d be the size of the dominating Jordan block of A. Then $A^m \in \Theta(r^m m^{d-1})$.*

Proof. The theorem trivially holds for $r = 0$. Assume $r \geq 1$. Without loss of generality, we can assume that A does not have an eigenvalue λ such that $\lambda \neq r$ and $|\lambda| = r$; if such an eigenvalue exists, replace A by A^h. Let J be the Jordan canonical form of A, i.e., $A = SJS^{-1}$, where S is a nonsingular matrix, and J is a diagonal block matrix of Jordan blocks. We use the following notation: $J_{\lambda,e}$ is a Jordan block of order e corresponding to eigenvalue λ, and O_x is a square matrix, where all entries are zero, except for x at the top-right corner. Let $J_{r,d}$ be the dominating Jordan block of A. It can be verified by induction that

$$
J_{r,d}^m = \begin{pmatrix}
r^m & \binom{m}{1}r^{m-1} & \binom{m}{2}r^{m-2} & \cdots & \binom{m}{d-2}r^{m-d+2} & \binom{m}{d-1}r^{m-d+1} \\
0 & r^m & \binom{m}{1}r^{m-1} & \cdots & \binom{m}{d-3}r^{m-d+3} & \binom{m}{d-2}r^{m-d+2} \\
\vdots & \vdots & \vdots & & \vdots & \vdots \\
0 & 0 & 0 & \cdots & r^m & \binom{m}{1}r^{m-1} \\
0 & 0 & 0 & \cdots & 0 & r^m
\end{pmatrix}.
$$

Thus the first row of $J_{r,d}^m$ has the form

$$
r^m \left[1 \quad \frac{m}{r} \quad \frac{m(m-1)}{2!r^2} \quad \cdots \quad \frac{m(m-1)\cdots(m-(d-2))}{(d-1)!r^{d-1}} \right],
$$

and so

$$
\lim_{m\to\infty} \frac{J_{r,d}^m}{r^m m^{d-1}} = O_\alpha, \quad \text{where } \alpha = \frac{1}{(d-1)!r^{d-1}}.
$$

All Jordan blocks other than the dominating block converge to zero blocks. and

$$
\lim_{m\to\infty} \frac{A^m}{r^m m^{d-1}} = S \lim_{m\to\infty} \frac{J^m}{r^m m^{d-1}} S^{-1}.
$$

The result follows.   □

A.2   Proofs Omitted from Section 3

Lemma 7. *Given a context free grammar, we can calculate for each nonterminal A the value of* length(A) *in polynomial time.*

Proof. We calculate the values of length$_i(A)$ defined as follows:

$$
\text{length}_1(A) := \begin{cases} 1 & A \to a \text{ is a production for some } a \\ \infty & \text{otherwise} \end{cases}
$$

$$
\text{length}_{i+1}(A) := \min\{\text{length}_i(B) + \text{length}_i(C) : A \to BC \text{ is a production }\}
$$

for every nonterminal A and $i = 1, 2, \ldots, n$, where n is the number of all nonterminals. This obviously takes polynomial time as we have n^2 values to compute and each length$_i(A)$ is either ∞ or at most 2^i. Observe that length$_i(A)$ is in fact the smallest possible length of a word that can be derived from A in such a way that the derivation tree is of depth at most i. If we consider a shortest word that can be derived from A, its derivation tree is of depth at most n, so we can take length$(A) = \text{length}_n(A)$.   □

In the following two proofs we show that the order of polynomial counting all words of length at most N does not change. It is equivalent to checking that the growth order does not change.

Lemma 10. *Given a generalized star expression, we can remove a constraint* $(v_i, v_j) \in \text{context}(A)$, *where A is an independent nonterminal, without changing the growth order of the language in question.*

Proof. By joining maximum segments of fake indices we can show that the number of words of length at most N is exactly the number of different vectors $\vec{x} = [\vec{a}_1 \vec{v}, \vec{a}_2 \vec{v}, \ldots, \vec{a}_k \vec{v}]$ where all $\vec{a}_k$ have nonnegative coordinates and the i-th coordinate of $\vec{v}$ is simply the value assigned to the i-th variable, for all possible assignments consistent with constraints such that $\vec{x}\vec{c} \leq N$ where $\vec{c}$ is some fixed vector with strictly positive coordinates. If all $\vec{a}_k^{(i)}$ are zero or all $\vec{a}_k^{(j)}$ are zero we can safely remove the constraint because the value of one variable does not matter. Otherwise there is exactly one k such that $\vec{a}_k^i$ is nonzero and exactly one k such that $\vec{a}_k^{(j)}$ is nonzero. If they are the same we can also safely remove the constraint because only the sum of those variables matters. Thus by reordering the coordinates of $\vec{x}$ we can assume that $\vec{a}_{k-1}^{(i)}, \vec{a}_k^{(j)} > 0$. Now the number of words of length N when we ignore the constraint $(v_i, v_j) \in \text{context}(A)$ is at most N^2 multiplied by the number of different vectors $[\vec{a}_1 \vec{v}, \vec{a}_2 \vec{v} \ldots \vec{a}_{k-2} \vec{v}]$ such that $\sum_j \vec{a}_j \vec{v} c_j \leq N$. On the other hand, we know that $(\alpha, \beta_1), (\alpha, \beta_2) \in \text{context}(A)$ for some $\alpha, \beta_1 < \beta_2$ and it is possible to show that the number of words of length tN when we keep all constraints is at least N^2 multiplied by the number of different vectors $[\vec{a}_1 \vec{v}, \vec{a}_2 \vec{v} \ldots \vec{a}_{k-2} \vec{v}]$ such that $\sum_j \vec{a}_j \vec{v} c_j \leq N$, where t is some constant ($t = 2\alpha c_{k-1} + 3\beta_2 c_k$ is big enough). This shows that ignoring the constraint $(v_i, v_j) \in \text{context}(A)$ does not increase the growth order. □

Lemma 11. *Given a generalized star expression, we can replace each constraint* $(v_i, v_j) \in \text{context}(A)$, *where A is a dependent nonterminal, by either* $(v_i, v_j) \in \mathbb{N} \times \{1\}$ *or* $(v_i, v_j) \cup \{1\} \times \mathbb{N}$ *without changing the growth order.*

Proof. As in the previous lemma, we join maximum segments of fake indices. If two constrained variables occur in the same segment, we can treat them as one unconstrained variable. If some segment contains an unconstrained variable, it contributes 1 to the final order. We can delete such segment, removing all constraints concerning variables it contains. Thus w.l.o.g. we can assume that each variable is constrained by some context(A) where A is a dependent nonterminal and each two constrained variables occur in different segments.

Again, the number of words of length at most N is exactly the number of different vectors $\vec{x} = [\vec{a}_1 \vec{v}, \vec{a}_2 \vec{v}, \ldots, \vec{a}_k \vec{v}]$ for all possible consistent assignments such that $\vec{x}\vec{c} \leq N$. Imagine a multigraph on k vertices, each of them corresponding to one coordinate of $\vec{x}$, in which we put an edge (x, y) for each constraint (v_i, v_j) such that $\vec{a}_x^{(i)}, \vec{a}_y^{(j)} > 0$. Now observe that coordinates corresponding to vertices from different connected components are completely independent; the resulting order is simply the sum of orders corresponding to different components. Thus it is enough to consider one such component. There are two possibilities:

Case 1: this component is a tree. Then the number of vectors $\vec{x}$ such that $\vec{x}\vec{c} \leq N$ is $\Theta(N^{k-1})$. To see why, choose any vertex to be the root of this tree and then fix coordinates starting from the leaves in a bottom-up fashion. After choosing the values of all coordinates but the root, the value of the coordinate corresponding to the root is uniquely determined. Thus the number of vectors is $O(N^{k-1})$. To see that it is $\Omega(N^{k-1})$, observe that there are constants $\alpha_v < \beta_v$ such that for each coordinate corresponding to a non-root v we can assign any value from $[\alpha_v N, \beta_v N]$.

Case 2: this component contains a cycle. Then the number of vectors $\vec{x}$ such that $\vec{x}\vec{c} \leq N$ is $\Theta(N^k)$. Obviously, it is $O(N^k)$. Take any spanning tree of this component and choose the root to be a vertex having an incident edge outside this spanning tree. As in the previous case, we can choose any value from some $[\alpha_v N, \beta_v N]$ for each coordinate corresponding to a non-root v. Then we can modify the coordinate corresponding to the root by using the variable corresponding to the additional edge. An appropriate choice of the constants $\alpha_v < \beta_v$ gives us that there are $\Omega(N^k)$ possible choices.

This shows how to replace each constraint by either $\mathbb{N} \times \{1\}$ or $\{1\} \times \mathbb{N}$ without decreasing the growth order. More specifically, we replace $(v_i, v_j) \in context(A)$ by $\mathbb{N} \times \{1\}$ if we want to fix the value of v_i and the value of v_j is not fixed yet.

It can be also checked that no replacement will result in an increased growth order.   □

Lemma 12. *Given a context-free grammar, we can check in polynomial time if a given nonterminal A is independent.*

Proof. First we check if for each nonterminal B that can appear in some derivation $A \overset{*}{\Rightarrow} u_1 B u_2 A v$ or $A \overset{*}{\Rightarrow} u A v_1 B v_2$, it holds that $|\{|w| : B \overset{*}{\Rightarrow} w\}| = 1$. This can be done in polynomial time by checking that for any production $C \to DE$ that can appear in some derivation $B \overset{*}{\Rightarrow} w$, $length(C) = length(D) + length(E)$. If it does not hold then clearly A is independent. Otherwise we build a graph $H = (V, E)$ where V is the set of all nonterminals and E contains edges $B \to C$ of weight $(0, length(D))$ and $B \to D$ of weight $(length(C), 0)$ for any production $B \to CD$ and take the strongly connected component containing A. We define the weight of a path to be the component-wise sum of weights of its edges. It is clear that to check if A is independent, we should check if there are two paths from A to A having different weights (α, β_1) and (α, β_2). Take any path from A to A with a weight of (x, y) where $x, y > 0$ (if there is no such path, we are done) and for each edge replace its weight (α, β) by $y\alpha - x\beta$. Now it is clear that we should check if there is a path from A to A with a nonzero modified weight. This can be done in polynomial time using a similar method to the one from Lemma 8.   □

More Concise Representation
of Regular Languages
by Automata and Regular Expressions*

Viliam Geffert[1], Carlo Mereghetti[2], and Beatrice Palano[2]

[1] Department of Computer Science – P. J. Šafárik University
Jesenná 5 – 04154 Košice – Slovakia
`viliam.geffert@upjs.sk`
[2] Dipartimento di Scienze dell'Informazione – Università degli Studi di Milano
via Comelico 39 – 20135 Milano – Italy
`{mereghetti,palano}@dsi.unimi.it`

Abstract. We consider two formalisms for representing regular languages: constant height pushdown automata and straight line programs for regular expressions. We constructively prove that their sizes are polynomially related. Comparing them with the sizes of finite state automata and regular expressions, we obtain optimal exponential and double exponential gaps, i.e., a more concise representation of regular languages.

Keywords: Pushdown automata; regular expressions; straight line programs; descriptional complexity.

1 Introduction

Several systems for representing *regular languages* have been presented and studied in the literature. For instance, for the original model of finite state automaton [10], a lot of modifications have been introduced: nondeterminism [10], alternation [4], probabilistic evolution [9], two-way input head motion [11], etc. Other important formalisms for defining regular languages are, e.g., regular grammars [6] and regular expressions [7]. All these models have been proved to share the same expressive power by exhibiting simulation results.

However, representation of regular languages may be much more "economical" in one system than another. This consideration has lead to a consolidated line of research — sometimes referred to as *descriptional complexity* — aiming to compare formalisms by comparing their *size*. The oldest and most famous result in this sense is the optimal exponential gap between the size of a deterministic (dfa) and nondeterministic (nfa) finite state automaton [8,10].

In this paper, we study the size of two formalisms for specifying regular languages, namely: a *constant height pushdown automaton* (h-pda) and a *straight line program for a regular expression* (slp).

---

* This work was partially supported by the Slovak Grant Agency for Science (VEGA) under contract "Combinatorial Structures and Complexity of Algorithms".

M. Ito and M. Toyama (Eds.): DLT 2008, LNCS 5257, pp. 359–370, 2008.

First, it is well known that the languages recognized by nondeterministic pdas (npdas) form the class of context-free languages, a proper superclass of the languages accepted by deterministic pdas (dpdas), which in turn is a proper superclass of regular languages [6]. However, if the maximum height of the push-down store is bounded by a constant, i.e., if it does not depend on the input length, it is a routine exercise to show that such machine accepts a regular language. (In general, it is not possible to bound the pushdown height by a constant.) Nevertheless, a representation by constant height pdas can potentially be more succinct than by the standard finite state automata, both for deterministic and nondeterministic machines. Here we prove *optimal exponential* and *optimal double exponential simulation costs* of constant height pdas by finite state automata. We also get an *exponential lower bound* for eliminating nondeterminism in constant height pdas. (Some related results on pdas accepting regular languages vs. finite state automata can be found in [12].)

Second, a natural counterpart to constant height pdas turned out to be straight line programs (slps), in perfect analogy to the relation of finite state automata vs. regular expressions. An slp is a loopless program representing a directed acyclic graph, the internal nodes of which represent the basic regular operations (i.e., union, concatenation, and star). Compared with the size of the standard regular expression represented by a binary tree, an slp can be more succinct by using the fact that replicated subtrees are shared. Here we prove an *optimal exponential gap* between the sizes of regular expressions and slps.

Moreover, we design conversions which construct equivalent constant height pdas from slps, and vice versa, in such a way that *the sizes of these two formalisms are polynomially related*. This should be compared with the relation between the standard nfas and regular expressions: the cost for the "←" conversion is linear [2,3], but it is exponential for the opposite conversion [5].

2   Preliminaries

In this section, we present the formalism we shall be dealing with. We assume the reader is familiar with the basic notions on formal language theory (see, e.g., [6]). The set of natural numbers, including zero, is denoted here by $\mathbf{N}$.

2.1   Straight Line Programs for Regular Expressions

A *regular expression*, defined over a given alphabet Σ, is: (i) $\emptyset$, ε, or any symbol $a \in \Sigma$, (ii) $r_1 + r_2$, $r_1 \cdot r_2$, or r_1^*, if r_1 and r_2 are regular expressions. The language represented by a given regular expression r, denoted by $L(r)$, is defined in the usual way [6]. With a slight abuse of terminology, we often identify a regular expression with the language it represents. Thus, Σ^* is the set of words on Σ, including the empty word ε. By $|w|$, we denote the length of a word $w \in \Sigma^*$ and by Σ^i the set of words of length $i \in \mathbf{N}$, with $\Sigma^0 = \{\varepsilon\}$ and $\Sigma^{\leq m} = \bigcup_{i=0}^{m} \Sigma^i$. By $|\Sigma|$, we denote the cardinality of the set Σ.

Definition 1. *The* size *of a regular expression r on Σ, denoted by* size(r), *is the number of occurrences of symbols of $\{\emptyset, \varepsilon\} \cup \Sigma$ plus the number of occurrences of operations $+, \cdot$ and $*$ inside r.*

Example 1. For $r = a \cdot (a + b)^* + (a + b)^* \cdot b \cdot a^*$, we have size$(r) = 16$.

A convenient way for representing regular expressions is given by straight line programs (see, e.g., [1]). Given a set of variables $X = \{x_1, \ldots, x_\ell\}$, a *straight line program for regular expressions* (slp) on Σ is a finite sequence of instructions

$$P \equiv \text{instr}_1 ; \ \ldots \ \text{instr}_i ; \ \ldots \ \text{instr}_\ell ,$$

where the i-th instruction instr$_i$ has one of the following forms:

(i) $x_i := \emptyset$, $x_i := \varepsilon$, or $x_i := a$ for any symbol $a \in \Sigma$,
(ii) $x_i := x_j + x_k$, $x_i := x_j \cdot x_k$, or $x_i := x_j^*$, for $1 \leq j, k < i$.

Such program P expands to the regular expression reg-exp$(P) = x_\ell$, obtained by nested macro-expansion of the variables $x_1, \ldots, x_{\ell-1}$, using the right parts of their instructions. Notice that a variable may be reused several times in the right parts. Such a number of occurrences is called a *fan-out* of the variable. The fan-out of x_ℓ is 0, while the fan-out of any other variable is at least 1, since, with the exception of x_ℓ, we can remove instructions defining variables not used at least once in some right part.

Definition 2. *The* size *of a straight line program P is the ordered pair* size$(P) = (\text{length}(P), \text{fan-out}(P))$, *where* length$(P)$ *denotes the number of instructions in P, and* fan-out(P) *the maximum fan-out of its variables.*

Example 2. Let us construct an slp for the regular expression in the Example 1:

$$P \equiv x_1 := a; \quad x_2 := b; \quad x_3 := x_1 + x_2; \quad x_4 := x_1^*; \quad x_5 := x_3^*;$$
$$x_6 := x_2 \cdot x_4; \quad x_7 := x_1 \cdot x_5; \quad x_8 := x_5 \cdot x_6; \quad x_9 := x_7 + x_8 .$$

Clearly, reg-exp$(P) = x_9 = a \cdot (a+b)^* + (a+b)^* \cdot b \cdot a^*$, and size$(P) = (9, 3)$.

Each slp P can be associated with a *vertex-labeled directed acyclic graph* (dag) $D_P = (V, E)$, where the vertices in $V = \{v_1, \ldots, v_\ell\}$ correspond to the respective variables in $X = \{x_1, \ldots, x_\ell\}$. That is, a vertex v_i is labeled by $e \in \{\emptyset, \varepsilon\} \cup \Sigma$, whenever the i-th instruction is $x_i := e$, and by '+' or '·', whenever this instruction is $x_i := x_j + x_k$ or $x_i := x_j \cdot x_k$, respectively. In the case of a binary operation, the directed arcs (v_j, v_i) and (v_k, v_i) are included in E, to connect v_i with its left and right sons, respectively. Similarly, v_i is labeled by '$*$', if the i-th instruction is $x_i := x_j^*$, with (v_j, v_i) included in E. (This idea is illustrated by Figure 1.)

From the definition of P, it is easy to see that D_P does not contain any directed cycle and that the fan-out of a variable establishes the out-degree of the corresponding vertex. So, there exists a unique *sink* v_ℓ (vertex without outgoing arcs) and some *sources* (vertices without ingoing arcs) labeled by $e \in \{\emptyset, \varepsilon\} \cup \Sigma$. We define the *depth* of D_P, depth(D_P), as the maximum length of a path from a source to the sink.

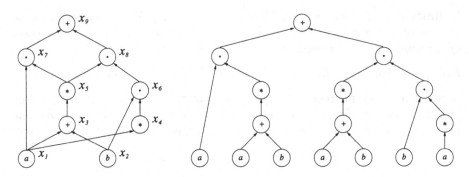

Fig. 1. On the left, the directed acyclic graph D_P associated with the slp P introduced in Example 2. The corresponding classical representation of the regular expression from Example 1 as a binary tree on the right.

In what follows, we point out some relations between the sizes of an slp and its regular expression. Clearly, an slp with fan-out bounded by 1 is just an ordinary regular expression written down in a slightly different way:

Proposition 1. *For each slp P, length(P) = size(reg-exp(P)) if and only if fan-out$(P) = 1$.*

In general, however, straight line programs can be exponentially more succinct than regular expressions. The following example shows that, even with only fan-out 2, we get an exponential gap.

Example 3. Consider the slp P_ℓ on $\Sigma = \{a\}$:

$$P_\ell \equiv \quad x_1 := a; \quad x_2 := x_1 \cdot x_1; \quad x_3 := x_2 \cdot x_2; \quad \ldots \quad x_\ell := x_{\ell-1} \cdot x_{\ell-1}.$$

It can be immediately seen that fan-out$(P_\ell) = 2$ and reg-exp$(P_\ell) = a^{2^{\ell-1}}$. Thus, for any $\ell \geq 1$, we obtain size(reg-exp(P_ℓ)) = $2^{\text{length}(P_\ell)} - 1$.

This latter example establishes the *optimality* of the following general result:

Proposition 2. *Let P and P' be two equivalent slps such that fan-out$(P') = 1$. Then, length$(P') \leq 2^{\text{depth}(P)}$.*

2.2   Constant Height Pushdown Automata

It is well known that the regular expressions (hence, slps as well) represent the class of *regular languages* [6]. This class can also be represented by automata.

A *nondeterministic finite state automaton* (nfa, for short) is a quintuple $A = \langle Q, \Sigma, H, q_0, F \rangle$, where Q is the finite set of states, Σ the finite input alphabet, $H \subseteq Q \times (\Sigma \cup \{\varepsilon\}) \times Q$ the transition relation, $q_0 \in Q$ the initial state, and $F \subseteq Q$ the set of final (accepting) states. An input string is *accepted*, if there exists a computation beginning in the state q_0 and ending in some final state $q \in F$ after reading this input. The set of all inputs accepted by A is denoted by $L(A)$. The automaton A is *deterministic* (dfa), if there are no ε-transitions in H and, for every $q \in Q$ and $a \in \Sigma$, there exists at most one $p \in Q$ such that $(q, a, p) \in H$.

In the literature, a *nondeterministic pushdown automaton* (npda) is usually obtained from an nfa by adding a pushdown store, containing symbols from Γ, the pushdown alphabet. At the beginning, the pushdown contains a single initial symbol $Z_0 \in \Gamma$. The transition relation is usually given in the form of δ, a mapping from $Q \times (\Sigma \cup \{\varepsilon\}) \times \Gamma$ to finite subsets of $Q \times \Gamma^*$. Let $\delta(q, x, X) \ni (p, \omega)$. Then A, being in the state q, reading x on the input and X on the top of the pushdown, can reach the state p, replace X by ω and finally, if $x \neq \varepsilon$, advance the input head one symbol. Its *deterministic* version (dpda) is obtained in the usual way. (For more details, see, e.g., [6].)

For technical reasons, we shall introduce the npdas in the following form, where moves manipulating the pushdown store are clearly distinguished from those reading the input tape: a npda is a 6-tuple $A = \langle Q, \Sigma, \Gamma, H, q_0, F \rangle$, where $Q, \Sigma, \Gamma, q_0, F$ are defined as above, while $H \subseteq Q \times (\{\varepsilon\} \cup \Sigma \cup \{+, -\} \cdot \Gamma) \times Q$ is the *transition relation* with the following meaning:

 (i) $(p, \varepsilon, q) \in H$: A reaches the state q from the state p without using the input tape or the pushdown store,
 (ii) $(p, a, q) \in H$: A reaches the state q from the state p by reading the symbol a from the input, not using the pushdown store,
(iii) $(p, -X, q) \in H$: if the symbol on top of the pushdown is X, A reaches the state q from the state p by popping X, not using the input tape,
(iv) $(p, +X, q) \in H$: A reaches the state q from the state p by pushing the symbol X onto the pushdown, not using the input tape.

Such machine does not use any initial pushdown symbol: an accepting computation begins in the state q_0 with the empty pushdown store and input head at the beginning, and ends in a final state $q \in F$ after reading the entire input.

A *deterministic pushdown automaton* (dpda) is obtained from npda by claiming that it can never get into a situation in which more than one instruction can be executed. (As an example, a dpda cannot have a pair of instructions of the form (q, ε, p_1) and (q, a, p_2).)

It is not hard to see that any npda in the classical form can be turned into this latter form and vice versa, preserving determinism in the case of dpdas.

At the cost of one more state, we can transform our npdas so that they accept by entering a *unique* final state at the end of input processing, with *empty* pushdown store. Notice however that the following transformation does not preserve determinism.

Lemma 1. *For any npda $A = \langle Q, \Sigma, \Gamma, H, q_0, F \rangle$, there exists an equivalent npda $A' = \langle Q \cup \{q_f\}, \Sigma, \Gamma, H', q_0, \{q_f\} \rangle$, where A' accepts by entering the unique final state $q_f \notin Q$ with empty pushdown store at the end of the input.*

Given a constant $h \in \mathbf{N}$, we say that the npda A is of pushdown height h if, for any word in $L(A)$, there exists an accepting computation along which the pushdown store never contains more than h symbols.

From now on, we shall consider *constant height npdas* only. Such machine will be denoted by a 7-tuple $A = \langle Q, \Sigma, \Gamma, H, q_0, F, h \rangle$, where $h \in \mathbf{N}$ is a constant denoting the pushdown height, and all other elements are defined as above. By definition, the meaning of the transitions in the form (iv) is modified as follows:

(iv') $(p, +X, q) \in H$: *if the current pushdown store height is smaller than h*, then A reaches the state q from the state p by pushing the symbol X onto the pushdown, not using the input tape.

Thus, this kind of transitions is disabled, if the current pushdown height is equal to h. A constant height npda can be replaced by an equivalent standard npda (without a built-in limit h on the pushdown) by storing, in the finite control states, a counter recording permitted pushdown heights (i.e., a number ranging within $\{0, \ldots, h\}$), hence, paying by a constant increase in the number of states.

Note that, for $h = 0$, the definition of constant height npda exactly coincides with that of an nfa, as one may easily verify. Moreover, Lemma 1 holds for constant height npdas as well, which enables us to consider acceptance by a *single final state* and, at the same time, with *empty pushdown store*. Therefore, from now on, a constant height npda will assume the form $A = \langle Q, \Sigma, \Gamma, H, q_0, \{q_f\}, h \rangle$.

Definition 3. *The* size *of a constant height npda* $A = \langle Q, \Sigma, \Gamma, H, q_0, \{q_f\}, h \rangle$ *is the ordered triple* $\mathrm{size}(A) = (|Q|, |\Gamma|, h)$.

Observe that this definition immediately gives that the size of an nfa is completely determined by the number of its states, since it is $(|Q|, 0, 0)$.

3   From a Constant Height npda to an slp

In this section, we show how to convert a constant height npda into an equivalent slp. Yet, we focus on the cost of such a conversion and prove that the size of the resulting slp is polynomial in the size of the original npda.

In what follows, to simplify our notation, a "long" regular expression $r_1 + r_2 + \cdots + r_n$ will also be written as $\sum_{i=1}^{n} r_i$. We define an "empty sum" as the regular expression $\emptyset$.

Let $A = \langle \{q_1, \ldots, q_k\}, \Sigma, \Gamma, H, q_1, \{q_k\}, h \rangle$ be a constant height npda. For each $i, j \in \{1, \ldots, k\}$, $s \in \{0, \ldots, k\}$, and $t \in \{0, \ldots, h\}$, we define $[q_i, s, t, q_j]$ to be the set of strings $x \in \Sigma^*$ such that, for each of them, there exists at least one computation with the following properties.

- The computation begins in the state q_i, with the pushdown empty.
- After reading the entire string x from the input, the computation ends in the state q_j, with the pushdown empty again.
- Any time the pushdown is empty during this computation, the current finite control state is from the set $\{q_1, \ldots, q_s\}$. (This restriction does not apply to q_i, q_j themselves.) For $s = 0$, the pushdown is never empty in the meantime.
- During this computation, the pushdown height never exceeds t.

We are now going to give an algorithm, consisting of two PHASES, which dynamically constructs regular expressions describing all sets $[q_i, s, t, q_j]$. The ultimate goal is to obtain $[q_1, k, h, q_k]$, the regular expression for the language $L(A)$. First, we easily construct $[q_i, 0, 0, q_j]$ for each q_i, q_j, basically by a direct inspection of the transitions in H. After that, we gradually increment the parameter s from 1 to k, thus obtaining $[q_i, k, 0, q_j]$ for each q_i, q_j. Second, we show how to upgrade from the parameters $k, t-1$ to parameters $0, t$, and then from parameters s, t to $s+1, t$, which leads up to $[q_i, k, h, q_j]$ for each q_i, q_j.

For any $1 \leq i, j \leq k$, we let $\Psi_0(q_i, q_j) = \{\alpha \in \Sigma \cup \{\varepsilon\} : (q_i, \alpha, q_j) \in H\} \cup \Delta_{i,j}$, where $\Delta_{i,j}$ is $\emptyset$ if $i \neq j$, or $\{\varepsilon\}$ if $i = j$. Thus, $\Psi_0(q_i, q_j)$ consists of the input symbols, possibly with ε, taking A from q_i to q_j in at most one computation step, without involving the pushdown.

PHASE I:

for each $1 \leq i, j \leq k$ **do**
$\quad [q_i, 0, 0, q_j] = \sum_{\alpha \in \Psi_0(q_i, q_j)} \alpha$;
for $s = 0$ **to** $k-1$ **do**
$\quad$ **for each** $1 \leq i, j \leq k$ **do**
$\quad\quad [q_i, s+1, 0, q_j] = [q_i, s, 0, q_j] + [q_i, s, 0, q_{s+1}] \cdot [q_{s+1}, s, 0, q_{s+1}]^* \cdot [q_{s+1}, s, 0, q_j]$;

Actually, $[q_i, 0, 0, q_j]$ is the regular expression representing the set $\Psi_0(q_i, q_j)$ by a formal sum of its elements, if any, otherwise by $\emptyset$. After that, this phase computes the regular expression $[q_i, s+1, 0, q_j]$ by adding to $[q_i, s, 0, q_j]$ the strings which enable A to use also the state q_{s+1}, without using the pushdown. This is exactly the Kleene's recursion for computing regular expressions from nfas [6].

Then the second phase starts, for which we need some more notation. For any $1 \leq i, j \leq k$ and $X \in \Gamma$, we let: $\Psi_+(q_i, X) = \{q \in Q : (q_i, +X, q) \in H\}$, and $\Psi_-(X, q_j) = \{q \in Q : (q, -X, q_j) \in H\}$. Thus, $\Psi_+(q_i, X)$ is the set of states reachable by A from q_i while pushing the symbol X onto the stack, and $\Psi_-(X, q_j)$ is the set of states from which A reaches q_j while popping X.

PHASE II:

for $t = 1$ **to** h **do begin**
$\quad$ **for each** $1 \leq i, j \leq k$ **do**
$(\diamond) \quad [q_i, 0, t, q_j] = [q_i, 0, t-1, q_j] + \sum_{X \in \Gamma} \sum_{p \in \Psi_+(q_i, X)} \sum_{q \in \Psi_-(X, q_j)} [p, k, t-1, q]$;
$\quad$ **for** $s = 0$ **to** $k-1$ **do**
$\quad\quad$ **for each** $1 \leq i, j \leq k$ **do**
$\quad\quad\quad [q_i, s+1, t, q_j] = [q_i, s, t, q_j] + [q_i, s, t, q_{s+1}] \cdot [q_{s+1}, s, t, q_{s+1}]^* \cdot [q_{s+1}, s, t, q_j]$;
end;
return $[q_1, k, h, q_k]$

For this phase, the construction marked by $(\diamond)$ is worth explaining. By definition, $[q_i, 0, t, q_j]$ consists of $[q_i, 0, t-1, q_j]$ plus the set of all strings x such that:

- starting in the state q_i with empty pushdown, A pushes some symbol X on the stack in the first move — the first two summands in $(\diamond)$,
- pops this symbol X in the last move only, by entering the state q_j with empty pushdown — third summand in $(\diamond)$ — and,

– during the computation processing the input string x, A pushes no more than $t-1$ symbols over X, so that globally the pushdown height is at least 1 and never exceeds t — variables $[p, k, t-1, q]$ in $(\diamond)$.

Notice that our algorithm can be easily regarded to as a straight line program P_A for $L(A)$. Informally, we can consider all $[q_i, s, t, q_j]$'s as the variables of P_A. Before PHASE I, we need to define some input variables, by instructions $x_\alpha := \alpha$ of the form (i), at most one per each $\alpha \in \Sigma \cup \{\varepsilon, \emptyset\}$. (See the definition of slps in Section 2.1.) Then, we easily unroll the for-cycles and sums, which translates the two phases into a finite sequence of instructions. To keep all such instructions in the form (ii), we only have to introduce some new auxiliary variables. Clearly, the output variable is $[q_1, k, h, q_k]$. The correctness can formally be proved by a double-induction on parameters s, t in $[q_i, s, t, q_j]$.

Concerning the length of P_A, there exist no more than $|\Sigma|+2$ input instructions, while those of PHASE I do not exceed $k^2 \cdot |\Sigma| + k^3 4$. Finally, PHASE II requires no more than $h \cdot k^3 \cdot (k \cdot |\Gamma| + 4)$ instructions, using also the fact that the number of variables involved in the triple sum of $(\diamond)$ is bounded by $|\Gamma| \cdot k^2$. Summing up, we get $\text{length}(P_A) \leq \mathcal{O}(h \cdot k^4 \cdot |\Gamma| + k^2 \cdot |\Sigma|)$. Moreover, the fan-out of any variable in P_A does not exceed $k^2 + 1$. Formally, we have shown:

Theorem 1. *Let $A = \langle Q, \Sigma, \Gamma, H, q_0, \{q_f\}, h \rangle$ be a constant height npda. Then there exists an slp P_A such that $\text{reg-exp}(P_A)$ denotes $L(A)$, with $\text{length}(P_A) \leq \mathcal{O}(h \cdot |Q|^4 \cdot |\Gamma| + |Q|^2 \cdot |\Sigma|)$ and $\text{fan-out}(P_A) \leq |Q|^2 + 1$. That is, for regular languages over a fixed alphabet, the size of P_A is polynomial in the size of A.*

4 From an slp to a Constant Height npda

Let us now show the converse result, namely, that any slp can be turned into an equivalent constant height npda whose size is polynomial in the size of the slp.

Let P be an slp with variables $\{x_1, \ldots, x_\ell\}$ on Σ. We consider the associated dag D_P, as described in Section 2.1, where the vertex v_i corresponds to the variable x_i. The enumeration of the variables in P induces a topological ordering on the vertices of D_P. Now we proceed as follows.

For $i = 1, \ldots, \ell$, we construct an npda $A_i = \langle Q_i, \Sigma, \Gamma_i, H_i, q_{0,i}, \{q_{f,i}\} \rangle$ such that $L(A_i)$ is exactly the language denoted by $\text{reg-exp}(x_i)$, a regular expression obtained by expanding the dag rooted in v_i. For a source node v_i, we define an "elementary" npda without a pushdown store — actually an nfa. An npda for an inner node is constructed inductively, using, as subprograms, npdas for vertices that are topologically smaller. The desired npda is A_ℓ. We start with the construction for sources.

SOURCES: let the source node v_i be labeled by $\alpha \in \Sigma \cup \{\varepsilon, \emptyset\}$. If $\alpha \neq \emptyset$, the single-transition npda recognizing α is defined as

$$A_i = \langle \{q_{0,i}, q_{f,i}\}, \Sigma, \emptyset, \{(q_{0,i}, \alpha, q_{f,i})\}, q_{0,i}, \{q_{f,i}\} \rangle.$$

For $\alpha = \emptyset$, we define the transition-free npda

$$A_i = \langle \{q_{0,i}, q_{f,i}\}, \Sigma, \emptyset, \emptyset, q_{0,i}, \{q_{f,i}\} \rangle.$$

In this latter case, the final state $q_{f,i}$ is settled only for technical reasons, but actually it cannot be reached. This is the basis of our inductive construction.

Now, let us define the inductive step. Let v_i be an internal node in the dag D_P. The construction of A_i depends on the label of v_i, and so we have the following cases:

LABEL '+': v_i is a vertex labeled by '+', with two ingoing arcs from vertices v_a and v_b, for $1 \leq a, b < i$, representing the instruction $x_i := x_a + x_b$. Define

$$A_i = \langle Q_a \cup Q_b \cup \{q_{0,i}, q_{f,i}\}, \Sigma, \Gamma_a \cup \Gamma_b \cup \{X_i\}, H_i, q_{0,i}, \{q_{f,i}\}\rangle, \quad \text{with}$$
$$H_i = H_a \cup H_b \cup$$
$$\{(q_{0,i}, +X_i, q_{0,a}), (q_{f,a}, -X_i, q_{f,i}), (q_{0,i}, +X_i, q_{0,b}), (q_{f,b}, -X_i, q_{f,i})\}.$$

Basically, A_i nondeterministically chooses to activate either the npda A_a or the npda A_b. Before activation, A_i pushes the symbol X_i onto the pushdown, and pops it right at the end of the processing of the activated npda.

LABEL '·': v_i is a vertex labeled by '·', with two ingoing arcs from v_a and v_b, for $1 \leq a, b < i$, representing the instruction $x_i := x_a \cdot x_b$. Define

$$A_i = \langle Q_a \cup Q_b \cup \{q_{0,i}, q_{m,i}, q_{f,i}\}, \Sigma, \Gamma_a \cup \Gamma_b \cup \{L_i, R_i\}, H_i, q_{0,i}, \{q_{f,i}\}\rangle, \quad \text{with}$$
$$H_i = H_a \cup H_b \cup$$
$$\{(q_{0,i}, +L_i, q_{0,a}), (q_{f,a}, -L_i, q_{m,i}), (q_{m,i}, +R_i, q_{0,b}), (q_{f,b}, -R_i, q_{f,i})\}.$$

Here A_i sequentially activates A_a and A_b. Before activating A_a, it pushes the symbol L_i onto the pushdown, and pops it out at the end of A_a-processing, by reaching the state $q_{m,i}$. From this state, A_i pushes another symbol R_i onto the pushdown, thus activating A_b, and pops it out at the end of A_b-processing.

LABEL '*': v_i is a vertex labeled by '*', with a single ingoing arc from the vertex v_a, for $1 \leq a < i$, representing the instruction $x_i := x_a^*$. Define

$$A_i = \langle Q_a \cup \{q_{0,i}, q_{f,i}\}, \Sigma, \Gamma_a \cup \{X_i\}, H_i, q_{0,i}, \{q_{f,i}\}\rangle, \quad \text{with}$$
$$H_i = H_a \cup \{(q_{0,i}, \varepsilon, q_{f,i}), (q_{0,i}, +X_i, q_{0,a}), (q_{f,a}, -X_i, q_{0,i})\}.$$

Here A_i nondeterministically chooses to activate A_a a certain number of times, including zero. In case A_a is going to be activated, A_i pushes the symbol X_i onto the pushdown, and pops it out at the end of A_a-processing by returning to the state $q_{0,i}$. In the state $q_{0,i}$, A_i can also terminate the iteration by reaching the state $q_{f,i}$ with an ε-move.

Informally, for parsing an input string, A_ℓ verifies matching with reg-exp(P) by starting from the source node of D_P and traveling along the arcs. When traveling towards sources, one symbol per each visited vertex is pushed onto the pushdown. Vice versa, when traveling back towards the sink, pushdown symbols are popped. These operations are needed to record the sequence of visited vertices, since some of them are shared (i.e., their fan-out is greater than 1). By induction on the depth of D_P, one may formally prove that $L(A_\ell)$ is the language denoted by reg-exp(P).

Let us measure the size of the npda A_ℓ. In the construction of each A_i, at most 3 new states and 2 new pushdown symbols are used, if v_i is an inner node, but only 2 states with no pushdown symbols, if it is a source node. Hence, $|Q_\ell| < 3\ell$ and $|\Gamma_\ell| < 2\ell$. Finally, the pushdown height of A_ℓ is easily seen to be equal to $\text{depth}(D_P) < \ell$.

Actually, some improvements on the size of A_ℓ can be obtained. Given the dynamics of A_ℓ above described, the use of the pushdown turns out to be necessary only for shared vertices, so that the machine can identify the proper ancestor by popping a symbol from the pushdown. Thus, by renaming the pushdown symbols, we can reduce the size of the pushdown alphabet to fan-out(P). Moreover, for vertices with fan-out equal to 1, the moves involving the pushdown can be transformed into ε-moves, thus reducing the pushdown height. Clearly, one may also eliminate ε-moves, possibly reducing the number of states. In conclusion

Theorem 2. *Let P be an slp. Then there exists a constant height npda $A_P = \langle Q, \Sigma, \Gamma, H, q_0, \{q_f\}, h \rangle$ such that $L(A_P)$ is denoted by reg-exp(P) and the size of A_P is linear in the size of P. In particular, $|Q| < 3 \cdot \text{length}(P)$, $|\Gamma| = $ fan-out(P), and $h < \text{length}(P)$. More precisely, h equals to the maximum number of vertices with fan-out greater than 1 along paths from sources to the sink.*

5   Constant Height pdas Versus Finite State Automata

Here we compare the sizes of constant height pushdown automata and the standard finite state automata. In what follows, npdas (but not dpdas) are in the form stated in Lemma 1, i.e., they accept by entering a unique final state with empty pushdown. First of all, we prove an exponential upper bound on the size of nfas (dfas) simulating constant height npdas (dpdas, respectively).

Proposition 3. *For each constant height npda $A = \langle Q, \Sigma, \Gamma, H, q_0, \{q_f\}, h \rangle$, there exists an equivalent nfa $A' = (Q', \Sigma, H', q_0', \{q_f'\})$ with $|Q'| \le |Q| \cdot |\Gamma^{\le h}|$. If $B = \langle Q, \Sigma, \Gamma, H, q_0, F, h \rangle$ is a constant height dpda, we can construct an equivalent dfa with no more than $|Q| \cdot |\Gamma^{\le h}|$ states.*

By Proposition 3 and the usual subset construction, one immediately gets:

Corollary 1. *Let $A = \langle Q, \Sigma, \Gamma, H, q_0, \{q_f\}, h \rangle$ be a constant height npda. Then there exists an equivalent dfa with no more than $2^{|Q| \cdot |\Gamma^{\le h}|}$ states.*

We are now going to show that the simulation costs in Proposition 3 and Corollary 1 are *optimal* by exhibiting two witness languages with matching exponential and double exponential gaps. For a string $x = x_1 \cdots x_n$, let $x^R = x_n \cdots x_1$ denote its reverse. Given an $h > 0$, an alphabet Γ, and two separator symbols $\sharp, \$ \notin \Gamma$, we define the language

$$L_{\Gamma,h} = \{\sharp w_1 \sharp w_2 \sharp \cdots \sharp w_m \$ w : w_1, \ldots, w_m \in \Gamma^*, w \in \Gamma^{\le h}, \text{ and } w \in \bigcup_{i=1}^m \{w_i^R\}\}.$$

We begin by providing upper bounds on the size of machines accepting $L_{\Gamma,h}$:

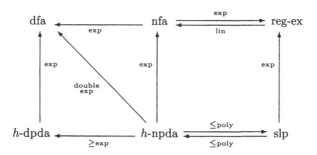

Fig. 2. Costs of simulations among different types of formalisms defining regular languages. Here h-dpda (h-npda) denotes constant height dpda (npda, respectively). An arc labeled by lin (poly, exp, double exp) from a vertex A to a vertex B means that, given a representation of type A, we can construct an equivalent representation of type B, paying by a linear (polynomial, exponential, double exponential, respectively) increase in the size. For clarity, some trivial linear conversions are omitted.

Lemma 2. *For each $h > 0$ and each alphabet Γ: (i) The language $L_{\Gamma,h}$ can be accepted by an npda with $\mathcal{O}(1)$ states, pushdown alphabet of size $|\Gamma|$, and constant height h. (ii) The language $L_{\Gamma,h}$ can also be accepted by an nfa (or dfa) with $\mathcal{O}(|\Gamma^{\leq h}|)$ states (or $2^{\mathcal{O}(|\Gamma^{\leq h}|)}$ states, respectively).*

Now we show that the sizes for $L_{\Gamma,h}$ stated in Lemma 2 (ii) are optimal.

Lemma 3. *For each $h > 0$ and each alphabet Γ, any dfa (or nfa) accepting the language $L_{\Gamma,h}$ must use at least $2^{|\Gamma^{\leq h}|}$ states (or $|\Gamma^{\leq h}|$ states, respectively).*

Let us now show the optimality of the exponential simulation cost of constant height dpdas by dfas presented in Proposition 3. Consider the following witness language: given an $h > 0$, an alphabet Γ, and a separator symbol $\sharp \notin \Gamma$, let

$$D_{\Gamma,h} = \{w \sharp w^R : w \in \Gamma^{\leq h}\}.$$

Lemma 4. *For each $h > 0$ and each alphabet Γ: (i) The language $D_{\Gamma,h}$ is accepted by a dpda with $\mathcal{O}(1)$ states, pushdown alphabet Γ and constant height h, and also by a dfa with $2 \cdot |\Gamma^{\leq h}| + 1$ states. (ii) Any dfa accepting the language $D_{\Gamma,h}$ must have at least $|\Gamma^{\leq h}|$ states.*

6 The Final Picture

In conclusion, in Figure 2, we sum up the main relations on the sizes of the different types of formalisms defining regular languages we considered in this paper. Let us briefly discuss the simulation costs displayed in this figure. The costs of the following simulations are asymptotically optimal:

- h-dpda $\rightarrow$ dfa: the exponential cost comes from Proposition 3, while its optimality follows from Lemma 4.
- h-npda $\rightarrow$ nfa: exponential cost, by Proposition 3 and Lemmas 2 (i) and 3.

- slp → reg-ex: the exponential cost comes from Proposition 2, while its optimality follows, e.g., from Example 3 in Section 2.1.
- h-npda → dfa: the double exponential cost was presented by Corollary 1, its optimality follows from Lemmas 2 (i) and 3.
- nfa → dfa: the exponential cost is known from [10], its optimality from [8].
- nfa ↔ reg-ex: the linear cost for the "←" conversion comes directly from the Kleene's Theorem (see also [2,3] for more sophisticated translations), while its optimality follows trivially by considering, e.g., the regular expression a^n, for a fixed $n > 0$. The exponential cost for the converse direction and its optimality is from [5].

The costs of the following simulations are not yet known to be optimal:
- h-npda ↔ slp: Theorems 1 and 2 prove polynomial upper bounds for both directions.
- h-npda → h-dpda: the exponential lower bound comes from the following consideration: a sub-exponential cost of h-npda → h-dpda conversion together with the optimal exponential cost for h-dpda → dfa would lead to a sub-double exponential cost of h-npda → dfa, thus contradicting the optimality of the double exponential cost. In general, for h-npda → h-dpda conversion, we conjecture a double exponential optimal cost.

Acknowledgements. The authors wish to thank the anonymous referees for useful and kind comments.

References

1. Aho, A.V., Hopcroft, J.E., Ullman, J.D.: The Design and Analysis of Computer Algorithms. Addison-Wesley, Reading (1974)
2. Brüggemann-Klein, A.: Regular expressions into finite automata. Theoretical Computer Science 120, 197–213 (1993)
3. Caron, P., Ziadi, D.: Characterization of Glushkov automata. Theoretical Computer Science 233, 75–90 (2000)
4. Chandra, A., Kozen, D., Stockmeyer, L.: Alternation. J. ACM 28, 114–133 (1981)
5. Ehrenfeucht, A., Zieger, P.: Complexity measures for regular expressions. J. Computer and System Sciences 12, 134–146 (1976)
6. Hopcroft, J.E., Motwani, R., Ullman, J.D.: Introduction to Automata Theory, Languages, and Computation. Addison-Wesley, Reading (2001)
7. Kleene, S.: Representation of events in nerve nets and finite automata. In: Shannon, C., McCarthy, J. (eds.) Automata Studies, pp. 3–42. Princeton University Press, Princeton (1956)
8. Meyer, A.R., Fischer, M.J.: Economy of description by automata, grammars, and formal systems. In: IEEE 12th Symp. Switching and Automata Theory, pp. 188–191 (1971)
9. Rabin, M.: Probabilistic automata. Information and Control 6, 230–245 (1963)
10. Rabin, M., Scott, D.: Finite automata and their decision problems. IBM J. Res. Develop. 3, 114–125 (1959)
11. Shepherdson, J.C.: The reduction of two–way automata to one–way automata. IBM J. Res. Develop. 3, 198–200 (1959)
12. Valiant, L.G.: Regularity and related problems for deterministic pushdown automata. J. ACM 22, 1–10 (1975)

A Taxonomy of
Deterministic Forgetting Automata

Jens Glöckler

Institut für Informatik, Universität Giessen
Arndtstr. 2, 35392 Giessen, Germany
Jens.Gloeckler@math.uni-giessen.de

Abstract. We investigate deterministic forgetting automata, i.e., deterministic linear bounded automata which can only use the operations 'move', 'erase' (rewrite with a blank symbol) and 'delete' (remove completely). We give a taxonomy of deterministic forgetting automata and draw comparisons to other kinds of automata (namely deterministic one-turn pushdown automata and one-way one-counter automata).

1 Introduction

Forgetting automata were introduced by Jančar, Mráz, and Plátek in [1] and have been studied further in a number of papers ([2,3,4,5,6]). They were introduced to model certain strategies from linguistics, e.g., the *analysis by reduction*: An input string is shortened repeatedly in order to finally decide whether the obtained short string (and consequently the original string) is syntactically correct or not.

In order to model this reduction, the operations *erase* and *delete* – as well as the operation *move* – were utilized. The erase operation originates from the work on *erasing automata* that was initiated in [7] (as a special case of *finite change automata*) and continued in [8,5]. The erase operation allows the automaton to rewrite the content of a tape field with an auxiliary blank symbol and thereby to irreversibly destroy the original information stored there. Nevertheless this operation turns out to be surprisingly powerful.

The delete operation, on the other hand, originates from the work on *list automata* that have been investigated in [9]. List automata work on a doubly linked list and can move around on the input list (this operation is called *move*), rewrite input symbols (*write*), remove entries from the list (*delete*) and insert new elements into the list (*insert*). The data structure of a doubly linked list and the associated operations seem very natural from a computational point of view. Moreover, the four levels of the Chomsky hierarchy can be represented in a uniform machine model via list automata (and certain restrictions on the operations available, see [10]).

An automaton that is allowed to use one or more of the operations move (in the following denoted by MV), erase (ER) and delete (DL) is called *forgetting automaton*. If we now consider the possible directions left and right for each of the given operations, we will get the following six operations in total:

M. Ito and M. Toyama (Eds.): DLT 2008, LNCS 5257, pp. 371–382, 2008.

- MV_L, MV_R: move head to the left and right, resp.,
- ER_L, ER_R: erase current field with ␣ and move head to the left and right, resp.
- DL_L, DL_R: delete current field and move head to the left and right, resp. (i.e., completely remove the current field and let the head reach the field originally on the left and on the right side of the head, respectively).

As any subset (except the empty set) of this set of operations can be examined, we have $2^6 - 1 = 63$ different automata models (and language families, respectively) to consider. We use the possible operations to denote the type of automaton considered, e.g., an (MV_R, ER, DL_R)-automaton[1]. The language family accepted by a certain type of automaton is denoted accordingly, e.g., $\mathscr{L}(MV_R, ER, DL_R)$.

Many of these language families, however, coincide trivially as certain lacking operations can be directly simulated by consecutively performing other operations. The series of operations $ER_L \to MV_R \to MV_R$, for example, corresponds to ER_R and therefore $\mathscr{L}(MV, ER_L) = \mathscr{L}(MV, ER)$. Other types of automata like (DL_R)-automata and (MV)-automata characterize the family of regular languages as they are in fact nothing else but one-way and two-way finite automata, respectively.

In this paper we compare deterministic (ER_R, DL)- and (MV_R, DL)-automata to the class of deterministic linear languages characterized by linear $LR(1)$ grammars and deterministic one-turn pushdown automata. Moreover the inclusions of $\mathscr{L}_{det}(MV_R, DL)$ in $\mathscr{L}_{det}(MV_R, ER)$ and $\mathscr{L}_{det}(MV_R, ER_R, DL)$ are both shown to be strict – a question that has been left open in [1]. The main result, however, is the separation of $\mathscr{L}_{det}(ER, DL)$ and $\mathscr{L}_{det}(MV_L, ER, DL)$ which solves another open problem from [1].

Furthermore we revise known results from the nondeterministic case to hold in the deterministic case (if possible) and quote recent results on deterministic forgetting automata from [11] to complete the taxonomy.

The remainder of this paper is organized as follows: In Section 2 some basic notations are introduced and the types of automata and grammars used in the paper are defined. In Section 3 the taxonomy of deterministic forgetting automata is examined. Finally, in Section 4 some open problems are presented.

2   Preliminaries

Let A^* denote the set of all words over the finite alphabet A. The empty word is denoted by ε. The reversal of a word w is denoted by w^R and the length of w by $|w|$. The number of occurrences of an alphabet symbol $a \in A$ in a word $w \in A^*$ is denoted by $|w|_a$. Set inclusion and strict set inclusion are denoted by $\subseteq$ and $\subset$, respectively.

---

[1] The missing subscript for an operation indicates that both directions are allowed (i.e., ER stands for ER_L and ER_R).

We use the following notations of language families: REG (regular languages), DCFL (deterministic context-free languages) and DCSL (deterministic context-sensitive languages).

We write $\mathscr{L}(X)$ for the family of languages accepted by devices of type X and $\mathscr{L}_{det}(X)$ for the family of languages accepted by deterministic devices X. Furthermore let $\mathbb{N}$ denote the set of positive natural numbers and $\mathbb{N}_0$ the set of natural numbers including zero.

Forgetting Automata

A *forgetting automaton* is a system $\mathcal{A} = \langle S, A, \triangleright, \triangleleft, \sqcup, O, \delta, s_0, F \rangle$, where S is a finite set of states, A is the input alphabet, $\triangleright, \triangleleft \notin A$ are the left and the right sentinels, $\sqcup \notin A$ is the blank symbol used for erasing, O is a subset of the set $\{MV_L, MV_R, ER_L, ER_R, DL_L, DL_R\}$, $\delta : S \times (A \cup \{\triangleright, \triangleleft, \sqcup\}) \rightarrow 2^{S \times O}$ is the transition function, $s_0 \in S$ is the initial state and $F \subseteq S$ is the set of final states. If $\mathcal{A}$ reads $\triangleright$ (or $\triangleleft$, respectively), it always implies an MV_R-operation (or MV_L-operation, respectively), even if $MV_R, MV_L \notin O$. Generally, a forgetting automaton is nondeterministic. A forgetting automaton is deterministic if $|\delta(s, x)| \leq 1$ for all $s \in S$ and $x \in (A \cup \{\triangleright, \triangleleft, \sqcup\})$.

A configuration of a forgetting automaton $\mathcal{A}$ is a string $w_1 s w_2$, where the word $w_1 w_2 \in \triangleright(A \cup \{\sqcup\})^* \triangleleft$ is the content of the list, s is the current state and $\mathcal{A}$ reads the first symbol of w_2. By $\vdash$ we denote the relation which describes the change of configurations according to δ; $\vdash^*$ is the reflexive, transitive closure of $\vdash$. An input word w is accepted by $\mathcal{A}$ if there is a computation, starting in the initial configuration $s_0 \triangleright w \triangleleft$, which reaches a configuration with an accepting state.

In case O contains both versions X_L and X_R of an operation, we write X for short. A forgetting automaton with a certain set of operations, e.g., MV_R and DL, is called (MV_R, DL)-automaton. For the family of languages accepted by such automata we write $\mathscr{L}(MV_R, DL)$.

Counter Automata

A *one-way one-counter automaton (1CA)* is a pushdown automaton (PDA) which accepts by final state and is allowed to use one pushdown symbol only (except for the bottom marker).

Linear LR(1) Grammars and One-Turn Pushdown Automata

A *linear grammar* is a context-free grammar $G = \langle N, T, S, P \rangle$, where all productions in P are of the form

$$A \rightarrow u_1 B u_2 \mid u_3$$

for $A, B \in N$ and $u_1, u_2, u_3 \in T^*$.

A *linear LR(1) grammar* is a linear grammar $G = \langle N, T, S, P \rangle$ where for the sentential forms $v_1 y a w_1$ and $v_1 y a w_2$, with $v_1, v_2, w_1, w_2 \in T^*, y \in (N \cup T)^*$, and $a \in T$, the derivations

$$S \Rightarrow^* v_1 A a w_1 \Rightarrow v_1 y a w_1$$
$$S \Rightarrow^* v_2 B w \Rightarrow v_1 y a w_2$$

imply that

$$v_1 = v_2, A = B, w = a w_1.$$

Furthermore N contains no useless symbols and S does not appear on the right side of any rule.

A *deterministic one-turn pushdown automaton* is a deterministic pushdown automaton $M = \langle S, A, \Gamma, \delta, s_0, Z_0, F \rangle$ for which each series of instantaneous descriptions $(s_{i_1}, w_1, y_1) \vdash^* \cdots \vdash^* (s_{i_k}, w_k, y_k)$ on an accepted word has the following property: There exists an $i \in \{1, \ldots, k\}$ such that

$$|y_1| \leq \cdots \leq |y_{i-1}| \leq |y_i| > |y_{i+1}| \geq \cdots \geq |y_k|.$$

The class of languages generated by linear LR(1) grammars coincides with the class of languages accepted by deterministic one-turn pushdown automata (see [12]). In the following we will denote it by DetLIN.

3   The Hierarchy of Deterministic Forgetting Automata

For known results in the nondeterministic case see [2,11]; many of the results given there are also valid for the deterministic case. We will, however, state these results here for the sake of completeness.

At the bottom of the classification[2] of deterministic forgetting automata (for a preview, see Figure 2 at the end of the paper) we find the class of regular languages, characterized, for example, by automata with right-moving operations only or only with one type of operation (MV, ER or DL).

In [11] the first language class above REG was shown to coincide with the class of languages accepted by deterministic one-way one-counter automata:

Proposition 1. $\mathscr{L}_{det}(\mathsf{ER_R}, \mathsf{DL}) = \mathscr{L}_{det}(1CA)$

$\mathscr{L}_{det}(\mathsf{ER_R}, \mathsf{DL})$ is obviously a superset of the family of regular languages; witness languages for the strictness of the inclusion are, e.g., $\{a^n b^n \mid n \in \mathbb{N}\}$ and $\{w \in \{0,1\}^* \mid |w|_0 = |w|_1\}$.

When comparing the language classes between REG and DCFL to the class of deterministic linear languages we get results very similar to the nondeterministic case:

Lemma 1. *DetLIN is incomparable to* $\mathscr{L}_{det}(\mathsf{ER_R}, \mathsf{DL})$.

---

[2] We omit the trivial classes $\mathscr{L}_{det}(\mathsf{MV_L}), \mathscr{L}_{det}(\mathsf{ER_L})$ and $\mathscr{L}_{det}(\mathsf{MV_L}, \mathsf{ER_L})$ below REG.

Proof. On the one hand $\{w \in \{0,1\}^* \mid |w|_0 = |w|_1\} \in \mathscr{L}_{det}(\mathsf{ER_R}, \mathsf{DL})$ is not (deterministic) linear, on the other hand $\{wcw^R \mid w \in \{a,b\}^*\} \notin \mathscr{L}_{det}(\mathsf{ER_R}, \mathsf{DL})$ ([2]) is a deterministic linear language.   □

The next language family in the hierarchy however contains all deterministic linear languages:

Theorem 1. *DetLIN* $\subset \mathscr{L}_{det}(\mathsf{MV_R}, \mathsf{DL})$

Proof. Let a one-turn pushdown automaton $\mathcal{A}$ be given. First of all we convert $\mathcal{A}$ into a linear grammar $G = \langle N, T, S, P \rangle$ using the method described in [13, Theorem 5.7.1]. As pointed out in [12], this construction leads to a linear LR(1) grammar.

The processing of a given input word w by a deterministic $(\mathsf{MV_R}, \mathsf{DL})$-automaton $\mathcal{B}$ consists of two phases: In the first phase the pushdown automaton $\mathcal{A}$ is simulated up to its first pop move (i.e., the move that decreases the stack height for the first time); in the second phase the grammar rules are processed successively by deleting symbols from the inside to the outside according to the given linear productions.

The pushdown automaton $\mathcal{A}$ can easily be simulated by the forgetting automaton $\mathcal{B}$ up to the first pop move as $\mathcal{B}$ only needs to move right and remember the top symbol of the stack. When the first pop move is reached during the simulation, $\mathcal{B}$ switches to the second phase where the grammar rules of G are applied. As a result of the construction of G the pop move occurs at the position of the input word at which a string that consists only of terminal symbols is generated in the last derivation step.

From that point on $\mathcal{B}$ can stepwise determine the nonterminal occurring in the preceding sentential form as there is only one nonterminal at a time (due to the linearity) and the nonterminal used afore is uniquely determined (due to the LR(1) condition). In order to be able to use the LR(1) condition, $\mathcal{B}$ needs to read the following terminal symbol (with an $\mathsf{MV_R}$ and an $\mathsf{DL_L}$ step) and to store it in its internal states. As the right hand sides of the productions in P have a maximum length of two (cf. the above mentioned proof on the construction of G), $\mathcal{B}$ knows about the substring ya of the current sentential form $uyaw_1$ and thus can uniquely determine the preceding nonterminal A such that $uAaw_1 \Rightarrow uyaw_1$. In this way $\mathcal{B}$ can stepwise delete the appropriate symbols, backtrack the (possible) derivation of grammar G and finally accept the input if the start symbol S can be obtained.

Furthermore the language $\{w \in \{0,1\}^* \mid |w|_0 = |w|_1\} \in \mathscr{L}_{det}(\mathsf{MV_R}, \mathsf{DL})$ is not (deterministic) linear and therefore the inclusion is strict.   □

The next results from [2] likewise hold in the deterministic case as the given witness languages can equally be accepted by deterministic automata of the corresponding classes:

Proposition 2. $\mathscr{L}_{det}(\mathsf{ER_R}, \mathsf{DL}) \subset \mathscr{L}_{det}(\mathsf{MV_R}, \mathsf{DL})$

Proof. The proof of the inclusion in [2] also holds for the deterministic case. Furthermore the witness language $\{wcw^R \mid w \in \{a,b\}^*\}$ for the separation can likewise be accepted by a deterministic (MV_R, DL)-automaton.   □

Proposition 3. $\mathscr{L}_{det}(ER_R, DL) \subset \mathscr{L}_{det}(ER, DL)$

Proof. The inclusion holds in any case as merely the ER_L-operation is added. The strictness of the inclusion holds as the witness language $\{a^{2^n} \mid n \in \mathbb{N}\}$ given for the nondeterministic case in [2] can also be accepted by a deterministic (ER, DL)-automaton.   □

Lemma 2. $\mathscr{L}_{det}(MV_R, DL)$ *is incomparable to* $\mathscr{L}_{det}(ER, DL)$.

Proof. The language $L_1 = \{wcw^R \mid w \in \{a,b\}^*\} \in \mathscr{L}_{det}(MV_R, DL)$ cannot be accepted by any deterministic (ER, DL)-automaton ([1]). Furthermore the language $L_2 = \{a^{2^n} \mid n \in \mathbb{N}\} \in \mathscr{L}_{det}(ER, DL)$ cannot be accepted by any deterministic (MV_R, DL)-automaton as L_2 is not context-free and $\mathscr{L}_{det}(MV_R, DL)$ is contained in DCFL (see below).   □

Using the same languages as in the preceding proof we get:

Corollary 1. $\mathscr{L}_{det}(MV_R, ER)$, $\mathscr{L}_{det}(MV_R, ER_R, DL)$ *and DCFL, respectively, are incomparable to* $\mathscr{L}_{det}(ER, DL)$.

While the strict inclusion of $\mathscr{L}_{det}(MV_R, DL)$ in DCFL was shown in [9], the question whether the inclusions $\mathscr{L}_{det}(MV_R, ER) \subseteq DCFL$ and $\mathscr{L}_{det}(MV_R, ER_R, DL) \subseteq DCFL$ are strict, was left open in [8,1,6]. We here show that the inclusions $\mathscr{L}_{det}(MV_R, DL) \subseteq \mathscr{L}_{det}(MV_R, ER)$ and $\mathscr{L}_{det}(MV_R, DL) \subseteq \mathscr{L}_{det}(MV_R, ER_R, DL)$ are both strict.

Theorem 2. $\mathscr{L}_{det}(MV_R, DL) \subset \mathscr{L}_{det}(MV_R, ER)$.

Proof. In [9] it was shown that the deterministic context-free language

$$L = \{a^{n_1}ba^{n_1}a^{n_2}ba^{n_2}\cdots a^{n_l}ba^{n_l}cb^l \mid l, n_1, \ldots, n_l \in \mathbb{N}\}$$

cannot be recognized by a deterministic (MV_R, DL)-automaton. In the following we will show that L can, however, be recognized by a deterministic (MV_R, ER)-automaton.

A deterministic (MV_R, ER)-automaton $\mathcal{A}$ can accept L as follows: $\mathcal{A}$ first moves to the first b and thereby uses MV_R-steps for all but the second symbol, whereas the second symbol is erased and later serves as a delimiter. Subsequently $\mathcal{A}$ erases one a at a time on the right hand and left hand side of the position reached (initially marked with b). In order to be able to distinguish the delimiter constructed before, $\mathcal{A}$ performs the erasing of the symbols on the left hand side via ER_L-operations and then moves back with MV_R. $\mathcal{A}$ can thus detect whether the delimiter is reached or not and can therefore prevent the first field of the tape from being erased. As soon as the delimiter is reached, $\mathcal{A}$ moves rightwards and erases two more a's.

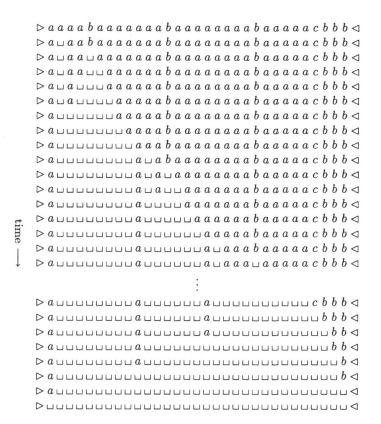

Fig. 1. Exemplary processing of the word $a^4ba^7ba^8ba^5cb^3 = a^4ba^4a^3ba^3a^5ba^5cb^3$ by the deterministic (MV_R, ER)-automaton from the proof of Theorem 2

At this point $\mathcal{A}$ has changed the first subword $a^{n_1}ba^{n_1}$ to $a\sqcup^{2n_1}$ and can continue with processing the second block. By iterating this procedure, every subword $a^{n_i}ba^{n_i}$ is changed to $a\sqcup^{2n_i}$. Figure 1 gives an example of how $\mathcal{A}$ processes the word $a^4ba^7ba^8ba^5cb^3 = a^4ba^4a^3ba^3a^5ba^5cb^3$; here every line shows the contents of the tape after the erasure of another field. As soon as $\mathcal{A}$ reaches the first c, the tape contains one a for each block $a^{n_i}ba^{n_i}$ on the left side of the head and $\mathcal{A}$ can therefore compare the a's with the b's on the right side (by alternately erasing one a and one b). $\mathcal{A}$ finally accepts the input if and only if all symbols have been erased after this comparison. □

Theorem 3. $\mathscr{L}_{det}(MV_R, DL) \subset \mathscr{L}_{det}(MV_R, ER_R, DL)$.

Proof. In analogy to the proof of Theorem 2, a deterministic (MV_R, ER_R, DL)-automaton $\mathcal{A}$ can compare the number of symbols to the left and to the right of each marker b and spare one a per block; here each block $a^{n_i}ba^{n_i}$ is rewritten into a single a. □

In [9] Plátek and Vogel showed that (MV, DL)-automata are able to recognize all deterministic context-free languages. Furthermore non-context-free languages like $\{a^{2^n} \mid n \in \mathbb{N}\}$ can be accepted; therefore we have:

Proposition 4. $DCFL \subset \mathscr{L}_{det}(MV, DL)$

In [11] an open problem from [2] was solved for the deterministic case, namely the question of an equality or strict inclusion of the families $\mathscr{L}_{det}(MV, DL)$ and $\mathscr{L}_{det}(MV, ER)$. In the nondeterministic case, however, the problem remains open.

Proposition 5. $\mathscr{L}_{det}(MV, DL) \subset \mathscr{L}_{det}(MV, ER)$

The next two results relate to (ER, DL)-automata, which in Proposition 3 were already shown to be more powerful than one-way one-counter automata:

Theorem 4. $\mathscr{L}_{det}(ER, DL) = \mathscr{L}_{det}(ER, DL_R)$

Proof. Although the proof of $\mathscr{L}(ER, DL) \subseteq \mathscr{L}(ER, DL_R)$ in [2] uses nondeterminism, the inclusion (and therefore the equality) holds in the deterministic case as well. The missing DL_L-operation (in both cases) can be simulated by consecutively performing ER_L and DL_R. □

Proposition 6. $\mathscr{L}_{det}(ER, DL) \subset \mathscr{L}_{det}(MV_R, ER, DL)$

Proof. The proof from [2] also holds in the deterministic case as the witness language $\{wcw^R \mid w \in \{a, b\}^*\} \notin \mathscr{L}(ER, DL)$ can also be accepted by a deterministic (MV_R, ER, DL)-automaton. □

When comparing the models of (ER, DL)-automata and (MV_L, ER, DL)-automata, one finds out that (MV_L, ER, DL)-automata by means of the additional MV_L-operation are merely able to leave the first input symbol (i.e., the first symbol not yet erased or deleted) to the left without changing it.

An (ER, DL)-automaton can leave an input symbol to the left only via ER_L or DL_L. If the processing of a certain subword v requires several movements back and forth on v, an (ER, DL)-automaton cannot perform these movements without deleting or erasing further symbols to the right of v. On the one hand, this is the case when some nonregular property needs to be checked. On the other hand such a property can only concern the length of a substring as an (ER, DL)-automaton can only read each symbol once before erasing or deleting it. We therefore require the length of the first subword of the input to be a power of two.

By adding another nonregular property on a second subword – separated by an otherwise unused marker symbol – we obtain a language suitable for fooling a deterministic (ER, DL)-automaton and we can thus separate $\mathscr{L}_{det}(ER, DL)$ and $\mathscr{L}_{det}(MV_L, ER, DL)$:

Theorem 5. $\mathscr{L}_{det}(ER, DL) \subset \mathscr{L}_{det}(MV_L, ER, DL)$

Proof. For the separation of the two families we use the following language:

$$L := \{a^{2^n} ca^m b^m \mid n, m \in \mathbb{N}\}$$

A deterministic (MV_L, ER, DL)-automaton can first check whether the length of the first string of a's is a power of two (by stepwise halving the number of symbols; here the MV_L-operation is crucial for not destroying the separator c), then the second block can be checked whether it has the form $a^n b^n$.

In order to show that $L \notin \mathscr{L}_{det}(ER, DL)$ holds, we assume the contrary and regard a deterministic (ER, DL)-automaton $\mathcal{A} = \langle S, A, \triangleright, \triangleleft, \sqcup, O, \delta, s_0, F \rangle$ with $L(\mathcal{A}) = L$.

Let $s = |S|$ and $w = a^{2^{2^n}} ca^n b^n \in L$ with $n > 2^s$ be an input word. If we observe $\mathcal{A}$'s computation on w, the following cases can be distinguished:

1. $\mathcal{A}$ reaches a loop of the form $\triangleright \sqcup^k xa^i ca^n b^n \triangleleft \vdash^* \triangleright \sqcup^k xa^i ca^n b^n \triangleleft$
   In this case $\mathcal{A}$ never reaches c and therefore accepts w if and only if it accepts $wb = a^{2^{2^n}} ca^n b^{n+1} \notin L$. This is a contradiction to the assumption $L(\mathcal{A}) = L$.
2. $\mathcal{A}$ reaches a loop of the form $\triangleright \sqcup^k xa^i ca^n b^n \triangleleft \vdash^* \triangleright \sqcup^k xa^j ca^n b^n \triangleleft$ with $i > j$ and $i - j \le s$
   Here $\mathcal{A}$ reaches c in a configuration $\triangleright \sqcup^{k'} x'ca^n b^n \triangleleft$ with $x' \in S$ and $k' < s$. $\mathcal{A}$ therefore accepts w if and only if it accepts $a^{i-j}w$.
3. $\mathcal{A}$ reaches a loop of the form $\triangleright \sqcup^k xa^i ca^n b^n \triangleleft \vdash^* \triangleright \sqcup^l xa^j ca^n b^n \triangleleft$ with $k < l, i > j$ and $k + i \ge l + j$
   We regard the minimal number of configuration changes for which this condition holds. Then $\mathcal{A}$ deletes $0 \le i + k - (j + l) \le s$ symbols in each loop and moves right, up to c. At this point, again three cases can be distinguished:

   (a) $\mathcal{A}$ reaches the first b before it reaches $\triangleright$ (or never reaches $\triangleright$)
       Here $\mathcal{A}$ enters a loop of the form $\triangleright \sqcup^{k'} ya^i b^n \triangleleft \vdash^* \triangleright \sqcup^{l'} ya^{j'} b^n \triangleleft$ with $i' > j'$. $\mathcal{A}$ therefore accepts w if and only if it accepts

       $$a^{2^{2^n} - (l'-k')(i-j)} ca^{n+(l-k)(i'-j')} b^n \notin L,$$

       because it reaches the first b in the same configuration for both words: The first subword $a^{2^{2^n}}$ is shortened by a multiple of $i - j$ and $\mathcal{A}$ reaches c in the same state as on w — whereas it leaves $(l' - k')(l - k)$ less blanks behind. The subword a^n is extended by a multiple of $i' - j'$ and $\mathcal{A}$ therefore reaches the first b in the same state as on w — whereas it leaves $(l - k)(l' - k')$ more blanks behind.
   (b) $\mathcal{A}$ never reaches a b or $\triangleright$ again
       In this case $\mathcal{A}$ accepts w if and only if it accepts $wb = a^{2^{2^n}} ca^n b^{n+1} \notin L$.
   (c) $\mathcal{A}$ reaches $\triangleright$ before it reaches the first b (or never reaches a b)
       In this case $\mathcal{A}$ can delete or erase at most s symbols near the first b before entering a loop that leads back to the left sentinel. After reaching $\triangleright$ again, the cases 1-3 need to be considered iteratedly. The cases 1, 2, 3 (a) and 3 (b) need only to be considered after at most s iterations of

case 3 (c), i.e., at most s movements between $\triangleright$ and the next a in loops that include at least one erase operation:

— If at some point the next 'a' will not be reached again (this corresponds to case 1), the above reasoning is still valid as $\mathcal{A}$ accepts w if and only if accepts wb.

— If $\mathcal{A}$ leaves less than s blanks on the tape before the first b is reached (for at most s iterations this corresponds to case 2), the first subword w can be extended in order to fool automaton $\mathcal{A}$: An input word leads $\mathcal{A}$ to the same state – before entering the loop that shortens the input as in case 2 – if it is contained in $N := \{a^\tau c a^n b^n \mid \tau \in \alpha \mathbb{N} + \beta\}$ (for some $\alpha, \beta \in \mathbb{N}$). Here $\alpha \mathbb{N} + \beta$ is the set of solutions of the simultaneous congruences that are defined by the 'sweeps' leading to the same state from $\triangleright$ to the first unerased symbol and vice versa. As one solution exists (namely for w), there exist infinitely many due to the Chinese remainder theorem.

As at most s iterations have been performed, α is smaller than 2^{2^n} and there exists a word $a^{n'} c a^n b^n \in N$ such that n' is not a power of two and $\mathcal{A}$ accepts $a^{n'} c a^n b^n$ if and only if it accepts w.

— If $\mathcal{A}$ reaches the first b after at most s iterations of case 3 (c) – this corresponds to case 3 (a) – there again exists some set $N' := \{a^\tau c a^n b^n \mid \tau \in \alpha' \mathbb{N} + \beta'\}$ of input words that lead $\mathcal{A}$ to the same state in which the loop directing to the first b is started. Therefore there exist $n' \in \mathbb{N}_0$ and $n'' \in \mathbb{N}$ such that $\mathcal{A}$ accepts w if and only if it accepts $a^{2^{2^n} - n'} c a^{n+n''} b^n$.

— If after at most s iterations of case 3 (c) $\mathcal{A}$ reaches the situation of case 3 (b), i.e., if it never reaches a b or $\triangleright$ again, $\mathcal{A}$ can still be fooled with the input word wb.

— If loops as in case 3(c) are iterated until the first b is reached, a series $z_1, z_2, \ldots$ of states – in which the first unerased a is reached in each loop – will occur. Due to the choice of n there exist p and q with $p < q$ such that $z_p = z_q$ holds. If only ER-operations are used in the loops (leading from $\triangleright$ to an a and vice versa) after the appearance of z_p, there exist $n', n'' \in \mathbb{N}$ such that $\mathcal{A}$ reaches the same configuration at the first b when processing w and $a^{2^{2^n} + n'} c a^{n-n''} b^n \notin L$. If, however, some DL-operations occur, there exist $n', n'' \in \mathbb{N}$ such that $\mathcal{A}$ reaches the same configuration at the first b when processing w and $a^{2^{2^n} + n'} c a^{n+n''} b^n \notin L$.

For all cases a contradiction to the assumption $L(\mathcal{A}) = L$ was reached and therefore the separation of $\mathscr{L}_{\text{det}}(\text{ER}, \text{DL})$ and $\mathscr{L}_{\text{det}}(\text{MV}_\text{L}, \text{ER}, \text{DL})$ follows.   □

The preceding proof also gives a solution to an open problem from [14] where all deterministic classes except for $\mathscr{L}_{\text{det}}(\text{ER}, \text{DL})$ were shown to be closed under marked concatenation; note that both $\{a^{2^n} \mid n \in \mathbb{N}\}$ and $\{a^m b^m \mid m \in \mathbb{N}\}$ can be accepted by a deterministic (ER, DL)-automaton:

Corollary 2. $\mathscr{L}_{det}(\text{ER}, \text{DL})$ *is not closed under marked concatenation.*

In [11] the following inclusion was shown to hold; it is another example of how surprisingly powerful the erase operation can be:

Proposition 7. $\mathscr{L}_{det}(MV_L, ER, DL) \subset \mathscr{L}_{det}(MV, ER)$

Finally, the last propositions deal with the family of languages accepted by deterministic forgetting automata equipped with all six possible operations:

Proposition 8. $\mathscr{L}_{det}(MV_R, ER, DL) \subset \mathscr{L}_{det}(MV, ER, DL)$

Proof. The proof from [2] also holds in the deterministic case as the witness language $\{wcw \mid w \in \{a,b\}^*\} \notin \mathscr{L}(MV_R, ER, DL)$ can also be accepted by a deterministic (MV, ER, DL)-automaton. $\qquad\square$

The integration into the Chomsky hierarchy at the top of the hierarchy was shown in [3]:

Proposition 9. $\mathscr{L}_{det}(MV, ER, DL) \subset DCSL$

For an overview of the classification of deterministic forgetting automata see Figure 2.

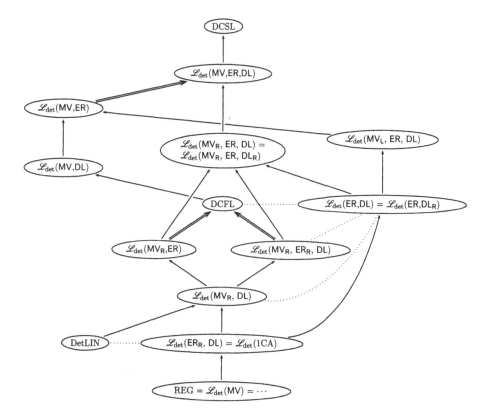

Fig. 2. The classification of deterministic forgetting automata. Here $\implies$ denotes inclusion and $\longrightarrow$ stands for strict inclusion, while dotted lines $\cdots$ mark incomparability.

4   Open Problems

The main open problems left concerning deterministic forgetting automata are probably the inclusions depicted in Figure 2 which are not known to be strict. As (MV, ER)-automata have turned out to be very powerful, there is no evidence that the inclusion $\mathscr{L}_{det}(MV, ER) \subseteq \mathscr{L}_{det}(MV, ER, DL)$ is strict.

In the nondeterministic case $\mathscr{L}(MV_R, ER)$ and $\mathscr{L}(MV_R, ER_R, DL)$ coincide with the family of context-free languages; the question whether the deterministic classes $\mathscr{L}_{det}(MV_R, ER)$ and $\mathscr{L}_{det}(MV_R, ER_R, DL)$ likewise coincide with the family of deterministic context-free languages (DCFL) remains open, however.

References

1. Jančar, P., Mráz, F., Plátek, M.: Forgetting automata and the Chomsky hierarchy. In: Proc. SOFSEM 1992, pp. 41–44. Masaryk University, Brno, Institute of Computer Science (1992)

2. Jančar, P., Mráz, F., Plátek, M.: A taxonomy of forgetting automata. In: Borzyszkowski, A.M., Sokolowski, S. (eds.) MFCS 1993. LNCS, vol. 711, pp. 527–536. Springer, Heidelberg (1993)

3. Jančar, P.: Nondeterministic forgetting automata are less powerful than deterministic linear bounded automata. Acta Mathematica et Informatica Universitatis Ostraviensis 1, 67–74 (1993)

4. Mráz, F., Plátek, M.: A remark about forgetting automata. In: Proc. SOFSEM 1993, pp. 63–66. Masaryk University, Brno, Institute of Computer Science (1993)

5. Mráz, F., Plátek, M.: Erasing automata recognize more than context-free languages. Acta Mathematica et Informatica Universitatis Ostraviensis 3, 77–85 (1995)

6. Jančar, P., Mráz, F., Plátek, M.: Forgetting automata and context-free languages. Acta Informatica 33, 409–420 (1996)

7. von Braunmühl, B., Verbeek, R.: Finite change automata. In: Weihrauch, K. (ed.) GI-TCS 1979. LNCS, vol. 67, pp. 91–100. Springer, Heidelberg (1979)

8. Jančar, P., Mráz, F., Plátek, M.: Characterization of context-free languages by erasing automata. In: Havel, I.M., Koubek, V. (eds.) MFCS 1992. LNCS, vol. 629, pp. 305–314. Springer, Heidelberg (1992)

9. Plátek, M., Vogel, J.: Deterministic list automata and erasing graphs. The Prague bulletin of mathematical linguistics 45, 27–50 (1986)

10. Chytil, M.P., Plátek, M., Vogel, J.: A note on the Chomsky hierarchy. Bulletin of the EATCS 27, 23–30 (1985)

11. Glöckler, J.: Forgetting automata and unary languages. International Journal of Foundations of Computer Science 18, 813–827 (2007)

12. Holzer, M., Lange, K.J.: On the complexities of linear LL(1) and LR(1) grammars. In: Ésik, Z. (ed.) FCT 1993. LNCS, vol. 710, pp. 299–308. Springer, Heidelberg (1993)

13. Harrison, M.A.: Introduction to Formal Language Theory. Addison-Wesley, Reading (1978)

14. Glöckler, J.: Closure properties of the families of languages accepted by forgetting automata. In: Proc. MEMICS 2007, pp. 51–58 (2007)

Provably Shorter Regular Expressions from Deterministic Finite Automata
(Extended Abstract)

Hermann Gruber[1] and Markus Holzer[2]

[1] Institut für Informatik, Ludwig-Maximilians-Universität München,
Oettingenstraße 67, D-80538 München, Germany
gruberh@tcs.ifi.lmu.de
[2] Institut für Informatik, Technische Universität München,
Boltzmannstraße 3, D-85748 Garching bei München, Germany
holzer@in.tum.de

Abstract. We study the problem of finding good elimination orderings for the state elimination algorithm, which is one of the most popular algorithms for the conversion of finite automata into equivalent regular expressions. Based on graph separator techniques we are able to describe elimination strategies that remove states in large induced subgraphs that are "simple" like, e.g., independent sets or subgraphs of bounded treewidth, of the underlying automaton, that lead to regular expressions of moderate size. In particular, we show that there is an elimination ordering such that every language over a binary alphabet accepted by an n-state *deterministic* finite automaton has alphabetic width at most $O(1.742^n)$, which is, to our knowledge, the algorithm with currently the best known performance guarantee. Finally, we apply our technique to the question on the effect of language operations on regular expression size. In case of the intersection operation we prove an upper bound which matches, up to a small factor, a lower bound recently obtained in [9,10], and thus settles an open problem stated in [7].

1 Introduction

One of the most basic theorems in formal language theory is that every regular expression can be effectively converted into an equivalent finite automaton, and *vice versa* [14], and algorithms accomplishing these tasks have been known since the beginning of automata theory, see, e.g., [17]. While regular expressions can be converted efficiently into nondeterministic finite automata, the other direction necessarily leads to an exponential blow-up in size [6]. Some very recent results on this problem imply an increase of $2^{\Omega(n)}$ in size, even given a deterministic finite automaton over a binary alphabet [9,10,11]. In spite of these strong negative results, already early authors noticed that, at least in many cases, it may be possible to improve the standard state elimination algorithm: The authors of the seminal work [17] noticed that the ordering in which the states of the given automaton are processed can greatly influence the size of the resulting regular expression, and an implementation study appearing in the 1960s notes [16]:

M. Ito and M. Toyama (Eds.): DLT 2008, LNCS 5257, pp. 383–395, 2008.
© Springer-Verlag Berlin Heidelberg 2008

"... a basic fault of the method is that it generates such cumbersome and so numerous expressions initially." ...

But only the last few years have seen a renewed interest in heuristic algorithms that produce, at least in some cases, shorter regular expressions than the standard, non-optimized textbook procedure, see, e.g., [3,7,12,18]. However, none of the mentioned algorithms is known to have a better performance guarantee than $O(4^n)$ in the worst case, which is (roughly) the guarantee of the standard textbook algorithms. It is worth mentioning that in [7] a recursive algorithm for converting planar n-state finite automata into regular expressions with a nontrivial performance guarantee of $2^{O(\sqrt{n})}$ was presented. As proved in [10], this bound is asymptotically optimal for the planar case. The mentioned algorithm exploits the separator theorem for planar graphs [15]. This was the starting point of our investigations.

The main idea underlying the graph separator technique is to identify large induced substructures that are "simple" that lead to regular expressions of moderate size or alphabetic width. Such a procedure is seemingly more difficult to implement than a mere state elimination strategy, but we will show how the idea of using separators can be generalized and implemented simply in a divide-and-conquer fashion. The difficulty when applying this idea is that on the one hand large or omnipresent substructures are needed, such that the algorithm can be applied successfully, and on the other hand, these substructures have to produce small regular expressions. These two conditions seem to clash at first thought. Nevertheless, we present two algorithms, one that uses independent sets, the other one induced subgraphs of bounded undirected treewidth, as basic building blocks for a strategy computing a good ordering on the states for the state elimination scheme. Both algorithms when applied to an n-state *deterministic* finite automata attain regular expressions with a performance guarantee of $O(c^n)$, for constants $c < 2.602$ and $c < 1.742$, respectively. As a side result, we identify a structural restriction, namely bounded treewidth, on the transition structure of the given finite automata that guarantees a polynomial upper bound on the resulting regular expression. These new insights on the conversion problem can be applied to some questions regarding the effect of language operations on regular expression size, too. Namely, we present a new algorithm computing a regular expression denoting the intersection of two regular languages. The performance guarantee is proved to be $2^{O(n \log \frac{m}{n})}$, where m and $n \leq m$ are sizes of the given regular expressions. This matches, up to a small factor,[1] a lower bound of $2^{\Omega(n)}$ recently established in [10]. We thus settle a question stated in [7] and complement previous lower bounds from [8,9,10]. We also prove a nontrivial upper bound for the alphabetic width of the language operation of half-removal, whose descriptional complexity in terms of finite automata was studied recently in [4].

---

[1] For example, assuming that storing a regular expression of alphabetic width k takes k bytes, and the larger expression is stored in an enormous plain text file taking 1 MByte = 2^{10} KByte disk space, while the smaller one needs only 1 KByte, we still have $\log \frac{m}{n} = 10$.

2 Basic Definitions

We introduce some basic notions in formal language and automata theory—for a thorough treatment, the reader might want to consult a textbook such as [21]. In particular, let Σ be a finite alphabet and Σ^* the set of all words over the alphabet Σ, including the empty word ϵ. The length of a word w is denoted by $|w|$, where $|\epsilon| = 0$.

A *nondeterministic finite automaton* (NFA) is a 5-tuple $A = (Q, \Sigma, \delta, q_0, F)$, where Q is a finite set of states, Σ is a finite set of input symbols, $\delta : Q \times \Sigma \to 2^Q$ is the transition function, $q_0 \in Q$ is the initial state, and $F \subseteq Q$ is the set of accepting states. The *language accepted* by the finite automaton A is defined as $L(A) = \{ w \in \Sigma^* \mid \delta(q_0, w) \cap F \neq \emptyset \}$, where δ is naturally extended to a function $Q \times \Sigma^* \to 2^Q$. A NFA $A = (Q, \Sigma, \delta, Q_0, F)$ is *deterministic*, for short a DFA, if $|\delta(q, a)| \leq 1$, for every $q \in Q$ and $a \in \Sigma$. In this case we simply write $\delta(q, a) = p$ instead of $\delta(q, a) = \{p\}$. Two finite automata are *equivalent* if they accept the same language. Without loss of generality we assume throughout this paper, that every finite automaton accepting a nonempty language has useful states only, i.e., every state is accessible from the initial state and co-accessible from some accepting state—this assumption is compatible with the definition of deterministic finite automata given above.

It is well known that finite automata and regular expressions are equally powerful, i.e., for every finite automaton on can construct an equivalent regular expression. Let Σ be an alphabet. The regular expressions over Σ are defined recursively in the usual way:[2] $\emptyset$, ϵ, and every letter a with $a \in \Sigma$ is a regular expression, and if r_1 and r_2 are regular expressions, then $(r_1 + r_2)$, $(r_1 \cdot r_2)$, and $(r_1)^*$ are also regular expressions. The language defined by a regular expression r, denoted by $L(r)$, is defined as follows: $L(\emptyset) = \emptyset$, $L(\epsilon) = \{\epsilon\}$, $L(a) = \{a\}$, $L(r_1 + r_2) = L(r_1) \cup L(r_2)$, $L(r_1 \cdot r_2) = L(r_1) \cdot L(r_2)$, and $L(r_1^*) = L(r_1)^*$. The *size* or *alphabetic width* of a regular expression r over the alphabet Σ, denoted by $\mathrm{alph}(r)$, is defined as the total number of occurrences of alphabet symbols of Σ in r. For a regular language L, we define its alphabetic width, $\mathrm{alph}(L)$, as the minimum alphabetic width among all regular expressions describing L. As with finite automata, the notion of equivalence is defined based on equality of the described language.

In the remainder of this section we fix some basic notations from graph theory. A *directed graph*, or *digraph*, $G = (V, E)$ consists of a finite set of vertices V with an associated set $E \subseteq V \times V$ of edges. If the edge relation E is symmetric, the digraph is said to be *symmetric*. Intuitively, a symmetric digraph is obtained by forgetting the orientation of the original edges in G. A digraph $H = (U, F)$ is a *subdigraph*, or simply *subgraph*, of a digraph $G = (V, E)$, if $U \subseteq V$ and for

---

[2] For convenience, parentheses in regular expressions are sometimes omitted and the concatenation is simply written as juxtaposition. The priority of operators is specified in the usual fashion: Concatenation is performed before union, and star before both product and union.

each edge $(u, v) \in F$ with $u, v \in U$, the pair (u, v) is an edge in E. For a subset $U \subseteq V$, the *subgraph induced by* U is the subgraph $G[U] = (U, E \cap (U \times U))$.

Finally, a *hammock* is a digraph $G = (V, E)$ having two distinguished vertices s and t satisfying the properties (1) that the indegree of s and the outdegree of t is zero, and (2) for every vertex v in G, there is both a path from s to v and a path from v to t. Here s is referred to as the *start vertex* and t as the *terminal vertex* of the hammock G. The remaining set of vertices $Q = V \setminus \{s, t\}$ is called the set of *internal vertices*. It is thus convenient to specify a hammock as a 4-tuple $H = (Q, E, s, t)$. With the finite automaton $A = (Q, \Sigma, \delta, q_0, F)$ we naturally associate a hammock $H(A) = (Q, E, s, t)$, where s and t are designated vertices not appearing in Q that play the role of the initial and a single final state, and $E = \{(s, q_0)\} \cup \{(q, t) \mid q \in F\} \cup \{(p, q) \in Q^2 \mid q \in \delta(p, a), \text{ for some } a \in \Sigma\}$. Due to the "dualism" of computations in A and walks in $H(A)$ one can reconstruct the language accepted by A from the walks in $H(A)$—a walk in $H(A)$ is a (possibly empty) sequence of edges along a path, with repeated edges allowed. To this end define the substitution $\sigma : E \rightarrow \Sigma^*$ by $(p, q) \mapsto \{a \in \Sigma \mid q \in \delta(p, a)\}$, if $p, q \in Q$, $(s, q_0) \mapsto \{\epsilon\}$, and $(q, t) \mapsto \{\epsilon\}$, for $q \in F$, which naturally extends to words and languages over E. It is easy to see that $L(A) = \sigma(L^Q_{st})$; here L^Z_{xy}, for $x, y \in Q \cup \{s, t\}$ and $Z \subseteq Q$, refers to the set of all walks in $H(A)$ from vertex x to vertex y whose *internal* vertices are all in Z—the internal vertices of a walk denote those that are visited by the walk after the leaving x and before entering y. This notion naturally extends to L^Z_{XY} for sets $X, Y \subseteq Q \cup \{s, t\}$ and $Z \subseteq Q$. The above definitions are particularly useful in connection with regular expressions because of the following well-known fact (see also [21]).

Lemma 1. *Let Γ and Σ be finite alphabets and r be a regular expression over Γ. Moreover, let $\rho : \Gamma \rightarrow 2^{\Sigma^*}$ be a regular substitution, i.e., a substitution satisfying $\rho(a) = L(r_a)$, for some regular expression r_a, for each $a \in \Gamma$. Then a regular expression describing $\rho(L(r))$ is obtained from r by substituting r_a for each letter $a \in \Sigma$.* $\square$

Thus, it suffices to describe the conversion to regular expressions from finite automata on the basis of the associated digraphs, which we will do in the forthcoming. Because our proofs and algorithmic ideas are mainly drawn from graph theory, this proves to be notationally more convenient.

3   Choosing a Good Elimination Ordering for the State Elimination Technique

The state elimination technique is an optimized version of the McNaughton-Yamada algorithm, avoiding the unnecessary computation of subexpressions. A detailed description of the state elimination algorithm can be found in [21]. Here we only introduce the necessary background and notations. For the hammock $H(A) = (Q, E, s, t)$ that is associated to a finite automaton $A = (Q, \Sigma, \delta, q_0, F)$ both algorithms compute regular expressions r^S_{jk} from the *regular expression matrix* $R^S = (r^S_{jk})_{j,k \in Q \cup \{s,t\}}$ satisfying $L(r^S_{jk}) = L^S_{jk}$, for every $j, k \in Q \cup \{s, t\}$

and a fixed ordering $S \subseteq Q$; it is convenient to write a total order on a finite set as a word, where the relative positions of the letters specify the order. Since any path from vertex j to k whose internal vertices are in $S \cup \{i\}$ can be written as $L_{jk}^{S \cup \{i\}} = L_{jk}^S + L_{ji}^S \cdot (L_{ii}^S)^* \cdot L_{ik}^S$, we are led to define the identity $r_{jk}^{S \cdot i} = r_{jk}^S + r_{ji}^S \cdot (r_{ii}^S)^* \cdot r_{ik}^S$ on regular expressions, for every $j, k \in Q \cup \{s, t\}$ and $S \cdot i$ prefix of the ordered set Q, which is the basic recurrence of both algorithms. After applying these rules for all pairs (j, k) with $j, k \neq i$ for an inner vertex i it becomes isolated and thus can safely be eliminated. This explains the term *state elimination algorithm*, because during the computation of the expressions r_{jk}^S we are led to the hammock $H^S(A) = (Q \setminus S, E^S, s, t)$, with $E^S = \{ (j, k) \mid$ there is a path from vertex j to k in $H(A)$ with internal vertices from $S \}$. Observe that the choice of the elimination order on Q can greatly influence the size of the resulting regular expression r_{st}^Q. A further slight enhancement on the algorithm concerns the usage of the similarity relation.[3] With a straightforward implementation one can ensure that $r_{jk}^S = \emptyset$ if and only if $L_{jk}^S = \emptyset$.

The size of the regular expression resulting from applying the McNaughton-Yamada algorithm has been analyzed in [7]. There it was shown that the algorithm produces a regular expression of alphabetic width at most $|\Sigma| \cdot n \cdot 4^n$. Here, state elimination is better by a factor of n, paradoxically because we *enlarged* the automaton, in adding an $(n + 1)$th state as single final state.

Theorem 2. *Let A be an n-state finite automaton. Then the state elimination algorithm produces for any ordering on the states a regular expression describing $L(A)$ of alphabetic width at most $|\Sigma| \cdot 4^n$.* $\Box$

Previous accounts on choosing elimination orderings can be naturally put into two groups: In the first group, we find algorithms that have a tail-recursive specification, and are most easily implemented by an iterative program [3,6,7,11,17,18], the others are based on the divide-and conquer paradigm [7,12], suggesting a recursive implementation. We present a lemma that proves useful for designing algorithms in both groups. The lemma gives rise to two algorithms for choosing good elimination orderings yielding nontrivial performance guarantees for deterministic finite automata, one of which gives polynomial-size regular expressions for a restricted yet large class of finite automata.

3.1   The Main Lemma

As before, for a finite automaton A, let $H(A) = (Q, E, s, t)$ be the hammock associated with A, and let S denote a subset of Q. We begin with an observation on the

---

[3] Two regular expressions r and s are called *similar*, in symbols $r \cong s$, if r and s can be transformed into each other by repeatedly applying one of the following rules to their subexpressions: (1) $r + r \cong r$, (2) $(r + s) + t \cong r + (s + t)$, (3) $r + s \cong s + r$, (4) $r + \emptyset \cong r \cong \emptyset + r$, (5) $r \cdot \emptyset \cong \emptyset \cong \emptyset \cdot r$, (6) $r \cdot \epsilon \cong r \cong \epsilon \cdot r$, and (7) $\emptyset^* \cong \epsilon \cong \epsilon^*$. The first three rules above define the notion of similarity introduced by Brzozowski [1], and the remaining three have been added because of their usefulness in the context of converting regular expressions into finite automata.

expressions r_{jk}^S resulting from eliminating S in case the induced subgraph $H[S]$ falls apart into mutually disjoint components.

Lemma 3. *Let $H = (Q, E, s, t)$ be a hammock. Assume $S \subseteq Q$ can be partitioned into two sets T_1 and T_2 such that the induced subgraph $H[S]$ falls apart into mutually disconnected components $H[T_1]$ and $H[T_2]$. Let j and k be vertices with $j, k \in (Q \setminus (T_1 \cup T_2)) \cup \{s, t\}$. Then for the expression obtained by elimination of the the vertices in T_1 followed by elimination of the vertices in T_2 holds $r_{jk}^{T_1 \cdot T_2} \cong r_{jk}^{T_1} + r_{jk}^{T_2}$.*

Proof. We prove the statement by induction on $|T_1| + |T_2|$. The induction is rooted at $|T_1| + |T_2| = 0$. For the case T_2 is empty, we have in general $r_{jk}^{T_1 T_2} = r_{jk}^{T_1} \cong r_{jk}^{T_1} + r_{jk}^{\epsilon}$, as desired. For the induction step, let $|T_1| + |T_2| = n$, with $T_2 \neq \emptyset$. Let t be the last element in T_2, that is, $T_2 = Tt$ for some prefix T of T_2. Then $r_{jk}^{T_1 T_2} \cong r_{jk}^{T_1 T} + r_{jt}^{T_1 T} \cdot (r_{tt}^{T_1 T})^* \cdot r_{tk}^{T_1 T}$. Since $|T_1| + |T| = n - 1$, for the first of the four subexpressions on the right-hand side the induction hypothesis applies: $r_{jk}^{T_1 T} \cong r_{jk}^{T_1} + r_{jk}^{T}$. For the last three subexpressions, we claim that $r_{jt}^{T_1 T} \cong r_{jt}^{T}$, as well as $(r_{tt}^{T_1 T})^* \cong (r_{tt}^{T})^*$, and $r_{tk}^{T_1 T} = r_{tk}^{T}$. We only prove the first congruence, the others are dealt with in a similar manner. It suffices to prove $r_{jt}^{T_1} \cong r_{jt}^{\epsilon}$, since the both sides of the former congruence are obtained from the latter by eliminating T, and state elimination preserves similarity of expressions. If there is an edge $(j, t) \in E$, then it is already described by r_{jt}^{ϵ}. It only remains to show that no further words are introduced by eliminating T_1. So we may as well assume that $(j, t) \notin E$ and prove the congruence for this case. This can be done as follows: Consider the subgraph $H[S]$. By assumption of the lemma, $t \in T_2$ is not reachable from any vertex in T_1, thus no walk from j to t can visit a vertex in T_1, and since there is no direct connection from j to t, the language $L_{jt}^{T_1}$ is empty. Every regular expression describing the empty set is similar to $\emptyset$, hence $r_{jt}^{T_1} \cong \emptyset$, provided $(j, t) \notin E$. This completes the proof of the congruence for this subexpression. Plugging in the four subexpression congruences we just found that $r_{jk}^{T_1 T_2} \cong r_{jk}^{T_1} + r_{jk}^{T} + r_{jt}^{T}(r_{tt}^{T})^* r_{tk}^{T} = r_{jk}^{T_1} + r_{jk}^{T_2}$.   □

3.2   Eliminating Independent Sets

The following theorem shows that eliminating an independent set from the vertex set before eliminating the remaining vertices produces intermediate regular expressions which are short and easy to understand.

Lemma 4. *Let $H = (Q, E, s, t)$ be a hammock. Assume $I \subseteq Q$ is an independent set in H. Let j and k be vertices with $j, k \in (Q \setminus I) \cup \{s, t\}$. Then for the regular expression r_{jk}^I obtained after elimination of I holds $r_{jk}^I \cong r_{jk}^{\epsilon} + \sum_{i \in I} r_{ji}^{\epsilon} \cdot (r_{ii}^{\epsilon})^* \cdot r_{ik}^{\epsilon}$.*

Proof. By induction on $|I|$, making repeated use of Lemma 3: The statement holds true in the case $|I| = 1$. For $|I| > 1$, in the notation of Lemma 3, set

$S = I$, let t be the last element in I, and assume that T is a suitable prefix such that $I = Tt$. Then $H[I]$ falls apart into mutually disjoint components $H[T]$ and $H[\{t\}]$. Thus, Lemma 3 is applicable, and $r_{jk}^I \cong r_{jk}^T + r_{jk}^t = r_{jk}^T + r_{jk}^\epsilon + r_{jt}^\epsilon \cdot (r_{tt}^\epsilon)^* \cdot r_{tk}^\epsilon$. By induction hypothesis, $r_{jk}^T \cong r_{jk}^\epsilon + \sum_{i \in T} r_{ji}^\epsilon \cdot (r_{ii}^\epsilon)^* \cdot r_{ik}^\epsilon$. Since the notion of similarity allows to suppress the multiple appearance of r_{jk}^ϵ in a sum of subexpressions, the expression $r_{jk}^T + r_{jk}^\epsilon + r_{jt}^\epsilon \cdot (r_{tt}^\epsilon)^* \cdot r_{tk}^\epsilon$ is similar to the right hand side of the congruence in the statement of the lemma.   □

The next observation is that we can use Lemma 4 repeatedly.

Lemma 5. *Let $H = (Q, E, s, t)$ be a hammock, and let S be an ordered subset of Q Assume $I \subseteq Q \setminus S$ is an independent set in H^S. Let j and k be vertices with $j, k \in (Q \setminus (S \cup I)) \cup \{s, t\}$. Then for the regular expression r_{jk}^{SI} obtained after elimination of SI holds $r_{jk}^{SI} \cong r_{jk}^S + \sum_{i \in I} r_{ji}^S \cdot (r_{ii}^S)^* \cdot r_{ik}^S$.*   □

This gives an algorithm for computing a good elimination ordering as follows: Choose a large independent set I_1 in $H = H(A)$, then choose an independent set I_2 in H^{I_1}, choose an independent set I_3 in $H^{I_1 I_2}$, and so on. To estimate the performance of the independent set elimination approach, we have to find a large independent set I_{k+1} in the hammock $H^{I_1 I_2 \ldots I_k}$. The cardinality of the maximum independent set in some intermediate graph G^S obtained after eliminating $S = I_1 I_2 \ldots I_k$ can be estimated using Turán's Theorem from graph theory [20]. The latter gives an estimate in terms of the *average degree* of a symmetric digraph $G = (V, E)$, the latter being defined as $\bar{d}(G) = |E|/|V|$—recall that each unordered pair $\{u, v\}$ forming an "undirected edge" is counted as two edges in E.

Theorem 6 (Turán). *If G is a symmetric digraph of average degree $\bar{d}$ with n vertices, then G has an independent set of size at least $n/(\bar{d} + 1)$.*

In spite of the well known fact that finding a maximum independent set is computationally hard, the proof of the above theorem implies that such a large independent set can also be found efficiently using a simple greedy algorithm. Due to lack of space we have to omit the proof of the following theorem.

Theorem 7. *Let A be an n-state deterministic finite automaton with input alphabet Σ. Then there exists an ordering on the states such that the state elimination algorithm produces a regular expression describing $L(A)$ of alphabetic width at most $|\Sigma| \cdot n^{O(1)} \cdot 4^{c \cdot n}$, where $c = \frac{2|\Sigma| \cdot 2|\Sigma|^2 \cdot 2|\Sigma|^4}{(2|\Sigma| + 1)(2|\Sigma|^2 + 1)(2|\Sigma|^4 + 1)}$.*   □

For the case of a binary alphabet, we have $c = \frac{1024}{1485}$ and $4^c \doteq 2.601$, thus giving a worst-case upper bound of, say, $O(2.602^n)$. This appears reasonable at once in presence of a worst-case lower bound of γ^n for the case of deterministic finite automata over binary alphabets, proved recently in [10]. Here, $\gamma > 1$ is a fixed constant[4] that is independent of n.

3.3   From Automata of Small Treewidth to Regular Expressions

We show that finite automata whose transition structure forms a graph of bounded undirected treewidth can be converted into regular expressions of polynomial size.

Definition 8. *Let $G = (V, E)$ be a digraph, and let $S \subseteq V$ be a set of vertices. A set of vertices X is a* balanced k-way separator *for S if the induced subgraph $G[S \setminus X]$ falls apart into k mutually disjoint subgraphs $G[T_i]$, for $1 \leq i \leq k$, with $0 \leq |T_i| \leq \frac{1}{2}|S \setminus X|$.*

It is known that for digraphs of undirected treewidth w, every nontrivial subset of the vertex set admits a small balanced k-way separator of size at most $w + 1$, for some k [19]. An elementary observation on sums of integers shows that we can always set $k = 3$, by grouping the disjoint subgraphs together in a suitable manner. Together with the mentioned result from [19], we thus have:

Lemma 9. *Let $G = (V, E)$ be a digraph of undirected treewidth at most w. Then for every subset S of V, there exists a balanced 3-way separator of size at most $w + 1$.*   □

This separation property can be used to convert finite automata of small undirected treewidth into relatively short regular expressions:

Theorem 10. *Let $A = (Q, \Sigma, \delta, q_0, F)$ be an n-state nondeterministic finite automaton, H its associated hammock, and let w denote the undirected treewidth of $H[Q]$. Then the there exists a ordering on the states such that the state elimination algorithm produces a regular expression describing language $L(A)$ of alphabetic width at most $|\Sigma| \cdot n^{2w+2+\log 3}$.*

Proof. We devise a recursive algorithm for finding an elimination ordering such that the size of the resulting regular expression obeys the desired bound as follows: By Lemma 9 for each set of states $S \subseteq Q$, we can find a balanced 3-way

---

[4] By tracking the size of the constants used in the chain of reductions used in that proof, one can deduce a concrete value for the constant γ. For alphabets of size $\ell \geq 3$, we get expression size at least $2^{\frac{\sqrt{\ell}(n-1)}{3 \cdot 2 \cdot (\ell+1)^2}}$, for infinitely many values of n. Here we exploited the fact from spectral graph theory that, using definitions and notation from [2], for the vertex expansion of ℓ-regular Ramanujan graphs G holds $g_G \geq h_G \geq \lambda_1/2 \geq \frac{\sqrt{\ell}}{2(\ell+1)}$, in particular for $\ell = 3$. Using a binary encoding that increases the size of the input deterministic finite automaton to $m = 10n$ whilst preserving star-height, the very same lower bound (but still in terms of $n = \frac{1}{10}m$) is proved for binary alphabets. Thus we obtain $\gamma \doteq 1.013$ for alphabet size at least 3, and $\gamma \doteq 1.001$ for binary alphabets. This estimate is most likely very loose, since the main goal in in [10] was merely to bound the value of γ away from 1.

separator X, such that $|X| \leq w + 1$, and the induced subgraph $H[S \setminus X]$ falls apart into three mutually disjoint subgraphs $H[T_i]$, for $1 \leq i \leq 3$. For each of the individual sets T_i, Lemma 3 ensures that for *every ordering*, $r_{jk}^{T_1 T_2 T_3} \cong \sum_{i=1}^{3} r_{jk}^{T_i}$, for all $j, k \in (Q \setminus (T_1 \cup T_2 \cup T_3)) \cup \{s, t\}$. Then we recursively compute an ordering for each T_i, placing a separator for $H[T_i]$ at the end of that ordering, and so on.

Since for each $S \subseteq Q$ the alphabetic width of r_{jk}^{S} is at most $4^{|X|}$ $\sum_{i=1}^{3} \text{alph}(r_{jk}^{T_i})$, for some X, T_1, T_2, and T_3 with $|X| \leq w+1$ and $|T_i| \leq \frac{1}{2}|S|$, for $1 \leq i \leq 3$. Moreover, the alphabetic width of the expression r_{jk}^{S} is bounded above by the recurrence $R(1) \leq 1$ and $R(n) \leq 4^{w+1} \cdot 3 \cdot R\left(\frac{n}{2}\right)$, for $n \geq 2$. We obtain $R(n) \leq 4^{(w+1)\log n} 3^{\log n}$. Applying the substitution σ increases the expression size by a factor of at most $|\Sigma|$. Thus we have an expression of alphabetic width $|\Sigma| \cdot n^{2w+2+\log 3}$ for the language $L(A)$. □

3.4   Eliminating Subgraphs of Small Treewidth

Now we present a fusion of our previous ideas: Instead of an independent set, we look for a large induced subgraph whose structure is "simple" in the sense that eliminating the states in the subgraph leads to a regular expression of moderate size. As we have seen in the previous section, one such example are induced subgraphs of small undirected treewidth. A very recent result states that every graph with bounded average degree has a large induced subgraph of treewidth at most two [5]:

Theorem 11. *Let G be a connected graph with average degree at most $\bar{d} \geq 2$. Then there is a polynomial-time algorithm which finds an induced subgraph with undirected treewidth at most two of size at least $\frac{3n}{d+1}$.* □

This gives rise to an algorithm with improved performance guarantee.

Theorem 12. *Let A be an n-state deterministic finite automaton with input alphabet Σ. Then there exists a ordering on the states such that the state elimination algorithm produces a regular expression describing $L(A)$ of alphabetic width at most $|\Sigma| \cdot n^{O(1)} \cdot 4^{c \cdot n}$, where $c = \frac{2|\Sigma|-2}{2|\Sigma|+1}$.*

Proof. Let $H = H(A) = (Q, E, s, t)$ be the hammock associated with the automaton A. Note that the average outdegree of $H[Q]$ is at most $|\Sigma|$ so the average degree of its undirected version is at most $2|\Sigma|$. By Theorem 11, we can find a subset S of Q having size $\frac{3n}{2|\Sigma|+1} = (1 - c) \cdot n$, for suitably chosen c, such that the induced subgraph $H[S]$ has undirected treewidth at most 2. The remaining states in $Q \setminus S$ are placed at the end of the elimination ordering. We set up a regular expression matrix $(r_{jk}^{\epsilon})_{j,k}$, whose rows j and columns k range over the set $(Q \setminus S) \cup \{s, t\}$. The algorithm from the proof of Theorem 10 can be used to compute an elimination ordering for S such that the set of walks from j to k using internal states only from $H[S]$ is described by the regular expression r_{jk}^{S}. One observes that, since this ordering does not depend on j or k, that the same result is obtained by eliminating S from the larger graph H by using that very ordering. As the size of the intermediate expressions after this phase is bounded by $|S|^{6+\log 3}$, and eliminating the remaining states in $Q \setminus S$ incurs a blow-up by a factor of $4^{c \cdot n}$, we obtain that the alphabet with of r_{st}^Q is at most $n^{O(1)} \cdot 4^{c \cdot n}$. Finally we apply the substitution σ to obtain a regular expression for $L(A)$ that has alphabetic width at most $|\Sigma| \cdot n^{O(1)} \cdot 4^{c \cdot n}$, for $c = (2|\Sigma| - 2)/(2|\Sigma| + 1)$. $\square$

In the case of a binary input alphabet, we obtain that the maximum blow-up arising in the conversion from deterministic finite automata to regular expressions is at most $n^{O(1)} \cdot 4^{2/5 \cdot n}$, where $4^{2/5} \doteq 1.741$.

4   Language Operations and Regular Expression Size

Studying descriptional complexity of language operations on regular expressions was first suggested in [7]. Lower bounds for the intersection and shuffle, and a tight lower bound for complementation were found recently in [8,9,10]. We are able to contrast these negative results with a comparable upper bound for intersection. A similar approach works for the half-removal operation.

Theorem 13. *Let $L_1, L_2 \subseteq \Sigma^*$ be regular languages with alphabetic width at most m and n, respectively. Then* $\mathrm{alph}(L_1 \cap L_2) \leq |\Sigma| \cdot 2^{O\left(1 + \log \frac{m}{n}\right) \min\{m,n\}}$. *Note that this bound is best possible for the case $m = \Theta(n)$ and $|\Sigma| = O(1)$.*

Proof. A regular expression of size m can be converted into an equivalent nondeterministic finite automaton A with at most $m+1$ states such that the digraph of the underlying transition structure has undirected treewidth at most two [13]—this nondeterministic finite automaton will in general have ϵ-transitions, but these do not cause any trouble when we treat them just like normal transitions. The construction ensures that the transition structure of the that automaton is a hammock with at most $m-1$ internal vertices.

Let A_1 and A_2 be finite automata thus obtained from suitable regular expressions of alphabetic width m and n describing the languages L_1 and L_2, respectively. Moreover, let Q_1 and Q_2 denote their respective state sets of A_1 and A_2, respectively. By applying the standard product construction for the intersection of regular languages, we obtain a nondeterministic finite automaton $A_1 \times A_2$ with $(m+1)(n+1)$ states accepting the language $L_1 \cap L_2$, by appropriately defining the initial state of $A_1 \times A_2$ and the accepting states of the product automaton. With G_1 and G_2 denoting the digraphs underlying each transition structure of the automata, the digraph underlying $A_1 \times A_2$ is (a subgraph of) the categorical product $G_1 \times G_2$. Let $H = H(A_1 \times A_2)$ denote the hammock associated with finite automaton $A_1 \times A_2$, where s and t are the distinguished vertices of H. The following claim is immediate from the definition of balanced 3-way separators (Definition 8) and the definition of categorical product:

Claim. Let $G_1 = (V_1, E_1)$ and $G_2 = (V_2, E_2)$ be digraphs, and $S_1 \subseteq V_1$, $S_2 \subseteq V_2$. Assume X is a balanced separator for S_1, such that the digraph $G[S_1 \setminus X]$ falls apart into the mutually disjoint subgraphs $G[T_i]$, for $1 \leq i \leq 3$, with

$0 \leq |T_i| \leq \frac{1}{2}|S_1 \setminus X|$. Then $X \times S_2$ is a balanced 3-way separator for $S_1 \times S_2$ in the product graph $G_1 \times G_2$, and the digraph $(G_1 \times G_2)[(S_1 \setminus X) \times S_2]$ falls apart into the mutually disjoint subgraphs $G[T_i \times S_2]$, for $1 \leq i \leq 3$, with $0 \leq |T_i \times S_2| \leq \frac{1}{2}|(S_1 \setminus X) \times S_2|$. $\square$

We proceed in a similar way as in the proof of Theorem 10, by recursively computing regular expressions $r_{jk}^{S_1 \times S_2}$ for $S_1 \subseteq Q_1$, $S_2 \subseteq Q_2$ and all $j, k \in ((Q_1 \times Q_2) \setminus (S_1 \times S_2)) \cup \{s, t\}$. This time we always choose a suitable separator according to the above stated claim. This is done as follows: If $|S_1| < |S_2|$, then exchange the roles of G_1 and G_2, and of S_1 and S_2, respectively. This is admissible by the symmetry of the categorical product. Afterwards, choose a 3-way separator X for $G_1[S_1]$ of size at most 3—recall that, by Lemma 9 such a separator exists, since both factor graphs have undirected treewidth at most 2. Let T_1, T_2, and T_3 be the disjoint subgraphs constituting $G_1[S_1 \setminus X]$ as given by Definition 8. Eliminating $(S_1 \setminus X) \times S_2$ gives regular expressions $r_{jk}^{(S_1 \setminus X) \times S_2}$, with j and k ranging over all states not in $(S_1 \setminus X) \times S_2$.

By the above claim and Lemma 3, we have $r_{jk}^{(S_1 \setminus X) \times S_2} \cong \sum_{i=1}^{3} r_{jk}^{T_i \times S_2}$. To recursively assign an elimination ordering to each of the subsets $T_i \times S_2$, we find next a balanced 3-way separator for the larger of the two graphs $G_1[T_i]$ and $G_2[S_2]$, which amounts to a corresponding separator in the product graph $G_1 \times G_2[T_i \times S_2]$, and recursively proceed to assign elimination orderings to such subsets until the subset sizes reach the value 1.

In order to get an upper bound on $\text{alph}(L_1 \cap L_2) \leq \text{alph}(r_{st}^{S_1 \times S_2})$, define

$$A(\beta, \eta) = \max_{\substack{S_1 \subseteq V_1, |S_1| \leq \beta \\ S_2 \subseteq V_2, |S_2| \leq \eta}} \{ \text{alph}(r_{jk}^{S_1 \times S_2}) \mid j, k \in ((Q_1 \times Q_2) \setminus (S_1 \times S_2)) \cup \{s, t\} \}.$$

An easy observation is that for the degenerate case, where S_1 and S_2 have both at most one element, we have $A(1, 1) \leq 4$. An upper bound is obtained thus by solving the recurrence $A(\beta, \eta) = A(\eta, \beta)$, if $1 < \beta < \eta$, $A(\beta, \eta) = 4$, if $\beta = \eta = 1$, and $A(\beta, \eta) = 3 \cdot A\left(\left\lfloor \frac{\beta}{2} \right\rfloor, \eta\right) \cdot 4^{3\eta}$, otherwise. This leads to the stated bound on $\text{alph}(L_1 \cap L_2)$. The analysis of the recurrence is omitted. $\square$

Basically the same technique can be used for the half-removal operation, defined as $\frac{1}{2}L = \{x \in \Sigma^* \mid \text{there exists } y \in \Sigma^* \text{ with } |x| = |y| \text{ such that } xy \in L\}$. The state complexity of this operation was studied in [4]. The theorem reads as follows—due to lack of space we omit the proof.

Theorem 14. Let $L \subseteq \Sigma^*$ be a regular language of alphabetic width at most n. Then $\text{alph}\left(\frac{1}{2}L\right) \leq |\Sigma| \cdot 2^{O(n)}$. $\square$

Thus, the technique used above is applicable for certain language operations that can be implemented on nondeterministic finite automata using a special kind of product construction. But there are also limitations: For instance, the authors failed to use the above technique to produce a nontrivial upper bound for the shuffle of two regular languages.

References

1. Brzozowski, J.A.: Derivatives of regular expressions. Journal of the ACM 11(4), 481–494 (1964)
2. Chung, F.R.K.: Spectral Graph Theory. In: CBMS Regional Conference Series in Mathematics, vol. 92. American Mathematical Society (1997)
3. Delgado, M., Morais, J.: Approximation to the smallest regular expression for a given regular language. In: Domaratzki, M., Okhotin, A., Salomaa, K., Yu, S. (eds.) CIAA 2004. LNCS, vol. 3317, pp. 312–314. Springer, Heidelberg (2005)
4. Domaratzki, M.: State complexity of proportional removals. Journal of Automata, Languages and Combinatorics 7(4), 455–468 (2002)
5. Edwards, K., Farr, G.E.: Planarization and fragmentability of some classes of graphs. Discrete Mathematics 308(12), 2396–2406 (2008)
6. Ehrenfeucht, A., Zeiger, H.P.: Complexity measures for regular expressions. Journal of Computer and System Sciences 12(2), 134–146 (1976)
7. Ellul, K., Krawetz, B., Shallit, J., Wang, M.: Regular expressions: New results and open problems. Journal of Automata, Languages and Combinatorics 10(4), 407–437 (2005)
8. Gelade, W.: Succinctness of regular expressions with interleaving, intersection and counting. In: Proceedings of the 33rd International Symposium on Mathematical Foundations of Computer Science, Toruń, Poland, August 2008. LNCS. Springer, Heidelberg (to appear, 2008)
9. Gelade, W., Neven, F.: Succinctness of the complement and intersection of regular expressions. In: Albers, S., Weil, P. (eds.) Proceedings of the 25th Symposium on Theoretical Aspects of Computer Science, Bordeaux, France, February 2008. Dagstuhl Seminar Proceedings, vol. 08001, pp. 325–336. Internationales Begegnungs- und Forschungszentrum fuer Informatik (IBFI), Schloss Dagstuhl, Germany (2008)
10. Gruber, H., Holzer, M.: Finite automata, digraph connectivity, and regular expression size. In: Aceto, L., Damgaard, I., Goldberg, L.A., Halldórsson, M.M., Ingólfsdóttir, A., Walkuwiewicz, I. (eds.) Proceedings of the 35th International Colloquium on Automata, Languages and Programming, Reykjavik, Iceland, July 2008. Springer, Heidelberg (2008)
11. Gruber, H., Johannsen, J.: Optimal lower bounds on regular expression size using communication complexity. In: Amadio, R. (ed.) FOSSACS 2008. LNCS, vol. 4962, pp. 273–286. Springer, Heidelberg (2008)
12. Han, Y.-S., Wood, D.: Obtaining shorter regular expressions from finite-state automata. Theoretical Computer Science 370(1-3), 110–120 (2007)
13. Ilie, L., Yu, S.: Follow automata. Information and Computation 186(1), 140–162 (2003)
14. Kleene, S.C.: Representation of events in nerve nets and finite automata. In: Shannon, C.E., McCarthy, J. (eds.) Automata Studies, Annals of Mathematics Studies, pp. 3–42. Princeton University Press, Princeton (1956)
15. Lipton, R.J., Tarjan, R.E.: A separator theorem for planar graphs. SIAM Journal on Applied Mathematics 36(2), 177–189 (1979)
16. McIntosh, H.V.: REEX: A CONVERT program to realize the McNaughton-Yamada analysis algorithm. Technical Report AIM-153, MIT Artificial Intelligence Laboratory (January 1968)
17. McNaughton, R., Yamada, H.: Regular expressions and state graphs for automata. IRA Transactions on Electronic Computers 9(1), 39–47 (1960)

18. Morais, J.J., Moreira, N., Reis, R.: Acyclic automata with easy-to-find short regular expressions. In: Farré, J., Litovsky, I., Schmitz, S. (eds.) CIAA 2005. LNCS, vol. 3845, pp. 349–350. Springer, Heidelberg (2006)
19. Robertson, N., Seymour, P.D.: Graph minors. II. Algorithmic aspects of tree-width. Journal of Algorithms 7(3), 309–322 (1986)
20. Turán, P.: On an extremal problem in graph theory (in Hungarian). Matematicko Fizicki Lapok 48, 436–452 (1941)
21. Wood, D.: Theory of Computation. John Wilet & Sons (1987)

Large Simple Binary Equality Words

Jana Hadravová* and Štěpán Holub**

Faculty of Mathematics and Physics, Charles University
186 75 Praha 8, Sokolovská 83, Czech Republic
holub@karlin.mff.cuni.cz, hadravova@ff.cuni.cz

Abstract. Let w be an equality word of two nonperiodic binary morphisms $g, h : \{a, b\}^* \to \Delta^*$. Suppose that no overflow occurs twice in w and that w contains at least 9 occurrences of a and at least 9 occurrences of b.

Then either $w = (ab)^i a$, or $w = a^i b^j$ with $\gcd(i, j) = 1$, up to the exchange of letters a and b.

1   Introduction

An equality word, also called a solution, of morphisms $g, h : \Sigma^* \to \Delta^*$ is a word satisfying $g(w) = h(w)$. All equality words of the morphisms g, h constitute the set $\mathrm{Eq}(g, h)$, which is called the equality language of g and h. Natural concept of equality languages was introduced in [1], and since then it has been widely studied. It turns out that the equality languages are very rich objects; for example, each recursively enumerable language can be obtained as a morphic image of generating words of a set $\mathrm{Eq}(g, h)$, see [2].

It is also well known, due to [3], that it is undecidable whether an equality language contains a nonempty word (an algorithmic problem known as the Post Correspondence Problem, or the PCP).

A lot of attention has been paid to the binary case, that is, when $|\Sigma| = 2$. This is the smallest domain alphabet for which the structure of $\mathrm{Eq}(g, h)$ is not completely trivial, and in the same time the largest for which there is any reasonable knowledge about the structure of the equality set. For $|\Sigma| = 3$ it is already a long-standing open problem whether the equality set has to be regular, see [4] and [5].

The structure of binary equality languages has been first studied in [6] and [7] and later in a series of papers [8,9,10]. It has been shown that binary equality languages are always generated by at most two words, provided that both morphisms are nonperiodic (the periodic case being rather easy). It is also known that if the set $\mathrm{Eq}(g, h)$ is generated by two distinct generators, then these generators are of the form ba^i and $a^i b$. Bi-infinite binary words were studied for example in [11]. It should be also mentioned that the binary case of the PCP is decidable, even in polynomial time ([12,13]).

* Supported by Hlavka's Foundation.
** Supported by the research project MSM 0021620839.

M. Ito and M. Toyama (Eds.): DLT 2008, LNCS 5257, pp. 396–407, 2008.
© Springer-Verlag Berlin Heidelberg 2008

However, very little is known so far about words which are single generators of binary equality languages. In this paper we make a step towards a characterization of such words. Our research will be limited only to so-called simple solutions, that is, to solutions that do not have the same overflow twice.

It is well known, since the proof of the decidability of the binary PCP, that each binary equality word can be divided into a sequence of so-called blocks, which are simple in the aforementioned sense. Simple solutions therefore represent a natural starting point of the research. We characterize all simple solutions that are long enough, more precisely all such solutions that contain each of the letters a and b at least nine times. Due to space limits we do not prove all details, we rather explain the main ideas, and include proofs that, instead of being purely technical, illustrate the underlying concepts.

2   Basic Concepts and Ideas

We shall mostly use standard notation and terminology of combinatorics of words (see for example [14] and [15]). We suppose that the reader is familiar with basic folklore facts concerning periods and primitive words. In particular, let us recall the Periodicity lemma, which can be formulated in the following way. If p and q are two primitive words such that the words p^ω and q^ω have a common factor of length at least $|p| + |q| - 1$, then p and q are conjugate.

We shall write $u \leq_p w$ to denote that u is a prefix of w. If, in addition, $u \neq w$, then we write $u <_p w$. Similarly, we use $u \leq_s w$ and $u <_s w$ for suffixes.

Let two binary morphisms $g, h : \{a, b\}^* \to \Delta^*$ be given. We suppose that both morphisms are nonperiodic, that is, $g(a)$ and $g(b)$ ($h(a)$ and $h(b)$ resp.) do not commute.

A word w is called a *solution* of g and h if $g(w) = h(w)$. A solution w is called *simple* if whenever w_1, w_1u, w_2 and w_2u' are prefixes of w^ω such that

$$g(w_1)z = h(w_2), and\ g(w_1u)z = h(w_2u')$$

for some word z, then $|u| = |u'| = k|w|$, for some $k \in \mathbb{N}_+$. We shall be interested only in simple solutions.

It is easy to see that if w is a simple solution, then it is a primitive word, that is, it is not a power of a shorter word.

Example 1. Trivial examples of non-simple solutions are words composed of shorter solutions. Apart from these, we can also find non-simple solutions that are minimal, that is, they cannot be decomposed into shorter solutions. As an example, consider morphisms

$$g(a) = bba, \qquad\qquad g(b) = bb,$$
$$h(a) = b, \qquad\qquad h(b) = abbabb.$$

They have a solution aab, which is not simple since

$$g(\varepsilon)bb = h(aa) \text{ and } g(aa)bb = h(aab).$$

We now formulate our main result.

Theorem 1. *Let* $g, h : \{a, b\}^* \to \Delta^*$ *be nonperiodic morphisms, and let* w *be their simple solution. If* $|w|_b \geq 9$ *and* $|w|_a \geq 9$, *then, up to the exchange of the letters* a *and* b, *either*

$$w = (ab)^i a$$

or

$$w = a^j b^i$$

with $\gcd(i, j) = 1$.

Example 2. Each word mentioned in Theorem 1 is indeed a simple solution for a pair of morphisms g and h. The word $w = (ab)^i a$ is a simple solution for example of morphisms:

$$g(a) = (ab)^i a, \qquad\qquad g(b) = b,$$
$$h(a) = a, \qquad\qquad h(b) = (ba)^{i+1} b.$$

The word $a^j b^i$ is a simple solution for example of morphisms:

$$g(a) = p^l, \qquad\qquad g(b) = a,$$
$$h(a) = a^i b a^i, \qquad\qquad h(b) = s^m$$

where

$$p = (a^i b a^i)^{j-1} a^i b, \qquad\qquad s = b a^i (a^i b a^i)^{j-1}$$

and $lj - mi = 1$.

It turns out that a lot of technical complications can be avoided if we work with cyclic words and cyclic solutions instead of ordinary ones. This motivates the following terminology.

Let $u = u_0 \ldots u_{n-1}$ be a finite word of length n, and let $(i, j) \in \mathbb{Z}_n \times \mathbb{Z}_n$, $i \neq j$, be an ordered pair. We define an *interval* $u[i, j]$ by

$$u[i, j] = \prod_{k=0}^{(j-i-1) \bmod n} u_{(i+k) \bmod n}.$$

Note that $u_i = u[i, i+1]$, and $u[i, i]$ is a word conjugate with u.

We denote an infinite word starting at the i-th position of u by

$$u[i, \infty] = u_i u_{i+1} \ldots u_{n-1} u_0 u_1 \ldots.$$

We have the following crucial definition.

Definition. Let $g, h : \{a, b\}^* \to \Delta^*$ be morphisms. A *cyclic solution* of g, h is an ordered quadruple $(w, \mathfrak{c}, G, H)$ where $w = w_0 w_1 \cdots w_{|w|-1} \in \{a, b\}^+$, $\mathfrak{c} \in \Delta^+$, $|\mathfrak{c}| = |g(w)| = h(w)$ and $G, H : \mathbb{Z}_{|w|} \to \mathbb{Z}_{|\mathfrak{c}|}$ are injective mappings such that

$$\mathfrak{c}[G(i), G(i+1)] = g(w_i) \text{ and } \mathfrak{c}[H(i), H(i+1)] = h(w_i),$$

for all $i \in \mathbb{Z}_{|w|}$.

The concept of a simple solution is extended to cyclic solutions in the following definition.

Definition. Let $(w, \mathfrak{c}, G, H)$ be a cyclic solution of g, h. We say that $(w, \mathfrak{c}, G, H)$ is *simple* if

$$\mathfrak{c}[G(r_1), H(t_1)] = \mathfrak{c}[G(r_2), H(t_2)]$$

implies $(r_1, t_1) = (r_2, t_2)$.

The prior definitions can be better understood if we use the informal concept of an overflow. Given two prefix comparable words u and v, we have either an overflow $v^{-1}u$ of u, or an overflow $u^{-1}v$ of v, depending on whether v is prefix of u, or the other way round. Since the role of an overflow is played by the word z in the definition of a simple solution and by the word $\mathfrak{c}[G(r_1), H(t_1)]$ in the definition of a simple cyclic solution, one can see that both definitions are in fact expressing the same thing: the solution does not contain the same overflow twice. Notice also that if $(w, \mathfrak{c}, G, H)$ is a simple cyclic solution, then w has to be primitive, similarly as in the case of an (ordinary) simple solution.

We now wish to define *p-synchronized overflows*. We have already mentioned that overflows in a cyclic solution $(w, \mathfrak{c}, G, H)$ are words $\mathfrak{c}[G(r), H(t)]$ given uniquely by pairs $(r, t) \in \mathbb{Z}_{|w|}$. Therefore, p-synchronized overflows will be k-tuples of overflows with some additional properties. Although our definition is slightly technical, we will see later on that this concept plays very important role in the proof of the theorem.

Definition. We say that a cyclic solution $(w, \mathfrak{c}, G, H)$ of morphisms g, h has k *p-synchronized overflows* if there is a k-tuple

$$((r_1, t_1), \ldots, (r_k, t_k)) \in (\mathbb{Z}_{|w|} \times \mathbb{Z}_{|w|})^k$$

which has the following properties:

1. for all $i \in \{1, \ldots, k-1\}$ there is $l_i \in \mathbb{N}^+$ such that

$$\mathfrak{c}[G(r_i), H(t_i)] = p^{l_i} \mathfrak{c}[G(r_{i+1}), H(t_{i+1})];$$

2. r_i are pairwise distinct and t_i are pairwise distinct;
3. the word $\mathfrak{c}[G(r_k), H(t_k)]$ is a nonempty prefix of p^ω;
4. for each $i \in \{1, \ldots, k\}$ there is some $0 \leq m < |h(b)|$ such that

$$G(r_i) = H(t_i - 1) + m \mod |\mathfrak{c}|,$$

and $w_{t_i - 1} = b$.

The following example illustrates the previous definitions.

Example 3. Let g, h be morphisms given by:

$$g(a) = (aab)^2 a, \qquad\qquad g(b) = ab,$$
$$h(a) = a, \qquad\qquad h(b) = (baa)^3 ba.$$

They have a simple cyclic solution $((ab)^2 a, \mathfrak{c}, G, H)$ where $\mathfrak{c} = (aab)^8 a$, and the mappings $G, H : \mathbb{Z}_5 \to \mathbb{Z}_{25}$ are given by:

$$G(0) = 0, \quad G(1) = 7, \quad G(2) = 9 \quad G(3) = 16, \quad G(4) = 18,$$
$$H(0) = 1, \quad H(1) = 2, \quad H(2) = 13, \quad H(3) = 14, \quad H(4) = 0.$$

The cyclic solution is depicted in Figure 1.

It is possible to verify that g and h have no equality word. Notice, on the other hand, that if w is an equality word for some morphisms g' and h', then we can find mappings G' and H' with $G'(0) = H'(0) = 0$ such that $(w, g'(w), G', H')$ is a cyclic solution. This example therefore shows that the concept of a cyclic solution generalizes nontrivially the concept of an equality word.

The example also features two aab-synchronized overflows, which are emphasized in Figure 1. They are given by pairs $(2, 2)$ and $(4, 4)$ since

$$\mathfrak{c}[G(2), H(2)] = \mathfrak{c}[9, 13] = (aab)a \quad \text{and} \quad \mathfrak{c}[G(4), H(4)] = \mathfrak{c}[18, 0] = (aab)(aab)a.$$

Notice that the cyclicity of the solution allows to speak easily for example about the overflow $(aab)^2 a(aab)^4 a$, which is given as $\mathfrak{c}[G(4), H(2)]$. One of the main advantages of simple cyclic solutions in comparison with (ordinary) simple solutions is that the definition of a simple cyclic solution does not need to employ infinite words.

It is not difficult to see the following properties of p-synchronized overflows. First, we have either

$$p \leq_\mathrm{p} \mathfrak{c}[G(r_i), H(t_i)], \qquad\qquad \text{or} \qquad\qquad p \leq_\mathrm{s} \mathfrak{c}[H(t_i), G(r_i)], \qquad (*)$$

for all $i \in \{1, \ldots, k\}$.

Second, if we define s by

$$s = \mathfrak{c}[G(r_1), H(t_1)]^{-1} p \, \mathfrak{c}[G(r_1), H(t_1)], \qquad\qquad (**)$$

then the following equations hold for all $i \in \{1, \ldots, k\}$:

$$\mathfrak{c}[G(r_i), \infty] \wedge p^\omega = \mathfrak{c}[G(r_i), H(t_i)](\mathfrak{c}[H(t_i), \infty] \wedge s^\omega). \qquad\qquad (***)$$

A morphism g is called *marked* if the first letter of $g(x)$ is distinct from the first letter of $h(y)$ as long as x, y are two distinct letters. Advantages of marked morphisms are well known in the theory of equality languages, as well as of the PCP. The crucial advantage is that if both morphisms are marked, the continuation of a solution is uniquely determined by any nonempty overflow.

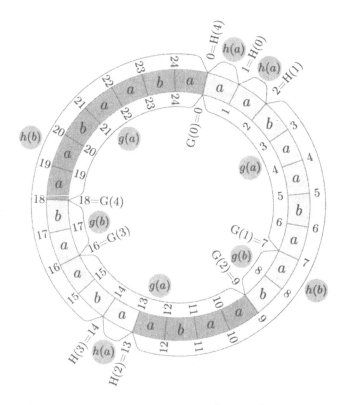

Fig. 1. Simple cyclic solution $((ab)^2a, (aab)^8a, G, H)$

Fortunately, each binary morphism has a so-called *marked version*, defined by:

$$g_{\mathbf{m}}(x) = z_g^{-1}g(x)z_g, \tag{1}$$

for each $x \in \Sigma$ with

$$z_g = g(ab) \wedge g(ba).$$

It is an important property of binary morphisms that $g_{\mathbf{m}}$ is well defined by (1), which, moreover, holds for any word $x \in \Sigma^*$.

It is not difficult to see that marked morphisms have the following property.

Lemma 1. *Let g be a marked morphism and u, v, w be words satisfying*

$$g(u) \wedge w <_{\mathbf{p}} g(v) \wedge w.$$

Then $g(u) \wedge w = g(u \wedge v)$.

Working with the cyclic solution allows to switch easily between any of the given morphisms and its marked version, which is another very convenient property of cyclic solutions.

3   Properties of Cyclic Solutions

3.1   Many bs Induce Rich Synchronized Overflows

The first step of the proof of Theorem 1 is to show that long words have to contain many synchronized overflows.

Let us adopt a convention. We use the symmetry of g and h, and a and b, and henceforth we shall assume that $h(b)$ is the longest of all four image words, that is,

$$|g(a)| \le |h(b)|, \quad |g(b)| \le |h(b)|, \quad \text{and} \quad |h(a)| \le |h(b)|.$$

A complicated combinatorial analysis, which we omit, yields that nine occurrences of the letter b are enough to enforce five p-synchronized overflows. This is formulated in the following lemma. Notice that we will be working with marked morphisms.

Lemma 2. *Let $(w, \mathfrak{c}, G, H)$ be a simple cyclic solution of marked morphisms $g, h : \{a, b\}^* \to \Sigma^*$. If $|w|_b \ge 9$, then there is a primitive word p such that*

- *$(w, \mathfrak{c}, G, H)$ has five p-synchronized overflows;*
- *$h(b)$ is a factor of p^ω; and*
- *at least one of the words $g(a)$ or $g(b)$ is longer than p.*

To give here just a basic hint of how the lemma is proved, we sketch the proof for a much more generous bound, namely $|w|_b \ge 25$.

We shall study the occurrences of $h(b)$ in $\mathfrak{c}$, which are of the form $\mathfrak{c}[H(i), H(i+1)]$, with $w_i = b$. We call them *true h-occurrences* of b. True g-occurrences are defined similarly.

Consider now the way a given true h-occurrence of b is covered by true g-occurrences of a and b. Since we are working with a simple cyclic solution, it is easy to see that if there are five distinct h-occurrences of b that are covered by the same pattern of g-occurrences of as and bs, then they produce the desired five p-synchronized overflows for a primitive word p.

It remains to show that only the following six types of covers are possible:

$$a^+ \qquad b^+ \qquad a^+b^+ \qquad b^+a^+ \qquad a^+b^+a^+ \qquad b^+a^+b^+. \qquad (2)$$

The desired result is then obtained easily by the pigeonhole principle.

In order to prove the remaining part, we look at the starting and ending positions of true g-occurrences of b. We are interested in situations when these occurrences start (end resp.) in some true h-occurrence of b.

Suppose, for a contradiction, that there is a true h-occurrence of b that is covered by a sequence of $g(a)$s and $g(b)$s that is not listed in (2). Inspection of the list shows that in such case there is a true h-occurrence of b in which at least two true g-occurrences of b start, or end. Let us discuss the first case, the second being similar.

Since the number of true g-occurrences of b equals the number of true h-occurrences of b, we deduce that there is a true h-occurrence of b in which no true g-occurrence of b starts. That occurrence is then covered either by a^+ or by ba^+, which implies that a word from $g(a^+)\mathrm{pref}_1(g(b))$ is a factor of $g(a)^\omega$. We get a contradiction with g marked.   □

It should not be too surprising that a much more detailed analysis of covers is possible, which leads eventually to the bound 9.

3.2   Impact of Five Synchronized Overflows

The next step is to employ the existence of five synchronized overflows in order to obtain information about the word w. Its structure is revealed in the following three lemmas.

Lemma 3. *Let* (w, c, G, H) *be a simple cyclic solution that has five p-synchronized overflows. Then the primitive root of c is not conjugate with p.*

Proof. Suppose, for a contradiction, that the primitive root of c and p are conjugate. Note that $h(b)$ is a factor of p^ω greater than $|p|$ by the existence of the synchronized overflows. It is not difficult to see that if $ba^i b$ and $ba^j b$ are two intervals in w, then $i = j$, unless $h(a)$ commutes with p. But if $h(a)$ commutes with p, then also $h(b)$ does, a contradiction. Therefore w is a power of $a^{i_1} ba^{i_2}$. This is a contradiction since w has to be primitive because (w, c, G, H) is simple.   □

The next lemma is a consequence of Lemma 1 and is presented without proof.

Lemma 4. *Let* (w, c, G, H) *be cyclic solution of binary marked morphisms g, h that has three p-synchronized overflows via $((r_1, t_1), (r_2, t_2), (r_3, t_3))$. Suppose that*

$$c[G(r_1), \infty] \wedge p^\omega = c[G(r_2), \infty] \wedge p^\omega = c[G(r_3), \infty] \wedge p^\omega. \tag{3}$$

Then (w, c, G, H) *is not simple.*

The following characterization of w is already quite strong.

Lemma 5. *Let* (w, c, G, H) *be a simple cyclic solution of binary marked morphisms g, h that has five p-synchronized overflows. Then there are words e and f conjugate with w, and primitive words u and v such that*

1. $g(e) = h(f)$;
2. u is conjugate with a suffix of e and $g(u) \in p^+$; and
3. v is conjugate with a suffix of f and $h(v) \in s^+$, where s is given by (**).

Proof. Let $((r_1, t_1), \ldots, (r_5, t_5))$ be a pentuple inducing p-synchronized overflows.

(1) Let $m \in \{1, \ldots, 5\}$ be chosen such that

$$|c[G(r_m), \infty] \wedge p^\omega| = \max_{k \in \{1, \ldots, 5\}} \{|c[G(r_k), \infty] \wedge p^\omega|\}.$$

According to Lemma 4, each three words $c[G(r_k), \infty] \wedge p^\omega$ are of different lengths. Then, by the pigeonhole principle, we obtain inequalities

$$c[G(r_{k_j}), \infty] \wedge p^\omega <_p c[G(r_m), \infty] \wedge p^\omega$$

for three different indices $k_1, k_2, k_3 \in \{1, \ldots, 5\}$; indeed, in the "maximal length hole" just two out of five lengths can be placed by Lemma 4.

Observe that $|c[G(r_{k_j}), \infty] \wedge p^\omega| < |c|$, otherwise p and the primitive root of c are conjugate, which we excluded by Lemma 3. By Lemma 1, we can find $\ell_1, \ell_2, \ell_3 \in \mathbb{Z}_{|w|}$ such that

$$c[G(r_{k_j}), \infty] \wedge p^\omega = c[G(r_{k_j}), G(\ell_j)],$$

for all $j \in \{1, 2, 3\}$.

Since the cyclic solution is simple, words $c[H(t_{k_j}), G(\ell_j)]$, $j \in \{1, 2, 3\}$, are all of different lengths, and are prefix comparable, see (***). We can suppose that

$$c[H(t_{k_1}), G(\ell_1)] <_p c[H(t_{k_2}), G(\ell_2)] <_p c[H(t_{k_3}), G(\ell_3)].$$

Consequently, by Lemma 1, there are n_1, n_2 such that $H(n_1) = G(\ell_1)$ and $H(n_2) = G(l_2)$. Thus

$$g(w[\ell_1, \ell_1]) = h(w[n_1, n_1]).$$

The first part of the lemma has been proved.

(2) Since $H(n_1) = G(\ell_1)$ and $H(n_2) = G(l_2)$, we have from the definition of p-synchronized overflow $n_1 = n_2$ and $l_1 = l_2$. Therefore, $c[G(r_{k_1}), G(\ell_1)]$ and $c[G(r_{k_2}), G(\ell_1)]$ are both prefixes of p^ω. Since they are also suffix comparable, it can be inferred from primitivity of p and (*) that

$$c[G(r_{k_1}), G(r_{k_2})] \in p^+.$$

Consequently, $g(u) \in p^+$ where u is found as the primitive root of the word $w[r_{k_1}, r_{k_2}]$. Since the morphism g is marked, there is a word $u_1 \leq_p u$ and $j \in \mathbb{N}$ such that

$$u \leq_p w[r_{k_1}, \ell_1] = u^j u_1.$$

The word $u_1^{-1} u u_1$ is then a suffix of $w[\ell_1, \ell_1]$, which completes the proof of the second part.

(3) The proof of the third part can be approached in a similar way.   □

In view of the previous lemma it is reasonable to investigate the structure of words (e, f) since the word w is their conjugate. The claims 2 and 3 of the lemma imply that there is a suffix $\tilde{u}$ of e and a suffix $\tilde{v}$ of f, such that $g(\tilde{u})$ and $h(\tilde{v})$ commute and their common primitive root is conjugate with p.

It is interesting to note that, in particular, there are positive integers i and j such that

$$g(\tilde{u}^i) = h(\tilde{v}^j).$$

However, the pair $(\tilde{u}^i, \tilde{v}^j)$ is not the one we are looking for, because the primitive root of c is not conjugate with p, as shown in Lemma 3.

We now have a piece of powerful information about the structure of w, which leads to the following claim.

Lemma 6. *Let $(w, \mathfrak{c}, G, H)$ be a simple cyclic solution of binary marked morphisms $g, h : \{a, b\}^* \to \Delta^*$. If $|w|_b \geq 9$, then there are words e, f conjugate with w such that $g(e) = h(f)$ and*

$$e = f = (ab)^i a \quad or \quad e = f = (ba)^i b \quad or \quad e = f = ab^i \quad or \quad (e, f) = (b^i a^j, a^j b^i)$$

with $\gcd(i, j) = 1$ and $j > i$.

Notice that in the foregoing lemma the condition $|w|_a \geq 9$ of Theorem 1 is missing. This is due to the fact that $h(b)$ is supposed to have the maximal length among the words $g(a)$, $g(b)$, $h(a)$ and $h(b)$. This distinguishes letters a and b and allows to drop the assumption on $|w|_a$.

Relaxing the assumptions of the theorem has impact on the final set of solutions. We can see from the previous lemma that the words conjugate with ab^i, that is, words $b^{i-j}ab^j$ are brought into question in the case that we do not suppose that $|w|_a \geq 9$.

Example 4. The word $w = b^{i-j}ab^j$, $j \leq i \in \mathbb{N}$, is a solution for example of

$$g(a) = b^{i-j}ab^j, \qquad\qquad g(b) = b^i,$$
$$h(a) = a, \qquad\qquad h(b) = b^{i+1}.$$

The proof of the lemma, which we omit, is achieved by a combinatorial analysis, which is not very deep, but rather complicated and tedious.

3.3 From Marked Morphisms to Ordinary Morphisms

We will finally proceed to prove Theorem 1. With help of Lemma 6 it should not be difficult. Note that there are two differences between Theorem 1 and Lemma 6, which are counterparts of each other:

- The lemma requires that the morphisms are marked, while the theorem speaks about general morphisms.
- The theorem requires that the morphisms agree on the same word, while the lemma only guarantees that e and f are conjugate.

Suppose that we are given a pair of (not necessarily marked) morphisms g and h, with a simple solution w. Consider marked versions g_m and h_m of g and h. Clearly, w can be seen as a cyclic solution $(w, g(w), G, H)$ satisfying in addition that $G(0) = H(0)$. Morphisms (g_m, h_m) now have a cyclic solution $(w, g(w), G_m, H_m)$ given by

$$G_m(j) = (G(j) + |z_g|) \bmod |g(w)|,$$
$$H_m(j) = (H(j) + |z_h|) \bmod |g(w)|.$$
(4)

Notice that $(w, g(w), G_m, H_m)$ is a simple cyclic solution.

Lemma 6 yields that if $|w|_a \geq 9$ and $|w|_b \geq 9$, then w is a conjugate (up to the exchange of letters of alphabet) with $(ab)^i a$ or $a^i b^j$ with $\gcd(i,j) = 1$. It remains to exclude all conjugate words other than trivial. Therefore, in order to complete the proof, we need the following two claims.

Claim 1. If $e = f = (ab)^i a$, $i \geq 9$, then $w = (ab)^i a$.

Claim 2. If $(e, f) = (b^i a^j, a^j b^i)$ with $i, j \geq 9$, then $w = a^j b^i$ or $w = b^i a^j$.

We prove only the former one.

Proof (of Claim 1). Since $i \geq 9$, the Periodicity lemma together with

$$g_{\mathbf{m}}((ab)^i a) = h_{\mathbf{m}}((ab)^i a)$$

implies that the words $g_{\mathbf{m}}(ab)$ and $h_{\mathbf{m}}(ab)$ have the same primitive root t. Hence there are nonempty words t_1, t_2 such that $t_1 t_2 = t$ and

$$\begin{aligned} g_{\mathbf{m}}(a) &= t^{i_1} t_1, & g_{\mathbf{m}}(b) &= t_2 t^{i_2}, \\ h_{\mathbf{m}}(a) &= t^{j_1} t_1, & h_{\mathbf{m}}(b) &= t_2 t^{j_2}. \end{aligned} \tag{5}$$

Primitivity of t implies that the longest common suffix of $g_{\mathbf{m}}(ab)$ and $g_{\mathbf{m}}(ba)$ is shorter than $|t|$. Since $g_{\mathbf{m}}$ is by definition equal to $z_g^{-1} g z_g$, we obtain that $|z_g| < |t|$. Similarly $|z_h| < |t|$.

Suppose that the word w is conjugate with $(ab)^i a$ in a nontrivial way. Therefore $(ab)^i a = e_1 e_2 = f_1 f_2$ such that $w = e_2 e_1 = f_2 f_1$. It is obvious that $e_1 = f_1$ and $e_2 = f_2$ since $(ab)^i a$ is a primitive word. Then $G_{\mathbf{m}}(k) = H_{\mathbf{m}}(k)$, where $k = |e_2| = |f_2|$. Equalities (4) imply that

$$G(k) - H(k) = |z_g| - |z_h| \mod |g(w)|,$$

and therefore

$$G(k) - H(k) < |t| \mod |g(w)|.$$

However, from (5) it is easy to infer that $G(k) - H(k)$ is a multiple of $|t|$, a contradiction. (Note that if $G(k) - H(k) = 0$ we obtain a contradiction as well since w is a simple solution.) □

4   Towards a Complete Characterization

The main obstacle for the generality of our result is the assumption that the solution is simple. As noted in the introduction, a general solution is composed of blocks, which are simple. Blocks of marked morphisms are pairs (e, f) that satisfy $g(e) = h(f)$ where e is not necessarily equal to f. The techniques used in this paper can be applied also for blocks. The missing assumption that $e = f$ or, more precisely, that e and f are conjugate, makes the classification more complicated, but not essentially different. Investigation of blocks is therefore a necessary further step towards a complete characterization of binary equality words.

Another missing part are the words with small number of one of the letters. This will probably require some ad hoc case analysis. It should be noted in this respect, that our proof requires essentially only $|w|_b \geq 9$ as soon as b is identified as the letter with the image of the maximal length, that is, as soon as $g(b)$ or $h(b)$ is the longest of the words $g(a)$, $g(b)$, $h(a)$ and $h(b)$. This makes the necessary case analysis of the short solutions a bit easier.

References

1. Salomaa, A.: Equality sets for homomorphisms of free monoid. Acta Cybern. 4, 127–139 (1980)
2. Čulík II, K.: A purely homomorphic characterization of recursively enumerable sets. J. ACM 26(2), 345–350 (1979)
3. Post, E.: A variant of a recursively unsolvable problem. Bulletin of the American Mathematical Society 52, 264–268 (1946)
4. Karhumäki, J.: On recent trends in formal language theory. In: Ottmann, T. (ed.) ICALP 1987. LNCS, vol. 267, pp. 136–162. Springer, Heidelberg (1987)
5. Karhumäki, J.: Open problems and exercises on words and languages (invited talk). In: Proceedings of Conference on Algebraic Information, Aristotle University of Thessaloniki, pp. 295–305 (2005)
6. Čulík, I.K., Karhumäki, J.: On the equality sets for homomorphisms on free monoids with two generators. ITA 14(4), 349–369 (1980)
7. Ehrenfeucht, A., Karhumäki, J., Rozenberg, G.: On binary equality sets and a solution to the test set conjecture in the binary case. J. Algebra 85(1), 76–85 (1983)
8. Holub, Š.: Binary equality sets are generated by two words. Journal of Algebra 259, 1–42 (2003)
9. Holub, Š.: A unique structure of two-generated binary equality sets. In: Developments in Language Theory, pp. 245–257 (2002)
10. Holub, Š.: Binary equality languages for periodic morphisms. In: Huang, C.-H., Sadayappan, P., Sehr, D. (eds.) LCPC 1997. LNCS, vol. 1366, pp. 52–54. Springer, Heidelberg (1998)
11. Maňuch, J.: Defect effect of bi-infinite words in the two-element case. Discrete Mathematics and Theoretical Computer Science 4(2), 273–290 (2001)
12. Ehrenfeucht, A., Karhumäki, J., Rozenberg, G.: The (generalized) Post Correspondence Problem with lists consisting of two words is decidable. Theor. Comput. Sci. 21, 119–144 (1982)
13. Halava, V., Holub, Š.: Binary (generalized) Post Correspondence Problem is in P. Technical Report 785, TUCS (September 2006)
14. Lothaire, M.: Combinatorics on words. Addison-Wesley, Reading (1983)
15. Rozenberg, G., Salomaa, A. (eds.): Handbook of formal languages. word, language, grammar, vol. 1. Springer, New York (1997)

On the Relation between Periodicity and Unbordered Factors of Finite Words*

Štěpán Holub[1] and Dirk Nowotka[2]

[1] Department of Algebra, Charles University in Prague, Czech Republic
holub@karlin.mff.cuni.cz
[2] Institute for Formal Methods in Computer Science,
University of Stuttgart, Germany
nowotka@fmi.uni-stuttgart.de

Abstract. Finite words and their overlap properties are considered in this paper. Let w be a finite word of length n with period p and where the maximum length of its unbordered factors equals k. A word is called unbordered if it possesses no proper prefix that is also a suffix of that word. Suppose $k < p$ in w. It is known that $n \leq 2k - 2$, if w has an unbordered prefix u of length k. We show that, if $n = 2k - 2$ then u ends in ab^i, with two different letters a and b and $i \geq 1$, and b^i occurs exactly once in w. This answers a conjecture by Harju and the second author of this paper about a structural property of maximum Duval extensions. Moreover, we show here that $i < k/3$, which in turn leads us to the solution of a special case of a problem raised by Ehrenfeucht and Silberger in 1979.

1 Introduction

Overlaps are one of the central combinatorial properties of words. Despite the simplicity of this concept, its nature is not very well understood and many fundamental questions are still open. For example, problems on the relation between the period of a word, measuring the self-overlap of a word, and the lengths of its unbordered factors, representing the absence of overlaps, are unsolved. The focus of this paper is on the investigation of such questions. In particular, we consider so called Duval extensions by solving a conjecture [6,4] about the structure of maximum Duval extensions. This result leads us to a partial answer of a problem raised by Ehrenfeucht and Silberger [5] in 1979.

When repetitions in words are considered then two notions are central: the *period*, which gives the least amount by which a word has to be shifted in order to overlap with itself, and the shortest *border*, which denotes the least (nonempty) overlap of a word with itself. Both notions are related in several ways, for example, the length of the shortest border of a word w is not larger than the period of w, and hence, the period of an unbordered word is its length, moreover, the

---

* The work on this article has been supported by the research project MSM 0021620839.

M. Ito and M. Toyama (Eds.): DLT 2008, LNCS 5257, pp. 408–418, 2008.

shortest border itself is always unbordered. Deeper dependencies between the period of a word and its unbordered factors have been investigated for decades; see also the references to related work below.

Let a word w be called a *Duval extension* of u, if $w = uv$ such that u is unbordered and for every unbordered factor x of w holds $|x| \leq |u|$. Let $\pi(w)$ denote the shortest period of a word w. A Duval extension is called *nontrivial* if $|u| < \pi(w)$. It is known that $|v| \leq |u| - 2$ for any nontrivial Duval extension uv [8,9,10]. This bound is tight, that is, Duval extensions with $|v| = |u| - 2$ exist. Let those be called *maximum Duval extensions*. The following conjecture has been raised in [6]; see also [4].

Conjecture 1. Let uv be a maximum Duval extension of $u = u'ab^i$ where $i \geq 1$ and a and b are different letters. Then b^i occurs only once in uv.

This conjecture is answered positively by Theorem 3 in this paper. Moreover, we show that $i < |u|/3$ in Theorem 4, which leads us to the result that a word z with unbordered factors of length at most k and $\pi(z) > k$ that contains a maximum Duval extension uv with $|u| = k$ is of length at most $7k/3 - 2$. This solves a special case of a conjecture in [5,1].

Previous Work. In 1979 Ehrenfeucht and Silberger [5] raised the problem about the maximum length of a word w, w.r.t. the length k of its longest unbordered factor, such that k is shorter than the period $\pi(w)$ of w. They conjectured that $|w| \geq 2k$ implies $k = \pi(w)$ where $|w|$ denotes the length of w. That conjecture was falsified shortly thereafter by Assous and Pouzet [1] by the following example:

$$w = a^n b a^{n+1} b a^n b a^{n+2} b a^n b a^{n+1} b a^n$$

where $n \geq 1$ and $k = 3n + 6$ and $\pi(w) = 4n + 7$ and $|w| = 7n + 10$, that is, $k < \pi(w)$ and $|w| = 7k/3 - 4 > 2k$. Assous and Pouzet in turn conjectured that $3k$ is the bound on the length of w for establishing $k = \pi(w)$. Duval [3] did the next step towards solving the problem. He established that $|w| \geq 4k - 6$ implies $k = \pi(w)$ and conjectures that, if w possesses an unbordered prefix of length k, then $|w| \geq 2k$ implies $k = \pi(w)$. Note that a positive answer to Duval's conjecture yields the bound $3k$ for the general question. Despite some partial results [11,4,7] towards a solution, Duval's conjecture was only solved in 2004 [8,9] with a new proof given in [10]. The proof of (the extended version of) Duval's conjecture lowered the bound for Ehrenfeucht and Silberger's problem to $3k - 2$ as conjectured by Assous and Pouzet [1]. However, there remains a gap of $k/3$ between that bound and the largest known example, which is given above. With this paper we take the next step towards the solution of the problem by Ehrenfeucht and Silberger by establishing the optimal bound of $7k/3$ for a special case.

2   Notation and Basic Facts

Let us fix a finite set A, called alphabet, of letters. Let A^* denote the monoid of all finite words over A including the *empty word* denoted by ε. In general, we denote variables over A by a, b, c, d and e and variables over A^* are usually denoted by f, g, h, r through z, and α, β, and γ including their subscripted and primed versions. The letters i through q are to range over the set of nonnegative integers.

Let $w = a_1 a_2 \cdots a_n$. The word $a_n a_{n-1} \cdots a_1$ is called the *reversal* of w denoted by $\overline{w}$. We denote the length n of w by $|w|$, in particular $|\varepsilon| = 0$. If w is not empty, then let ${}^\bullet w = a_2 \cdots a_{n-1} a_n$ and $w^\bullet = a_1 a_2 \cdots a_{n-1}$. We define ${}^\bullet \varepsilon = \varepsilon^\bullet = \varepsilon$. Let $0 \leq i \leq n$. Then $u = a_1 a_2 \cdots a_i$ is called a *prefix* of w, denoted by $u \leq_{\mathrm{p}} w$, and $v = a_{i+1} a_{i+2} \cdots a_n$ is called a *suffix* of w, denoted by $v \leq_{\mathrm{s}} w$. A prefix or suffix is called proper when $0 < i < n$. An integer $1 \leq p \leq n$ is a *period* of w if $a_i = a_{i+p}$ for all $1 \leq i \leq n - p$. The smallest period of w is called *the period* of w, denoted by $\pi(w)$. A nonempty word u is called a *border* of a word w, if $w = uy = zu$ for some words y and z. We call w *bordered*, if it has a border that is shorter than w, otherwise w is called *unbordered*. Note that every bordered word w has a minimum border u such that $w = uvu$, where u is unbordered.

Let $\lhd$ be a total order on A. Then $\lhd$ extends to a *lexicographic order*, also denoted by $\lhd$, on A^* with $u \lhd v$ if either $u \leq_{\mathrm{p}} v$ or $xa \leq_{\mathrm{p}} u$ and $xb \leq_{\mathrm{p}} v$ and $a \lhd b$. Let $\overline{\lhd}$ denote a lexicographic order on the reversals, that is, $u \; \overline{\lhd} \; v$ if $\overline{u} \lhd \overline{v}$. Let $\lhd^a$ and $\lhd_b$ and $\lhd_b^a$ denote lexicographic orders where the maximum letter or the minimum letter or both are fixed in the respective orders on A. We establish the following convention for the rest of this paper: in the context of a given order $\lhd$ on A, we denote the inverse order of $\lhd$ by $\blacktriangleleft$. A $\lhd$-maximal prefix (suffix) α of a word w is defined as a prefix (suffix) of w such that $v \; \overline{\lhd} \; \alpha$ ($v \lhd \alpha$) for all $v \leq_{\mathrm{p}} w$ ($v \leq_{\mathrm{s}} w$).

The notion of maximum pre- and suffix are symmetric. It is general practice that facts involving the maximum ends of words are mostly formulated for maximum suffixes. The analogue version involving maximum prefixes is tacitly assumed.

Remark 1. Any maximum suffix of a word w is longer than $|w| - \pi(w)$ and occurs only once in w.

Indeed, let α be the $\lhd$-maximal suffix of u for some order $\lhd$. Then $u = x\alpha y$ and $\alpha \lhd \alpha y$ implies $y = \varepsilon$ by the maximality of α. If $w = uv\alpha$ with $|v| = \pi(w)$, then $u\alpha \leq_{\mathrm{p}} w$ gives a contradiction again.

Let an integer q with $0 \leq q < |w|$ be called *point* in w. A nonempty word x is called a *repetition word* at point q if $w = uv$ with $|u| = q$ and there exist words y and z such that $x \leq_{\mathrm{s}} yu$ and $x \leq_{\mathrm{p}} vz$. Let $\pi(w, q)$ denote the length of the shortest repetition word at point q in w. We call $\pi(w, q)$ the *local period* at point q in w. Note that the repetition word of length $\pi(w, q)$ at point q is necessarily unbordered and $\pi(w, q) \leq \pi(w)$. A factorization $w = uv$, with $u, v \neq \varepsilon$ and $|u| = q$, is called *critical*, if $\pi(w, q) = \pi(w)$, and if this holds, then q is called a *critical point*.

Let $\lhd$ be an order on A. Then the shorter of the $\lhd$-maximal suffix and the $\blacktriangleleft$-maximal suffix of some word w is called a *critical suffix* of w. Similarly, we define a *critical prefix* of w by the shorter of the two maximum prefixes resulting from some order and its inverse. This notation is justified by the following formulation of the so called critical factorization theorem (CFT) [2], which relates maximum suffixes and critical points.

Theorem 1 (CFT). *Let $w \in A^*$ be a nonempty word and γ be a critical suffix of w. Then $|w| - |\gamma|$ is a critical point.*

Let uv be a *Duval extension* of u if u is an unbordered word and every factor in uv longer than $|u|$ is bordered. A Duval extension uv of u is called trivial if $v \leq_p u$. The following fact was conjectured in [3] and proven in [8,9,10].

Theorem 2. *Let uv be a nontrivial Duval extension of u. Then $|v| \leq |u| - 2$.*

Following Theorem 2 let a *maximum Duval extension* of u be a nontrivial Duval extension uv with $|v| = |u| - 2$. This length constraint on v will often tacitly be used in the rest of this paper.

Let wuv be an *Ehrenfeucht-Silberger extension* of u if both uv and $\overline{wu}$ are Duval extensions of u and $\overline{u}$, respectively, moreover, uv and $\overline{wu}$ are called the Duval extensions corresponding to the Ehrenfeucht-Silberger extension of u.

Ehrenfeucht and Silberger were the first to investigate the bound on the length of a word w, w.r.t. the length k of its longest unbordered factors, such that $k < \pi(w)$. Some bounds have been conjectured. The latest such conjecture is taken from [9].

Conjecture 2. Let wuv be a nontrivial Ehrenfeucht-Silberger extension of u. Then $|wv| < \frac{4}{3}|u|$.

3   Periods and Maximum Suffixes

Note the following simple but noteworthy fact.

Lemma 1. *Let u be an unbordered word, and let v be a word that does not contain u. Let α be the $\lhd$-maximal suffix of u. Then any prefix w of uv such that α is a suffix of w, is unbordered.*

Proof. Certainly, $|w| \geq |u|$ by Remark 1. Suppose that w has a shortest border h. Then $|h| < |u|$ otherwise $u \leq_p h$ and u occurs in v since h is the shortest border; a contradiction. But now, h is a border of u; again a contradiction.   □

This implies immediately the following version of Lemma 1 for Duval extensions, which will be used frequently further below.

Lemma 2. *Let uv be a nontrivial Duval extension of u, and let α be the $\lhd$-maximal suffix of u. Then uv contains just one occurrence of α.*

The next lemma highlights an interesting fact about borders involving maximum suffixes. It will mostly be used on maximum prefixes of words, the dual to maximum suffixes, in later proofs. However, it is general practice to reason about ordered factors of words by formulating facts about suffixes rather than prefixes. Both ways are of course equivalent. We have chosen to follow general practice here despite its use on prefixes later in this paper.

Lemma 3. *Let αa be the $\lhd$-maximal suffix of a word wa where a is a letter. Let u be a word such that αa is a prefix of u and wb is a suffix of u, with $b \neq a$ and $b \lhd a$. Then u is either unbordered, or its shortest border has the length at least $|w| + 2$.*

Proof. Suppose that u has a shortest border hb. If $|h| < |\alpha|$ then $hb \leq_p \alpha$ and $h \leq_s \alpha$ and $hb \lhd ha$ contradict the maximality of αa. Note that $|h| \neq |\alpha|$ since $a \neq b$. If $|\alpha| < |h| \leq |w|$ then $\alpha a \leq_p h$, and hence, αa occurs in w contradicting the maximality of αa again; see Remark 1. Hence, $|hb| \geq |w| + 2$. □

The next lemma is taken from a result in [7] about so called minimal Duval extensions. However, the shorter argument given here (including the use of Lemma 3) gives a more concise proof than the one in [7].

Lemma 4. *Let uv be a nontrivial Duval extension of u where $u = xazb$ and $xc \leq_p v$ and $a \neq c$. Then bxc occurs in u.*

Proof. Let ya be the $\lhd^a$-maximal suffix of xa. Consider the factor $yazbxc$ of uv, which is longer than u and therefore bordered with a shortest border r. Now, Lemma 3 implies that $|r| > |xc|$, and hence, $bxc \leq_s r$ occurs in u. □

4   Some Facts about Certain Suffixes of a Word

This section is devoted to the foundational proof technique used in the remainder of this paper. The main idea is highlighted in Lemma 5, which identifies a certain unbordered factor of a word.

Lemma 5. *Let α be the $\lhd$-maximal suffix and β be the $\blacktriangleleft$-maximal suffix of a word u, and let v be such that neither α nor β occur in uv more than once. Let a be the last letter of v and b be the first letter of x where $x \leq_s \alpha v^\bullet$ and $|x| = \pi(\alpha v^\bullet)$.*

If $\pi(\alpha v) > \pi(\alpha v^\bullet)$, then αv is unbordered, in case $a \lhd b$, and βv is unbordered, in case $b \lhd a$.

Proof. Let γ be the longest border of $\alpha v^\bullet$. Note that $|\gamma| < |\alpha|$ since $^\bullet \alpha v$ does not contain the critical suffix of u, by assumption. We have $\alpha = \gamma b a'$ and $\alpha v = v' \gamma a$. Note that $\pi(\alpha v^\bullet) = |v'|$, and the inequality $\pi(\alpha v) > \pi(\alpha v^\bullet)$ means $a \neq b$.

Suppose that $a \lhd b$. We claim that αv is unbordered in this case. Suppose the contrary, and let αv have a shortest border ha. Then $|h| < |\gamma|$ otherwise either $a = b$, if $|h| = |\gamma|$, or γ is not the longest border of $\alpha v^\bullet$, if $|h| > |\gamma|$; a contradiction in both cases. But now $\alpha \lhd hba'$ since $ha \leq_p \alpha$ and $a \lhd b$ contradicting the maximality of α because $hba' \leq_s \alpha$.

Suppose that $b \lhd a$. In this case the word βv is unbordered. To see this suppose that βv has a shortest border ha. The assumption that uv contains just one occurrence of the maximal suffixes implies that ha is a proper prefix of β. If $|h| \geq |\gamma|$ then γa occurs in u contradicting the maximality of α since $\gamma b \leq_p \alpha \lhd \gamma a$. But now $ha \leq_p \beta \blacktriangleleft hba'$ (since $b \lhd a$) contradicting the maximality of β. □

Proposition 1. *Let uv be a nontrivial Duval extension of u, and let α be a critical suffix w.r.t. an order $\lhd$. Then $|v| < \pi(\alpha v) \leq |u|$.*

Proof. If $|v| \geq \pi(\alpha v)$ then α occurs twice in αv contradicting Lemma 2. Suppose that $\pi(\alpha v) > |u|$, and let z be the shortest prefix of v such that already $\pi(\alpha z) > |u|$. Then $\pi(\alpha z) > \pi(\alpha z^\bullet)$, and Lemma 5 implies that either αz or βz is unbordered, where β is the $\blacktriangleleft$-maximal suffix of u. This contradicts the assumption that uv is a Duval extension, since both the candidates are longer than u, which follows from $\pi(\alpha z) > |u|$ and $|\beta| > |\alpha|$. □

5 About Maximum Duval Extensions

In this section we consider the general results of the previous section for the special case of Duval extensions, which leads is to the main results, Theorem 3 and 4. Theorem 3 confirms a conjecture in [6]. Theorem 4 constitutes a further step to answer Conjecture 2.

Definition 1. *Let uv be a Duval extension of u. The suffix s of uv is called a trivial suffix if $\pi(s) = |u|$ and s is of maximum length.*

Note that $s = uv$, if uv is a trivial Duval extension, and $as \leq_s uv$ with $\pi(as) > |u|$, if uv is a nontrivial Duval extension. Moreover, Proposition 1 implies that $|s| \geq |\alpha v|$ where α is any critical suffix of u.

Let us begin with considerations about the periods of suffixes of maximum Duval extensions.

Lemma 6. *Let uv be a maximum Duval extension of u, and let $\lhd$ be an order such that the $\lhd$-maximal suffix α is critical. Then $\pi(\alpha v) = |u|$.*

Proof. It follows from Proposition 1 that $|u| - 1 \leq \pi(\alpha v) \leq |u|$ since $|v| = |u| - 2$. Suppose $\pi(\alpha v) = |u| - 1$. Let $w\alpha$ be the longest suffix of u such that $\pi(w\alpha v) = |u| - 1$. We have $w\alpha \neq u$ since u is unbordered. We can write $w\alpha v = w\alpha v'w\alpha^\bullet$, where v' is a prefix of v such that $|w\alpha v'| = |u| - 1$. The maximality of $w\alpha$ implies that $aw\alpha$ is a suffix of u, and $bw\alpha^\bullet$ is a suffix of αv, with $a \neq b$.

Choose a letter c in $w\alpha^\bullet$ such that $c \neq a$. Such a letter exists for otherwise $aw\alpha^\bullet \in a^+$ and α is just a letter, different from a. But this implies $u \in a^+\alpha$ and $v \notin a^+$ for uv to be nontrivial, that is, $v'd \leq_p v$ with $d \neq a$; a contradiction since $uv'd$ is unbordered in this case.

Consider the $\overline{\lhd}^c$-maximal prefix of $bw\alpha^\bullet$ denoted by bt. Note that $|t| \geq 1$. We claim that $aw\alpha v't$ is unbordered. Suppose on the contrary that r is the shortest border of $aw\alpha v't$. By Lemma 3 applied to the reversal of $aw\alpha v't$, the border r is longer than $bw\alpha^\bullet$. Hence, r contains α contradicting Lemma 2. But now, since $|w\alpha v'| = |u| - 1$ and $|t| \geq 1$, the unbordered factor $aw\alpha v't$ is longer than u; a contradiction. □

Lemma 7. *Let uv be a maximum Duval extension of u, let a be the last letter of u, and let xv be the trivial suffix of uv. Then $|\alpha| \leq |x|$ for the $\lhd^a$-maximal suffix α of any order $\lhd^a$.*

Proof. Suppose on the contrary that $|\alpha| > |x|$, which implies that the $\blacktriangleleft^a$-maximal suffix β is critical and $\beta \leq_s x$ by Lemma 6. Since uv is nontrivial, we can write $u = u'cwba$ and $v = v'dw$ where $wba = x$.

Consider the maximum prefix t of dw with respect to any order on the reversals where d is maximal. Note that $d \leq_s t$. The word $cwbav't$ is longer than u, therefore it is bordered. Let r be its shortest border. By Lemma 3, we have $|cw| < |r|$. Lemma 2 implies that $r = cwb$, and we have $d = b$ since $d \leq_s t$. Note that $|t| < |bw|$ otherwise $t = bw = wb$, which implies $|u| = \pi(xv) = \pi(wbav'bw) = \pi(bwav'bw) \leq |v| + 1 < |u|$; a contradiction. Hence, $te \leq_p bw$ for some letter $e \neq b$. Moreover, $e \neq a$ since $\beta^\bullet \leq_s r$ and β does occur only once in βv by Lemma 2.

Consider the factor $\alpha v'te$, which is longer than u, and hence, bordered. Let s be the shortest border of $\alpha v'te$. Note that $|s| < |\beta|$ otherwise $\beta^\bullet e \leq_s s$ contradicting the maximality of β since $\beta = \beta^\bullet a \blacktriangleleft^a \beta^\bullet e$. Let $s = \beta'e$ where $\beta' \leq_s \beta^\bullet$. But then $\beta'e \leq_p \alpha \lhd^a \beta'a$ and $\beta'a \leq_s u$ contradicting the maximality of α. $\square$

Lemma 8. *Let uv be a maximum Duval extension of $u = u'ab$ where a and b are letters. Then a occurs in u'.*

Proof. Suppose on the contrary that a does not occur in u'. Note that b occurs in u' by Lemma 4. So, we may assume that $a \neq b$. Moreover, we have that also a letter c different from a and b has to occur in u' otherwise $u = b^iab$ and $v = b^jdv'$ for some $d \neq b$ and $j < i$, but then ub^jd is unbordered; a contradiction.

Let β be the maximum suffix of u w.r.t. some order $\lhd_c^b$, and let α be a maximum suffix of u w.r.t. the order $\blacktriangleleft_c^b$. Let γ be the shorter of the two suffixes α and β, and note that $|\gamma| > 2$.

Lemma 6 implies $\pi(\gamma v) = |u|$. Let $w\gamma v$ be the trivial suffix of uv. We have that $u \neq w\gamma$ since uv is a nontrivial Duval extension of u. Therefore, we can write $u = u'dw\gamma$ and $v = v'ew\gamma^{\bullet\bullet}$ where d and e are different letters and $|w\gamma v'e| = |u|$. Note that e occurs in $u^{\bullet\bullet}$ otherwise $uv'e$ is unbordered; a contradiction. Consider an order $\lhd^e$ and let t be the $\overline{\lhd}^e$-maximal prefix of $ew\gamma^{\bullet\bullet}$.

The word $dw\gamma v't$ is longer than u, therefore it is bordered. Let r be its shortest border. By Lemma 3, we have $|dw\gamma| - 2 < |r|$. Lemma 2 implies that $|r|$ is exactly $|dw\gamma| - 1$, whence $r = dw\gamma^\bullet$. Clearly, the letter e is a suffix of t, and thus also of r, which implies that e is a suffix of $u^\bullet$; a contradiction since $e \neq a$. $\square$

The following example shows that the requirement of a maximum Duval extension is indeed necessary in Lemma 8.

Example 1. Let a, b, and c be different letters, and consider $u = a^iba^{i+j}bcb$ and $v = a^{i+j}ba^{i-1}$ with $i, j \geq 1$. Then $u.v = a^iba^{i+j}bcb.a^{i+j}ba^{i-1}$ is a nontrivial Duval extension of length $2|u| - 4$ such that c occurs only in the second last position of u. However, a maximum Duval extension of a word $|u|$ has length $2|u| - 2$.

The next lemma highlights a relation between the trivial suffix of a maximum Duval extension uv and the set $alph(u)$ of all letters occurring in u.

Lemma 9. *Let uv be a maximum Duval extension of u and wxw be the trivial suffix of uv where $|wx| = |u|$. Then either* $\mathrm{alph}(w) = \mathrm{alph}(u)$ *or there exists a letter b such that* $\mathrm{alph}(w) = \mathrm{alph}(u) \setminus \{b\}$ *and $u = u'bb$ and bb does not occur in u'.*

Proof. Suppose contrary to the claim that $|\mathrm{alph}(w)| < |\mathrm{alph}(u)|$ and for any $b \in \mathrm{alph}(u) \setminus \mathrm{alph}(w)$ we have bb is not a suffix of u or bb occurs in $u^{\bullet\bullet}$.

Let $btwac \leq_\mathrm{s} u$ where $a, b, c \in \mathrm{alph}(u)$ and b does not occur in tw. Consider $btwxw$, which is longer than u and therefore has to be bordered. Let r be the shortest border of $btwxw$. Certainly, $|w| < |r|$ since $b \leq_\mathrm{p} r$ and $b \notin \mathrm{alph}(w)$. Moreover, $btw \leq_\mathrm{p} r$ implies $\pi(btwxw) \leq |u|$ contradicting the maximality of wxw. So, we note that $|w| < |r| < |btw|$.

Suppose $a \neq b$. Let $v = v'r$ and consider the factor $twacv'b$, which has to be bordered since $|twacv'b| = |twacv| - |r| + 1 > |acv| = |u|$. Let s be the shortest border of $twacv'b$. We have $|s| > |twa|$ because b is a suffix of s and does not occur in tw and $a \neq b$ by assumption. But now, $twac \leq_\mathrm{p} s$ contradicting Lemma 2 since wac contains a maximum suffix of u.

Suppose $a = b$. This is the only case where we need to consider that either $bb \not\leq_\mathrm{s} u$ or bb occurs at least twice in u. Let $d \in \mathrm{alph}(u)$ be such that $d = c$, if $c \neq b$, and d be an arbitrary letter different from b otherwise. Consider an order $\lhd_d^b$ on $\mathrm{alph}(u)$. Let α be the $\lhd_d^b$-maximal suffix of u. Note that $|\alpha| > |wbc|$ since either $c = b$ or $c = d$. If $c = b$ then $bb \leq_\mathrm{p} \alpha$ occurs in $u^\bullet$ by assumption. If $c = d$ then be occurs in $u^\bullet$ for some letter e by Lemma 8 where we have $be \leq_\mathrm{p} \alpha$ since either $d \lhd_d^b e$ or $e = d$. Since every critical suffix of u is a suffix of wbc by Lemma 6 and $\alpha \not\leq_\mathrm{s} wbc$, we have that the $\blacktriangleleft_d^b$-maximal suffix β is critical and $\beta \leq_\mathrm{s} wbc$. Moreover, $|\beta| > 2$ since $bc \leq_\mathrm{s} u$ and d occurs in $u^\bullet$ by Lemma 4. We have that $\beta^{\bullet\bullet} \leq_\mathrm{s} w$, and hence, $\beta^{\bullet\bullet} \leq_\mathrm{s} r$. From $|r| < |btw|$ follows that $\beta^{\bullet\bullet}c'$ occurs in tw where c' is a letter in tw, and therefore $c' \neq b$. But this contradicts the maximality of β since $\beta^{\bullet\bullet}b \blacktriangleleft_d^b \beta^{\bullet\bullet}c'$. $\square$

The next two results, Lemma 10 and 11, constitute a case split of the proof of Theorem 3. Namely, the cases when exactly two or more than two letters occur in a maximum Duval extension.

Lemma 10. *Let uv be a maximum Duval extension of $u = u'ab^i$ where $i \geq 1$ and $|\mathrm{alph}(u)| > 2$ and $a \neq b$. Then u' does not contain the factor b^i.*

Proof. Suppose, contrary to the claim, that b^i occurs in u'. Consider the trivial suffix $wcbv'dw$ of uv where $|cbv'dw| = |u|$ and $c \in \{a, b\}$. We can write $u = u'ewcb$ with $d \neq e$ since $|u| > |wcb|$. We have that $\mathrm{alph}(w) = \mathrm{alph}(u)$ by Lemma 9. Choose a letter f in dw such that $f \neq e$ and $f \neq c$. Let $\lhd_e^f$ be an order. Let dt be the $\overline{\lhd}_e^f$-maximal prefix of dw. The word $ewcbv't$ is longer than u, therefore it is bordered. Let r be its shortest border. By Lemma 3, we have $|dw| < |r|$. Lemma 2 implies that $|r|$ is exactly $|dwc|$, and hence, $r = ewc$. Clearly, the letter f is a suffix of t, and thus also of r, which implies that $f = c$; a contradiction. $\square$

Lemma 11. *Let uv be a maximum Duval extension of $u = u'ab^i$ over a binary alphabet where $i \geq 1$ and $a \neq b$. Then u' does not contain the factor b^i and $awbb \leq_s u$ and $v = v'bw$ where $wbbv$ is the trivial suffix of uv.*

Proof. Let s be the trivial suffix of uv, and let $u = u_0cwdb$ and $v = v'ew$ where $wdbv'ew = s$. Note that $c \neq e$ by the maximality of s. Let $\lhd$ be the order such that $a \lhd b$.

Suppose $c = b$ and $e = a$. Let t be the $\overline{\lhd}$-maximal prefix of aw. Consider the factor $bwdbv't$, which is longer than $|u|$ and hence bordered. Lema 3 implies that $|bw| < |r|$. Lemma 2 implies that $r = bwd$, in fact, $r = bwa$ since $a \leq_s t$. Note that $|t| \leq |w|$ otherwise $r = bwa = baw = ba^{|w|+1}$ contradicting Lemma 9. So, we have $tb \leq_p aw$ by the maximality of t. But now wab occurs in v, and hence, the critical suffix of u occurs in v by Lemma 6 contradicting Lemma 2.

We conclude that $c = a$ and $e = b$. Consider the $\lhd$-maximal suffix β of u. Suppose contrary to the claim that b^i occurs in u'. Then $b^ja \leq_p \beta$ for some $j \geq i$.

Let t be the $\overline{\lhd}$-maximal prefix of bw. Similarly to the reasoning above, we consider the factor $awdbv't$ and conclude that it has the border $r = awb$ and $d = b$ and $ta \leq_p bw$. Lemma 7 implies that $\beta \leq_s wbb$. Note that b^j is a power of b in u of maximum size and occurs in w by assumption, and hence, $b^j \leq_s t$. But now, $b^j \leq_s r$ and $b^{j+1} \leq_s u$; a contradiction. $\qquad\square$

The main result follows directly from the previous two lemmas.

Theorem 3. *Let uv be a maximum Duval extension of $u = u'ab^i$ where $i \geq 1$ and $a \neq b$. Then b^i occurs only once in uv.*

Indeed, b^i does not occur in u' by Lemma 10 and 11. If b^i occurs in $b^{i-1}v$, that is, $b^{i-1}v = wb^iv'$, then $u'abwb^i$ is unbordered; a contradiction.

Let us consider the results obtained so far for the special case of a binary alphabet in the following remark.

Remark 2. Let $uv \in \{a,b\}^+$ be a maximum Duval extension with $b \leq_s u$, and let wv be the trivial suffix of uv.

Theorem 3 implies that the $\lhd_a^b$-maximal suffix of u is critical and equal to b^i. Lemma 4 implies that $i \geq 2$. Lemma 9 implies that a occurs in w, and in particular, $w \in a^+bb$, if $i = 2$. Lemma 11 implies that $axb^i \leq_s u$ and $bxb^{i-2} \leq_s v$, where $w = xb^i$.

Theorem 4. *Let uv be a maximum Duval extension of $u = u'ab^i$ where $i \geq 1$ and $a \neq b$. Then $3i \leq |u|$.*

Proof. The shortest possible maximum Duval extension of a word u is of the form uv with $u = abaabb$ and $v = aaba$. This proves the claim for $i \leq 2$. Assume $i > 2$ in the following.

Let $cb^k \leq_s v$ with $c \neq b$. Lemma 6 implies that $k \geq i-2$, and Lemma 2 yields $k \leq i-1$. Consider the shortest border h of uv. Then $|h| < |u| - 2$ otherwise uv

is trivial. Let $h = gb^k$, and let j be the maximum integer such that $gb^j \leq_p u$. Clearly, $k \leq j \leq i - 1$ since b^i occurs only as a suffix of u. Let $u = gb^j fb^i$. Note that

$$b \notin \{\mathrm{pref}_1(g), \mathrm{pref}_1(f), \mathrm{suff}_1(g), \mathrm{suff}_1(f)\} . \tag{1}$$

Next we show that b^k occurs in g or f. Suppose the contrary, that is, neither g nor f contains b^k. Consider the shortest border x of $fb^i v$. We have $|x| < |fb^i|$, since b^i does not occur in v. Property (1) and the assumption that b^k does not occur in f imply that $x = fb^k$. Let $v = v'fb^k$. Consider the shortest border y of $b^j fb^i v' f$. Again, we have $|y| < |b^j fb^i|$ since b^i does not occur in v, and property (1) implies that $y = b^j h$. Let $v = v''b^j fb^k$. Finally, consider the shortest border z of $uv''b^j$. Property (1) and the assumption that b^k does not occur in g or f imply that either $z = gb^j$ or $z = gb^j fb^i$. The former implies that $uv = gb^j fb^i gb^j fb^k$ is a trivial Duval extension, and the latter implies that $|u| < |v|$; a contradiction in both cases.

We conclude that b^k occurs in g or f. Let $u = u_1 b^m u_2 b^n u_3 b^i$ where u_1, u_2, and u_3 are not empty and neither begin nor end with b and $k \leq m, n \leq i - 1$. The claim is proven if $|u_1 u_2 u_3| > 3$ or $m = i - 1$ or $n = i - 1$. Suppose the contrary, that is, u_1, u_2, and u_3 are letters and $m = i - 2$ and $n = i - 2$ and $k = i - 2$.

Let us consider the shape of v next. Note that every factor of length 2 in v contains b otherwise there exists a prefix w of v that ends in two letters not equal to b and uw is unbordered; a contradiction. Moreover, for every power $b^{k'}$ in v holds $i - 1 \leq k'$ otherwise $w'cb^{k'}d$ is a prefix of v where c and d are letters different from b and $b^m u_2 b^n u_3 b^i w' cb^{k'}d$ is unbordered; a contradiction. Considering possible borders of words $uv_1 b^{i-2}$, $uv_1 b^{i-2}v_2 b^{i-2}$, $u_2 b^{i-2}u_3 b^i v_1 b^{i-2}v_2 b^{i-2}$ and $u_3 b^i v_1 b^{i-2}v_2 b^{i-2}v_3 b^{i-2}$ we deduce that $v_1 = u_1$, $v_2 = u_2$ and $v_3 = u_3$; a contradiction since uv is assumed to be nontrivial. This proves the claim. $\qquad \square$

Corollary 1. *Let w be a nontrivial Ehrenfeucht-Silberger extension of u such that one of its corresponding Duval extensions is of maximum length. Then $|w| < \frac{7}{3}|u| - 2$.*

Indeed, suppose on the contrary that $w = xuv$ and uv is a maximum Duval extension with $ab^i \leq_s u$ and $|x| \geq i$ where $a \neq b$. The case where $\overline{x}u$ is a maximum Duval extension is symmetric. Now, either $b^i \leq_s x$ or $eb^j \leq_s x$ with $j < i$ and $e \neq b$. If $eb^j \leq_s x$ with $j < i$ and $e \neq b$, then $eb^j u$ is unbordered; a contradiction. If $b^i \leq_s x$ then $b^i ub^{-i}$ is unbordered by Theorem 3, and its Duval extension $b^i uv$ is trivial, since it is too long; a contradiction.

The following example is taken from [1].

Example 2. Consider the following word xuv where we separate the factors x, u, and v for better readability

$$x.u.v = b^{i-2}.ab^{i-1}ab^{i-2}ab^i.ab^{i-2}ab^{i-1}ab^{i-2}$$

where $i > 2$. We have that the largest unbordered factors of xuv are of length $3i$, namely the factors $u = ab^{i-1}ab^{i-2}ab^i$ and $b^iab^{i-2}ab^{i-1}a$, and $\pi(xuv) = 4i - 1$, and hence, xuv is a nontrivial Ehrenfeucht-Silberger extension of u. Note that uv is a maximum Duval extension. We have $|xuv| = 7i - 4 = \frac{7}{3}|u| - 4$.

References

1. Assous, R., Pouzet, M.: Une caractérisation des mots périodiques. Discrete Math. 25(1), 1–5 (1979)
2. Crochemore, M., Perrin, D.: Two-way string-matching. J. ACM 38(3), 651–675 (1991)
3. Duval, J.-P.: Relationship between the period of a finite word and the length of its unbordered segments. Discrete Math. 40(1), 31–44 (1982)
4. Duval, J.-P., Harju, T., Nowotka, D.: Unbordered factors and Lyndon words. Discrete Math. 308(11), 2261–2264 (2008)
5. Ehrenfeucht, A., Silberger, D.M.: Periodicity and unbordered segments of words. Discrete Math. 26(2), 101–109 (1979)
6. Harju, T., Nowotka, D.: Duval's conjecture and Lyndon words. TUCS Tech. Rep. 479, Turku Centre of Computer Science, Finland (2002)
7. Harju, T., Nowotka, D.: Minimal Duval extensions. Internat. J. Found. Comput. Sci. 15(2), 349–354 (2004)
8. Harju, T., Nowotka, D.: Periodicity and unbordered words. In: Diekert, V., Habib, M. (eds.) STACS 2004. LNCS, vol. 2996, pp. 294–304. Springer, Heidelberg (2004)
9. Harju, T., Nowotka, D.: Periodicity and unbordered words: A proof of the extended Duval conjecture. J. ACM 54(4) (2007)
10. Holub, Š.: A proof of the extended Duval's conjecture. Theoret. Comput. Sci. 339(1), 61–67 (2005)
11. Mignosi, F., Zamboni, L.Q.: A note on a conjecture of Duval and Sturmian words. Theor. Inform. Appl. 36(1), 1–3 (2002)

Duplication in DNA Sequences*

Masami Ito[1], Lila Kari[2], Zachary Kincaid[2], and Shinnosuke Seki[2]

[1] Department of Mathematics, Faculty of Science, Kyoto Sangyo University,
Kyoto, Japan, 603-8555
ito@ksuvx0.kyoto-su.ac.jp

[2] Department of Computer Science, University of Western Ontario, London,
Ontario, Canada, N6A 5B7
{lila,sseki}@csd.uwo.ca, zkincaid@uwo.ca

Abstract. Duplication and repeat-deletion are the basic models of errors occurring during DNA replication from the viewpoint of formal languages. During DNA replication, subsequences of a strand of DNA may be copied several times (duplication) or skipped (repeat-deletion). Iterated duplication and repeat-deletion have been well-studied, but little is known about single-step duplication and repeat-deletion. In this paper, we investigate properties of these operations, such as closure properties of language families in the Chomsky hierarchy, language equations involving these operations. We also make progress towards a characterization of regular languages that are generated by duplicating a regular language.

1 Introduction

Duplication grammars and duplication languages have recently received a great deal of attention in the formal language theory community. Duplication grammars, defined in [12], model duplication using string rewriting systems. Several properties of languages generated by duplication grammars were investigated in [12] and [13]. Another prevalent model for duplication is a unary operation on words [1], [2], [5], [7], [8], [9]. The biological phenomenon which motivates the research on duplication is a common error occurring during DNA replication: the insertion or deletion of repeated subsequences in DNA strands, [3].

A DNA single strand is a string over the DNA alphabet of bases $\{A, C, G, T\}$. Due to the Watson-Crick complementarity $A-T$, $C-G$, two complementary DNA single strands of opposite orientation can bind to each other to form a DNA double strand. DNA replication is the process by which given a "template" DNA strand, an enzyme called DNA polymerase creates a new "nascent" DNA strand that is a complement of the template. To be more precise, a special short DNA single strand called a "primer" is attached to the template as a toe-hold, and then DNA polymerase adds complementary bases to the template strand, one by one, until the entire template strand becomes double-stranded.

* This research was supported by Grant-in-Aid for Scientific Research No. 19-07810 by Japan Society for the Promotion of Sciences and Research Grant No. 015 by Kyoto Sangyo University to M. I., and The Natural Sciences and Engineering Council of Canada Discovery Grant and Canada Research Chair Award to L.K.

M. Ito and M. Toyama (Eds.): DLT 2008, LNCS 5257, pp. 419–430, 2008.

It has been observed that errors can happen during this process, the most common of them being repeat insertions and deletions of bases. The "strand slippage model" that was proposed as an explanation of these phenomena suggests that these errors are caused by misalignments between the template and nascent strands during replication. DNA polymerase is not known to have any "memory" to remember which base on the template has been just copied onto the nascent strand, and hence the template and nascent strands can *slip*. As such, the DNA polymerase may copy a part of the template twice (resulting in an insertion) or forget to copy it (deletion). These errors occur most frequently on repeated sequences so that they are appropriately modelled by the rewriting rules $u \rightarrow uu$ and $uu \rightarrow u$.

The rule $u \rightarrow uu$ is a natural model for duplication, and the rule $uu \rightarrow u$ models the dual of duplication, which we call *repeat-deletion*. Since strand slippage is responsible for both these operations, it is natural to study both duplication and repeat-deletion. Repeat-deletion has already been extensively studied, e. g. , in [6]. However, the existing literature addresses mainly the iterated application of both repeat-deletion and duplication. This paper investigates the effects of a *single* duplication or repeat-deletion. This restriction introduces subtle new complexities into languages that can be obtained as a duplication or repeat-deletion of a language.

The paper is organized as follows: In Section 2 we define the terminology and notations we use. Section 3 is dedicated to the closure properties of language families of the Chomsky hierarchy under duplication and repeat-deletion. In Section 4, we present and solve language equations based on these operations, and give a constructive method for obtaining maximal solutions. In Section 5 we introduce a generalization of duplication, namely controlled duplication, and investigate characterizations of regular languages that can be obtained by the duplication of a regular language. Lastly, we present some results on the relationship between duplication and primitive words.

2   Preliminaries

We provide definitions for terms and notations to be used throughout the paper. For basic concepts in formal language theory, we refer the reader to [4], [16], [18].

Let Σ be a finite alphabet, Σ^* be the set of words over Σ, and $\Sigma^+ = \Sigma^* \setminus \{\lambda\}$, where λ is the empty word. The length of a word $w \in \Sigma^*$ is denoted by $|w|$. For a non-negative integer $n > 0$ let $\Sigma^n = \{w \in \Sigma^* \mid |w| = n\}$ and $\Sigma^{\leq n} = \bigcup_{i=1}^{n} \Sigma^i$

For a finite automaton $A = (Q, \Sigma, \delta, s, F)$ (where Q is a state set, $\delta : Q \times \Sigma \to 2^Q$ is a transition function, $s \in Q$ is the start state, and $F \subseteq Q$ is a set of final states), let $\mathcal{L}(A)$ denote the language accepted by A. We extend δ to $\hat{\delta} : Q \times \Sigma^* \to 2^Q$ as follows: (1) $\hat{\delta}(q, \lambda) = \{q\}$ for $q \in Q$ and (2) $\hat{\delta}(q, wa) = \cup_{p \in \hat{\delta}(q,w)} \delta(p, a)$ for $q \in Q$, $w \in \Sigma^*$, and $a \in \Sigma$. For $P_1, P_2 \subseteq Q$, we define an automaton $A_{(P_1,P_2)} = (Q \cup s_0, \Sigma, \delta', s_0, P_2)$, where $s_0 \notin Q$ is a new start state and $\delta' = \delta \cup (s_0, \lambda, P_1)$. Hence, $\mathcal{L}(A_{(P_1,P_2)}) = \{w \mid \hat{\delta}(p_1, w) \cap P_2 \neq \emptyset \text{ for some } p_1 \in P_1\}$. If P_i is the singleton set $\{p_i\}$, then we may simply write p_i for $i \in \{1, 2\}$.

The aim of this paper is to investigate two operations that are defined on words and languages: *duplication* and *repeat-deletion*. The unary duplication operation is defined for a word $u \in \Sigma^*$ as follows:

$$u^\heartsuit = \{v \mid u = xyz, v = xyyz \text{ for some } x, z \in \Sigma^*, y \in \Sigma^+\}.$$

The duplication operation is extended to a language $L \subseteq \Sigma^*$ as $L^\heartsuit = \bigcup_{u \in L} u^\heartsuit$. Some authors, e.g., [2] require the duplicated factor y to be in a finite set of words called the *duplication scheme*. We discuss a generalization of duplication schemes which we call *controlled duplication* in Section 5.

We also define another unary operation based on the dual of the $\heartsuit$ operation. We call this operation *repeat-deletion* and denote it by $\spadesuit$, which is defined for a word $v \in \Sigma^*$ as follows:

$$v^\spadesuit = \{u \mid v = xyyz, u = xyz \text{ for some } x, z \in \Sigma^*, y \in \Sigma^+\}.$$

As above, for a given language $L \subseteq \Sigma^*$, we define $L^\spadesuit = \bigcup_{v \in L} v^\spadesuit$.

Previous work focused on the reflexive transitive closure of the duplication operation, which we will refer to as duplication closure. In this paper, all occurrences of $\heartsuit$ and $\spadesuit$ refer to the *single step* variations of the duplication and repeat-deletion, respectively.

3   Closure Properties

Much of the work on duplication closure has been concerned with determining which of the families of languages on the Chomsky hierarchy are closed under this operation. It is known that on a binary alphabet the family of regular languages is closed under duplication closure. In contrast, on a bigger alphabet it is still closed under n-bounded duplication closure for $n \leq 2$ but not closed under n-bounded operation closure for any $n \geq 4$. The family of context-free languages is closed under (uniformly) bounded duplication closure. The readers are referred to [5] for these results.

It is a natural first step to determine these closure properties under (single step) duplication. In this section, we show that the family of regular languages is closed under repeat-deletion but not duplication, the family of context-free languages is not closed under either operation, and the family of context-sensitive languages is closed under both operations.

The following two propositions are due to [17] (without proofs).

Proposition 1. REG *is not closed under duplication.*

Proposition 2. CFL *is not closed under duplication.*

The proof of Proposition 1 requires an alphabet that is at least binary. As we shall see in Section 5, this bound is strict. That is, the family of regular languages over a unary alphabet is closed under duplication. In addition, we have:

Proposition 3. CSL *is closed under duplication.*

In the following, we consider the closure properties of the language families on the Chomsky hierarchy under repeat-deletion. Our first goal is to prove that the family of regular languages is closed under repeat-deletion. For this purpose, we define the following binary operation $\natural$ on languages $L, R \subseteq \Sigma^*$:

$$L \natural R = \{xyz \mid xy \in L, yz \in R, y \neq \lambda\}.$$

Proposition 4. REG *is closed under* $\natural$.

Proof. Let $L_1, L_2 \in$ REG. Let $\# \notin \Sigma$ and let h be defined by $h(a) = a$ for $a \in \Sigma^*$ and $h(\#) = \lambda$. Let $L_1' = L_1 \leftarrow \{\#\} = \{u\#v \mid uv \in L_1\}$ ($\leftarrow$ denotes the insertion operation) and $L_2' = L_2 \leftarrow \{\#\}$. Moreover, let $\overline{L_1} = L_1' \# \Sigma^*$ and let $\overline{L_2} = \Sigma^* \# L_2'$. Then $L_1 \natural L_2 = h(\overline{L_1} \cap \overline{L_2})$. Since REG is closed under insertion, concatenation, intersection, and homomorphism, $L_1 \natural L_2$ is regular. $\square$

For a regular language L, there is a finite automaton $A = (Q, \Sigma, \delta, s, F)$ such that $\mathcal{L}(A) = L$. Recall that for any state $q \in Q$, $\mathcal{L}(A_{(s,q)}) = \{w \mid q \in \hat{\delta}(s, w)\}$ and $\mathcal{L}(A_{(q,F)}) = \{w \mid \hat{\delta}(q, w) \cap F \neq \emptyset\}$. Intuitively, $\mathcal{L}(A_{(s,q)})$ is the set of words accepted "up to q", and $\mathcal{L}(A_{(q,F)})$ is the set of words accepted "after q" so that $\mathcal{L}(A_{(s,q)})\mathcal{L}(A_{(q,F)}) \subseteq L$ is the set of words in L which have a derivation that passes through state q.

Lemma 1. *Let L be a regular language and $A = (Q, \Sigma, \delta, s, F)$ be a finite automaton accepting L. Then $L^{\spadesuit} = \bigcup_{q \in Q} \mathcal{L}(A_{(s,q)}) \natural \mathcal{L}(A_{(q,F)})$.*

Proof. Let $L' = \bigcup_{q \in Q} \mathcal{L}(A_{(s,q)}) \natural \mathcal{L}(A_{(q,F)})$. First we prove that $L^{\spadesuit} \subseteq L'$. Let $\alpha \in L^{\spadesuit}$. Then there exists a decomposition $\alpha = xyz$ for some $x, y, z \in \Sigma^*$ such that $xyyz \in L$ and $y \neq \lambda$. Since A accepts $xyyz$, there exists some $q \in Q$ such that $q \in \hat{\delta}(s, xy)$ and $\hat{\delta}(q, yz) \cap F \neq \emptyset$. By construction, $xy \in \mathcal{L}(A_{(s,q)})$ and $yz \in \mathcal{L}(A_{(q,F)})$. This implies that $xyz \in \mathcal{L}(A_{(s,q)}) \natural \mathcal{L}(A_{(q,F)})$, from which we have $L^{\spadesuit} \subseteq L'$.

Conversely, if $\alpha \in L'$, then there exists $q \in Q$ such that $\alpha \in \mathcal{L}(A_{(s,q)}) \natural \mathcal{L}(A_{(q,F)})$. We can decompose α into xyz for some $x, y, z \in \Sigma^*$ such that $xy \in \mathcal{L}(A_{(s,q)})$, $yz \in \mathcal{L}(A_{(q,F)})$, and $y \neq \lambda$. Since $\mathcal{L}(A_{(s,q)})\mathcal{L}(A_{(q,F)}) \subseteq L$, we have that $xyyz$ belongs to L. It follows that $\alpha = xyz \in L^{\spadesuit}$ and $L' \subseteq L^{\spadesuit}$. $\square$

The following is an immediate consequence of Proposition 4 and Lemma 1.

Proposition 5. REG *is closed under repeat-deletion.*

In contrast, the family of context-free languages is not closed under repeat-deletion, despite the following proposition.

Proposition 6. CFL *is closed under* ♮ *with regular languages.*

Lemma 2. CFL *is not closed under* ♮.

Proof. Let $L_1 = \{a^i \# b^i \$ \mid i \geq 0\}$ and $L_2 = \{\# b^j \$ c^j \mid j \geq 0\}$. Although L_1 and L_2 are CFLs, $L_1 ♮ L_2 = \{a^i \# b^i \$ c^i \mid i \geq 0\}$, which is not context-free.   □

Proposition 7. CFL *is not closed under repeat-deletion.*

Proof. Let $L = \{a^i \# b^i \# b^j c^j \mid i, j \geq 0\}$, which is context-free. Then $L^{\spadesuit} \cap a^* \# b^* c^* = \{a^i \# b^j c^j \mid i, j \geq 0, i \leq j\}$, which is not context free. Since CFL is closed under intersection with regular languages, and since $L^{\spadesuit} \cap a^* \# b^* c^*$ is not context-free, we conclude that $L^{\spadesuit}$ is not context-free.   □

Proposition 8. CSL *is closed under repeat-deletion.*

In summary, the following closure properties of duplication, repeat-deletion, and the ♮ operation hold:

	$\heartsuit$	$\spadesuit$	♮	♮ with regular
FIN	Y	Y	Y	N
REG	N	Y	Y	Y
CFL	N	N	N	Y
CSL	Y	Y	Y	Y

4   Language Equations

We now consider the language equation problem posed by duplication: for a given language $L \subseteq \Sigma^*$, can we find a language $X \subseteq \Sigma^*$ such that $X^{\heartsuit} = L$? In the following, we show that if L is a regular language and there exists a solution to $X^{\heartsuit} = L$, then we can compute a maximal solution. We note that the solution to the language equation is not unique in general.

Example 1. $\{aaa, aaaa, aaaaa\}^{\heartsuit} = \{aaa, aaaaa\}^{\heartsuit} = \{a^i : 4 \leq i \leq 10\}$

In view of the fact that a language equation may have multiple solutions, we define an equivalence relation $\sim_{\heartsuit}$ on languages as follows:

$$X \sim_{\heartsuit} Y \Leftrightarrow X^{\heartsuit} = Y^{\heartsuit}.$$

For the same reason, we define an equivalence relation $\sim_{\spadesuit}$ as follows:

$$X \sim_{\spadesuit} Y \Leftrightarrow X^{\spadesuit} = Y^{\spadesuit}.$$

Lemma 3. *The equivalence classes of $\sim_\heartsuit$ are closed under arbitrary unions. That is, if $[X] \in 2^{\Sigma^*}/\sim_\heartsuit$ and if $\Xi \subseteq [X]$ ($\Xi \neq \emptyset$), then $\bigcup_{L \in \Xi} L \in [X]$.*

Corollary 1. *For an equivalence class $[X] \in 2^{\Sigma^*}/\sim_\heartsuit$, there exists a unique maximal element $X_{\max}$ with respect to the set inclusion partial order defined as follows:*

$$X_{\max} = \bigcup_{L \in [X]} L.$$

We provide a way to construct the maximum element of a given equivalence class. First, we prove a more general result.

Proposition 9. *Let $L \subseteq \Sigma^*$, and let $f, g : \Sigma^* \to 2^{\Sigma^*}$ be any functions such that $u \in g(v) \Leftrightarrow v \in f(u)$ for all $u, v \in \Sigma^*$. If a solution to the language equation $\bigcup_{x \in X} f(x) = L$ exists, then the maximum solution (with respect to the set inclusion partial order) is given by $X_{\max} = \left(\bigcup_{y \in L^c} g(y)\right)^c$.*

Proof. For two languages $X, Y \subseteq \Sigma^*$ such that $\bigcup_{x \in X} f(x) = L$ and $\bigcup_{y \in Y} f(y) = L$, $\bigcup_{z \in X \cup Y} f(z) = L$ holds. Hence the assumption implies the existence of $X_{\max}$. ($\subseteq$) Suppose $\exists w \in g(v) \cap X_{\max}$ for some $v \in L^c$. This means that $v \in f(w)$. However, $f(w) \subseteq \bigcup_{x \in X_{\max}} f(x) = L$, and hence $v \in L$, a contradiction. ($\supseteq$) Suppose that $\exists w \in X_{\max}^c \cap \left(\bigcup_{y \in L^c} g(y)\right)^c$. If $f(w) \subseteq L$, then $w \in X_{\max}$ (by the maximality of $X_{\max}$). Otherwise, $\exists v \in f(w) \cap L^c$. This implies that $w \in g(v) \subseteq \bigcup_{y \in L^c} g(y)$. In both cases, we have a contradiction. Therefore, we have $X_{\max}^c = \bigcup_{y \in L^c} g(y)$, i.e., $X_{\max} = \left(\bigcup_{y \in L^c} g(y)\right)^c$. $\qquad\square$

Lemma 4. *Let $u, v \in \Sigma^*$. Then $u \in v^\heartsuit$ if and only if $v \in u^\spadesuit$.*

Proof. ($\Rightarrow$) If $u \in v^\heartsuit$, then there exist $x, z \in \Sigma^*$ and $y \in \Sigma^+$ such that $v = xyz$ and $u = xyyz$. Then $u^\spadesuit$ contains $xyz = v$. ($\Leftarrow$) If $v \in u^\spadesuit$, then there exist $x', z' \in \Sigma^*$ and $y' \in \Sigma^+$ such that $v = x'y'z'$ and $u = x'y'y'z'$. Then $x'y'y'z' = u \in v^\heartsuit$. $\qquad\square$

Proposition 9 and Lemma 4 imply the following corollaries.

Corollary 2. *Let $L \subseteq \Sigma^*$. If there exists a language $X \subseteq \Sigma^*$ such that $X^\spadesuit = L$, then the maximum element $X_{\max}$ of $[X]_{\sim_\spadesuit}$ is given by $((L^c)^\heartsuit)^c$.*

Corollary 3. *Let $L \subseteq \Sigma^*$. If there exists a language $X \subseteq \Sigma^*$ such that $X^\heartsuit = L$, then the maximum element $X_{\max}$ of $[X]_{\sim_\heartsuit}$ is given by $((L^c)^\spadesuit)^c$.*

Proposition 10. *Let L, X be regular languages satisfying $X^\heartsuit = L$. Then it is decidable whether X is the maximal solution for this language equation.*

Proof. Since REG is closed under repeat-deletion and complement, the maximum solution of $X^\heartsuit = L$ given in Corollary 3, $((L^c)^\spadesuit)^c$, is regular. Since the equivalence problem for regular languages is decidable, it is decidable whether a given solution to the duplication language equation is maximal. $\qquad\square$

Due to the fact that the family of regular languages is not closed under duplication, we cannot obtain a similar decidability result for the repeat-deletion language equation, $X^{\spadesuit} = L$. This motivates our investigation in the next section of a necessary and sufficient condition for the duplication of a regular language to be regular.

5   Controlled Duplication

In Section 4 we showed that for a given language $L \subseteq \Sigma^*$, the maximal solution of the repeat-deletion language equation $X^{\spadesuit} = L$ is given by $((L^c)^{\heartsuit})^c$. However, unlike the duplication language equation, we do not have an efficient algorithm to compute this language due to the fact that the family of regular languages is not closed under duplication. This motivates "controlling" the duplication in such a manner that duplications can occur only for some specific words. In this section, we first introduce a *controlled duplication*, together with some of its basic properties. Then we propose a possible way of characterizing regular languages whose duplication can be controlled so as to generate regular languages, and give partial answers in several particular cases.

For languages $L, C \subseteq \Sigma^*$, we define the duplication of L using the control set C as follows:

$$L^{\heartsuit(C)} = \{xyyz \mid xyz \in L, y \in C\}.$$

Note that this "controlled" duplication operation can express two variants of duplication that appear in previous literature ([8], [9]), namely uniform and length-bounded duplication. Indeed, using the notations in [8], we have $D^1_{\{n\}}(L) = L^{\heartsuit(\Sigma^n)}$ and $D^1_{\{0,1,\dots,n\}}(L) = L^{\heartsuit(\Sigma^{\leq n})}$.

The following two lemmata are basic properties of *controlled duplication*.

Lemma 5. *Let* $L, C_1, C_2 \subseteq \Sigma^*$. *If* $C_1 \subseteq C_2$, *then* $L^{\heartsuit(C_1)} \subseteq L^{\heartsuit(C_2)}$.

Lemma 6. *Let* $L, C_1, C_2 \subseteq \Sigma^*$. *Then* $L^{\heartsuit(C_1 \cup C_2)} = L^{\heartsuit(C_1)} \cup L^{\heartsuit(C_2)}$.

Let L be a language and C be a control set. We say that a word $w \in C$ is *useful with respect to* L if $w \in F(L)$; otherwise, it is called *useless with respect to* L. The control set C is said to *contain an infinite number of useful words with respect to* L if $|F(L) \cap C| = \infty$.

Lemma 7. *Let* $L \subseteq \Sigma^*$ *be a language,* $C \subseteq \Sigma^*$ *be a control set, and* C' *be the set of all useless words in* C *with respect to* L. *Then* $L^{\heartsuit(C)} = L^{\heartsuit(C \setminus C')}$.

Proposition 11. *For a regular language* $L \subseteq \Sigma^*$ *and a regular control set* $C \subseteq \Sigma^*$, *it is decidable whether* C *contains an infinite number of useful words with respect to* L.

For a regular language L and a control set C, we now investigate a necessary and sufficient condition for $L^{\heartsuit(C)}$ to be regular. A sufficient condition is a corollary of the following result in [2]. A family of languages is called a *trio* if it is closed under λ-free homomorphism, inverse homomorphisms, and intersections with regular languages. Note that both the families of regular languages and of context-free languages are trio.

Theorem 1 ([2]). *Any trio is closed under duplication with a finite control set.*

Corollary 4. *Let $L \subseteq \Sigma^*$ be a regular language and $C \subseteq \Sigma^*$. If there exists a finite control set $C' \subseteq \Sigma^*$ such that $L^{\heartsuit(C)} = L^{\heartsuit(C')}$, then $L^{\heartsuit(C)}$ is regular.*

Results in [15] that state that infinite repetitive languages cannot be even context-free indicate that the converse of Corollary 4 may also be true. Hence, in the remainder of this section we shall investigate the following claim:

Claim. Let $L \subseteq \Sigma^*$ be a regular language and $C \subseteq \Sigma^*$ be a control set. If $L^{\heartsuit(C)}$ is regular then there exist a finite control set $C' \subseteq \Sigma^*$ such that $L^{\heartsuit(C)} = L^{\heartsuit(C')}$.

As shown in the following example, this claim generally does not hold.

Example 2. Let $\Sigma = \{a, b\}$, $L = ba^+b$, and $C = ba^+ \cup a^+b$. We can duplicate a prefix ba^i of a word $ba^jb \in L$ ($i \leq j$) to obtain a word $ba^iba^jb \in L^{\heartsuit(C)}$. In the same way, the duplication of a suffix $a^\ell b$ of a word ba^kb ($k \geq \ell$) results in a word $ba^kba^\ell b \in L^{\heartsuit(C)}$. Thus $L^{\heartsuit(C)} = ba^+ba^+b$. Note that L and $L^{\heartsuit(C)}$ are regular. However there exists no finite control set C' satisfying $L^{\heartsuit(C)} = L^{\heartsuit(C')}$. This is because ba^+ba^+b can have arbitrary long repetitions of a's, and hence arbitrary long control factors are required to generate it.

Nevertheless this claim holds for several interesting cases: the case where L is finite or C contains at most a finite number of useful words with respect to L, the case of a unary alphabet $\Sigma = \{a\}$, the case $L = \Sigma^*$, and the case where the control set is "marked", i.e. there exists $a \in \Sigma$ such that $C \subseteq a(\Sigma \setminus \{a\})^*a$. In the following, we prove the direct implication of the claim for these cases (the reverse one is clear from Corollary 4).

The first case we consider is when L is finite. Then $L^{\heartsuit(C)}$ is finite and hence regular. Since $F(L)$ is finite, by letting $C' = C \cap F(L)$, $L^{\heartsuit(C)} = L^{\heartsuit(C')}$. Thus the claim holds in this case. Moreover, even for an infinite L, we can reach the same conclusion if C contains at most a finite number of useful words with respect to L because C', defined as above, is finite.

Next, we consider the case of a unary alphabet. We omit the proof that is mainly based on number theory arguments.

Proposition 12. *Let $\Sigma = \{a\}$ be a unary alphabet, $L \subseteq \Sigma^*$ be a regular language, and $C \subseteq \Sigma^*$ be an arbitrary language. Then $L^{\heartsuit(C)}$ is regular, and there exists a finite control set $C' \in FIN$ such that $L^{\heartsuit(C)} = L^{\heartsuit(C')}$.*

By letting $C = \Sigma^*$, Proposition 12 implies that the family of regular languages is closed under duplication when Σ is unary.

Thirdly we prove that the claim holds for the case when $L = \Sigma^*$ (Corollary 5). This requires the following known two lemmata.

Lemma 8 ([10]). *For a primitive word p, any conjugate word of p is primitive.*

Lemma 9 ([11]). *Let p and q be primitive words with $p \neq q$ and let $i, j \geq 2$. Then $p^i q^j$ is primitive.*

For a language $C \subseteq \Sigma^*$, we define $\mathrm{Dup}(C) = \{ww \mid w \in C\}$.

Proposition 13. *Let $C \subseteq \Sigma^*$. Then $\Sigma^*\mathrm{Dup}(C)\Sigma^*$ is regular if and only if there exists a finite language C' such that $\Sigma^*\mathrm{Dup}(C')\Sigma^* = \Sigma^*\mathrm{Dup}(C)\Sigma^*$.*

Proof. The proof of 'if'-part is obvious since $\Sigma^*\mathrm{Dup}(C')\Sigma^*$ is regular. Now consider the proof of 'only if'-part. Assume $L = \Sigma^*\mathrm{Dup}(C)\Sigma^*$ is regular and consider the regular language $L \cap (\Sigma^* \setminus L\Sigma^+) \cap (\Sigma^* \setminus \Sigma^+ L)$. All words in this language have a representation ww for some $w \in C$. Hence there exists $C' \subseteq C$ such that $\mathrm{Dup}(C') = L \cap (\Sigma^* \setminus L\Sigma^+) \cap (\Sigma^* \setminus \Sigma^+ L)$. Notice that for any $w \in C$ there exist $w' \in C'$ and $x, y \in \Sigma^*$ such that $ww = xw'w'y$. Therefore, $\Sigma^*\mathrm{Dup}(C)\Sigma^* = \Sigma^*\mathrm{Dup}(C')\Sigma^*$.

Suppose C' is infinite. Then there exists a word $uu \in \mathrm{Dup}(C')$ with length twice that of the pumping lemma constant for $\mathrm{Dup}(C')$. So by the pumping lemma, there exists a decomposition $uu = u_1 u_2 u_3 u_1 u_2 u_3$, of uu such that $u_1, u_3 \in \Sigma^*$, $u_2 \in \Sigma^+$ and $u_1 u_2^i u_3 u_1 u_2 u_3 \in \mathrm{Dup}(C')$ for any $i \in \mathbb{N}$. Notice that for any $i \in \mathbb{N}$, $u_1 u_2^i u_3 u_1 u_2 u_3$ is not primitive because it is in $\mathrm{Dup}(C')$. Consider the case $i \geq 3$. By Lemma 8, $u_2^{i-1}(u_2 u_3 u_1)^2$ is not primitive. Then Lemma 9 implies that u_2 and $u_2 u_3 u_1$ share a primitive root, say $p \in \Sigma^+$. We may now write $u_2 = p^n$ and $u_2 u_3 u_1 = p^m$ for some $n, m \geq 1$. Hence $u_2^{i-1}(u_2 u_3 u_1)^2 = p^{n(i-1)+2m}$. From Lemma 8, it follows that $u_1 u_2^i u_3 u_1 u_2 u_3 = q^{n(i-1)+2m}$, where q is a conjugate word of p. Now we have that $u_1 u_2^i u_3 u_1 u_2 u_3 = q^{n(i-1)+2m}$ is a proper prefix (and suffix) of $u_1 u_2^{i+1} u_3 u_1 u_2 u_3 = q^{ni+2m}$, which contradicts the definition of $\mathrm{Dup}(C')$. Thus C' must be finite. □

Lemma 10. *Let $C \subseteq \Sigma^*$. Then $(\Sigma^*)^{\heartsuit(C)} = \Sigma^*\mathrm{Dup}(C)\Sigma^*$.*

Proof. Let $w \in (\Sigma^*)^{\heartsuit(C)}$. Then there exist $x, y, z \in \Sigma^*$ such that $y \in C$ and $w = xyyz$. Thus, $w \in \Sigma^*\mathrm{Dup}(C)\Sigma^*$. Conversely, let $v \in \Sigma^*\mathrm{Dup}(C)\Sigma^*$. Then v is of the form $xyyz$ such that $x, z \in \Sigma^*$ and $yy \in \mathrm{Dup}(C)$ (so, $y \in C$). The duplication of y in $xyz \in \Sigma^*$ results in $xyyz = v$, and hence $v \in (\Sigma^*)^{\heartsuit(C)}$. □

The following corollary is a consequence of Proposition 13 and Lemma 10. In fact, this corollary asserts the claim in the case when $L = \Sigma^*$.

Corollary 5. *Let $C \subseteq \Sigma^*$. Then $(\Sigma^*)^{\heartsuit(C)}$ is regular if and only if there exists a finite subset $C' \subseteq C$ such that $(\Sigma^*)^{\heartsuit(C')} = (\Sigma^*)^{\heartsuit(C)}$.*

The last case we consider is that of marked duplication, where given a word w in $L^{\heartsuit(C)}$, we can deduce or at least guess the factor whose duplication generates w from a word in L according to some mark of a control set C. Here we consider a mark which shows the beginning and end of a word in C, that is, $C \subseteq \#(\Sigma \setminus \{\#\})^*\#$ for some character $\#$. For a strongly-marked duplication, where $\# \notin \Sigma$ and $L \subseteq \Sigma^*\#\Sigma^*\#\Sigma^*$, we can easily show that the existence of a finite control set provided $L^{\heartsuit(C)}$ is regular using the pumping lemma for the regular language. Hence we consider the case when the mark itself is a character in Σ, say $\# = a$ for some $a \in \Sigma$.

We introduce several needed notions related to controlled duplication. Let $L \subseteq \Sigma^*$ be a language and $C \subseteq \Sigma^*$ be a control set. For a word $w \in L^{\heartsuit(C)}$, we call a tuple (x, y, z) a *dup-factorization of w with respect to L and C* if $w = xyyz$, $xyz \in L$, and $y \in C$. When L and C are clear from the context, we simply say that (x, y, z) is a dup-factorization of w. For $y \in C$, if there are $x, z \in \Sigma^*$ such that (x, y, z) is a dup-factorization of w, then we call y a *dup-factor* of w.

Proposition 14. *Let Σ be a finite alphabet of more than one character, $L \subseteq \Sigma^*$ be a regular language, and $C \subseteq a(\Sigma \setminus \{a\})^*a$ for some $a \in \Sigma$. Then $L^{\heartsuit(C)}$ is regular if and only if there exists a finite language C' such that $L^{\heartsuit(C)} = L^{\heartsuit(C')}$.*

Proof. We consider following two syntactic equivalence relations:

$$\equiv_L = \{(u, v) \mid \forall x, y \in \Sigma^*, xuy \in L \Leftrightarrow xvy \in L\},$$
$$\equiv_\heartsuit = \{(u, v) \mid \forall x, y \in \Sigma^*, xuy \in L^{\heartsuit(C)} \Leftrightarrow xvy \in L^{\heartsuit(C)}\},$$

and define $\equiv \, = \, \equiv_L \cap \equiv_\heartsuit$. Since both L and $L^{\heartsuit(C)}$ are regular, $C/\equiv$ is finite. Let $\Gamma_2 = \{[c] \in C/\equiv \, \mid \, |[c]| \leq 2\}$. Using induction on the number of dup-factorizations, we prove that (i) $\Gamma_2 \neq \emptyset$, and (ii) any word in $L^{\heartsuit(C)}$ has a dup-factor which is in an equivalence class in Γ_2.

Firstly, we consider a word w in $L^{\heartsuit(C)}$ which has the smallest number of dup-factorizations among the elements of $L^{\heartsuit(C)}$. Suppose that no dup-factor of w is in equivalence classes in Γ_2. Let (x, aya, z) be a dup-factorization of w for some $x, y, z \in \Sigma^*$. Then there exists $ay'a \in C$ such that $ay'a \equiv aya$, $y' \neq y$, and $ay'a \notin \text{Suff}(x)$. Let $w' = xay'aayaz$. This is in $L^{\heartsuit(C)}$, and hence w' must have a dup-factorization, say $(\alpha, a\beta a, \gamma)$ for some $\alpha, \beta, \gamma \in \Sigma^*$. Due to the fact that y', y, β do not contain any a, $(a\beta a)^2$ is either (1) a factor of x, (2) a factor of z, or (3) $\beta = y$ and $a\beta a \in \text{Pref}(z)$. Here we consider only the case (1), and let $x = \alpha(a\beta a)^2\gamma'$, $\gamma = \gamma'ay'aayaz$. Then $w' = \alpha(a\beta a)^2\gamma' \in L^{\heartsuit(C)} \Rightarrow \alpha a\beta a\gamma'ay'aayaz \in L \Rightarrow \alpha a\beta a\gamma'(aya)^2z \in L \Rightarrow \alpha(a\beta a)^2\gamma'(aya)^2z = w \in L^{\heartsuit(C)}$, and hence $\alpha, a\beta a, \gamma'(aya)^2z)$ is a dup-factorization of w. This means that a dup-factorization $(\alpha, a\beta a, \gamma)$ of w' induces a dup-factorization $(\alpha_0, a\beta a, \gamma_0)$ of w, where a single occurrence of y' in either α or γ is replaced by y to obtain α_0 and γ_0. The original dup-factorization (x, aya, z) of w cannot be obtained this way. Hence w' has a smaller number of dup-factorizations than w, a contradiction. Thus w has a dup-factor which is in an equivalence class in Γ_2, and hence $\Gamma_2 \neq \emptyset$.

Now we assume that all words in $L^{\heartsuit(C)}$ with at most n dup-factorizations have a dup-factor which is in an equivalence class in Γ_2. Suppose that there were $v \in L^{\heartsuit(C)}$ with $n+1$ dup-factorizations and without any dup-factor which is in the equivalence class of size at most 2. Then we can construct a word v' as above which has at most n dup-factorizations but does not satisfy the assumption, which is a contradiction. □

Note that the property of a control set required in this proof is that none of its elements "overlap" with each other. That is, we can use a similar proof to settle a more general case where a control set is non-overlapping and an infix code. (See [18] for definitions.)

Corollary 6. *Let L be a regular language and C be a control set such that $L^{\heartsuit(C)}$ is regular. If C is non-overlapping and an infix code, then there exists a finite control set C' such that $L^{\heartsuit(C)} = L^{\heartsuit(C')}$.*

Moreover, the proof of Proposition 14 shows that if we let $m = |C/ \equiv |$, the size of finite control set C' given there is at most $2|\Gamma_2|$, which is not bigger than $2(m-1)$ because at least one equivalence class in $C/\equiv$ must have infinite cardinality. Finally, we provide a result slightly stronger than Corollary 6.

Corollary 7. *Let L be a regular language and C be a control set. If there exists a finite set $C_1 \subset C$ such that $C \setminus C_1$ is non-overlapping and an infix code, then the regularity of $L^{\heartsuit(C)}$ implies the existence of a finite control set C' such that $L^{\heartsuit(C)} = L^{\heartsuit(C')}$.*

6 Duplication and Primitivity

There is evidently a connection between duplication, repeat-deletion, and primitive words, but the nature of this relationship is unclear. This section elucidates some of the properties of this relationship.

Proposition 15 (see, for instance, [14]). *Let $u, v \in \Sigma^+$ such that uv is primitive. Then both $u(uv)^n$ and $v(uv)^n$ are primitive for any $n \geq 2$.*

Proposition 16. *Let $w \in \Sigma^*$ be a non-primitive word. If we duplicate a factor of w which is properly shorter than the primitive root of w, then the resulting word is primitive.*

We can derive the following proposition from Lemma 9.

Proposition 17. *Let $x, y, z \in \Sigma^*$. If xyz is primitive and $xyyz$ is not primitive, then xz is primitive.*

7 Discussion and Future Work

In this paper, we studied duplication and repeat-deletion, two formal language theoretic models of insertion and deletion errors occurring during DNA replication. Specifically, we obtained the closure properties of the families of languages in the Chomsky hierarchy under these operations, the language equations of the form $X^{\heartsuit} = L$ and $X^{\spadesuit} = L$ for a given language L, and the operation of controlled duplication. In addition, we made steps towards finding a necessary and sufficient condition for a controlled duplication of a regular language to be regular.

Two problems for further investigation are: the problem of how to decide for a given language L whether the language equation $X^{\heartsuit} = L$ has a solution, and the problem of finding a necessary condition for the controlled duplication of a regular language to be regular in the general case.

Acknowledgements

We wish to express our gratitude to Dr. Zoltán Ésik for the concise proof of Proposition 4. We would also like to thank Dr. Helmut Jürgensen for discussions about the claim and Dr. Kathleen Hill for extended discussions on the biological motivation for duplication and repeat-deletion.

References

1. Dassow, J., Mitrana, V., Păun, G.: On the regularity of duplication closure. Bull. EATCS 69, 133–136 (1999)
2. Dassow, J., Mitrana, V., Salomaa, A.: Operations and language generating devices suggested by the genome evolution. Theoretical Computer Science 270, 701–738 (2002)
3. Garcia-Diaz, M., Kunkel, T.A.: Mechanism of a genetic glissando: structural biology of indel mutations. Trends in Biochemical Sciences 31(4), 206–214 (2006)
4. Ito, M.: Algebraic Theory of Automata and Languages. World Scientific Pub. Co. Inc., Singapore (2004)
5. Ito, M., Leupold, P., S-Tsuji, K.: Closure of language classes under bounded duplication. In: Ibarra, O.H., Dang, Z. (eds.) DLT 2006. LNCS, vol. 4036, pp. 238–247. Springer, Heidelberg (2006)
6. Leupold, P.: Duplication roots. In: Harju, T., Karhumäki, J., Lepistö, A. (eds.) DLT 2007. LNCS, vol. 4588, pp. 290–299. Springer, Heidelberg (2007)
7. Leupold, P.: Languages generated by iterated idempotencies and the special case of duplication. Ph.D. thesis, Department de Filologies Romaniques, Facultat de Lletres, Universitat Rovira i Virgili, Tarragona, Spain (2006)
8. Leupold, P., Mitrana, V., Sempere, J.: Formal languages arising from gene repeated duplication. In: Jonoska, N., Păun, G., Rozenberg, G. (eds.) Aspects of Molecular Computing. LNCS, vol. 2950, pp. 297–308. Springer, Heidelberg (2003)
9. Leupold, P., M-Vide, C., Mitrana, V.: Uniformly bounded duplication languages. Discrete Applied Mathematics 146(3), 301–310 (2005)
10. Lothaire, M.: Combinatorics on Words, Encyclopedia of Mathematics and its Applications 17. Addison-Wesley Publishing Co., Reading (1983)
11. Lyndon, R.C., Schützenberger, M.P.: On the equation $a^M = b^N c^P$ in a free group. Michigan Mathematical Journal 9, 289–298 (1962)
12. M-Vide, C., Păun, G.: Duplication grammars. Acta Cybernetica 14, 151–164 (1999)
13. Mitrana, V., Rozenberg, G.: Some properties of duplication grammars. Acta Cybernetica 14, 165–177 (1999)
14. Reis, C.M., Shyr, H.J.: Some properties of disjunctive languages on a free monoid. Information and Control 37, 334–344 (1978)
15. Ross, R., Winklmann, K.: Repetitive strings are not context-free. R.A.I.R.O informatique théorique / Theoretical Informatics 16(3), 191–199 (1982)
16. Rozenberg, G., Salomaa, A. (eds.): Handbook of Formal Languages. Springer, Heidelberg (1997)
17. Searls, D.B.: The computational linguistics of biological sequences. In: Hunter, L. (ed.) Artificial Intelligence and Molecular Biology, pp. 47–120. AAAI Press, The MIT Press (1993)
18. Yu, S.S.: Languages and Codes. Lecture Notes, Department of Computer Science, p. 402. National Chung-Hsing University, Taichung (2005)

On the State Complexity of Complements, Stars, and Reversals of Regular Languages*

Galina Jirásková

Mathematical Institute, Slovak Academy of Sciences, Grešákova 6, 040 01 Košice, Slovakia
jiraskov@saske.sk

Abstract. We examine the deterministic and nondeterministic state complexity of complements, stars, and reversals of regular languages. Our results are as follows:

1. The nondeterministic state complexity of the complement of an n-state NFA language over a five-letter alphabet may reach each value in the range from $\log n$ to 2^n.
2. The state complexity of the star (reversal) of an n-state DFA language over a growing alphabet may reach each value in the range from 1 to $\frac{3}{4}2^n$ (from $\log n$ to 2^n, respectively).
3. The nondeterministic state complexity of the star (reversal) of an n-state NFA binary language may reach each value in the range from 1 to $n + 1$ (from $n - 1$ to $n + 1$, respectively).

We also obtain some partial results on the nondeterministic state complexity of the complements of binary regular languages. As a bonus, we get an exponential number of values that are non-magic, which improves a similar result of Geffert (*Proc. 7th DCFS*, Como, Italy, 23–37).

1 Introduction

Regular languages and finite automata are among the oldest and simplest topics in formal language theory. They have been intensively studied since the forties. Nevertheless, some important problems are still open. The most famous is the question of how many states are sufficient and necessary for two-way deterministic finite automata to simulate two-way nondeterministic finite automata [1,17].

Recently, there have been a new interest in automata theory; for a discussion, we refer to [10,20]. Many researchers have investigated various problems concerning descriptional complexity which studies the costs of description of languages by different formal systems. Here we focus on the deterministic and nondeterministic state complexity of complements, stars, and reversals of regular languages.

In 1997, at the 3rd Conference on Developments in Language Theory, Iwama at al. [11] stated the question of whether there always exists a minimal nondeterministic finite automaton (NFA) of n states whose equivalent minimal deterministic finite automaton (DFA) has exactly α states for all integers n and α

* Research supported by the VEGA grant 2/6089/26.

M. Ito and M. Toyama (Eds.): DLT 2008, LNCS 5257, pp. 431–442, 2008.

satisfying that $n \leqslant \alpha \leqslant 2^n$. The question has also been considered in [12], where an integer Z with $n < Z < 2^n$ is called a "magic number" if no DFA of Z states can be simulated by any NFA of n states. In [13] it has been shown that there are no magic numbers, that is, appropriate automata have been described for all integers n and α. However, the constructions have used a growing alphabet of size $2^{n-1} + 1$. Later, in [5], the size of the alphabet has been decreased to $n + 2$, and finally, in [16], the result has been proved for a fixed four-letter alphabet. On the other hand, there are a lot of magic numbers in a unary case [6]. The problem remains open for binary and ternary alphabets.

A similar question for complements of regular languages has been examined in [15]. Using a growing alphabet of size 2^{n+1} it has been proved that all values in the range from $\log n$ to 2^n can be obtained as the nondeterministic state complexity of an n-state NFA language. Here we improve this result by showing that it still holds for a fixed five-letter alphabet. We also consider a binary case, and, as a bonus, we get an exponential number of so called "non-magic" values.

We next investigate the deterministic and nondeterministic state complexity of stars and reversals of regular languages. In all cases, we show that the whole range of complexities up to the known upper bounds can be obtain. To prove the results on state complexity we use growing alphabets. In the nondeterministic case, a binary alphabet is enough to describe appropriate automata.

To conclude this section let us mention some other related works. Magic numbers for symmetric difference NFAs have been studied by Zijl [22]. In [9], it has been shown that the deterministic and nondeterministic state complexity of union and intersection of regular languages may reach each value from 1 up to the upper bounds mn or $m + n + 1$. Similar results for the nonterminal complexity of some operations on context-free languages have been recently obtained by Dassow and Stiebe [4].

2   Preliminaries

In this section, we give some basic definitions, notations, and preliminary results used throughout the paper. For further details, we refer to [18,19].

Let Σ be a finite alphabet and Σ^* the set of all strings over the alphabet Σ including the empty string ε. The length of a string w is denoted by $|w|$. A language is any subset of Σ^*. The complement of a language L is denoted by L^c, its star by L^*, and it reversal by L^R. We denote the cardinality of a finite set A by $|A|$ and its power-set by 2^A.

A *deterministic finite automaton* (DFA) is a 5-tuple $M = (Q, \Sigma, \delta, q_0, F)$, where Q is a finite set of states, Σ is a finite input alphabet, δ is the transition function that maps $Q \times \Sigma$ to Q, q_0 is the initial state, $q_0 \in Q$, and F is the set of accepting states, $F \subseteq Q$. In this paper, all DFAs are assumed to be complete, that is, the next state $\delta(q, a)$ is defined for each state q in Q and each symbol a in Σ. The transition function δ is extended to a function from $Q \times \Sigma^*$ to Q in a natural way. A string w in Σ^* is accepted by the DFA M if the state $\delta(q_0, w)$ is an accepting state of the DFA M. The language accepted by the DFA M, denoted $L(M)$, is the set of strings $\{w \in \Sigma^* \mid \delta(q_0, w) \in F\}$.

A *nondeterministic finite automaton* (NFA) is a 5-tuple $M = (Q, \Sigma, \delta, q_0, F)$, where Q, Σ, q_0 and F are defined in the same way as for a DFA, and δ is the nondeterministic transition function that maps $Q \times \Sigma$ to 2^Q. The transition function can be naturally extended to the domain $Q \times \Sigma^*$. A string w in Σ^* is accepted by the NFA M if the set $\delta(q_0, w)$ contains an accepting state of the NFA M. The *language accepted by* the NFA M is the set of strings $L(M) = \{w \in \Sigma^* \mid \delta(q_0, w) \cap F \neq \varnothing\}$.

Two automata are said to be *equivalent* if they accept the same language. A DFA (an NFA) M is called *minimal* if all DFAs (all NFAs, respectively) that are equivalent to M have at least as many states as M. It is well-known that a DFA $M = (Q, \Sigma, \delta, q_0, F)$ is minimal if *(i)* all its states are reachable from the initial state, and *(ii)* no two its different states are equivalent (states p and q are said to be equivalent if for all strings w in Σ^*, the state $\delta(p, w)$ is accepting iff the state $\delta(q, w)$ is accepting). Each regular language has a unique minimal DFA, up to isomorphism. However, the same result does not hold for NFAs.

The *(deterministic) state complexity* of a regular language is the number of states in its minimal DFA. The *nondeterministic state complexity* of a regular language is defined as the number of states in a minimal NFA accepting this language. A regular language with deterministic (nondeterministic) state complexity n is called an n-state DFA language (an n-state NFA language, respectively).

Every nondeterministic finite automaton $M = (Q, \Sigma, \delta, q_0, F)$ can be converted to an equivalent deterministic finite automaton $M' = (2^Q, \Sigma, \delta', q_0', F')$ using an algorithm known as the "subset construction" in the following way. Every state of the DFA M' is a subset of the state set Q. The initial state of the DFA M' is the set $\{q_0\}$. The transition function δ' is defined by $\delta'(R, a) = \bigcup_{r \in R} \delta(r, a)$ for each state R in 2^Q and each symbol a in Σ. A state R in 2^Q is an accepting state of the DFA M' if it contains at least one accepting state of the NFA M. The DFA M' need not be minimal since some states may be unreachable or equivalent. Sometimes, also NFAs with a set of initial states are considered. In such a case, the subset construction starts with this set being the initial state of an equivalent DFA.

To prove that an NFA is minimal we use a fooling-set lower-bound technique [2,3,7]. After defining a fooling set, we recall the lemma from [2] describing this lower-bound technique.

Definition 1. *A set of pairs of strings $\{(x_i, y_i) \mid i = 1, 2, \ldots, n\}$ is said to be a fooling set for a regular language L if for every i and j in $\{1, 2, \ldots, n\}$,*
(1) the string $x_i y_i$ is in the language L, and
(2) if $i \neq j$, then at least one of the strings $x_i y_j$ and $x_j y_i$ is not in L.

Lemma 1 (Birget [2]). *Let a set of pairs of strings $\{(x_i, y_i) \mid i = 1, 2, \ldots, n\}$ be a fooling set for a regular language L. Then every NFA for the language L needs at least n states.* □

3   Complements

We start with the complements of regular languages. In the deterministic case, there is not much to say. The state complexity of a language and its complement is the same since to get a DFA for the complement we can simply exchange the accepting and the rejecting states in a DFA for the given language. The nondeterministic case is completely different. Given an n-state NFA we can apply the subset construction, and then exchange the accepting and the rejecting states, which gives an upper bound 2^n on the size of an NFA for the complement. This upper bound is known to be tight [17,3], and can be reached by the complement of a binary regular language [14].

Here we deal with the question of what values can be reached as the size of a minimal NFA accepting the complement of an n-state NFA language. In [15] it has been shown that all values from $\log n$ to 2^n can be reached, however, appropriate automata have been defined over a growing alphabet of size 2^{n+1}. In this section, we prove that this result still holds for a fixed five-letter alphabet. For each α with $\log n \leqslant \alpha \leqslant 2^n$, we describe a minimal n-state NFA M with a five-letter input alphabet such that every minimal NFA for the complement of the language $L(M)$ has exactly α states. In the second part of this section, we study a binary case, and show that here the whole range of complexities from $3 \log n$ to $n + 2^{n/3}$ can be obtained. As a bonus, we get an exponential number of so called non-magic values, which improves a similar result of Geffert [5].

The first two lemmata solve special cases of $\alpha = n$ and $\alpha = 2^n$. The next one has been recently proved in [16].

Lemma 2 ([15]). *For every $n \geqslant 1$, there exists a minimal binary NFA M of n states such that every minimal NFA for the complement of the language $L(M)$ has n states.*   □

Lemma 3 ([14]). *For every $n \geqslant 1$, there exists a minimal binary NFA M of n states such that every minimal NFA for the complement of the language $L(M)$ has 2^n states.*   □

Lemma 4 ([16], Theorem 1). *For all integers n and α with $n < \alpha < 2^n$, there exists a minimal NFA of n states with a four-letter input alphabet whose equivalent minimal DFA has exactly α states.*   □

We use the automata from the lemma above to prove the next result which shows that the nondeterministic state complexity of the complement of an n-state NFA language over a five-letter alphabet may reach an arbitrary value from $n + 1$ to $2^n - 1$.

Lemma 5. *For all integers n and α with $n < \alpha < 2^n$, there exists a minimal NFA M of n states with a five-letter input alphabet such that every minimal NFA for the complement of the language $L(M)$ has α states.*

Proof. Let $n < \alpha < 2^n$. Then there is an integer k such that $1 \leqslant k \leqslant n-1$ and $n - k + 2^k \leqslant \alpha < n - (k+1) + 2^{k+1}$. It follows that $\alpha = n - (k+1) + 2^k + m$, where m is an integer such that $1 \leqslant m < 2^k$.

Let $C = C_{n,k,m} = (Q, \{a, b, c, d\}, \delta_C, q_0, \{k\})$, where $Q = \{0, 1, \ldots, n-1\}$, be the n-state NFA from Lemma 4 whose minimal DFA has α states.

Now, let $M = M_{n,k,m} = (Q, \{a, b, c, d, f\}, \delta, q_0, \{k\})$ be an n-state NFA obtained from the NFA C by adding transitions on a new symbol f so that by f, state i with $0 \leqslant i \leqslant k-1$ goes to $\{i+1\}$, state k goes to $\{0, 1, \ldots, k\}$, and each other state goes to the empty set.

Let M' be the DFA obtained from the NFA M by the subset construction. It can be shown that the DFA M' has α reachable states. After exchanging the accepting and the rejecting states we get a DFA of the same number of states for the language $L(M)^c$. To prove the lemma it is sufficient to show that every NFA for the language $L(M)^c$ needs at least α states. This can be shown by describing a fooling set for the language $L(M)^c$ of size α. $\square$

As a corollary of Lemmata 2, 3, and 5, and taking into account that $(L^c)^c = L$, we get the following result.

Theorem 1. *For all integers n and α with $\log n \leqslant \alpha \leqslant 2^n$, there exists a minimal nondeterministic finite automaton M of n states with a five-letter input alphabet such that every minimal nondeterministic finite automaton for the complement of the language $L(M)$ has exactly α states.* $\square$

The second part of this section is devoted to the nondeterministic state complexity of the complements of binary regular languages. The first lemma deals with values from $n+4$ up to $2^{\lfloor n/3 \rfloor} - 1$, the second one covers the remaining cases.

Lemma 6. *For all integers n and α with $n + 4 \leqslant \alpha < n + 2^{\lfloor n/3 \rfloor}$, there exists a minimal binary NFA M of n states such that every minimal NFA for the complement of the language $L(M)$ has α states.*

Proof. Let $n+4 \leqslant \alpha < n+2^{\lfloor n/3 \rfloor}$ and let $k = \lfloor n/3 \rfloor$. Then α can be expressed as $\alpha = n + \sum_{i=0}^{k-1} c_i \cdot 2^i$, where $c_i \in \{0, 1\}$ for $i = 0, 1, \ldots, k-1$. Denote by $m = \max\{i \mid c_i = 1\}$ and $\ell = |\{i > 0 \mid c_i = 1\}|$. Since $\alpha \geqslant n + 4$, we have $m \geqslant 2$.

Define an n-state NFA $M = (Q, \{a, b\}, \delta, p_1, \{1\})$, where $Q = \{p_1, p_2, \ldots, p_k\} \cup \{s_1, s_2, \ldots, s_k\} \cup \{1, 2, \ldots, n - 2k\}$, and δ is defined as follows (see Fig. 1). If $1 \leqslant i < k$ and $c_i = 0$, then $\delta(p_i, a) = \{s_i\}$, $\delta(s_i, a) = \{p_{i+1}\}$, and $\delta(p_i, b) = \delta(s_i, b) = \varnothing$. If $1 \leqslant i < k$ and $c_i = 1$, then $\delta(p_i, a) = \{p_{i+1}\}$, $\delta(p_i, b) = \{s_i\}$, $\delta(s_i, a) = \{s_i, i\}$, and $\delta(s_i, b) = \{s_i\}$. Next, $\delta(p_k, a) = \{s_k\}$, $\delta(p_k, b) = \{\ell, \ell - 1\}$, $\delta(s_k, a) = \{n - 2k\}$, and $\delta(s_k, b) = \varnothing$ if $c_0 = 0$ and $\delta(s_k, b) = \{\ell + 1, \ell\}$ if $c_0 = 1$. Finally, $\delta(q, a) = \delta(q, b) = \{q - 1\}$ if $2 \leqslant q \leqslant n - 2k$, $\delta(1, a) = \{s_m\} \cup \{1, 2, \ldots, m + 1\}$, and $\delta(1, b) = \varnothing$.

Notice that there is a chain of a's going from state p_1 to state 1, which goes through all p_i's, those s_i's with $c_i = 0$, and states $n - 2k, n - 2k - 1, \ldots, 2, 1$. The length of this chain is $n - 1 - \ell$, i.e., the string $a^{n-1-\ell}$ is in $L(M)$. Next, for all i with $c_i = 1$, all strings with an a in the i-th position from the end are accepted by M from state s_i, and no string in b^* is accepted from s_i.

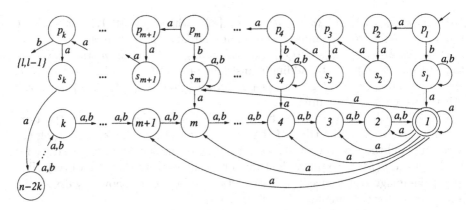

Fig. 1. The nondeterministic finite automaton M

It can be shown that the NFA M is minimal, the DFA M' obtained from the NFA M by the subset construction has α reachable states, and there is a fooling set for the language $L(M)^c$ of size α.   □

Lemma 7. *For all integers n and α with $n+1 \leqslant \alpha \leqslant 2n$, there exists a minimal binary NFA M of n states such that every minimal NFA for the complement of the language $L(M)$ has α states.*   □

By Lemmata 2, 6, 7, and the fact that $(L^c)^c = L$, we have the next result.

Theorem 2. *For all integers n and α with $3 \log n \leqslant \alpha < n + 2^{\lfloor n/3 \rfloor}$, there exists a minimal binary NFA M of n states such that every minimal NFA for the complement of the language $L(M)$ has exactly α states.*   □

As a corollary, we get an exponential number of non-magic values in a binary case, which improves the current number $2^{\Omega(n^{1/3} \ln^{2/3} n)}$ obtained by Geffert [5] using binary bounded languages.

Corollary 1. *For every $n \geqslant 1$, all values from n to $n + 2^{\lfloor n/3 \rfloor}$ are non-magic in a binary case, that is, for each integer α with $n \leqslant \alpha < n + 2^{\lfloor n/3 \rfloor}$, there exists a minimal binary NFA of n states whose equivalent minimal DFA has α states.*

Proof. Consider a binary NFA M described in Lemmata 2, 6, 7, for a given α. The DFA obtained from this NFA by the subset construction has α reachable sets. These sets must be inequivalent because otherwise we would have a smaller DFA for the language $L(M)$, and so, also a smaller DFA for the language $L(M)^c$. However, every NFA for the language $L(M)$ needs at least α states, a contradiction. Thus the minimal DFA for the language $L(M)$ has α states as desired.   □

4   Stars

This section deals with the deterministic and nondeterministic state complexity of stars of regular languages.

The upper bound on the state complexity of star operation is known to be $\frac{3}{4}2^n$ [21]. In the first part of this section, we show that each value from 1 to this upper bound can be reached as the state complexity of the star of an n-state DFA language. With an upper bound $n + 1$, we prove a similar result for the nondeterministic state complexity of stars in the second part of this section. To get the result in the deterministic case we use a growing alphabet. In the nondeterministic case, a binary alphabet is enough to describe appropriate automata.

Let us start with recalling binary languages that reach the upper bound on the state complexity of star operation. Let $k \geqslant 2$ and let A_k be the binary k-state DFA depicted in Fig. 2. The following result has been shown by Yu, Zhuang and Salomaa [21].

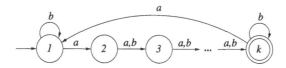

Fig. 2. The deterministic finite automaton A_k

Lemma 8 ([21]). *For every $k \geqslant 2$, the minimal DFA for the language $L(A_k)^*$ has $\frac{3}{4}2^k$ states.* □

Using automata A_k described above we prove the following lemma.

Lemma 9. *For all integers n and k with $2 \leqslant k \leqslant n$, there exists a minimal DFA $B_{n,k}$ of n states with a four-letter input alphabet such that the minimal DFA for the language $L(B_{n,k})^*$ has $n - k + \frac{3}{4}2^k$ states.*

Proof. If $k = n$, then take the DFA A_n from Lemma 8. Let $2 \leqslant k \leqslant n - 1$ and let $\Sigma = \{a, b, c, d\}$.

Let us construct an n-state DFA $B_{n,k}$ with the input alphabet Σ from the k-state DFA A_k by adding new states $k + 1, k + 2, \ldots, n$, which go to itself by a, b, c except for state $k + 1$ which goes to state 1 by a, b, c. Each of the states in $\{1, 2, \ldots, k\}$ goes to state $k + 1$ by c and to state $k + 2$ by d. By d, state n goes to state 1, and state q with $k + 1 \leqslant q \leqslant n - 1$ to state $q + 1$. The DFA $B_{n,k}$ is shown in Fig. 3 and is minimal since no two of its states are equivalent. If $k = n - 1$, then the DFA $B_{n,k}$ is defined over the alphabet $\{a, b, c\}$.

Construct an NFA B' be for the language $L(B_{n,k})^*$ from the DFA $B_{n,k}$ by adding a new initial (and accepting) state q_0 which goes to state 2 by a, to state 1 by b, to state $k + 1$ by c, and to state $k + 2$ by d. Next, add transitions by a and by b from state $k - 1$ to state 1.

Let B'' be the DFA obtained from the NFA B' by the subset construction. The DFA B'' has $n - k + \frac{3}{4}2^k$ reachable and pairwise inequivalent states, and the lemma follows. □

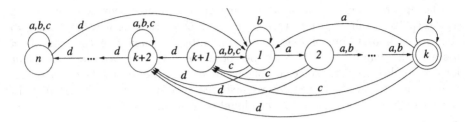

Fig. 3. The deterministic finite automaton $B_{n,k}$

Using automata $B_{n,k}$ we prove the following result showing that the state complexity of the star of an n-state DFA language may be arbitrary from $n+1$ to $\frac{3}{4}2^n$.

Lemma 10. *For all n and α with $n+1 \leqslant \alpha \leqslant \frac{3}{4}2^n$, there is a minimal DFA M of n states such that the minimal DFA for the language $L(M)^*$ has α states.*

Proof. If $\alpha = n - k + \frac{3}{4}2^k$, where $2 \leqslant k \leqslant n$, then take the n-state DFA $B_{n,k}$ from Lemma 9. Otherwise, let k be an integer such that $n - k + \frac{3}{4}2^k < \alpha < n - (k+1) + \frac{3}{4}2^{k+1}$. Then $\alpha = n - k + \frac{3}{4}2^k + m$ for some integer m with $1 \leqslant m \leqslant 2^{k-1} + 2^{k-2} - 2$.

Let $S_1, S_2, \ldots, S_\ell$, where $\ell = 2^{k-1} + 2^{k-2} - 2$, be all subsets of $\{1, 2, \ldots, k-1\}$ and all subsets $\{1, k\} \cup T$ with $T \subseteq \{2, 3, \ldots, k-1\}$, except for the empty-set and the set $\{1, 2, \ldots, k\}$, ordered in such a way that $S_1 = \{1\}$, and the sets of a smaller cardinality precede the sets with a larger cardinality. Now let $S_1, S_2, \ldots, S_m$ be the first m sets in the sequence.

Construct the DFA $M = M_{n,k,m}$ from the DFA $B_{n,k}$ by adding transitions on m new symbols $f_1, f_2, \ldots, f_m$ so that by symbol f_i $(1 \leqslant i \leqslant m)$, each state q in S_i goes to itself, and each state q in $\{1, 2, \ldots, n\} \setminus S_i$ goes to state $k + 1$.

Let M' be an NFA for the language $L(M)^*$ obtained from the DFA M by adding a new initial (and accepting) state q_0 as in Lemma 9. By f_i $(1 \leqslant i \leqslant m)$, state q_0 goes to state 1 if $1 \in S_i$, and to state $k + 1$ if $1 \notin S_i$. If the accepting state k is in S_i, then we add the transition by f_i from state k to state 1.

Let M'' be the DFA obtained from the NFA M' by the subset construction. The DFA M'' has $n - k + \frac{3}{4}2^k + m$ reachable and pairwise inequivalent states, which proves the lemma. □

The next lemma shows that sometimes even less than n states are sufficient to accept the star of an n-state DFA language. To describe appropriate automata it uses unary or binary alphabets.

Lemma 11. *For all integers n and k with and $1 \leqslant k \leqslant n$, there exists a minimal binary DFA M of n states such that the minimal DFA for the language $L(M)^*$ has k states.* □

Let us summarize the above results in the following theorem.

Theorem 3. *For all integers n and α with either $1 = n \leqslant \alpha \leqslant 2$, or $n \geqslant 2$ and $1 \leqslant \alpha \leqslant \frac{3}{4}2^n$, there exists a minimal DFA M of n states with a 2^n-letter input alphabet such that the minimal DFA for the star of the language $L(M)$ has exactly α states.*   □

The upper bound on the nondeterministic state complexity of stars of n-state NFA languages is known to be $n+1$ [8]. The next theorem shows that each value from 1 to $n + 1$ can be reached as the nondeterministic state complexity of the star of an n-state binary NFA language.

Theorem 4. *The nondeterministic state complexity of the star of each 1-state NFA language is 1. If $n \geqslant 2$, then for every k with $1 \leqslant k \leqslant n + 1$, there exists a minimal NFA M of n states with a binary input alphabet such that every minimal NFA for the star of the language $L(M)$ has exactly k states.*   □

5   Reversals

This section studies the deterministic and nondeterministic state complexity of reversals of regular languages.

If a regular language is accepted by an n-state DFA, then an n-state NFA for its reversal can be obtained from this DFA by interchanging the initial and the accepting states, and by reversing all transitions. By applying the subset construction to this NFA, we get a DFA for the reversal of at most 2^n states. Since the reversal of the reversal of a language is the same language, the lower bound on the size of the minimal DFA for the reversal of an n-state DFA language is $\log n$ (whenever $n \geqslant 3$; note that the reversal of an 1-state DFA language is the same language). In this section, we show that each value from $\log n$ to 2^n can be reached as the state complexity of the reversal of an n-state DFA language. In the second part of this section, we deal with the nondeterministic state complexity of reversals.

We start with the following lemma showing that all values from n to $2n$ can be reached as the state complexity of the reversal of an n-state DFA binary language.

Lemma 12. *For all integers n and α with $2 \leqslant n \leqslant \alpha \leqslant 2n$, there exists a minimal binary DFA A of n states such that the minimal DFA for the language $L(A)^R$ has α states.*   □

The next lemma describes an n-state DFA language over an $(n-1)$-letter alphabet such that the state complexity of its reversal is $2n+1$. We use this automaton later in our constructions.

Lemma 13. *Let $n \geqslant 3$ and let $\Sigma = \{a_1, \ldots, a_{n-1}\}$ be an $(n-1)$-letter alphabet. There exists a minimal DFA B of n states with the input alphabet Σ such that the minimal DFA for the language $L(B)^R$ has $2n + 1$ states.*

Proof. Define an n-state DFA $B = (Q, \Sigma, \delta, n, \{1\})$, where $Q = \{1, 2, \ldots, n\}$, and for all $q = 1, 2, \ldots, n$ and all $i = 1, 2, \ldots, n-1$, $\delta(i+1, a_i) = i$ and $\delta(q, a_i) = n$ if $q \neq i+1$, that is, by a_i, state $i+1$ goes to state i and each other state goes to state n. The DFA B is minimal since if $1 \leqslant i < j \leqslant n$, then the string $a_{i-1}a_{i-2} \cdots a_1$ is accepted by the DFA B from state i but not from state j.

Let B' be the NFA for the language $L(B)^R$ obtained from the DFA B by interchanging the accepting and the rejecting state and by reversing all transitions. Let B'' be the DFA obtained from the NFA B' by the subset construction. The DFA B'' has $2n+1$ reachable and pairwise inequivalent states, which proves the lemma. $\square$

The next lemma deals with the case, when α is between $2n+2$ and 2^n, and uses a growing alphabet of size $n + \lfloor \alpha/2 \rfloor$ to describe appropriate automata.

Lemma 14. *For all integers n and α with $n \geqslant 3$ and $2n + 2 \leqslant \alpha \leqslant 2^n$, there exists a minimal DFA C of n states such that the minimal DFA for the language $L(C)^R$ has α states.*

Proof. Let $\alpha = 2n + 1 + m$, where $1 \leqslant m \leqslant 2^n - 2n - 1$.

Let $k = \lfloor m/2 \rfloor$ and let $\Sigma_m = \{a_1, a_2, \ldots, a_{n-1}, b_1, b_2, \ldots, b_k\}$ if m is even, and $\Sigma_m = \{a_1, a_2, \ldots, a_{n-1}, b_1, b_2, \ldots, b_k, c\}$ if m is odd.

Let $Q = \{1, 2, \ldots, n\}$ and $T = \{2, 3, \ldots, n\}$. Now take all subsets of Q with cardinality more than 1, and order them in a sequence

$$S_1, Q \setminus S_1, S_2, Q \setminus S_2, \ldots, S_k, Q \setminus S_k, \ldots, S_{2^{n-1}-n-1}, Q \setminus S_{2^{n-1}-n-1},$$

(that is, each odd set of size at least two is followed by its complement in Q).

Define an n-state DFA $C = (Q, \Sigma_m, \delta, n, \{1\})$, in which for all $i = 1, \ldots, n-1$, the transitions by symbol a_i are the same as in the DFA B described in the proof of Lemma 13. Next, for all $j = 1, 2, \ldots, k$, by symbol b_j, each state in S_j goes to state 1, and each state in $Q \setminus S_j$ goes to state n. If m is odd, then, moreover, by symbol c, state 1 goes to state n, and each other state goes to state 1.

Let C' be the NFA for the language $L(C)^R$ obtained from the DFA C by interchanging the accepting and the rejecting state and by reversing all transitions. Let C'' be the DFA obtained from the NFA C' by the subset construction. The DFA C'' has $2n + 1 + m$ reachable and pairwise inequivalent states, which completes the proof of the lemma. $\square$

As a corollary of the three lemmata above and using the fact that $(L^R)^R = L$ we get the following result.

Theorem 5. *For all integers n and α with $n \geqslant 3$ and $\log n \leqslant \alpha \leqslant 2^n$, there exists a minimal DFA M of n states with a 2^n-letter input alphabet such that the minimal DFA for the reversal of the language $L(M)$ has exactly α states. The minimal DFA for the reversal of a 2-state DFA language may have 2, 3, or 4 states, and the reversal of a 1-state DFA language is a 1-state DFA language.* $\square$

We now turn our attention to the nondeterministic state complexity of reversals of regular languages represented by NFAs. The reversal of each 1-state NFA language is the same language. For $n \geqslant 2$, the upper bound on the nondeterministic state complexity of an n-state NFA language is known to be $n+1$ [8], and can be reached by the reversal of a binary language [14]. By the reversal of this binary language, in the case of $n \geqslant 3$, the lower bound $n-1$ is reached. The reversal of the n-state NFA language $\{w \in \{a,b\} \mid |w| \equiv 0 \mod n\}$ is the same language. Thus, the nondeterministic state complexity of a 2-state NFA language is 2 or 3, and for $n \geqslant 3$, we get the following result.

Theorem 6. *Let $n \geqslant 3$. Then the nondeterministic state complexity of the reversal of an n-state NFA binary language is either $n-1$, or n, or $n+1$.* □

6 Conclusions

We have investigated the deterministic and nondeterministic state complexity of complements, stars, and reversals of regular languages. In all cases, we have shown that the whole ranges of complexities up to the known upper bounds can be obtained. Our results are summarized in the following tables (where $[r \mathrel{..} s]$ denotes the set of all integers α with $r \leqslant \alpha \leqslant s$).

	State Complexity			Alphabet Size
L^c	$\{n\}$		trivial	arbitrary
L^*	$[1 \mathrel{..} \frac{3}{4}2^n]$		Theorem 3	2^n
L^R	$[\log n \mathrel{..} 2^n]$		Theorem 5	2^n

	Nondeterministic State Complexity			Alphabet Size
L^c	$[\log n \mathrel{..} 2^n]$		Theorem 1	5
L^*	$[1 \mathrel{..} n+1]$		Theorem 4	2
L^R	$\{n-1, n, n+1\}$		Theorem 6	2

To prove the results on nondeterministic state complexity we have used a fixed five-letter alphabet in the case of complements, and a binary alphabet in the case of stars and reversals. The results on the state complexity of stars and reversals have been shown for a growing alphabet. Whether or not they still hold for a fixed alphabet remains open. We also have proved some partial results on complements in a binary case, and, as a corollary, we have obtained exponentially many "non-magic" numbers, which improves a similar result of Geffert [5].

References

1. Berman, P., Lingas, A.: On the complexity of regular languages in terms of finite automata. Technical Report 304, Polish Academy of Sciences (1977)
2. Birget, J.C.: Intersection and union of regular languages and state complexity. Inform. Process. Lett. 43, 185–190 (1992)

3. Birget, J.C.: Partial orders on words, minimal elements of regular languages, and state complexity. Theoret. Comput. Sci. 119, 267–291 (1993)
4. Dassow, J., Stiebe, R.: Nonterminal complexity of some operations on context-free languages. In: Geffert, V., Pighizzini, G. (eds.) 9th International Workshop on Descriptional Complexity of Formal Systems, pp. 162–169. P. J. Šafárik University of Košice, Slovakia (2007)
5. Geffert, V. (Non)determinism and the size of one-way finite automata. In: Mereghetti, C., Palano, B., Pighizzini, G., Wotschke, D. (eds.) 7th International Workshop on Descriptional Complexity of Formal Systems, pp. 23–37. University of Milano, Italy (2005)
6. Geffert, V.: Magic numbers in the state hierarchy of finite automata. In: Královič, R., Urzyczyn, P. (eds.) MFCS 2006. LNCS, vol. 4162, pp. 412–423. Springer, Heidelberg (2006)
7. Glaister, I., Shallit, J.: A lower bound technique for the size of nondeterministic finite automata. Inform. Process. Lett. 59, 75–77 (1996)
8. Holzer, M., Kutrib, M.: Nondeterministic descriptional complexity of regular languages. Internat. J. Found. Comput. Sci. 14, 1087–1102 (2003)
9. Hricko, M., Jirásková, G., Szabari, A.: Union and intersection of regular languages and descriptional complexity. In: Mereghetti, C., Palano, B., Pighizzini, G., Wotschke, D. (eds.) 7th International Workshop on Descriptional Complexity of Formal Systems, pp. 170–181. University of Milano, Italy (2005)
10. Hromkovič, J.: Descriptional complexity of finite automata: Concepts and open problems. J. Autom. Lang. Comb. 7, 519–531 (2002)
11. Iwama, K., Kambayashi, Y., Takaki, K.: Tight bounds on the number of states of DFAs that are equivalent to n-state NFAs. Theoret. Comput. Sci. 237, 485–494 (2000); Preliminary version In:Bozapalidis, S. (ed.) 3rd International Conference on Developments in Language Theory. Aristotle University of Thessaloniki (1997)
12. Iwama, K., Matsuura, A., Paterson, M.: A family of NFAs which need $2^n - \alpha$ deterministic states. Theoret. Comput. Sci. 301, 451–462 (2003)
13. Jirásková, G.: Note on minimal finite automata. In: Sgall, J., Pultr, A., Kolman, P. (eds.) MFCS 2001. LNCS, vol. 2136, pp. 421–431. Springer, Heidelberg (2001)
14. Jirásková, G.: State complexity of some operations on binary regular languages. Theoret. Comput. Sci. 330, 287–298 (2005)
15. Jirásek, J., Jirásková, G., Szabari, A.: State complexity of concatenation and complementation. Internat. J. Found. Comput. Sci. 16, 511–529 (2005)
16. Jirásek, J., Jirásková, G., Szabari, A.: Deterministic blow-ups of minimal nondeterministic finite automata over a fixed alphabet. In: Harju, T., Karhumäki, J., Lepistö, A. (eds.) DLT 2007. LNCS, vol. 4588, pp. 254–265. Springer, Heidelberg (2007)
17. Sakoda, W.J., Sipser, M.: Nondeterminism and the size of two-way finite automata. In: 10th Annual ACM Symposium on Theory of Computing, San Diego, California, USA, pp. 275–286 (1978)
18. Sipser, M.: Introduction to the theory of computation. PWS Publishing Company, Boston (1997)
19. Yu, S.: Regular languages. In: Rozenberg, G., Salomaa, A. (eds.) Handbook of Formal Languages, ch. 2, vol. I, pp. 41–110. Springer, Heidelberg (1997)
20. Yu, S.: A renaissance of automata theory? Bull. Eur. Assoc. Theor. Comput. Sci. 72, 270–272 (2000)
21. Yu, S., Zhuang, Q., Salomaa, K.: The state complexity of some basic operations on regular languages. Theoret. Comput. Sci. 125, 315–328 (1994)
22. Zijl, L.: Magic numbers for symmetric difference NFAs. Internat. J. Found. Comput. Sci. 16, 1027–1038 (2005)

On the State Complexity of Operations on Two-Way Finite Automata*

Galina Jirásková[1] and Alexander Okhotin[2,3]

[1] Mathematical Institute, Slovak Academy of Sciences, Košice, Slovakia
jiraskov@saske.sk
[2] Academy of Finland
[3] Department of Mathematics, University of Turku, Finland
alexander.okhotin@utu.fi

Abstract. The number of states in two-way deterministic finite automata (2DFAs) is considered. It is shown that the state complexity of basic operations is: at least $m + n - o(m + n)$ and at most $4m + n + 1$ for union; at least $m+n-o(m+n)$ and at most $m+n+1$ for intersection; at least n and at most $4n$ for complementation; at least $\Omega(\frac{m}{n}) + \frac{2^{\Omega(n)}}{\log m}$ and at most $2m^{m+1} \cdot 2^{n^{n+1}}$ for concatenation; at least $\frac{1}{n}2^{\frac{n}{2}-1}$ and at most $2^{O(n^{n+1})}$ for both star and square; between n and $n + 2$ for reversal; exactly $2n$ for inverse homomorphism. In each case m and n denote the number of states in 2DFAs for the arguments.

1 Introduction

State complexity of one-way deterministic finite automata (1DFA) has been studied very well. In particular, the state complexity of virtually all reasonable operations on regular languages has now been determined, and techniques for obtaining such results have been perfected. However, by now, most problems of interest have been researched, and there seem to be no more applications for these techniques apart from formulating and solving artificial problems.

The goal of this paper is to apply the methods for dealing with 1DFAs to begin the study of the state complexity of another important model: the *two-way deterministic finite automaton* (2DFA). These automata, in which the head may move over the tape in both directions, were introduced by Rabin and Scott [11], who proved that every n-state 2DFA can be simulated by a 1DFA with $(n + 1)^{n+1}$ states. The exact tradeoff, recently established by Kapoutsis [7], is $n(n^n - (n - 1)^n)$ states, which was proved by a precise construction of a 1DFA and a matching lower bound argument for 1DFAs.

Another important recent result on 2DFAs is the theorem by Geffert et al. [5] stating that every n-state 2DFA can be converted to a 2DFA with $4n$ states recognizing the same language, which halts on every input, that is, never goes into

---

* Supported by VEGA grant 2/6089/26 and by Academy of Finland grant 118540.

an infinite loop. This result implies an upper bound of $4n$ on the state complexity of complementation for 2DFAs. Unfortunately, no lower bound arguments applicable to 2DFAs are known yet, and probably this is why no study of the state complexity of operations on 2DFAs has been undertaken.

This paper proposes a straightforward technique of proving lower bounds on the number of states in 2DFAs recognizing a language L: first establish a lower bound on the number of states in a 1DFA recognizing L, and then apply the inverse of the function $f(n) = n(n^n - (n-1)^n)$ to this number. This gives a lower bound on the size of any 2DFA recognizing L. The effectiveness of this method relies on having really high lower bounds for 1DFAs, and this is where the methods developed in the recent years come to use.

The above approach is used to prove lower bounds on the state complexity of several operations on 2DFAs: union, intersection, concatenation, star and square. For intersection, the resulting lower bound $m+n-o(m+n)$ is asymptotically the same as the straightforward upper bound $m + n + 1$; for union, the lower bound $m+n-o(m+n)$ is of the same order of magnitude as the upper bound $4m+n+1$. For concatenation and related operations the lower bound is exponential, while the upper bound is double exponential.

A few more operations are handled using different methods. For complementation and reversal, lower bounds of n are obtained by straightforward arguments based on the size of 1DFAs. For the inverse homomorphism, a different argument based upon 1DFAs establishes that its state complexity is exactly $2n$.

2   Finite Automata and Tradeoffs between Them

Definition 1. *A two-way deterministic finite automaton (2DFA) is a quintuple $(Q, \Sigma, \delta, q_0, F)$, in which Q is a finite set of states, Σ is a finite alphabet with $\vdash, \dashv \notin \Sigma$, $q_0 \in Q$ is the initial state, $\delta \colon Q \times (\Sigma \cup \{\vdash, \dashv\}) \to Q \times \{-1, +1\}$ is a partially defined transition function, and $F \subseteq Q$ is the set of accepting states (effective at the right end marker).*

For an input string $w = a_1 \ldots a_\ell \in \Sigma^*$, let $a_0 = \vdash$ and let $a_{\ell+1} = \dashv$. The computation of a 2DFA on w is the longest, possibly infinite, sequence $(p_0, i_0), \ldots, (p_j, i_j), \ldots$, in which $p_j \in Q$, $0 \leqslant i_j \leqslant \ell + 1$, and

- *$(p_0, i_0) = (q_0, 0)$;*
- *$\delta(p_{j-1}, a_{i_{j-1}}) = (p_j, d_j)$ and $i_j = i_{j-1} + d_j$.*

If the sequence is infinite, the automaton is said to loop on w. If the sequence is finite, ending with (p_j, i_j), then the automaton is said to accept w if $i_j = \ell + 1$ and $p_j \in F$, otherwise, the automaton is said to reject w. Define $L(A) = \{w \mid A$ accepts $w\}$.

This definition of a computation has a convenient interpretation it terms of directed graphs with out-degree 1, which was proposed by Sakoda and Sipser [13]. Consider a graph (V, E), with $V = Q \times \{0, \ldots, \ell+1\}$ and $E = \big\{ \langle (q, i), (q', i + d) \rangle \mid \delta(q, a_i) = (q', d) \big\}$. Every vertex (q, i) represents the automaton being in state

q, with its head in position i over the input string. The initial vertex is $(q_0, 0)$, and at every step the automaton follows the unique outgoing arc. The accepting vertices are $(q, \ell + 1)$ with $q \in F$.

Note that the transition table by each symbol a_i produces the subgraph of nodes $Q \times \{i\}$ and their outgoing arc. The whole graph is a composition of such subgraphs, and it can be said that every 2DFA solves the reachability problem in a graph induced by the subgraphs in its alphabet. This graph interpretation will be used in the following to explain several constructions.

A 2DFA is *one-way* (1DFA; that is, a standard deterministic finite automaton) if $\delta(q_0, \vdash) = q_0$, $\delta(q, a) = (q', +1)$ and $\delta(q, \dashv)$ is undefined for all q. Nondeterministic variants of these models (2NFA and 1NFA) can be defined by letting $\delta \colon Q \times (\Sigma \cup \{\vdash, \dashv\}) \to 2^{Q \times \{-1, +1\}}$; then there may be multiple computations for a single input string, and if at least one of them is accepting, the string is said to be accepted.

All four models recognize only regular languages. However, the number of states required to recognize the same language by different types of finite automata may be significantly different. In particular, simulating 2DFAs by 1DFAs involves a superexponential blowup in the worst case.

Lemma 1. *For every language L recognized by a 2DFA with n states there exists a partial 1DFA with $n(n^n - (n-1)^n)$ states recognizing L, as well as a partial 1DFA with $(n+2)^n - (n+1)^n$ states recognizing L^R.*

The first upper bound is known from Kapoutsis [7], and the construction is by having states of the form (q, φ), where $q \in Q$ is a state of the 2DFA and $\varphi \colon Q \to Q$ is a function mapping states to states. The constructed 1DFA reaches a state (q, φ) on a string w, for which (1) the 2DFA executed on $\vdash w$ first comes to the right of the last symbol of w in state q; (2) if the 2DFA starts on the rightmost symbol of $\vdash w$ in state $\widetilde{q}$, then it eventually goes to the right of this rightmost symbol in state $\varphi(\widetilde{q})$. By the construction, the image of φ must contain q, which accounts for $-(n-1)^n$ in the expression.

The second upper bound can be established by a similar construction. This time the states will be functions $\psi \colon Q \to Q \cup \{Acc, Rej\}$ whose image contains Acc. The 1DFA for L^R reaches a state ψ on w if, for every state $\widetilde{q}$, the 2DFA, having started on the leftmost symbol of $w^R \dashv$ in state $\widetilde{q}$ eventually goes to the left of this leftmost symbol in state $\psi(\widetilde{q})$ (or, depending on the value of ψ, accepts or rejects).

Kapoutsis [7] has also proved that his $n(n^n - (n-1)^n)$ upper bound of the 2DFA–1DFA tradeoff is precise. Our upper bound $(n+2)^n - (n+1)^n$ on the size of L^R is precise as well, and furthermore, both bounds are reached on a single witness automaton.

Lemma 2. *For every n there exists an alphabet Σ_n and a language $L \subseteq \Sigma_n^*$ recognized by an n-state 2DFA, such that every partial 1DFA for L must have at least $n(n^n - (n-1)^n)$ states, while every partial 1DFA for L^R must have at least $(n+2)^n - (n+1)^n$ states.*

The following tradeoff between 1DFAs and 2DFAs is thus established:

Theorem 1 (Kapoutsis [7]). *For every $n \geqslant 2$, a language recognized by an n-state 2DFA can be represented by a 1DFA with $n(n^n - (n-1)^n)$ states. This size is in the worst case necessary, with witness languages given over a growing alphabet of size $\Theta(n^n)$.*

The following lemma represents a reverse application of this tradeoff. It is based upon the fact that $\frac{\log k}{\log \log k}$ is a lower bound on the inverse of the function $f(n) = n(n^n - (n-1)^n)$.

Lemma 3. *Let L be a regular language, let $k \geqslant 4$. Then, if L requires a 1DFA with at least k states, then L requires a 2DFA with at least $\frac{\log k}{\log \log k}$ states.*

Proof (sketch). If there is a 2DFA with $n < \frac{\log k}{\log \log k}$ states for such a language, then, by Lemma 1, there is a 1DFA with $n(n^n - (n-1)^n)$ states recognizing L. It can be shown that $n(n^n - (n-1)^n) < k$, which contradicts the assumption that k states are necessary for a 1DFA to accept L. $\square$

Although two-way automata are generally much more succinct than their one-way counterparts, for some languages two-way motion is useless, and the size of 1DFAs and 2DFAs is the same:

Proposition 1 (Geffert [3,4]). *For every $n \geqslant 1$, the singleton language $\{a^{n-1}\}$ requires a 2DFA of n states.*

Finally, let us note the following property of the state complexity of 1DFAs under injective homomorphisms (codes).

Lemma 4. *Let $h : \Sigma^* \to \Gamma^*$ be a code. Then, if a language $L \subseteq \Sigma^*$ requires a 1DFA of at least n states, then each language $K \subseteq \Gamma^*$ with $K \cap h(\Sigma^*) = h(L)$ requires a 1DFA of at least n states.*

3   Complementation

Complementation is a trivial operation for 1DFAs, where it can be performed by inverting the set of accepting states. In contrast, the complement of a language recognized by an n-state 1NFA may require as many as 2^n states [1]. For 2DFAs this operation is also non-trivial, because a 2DFA may reject by an undefined transition or, more importantly, by looping.

Accordingly, a 2DFA for the complement of a language recognized by a given 2DFA can be constructed by first reconstructing the given 2DFA to make it halt on every input. The best known construction is the following:

Theorem 2 (Geffert et al. [5]). *For every n-state 2DFA there exists an equivalent 2DFA with 4n states that halts on every input.*

This implies that the complement can always be recognized by a 2DFA with $4n$ states. A straightforward lower bound of n states can also be proved, which gives the following theorem:

Theorem 3. *For every $n \geqslant 2$, the state complexity of complementation for n-state 2DFAs is at least n and at most $4n$.*

Proof (sketch). If for any single value of n, the complement of every n-state 2DFA language could be represented with $n - 1$ states, then the size of the lower bound 1DFAs in Theorem 1 could be reduced for this n, contradicting the theorem. $\square$

4   Union and Intersection

With respect to 1DFAs, both union and intersection have state complexity mn [9]. The straightforward upper bounds for the 2DFA state complexity of these operations are the following:

Proposition 2. *For every 2DFA A with m states and 2DFA B with n states there exists a 2DFA for the language $L(A) \cap L(B)$ with $m + n + 1$ states and a 2DFA for $L(A) \cup L(B)$ with $4m + n + 1$ states.*

A 2DFA C recognizing the intersection will first simulate A (if A does not accept the input, neither will C), and once A accepts, C will return to the left end marker in a special state and then proceed with simulating B. This gives $m+n+1$ states.

For union, the first step is to convert A to an equivalent 2DFA A' with $4m$ states that halts on every input, which can be done by Theorem 2. Now a 2DFA D recognizing the union will first simulate A', and if A' accepts, it will accept as well, and otherwise, if A' halts and rejects, then D returns to the beginning of the tape and starts simulating B. So $4m + n + 1$ states are enough.

In order to establish lower bounds on the size of a 2DFA recognizing union or intersection, let us first obtain lower bounds on the size of 1DFAs for these languages.

Lemma 5. *For every $m, n \geqslant 2$, there exists an alphabet $\Sigma_{m,n}$ and languages $K, L \subseteq \Sigma_{m,n}^*$ recognized by m-state and n-state 2DFAs, respectively, such that the smallest 1DFA recognizing $K \cup L$ ($K \cap L$, respectively) has at least $m^m \cdot n^n$ states.*

The proof proceeds by taking alphabets Σ_m and Σ_n from Lemma 2 and considering the alphabet $\Sigma_{m,n} = \Sigma_m \times \Sigma_n$. Then two automata A_m and B_n are defined similarly to the automata from Lemma 2, which require 1DFAs of size m^m and n^n, respectively. When reading a symbol from $\Sigma_{m,n}$, A_m considers only its first component, while B_n works on the second components. Then a 1DFA for the union $L(A_m) \cup L(B_n)$ or for the intersection $L(A_m) \cap L(B_n)$ basically has to simulate both 2DFAs in parallel, which gives the lower bound.

It remains to infer a lower bound on the number of states in a 2DFA recognizing union and intersection from the above lower bound on the size of a 1DFA. Using Lemma 3, this gives $m + n - o(m + n)$ states, so the following theorem can be stated:

Theorem 4. *The state complexity of union (intersection) with respect to 2DFAs is at least $m + n - o(m + n)$ and at most $4m + n$ ($m + n + 1$, respectively), with the witness languages given over a growing alphabet.*

5   Concatenation

The next operation to consider is concatenation of two regular languages represented by 2DFAs. A string w is in $L(A) \cdot L(B)$ if and only if it can be factorized as $w = uv$ with $u \in L(A)$ and $v \in L(B)$. A 2DFA recognizing $L(A) \cdot L(B)$ should somehow simulate A on the first part and B on the second part. However, there is no evident way for these simulated computations to detect the boundary between u and v. Because of this, the only known way of recognizing such a concatenation by a 2DFA is by first converting both A and B to 1DFAs and then using the known construction for a concatenation of two 1DFAs, which yields a 1DFA with $(2m - 1)2^{n-1}$ states. This results in the following huge 1DFA for the concatenation:

Proposition 3. *For every 2DFAs A with m states and 2DFA B with n states there exists a 1DFA for the language $L(A) \cdot L(B)$ containing the following number of states:*

$$(2m(m^m - (m-1)^m) - 1) \cdot 2^{n(n^n - (n-1)^n) - 1} = 2^{O(m \log m + n^{n+1})}.$$

The goal of this section is to establish a comparable lower bound on the size of a 1DFA recognizing a concatenation of two 2DFAs. Such a lower bound is first established for the special case of a fixed A.

Lemma 6. *For every $n \geqslant 2$, there exists a language L over an alphabet $\Sigma_n = \{a_j \mid 1 \leqslant j \leqslant \lfloor \frac{n-1}{2} \rfloor\} \cup \{c\}$ recognized by an n-state 2DFA, such that every 1DFA for $\Sigma_n^* c \cdot L$ requires at least $2^{2^{\frac{n-2}{2}}}$ states.*

Proof. Let $k = \lfloor \frac{n-1}{2} \rfloor \geqslant \frac{n-2}{2}$. The language L is defined as $\{w \mid \text{for every } j \in \{1, \ldots, k\}, |w|_{a_j} \text{ is even}\}$. A 2DFA can recognize it by making k passes over the string. This is done by a 2DFA $B_n = (\{1, 2, \ldots, 2k - 1, 2k, 2k + 1\}, \Sigma_n, \delta_B, 1, \{2k + 1\})$, in which the states are arranged into pairs $\{2i - 1, 2i\}$ with $1 \leqslant i \leqslant k$, and there is one extra state $2k + 1$. In each j-th pair of states B_n checks the number of a_j modulo 2; if j is odd, the direction of its motion is from left to right, and if j is even, the automaton goes from right to left. Accordingly, denote $d(j) = +1$ for j odd and $d(j) = -1$ for j even. The transitions are defined as follows. Over the left end marker,

$$\delta_B(1, \vdash) = (1, +1)$$
$$\delta_B(2i - 1, \vdash) = (2(i + 1) - 1, +1) \quad (1 \leqslant i \leqslant k, \, i \text{ is even})$$

and the rest of transitions at $\vdash$ are undefined. The transitions by c are:

$$\delta_B(2i - t, c) = (2i - t, d(i)) \quad (1 \leqslant i \leqslant k, \, t \in \{0, 1\})$$

For each symbol a_j, most transitions are the same as for c, with the exception of the j-th pair of wires, which gets crossed:

$$\delta_B(2j-1, a_j) = (2j, d(j))$$
$$\delta_B(2j, a_j) = (2j-1, d(j))$$
$$\delta_B(2i-t, a_j) = (2i-t, d(i)) \quad (1 \leqslant i \leqslant k,\ i \neq j,\ t \in \{0,1\})$$

The following transitions are defined at the right end marker:

$$\delta_B(2i-1, \dashv) = (2(i+1)-1, -1) \quad (1 \leqslant i \leqslant k,\ i \text{ is odd})$$

Finally, in state $2k+1$ the automaton always goes to the right: $\delta_B(2k+1, s) = (2k+1, +1)$ for all $s \in \Sigma_n$.

These transitions are illustrated in the graph representation of a sample string given in Figure 1. Symbols from Σ_n form subgraphs, and the string w in Σ_n^* is in the language $L(B_n)$ if the graph formed by $\vdash w \dashv$ has a path from the node 1 in the leftmost column to the node $2k+1$ in the rightmost column. The symbol c effectively acts as an identity, since it preserves the connectivity of the graph.

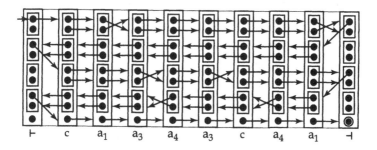

Fig. 1. $ca_1a_3a_4a_3ca_4a_1 \in L(B_9)$

A 1DFA B_n' simulating the 2DFA B_n needs to count the parity of each symbol at once, which requires storing k bits. It can be defined with a set of states $\mathbb{B}^k$, the set of all Boolean vectors of length k. The initial and the sole accepting state is $(0, 0, \ldots, 0)$. Each state $(b_1, b_2, \ldots, b_k)$ goes to state $(b_1, b_2, \ldots, b_{j-1}, \neg b_j, b_{j+1}, \ldots, b_k)$ on a_j, and to itself on c.

The difficulty of recognizing $\Sigma_n^* c \cdot L$ by an 1DFA lies with the uncertainty of *when* to start counting the parity of the number of letters. This can be guessed nondeterministically. Construct a 1NFA M' for the language $\Sigma_n^* c \cdot L(B_n)$ from the 1DFA B_n' by adding a new initial state q_0, which goes to itself on a_j $(1 \leqslant j \leqslant k)$, and goes to $\{q_0, (0, 0, \ldots, 0)\}$ on c.

Let $M = (Q, \Sigma_n, \delta, \{q_0\}, F)$, where $Q = 2^{\{q_0\} \cup \mathbb{B}^k}$, be the 1DFA obtained from the 1NFA M' by the subset construction. It will be shown that the 1DFA M has 2^{2^k} reachable states, which are pairwise inequivalent.

The first claim is that for every set of Boolean vectors $S \subseteq \mathbb{B}^k$, the set $\{q_0\} \cup S$ is a reachable state in the 1DFA M. The proof is by induction on the size of a set.

The basis, $S = \varnothing$, holds true since $\{q_0\}$ is the initial state of M. For the induction step, let $S \cup \{(\sigma_1, \sigma_2, \ldots, \sigma_k)\}$ be any subset with $(\sigma_1, \sigma_2, \ldots, \sigma_k) \notin S$. For each vector $x = (x_1, x_2, \ldots, x_k)$ in S, construct the vector $x' = (x'_1, x'_2, \ldots, x'_k)$, such that $x'_i = x_i$ if $\sigma_i = 0$, and $x'_i = \neg x_i$ if $\sigma_i = 1$. Let $S' = \{x' \mid x \in S\}$. Then the set $\{q_0\} \cup S'$ is reachable by induction. Now take the string $w = ca_{i_1} a_{i_2} \cdots a_{i_r}$, where $1 \leqslant i_1 < i_2 < \cdots < i_r \leqslant k$ are all numbers, for which $\sigma_{i_j} = 1$. Each x' in S' goes to x by w, while state q_0 goes to $\{q_0, (0, 0, \ldots, 0)\}$ by c and then to $\{q_0, (\sigma_1, \sigma_2, \ldots, \sigma_k)\}$ by $a_{i_1} a_{i_2} \cdots a_{i_r}$. Thus, the set $\{q_0\} \cup S'$ goes to $\{q_0\} \cup S \cup \{(\sigma_1, \ldots, \sigma_k)\}$ by w, which concludes the proof of reachability.

To prove inequivalence, for each state $\sigma = (\sigma_1, \sigma_2, \ldots, \sigma_k)$ of the 1NFA M', consider the string $w(\sigma) = a_{i_1} a_{i_2} \cdots a_{i_r}$, where $1 \leqslant i_1 < i_2 < \cdots < i_r \leqslant k$ are all numbers, for which $\sigma_{i_j} = 1$. The string $w(\sigma)$ is accepted by the 1NFA M' from state σ, but is not accepted by M' from any other state. Now, if $\{q_0\} \cup S$ and $\{q_0\} \cup T$ are two different states of the 1DFA M, they must differ in a state $\sigma \in \mathbb{B}^k$, and so the string $w(\sigma)$ distinguishes them.   □

For the next step, let us prove a stronger statement, where the language L is defined over a fixed 5-letter alphabet. This is done by encoding the symbols used in the previous construction over these five symbols.

Lemma 7. *Let $\Sigma = \{a, b, a', b', c\}$. For every $n \geqslant 1$, there exists a language $L \subseteq \Sigma^*$ recognized by an n-state 2DFA, such that every 1DFA for $\Sigma^* c \cdot L$ requires at least $2^{2^{\frac{n-2}{2}}}$ states.*

Proof (sketch). Let $k = \lfloor \frac{n-1}{2} \rfloor$. The automaton B'_n to be constructed is very similar to the one in Lemma 6 and will be explained in terms of the previous construction. It again has $2k + 1 \leqslant n$ states, and its transitions by c and by both markers remain the same as in the proof of Lemma 6.

The transitions by a are the same as the transitions for a_1 in the above construction: they cross the first pair of wires going forward. The transitions by a' cross the second pair of wires (which go backward), that is, a' is the same as a_2 in the previous automaton.

The symbols b and b' have the following transitions, for all $i \in \{1, \ldots, k\}$ and for all $j \in \{1, 2\}$:

$$\delta_{B'}(2(i-1)+j, b) = \begin{cases} (2(i+1)+j, +1), & \text{if } i+2 < k \\ (j, +1), & \text{if } i+2 \geqslant k \end{cases} \quad \text{(for odd } i\text{)}$$

$$\delta_{B'}(2(i-1)+j, b) = (2(i-1)+j, -1) \quad \text{(for even } i\text{)}$$

$$\delta_{B'}(2(i-1)+j, b') = (2(i-1)+j, +1) \quad \text{(for odd } i\text{)}$$

$$\delta_{B'}(2(i-1)+j, b') = \begin{cases} (2(i+1)+j, -1), & \text{if } i+2 < k \\ (j, -1), & \text{if } i+2 \geqslant k \end{cases} \quad \text{(for even } i\text{)}$$

$$\delta_{B'}(2k+1, b) = \delta_B(2k+1, b') = (2k+1, +1)$$

Basically, b performs a cyclic shift of all pairs of forward wires, while b' does the same for backward wires. This allows putting an arbitrary pair in the topmost position, and thus each symbol a_{2i-1} in the above construction is simulated by

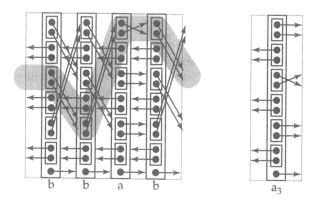

Fig. 2. Simulation of a_3 by $bbab$

a string $b^j a b^i$ (for an appropriate j), as shown in Figure 2, while each symbol a_{2i} is simulated by $(b')^j a' (b')^i$, again for some appropriate j.

Formally, define a code $h : \Sigma_n \to \Sigma$ by $h(a_{2i-1}) = b^j a b^i$ and $h(a_{2i}) = (b')^j a' (b')^i$ as above, and with $h(c) = c$. Since each codeword $h(a_i)$ forms a subgraph with the same connectivity as a_i in the original automaton, it follows that $L(B') \cap h(\Sigma_n^*) = h(L(B_n))$. Then the same lower bound 2^{2^k} follows by Lemma 4. □

Now, once the methods of extracting many states from the second argument of concatenation have been developed, let us allow its first argument to be an arbitrary 2DFA. Then the following lower bound on the size of a 1DFA can be established.

Lemma 8. *For every $m, n \geqslant 1$ there exist languages K and L over a 9-letter alphabet, recognized by m-state and n-state 2DFAs, respectively, such that every 1DFA for $K \cdot L$ requires at least $\left(\frac{m-3}{2}\right)^{\frac{m-3}{2}} \cdot \left(2^{2^{\frac{n-2}{2}}} - 1\right)$ states.*

The alphabet is $\Sigma = \{a, b, a', b', c, d, f_1, f_2, f_3\}$, where the first five symbols are the same as in the previous construction, and the second automaton operates in exactly the same way. The new symbols d, f_1, f_2, f_3 are identities for the second automaton in the same way as the symbol c. The first automaton, which replaces the language $\Sigma^* c$, accepts a string if the graph formed by symbols d, f_1, f_2, f_3 contains a certain path; this condition requires $\left(\frac{m-3}{2}\right)^{\frac{m-3}{2}}$ states in a 1DFA to check.

This lower bound on the size of a 1DFA representing the concatenation implies the following lower bound on the size of a 2DFA:

Theorem 5. *The state complexity of concatenation of 2DFAs is at least $\Omega(\frac{m}{n}) + \frac{2^{\Omega(n)}}{\log m}$ and at most $2m^{m+1} \cdot 2^{n+1}$ with the lower bound languages defined over a 9-letter alphabet.*

6   Square and Star

Let us consider two more operations related to concatenation. One of them is Kleene star, which has state complexity of $\frac{3}{4}2^n$ for 1DFAs [9], while for 1NFAs it requires exactly $n+1$ states [6]. The other operation, the so-called *square*, is the concatenation of a language with itself. It has state complexity of $n \cdot 2^n - 2^{n-1}$ for 1DFAs [12] and $2n$ for 1NFAs [2].

For 2DFAs, like in the case of concatenation, the only evident method of constructing a 2DFA for square or star of a given language is by first converting the given 2DFA to a 1DFA. This gives the following upper bound:

Proposition 4. *For every 2DFA A with n states there exists a 1DFA for the language $L(A)^2$ with*

$$(2n(n^n - (n-1)^n) - 1) \cdot 2^{n(n^n - (n-1)^n)-1} = 2^{2^{\Theta(n \log n)}}$$

states, as well as a 1DFA for the language $L(A)^*$ containing the following number of states:

$$\frac{3}{4}2^{n(n^n-(n-1)^n)} = 2^{2^{\Theta(n \log n)}}.$$

Let us now establish a lower bound on the number of states in this 1DFA.

Lemma 9. *For every $n \geqslant 1$ there exists a language $L \subseteq \{a, b, a', b', c, d, e\}^*$ recognized by an n-state 2DFA, such that every 1DFA for L^2 and every 1DFA for L^* requires at least $2^{2^{\frac{n-4}{2}}}$ states.*

The proof is similar to that of Lemma 7, though the construction of a 2DFA has to be elaborated. For $L \cdot L$, the same 2DFA is naturally used twice: as the first argument of concatenation, it is used to recognize strings of the form $d\{a, b, a', b', cc\}^*c$; as the second argument, it works basically as the 2DFA in Lemma 7. In this way the general outline of the proof is preserved. It is extended to the case of star by preventing a concatenation of more than two strings using a special symbol e.

This lower bound on the size of a 1DFA implies a lower bound on the number of states in a 2DFA according to Lemma 3. Thus the following bounds are obtained:

Theorem 6. *The state complexity of square and star of a 2DFAs is at least $\frac{1}{n}2^{\frac{n}{2}-1}$ and at most $2^{O(n^{n+1})}$, with the lower bound languages defined over a 7-letter alphabet.*

7   Reversal

Consider the operation of reversal of a regular language. Unlike complementation, it is expensive for 1DFAs, where it requires 2^n states [8]. For 1NFAs, only $n + 1$ states are needed [6]. For 2DFAs, the reversal of a given language can be recognized by adding two extra states:

Proposition 5. *For every 2DFAs A with n states there exists an $(n+2)$-state 2DFA recognizing the language $L(A)^R$.*

A 2DFA B recognizing this language first goes to the end of the input string in a special state and then simulates A with the direction of its movement reversed. The simulated A accepts over the left end marker, and once this happens, B proceeds to the right end marker in another special state.

A close lower bound follows from Proposition 1, which states that the language $\{a^{n-1}\}$ requires exactly n states of a 2DFA. Since $\{a^{n-1}\}^R = \{a^{n-1}\}$, this gives a lower bound of n states for the reversal. In overall, the state complexity of reversal for 2DFAs lies within the following bounds:

Theorem 7. *For every $n \geqslant 4$, the state complexity of reversal of an n-state 2DFA is between n and $n+2$, with the witness languages given over a unary alphabet.*

8   Inverse Homomorphism

Let Σ and Γ be two alphabets. A *homomorphism* is a mapping $h : \Sigma^* \to \Gamma^*$ that satisfies the conditions $h(uv) = h(u)h(v)$ for all $u, v \in \Sigma^*$ and $h(\varepsilon) = \varepsilon$. It is completely defined by the images of letters from Σ.

For every regular language $L \subseteq \Gamma^*$, the pre-image $\{w \mid h(w) \in L\} \subseteq \Sigma^*$ with respect to h is regular as well. Its state complexity for 1DFAs is known to be n. It will now be proved that with respect to 2DFAs its state complexity is exactly $2n$.

Lemma 10. *For every homomorphism $h : \Sigma^* \to \Gamma^*$ and for every n-state 2DFA A over Γ, there exists a $2n$-state 2DFA for $h^{-1}(L(A))$.*

The construction uses twice as many states in order to remember, besides a state of the original automaton, the current direction of its motion.

It turns out that $2n$ states are necessary in the worst case. To see this, it will be established that a smaller 2DFA for an inverse homomorphic image of a language L allows one to construct a 1DFA for the language L or for the language L^R that has fewer states than the worst-case bound established in Lemma 2.

Lemma 11. *Let Σ be an alphabet, let $c \notin \Sigma$ and define a homomorphism $h : (\Sigma \cup \{c\})^* \to \Sigma^*$ by $h(a) = a$ for all $a \in \Sigma$ and $h(c) = \varepsilon$. Then for every language $L \subseteq \Sigma$ and for every $n \geqslant 3$, if $h^{-1}(L)$ is recognized by a 2DFA with $2n-1$ states, then L is recognized by a 1DFA with n^n states or L^R is recognized by a 1DFA with $(n+1)^{n-1}$ states.*

Now the state complexity of h^{-1} can be established as follows:

Theorem 8. *The state complexity of inverse homomorphisms with respect to 2DFAs is $2n$, with the lower bound defined over a growing alphabet.*

Proof. To show the lower bound, for every $n \geqslant 3$, let L be as defined in Lemma 2 and suppose $h^{-1}(L)$ is recognized by a $(2n-1)$-state 2DFA. Then, by Lemma 11, there is an n^n-state 1DFA for L or an $(n+1)^{n-1}$-state 1DFA for L^R. In each case this contradicts Lemma 2. $\square$

9 Summary

The state complexity of basic operations on regular languages with respect to 1DFAs, 1NFAs and 2DFAs is compared in the following table:

	1DFA	1NFA	2DFA
$\cup$	mn [9]	$m+n+1$ [6]	$m+n-o(m+n) \leqslant \cdot \leqslant 4m+n+1$
$\cap$	mn [9]	mn [6]	$m+n-o(m+n) \leqslant \cdot \leqslant m+n+1$
$\sim$	n	2^n [1]	$n \leqslant \cdot \leqslant 4n$
$\cdot$	$m \cdot 2^n - 2^{n-1}$ [9]	$m+n$ [6]	$\Omega(\frac{m}{n}) + \frac{2^{\Omega(n)}}{\log m} \leqslant \cdot \leqslant 2m^{m+1} \cdot 2^{n^{n+1}}$
2	$n \cdot 2^n - 2^{n-1}$ [12]	$2n$ [2]	$\frac{1}{n}2^{\frac{n}{2}-1} \leqslant \cdot \leqslant 2^{O(n^{n+1})}$
$*$	$\frac{3}{4}2^n$ [9]	$n+1$ [6]	$\frac{1}{n}2^{\frac{n}{2}-1} \leqslant \cdot \leqslant 2^{O(n^{n+1})}$
R	2^n [8]	$n+1$ [6]	$n \leqslant \cdot \leqslant n+2$
h^{-1}	n	?	$2n$

References

1. Birget, J.C.: Partial orders on words, minimal elements of regular languages, and state complexity. Theoretical Computer Science 119, 267–291 (1993)
2. Domaratzki, M., Okhotin, A.: State complexity of power, TUCS Technical Report No 845, Turku Centre for Computer Science, Turku, Finland (January 2007)
3. Geffert, V.: Nondeterministic computations in sublogarithmic space and space constructibility. SIAM Journal on Computing 20(3), 484–498 (1991)
4. Geffert, V.: Personal communication (March 2008)
5. Geffert, V., Mereghetti, C., Pighizzini, G.: Complementing two-way finite automata. Information and Computation 205(8), 1173–1187 (2007)
6. Holzer, M., Kutrib, M.: Nondeterministic descriptional complexity of regular languages. International Journal of Foundations of Computer Science 14, 1087–1102 (2003)
7. Kapoutsis, C.A.: Removing bidirectionality from nondeterministic finite automata. In: Jedrzejowicz, J., Szepietowski, A. (eds.) MFCS 2005. LNCS, vol. 3618, pp. 544–555. Springer, Heidelberg (2005)
8. Leiss, E.L.: Succinct representation of regular languages by Boolean automata. Theoretical Computer Science 13, 323–330 (1981)
9. Maslov, A.N.: Estimates of the number of states of finite automata. Soviet Mathematics Doklady 11, 1373–1375 (1970)
10. Moore, F.R.: On the bounds for state-set size in the proofs of equivalence between deterministic, nondeterministic, and two-way finite automata. IEEE Transactions on Computers 20, 1211–1214 (1971)
11. Rabin, M.O., Scott, D.: Finite automata and their decision problems. IBM Journal of Research and Development 3, 114–125 (1959)
12. Rampersad, N.: The state complexity of L^2 and L^k. Information Processing Letters 98, 231–234 (2006)
13. Sakoda, W.J., Sipser, M.: Nondeterminism and the size of two way finite automata. In: 10th ACM Symposium on Theory of Computing (STOC 1978), pp. 275–286 (1978)
14. Sipser, M.: Halting space-bounded computations. Theoretical Computer Science 10(3), 335–338 (1980)

On the Size Complexity of
Rotating and Sweeping Automata*

Christos Kapoutsis, Richard Královič, and Tobias Mömke

Department of Computer Science, ETH Zürich

Abstract. We examine the succinctness of *one-way, rotating, sweeping,* and *two-way* deterministic finite automata (1DFAs, RDFAs, SDFAs, 2DFAs). Here, a SDFA is a 2DFA whose head can change direction only on the endmarkers and a RDFA is a SDFA whose head is reset on the left end of the input every time the right endmarker is read. We introduce a list of language operators and study the corresponding closure properties of the size complexity classes defined by these automata. Our conclusions reveal the logical structure of certain proofs of known separations in the hierarchy of these classes and allow us to systematically construct alternative problems to witness these separations.

1 Introduction

One of the most important open problems in the study of the size complexity of finite automata is the comparison between determinism and nondeterminism in the two-way case: Does every two-way nondeterministic finite automaton (2NFA) with n states have a deterministic equivalent (2DFA) with a number of states polynomial in n? [6,5] Equivalently, if 2N is the class of families of languages that can be recognized by families of polynomially large 2NFAs and 2D is its deterministic counterpart, is it 2D = 2N? The answer is conjectured to be negative, even if all 2NFAs considered are actually one-way (1NFAs). That is, even 2D $\not\supseteq$ 1N is conjectured to be true, where 1N is the one-way counterpart of 2N.

To confirm these conjectures, one would need to prove that some n-state 2NFA or 1NFA requires superpolynomially (in n) many states on every equivalent 2DFA. Unfortunately, such lower bounds for arbitrary 2DFAs are currently beyond reach. They have been established only for certain restricted special cases. Two of them are the *rotating* and the *sweeping* 2DFAs (RDFAs and SDFAs, respectively).

A SDFA is a 2DFA that changes the direction of its head only on the input endmarkers. Thus, a computation is simply an alternating sequence of rightward and leftward one-way scans. A RDFA is a SDFA that performs no leftward scans: upon reading the right endmarker, its head jumps directly to the left end. The subsets of 2D that correspond to these restricted 2DFAs are called SD and RD.

Several facts about the size complexity of SDFAs have been known for quite a while (e.g., 1D $\not\supseteq$ SD [7], SD $\not\supseteq$ 2D [7,1,4], SD $\not\supseteq$ 1N [7], SD $\not\supseteq$ 1N ∩ co-1N [3]) and, often, at the core of their proofs one can find proofs of the corresponding

---

* Work supported by the Swiss National Science Foundation grant 200021-107327/1.

M. Ito and M. Toyama (Eds.): DLT 2008, LNCS 5257, pp. 455–466, 2008.

facts for RDFAs (e.g., 1D $\not\supseteq$ RD, RD $\not\supseteq$ 2D, etc.). Overall, though, our study of these automata has been fragmentary, exactly because they have always been examined only on the way to investigate the 2D vs. 2N question.

In this article we take the time to make the hierarchy 1D $\subseteq$ RD $\subseteq$ SD $\subseteq$ 2D itself our focus. We introduce a list of language operators and study the closure properties of our complexity classes with respect to them. Our conclusions allow us to reprove the separation of [3], this time with a new witness language family, which (i) is constructed by a sequence of applications of our operators to a single, minimally hard, 'core' family and (ii) mimicks the design of the original witness. This uncovers the logical structure of the original proof and explains how hardness propagates upwards in our hierarchy of complexity classes when appropriate operators are applied. It also enables us to construct many other witnesses of the same separation, using the same method but a different sequence of operators and/or a different 'core' family. Some of these witnesses are both *simpler* (produced by operators of lower complexity) and *more effective* (establish a greater exponential gap) than the one of [3].

More generally, our operators provide a systematic way of proving separations by building witnesses out of simpler and easier 'core' language families. For example, given any family $\mathcal{L}$ which is hard for 1DFAs reading from left to right (as usual) but easy for 1DFAs reading from right to left ($\mathcal{L} \notin$ 1D but $\mathcal{L}^R \in$ 1D), one can build a family $\mathcal{L}'$ which is hard for SDFAs but easy for 1NFAs, easy for 1NFAs recognizing the complement, and easy for 2DFAs ($\mathcal{L}' \in (1N \cap co\text{-}1N \cap 2D) \setminus$ SD), a simultaneous witness for the theorems of [7,3,1,4]. We believe that this operator-based reconstruction or simplification of witnesses deepens our understanding of the relative power of these automata.

The next section defines the objects that we work with. Section 3 introduces two important tools for working with parallel automata and uses them to prove hardness propagation lemmata. These are then applied in Sect. 4 to establish the hierarchy and closures map of Fig. 1. Section 5 lists our final conclusions.

2 Preliminaries

Let Σ be an alphabet. If $z \in \Sigma^*$ is a string, then $|z|$, z_t, z^t, and z^R are its length, t-th symbol (if $1 \leq t \leq |z|$), t-fold concatenation with itself (if $t \geq 0$), and reverse. If $P \subseteq \Sigma^*$, then $P^R := \{z^R \mid z \in P\}$.

A *(promise) problem* over Σ is a pair $L = (L_Y, L_N)$ of disjoint subsets of Σ^*. The *promise* of L is $L_P := L_Y \cup L_N$. If $L_P = \Sigma^*$, then L is a *language*. If $L_Y, L_N \neq \emptyset$, then L is *nontrivial*. We write $w \in L$ iff $w \in L_Y$, and $w \notin L$ iff $w \in L_N$. (Note that "$x \notin L$" is equivalent to the negation of "$x \in L$" only when $x \in L_P$.) To *solve* L is to accept all $w \in L$ but no $w \notin L$ (and decide arbitrarily on $w \notin L_P$).

A *family of automata* $\mathcal{M} = (M_n)_{n \geq 1}$ solves a *family of problems* $\mathcal{L} = (L_n)_{n \geq 1}$ iff, for all n, M_n solves L_n. The automata of $\mathcal{M}$ are 'small' iff, for some polynomial p and all n, M_n has at most $p(n)$ states.

Problem Operators. Fix a delimiter # and let L, L_1, L_2 be arbitrary problems. If $\#x_1\#\cdots\#x_l\#$ denotes strings from $\#(L_P\#)^*$ and $\#x\#y\#$ denotes strings from $\#(L_1)_P\#(L_2)_P\#$, then the following pairs are easily seen to be problems, too:

$$\neg L := (L_N, L_Y) \qquad L^R := (L_Y^R, L_N^R)$$
$$L_1 \wedge L_2 := (\ \{\#x\#y\# \mid x \in L_1 \wedge y \in L_2\},\ \{\#x\#y\# \mid x \notin L_1 \vee y \notin L_2\}\ )$$
$$L_1 \vee L_2 := (\ \{\#x\#y\# \mid x \in L_1 \vee y \in L_2\},\ \{\#x\#y\# \mid x \notin L_1 \wedge y \notin L_2\}\ )$$
$$L_1 \oplus L_2 := (\ \{\#x\#y\# \mid x \in L_1 \Leftrightarrow y \notin L_2\},\ \{\#x\#y\# \mid x \in L_1 \Leftrightarrow y \in L_2\}\ )$$
$$\bigwedge L := (\ \{\#x_1\#\cdots\#x_l\# \mid (\forall i)(x_i \in L)\},\ \{\#x_1\#\cdots\#x_l\# \mid (\exists i)(x_i \notin L)\}\ ) \tag{1}$$
$$\bigvee L := (\ \{\#x_1\#\cdots\#x_l\# \mid (\exists i)(x_i \in L)\},\ \{\#x_1\#\cdots\#x_l\# \mid (\forall i)(x_i \notin L)\}\ )$$
$$\bigoplus L := (\ \{\#x_1\#\cdots\#x_l\# \mid \text{the number of } i \text{ such that } x_i \in L \text{ is odd}\},$$
$$\{\#x_1\#\cdots\#x_l\# \mid \text{the number of } i \text{ such that } x_i \in L \text{ is even}\}\ )$$

over the promises, respectively: L_P, $(L_P)^R$, $\#(L_1)_P\#(L_2)_P\#$ (for $L_1 \wedge L_2$, $L_1 \vee L_2$, $L_1 \oplus L_2$) and $\#(L_P\#)^*$ (for the rest). We call these problems, respectively: the *complement* and *reversal* of L; the *conjunctive*, *disjunctive*, and *parity concatenation* of L_1 with L_2; the *conjunctive*, *disjunctive*, and *parity star* of L. By the definitions, we easily have $\neg(L^R) = (\neg L)^R$, and also:

$$\neg(L_1 \wedge L_2) = \neg L_1 \vee \neg L_2 \qquad \neg(\bigwedge L) = \bigvee \neg L \qquad (L_1 \wedge L_2)^R = L_2^R \wedge L_1^R$$
$$\neg(L_1 \vee L_2) = \neg L_1 \wedge \neg L_2 \qquad \neg(\bigvee L) = \bigwedge \neg L \qquad (L_1 \vee L_2)^R = L_2^R \vee L_1^R \tag{2}$$
$$\neg(L_1 \oplus L_2) = \neg L_1 \oplus L_2 \qquad (\bigwedge L)^R = \bigwedge L^R \qquad (L_1 \oplus L_2)^R = L_2^R \oplus L_1^R.$$
$$(\bigvee L)^R = \bigvee L^R$$

Our definitions extend naturally to families of problems: we just apply the problem operator to (corresponding) components. E.g., if $\mathcal{L}, \mathcal{L}_1, \mathcal{L}_2$ are families of problems, then $\neg\mathcal{L} = (\neg L_n)_{n \geq 1}$ and $\mathcal{L}_1 \vee \mathcal{L}_2 = (L_{1,n} \vee L_{2,n})_{n \geq 1}$. Clearly, the identities of (2) remain true when we replace L, L_1, L_2 with $\mathcal{L}, \mathcal{L}_1, \mathcal{L}_2$.

Finite Automata. Our automata are *one-way*, *rotating*, *sweeping*, or *two-way*. We refer to them by the naming convention bDFA, where $b = 1, R, S, 2$. E.g., RDFAs are rotating (R) deterministic finite automata (DFA). We assume the reader is familiar with all these machines. This section simply fixes some notation.

A SDFA [7] over an alphabet Σ and a set of states Q is a triple $M = (q_s, \delta, q_a)$ of a *start* state $q_s \in Q$, an *accept* state $q_a \in Q$, and a *transition function* δ which partially maps $Q \times (\Sigma \cup \{\vdash, \dashv\})$ to Q, for some endmarkers $\vdash, \dashv \notin \Sigma$. An input $z \in \Sigma^*$ is presented to M surrounded by the endmarkers, as $\vdash z \dashv$. The computation starts at q_s and on $\vdash$. The next state is always derived from δ and the current state and symbol. The next position is always the adjacent one in the direction of motion; except when the current symbol is $\dashv$ and the next state is not q_a or when the current symbol is $\vdash$, in which two cases the next position is the adjacent one towards the other endmarker. Note that the computation can either loop, or hang, or fall off $\dashv$ into q_a. In this last case, we say M *accepts* z.

More generally, for any input string $z \in \Sigma^*$ and state p, the *left computation of M from p on z* is the unique sequence $\text{LCOMP}_{M,p}(z) := (q_t)_{1 \leq t \leq m}$ where:

$q_1 := p$; every next state is $q_{t+1} := \delta(q_t, z_t)$, provided that $t \leq |z|$ and the value of δ is defined; and m is the first t for which this provision fails. If $m = |z| + 1$, we say the computation *exits* z into q_m or *results in* q_m; otherwise, $1 \leq m \leq |z|$ and the computation *hangs* at q_m and *results in* $\perp$; the set $Q_\perp := Q \cup \{\perp\}$ contains all possible *results*. The *right computation of M from p on z* is denoted by $\text{RCOMP}_{M,p}(z)$ and defined symmetrically, with $q_{t+1} := \delta(q_t, z_{|z|+1-t})$.

We say M is a RDFA if its next position is decided differently: it is always the adjacent one to the right, except when the current symbol is $\dashv$ and the next state is not q_a, in which case it is the one to the right of $\vdash$.

We say M is a 1DFA if it halts immediately after reading $\dashv$: the value of δ on any state q and on $\dashv$ is always either q_a or undefined. If it is q_a, we say q is a *final* state; if it is undefined, we say q is *nonfinal*. The state $\delta(q_s, \vdash)$, if defined, is called *initial*. If M is allowed more than one next move at each step, we say it is *nondeterministic* (a 1NFA).

Parallel Automata. The following additional models will also be useful.

A *(two-sided) parallel automaton* (P$_2$1DFA) [7] is any triple $M = (\mathfrak{L}, \mathfrak{R}, F)$ where $\mathfrak{L} = \{C_1, \ldots, C_k\}$, $\mathfrak{R} = \{D_1, \ldots, D_l\}$ are disjoint families of 1DFAs, and $F \subseteq Q_\perp^{C_1} \times \cdots \times Q_\perp^{C_k} \times Q_\perp^{D_1} \times \cdots \times Q_\perp^{D_l}$, where Q^A is the state set of automaton A. To run M on z means to run each $A \in \mathfrak{L} \cup \mathfrak{R}$ on z from its initial state and record the result, but with a twist: each $A \in \mathfrak{L}$ reads from left to right (i.e., reads z), while each $A \in \mathfrak{R}$ reads from right to left (i.e., reads z^R). We say M accepts z iff the tuple of the results of these computations is in F. When $\mathfrak{R} = \emptyset$ or $\mathfrak{L} = \emptyset$, we say M is *left-sided* (a P$_L$1DFA) or *right-sided* (a P$_R$1DFA), respectively.

A *parallel intersection automaton* ($\cap_2$1DFA, $\cap_L$1DFA, or $\cap_R$1DFA) [5] is a parallel automaton whose F consists of the tuples where *all* results are final states. If F consists of the tuples where *some* result is a final state, the automaton is a *parallel union automaton* ($\cup_2$1DFA, $\cup_L$1DFA, or $\cup_R$1DFA) [5]. So, a $\cap_2$1DFA accepts its input iff all components accept it; a $\cup_2$1DFA accepts iff any component does.

We say that a family of parallel automata $\mathcal{M} = (M_n)_{n \geq 1}$ are 'small' if for some polynomial p and all n, each component of M_n has at most $p(n)$ states. Note that this restricts only the *size* of the components—not their *number*.

Complexity Classes. The size-complexity class 1D consists of every family of problems that can be solved by a family of small 1DFAs. The classes RD, SD, 2D, $\cap_L$1D, $\cap_R$1D, $\cap_2$1D, $\cup_L$1D, $\cup_R$1D, $\cup_2$1D, P$_L$1D, P$_R$1D, P$_2$1D, and 1N are defined similarly, by replacing 1DFAs with RDFAs, SDFAs, etc. The naming convention is from [5]; there, however, 1D, 1N, and 2D contain families of *languages*, not problems.

If $\mathcal{C}$ is a class, then re-$\mathcal{C}$ consists of all families of problems whose reversal is in $\mathcal{C}$ and co-$\mathcal{C}$ consists of all families of problems whose complement is in $\mathcal{C}$. Of special interest to us is the class 1N $\cap$ co-1N; we also denote it by 1Δ.

The following inclusions are easy to verify, by the definitions and by [7, Lemma 1], for every side mode $\sigma = $ L, R, 2 and every parallel mode $\pi = \cap, \cup, \text{P}$:

$$
\begin{array}{llll}
\text{co-}\cap_\sigma 1\text{D} = \cup_\sigma 1\text{D} & \cap_\sigma 1\text{D}, \cup_\sigma 1\text{D} \subseteq \text{P}_\sigma 1\text{D} & 1\text{D} \subseteq \cap_L 1\text{D}, \cup_L 1\text{D}, \text{RD} \subseteq \text{P}_L 1\text{D} & \\
\text{re-}\pi_L 1\text{D} = \pi_R 1\text{D} & \pi_L 1\text{D}, \pi_R 1\text{D} \subseteq \pi_2 1\text{D} & \text{RD} \subseteq \text{SD} \subseteq \text{P}_2 1\text{D}, 2\text{D}. &
\end{array} \tag{3}
$$

A Core Problem. Let $[n] := \{1, \ldots, n\}$. All witnesses in Sect. 4 will be derived from the operators of (1) and the following 'core' problem: "Given two symbols describing a set $\alpha \subseteq [n]$ and a number $i \in [n]$, check that $i \in \alpha$." Formally,

$$J_n := (\, \{\alpha i \mid \alpha \subseteq [n] \,\&\, i \in \alpha\}, \{\alpha i \mid \alpha \subseteq [n] \,\&\, i \in \overline{\alpha}\} \,). \tag{4}$$

Lemma 1. $\mathcal{J} := (J_n)_{n \geq 1}$ *is not in* 1D *but is in* re-1D, 1N, co-1N, $\cap_{\!l}$1D, $\cup_{\!l}$1D.

3   Basic Tools and Hardness Propagation

To draw the map of Fig. 1, we need several lemmata that explain how the operators of (1) can increase the hardness of problems. In turn, to prove these lemmata, we need two basic tools for the construction of hard inputs to parallel automata: the *confusing* and the *generic* strings. We first describe these tools and then use them to prove the hardness propagation lemmata.

Confusing Strings. Let $M = (\mathfrak{L}, \mathfrak{R})$ be a $\cap_2$1DFA and L a problem. We say a string y *confuses* M *on* L if it is a positive instance but some component hangs on it or is negative but every component treats it identically to a positive one:

$$\text{or} \quad \begin{array}{ll} y \in L & \& \quad (\exists A \in \mathfrak{L} \cup \mathfrak{R})(A(y) = \bot) \\ y \notin L & \& \quad (\forall A \in \mathfrak{L} \cup \mathfrak{R})(\exists \tilde{y} \in L)(A(y) = A(\tilde{y})) \end{array} \tag{5}$$

where $A(z)$ is the result of LCOMP$_A(z)$, if $A \in \mathfrak{L}$, or of RCOMP$_A(z)$, if $A \in \mathfrak{R}$. It can be shown that, if some y confuses M on L, then M does not solve L. Note, though, that (5) is independent of the selection of final states in the components of M. So, if $\mathfrak{F}(M)$ is the class of $\cap_2$1DFAs that may differ from M only in the selection of final states, then a y that confuses M on L confuses every $M' \in \mathfrak{F}(M)$, too, and thus no $M' \in \mathfrak{F}(M)$ solves L, either. The converse is also true.

Lemma 2. *Let* $M = (\mathfrak{L}, \mathfrak{R})$ *be a* $\cap_2$1DFA *and* L *a problem. Then, strings that confuse* M *on* L *exist iff no member of* $\mathfrak{F}(M)$ *solves* L.

Proof. [$\Rightarrow$] Suppose some y confuses M on L. Fix any $M' = (\mathfrak{L}', \mathfrak{R}') \in \mathfrak{F}(M)$. Since (5) is independent of the choice of final states, y confuses M' on L, too. If $y \in L$: By (5), some $A \in \mathfrak{L}' \cup \mathfrak{R}'$ hangs on y. So, M' rejects y, and thus fails. If $y \notin L$: If M' accepts y, it fails. If it rejects y, then some $A \in \mathfrak{L}' \cup \mathfrak{R}'$ does not accept y. Consider the $\tilde{y}$ guaranteed for this A by (5). Since $A(\tilde{y}) = A(y)$, we know $\tilde{y}$ is also not accepted by A. Hence, M' rejects $\tilde{y} \in L$, and fails again.

[$\Leftarrow$] Suppose no string confuses M on L. Then, no component hangs on a positive instance; and every negative instance is 'noticed' by some component, in the sense that the component treats it differently than all positive instances:

$$\text{and} \quad \begin{array}{l} (\forall y \in L)(\forall A \in \mathfrak{L} \cup \mathfrak{R})(A(y) \neq \bot) \\ (\forall y \notin L)(\exists A \in \mathfrak{L} \cup \mathfrak{R})(\forall \tilde{y} \in L)(A(y) \neq A(\tilde{y})). \end{array} \tag{6}$$

This allows us to find an $M' \in \mathfrak{F}(M)$ that solves L, as follows. We start with all states of all components of M unmarked. Then we iterate over all $y \notin L$. For each of them, we pick an A as guaranteed by (6) and, if the result $A(y)$ is a state, we mark it. When this (possibly infinite) iteration is over, we make all marked states nonfinal and all unmarked states final. The resulting $\cap_2 1\text{DFA}$ is our M'.

To see why M' solves L, consider any string y. If $y \notin L$: Then our method examined y, picked an A, and ensured $A(y)$ is either $\bot$ or a nonfinal state. So, this A does not accept y. Therefore, M' rejects y. If $y \in L$: Towards a contradiction, suppose M' rejects y. Then some component A^* does not accept y. By (6), $A^*(y) \neq \bot$. Hence, $A^*(y)$ is a state, call it q^*, and is nonfinal. So, at some point, our method marked q^*. Let $\hat{y} \notin L$ be the string examined at that point. Then, the selected A was A^* and $A(\hat{y})$ was q^*, and thus no $\tilde{y} \in L$ had $A^*(\tilde{y}) = q^*$. But this contradicts the fact that $y \in L$ and $A^*(y) = q^*$.   □

Generic Strings [7]. Let A be a 1DFA over alphabet Σ and states Q, and $y, z \in \Sigma^*$. The (*left*) *views of A on y* is the set of states produced on the right boundary of y by left computations of A:

$$\text{LVIEWS}_A(y) := \{q \in Q \mid (\exists p \in Q)[\text{LCOMP}_{A,p}(y) \text{ exits into } q]\}.$$

The (*left*) *mapping of A on y and z* is the partial function

$$\text{LMAP}_A(y, z) : \text{LVIEWS}_A(y) \rightarrow Q$$

which, for every $q \in \text{LVIEWS}_A(y)$, is defined only if $\text{LCOMP}_{A,q}(z)$ does not hang and, if so, returns the state that this computation exits into. It is easy to verify that this function is a *partial surjection* from $\text{LVIEWS}_A(y)$ to $\text{LVIEWS}_A(yz)$. This immediately implies Fact 1. Fact 2 is equally simple.

Fact 1. *For all A, y, z as above:* $|\text{LVIEWS}_A(y)| \geq |\text{LVIEWS}_A(yz)|$.

Fact 2. *For all A, y, z as above:* $\text{LVIEWS}_A(yz) \subseteq \text{LVIEWS}_A(z)$.

Now consider any $\text{P}_i 1\text{DFA}$ $M = (\mathcal{L}, \emptyset, F)$ and any problem L which is *infinitely right-extensible*, in the sense that every $u \in L$ can be extended into a $uu' \in L$. By Fact 1, if we start with any $u \in L$ and keep right-extending it ad infinitum into $uu', uu'u'', uu'u''u''', \cdots \in L$ then, from some point on, the corresponding sequence of tuples of sizes $(|\text{LVIEWS}_A(\cdot)|)_{A \in \mathcal{L}}$ will become constant. If y is any of the extensions after that point, then y satisfies

$$y \in L \quad \& \quad (\forall yz \in L)(\forall A \in \mathcal{L})\big(|\text{LVIEWS}_A(y)| = |\text{LVIEWS}_A(yz)|\big) \quad (7)$$

and is called L-*generic (for M) over L.* The next lemma uses such strings.

Lemma 3. *Suppose a $\text{P}_i 1\text{DFA}$ $M = (\mathcal{L}, \emptyset, F)$ solves $\bigwedge L$ and y is L-generic for M over $\bigwedge L$. Then, $x \in L$ iff $\text{LMAP}_A(y, xy)$ is total and injective for all $A \in \mathcal{L}$.*

Proof. [⇒] Let $x \in L$. Then $yxy \in \bigwedge L$ (since $y \in \bigwedge L$ and $x \in L$). So, yxy right-extends y inside $\bigwedge L$. Since y is L-generic, $|\text{LVIEWS}_A(y)| = |\text{LVIEWS}_A(yxy)|$, for all $A \in \mathcal{L}$. Hence, each partial surjection $\text{LMAP}_A(y, xy)$ has domain and codomain of the same size. This is possible only if the function is both total and injective.

[⇐] Suppose each partial surjection $\text{LMAP}_A(y, xy)$ is total and injective. Then it bijects the set $\text{LVIEWS}_A(y)$ into the set $\text{LVIEWS}_A(yxy)$, which is actually a subset of $\text{LVIEWS}_A(y)$ (Fact 2). Clearly, this is possible only if this subset is the set itself. So, $\text{LMAP}_A(y, xy)$ is a permutation π_A of $\text{LVIEWS}_A(y)$.

Now pick $k \geq 1$ so that each π_A^k is an identity, and let $z := y(xy)^k$. It is easy to verify that $\text{LMAP}_A(y, (xy)^k)$ equals $\text{LMAP}_A(y, xy)^k = \pi_A^k$, and is therefore the identity on $\text{LVIEWS}_A(y)$. This means that, reading through z, the left computations of A do not notice the suffix $(xy)^k$ to the right of the prefix y. So, no A can distinguish between y and z: it either hangs on both or exits both into the same state. Thus, M does not distinguish between y and z, either: it either accepts both or rejects both. But M accepts y (because $y \in \bigwedge L$), so it accepts z. Hence, every #-delimited infix of z is in L. In particular, $x \in L$.   □

If $M = (\mathfrak{L}, \mathfrak{R}, F)$ is a $\text{P}_2\text{1DFA}$, we can also work symmetrically with right computations and left-extensions: we can define $\text{RVIEWS}_A(y)$ and $\text{RMAP}_A(z, y)$ for $A \in \mathfrak{R}$, derive Facts 1, 2 for $\text{RVIEWS}_A(y)$ and $\text{RVIEWS}_A(zy)$, and define R-generic strings. We can then construct strings, called *generic*, that are simultaneously L- and R-generic, and use them in a counterpart of Lemma 3 for $\text{P}_2\text{1DFAs}$:

Lemma 4. *Suppose a* $\text{P}_2\text{1DFA}$ $M = (\mathfrak{L}, \mathfrak{R}, F)$ *solves* $\bigwedge L$ *and* y *is generic for* M *over* $\bigwedge L$. *Then,* $x \in L$ *iff* $\text{LMAP}_A(y, xy)$ *is total and injective for all* $A \in \mathfrak{L}$ *and* $\text{RMAP}_A(yx, y)$ *is total and injective for all* $A \in \mathfrak{R}$.

Hardness Propagation. We are now ready to show how the operators of (1) can allow us to build harder problems out of easier ones.

Lemma 5. *If no* m-*state* 1DFA *can solve problem* L, *then no* $\cap_{\text{L}}\text{1DFA}$ *with* m-*state components can solve problem* $\bigvee L$. *Similarly for* $\bigoplus L$.

Proof. Suppose no m-state 1DFA can solve L. By induction on k, we prove that no $\cap_{\text{L}}\text{1DFA}$ with k m-state components can solve $\bigvee L$ (the proof for $\bigoplus L$ is similar).

If $k = 0$: Fix any such $\cap_{\text{L}}\text{1DFA}$ $M = (\mathfrak{L}, \emptyset)$. By definition, # $\notin \bigvee L$. But M accepts #, because all components do (vacuously, since $\mathfrak{L} = \emptyset$). So M fails.

If $k \geq 1$: Fix any such $\cap_{\text{L}}\text{1DFA}$ $M = (\mathfrak{L}, \emptyset)$. Pick any $D \in \mathfrak{L}$ and remove it from M to get $M_1 = (\mathfrak{L}_1, \emptyset) := (\mathfrak{L} - \{D\}, \emptyset)$. By the inductive hypothesis, no member of $\mathfrak{F}(M_1)$ solves $\bigvee L$. So (Lemma 2), some y confuses M_1 on $\bigvee L$.

Case 1: $y \in \bigvee L$. Then some $A \in \mathfrak{L}_1$ hangs on y. Since $A \in \mathfrak{L}$, too, y confuses M as well. So, M does not solve $\bigvee L$, and the inductive step is complete.

Case 2: $y \notin \bigvee L$. Then every $A \in \mathfrak{L}_1$ treats y identically to a positive instance:

$$(\forall A \in \mathfrak{L} - \{D\})(\exists \tilde{y} \in \bigvee L)(A(y) = A(\tilde{y})). \tag{8}$$

Let M_2 be the single-component $\cap_{\text{L}}\text{1DFA}$ whose only 1DFA, call it D', is the one derived from D by changing its initial state to $D(y)$. By the hypothesis of the lemma, no member of $\mathfrak{F}(M_2)$ solves L. So (Lemma 2), some x confuses M_2 on L. We claim that yx# confuses M on $\bigvee L$. Thus, M does not solve $\bigvee L$, and the induction is again complete. To prove the confusion, we examine cases:

Case 2a: $x \in L$. Then $yx\# \in \bigvee L$, since $y \in (\bigvee L)_P$ and $x \in L$. And D' hangs on x (since x is confusing and D' is the only component), thus $D(yx\#) = D'(x\#) = \perp$. So, component D of M hangs on $yx\# \in \bigvee L$. So, $yx\#$ confuses M on $\bigvee L$.

Case 2b: $x \notin L$. Then $yx\# \notin \bigvee L$, because $y \notin \bigvee L$ and $x \in L_P$. And, since x is confusing, D' treats it identically to some $\tilde{x} \in L$: $D'(x) = D'(\tilde{x})$. Then, each component of M treats $yx\#$ identically to a positive instance of $\bigvee L$:

- D treats $yx\#$ as $y\tilde{x}\#$: $D(y\tilde{x}\#) = D'(\tilde{x}\#) = D'(x\#) = D(yx\#)$. And we know $y\tilde{x}\# \in \bigvee L$, because $y \in (\bigvee L)_P$ and $\tilde{x} \in L$.
- each $A \neq D$ treats $yx\#$ as $\tilde{y}x\#$, where $\tilde{y}$ the string guaranteed for A by (8): $A(\tilde{y}x\#) = A(yx\#)$. And we know $\tilde{y}x\# \in \bigvee L$, since $\tilde{y} \in \bigvee L$ and $x \in L_P$.

Overall, $yx\#$ is again a confusing string for M on $\bigvee L$, as required. □

Lemma 6. *If L_1 has no $\cap_L 1$DFA with m-state components and L_2 has no $\cap_R 1$DFA with m-state components, then $L_1 \vee L_2$ has no $\cap_2 1$DFA with m-state components. Similarly for $L_1 \oplus L_2$.*

Proof. Let $M = (\mathfrak{L}, \mathfrak{R})$ be a $\cap_2 1$DFA with m-state components. Let $M_1 := (\mathfrak{L}', \emptyset)$ and $M_2 := (\emptyset, \mathfrak{R}')$ be the $\cap_2 1$DFAs derived from the two 'sides' of M after changing the initial state of each $A \in \mathfrak{L} \cup \mathfrak{R}$ to $A(\#)$. By the lemma's hypothesis, no member of $\mathfrak{F}(M_1)$ solves L_1 and no member of $\mathfrak{F}(M_2)$ solves L_2. So (Lemma 2), some y_1 confuses M_1 on L_1 and some y_2 confuses M_2 on L_2. We claim that $\#y_1\#y_2\#$ confuses M on $L_1 \vee L_2$ and thus M fails. (Similarly for $L_1 \oplus L_2$.)

Case 1: $y_1 \in L_1$ or $y_2 \in L_2$. Assume $y_1 \in L_1$ (if $y_2 \in L_2$, we work similarly). Then $\#y_1\#y_2\# \in L_1 \vee L_2$ and some $A' \in \mathfrak{L}'$ hangs on y_1. The corresponding $A \in \mathfrak{L}$ has $A(\#y_1\#y_2\#) = A'(y_1\#y_2\#) = \perp$. So, $\#y_1\#y_2\#$ confuses M on $L_1 \vee L_2$.

Case 2: $y_1 \notin L_1$ and $y_2 \notin L_2$. Then $\#y_1\#y_2\# \notin L_1 \vee L_2$, and each component of M_1 treats y_1 identically to a positive instance of L_1, and same for M_2, y_2, L_2:

$$(\forall A' \in \mathfrak{L}')(\exists \tilde{y}_1 \in L_1)\big(A'(y_1) = A'(\tilde{y}_1)\big), \tag{9}$$

$$(\forall A' \in \mathfrak{R}')(\exists \tilde{y}_2 \in L_2)\big(A'(y_2) = A'(\tilde{y}_2)\big). \tag{10}$$

It is then easy to verify that every $A \in \mathfrak{L}$ treats $\#y_1\#y_2\#$ as $\#\tilde{y}_1\#y_2\# \in L_1 \vee L_2$ ($\tilde{y}_1$ as guaranteed by (9)), and every $A \in \mathfrak{R}$ treats $\#y_1\#y_2\#$ as $\#y_1\#\tilde{y}_2\# \in L_1 \vee L_2$ ($\tilde{y}_2$ as guaranteed by (10)). Therefore, $\#y_1\#y_2\#$ confuses M on $L_1 \vee L_2$, again. □

Lemma 7. *Let L' be nontrivial, $\pi \in \{\cap, \cup, P\}$, $\sigma \in \{L, R, 2\}$. If L has no $\pi_\sigma 1$DFA with m-state components, then neither $L \wedge L'$ has. Similarly for $\neg L$ and $L \oplus L'$.*

Proof. We prove only the first claim, for $\pi = \cap$ and $\sigma = L$. Fix any $y' \in L'$. Given a $\cap_L 1$DFA M' solving $L \wedge L'$ with m-state components, we build a $\cap_L 1$DFA M solving L with m-state components: We just modify each component A' of M' so that the modified A' works on y exactly as A' on $\#y\#y'\#$. Then, M accepts y $\Leftrightarrow$ M' accepts $\#y\#y'\#$ $\Leftrightarrow$ $y \in L$. The modifications are straightforward. □

Lemma 8. *If L has no $\cap_L 1$DFA with $\binom{m}{2}$-state components, then $\bigwedge L$ has no $P_L 1$DFA with m-state components.*

Proof. Let $M = (\mathfrak{L}, \emptyset, F)$ be a $\text{P}_{\text{L}}1\text{DFA}$ solving $\bigwedge L$ with m-state components. Let y be L-generic for M over $\bigwedge L$. We will build a $\cap_{\text{L}}1\text{DFA}$ M' solving L.

By Lemma 3, an arbitrary x is in L iff $\text{LMAP}_A(y, xy)$ is total and injective for all $A \in \mathfrak{L}$; i.e., iff for all $A \in \mathfrak{L}$ and every two distinct $p, q \in \text{LVIEWS}_A(y)$,

$$\text{LCOMP}_{A,p}(xy) \text{ and } \text{LCOMP}_{A,q}(xy) \text{ exit } xy, \text{ into different states.} \tag{11}$$

So, checking $x \in L$ reduces to checking (11) for each A and two-set of states of $\text{LVIEWS}_A(y)$. The components of M' will perform exactly these checks. To describe them, let us first define the following relation on the states of an $A \in \mathfrak{L}$:

$$r \asymp_A s \iff \text{LCOMP}_{A,r}(y) \text{ and } \text{LCOMP}_{A,s}(y) \text{ exit } y, \text{ into different states,}$$

and restate our checks as follows: for all $A \in \mathfrak{L}$ and all distinct $p, q \in \text{LVIEWS}_A(y)$,

$$\text{LCOMP}_{A,p}(x) \text{ and } \text{LCOMP}_{A,q}(x) \text{ exit } x, \text{ into states that relate under } \asymp_A. \tag{11'}$$

Now, building 1DFAs to perform these checks is easy. For each $A \in \mathfrak{L}$ and $p, q \in \text{LVIEWS}_A(y)$, the corresponding 1DFA has 1 state for each two-set of states of A. The initial state is $\{p, q\}$. At each step, the automaton applies A's transition function on the current symbol and each state in the current two-set. If either application returns no value or both return the same value, it hangs; otherwise, it moves to the resulting two-set. A state $\{r, s\}$ is final iff $r \asymp_A s$.   □

Lemma 9. *If L has no $\cap_2 1\text{DFA}$ with $\binom{m}{2}$-state components, then $\bigwedge L$ has no $\text{P}_2 1\text{DFA}$ with m-state components.*

4   Closure Properties and a Hierarchy

We are now ready to confirm the information of Fig. 1. We start with the positive cells of the table, continue with the diagram, and finish with the negative cells of the table. On the way, Lemma 11 proves a few useful facts.

Lemma 10. *Every '+' in the table of* Fig. 1b *is correct.*

Proof. Each closure can be proved easily, by standard constructions. We also use the fact that every m-state RDFA (resp., SDFA) can be converted into an equivalent one with $O(m^2)$ states that keeps track of the number of rotations (resp., sweeps), and thus never loops. Similarly for 2DFAs and $O(m)$ [2].   □

Lemma 11. *The following separations and fact hold:*

[I] $\cap_{\text{L}}1\text{D} \not\supseteq$ re-1D,   [III] $\cap_2 1\text{D} \not\supseteq \cup_{\text{L}}1\text{D} \cap \text{RD}$,   [V] there exists $\mathcal{L} \in \cup_{\text{L}}1\text{D} \cap \text{RD}$
[II] $\text{P}_{\text{L}}1\text{D} \not\supseteq$ re-1D,   [IV] $\cap_{\text{L}}1\text{D} \cup \cap_{\text{R}}1\text{D} \not\supseteq \cap_2 1\text{D} \cap \text{SD}$ such that $\bigwedge \mathcal{L} \notin \text{P}_2 1\text{D}$.

Proof. [I] Let $\mathcal{L} := \bigvee \mathcal{J}$. We prove $\mathcal{L}$ is a witness. *First*, $\mathcal{J} \notin 1\text{D}$ (Lemma 1) implies $\bigvee \mathcal{J} \notin \cap_{\text{L}}1\text{D}$ (Lemma 5). *Second*, $\mathcal{J}^{\text{R}} \in 1\text{D}$ (Lemma 1) implies $\bigvee \mathcal{J}^{\text{R}} \in 1\text{D}$ (by A7 of Fig. 1b), and thus $(\bigvee \mathcal{J})^{\text{R}} \in 1\text{D}$ (by (2)).

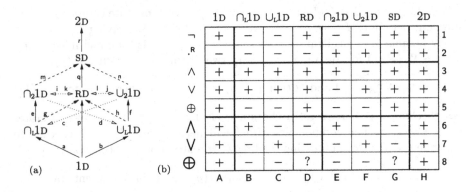

Fig. 1. (a) A hierarchy from 1D to 2D: a solid arrow $\mathcal{C} \to \mathcal{C}'$ means $\mathcal{C} \subseteq \mathcal{C}'$ & $\mathcal{C} \not\supseteq \mathcal{C}'$; a dashed arrow means the same, but $\mathcal{C} \subseteq \mathcal{C}'$ only for the part of $\mathcal{C}$ that can be solved with polynomially many components; a dotted arrow means only $\mathcal{C} \not\supseteq \mathcal{C}'$. (b) Closure properties: '+' means closure; '−' means non-closure; '?' means we do not know.

[II] Let $\mathcal{L} := \bigwedge\bigvee\mathcal{J}$. We prove $\mathcal{L}$ is a witness. *First*, $\bigvee\mathcal{J} \notin \cap_L 1D$ (by I) implies $\bigwedge\bigvee\mathcal{J} \notin P_L 1D$ (Lemma 8). *Second*, $(\bigvee\mathcal{J})^R \in 1D$ (by I) implies $\bigwedge(\bigvee\mathcal{J})^R \in 1D$ (by A6 of Fig. 1b), and thus $(\bigwedge\bigvee\mathcal{J})^R \in 1D$ (by (2)).

[III] Let $\mathcal{L} := (\bigvee\mathcal{J})\vee(\bigvee\mathcal{J}^R)$. We prove $\mathcal{L}$ is a witness. *First*, $\bigvee\mathcal{J} \notin \cap_L 1D$ (by I) implies $(\bigvee\mathcal{J})^R \notin \text{re-}\cap_L 1D$ or, equivalently, $\bigvee\mathcal{J}^R \notin \cap_R 1D$ (by (2), (3)). Overall, both $\bigvee\mathcal{J} \notin \cap_L 1D$ and $\bigvee\mathcal{J}^R \notin \cap_R 1D$, and thus $\mathcal{L} \notin \cap_2 1D$ (Lemma 6). *Second*, $\mathcal{J} \in \cup_L 1D$ via $\cup_L 1DFAs$ with few components (Lemma 1) and thus $\bigvee\mathcal{J} \in \cup_L 1D$ also via $\cup_L 1DFAs$ with few components (by C7); therefore $\bigvee\mathcal{J} \in RD$ via the RDFA that simulates these components one by one. Hence, $\bigvee\mathcal{J} \in \cup_L 1D \cap RD$. In adddition, $\mathcal{J}^R \in 1D$ (Lemma 1) implies $\bigvee\mathcal{J}^R \in 1D$ (by A7), and thus $\bigvee\mathcal{J}^R \in \cup_L 1D \cap RD$ as well (since $1D \subseteq \cup_L 1D, RD$). Overall, both $\bigvee\mathcal{J}$ and $\bigvee\mathcal{J}^R$ are in $\cup_L 1D \cap RD$. Hence, $\mathcal{L} \in \cup_L 1D \cap RD$ as well (by C4,D4).

[IV] Let $\mathcal{L} := (\bigvee\mathcal{J}) \wedge (\bigvee\mathcal{J}^R)$. We prove $\mathcal{L}$ is a witness. However, given that $\mathcal{L}^R = (\bigvee\mathcal{J}^R)^R \wedge (\bigvee\mathcal{J})^R = \bigvee(\mathcal{J}^R)^R \wedge \bigvee\mathcal{J}^R = \mathcal{L}$, we know $\mathcal{L} \in \cap_L 1D \iff \mathcal{L} \in \cap_R 1D$, and thus it enough to prove only that $\mathcal{L} \in (\cap_2 1D \cap SD) \setminus \cap_L 1D$. Here is how. *First*, $\bigvee\mathcal{J} \notin \cap_L 1D$ (by I) and $\bigvee\mathcal{J}^R$ is nontrivial, so $\mathcal{L} \notin \cap_L 1D$ (by Lemma 7). *Second*, $\bigvee\mathcal{J}^R \in 1D$ (by I) implies $(\bigvee\mathcal{J})^R \in \cap_2 1D \cap SD$ (since $1D \subseteq \cap_2 1D, SD$) and thus $\bigvee\mathcal{J} \in \cap_2 1D \cap SD$ as well (by E2,G2). Since both $\bigvee\mathcal{J}$ and $\bigvee\mathcal{J}^R$ are in $\cap_2 1D \cap SD$, the same is true of $\mathcal{L}$ (by E3,G3).

[v] Let $\mathcal{L} := (\bigvee\mathcal{J}) \vee (\bigvee\mathcal{J}^R)$. By III, $\mathcal{L} \in (\cup_L 1D \cap RD) \setminus \cap_2 1D$. By Lemma 9, $\mathcal{L} \notin \cap_2 1D$ implies $\bigwedge\mathcal{L} \notin P_2 1D$. □

Lemma 12. *Every arrow in the hierarchy of Fig. 1a is correct.*

Proof. All inclusions are immediate, either by the definitions or by easy constructions. Note that g, h, m, n refer only to the case of parallel automata with polynomially many components. The non-inclusions are established as follows.

[a,b] By Lemma 1. [d,k,m] By III. [c,l,n] By d,k,m, respectively, and (3),D1,G1. [g,p,h] By k,l. [e] By I, since re-1D $\subseteq \cap_2 1D$. [f] By e and (3). [i,q,j] By II and since

re-1D $\subseteq \cap_2$1D, $\cup_2$1D, SD and RD $\subseteq$ P$_1$1D. [r] Pick $\mathcal{L}$ as in V. Then $\bigwedge \mathcal{L} \notin$ SD (since SD $\subseteq$ P$_2$1D) but $\bigwedge \mathcal{L} \in$ 2D (by H6 and since $\mathcal{L} \in$ RD $\subseteq$ 2D). □

Lemma 13. *Every '−' in the table of Fig. 1b is correct.*

Proof. We examine the cells visiting them row by row, from top to bottom.

[B1] By C1 and (3). [C1] Pick $\mathcal{L}$ as in III. Then $\mathcal{L} \in \cup_1$1D but $\mathcal{L} \notin \cap_2$1D, so $\neg\mathcal{L} \notin \cup_2$1D and thus $\neg\mathcal{L} \notin \cup_1$1D. [E1] Pick $\mathcal{L}$ as in III. Then $\mathcal{L} \notin \cap_2$1D. But $\neg\mathcal{L} \in \cap_2$1D (because $\mathcal{L} \in \cup_1$1D, so $\neg\mathcal{L} \in \cap_1$1D). [F1] By E1 and (3).

[A2] By Lemma 1, $\mathcal{J}^R \in$ 1D but $\mathcal{J} \notin$ 1D. [B2] Pick $\mathcal{L}$ as in I. Then $\mathcal{L} \notin \cap_1$1D but $\mathcal{L}^R \in$ 1D $\subseteq \cap_1$1D. [C2] Pick $\mathcal{L}$ as in I. Since $\mathcal{L} \notin \cap_1$1D, we know $\neg\mathcal{L} \notin \cup_1$1D. Since $\mathcal{L}^R \in$ 1D, we know $\neg(\mathcal{L}^R) \in$ 1D (by A1) and thus $(\neg\mathcal{L})^R \in \cup_1$1D. [D2] Pick $\mathcal{L}$ as in II. Then $\mathcal{L}^R \in$ 1D $\subseteq$ RD but $\mathcal{L} \notin$ P$_1$1D $\supseteq$ RD.

[F3] Let $\mathcal{L}_1, \mathcal{L}_2$ be the witnesses for E4. Then $\mathcal{L}_1, \mathcal{L}_2 \in \cap_2$1D, hence $\neg\mathcal{L}_1, \neg\mathcal{L}_2 \in \cup_2$1D. But $\mathcal{L}_1 \vee \mathcal{L}_2 \notin \cap_2$1D, hence $\neg(\mathcal{L}_1 \vee \mathcal{L}_2) \notin \cup_2$1D or, equivalently $\neg\mathcal{L}_1 \wedge \neg\mathcal{L}_2 \notin \cup_2$1D. [E4] Pick $\mathcal{L}$ as in I. Then $\mathcal{L}^R \in$ 1D, hence $\mathcal{L}^R \in \cap_2$1D (since 1D $\subseteq \cap_2$1D), and thus $\mathcal{L} \in \cap_2$1D (by E2). But $\mathcal{L} \vee \mathcal{L}^R \notin \cap_2$1D (by III).

[B5,E5] Let $\mathcal{L}$ be the complement of the family of III. Then $\mathcal{L} \in \cap_1$1D $\subseteq \cap_2$1D. But $\neg\mathcal{L} \notin \cap_2$1D, and thus $\mathcal{L} \oplus \mathcal{L} \notin \cap_2$1D $\supseteq \cap_1$1D (Lemma 7). [C5,F5] Pick $\mathcal{L}$ as in III. Then $\mathcal{L} \in \cup_1$1D $\subseteq \cup_2$1D. But $\mathcal{L} \notin \cap_2$1D, hence $\neg\mathcal{L} \notin \cup_2$1D, and thus $\mathcal{L} \oplus \mathcal{L} \notin \cup_2$1D $\supseteq \cup_1$1D (Lemma 7).

[C6,D6,F6,G6] Pick $\mathcal{L}$ as in V. Then $\mathcal{L} \in \cup_1$1D $\cap$ RD $\subseteq \cup_2$1D, SD. But $\bigwedge \mathcal{L} \notin$ P$_2$1D and thus $\bigwedge \mathcal{L} \notin \cup_1$1D, RD, $\cup_2$1D, SD. [B7,D7,E7,G7] Let $\mathcal{L}$ be the complement of the family of V. Then $\mathcal{L} \in \cap_1$1D $\cap$ RD (by D1), and thus also $\mathcal{L} \in \cap_2$1D, SD. But $\neg\bigvee \mathcal{L} = \bigwedge \neg\mathcal{L} \notin$ P$_2$1D, so $\neg\bigvee \mathcal{L} \notin \cap_1$1D, RD, $\cap_2$1D, SD, and same for $\bigvee \mathcal{L}$ (by D1,G1).

[B8,C8,E8,F8] By B5,C5,E5,F5. The witnesses there, are problems of the form $\mathcal{L} \oplus \mathcal{L}$, for some $\mathcal{L}$. Such problems simply restrict the corresponding $\bigoplus \mathcal{L}$. □

5   Conclusions

For each $n \geq 1$, let S_n be the problem: "Given a set $\alpha \subseteq [n]$ and two numbers $i, j \in [n]$ exactly one of which is in α, check that the one in α is j." Formally:

$$S_n := (\{\alpha i j \mid \alpha \subseteq [n] \ \& \ i \in \bar{\alpha} \ \& \ j \in \alpha\}, \{\alpha i j \mid \alpha \subseteq [n] \ \& \ i \in \alpha \ \& \ j \in \bar{\alpha}\}).$$

For $\mathcal{S} := (S_n)_{n \geq 1}$ the corresponding family, consider the family

$$\mathcal{R} = (R_n)_{n \geq 1} := \bigwedge ((\bigoplus \mathcal{S}) \oplus (\bigoplus \mathcal{S}^R)).$$

It is easy to see that $\mathcal{S} \in$ 1Δ = 1N$\cap$co-1N and that 1Δ is closed under $\cdot^R, \bigoplus, \oplus, \bigwedge$. Hence, $\mathcal{R} \in$ 1Δ as well. At the same time, $\mathcal{S} \notin$ 1D (easily), so $\bigoplus \mathcal{S} \notin \cap_1$1D (Lemma 5) and $\bigoplus \mathcal{S}^R = (\bigoplus \mathcal{S})^R \notin \cap_R$1D, which implies $(\bigoplus \mathcal{S}) \oplus (\bigoplus \mathcal{S}^R) \notin \cap_2$1D (Lemma 6) and thus $\mathcal{R} \notin$ P$_2$1D (Lemma 9). Hence, $\mathcal{R} \notin$ SD either. Overall, $\mathcal{R}$ witnesses that 1$\Delta \nsubseteq$ SD. This separation was first proven in [3]. There, it was witnessed by a language family $(\Pi_n)_{n \geq 1}$ that restricted liveness [5].

We claim that, for all n, Π_n and R_n are '*essentially the same*': For each direction (left-to-right, right-to-left), there exists a $O(n)$-state one-way transducer that converts any well-formed instance u of Π_n into a string v in the promise of R_n such that $u \in \Pi_n \iff v \in R_n$. Conversely, it is also true that for each direction some $O(n)$-state one-way transducer converts any v from the promise of R_n into a well-formed instance u of Π_n such that $u \in \Pi_n \iff v \in R_n$.

Therefore, using our operators, we essentially 'reconstructed' the witness of [3] in a way that identifies the source of its complexity (the witness of $1\Delta \nsubseteq 1D$ at its core) and reveals how its definition used reversal, parity, and conjuction to propagate its deterministic hardness upwards from $1D$ to SD without increasing its hardness with respect to 1Δ.

At the same time, using our operators, we can easily show that the witness of [3] is, in fact, unnecessarily complex. Already from the proof of Lemma 11[v] (and the easy closure of 1Δ under $\cdot^R, \bigvee, \vee, \bigwedge$), we know that even

$$\mathcal{L} = (L_n)_{n \geq 1} := \bigwedge((\bigvee \mathcal{J}) \vee (\bigvee \mathcal{J}^R))$$

witnesses $1\Delta \nsubseteq SD$. Indeed, $\mathcal{L}$ is both *simpler* than $\mathcal{R}$ (uses $\mathcal{J}, \bigvee, \vee$ instead of $\mathcal{S}, \bigoplus, \oplus$) and *more effective* (we can prove it needs $O(n)$ states on 1NFAs and co-1NFAs and $\Omega(2^{n/2})$ states on SDFAs, compared to $\mathcal{R}$'s $O(n^2)$ and $\Omega(2^{n/2}/\sqrt{n})$ [3]).

Finally, using our operators, we can systematically produce many different witnesses for each provable separation. The following corollary is indicative.

Corollary 1. *Let $\mathcal{L}$ by any family of problems.*
- *If $\mathcal{L} \in 1\Delta \setminus 1D$, then $\bigwedge\bigvee\mathcal{L} \in 1\Delta \setminus RD$.*
- *If $\mathcal{L} \in 1\Delta \setminus (1D \cup \text{re-}1D)$, then $\bigwedge\bigvee\mathcal{L} \in 1\Delta \setminus SD$.*
- *If $\mathcal{L} \in \text{re-}1D \setminus 1D$, then $\bigwedge\bigvee(\mathcal{L} \vee \mathcal{L}^R) \in (1\Delta \cap 2D) \setminus SD$.*

Note how the alternation of $\bigwedge$ and $\bigvee$ (in 'conjunctive normal form' style) increases the hardness of a core problem; it would be interesting to further understand its role in this context. Answering the ?'s of Fig. 1b would also be very interesting—they seem to require tools other than the ones currently available.

References

1. Berman, P.: A note on sweeping automata. In: de Bakker, J.W., van Leeuwen, J. (eds.) ICALP 1980. LNCS, vol. 85, pp. 91–97. Springer, Heidelberg (1980)
2. Geffert, V., Mereghetti, C., Pighizzini, G.: Complementing two-way finite automata. Information and Computation 205(8), 1173–1187 (2007)
3. Kapoutsis, C., Královič, R., Mömke, T.: An exponential gap between LasVegas and deterministic sweeping finite automata. In: Hromkovič, J., Královič, R., Nunkesser, M., Widmayer, P. (eds.) SAGA 2007. LNCS, vol. 4665, pp. 130–141. Springer, Heidelberg (2007)
4. Micali, S.: Two-way deterministic finite automata are exponentially more succinct than sweeping automata. Information Processing Letters 12(2), 103–105 (1981)
5. Sakoda, W.J., Sipser, M.: Nondeterminism and the size of two way finite automata. In: Proceedings of the 10th Annual ACM Symposium on Theory of Computing, San Diego, California, USA, May 1-3, 1978. ACM, New York (1978)
6. Seiferas, J.I.: Untitled manuscript. Communicated to Michael Sipser (October 1973)
7. Sipser, M.: Lower bounds on the size of sweeping automata. Journal of Computer and System Sciences 21(2), 195–202 (1980)

An Analysis and a Reproof of Hmelevskii's Theorem*
(Extended Abstract)

Juhani Karhumäki and Aleksi Saarela

Department of Mathematics and Turku Centre for Computer Science TUCS,
University of Turku, 20014 Turku, Finland
karhumak@utu.fi, amsaar@utu.fi

Abstract. We analyze and reprove the famous theorem of Hmelevskii, which states that the general solutions of constant-free equations on three unknowns are finitely parameterizable, that is expressible by a finite collection of formulas of word and numerical parameters. The proof is written, and simplified, by using modern tools of combinatorics on words. As a new aspect the size of the finite representation is estimated; it is bounded by a double exponential function on the size of the equation.

1 Introduction

Theory of word equations is a fundamental part of combinatorics on words. It plays an essential role in a number of areas of mathematical research, such as in representation results of algebra, theory of algorithms and pattern matching. During the few last decades it has provided several challenging problems as well as fundamental, or even breakthrough, results in discrete mathematics.

Remarkable achievements of the topic are the decidability of the satisfiability problem for word equations, and the compactness result of systems of word equations, see [9] for the first and [1] and [4] for the second. The first result was reproved and sharpened to a PSPACE algorithm in [10]. For the latter one the question of bounding the size of an equivalent finite subset is still a challenge.

In the case of word equations with only three unknowns fundamental results have also been achieved. In one direction Hmelevskii [6] proved already in 1970 that any such constant-free equation is finitely parameterizable, that is the general solution can be expressed as a finite formula on word and numerical parameters. On other direction Spehner [11,12] classified all sets of relations a given solution, that is a triple of words, can satisfy. A remarkable thing is that both of these results have only very complicated proofs. This, if any, is a splendid example of a challenging nature of word problems. Indeed, even the basic question of finding any upper bound for the maximal size of independent system of word equations on three unknowns is still open, see [5] and [3].

The goal of this paper is to analyze the proof of Hmelevskii's theorem. The result itself is, of course, very well known, see e.g. [8]. However, a compact and

---

* Supported by the Academy of Finland under grant 8121419.

readable presentation of it seems to be lacking. We hope to fill this gap. In other words, we search for a self-contained proof using achievements and tools of combinatorics on words obtained over the last 30 years. The hope will be completed only in the full paper, but we do believe that already this presentation will give the reader justified impression of the proof. In addition, we conclude, for the first time, an upper bound for the size of the formula giving the general solution of a constant-free equation on three unknowns. Our bound is double exponential in terms of the length of the equation – and thus not likely to be even close to the optimal one.

In this extended abstract the proof is outlined in modern terms and tools of combinatorics on words, but many details are left to the final version of the full paper [7].

2   Definitions and Basic Results

In this section we fix the terminology and state the basic auxiliary results needed, for more see [2].

We consider word equations $U = V$, where $U, V \in \varXi^*$ and $\varXi$ is the alphabet of unknowns. A morphism $h : \varXi^* \to \varSigma^*$ is a solution of this equation, if $h(U) = h(V)$. We also consider *one-sided* equations $xU \rightrightarrows yV$. A morphism $h : \varXi^* \to \varSigma^*$ is a solution of this equation, if $h(xU) = h(yV)$ and $|h(x)| \geq |h(y)|$.

A solution h is *periodic*, if there exists such $t \in \varSigma^*$ that every $h(x)$, where $x \in \varXi$, is a power of t. Otherwise h is *nonperiodic*. Periodic solutions are easy to find and represent, so in many cases it is enough to consider nonperiodic ones.

If a word u is a *prefix* of a word v, that is $v = uw$ for some w, the notation $u \leq v$ is used. If also $u \neq v$, then u is a *proper prefix*; this is denoted by $u < v$.

Let $w = a_1 \dots a_n$. Its *reverse* is $w^R = a_n \dots a_1$, and its *length* is $|w| = n$. The number of occurrences of a letter a in w is denoted by $|w|_a$.

If $\varSigma = \{a_1, \dots, a_n\}$, then $U \in \varSigma^*$ can be denoted $U(a_1, \dots, a_n)$, and its image under a morphism h can be denoted $h(U) = U(h(a_1), \dots, h(a_n))$. If $u \in \varSigma^*$, then the morphism $a_1 \mapsto u$ means the morphism, which maps $a_1 \mapsto u$ and $a_i \mapsto a_i$, when $i = 2, \dots, n$.

The following theorems and lemmas are easy to prove by using standard methods for solving equation. They give solutions to some simple equations. These solutions will be the basis of parametric solutions of all equations with three unknowns. We start with the well known lemmata, see [2].

Theorem 2.1. *Let $U, V \in \{x, y\}^*$ and $U \neq V$. Assume that $|U|_x = a$, $|U|_y = b$, $|V|_x = c$ and $|V|_y = d$. The solutions of the equation $U = V$ are $x = t^i$, $y = t^j$, where $t \in \varSigma^*$, $ai + bj = ci + dj$ and $i, j \geq 0$.*

Theorem 2.2. *The solutions of the equation $xz = zy$ are $x = pq$, $y = qp$, $z = p(qp)^i$ or $x = y = 1$, $z = p$, where $p, q \in \varSigma^*$ and $i \geq 0$.*

Lemma 2.3. *The nonperiodic solutions of the equation $xyz = zyx$ are $x = (pq)^i p$, $y = q(pq)^j$, $z = (pq)^k p$, where $p, q \in \varSigma^*$, $i, j, k \geq 0$, $pq \neq qp$ and pq can be assumed to be primitive.*

Lemma 2.4. *The nonperiodic solutions of the equation* $xyz = zxy$ *are* $x = (pq)^i p$, $y = q(pq)^j$, $z = (pq)^k$, *where* $p, q \in \Sigma^*$, $i, j, k \geq 0$ *and* $pq \neq qp$.

Lemma 2.5. *Let* $a \geq 2$. *The nonperiodic solutions of the equation* $xzx = y^a$ *are* $x = (pq)^i p$, $y = (pq)^{i+1} p$, $z = qp((pq)^{i+1}p)^{a-2}pq$, *where* $p, q \in \Sigma^*$, $i \geq 0$ *and* $pq \neq qp$.

Lemma 2.6. *Let* $a \geq 2$. *The nonperiodic solutions of the equation* $xy^a z = zy^a x$ *are* $x = (pq^a)^i p$, $y = q$, $z = (pq^a)^j p$ *or*

$$\begin{cases} x &= qp((pq)^{k+1}p)^{a-2}pq(((pq)^{k+1}p)^{a-1}pq)^i, \\ y &= (pq)^{k+1}p, \\ z &= qp((pq)^{k+1}p)^{a-2}pq(((pq)^{k+1}p)^{a-1}pq)^j, \end{cases}$$

where $p, q \in \Sigma^*$, $i, j, k \geq 0$ *and* $pq \neq qp$.

Lemma 2.7. *The nonperiodic solutions of the equation* $xyxz \rightrightarrows zx^2 y$ *are* $x = (pq)^i p$, $y = qp((pq)^{i+1}p)^j pq$, $z = pq$, *where* $p, q \in \Sigma^*$, $i, j \geq 0$ *and* $pq \neq qp$.

Lemma 2.8. *Let* $a, b \geq 1$ *and* $U, V \in \Sigma^*$. *If* h *is a solution of the equation* $x^a y U = y^b x V$, *then* $h(x)$ *and* $h(y)$ *commute.*

The following corollary of the graph lemma is also useful. A proof can be found in [2]. This result simplifies the original proof of Hmelevskii in several places.

Theorem 2.9. *Let* $A, B, C, D \in \{x, y, z\}^*$. *If* h *is a solution of the pair of equations* $xA = yB$, $xC = zD$, *then* h *is periodic or one of* $h(x), h(y), h(z)$ *equals* 1.

3   Parametric Words

In this section, we define the central notions of this presentation, namely parametric words, parameterizability and parametric solutions.

Fix the alphabet of *word parameters* Δ and the set of *numerical parameters* Λ. Now *parametric words* are defined inductively as follows:

(i) if $a \in \Delta \cup \{1\}$, then (a) is a parametric word,
(ii) if α and β are parametric words, then so is $(\alpha\beta)$,
(iii) if α is a parametric word and $i \in \Lambda$, then (α^i) is a parametric word,

The set of parametric words is denoted by $\mathcal{P}(\Delta, \Lambda)$. The sets of parameters are always denoted by Δ and Λ.

When there is no danger of confusion, unnecessary parenthesis can be omitted and notations like $\alpha^i \alpha^j = \alpha^{i+j}$ and $(\alpha^i)^j = \alpha^{ij}$ can be used. Then parametric words form a monoid, if the product of α and β is defined to be $\alpha\beta$.

If f is a function $\Lambda \to \mathbb{N}_0$, we can abuse the notation and use the same symbol for the function, which maps parametric words by giving values for the numerical parameters with f: if $a \in \Delta \cup \{1\}$, then $f((a)) = a$; if $\alpha, \beta \in \mathcal{P}(\Delta, \Lambda)$, then $f((\alpha\beta)) = f(\alpha)f(\beta)$; if $\alpha \in \mathcal{P}(\Delta, \Lambda)$ and $i \in \Lambda$, then $f((\alpha^i)) = f(\alpha)^{f(i)}$. A parametric word is thus mapped by f to a word of Δ^*. This can be further mapped by a morphism $h : \Delta^* \to \Sigma^*$ to a word of Σ^*. The mapping $h \circ f$ is a *valuation* of a parametric word into Σ^*, and f is its valuation to the set Δ^*.

We define the *length* of a parametric word: the length of 1 is zero; if $a \in \Delta$, then the length of a is one; if $\alpha, \beta \in \mathcal{P}(\Delta, \Lambda)$, then the length of $\alpha\beta$ is the sum of lengths of α and β; if $\alpha \in \mathcal{P}(\Delta, \Lambda) \smallsetminus \{1\}$ and $i \in \Lambda$, then the length of α^i is the length of α plus one.

Next we define the *height* of a parametric word: if $a \in \Delta \cup \{1\}$, then the height of a is zero; if $\alpha, \beta \in \mathcal{P}(\Delta, \Lambda)$, then the height of $\alpha\beta$ is the maximum of heights of α and β; if $\alpha \in \mathcal{P}(\Delta, \Lambda) \smallsetminus \{1\}$ and $i \in \Lambda$, then the height of α^i is the height of α plus one. Parametric words of height zero can be considered to be words of Δ^*.

A *linear Diophantine relation* R is a disjunction of systems of linear Diophantine equations with lower bounds for the unknowns. For example,

$$((x + y - z = 0) \wedge (x \geq 2)) \vee ((x + y = 3) \wedge (x + z = 4))$$

is a linear Diophantine relation over the unknowns x, y, z. We are only interested in the nonnegative values of the unknowns. If $\Lambda = \{i_1, \ldots, i_k\}$, f is a function $\Lambda \to \mathbb{N}_0$ and $f(i_1), \ldots, f(i_k)$ satisfy R, then the notation $f \in R$ can be used.

Let S be a set of morphisms $\Xi^* \to \Sigma^*$, $\Lambda = \{i_1, \ldots, i_k\}$, h_j a morphism from Ξ^* to parametric words and R_j a linear Diophantine relation, when $j = 1, \ldots, m$. The set $\{(h_j, R_j) : 1 \leq j \leq m\}$ is a *parametric representation* of S, if

$$S = \{h \circ f \circ h_j : 1 \leq j \leq m, f \in R_j\},$$

where $h \circ f$ runs over all valuations to Σ^*. The linear Diophantine relations are not strictly necessary, but they make some proofs easier. A set can be *parameterized*, if it has a parametric representation. The *length* of the parametric representation is the sum of the lengths of all $h_j(x)$, where $j = 1, \ldots, m$ and $x \in \Xi$.

It follows immediately that if two sets can be parameterized, then also their union can be parameterized.

Let $S, S_1, \ldots, S_n$ be sets of morphisms $\Xi^* \to \Sigma^*$. The set S can be *parameterized in terms of the sets* $S_1, \ldots, S_n$, if there exists such morphisms $h_1, \ldots, h_n$ from Ξ^* to $\mathcal{P}(\Xi, \Lambda)$ that

$$S = \{g \circ f \circ h_j : 1 \leq j \leq n, g \in S_j\},$$

where f runs over functions $\Lambda \to \mathbb{N}_0$.

Again it is a direct consequence of the definitions that the parameterizability is preserved in compositions. Namely, if S can be parameterized in terms of the sets $S_1, \ldots, S_n$ and every S_i can be parameterized in terms of the sets $S_{i1}, \ldots, S_{in_i}$, then S can be parameterized in terms of the sets S_{ij}.

We conclude these definitions by saying that solutions of an equation can be *parameterized*, if the set of its all solutions can be parameterized. A parametric representation of this set is a *parametric solution* of the equation.

These definitions can be generalized in an obvious way for systems of equations. Theorems 2.1 and 2.2 and Lemmas 2.3 – 2.7 give parametric solutions for some equations. The following theorem states that the basic tool in solving equations, namely the cancellation of the first variable, preserves the parameterizability of solutions.

Theorem 3.1. *Let $U, V \in \Xi^*$, $x, y \in \Xi$ and $x \neq y$. Let $h : \Xi^* \to \Xi^*$ be the morphism $x \mapsto yx$. If the equation $xh(U) = h(V)$ has a parametric solution, then so does the equation $xU \rightrightarrows yV$.*

Let α and β be parametric words. The pair (α, β) can be viewed as an equation, referred to as an *exponential equation*. The *height* of this equation is the height of $\alpha\beta$. The solutions of this equation are the functions $f : \Lambda \to \mathbb{N}_0$ that satisfy $f(\alpha) = f(\beta)$.

If we know some parametric words, which give all solutions of an equation, but which also give some extra solutions, then often the right solutions can be picked by adding some constraints for the numerical parameters. These constraints can be found by exponential equations, and the following theorems prove that they are in our cases equivalent with linear Diophantine relations.

Theorem 3.2. *Let E be an exponential equation of height one. There exists a linear Diophantine relation R such that a function $f : \Lambda \to \mathbb{N}_0$ is a solution of E if and only if $f \in R$.*

In some cases Theorem 3.2 can be generalized for exponential equations of height two.

Theorem 3.3. *Let $\Lambda = \{i, j\}$ and let $s_0, \ldots, s_m, t_1, \ldots, t_m, u_0, \ldots, u_n$ and $v_1, \ldots, v_n$ be parametric words of height at most one, with no occurrences of parameter j. Assume that i occurs at least in the words $t_1, \ldots, t_m$ and $v_1, \ldots, v_n$. Let $\alpha = s_0 t_1^j s_1 \ldots t_m^j s_m$ and $\beta = u_0 v_1^j u_1 \ldots v_n^j u_n$. Now there exists a linear Diophantine relation R such that a function $f : \Lambda \to \mathbb{N}_0$ is a solution of the exponential equation $E : \alpha = \beta$ if and only if $f \in R$.*

The parametric words in the next theorem come from Lemma 2.6.

Theorem 3.4. *Let $\Delta = \{p, q\}$, $\Lambda = \{i, j, k\}$ and $a \geq 2$. Let $\alpha = (pq^a)^i p$, $\beta = q$, $\gamma = (pq^a)^j p$, or*

$$
\begin{cases}
\alpha & = qp((pq)^{k+1}p)^{a-2}pq(((pq)^{k+1}p)^{a-1}pq)^i, \\
\beta & = (pq)^{k+1}p, \\
\gamma & = qp((pq)^{k+1}p)^{a-2}pq(((pq)^{k+1}p)^{a-1}pq)^j.
\end{cases}
$$

Let $A, B \in \{x, y, z\}^*$ and let h be the morphism mapping $x \mapsto \alpha, y \mapsto \beta, z \mapsto \gamma$. Now there exists a linear Diophantine relation R such that a function $f : \Lambda \to \mathbb{N}_0$ is a solution of the exponential equation $E : h(A) = h(B)$ if and only if $f \in R$.

4   Basic Equations

From now on we only consider equations with three unknowns. The alphabet of unknowns is $\Xi = \{x, y, z\}$. The left-hand side of an equation can be assumed to begin with x. We can also assume that x occurs on the right-hand side, but not as the first letter.

Periodic solutions and solutions, where some unknown has the value 1, are called *trivial*. These are easy to parameterize by Theorem 2.1.

An equation is a *basic equation*, if it is a trivial equation $U = U$, where $U \in \Xi^*$, if it has only trivial solutions, or if it is of one of the following forms, where $a, b \geq 1$, $c \geq 2$ and $t \in \{x, z\}$:

B1. $x^a y \ldots = y^b x \ldots$

B2. $x^2 \ldots \rightrightarrows y^a x \ldots$

B3. $xyt \ldots \rightrightarrows zxy \ldots$

B4. $xyt \ldots \rightrightarrows zyx \ldots$

B5. $xyz \ldots = zxy \ldots$

B6. $xyz \ldots = zyx \ldots$

B7. $xy^c z \ldots = zy^c x \ldots$

B8. $xyt \ldots \rightrightarrows z^a xy \ldots$

B9. $xyxz \ldots \rightrightarrows zx^2 y \ldots$

The parameterizability of basic equations is easy to prove with the help of previous lemmas and theorems.

Theorem 4.1. *Every basic equation has a parametric solution of bounded length.*

Proof. For equations $U = U$ and for equations with only trivial solutions the claim is clear. We prove it for equations B1 – B9. First we reduce equations to other equations by Theorem 3.1. The equation B2 is reduced by the substitution $x \mapsto yx$ to the equation $xyx \ldots = y^a x \ldots$, which is of the form B1. The equations B3 and B4 are reduced by the substitution $x \mapsto zx$ to the equations $xyz \ldots = zxy \ldots$ and $xyz \ldots = yzx \ldots$, which are of the form B5. The equation B8 is reduced by the substitution $x \mapsto zx$ to the equation $xyzA = z^a xyB$ for some $A, B \in \Xi^*$. By Lemma 2.8, this is equivalent with the equation $xyzxyzA = zxyz^a xyB$, which is of the form B5.

Consider the equations B1, B5, B6, B7 and B9. Their solutions are also solutions of $xy = yx$, $xyz = zxy$, $xyz = zyx$, $xy^c z = zy^c x$ and $xyxz \rightrightarrows zx^2 y$, respectively. For B1 this follows from Lemma 2.8, otherwise by a length argument. By Lemmas 2.1, 2.4, 2.3, 2.6 and 2.7, these latter equations have parametric solutions over word parameters p, q and numerical parameters i, j, k. By substituting the parametric words from these solutions to the original basic equations, we get exponential equations, which are equivalent with linear Diophantine relations by Theorems 3.2, 3.3 and 3.4. The parametric solutions with these linear Diophantine relations, together with parametric representations for the periodic solutions, determine parametric solutions for these equations.   □

5   Images and θ-Images

In this section we define images and θ-images of equations and prove some results about them. If h is a solution of the equation $xU \rightrightarrows yV$, then $h(y) \leq h(x)$. This fact was already behind Theorem 3.1. This will be generalized.

Let $t_1, \ldots, t_n \in \{y, z\}$ and $V = t_1 \ldots t_n$. Let $t_{n+1} = t_1$. If a morphism h is a solution of the equation $E : xU \rightrightarrows VxW$, then

$$h(x) = h(V^k t_1 \ldots t_i)u \tag{1}$$

for some numbers k, i and word u satisfying $k \geq 0$, $0 < i \leq n$ and $h(t_{i+1}) \not\leq u$.

On the other hand, a morphism h satisfying (1) is a solution of E iff $uh(U) = h(t_{i+1} \ldots t_n t_1 \ldots t_i)uh(W)$. We can write $h = g \circ f$, where f is the morphism $x \mapsto V^k t_1 \ldots t_i x$ and g is the morphism for which $g(x) = u$, $g(y) = h(y)$ and $g(z) = h(z)$. Now h is a solution of E iff g is a solution of

$$xf(U) \Leftarrow f(t_{i+1} \ldots t_n t_1 \ldots t_i)xf(W).$$

An *image* of an equation $xU(x, y, z) \rightrightarrows V(y, z)xW(x, y, z)$ under the morphism $x \mapsto V^k Px$, where $k \geq 0$, $V = PQ$ and $Q \neq 1$, is

$$xU(V^k Px, y, z) \Leftarrow QPxW(V^k Px, y, z).$$

If V contains only one of y, z or if $P = 1$, the image is *degenerated*.

Images are needed in the most important reduction steps used in the proof of parameterizability of equations with three unknowns. The solutions of an equation are easily acquired from the solutions of its images, so it is enough to consider them. There are infinitely many images, but a finite number is enough, if one of them is turned from a one-sided equation to an ordinary equation.

Equation E is *reduced to the equations* $E_1, \ldots, E_n$ by an *n-tuple of substitutions*, if E is of the form $xU(x, y, z) \rightrightarrows t_1 \ldots t_k xV(x, y, z)$, where $1 \leq n \leq k$ and $t_1, \ldots, t_k \in \{y, z\}$, equation E_i is

$$xU(t_1 \ldots t_i x, y, z) \Leftarrow t_{i+1} \ldots t_k t_1 \ldots t_i xV(t_1 \ldots t_i x, y, z),$$

when $1 \leq i < n$, and equation E_n is

$$xU(t_1 \ldots t_n x, y, z) = t_{n+1} \ldots t_k t_1 \ldots t_n xV(t_1 \ldots t_n x, y, z).$$

By the above, Theorem 3.1 can be generalized.

Theorem 5.1. *Let E be an equation of length n. If E is reduced to the equations $E_1, \ldots, E_m$ by an m-tuple of substitutions, and if $E_1, \ldots, E_m$ have parametric solutions of length at most c, then E has a parametric solution of length $O(mn)c$.*

Reductions with n-tuples of substitutions are not sufficient. Other ways to restrict the considerations to a finite number of images are needed.

Equation

$$xU(x, y, z) \rightrightarrows V(y, z)xW(x, y, z)$$

is of *type I*, if both unknowns y, z occur in V. Equation

$$xy^b U(x,y,z) \rightrightarrows z^c x V(x,z) y W(x,y,z),$$

where $b, c \geq 1$, is of *type II*, if $b > 1$ or $V \neq 1$.

Theorem 5.2. *The solutions of an equation of type I of length n can be parameterized in terms of the solutions of $O(n^2)$ of its images of length $O(n^3)$.*

Theorem 5.2 can be generalized by defining θ-images.

A sequence of equations $E_0, \ldots, E_n$ is a *chain*, if E_i is an image of E_{i-1} for all i, $1 \leq i \leq n$. Then E_n is an *image of order n* of E_0. If every E_i is a degenerated image, then the chain is degenerated and E_n is a degenerated image of order n.

We define θ-*images* of equations of type I and II. For equations of type I all images are θ-images. For equations of type II the degenerated images of order 2 and nondegenerated images of order 3 are θ-images.

The proofs of the following three lemmas, and also the proof of Theorem 5.2, consist of examining images and their solutions and using exponential equations. Especially the proof of Lemma 5.4 is somewhat complicated. We consider an equation of type II

$$xy^b A(x,y,z) \rightrightarrows z^c x B(x,z) y C(x,y,z), \tag{2}$$

where $b, c \geq 1$ and $b > 1$ or $B \neq 1$. Its images are degenerated and of the form

$$xy^b A(z^i x, y, z) \Leftleftarrows z^c x B(z^i x, z) y C(z^i x, y, z).$$

Lemma 5.3. *The solutions h of (2) satisfying $|h(y)| \leq |h(z)|$ can be parameterized in terms of the solutions of $O(n^{17})$ of its θ-images of length $O(n^{18})$.*

Lemma 5.4. *If x occurs in B, then the nonperiodic solutions of (2), and some periodic solutions, can be parameterized in terms of the solutions of $O(n^{17})$ of its θ-images of length $O(n^{18})$.*

Lemma 5.5. *If $B = z^d$, where $d \geq 1$, then the solutions of (2) can be parameterized in terms of the solutions of $O(n^{26})$ of its θ-images of length $O(n^{27})$.*

We define a *complete set of θ-images* of an equation of type I or II. For equations of type I it is the set of Théorem 5.2. For equations of the form (2) it is the set of Lemma 5.3, if $B = 1$, the set of Lemma 5.4, if x occurs in B, and the set of Lemma 5.5, if $B = z^d$, $d \geq 1$. The next theorem follows immediately from this definition.

Theorem 5.6. *Every equation of type I or II of length n has a complete set of θ-images consisting of $O(n^{26})$ equations of length $O(n^{27})$.*

We assume that every complete set of θ-images satisfies the conditions of Theorem 5.6. The next theorem requires only little extra work.

Theorem 5.7. *Let E be a word equation of length n. If $\{E_1, \ldots, E_m\}$ is a complete set of θ-images of E and every E_i has a parametric solution of length at most c, then E has a parametric solution of length $O(mn^{26})c$.*

6 Trees of Equations

The proof of the parameterizability of equations with three unknowns consists mainly of reducing equations to other equations. This forms a tree-like structure. The intention is to make all leaf equations in this tree to be basic equations. The possible reduction steps are given in the definition of a neighborhood.

Lemma 6.1. *Let E_0 be the equation $xy^a zy^p s \ldots \rightrightarrows zy^b xy^q t \ldots$, where $s, t \in \{x, z\}$ and $a + p \neq b + q$. Let $k \geq 8 + |p - q|$ be even, E_k be the equation $xP \rightrightarrows zQ$ and $E_0, \ldots, E_k$ be a degenerated chain. Now the solutions of E_k satisfying $y \neq 1$ are also solutions of the equation $xy^a zy^b \rightrightarrows zy^b xy^a$.*

The equations $E_1, \ldots, E_n$ form a *neighborhood* of an equation E, if one of the following conditions holds:

N1. $E_1, \ldots, E_n$ form a complete set of θ-images of E,

N2. E reduces to $E_1, \ldots, E_n$ with an n-tuple of substitutions,

N3. E is the equation $U = V$, U and V begin with different letters, $n = 2$, and E_1 and E_2 are equations $U \rightrightarrows V$ and $V \rightrightarrows U$,

N4. $n = 1$ and E is the equation $U = V$ and E_1 is the equation $U^R = V^R$,

N5. E is the equation $SU = TV$, $|S|_t = |T|_t$ for all $t \in \Xi$, $n = 1$ and E_1 is the equation $US = VT$,

N6. $n = 1$ and E_1 is E reduced from the left or multiplied from the right,

N7. $n = 1$ and, with the assumptions of lemma 6.1, E is the equation $xP \rightrightarrows zQ$ and E_1 the equation $xy^a zy^b xP \rightrightarrows zy^b xy^a zQ$.

Theorem 6.2. *Let E be a word equation of length n and let $E_1, \ldots, E_m$ be its neighborhood. If each E_i has a parametric solution of length at most c, then E has a parametric solution of length $O(mn^{26})c$.*

Proof. For N1 this follows from Theorem 5.7, for N2 from Theorem 5.1 and for N7 from Lemma 6.1. The other cases are clear.   □

Directed acyclic graph, whose vertices are equations, is a *tree* of E, if the following conditions hold:

(i) only vertex with no incoming edges is E,

(ii) all other vertices have exactly one incoming edge,

(iii) if there are edges from E_0 to exactly $E_1, \ldots, E_n$, then these equations form a neighborhood of E.

Theorem 6.3. *Let E be a word equation of length n. If E has a tree of height k, then all equations in the tree are of length $O(n)^{27^k}$. If each leaf equation in this tree has a parametric solution of length at most c, then E has a parametric solution of length $O(n)^{52 \cdot 27^k} c$.*

Proof. In the case N1 the first claim follows directly from Theorem 5.6, and for the other cases the bound $O(n)^{27^k}$ is more than enough. Now, by Theorem 6.2, there exists a constant a such that E has a parametric solution of length

$$a(an)^{52} \cdot a((an)^{27})^{52} \cdot a((an)^{27^2})^{52} \cdots a((an)^{27^{k-1}})^{52} \cdot c$$

$$<a^k(an)^{52 \cdot 27^k} c = O(n)^{52 \cdot 27^k} c. \qquad \square$$

A tree in which all leaves are basic equations is a *basic tree*.

If every θ-image of an equation of type I or II has a basic tree, then the equation has a basic tree, because it has a complete set of θ-images. The rule N1 is used this way instead of explicitly selecting some complete set of θ-images.

The main theorem is proved by a sequence of lemmas. The lemmas are proved by using the rules of the definition of a neighborhood in various ways.

Lemma 6.4. *The equation $xyz^2A(x,y,z) = yz^2xB(x,y,z)$ has a basic tree.*

Lemma 6.5. *The equation $x^2yz \ldots \rightrightarrows zyxy \ldots$ has a basic tree.*

Lemma 6.6. *Let $s \neq x$ and $t \neq y$. The following equations have basic trees:*

(a) $xy^2z \ldots \rightrightarrows zx^2y \ldots,$
(b) $xyzs \ldots \rightrightarrows zx^2y \ldots,$
(c) $xy^2z \ldots \rightrightarrows zxyt \ldots,$
(d) $xyzt \ldots \rightrightarrows zy^2x \ldots,$
(e) $xyz \ldots \rightrightarrows zy^2x \ldots.$

Let $1 \leq a, b \leq 2, d \geq 1$ and $t \neq y$. The equations $x^a y^b t \ldots \rightrightarrows zyx \ldots$, $x^a y^b t \ldots \rightrightarrows zxy \ldots$ and $x^a y^b t \ldots \rightrightarrows z(yz)^d x \ldots$ are *supporting equations*.

Lemma 6.7. *Every supporting equation has a basic tree.*

Lemma 6.8. *The equation $xy^a zy^p s \ldots \rightrightarrows zy^b xy^q t \ldots$, where $a > 0$, $a+p = b+q$ and $s, t \neq y$, has a basic tree.*

The next proof contains maybe the most critical part of the construction, because very long chains of images are considered. Similar construction was needed also in the proof of Lemma 6.6.

Lemma 6.9. *The equation $xy^a z \ldots \rightrightarrows zy^b x \ldots$, where $a > 0$, has a basic tree.*

Proof. The equation can be written in the form $E_0 : xy^a zy^p u \ldots \rightrightarrows zy^b xy^q v \ldots$, where $u, v \neq y$. If $a+p = b+q$, then the claim follows from Lemma 6.8. Assume that $a+p \neq b+q$. Let $l \geq 8 + |p-q|$ be even. Form a complete set of θ-images of E_0, a complete set of θ-images of these, and so on l times. These θ-images form chains $E_0, \ldots, E_l$. We show that each chain has an equation with a basic tree; this proves the claim.

First, consider chains of degenerated θ-images. There is a corresponding chain of ordinary images and we can use the rule N7. The equation E_l is replaced by the equation $xy^a zy^b xP \rightrightarrows zy^b xy^a zQ$, which has a basic tree by Lemma 6.8.

Second, consider nondegenerated chains. Assume that the part $E_0, \ldots, E_{j-1}$ of the chain is degenerated and that E_j is a nondegenerated θ-image of E_{j-1}. If $b = 0$, the equation E_0 is of the form $xy^a z \ldots \Rightarrow zx \ldots$, and E_{j-1} is of the same form. The equation E_j can be seen to be a supporting equation and thus it has a basic tree. If $b > 0$, then E_0 is of the form $xy^a z \ldots \Rightarrow zy^b x \ldots$. Equation E_{j-1} is of the same form. Now E_j is of the form $y^c z y^d x \ldots \Rightarrow xy^a z \ldots$, where $c + d = a$ and $c \geq 1$. If $c > 1$, then E_j is basic of the form B2. If $c = 1$, then all θ-images of E_j can be seen to have basic trees by Lemmas 6.6 and 6.7.   □

Lemma 6.10. *The equation* $xy^a t \ldots \Rightarrow z^c x B(x, z) y \ldots$, *where* $a, c \geq 1$ *and* $t \neq y$, *has a basic tree.*

Lemma 6.11. *The equation* $x^n y^m t \ldots \Rightarrow zy A(y, z) x \ldots$, *where* $n, m \geq 1$ *and* $t \neq y$, *has a basic tree.*

The proof of the next theorem finally gathers the previous results together and gives the idea of how the height of the tree can be estimated.

Theorem 6.12. *Every equation of length n with three unknowns has a basic tree of height $O(n)$.*

Proof. The trivial equation $U = U$ is a basic equation. All other equations can be reduced from the left and split into one-sided equations. By multiplication from the right, every one-sided equation can be turned into one of the equations

$$x^2 \ldots \Rightarrow y^c x \ldots \tag{3}$$

$$xy \ldots \Rightarrow y^c x \ldots \tag{4}$$

$$xz^a t \ldots \Rightarrow y^c x B(x, y) z \ldots \tag{5}$$

$$x^a y^b s \ldots \Rightarrow y^c z B(y, z) x \ldots \tag{6}$$

$$x^a z^b t \ldots \Rightarrow yz B(y, z) x \ldots \tag{7}$$

$$x^a z^b t \ldots \Rightarrow y^d z B(y, z) x \ldots, \tag{8}$$

where $a, b, c \geq 1$, $d > 1$, $t \neq z$ and $s \neq y$. We prove that these have basic trees.

Equation (3) is basic of the form B2. Equation (4) is reduced by the substitution $x \mapsto yx$ to the equation $xy \ldots = y^c x \ldots$, which is basic of the form B1. Equation (5) is the equation of Lemma 6.10. Equation (7) is the equation of Lemma 6.11.

The equation (6) is of type I and its images are of the form $xy \ldots \Leftarrow Dx \ldots$, where D is a conjugate of $y^c z B$. If $y^2 \leq D$, then this is of the form (3), if $yz \leq D$, then of the form (5), and if $z \leq D$, then of the form (7). So every image of (6) and thus the equation itself has a basic tree.

The equation (8) is of type I and its images are of the form $x(y \ldots)^{a-1} z^b y \ldots \Leftarrow Dx \ldots$, where D is a conjugate of $y^d z B$. Again it is of the form (3), (5) or (7). So every image of (6) and thus the equation itself has a basic tree.

The constructions of trees in the lemmas produce trees of bounded height with two exceptions: Lemmas 6.6 and 6.9, where a tree with height of order $|p - q|$ is constructed for the equation

$$xy^a zy^p \ldots \rightrightarrows zy^b xy^q \ldots . \tag{9}$$

We prove that the powers of y here cannot be more than n, which proves this theorem. In the definition of neighborhood, the rules N1, N2, N5 and N6 can produce higher powers than those in the initial equation. There is no need to use N6 to generate high powers and N5 is only used in 6.4, 6.6 and 6.8, where it does not generate high powers. Consider N1 and N2. Here an equation $xU(x, y, z) \rightrightarrows y^a xV(x, y, z)$ can be turned into $xU(y^i x, y, z) \Leftarrow y^a xV(y^i x, y, z)$ for high values of i. But in order for y to be in the position of (9), the rules N1 or N2 must be used again. Then y is replaced by xuy for some $u \in \{x, z\}^*$ and the powers of y disappear. The claim is proved. $\square$

In the next theorem $\exp^2$ denotes the double exponential function $\exp \circ \exp$.

Theorem 6.13. *Every equation of length n with three unknowns has a parametric solution of length $\exp^2(O(n))$.*

Proof. By Theorem 6.12 every equation has a basic tree of height $O(n)$. By Theorem 4.1 the leaf equations have parametric solutions of bounded length. Now from Theorem 6.3 it follows that E has a parametric solution of length $O(n)^{52 \cdot 27^k}$, where $k = O(n)$, that is of length $\exp^2(O(n))$. $\square$

References

1. Albert, M.H., Lawrence, J.: A proof of Ehrenfeucht's Conjecture. Theoret. Comput. Sci. 41, 121–123 (1985)
2. Choffrut, C., Karhumäki, J.: Combinatorics of words. In: Rozenberg, G., Salomaa, A. (eds.) Handbook of Formal Languages. Springer, Heidelberg (1997)
3. Czeizler, E., Karhumäki, J.: On non-periodic solutions of independent systems of word equations over three unknowns. Internat. J. Found. Comput. Sci. 18, 873–897 (2007)
4. Guba, V.S.: Equivalence of infinite systems of equations in free groups and semigroups to finite subsystems. Mat. Zametki 40, 321–324 (1986)
5. Harju, T., Karhumäki, J., Plandowski, W.: Independent system of equations. In: Lothaire, M. (ed.) Algebraic Combinatorics on Words. Cambridge University Press, Cambridge (2002)
6. Hmelevskii, Y.I.: Equations in free semigroups. Proc. Steklov Inst. of Math. 107 (1971); Amer. Math. Soc. Translations (1976)
7. Karhumäki, J., Saarela, A.: A Reproof of Hmelevskii's Theorem (manuscript)
8. Lothaire, M.: Combinatorics on Words. Addison-Wesley, Reading (1983)
9. Makanin, G.S.: The problem of solvability of equations in a free semigroup. Mat. Sb. 103, 147–236 (1977); English transl. in Math. USSR Sb. 32, 129–198
10. Plandowski, W.: Satisfiability of word equations with constants is in PSPACE. J. ACM 51, 483–496 (2004)
11. Spehner, J.-C.: Quelques Problemes d'extension, de conjugaison et de presentation des sous-monoides d'un monoide libre. Ph.D. Thesis, Univ. Paris (1976)
12. Spehner, J.-C.: Les presentations des sous-monoides de rang 3 d'un monoide libre. In: Semigroups, Proc. Conf. Math. Res. Inst., pp. 116–155 (1978)

Hierarchies of Piecewise Testable Languages

Ondřej Klíma and Libor Polák*

Department of Mathematics, Masaryk University
Janáčkovo nám 2a, 662 95 Brno, Czech Republic

Abstract. The classes of languages which are boolean combinations of languages of the form

$$A^* a_1 A^* a_2 A^* \ldots A^* a_\ell A^*, \text{ where } a_1, \ldots, a_\ell \in A, \ \ell \leq k,$$

for a fixed $k \geq 0$, form a natural hierarchy within piecewise testable languages and have been studied in papers by Simon, Blanchet-Sadri, Volkov and others. The main issues were the existence of finite bases of identities for the corresponding pseudovarieties of monoids and generating monoids for these pseudovarieties.

Here we deal with similar questions concerning the finite unions and positive boolean combinations of the languages of the form above. In the first case the corresponding pseudovarieties are given by a single identity, in the second case there are finite bases for k equal to 1 and 2 and there is no finite basis for $k \geq 4$ (the case $k = 3$ remains open). All the pseudovarieties are generated by a single algebraic structure.

Keywords: varieties of languages, piecewise testable languages, syntactic monoid.

1 Introduction

A language L over an alphabet A is called *piecewise testable* if it is a finite boolean combination of languages of the form

$$A^* a_1 A^* a_2 A^* \ldots A^* a_\ell A^*, \text{ where } a_1, \ldots, a_\ell \in A, \ \ell \geq 0 . \quad (*)$$

A characterization of piecewise testable languages was given by Simon [18] who proved that a language L is piecewise testable if and only if its syntactic monoid is $\mathcal{J}$-trivial. Note that nowadays there exist several proofs of this deep result [1,7,19,22]. See survey papers [11,13] for more information and connections to concatenation hierarchies.

The Simon theorem was one of the first deep examples of Eilenberg's correspondence [5] between boolean varieties of languages and pseudovarieties of monoids. The correspondence uses the concept of the syntactic monoid of a language. Pin's modification [12,13] of Eilenberg's result gives a correspondence

---

* Both authors were supported by the Ministry of Education of the Czech Republic under the project MSM 0021622409 and by the Grant no. 201/06/0936 of the Grant Agency of the Czech Republic.

M. Ito and M. Toyama (Eds.): DLT 2008, LNCS 5257, pp. 479–490, 2008.
© Springer-Verlag Berlin Heidelberg 2008

between positive varieties of languages and pseudovarieties of finite ordered monoids. For example, finite unions of languages of the form (∗) form a positive variety of languages $\mathscr{P}$ which corresponds to the pseudovariety of all ordered monoids satisfying the identity $x \leq 1$ [14].

Later on Polák [15] presented another modification of Eilenberg's correspondence: conjunctive varieties are related to pseudovarieties of finite idempotent semirings. The difference is that conjunctive varieties of languages are not closed, in general, under complements and unions and one uses a stronger notion of preimages. Note that there is a dual version. Namely, the so-called disjunctive varieties which can be obtained from conjunctive varieties in the following way: for a conjunctive variety $\mathscr{V}$, we consider the class $\mathscr{V}^c$ of complements of languages from $\mathscr{V}$ (see [16] for more details).

Next steps were the literal varieties of languages by Ésik and Ito (see [6]), Straubing's C-varieties of languages (see [21]), and more generally Polák's D-varieties (see [16]). We make use of all mentioned modifications of Eilenberg's correspondence in this paper. We slightly modify the definition of a semiring (no zero element is postulated) and D-varieties of languages are present in our paper only implicitly.

We are interested in piecewise testable languages. If we fix a number k then we can consider boolean combinations of languages of the form (∗), where $\ell \leq k$. The resulting class is a boolean variety of languages, which is denoted $\mathscr{BV}_k$. These k-levels in the hierarchy of piecewise testable languages were considered by I. Simon [18], who found a simple characterization by identities for levels 1 and 2. A lot of work was done by Blanchet-Sadri [3,4] who found a basis of identities for the level 3 and proved that there is no finite basis of identities for $k > 3$.

In this paper we consider similar levels of positive varieties $\mathscr{PV}_k$ and disjunctive varieties $\mathscr{DV}_k$ of piecewise testable languages. Here $\mathscr{PV}_k$ is formed by finite intersections of finite unions of languages of the form (∗), where $\ell \leq k$, and $\mathscr{DV}_k$ is formed by finite unions of languages of the form (∗), where $\ell \leq k$.

After introductory Sections 1 and 2 we discuss the identity problem for the pseudovarieties corresponding to $\mathscr{BV}_k$, $\mathscr{PV}_k$ and $\mathscr{DV}_k$ in Section 3. In Section 4 we obtain a characterization of $\mathscr{DV}_k$ for arbitrary $k \geq 1$, $\mathscr{PV}_1$ and $\mathscr{PV}_2$ in terms of a basis of identities for the corresponding pseudovariety of semirings (resp. monoids) and we show that such a finite basis does not exist for $\mathscr{PV}_k$ for $k > 3$. The characterization of $\mathscr{PV}_3$ by a finite basis of identities is stated as an open problem. Note that we do not reprove the results of Simon or Blanchet-Sadri.

In [23] Volkov showed that the pseudovariety of monoids corresponding to $\mathscr{BV}_k$ is generated by a single monoid for each k. He proved that in order to get this result we can use any of the three different series of monoids described by Straubing [20] and Pin [10]. In Section 5 we present an alternative proof concerning one of the series and we show that also $\mathscr{PV}_k$ and $\mathscr{DV}_k$ are generated by a single ordered monoid and by a single idempotent semiring, respectively. Finally, in the last section we state among others that the pseudovarieties $\mathbf{BV}_k$ and $\mathbf{PV}_k$ for $k \geq 4$ have no finite bases of pseudoidentities.

Due to the space limitations certain parts of proofs are placed into Appendix (see [9]).

2   Eilenberg Type Correspondences

Let A be a finite alphabet, let A^* be the set of all words over A with the operation of concatenation $\cdot$, i.e. A^* is the free monoid over A. The empty word is denoted by λ. A language over an alphabet A is a subset of A^*. A language is *regular* if it is accepted by a finite automaton – see, for instance, [2] or [13]. We will work with certain classes of (regular) languages.

2.1   Boolean and Positive Varieties of Languages

We recall here the basics concerning the Eilenberg type theorems. The boolean case was invented by Eilenberg [5] and the positive case by Pin [12].

A *boolean variety of languages* $\mathcal{V}$ associates to every finite alphabet A a class $\mathcal{V}(A)$ of regular languages over A in such a way that

- $\mathcal{V}(A)$ is closed under boolean operations (in particular $\emptyset, A^* \in \mathcal{V}(A)$),
- $\mathcal{V}(A)$ is closed under derivatives, i.e.
  $L \in \mathcal{V}(A)$, $u, v \in A^*$ implies $u^{-1}Lv^{-1} = \{\, w \in A^* \mid uwv \in L \,\} \in \mathcal{V}(A)$,
- $\mathcal{V}$ is closed under preimages in morphisms, i.e.
  $f : B^* \to A^*$, $L \in \mathcal{V}(A)$ implies $f^{-1}(L) \in \mathcal{V}(B)$.

To get the notion of a *positive* variety of languages, we use in the first item only intersections and unions (not complements).

A *pseudovariety* of finite monoids is a class of finite monoids closed under submonoids, morphic images and products of finite families. Similarly for ordered monoids (see [13]).

For a regular language $L \subseteq A^*$, we define the relations $\sim_L$ and $\preceq_L$ on A^* as follows: for $u, v \in A^*$ we have

$$u \sim_L v \text{ if and only if } (\, \forall\, p, q \in A^* \,)\, (\, puq \in L \iff pvq \in L \,),$$

$$u \preceq_L v \text{ if and only if } (\, \forall\, p, q \in A^* \,)\, (\, pvq \in L \implies puq \in L \,).$$

The relation $\sim_L$ is a congruence on A^* of finite index (i.e. there are only finitely many classes) and the quotient structure $\mathsf{M}(L) = A^*/\!\sim_L$ is called the *syntactic monoid* of L.

The relation $\preceq_L$ is a preorder (i.e. a reflexive and transitive relation) on A^* and the corresponding equivalence relation is $\sim_L$. Hence $\preceq_L$ induces an order on $\mathsf{M}(L) = A^*/\!\sim_L$, namely: $u\!\sim_L \leq v\!\sim_L$ if and only if $u \preceq_L v$. Then we speak about a *syntactic ordered monoid* of L and we denote the structure by $\mathsf{O}(L)$.

Result 1. (Eilenberg [5], Pin [12]) *Boolean varieties (positive varieties) of languages correspond to pseudovarieties of finite monoids (ordered monoids). The correspondence, written $\mathcal{V} \longleftrightarrow \mathbf{V}$ ($\mathscr{P} \longleftrightarrow \mathbf{P}$), is given by the following relationship: for $L \subseteq A^*$ we have*

$$L \in \mathcal{V}(A) \iff \mathsf{M}(L) \in \mathbf{V} \quad (\, L \in \mathscr{P}(A) \iff \mathsf{O}(L) \in \mathbf{P} \,).$$

The pseudovarieties can be characterized by pseudoidentities (see e.g. [2], [13]). The classes we consider here are *equational* – they are given by identities. For a set X, an *identity* is a pair $u = v$ ($u \leq v$) of words over X, i.e. $u, v \in X^*$. An identity $u = v$ ($u \leq v$) is *satisfied* in a finite monoid M (ordered monoid $(M, \leq)$) if for each morphism $\phi : X^* \to M$ we have $\phi(u) = \phi(v)$ ($\phi(u) \leq \phi(v)$). In such a case we write $M \models u = v$, and for a set of identities Π, we define $\mathsf{Mod}(\Pi) = \{ M \mid (\forall \pi \in \Pi) \, M \models \pi \}$. For a class $\mathcal{M}$ of monoids, the meaning of $\mathcal{M} \models \Pi$ is that, for each $M \in \mathcal{M}$, we have $M \models \Pi$. Similarly for the ordered case.

2.2   Disjunctive Varieties of Languages

Conjunctive (and dually disjunctive) varieties of languages were introduced by the second author [15]. In the definition of such classes of languages union (or dually intersection) is omitted but morphisms are from a larger class. Motivated by Straubing [21] the second author [16] generalized conjunctive varieties to D-conjunctive varieties where semiring morphisms can be taken from a fixed class of morphisms D. As our application uses the non-killing morphisms from [8], we modify mentioned definitions without using a concept of D-conjunctive varieties. We even modify also the basic definition of idempotent semiring to make the presentation more clear, namely we omit a neutral element with respect to the second operation, and finally we will consider the dual version, i.e. disjunctive varieties.

An *idempotent semiring* is a structure $(S, \cdot, +, 1)$ where

- $(S, \cdot)$ is a monoid with the neutral element 1,
- $(S, +)$ is a semilattice,
- $(\forall a, b, c \in S)$ ($a(b + c) = ab + ac$, $(a + b)c = ac + bc$).

A *pseudovariety* of finite idempotent semirings is a class of finite idempotent semirings closed under sub-semirings, morphic images and finite products.

Let $A^{\cup}$ denote the set of all finite non-empty subsets of A^*. For $U, V \in A^{\cup}$, we define $U \cdot V = \{uv \mid u \in U, v \in V\}$. Then $(A^{\cup}, \cdot, \cup, \{\lambda\})$ is a free idempotent semiring over A. For sets A, B, a language $L \subseteq A^*$ and idempotent semiring morphism $f : B^{\cup} \to A^{\cup}$ we define $f^{-1}(L) = \{u \in B^* \mid f(u) \cap L \neq \emptyset\}$.

An *identity* is a pair $U = V$ of elements of $X^{\cup}$. An identity $U = V$ is *satisfied* in a finite idempotent semiring S if for each morphism $\phi : X^{\cup} \to S$ we have $\phi(U) = \phi(V)$. In such a case we write $S \models U = V$ and for a set of identities Π we define $\mathsf{Mod}(\Pi) = \{ S \mid (\forall \pi \in \Pi) \, S \models \pi \}$.

A *disjunctive variety* of languages $\mathscr{D}$ associates to every finite alphabet A a class $\mathscr{D}(A)$ of regular languages over A in such a way that

- $A^* \in \mathscr{D}(A)$,
- $\mathscr{D}(A)$ is closed under finite unions (in particular $\emptyset \in \mathscr{D}(A)$),
- $\mathscr{D}(A)$ is closed under derivatives,
- $\mathscr{D}$ is closed under preimages in semiring morphisms, i.e.
  $f : B^{\cup} \to A^{\cup}$, $L \in \mathscr{D}(A)$ implies $f^{-1}(L) \in \mathscr{D}(B)$.

Let $L \subseteq A^*$ be a regular language. We define the relation $\equiv_L$ on $A^{\cup}$ as follows: for $U, V \in A^{\cup}$ we have

$$U \equiv_L V \text{ if and only if } (\forall p, q \in A^*) (pUq \cap L \neq \emptyset \iff pVq \cap L \neq \emptyset).$$

This relation has a finite index and the quotient structure $S(L) = A^{\cup}/\equiv_L$ is called the *syntactic semiring* of L. Notice that we are using the syntactic semiring from [15,16] for the complement of L.

Result 2. (Polák [16]) *Disjunctive varieties of languages correspond to pseudovarieties of idempotent semirings. The correspondence, written $\mathscr{D} \longleftrightarrow \mathbf{D}$, is given by the following relationship: for $L \subseteq A^*$ we have*

$$L \in \mathscr{D}(A) \iff S(L) \in \mathbf{D}.$$

3   Hierarchies of Piecewise Testable Languages

For a word $u = a_1 a_2 \ldots a_n$, where $a_1, a_2, \ldots, a_n \in A$, we define the language

$$L_u = A^* a_1 A^* \ldots A^* a_n A^*$$

and we denote by $|u|$ the length of the word u, i.e. $|u| = n$. Note that the length of the empty word λ is $|\lambda| = 0$ and $L_\lambda = A^*$. For a fixed $k \geq 0$, we define the classes $\mathscr{DV}_k$, $\mathscr{PV}_k$ and $\mathscr{BV}_k$ as follows. For each finite alphabet A, we have

- $L \in \mathscr{DV}_k(A)$ if L is a finite union of languages of the form L_u, where $u \in A^*$, $|u| \leq k$;
- $L \in \mathscr{PV}_k(A)$ if L is a finite intersection of finite unions of languages of the form L_u, where $u \in A^*$, $|u| \leq k$;
- $L \in \mathscr{BV}_k(A)$ if L is a boolean combination of languages of the form L_u, where $u \in A^*$, $|u| \leq k$.

Proposition 1. *Let $k \geq 0$. Then*
(i) $\mathscr{DV}_k$ is a disjunctive variety of languages,
(ii) $\mathscr{PV}_k$ is a positive variety of languages,
(iii) $\mathscr{BV}_k$ is a boolean variety of languages.

Proof. All statements are straightforward.   □

We say that a word $u = b_1 b_2 \ldots b_m$, where $b_1, \ldots, b_m \in A$, is a *subword* of a word $v = c_1 c_2 \ldots c_n$, where $b_1, \ldots, c_n \in A$, if $b_1 = c_{i_1}, \ldots, b_m = c_{i_m}$ for some $1 \leq i_1 < i_2 < \cdots < i_m \leq n$. We write $u \triangleleft v$ in this case. Note that for $u \in A^*$, we have $L_u = \{ v \in A^* \mid u \triangleleft v \}$.

For $u \in A^*$, we define $\mathsf{Sub}_k(u)$ as the set of all subwords of u of length at most k and $\mathsf{Sub}(u) = \bigcup_{k \geq 0} \mathsf{Sub}_k(u)$. For $U \in A^{\cup}$, we define $\mathsf{Sub}_k(U) = \bigcup_{u \in U} \mathsf{Sub}_k(u)$.

Next, we define relations $\sim_k^A$, $\prec_k^A$ on A^* and $\equiv_k^A$ on $A^{\cup}$ as follows: for $u, v \in A^*$, $U, V \in A^{\cup}$, we have

$$u \sim_k^A v \quad \text{if and only if} \quad \mathsf{Sub}_k(u) = \mathsf{Sub}_k(v),$$

$$u \preceq_k^A v \quad \text{if and only if} \quad \mathsf{Sub}_k(v) \subseteq \mathsf{Sub}_k(u),$$

$$U \equiv_k^A V \quad \text{if and only if} \quad \mathsf{Sub}_k(U) = \mathsf{Sub}_k(V).$$

We write only $\sim_k$, $\preceq_k$ and $\equiv_k$ when the alphabet A is known from the context.

The first item of the following lemma is due to Simon [17] and this is a basic step in every paper concerning piecewise testable languages (see e.g. [1,18,23]).

Proposition 2. *Let $k \geq 0$. Then for $L \subseteq A^*$, we have*
(i) $L \in \mathscr{BV}_k$ if and only if $\sim_k \subseteq \sim_L$,
(ii) $L \in \mathscr{PV}_k$ if and only if $\preceq_k \subseteq \preceq_L$,
(iii) $L \in \mathscr{DV}_k$ if and only if $\equiv_k \subseteq \equiv_L$.

The proof is placed into Appendix (see [9]).

By Proposition 2 the quotient structures $A^*/\sim_k$, $A^*/\preceq_k$, $A^\cup/\equiv_k$ are free objects in the (equational) pseudovarieties $\mathbf{BV}_k, \mathbf{PV}_k$ and $\mathbf{DV}_k$ over the set A. Since the equivalence relations $\sim_k$ and $\equiv_k$ have finite indices the corresponding pseudovarieties are locally finite. This is not a surprise because for a given alphabet A there are only finitely many languages of the form $(*)$ with $\ell \leq k$, hence $\mathscr{BV}_k(A)$ ($\mathscr{PV}_k(A)$ and $\mathscr{DV}_k(A)$, respectively) are finite.

This proposition also solves the identity problem for the pseudovarieties $\mathbf{BV}_k$, $\mathbf{PV}_k$ and $\mathbf{DV}_k$. But the solution of the identity problem is not a solution of the membership problem. Only if we have a finite basis of identities we can test them. Our goal is to find such bases for the mentioned classes.

From another point of view the proposition solves the membership problem too. One can compute the finite free structure $A^*/\sim_k$ (or $A^*/\preceq_k$, or $A^\cup/\equiv_k$) and check whether the syntactic monoid (ordered monoid, semiring) of a given language L is a quotient of this free structure.

We already stated that the classes $\mathscr{BV}_k$ were studied in many contributions. The first two items of the following lemma are due to Simon [18], the third can be found in [3]. The last item is proved in [4].

Result 3. *Let $k \geq 0$. Then (i) $\mathbf{BV}_1 = \mathsf{Mod}(xy = yx, x^2 = x)$,*
(ii) $\mathbf{BV}_2 = \mathsf{Mod}((xy)^2 = (yx)^2, xyzx = xyxzx)$,
(iii) $\mathbf{BV}_3 = \mathsf{Mod}((xy)^3 = (yx)^3, xzyxvxwy = xzxyxvxwy, ywxvxyz = ywxvxyxzx)$,
(iv) $\mathbf{BV}_k$ is not finitely based for $k \geq 4$.

Up to our knowledge there are no similar results for the hierarchies $\mathscr{PV}_k$ and $\mathscr{DV}_k$. These will be established in the next section.

4   Bases of Identities for $\mathscr{PV}_k$ and $\mathscr{DV}_k$

4.1   Disjunctive Varieties $\mathscr{DV}_k$

It is not hard to see that the disjunctive variety $\mathscr{DV}_1$ corresponds to the pseudovariety of idempotent semirings $\mathsf{Mod}(xy = x + y)$. The next theorem establishes the result for an arbitrary k. We consider

$$x_1 x_2 \ldots x_{k+1} = \sum_{i=1}^{k+1} x_1 \ldots x_{i-1} x_{i+1} \ldots x_{k+1} . \qquad (\pi_k)$$

Theorem 1. *Let $k \geq 0$. Then $\mathbf{DV}_k = \mathsf{Mod}(\pi_k)$.*

Proof. It is easy to see that both sides of π_k have the same set of subwords of length at most k. With respect to Proposition 2 we have to show that each identity $U = V$ such that $\mathsf{Sub}_k(U) = \mathsf{Sub}_k(V)$ is a consequence of the identity π_k. If we put in the identity π_k all variables equal to 1 with exception of the variable x_1, then we obtain the identity $x_1 = x_1 + 1$. From this identity we have $xy = (x+1)(y+1) = xy + x + y + 1$ and more generally $u = \mathsf{Sub}(u)$ for each word u. Now using the identity π_k we can rewrite each word in $\mathsf{Sub}(u)$ by all its subwords of length at most k. Hence we obtain the identity $u = \mathsf{Sub}_k(u)$. The identity $U = \mathsf{Sub}_k(U)$ follows and from that we obtain each identity $U = V$ such that $\mathsf{Sub}_k(U) = \mathsf{Sub}_k(V)$. □

4.2 Positive Varieties $\mathscr{PV}_k$

We prove a certain analogue of the characterizations from Result 3. Recall that the positive variety $\mathscr{P} = \bigcup_{k \geq 0} \mathscr{PV}_k$ is characterized by the identity $x \leq 1$. When we study the positive varieties $\mathscr{PV}_k$ then it is natural to consider the classes $\mathscr{BV}_k \cap \mathscr{P}$ and ask whether the equality $\mathscr{PV}_k = \mathscr{BV}_k \cap \mathscr{P}$ holds. The inclusion $\mathscr{PV}_k \subseteq \mathscr{BV}_k \cap \mathscr{P}$ is trivial but the opposite one is much more delicate as the next result shows.

Theorem 2. *(i) $\mathscr{PV}_1 = \mathscr{BV}_1 \cap \mathscr{P}$, i.e. $\mathbf{PV}_1 = \mathsf{Mod}(xy = yx, x^2 = x, x \leq 1)$.*
(ii) $\mathscr{PV}_2 = \mathscr{BV}_2 \cap \mathscr{P}$, i.e. $\mathbf{PV}_2 = \mathsf{Mod}((xy)^2 = (yx)^2, xyzx = xyxzx, x \leq 1)$.
(iii) For $k \geq 3$, $\mathscr{PV}_k \neq \mathscr{BV}_k \cap \mathscr{P}$.
(iv) For $k \geq 4$, $\mathbf{PV}_k$ has no finite basis of identities.

Proof. **Part (i).** Let $u \leq v$ be an identity satisfied in $\mathbf{PV}_1$. This means that $u \preceq_1 v$, i.e. $\mathsf{Sub}_1(v) \subseteq \mathsf{Sub}_1(u)$. Then $uv \sim_1 u$ and $v \vartriangleleft uv$. Hence $\mathbf{BV}_1 \models u = uv$ and $\mathbf{P} \models uv \leq v$ and the identity $u \leq v$ is a consequence of identities satisfied in $\mathbf{BV}_1$ and $\mathbf{P}$. This implies $\mathscr{BV}_1 \cap \mathscr{P} \subseteq \mathscr{PV}_1$. The statement follows.

Part (ii). We start with a technical lemma which we use inductively afterwards.

Lemma 1. *Let $u, v \in A^*$ be such that $u \preceq_2 v$ and $u \npreceq_2 v$. Then there exists $w \in A^*$ such that $u \preceq_2 w \preceq_2 v$ and at least one of the following two conditions happens: i) $v \vartriangleleft w$ and $v \npreceq_2 w$ or ii) $w \vartriangleleft u$ and $w \npreceq_2 u$.*

The proof is placed into Appendix (see [9]).

Claim 1. *$\mathscr{PV}_2 = \mathscr{BV}_2 \cap \mathscr{P}$.*

Proof. We show that each identity $u \leq v$ which is satisfied in $\mathbf{PV}_2$ is a consequence of identities satisfied in $\mathbf{BV}_2$ and the identity $x \leq 1$. We show this by an induction with respect to the cardinality of the set $M = \mathsf{Sub}_2(u) \setminus \mathsf{Sub}_2(v)$.

If $M = \emptyset$ then $u \sim_2 v$ and the statement is clear.

If $M \neq \emptyset$ then the assumptions of Lemma 1 are valid. So, there exists w such that $u \preceq_2 w \preceq_2 v$, $v \lhd w$ and $v \not\sim_2 w$ (or $w \lhd u$ and $w \not\sim_2 u$ which can be proved in a similar way). Then $\mathsf{Sub}_2(v) \subset \mathsf{Sub}_2(w) \subseteq \mathsf{Sub}_2(u)$ and $\mathsf{Sub}_2(u) \setminus \mathsf{Sub}_2(w) \subset M$ follows. By an induction assumption the identity $u \leq w$ is a consequence of the identities satisfied in $\mathbf{BV}_2$ and the identity $x \leq 1$. Because $v \lhd w$, the identity $w \leq v$ is a consequence of the identity $x \leq 1$. This implies that $u \leq v$ is a consequence of the identities satisfied in $\mathbf{BV}_2$ and the identity $x \leq 1$.   □

Part (iii).

Claim 2. *For $k \geq 3$, it holds $\mathscr{PV}_k \neq \mathscr{BV}_k \cap \mathscr{P}$.*

Proof. We show that the identity $(xy)^{k-1} \leq x^{k-1}y^{k-1}$ is satisfied in $\mathbf{PV}_k$ but it is not satisfied in $\mathbf{BV}_k \cap \mathbf{P}$.

The first observation is clear since $\mathsf{Sub}_k(x^{k-1}y^{k-1}) \subseteq \mathsf{Sub}_k((xy)^{k-1})$.

For the second part, we assert first that there is no word v different from $(xy)^{k-1}$ such that $v \sim_k (xy)^{k-1}$. Indeed, if $\mathsf{Sub}_k(v) = \mathsf{Sub}_k((xy)^{k-1})$ then v contains exactly $k-1$ occurrences of variable x and the same number of occurrences of y. Now $yx^{k-1} \notin \mathsf{Sub}_k(v)$ and $y^{k-1}x \notin \mathsf{Sub}_k(v)$ hence the first letter of v is x and the last letter of v is y. Moreover, $x^i y x^{k-1-i} \in \mathsf{Sub}_k(v)$ for each $i = 1, \ldots, k-2$, so, between i-th and $(i+1)$-th occurrence of x in v has to be some y. We can conclude with $v = (xy)^{k-1}$.

Now we assert that there is no proper subword v of $(xy)^{k-1}$ such that $v \preceq_k x^{k-1}y^{k-1}$. Assume that there is some word v with this property. Then from $\mathsf{Sub}_k(x^{k-1}y^{k-1}) \subseteq \mathsf{Sub}_k(v)$ we can deduce that v contains exactly $k-1$ occurrences of variable x and the same number of occurrences of y, which is a contradiction.

Our two assertions imply the statement, since there is no proof of $(xy)^{k-1} \leq x^{k-1}y^{k-1}$ using the identity $x \leq 1$ and the identities which are satisfied in $\mathbf{BV}_k$.   □

Remark 1. The idea from our proof of Claim 2 can be also used for direct construction of a language L with the properties: $L \in \mathscr{BV}_k \cap \mathscr{P}$, $L \notin \mathscr{PV}_k$. We show such an example for the case $k = 3$.

We consider the following language L over $A = \{a, b\}$

$$L = L_{aaa} \cup L_{bbb} \cup \{aabb\} \tag{1}$$

$$= L_{aaa} \cup L_{bbb} \cup L_{aabb} \tag{2}$$

$$= L_{aaa} \cup L_{bbb} \cup (L_{aa} \cap L_{bb} \cap L_{ba}^{\mathsf{c}}) . \tag{3}$$

The fact $L \in \mathscr{P}$ follows from (2) and the fact $L \in \mathscr{BV}_3$ follows from (3). On the other hand, we can show that $L \notin \mathscr{PV}_3$. Assume, for a moment, that $L \in \mathscr{PV}_3$. Then $\preceq_3 \subseteq \preceq_L$ by Proposition 2. It is clear that $abab \preceq_3 aabb$, so we have $abab \preceq_L aabb$ which is a contradiction with $aabb \in L$ and $abab \notin L$.

Part (iv). This part of the theorem is proved for $k = 4$ first.

Claim 3. *There is no finite basis of identities for the pseudovariety $\mathbf{PV}_4$.*

Proof. Assume that $\mathbf{PV}_4$ has a finite basis Π of identities. Let n be the number of variables used in Π. We consider the identity $u \leq v$ where

$$u = x\,y\,x\,X\,y\,Y\,y \,, \quad v = x\,x\,y\,X\,y\,Y\,y \quad \text{with } X = z_1 z_2 \ldots z_n \text{ and } Y = z_n \ldots z_2 z_1 \,.$$

One can show that this identity is satisfied in $\mathbf{PV}_4$ and it is not a consequence of identities in Π. A full proof is placed into appendix (see [9]).   □

The previous proposition can be easily modified for every $k > 4$. The change is that we multiply the words u and v by x^{k-4} from left. Hence we have

$$u = x^{k-3}\,y\,x\,X\,y\,Y\,y \,, \quad v = x^{k-2}y\,X\,y\,Y\,y \,.$$

This ends the sketch of the proof of Theorem 2.   □

For the last case $k = 3$, the proof of Claim 3 does not work. The easiest example of the identity which is satisfied in $\mathbf{PV}_3$ but which is not a consequence of the identities from $\mathbf{BV}_3$ and the identity $x \leq 1$ is the identity $xz_1yxz_2y \leq xz_1xyz_2y$. It seems that this identity is strong enough as we did not find some identity which is not a consequence of this one. This leads us to the following conjecture about finite basis of identities for $\mathbf{PV}_3$.

Conjecture. $\mathbf{PV}_3 = \mathbf{BV}_3 \cap \mathbf{P} \cap \mathsf{Mod}(\,xz_1yxz_2y \leq xz_1xyz_2y\,)$.

5   Generating by a Single Monoid and Semiring

Volkov in [23] proved that each pseudovariety $\mathbf{BV}_k$ is generated by a single monoid. We show an alternative proof of this fact and we will prove the similar results concerning the pseudovarieties $\mathbf{PV}_k$ and $\mathbf{DV}_k$. The idea is that we will generate the varieties of languages $\mathscr{BV}_k$, $\mathscr{PV}_k$ and $\mathscr{DV}_k$ by a single language instead of generating the pseudovarieties of monoids and semirings.

Volkov used three types of monoids which were introduced by Straubing and Pin, namely the monoid $\mathcal{R}_k$ of all reflexive binary relations (viewed as a submonoid of the monoid of all $(k+1) \times (k+1)$ matrices over the Boolean semiring $\mathbf{B} = (\{0,1\}, \wedge, \vee)$), its submonoid $\mathcal{U}_k$ of all upper unitriangular matrices (i.e. there are only zeros under the main diagonal and all diagonal entries are 1), and the monoid $\mathcal{C}_k$ of all order preserving and extensive transformations of a chain with $k + 1$ elements. We identify such transformation ϕ with the matrix $C(\phi)$ having exactly one non-zero entry in each row, namely at the position $(i, \phi(i))$ for $i = 1, \ldots, k + 1$. Clearly, the composition of transformations corresponds to the multiplication of matrices.

The last monoid we will use, denoted $\mathcal{S}_k$, is the submonoid of $\mathcal{U}_k$ consisting of all *stair triangular* matrices, i.e. matrices satisfying: if $a_{i,j} = 1$, $i < j$ then

$$a_{i,i} = a_{i,i+1} = \cdots = a_{i,j} = 1, \ a_{i,j} = a_{i+1,j} = \cdots = a_{j,j} = 1 \,.$$

The monoids $\mathcal{R}_k, \mathcal{U}_k$ and $\mathcal{S}_k$ are idempotent semirings with respect to $\vee$ taken componentwise.

Notice that the mapping $\phi \mapsto S(\phi)$, $\phi \in \mathcal{C}_k$, where $(S(\phi))_{i,j} = 1$ if and only if $j \in \{i, i+1, \ldots, \phi(i)\}$, induces a monoid isomorphism of $\mathcal{C}_k$ onto $\mathcal{S}_k$.

For each k, we fix the k-element alphabet $B = \{b_1, b_2, \ldots, b_k\}$ and the language

$$L(k, B) = B^* b_1 B^* b_2 B^* \ldots B^* b_k B^* .$$

A crucial property of $L(k, B)$ is the following lemma.

Lemma 2. *For every finite alphabet A and a word $u \in A^*$ of length k, there exists a morphism $f : A^* \to B^*$ such that $f^{-1}(L(k, B)) = L_u$.*

Proof. Let $u = a_1 a_2 \ldots a_k$, where $a_i \in A$. For each $a \in A$ we consider the sequence of indices $i_1 < i_2 < \cdots < i_\ell$ such that $a_{i_1} = a_{i_2} = \cdots = a_{i_\ell} = a$ and define $f(a) = b_{i_\ell} \ldots b_{i_2} b_{i_1}$.

An example for a better understanding: if $k = 8$, $A = \{c_1, \ldots, c_4\}$ and $u = c_4 c_3 c_1 c_4 c_1 c_3 c_1 c_4$ then $f : c_1 \mapsto b_7 b_5 b_3$, $c_2 \mapsto \lambda$, $c_3 \mapsto b_6 b_2$, $c_4 \mapsto b_8 b_4 b_1$.

Note that $b_i b_{i+1} \ldots b_{i+j} \lhd f(a)$ implies $j = 0$; in other terms $\mathrm{Sub}(f(a)) \cap \mathrm{Sub}(b_1 b_2 \ldots b_k) \subseteq B$. So, we have defined $f : A^* \to B^*$ morphism and we have to check that $f^{-1}(L(k, B)) = L_u$.

"$\subseteq$" : Let $w \in f^{-1}(L(k, B))$, $w = c_1 c_2 \ldots c_m$, where $c_1, \ldots c_m \in A$. Then there exist indices $j_1 < j_2 < \cdots < j_k$ such that $f(c_{j_i})$ contains b_i for all $i = 1, \ldots, k$. Hence $c_{j_i} = a_i$ for all $i = 1, \ldots, k$. This means $u = a_1 a_2 \ldots a_k \lhd w$, i.e. $w \in L_u$.

"$\supseteq$" : Now, let $w \in L_u$. Then $f(u) \lhd f(w)$. From the definition of images of letters we have $b_i \lhd f(a_i)$ for all $i = 1, \ldots, k$ and we can conclude with $b_1 b_2 \ldots b_k \lhd f(w)$, i.e. $w \in f^{-1}(L(k, B))$. $\qquad\square$

Lemma 3. *For the language $L(k, B)$ over the alphabet B the following is true.*
(i) If a boolean variety of languages $\mathscr{B}$ satisfies $L(k, B) \in \mathscr{B}(B)$, then $\mathscr{B}\mathscr{V}_k \subseteq \mathscr{B}$.
(ii) If a positive variety of languages $\mathscr{V}$ satisfies $L(k, B) \in \mathscr{V}(B)$, then $\mathscr{P}\mathscr{V}_k \subseteq \mathscr{V}$.
(iii) If a disjunctive variety $\mathscr{D}$ satisfies $L(k, B) \in \mathscr{D}(B)$, then $\mathscr{D}\mathscr{V}_k \subseteq \mathscr{D}$.

Proof. In all cases the classes are closed under the preimages in morphisms. If we apply the previous lemma we see that for any alphabet A and the word u of length k we have $L_u \in \mathscr{B}(A)$ (and $L_u \in \mathscr{V}(A)$ and $L_u \in \mathscr{D}(A)$, respectively). The classes are also closed under derivatives since $a^{-1} L_{av} = L_v$ and $L_{va} a^{-1} = L_v$. Hence, for any alphabet A and the word u of length at most k, we have $L_u \in \mathscr{B}(A)$ (and $L_u \in \mathscr{V}(A)$ and $L_u \in \mathscr{D}(A)$, respectively). Now the statements are consequences of the definitions of the classes $\mathscr{B}\mathscr{V}_k$, $\mathscr{P}\mathscr{V}_k$, $\mathscr{D}\mathscr{V}_k$. $\qquad\square$

Proposition 3. *For each $k \geq 1$, we have:*
(i) $\mathbf{BV}_k$ is generated by the syntactic monoid $B^* / {\sim_{L(k,B)}}$.
(ii) $\mathbf{PV}_k$ is generated by the syntactic ordered monoid $B^* / {\preceq_{L(k,B)}}$.
(iii) $\mathbf{DV}_k$ is generated by the syntactic semiring $B^\cup / {\equiv_{L(k,B)}}$.

Proof. It is a direct consequence of Lemma 3. $\qquad\square$

Now we present natural models of the syntactic structures of the language $L(k, B)$. We define $\mu : B \to \mathcal{S}_k$ as follows: the only non-zero non-diagonal entry in the matrix $\mu(b_i)$ is $(\mu(b_i))_{i,i+1} = 1$ for $i = 1, \ldots, k$. This mapping naturally extends to B^* and $B^\cup$.

Proposition 4. *The structures* $(\mathcal{S}_k, \cdot), ((\mathcal{S}_k, \cdot, \leq)$ *and* $(\mathcal{S}_k, \cdot, \vee)$, *respectively) are isomorphic to the syntactic monoid (ordered syntactic monoid and syntactic semiring, respectively) of the language* $L(k, B)$.

Proof. Indeed, using the induction with respect to the lengths of words we see that the extension $\mu : B^* \to \mathcal{S}_k$ is given by $(\mu(u))_{i,j} = 1$ if and only if $i \leq j$ and $b_i \ldots b_{j-1} \lhd u$ for each $u \in A^*$.

For a matrix $S \in \mathcal{S}_k$ with non-zero entries $s_{1,1}, \ldots, s_{1,p_1}, \ldots, s_{k,k}, \ldots, s_{k,p_k}$, $s_{k+1.k+1}$, we see that $\mu(b_k \ldots b_{p_k-1} \ldots b_1 \ldots b_{p_1-1}) = S$ and thus μ is surjective.

Further, for each $u, v \in A^*$, $U, V \in A^\cup$, we have $u \sim_{L(k,B)} v$ if and only if $\mu(u) = \mu(v)$, $u \preceq_{L(k,B)} v$ if and only if $\mu(u) \geq \mu(v)$, and finally $U \equiv_{L(k,B)} V$ if and only if $\mu(U) = \mu(V)$. □

6 Final Remarks

Remark 2. [1] We know that the pseudovarieties $\mathbf{BV}_k$ and $\mathbf{PV}_k$, for $k \geq 4$, have no finite bases of identities. A natural question is whether there exist finite bases of pseudoidentities for these classes. (One can consult the background concerning pseudoidentities in Almeida's book [2].)

By Proposition 3 or by [23] each pseudovariety $\mathbf{BV}_k$ is generated by a single monoid and such a pseudovariety admits a finite basis of identities if and only if it admits a finite basis of pseudoidentities (see Corollary 4.3.8 in the book [2]). The same arguments can be used in the case of the pseudovarieties $\mathbf{PV}_k$. Therefore the pseudovarieties $\mathbf{BV}_k$ and $\mathbf{PV}_k$ have no finite bases of pseudoidentities.

Remark 3. Our goal was to get a better understanding of languages of level 1 in Straubing-Thérien hierarchy. We expect that some results from the present paper can be extended also to other hierarchies. For example, it could be interesting to study hierarchies based of locally testable languages, group languages or languages of the form

$$B_0^* a_1 B_1^* a_2 B_2^* \ldots a_\ell B_\ell^*, \text{ where } a_1, \ldots, a_\ell \in A, \ B_0, \ldots, B_\ell \subseteq A, \ \ell \leq k, \ k \text{ fixed }.$$

We also have formulated the conjecture that $\mathbf{PV}_3$ is finitely based.

References

1. Almeida, J.: Implicit operations on finite $\mathcal{J}$-trivial semigroups and a conjecture of I. Simon. J. Pure Appl. Algebra 69, 205–218 (1990)
2. Almeida, J.: Finite Semigroups and Universal Algebra. World Scientific, Singapore (1994)
3. Blanchet-Sadri, F.: Games, equations and the dot-depth hierarchy. Comput. Math. Appl. 18, 809–822 (1989)
4. Blanchet-Sadri, F.: Equations and monoids varieties of dot-depth one and two. Theoret. Comput. Sci. 123, 239–258 (1994)

---

[1] We express here our gratitude to Jorge Almeida for a discussion on the topic.

5. Eilenberg, S.: Automata, Languages and Machines, vol. B. Academic Press, New York (1976)
6. Ésik, Z., Ito, M.: Temporal logic with cyclic counting and the degree of aperiodicity of finite automata. Acta Cybernetica 16, 1–28 (2003)
7. Higgins, P.: A proof of Simon's Theorem on piecewise testable languages. Theoret. Comput. Sci. 178, 257–264 (1997)
8. Klíma, O., Polák, L.: Classes of meet automata. Theoret. Comput. Sci. (to appear)
9. Klíma, O., Polák, L.: Hierarchies of piecewise testable languages, a version containing also Appendix, http://www.math.muni.cz/~polak
10. Pin, J.-E.: Varieties of Formal Languages. North Oxford Academic, Plenum (1986)
11. Pin, J.-E.: Finite semigroups and recognizable languages: an introduction. In: Fountain, J. (ed.) NATO Advanced Study Institute Semigroups, Formal Languages and Groups, pp. 1–32. Kluwer Academic Publisher, Dordrecht (1995)
12. Pin, J.-E.: A variety theorem without complementation. Russian Mathem (Iz. VUZ) 39, 74–83 (1995)
13. Pin, J.-E.: Syntactic semigroups. In: Rozenberg, G., Salomaa, A. (eds.) Handbook of Formal Languages, ch. 10. Springer, Heidelberg (1997)
14. Pin, J.-E., Weil, P.: Polynomial closure and unambiguous product. Theory Comput. Syst. 30, 1–39 (1997)
15. Polák, L.: A classification of rational languages by semilattice-ordered monoids. Arch. Math. (Brno) 40, 395–406 (2004)
16. Polák, L.: On pseudovarieties of semiring morphisms. In: Fiala, J., Koubek, V., Kratochvíl, J. (eds.) MFCS 2004. LNCS, vol. 3153, pp. 635–647. Springer, Heidelberg (2004)
17. Simon, I.: Hierarchies of events of dot-depth one, Ph.D. thesis, University of Waterloo (1972)
18. Simon, I.: Piecewise testable events. In: Proc. ICALP 1975. LNCS, vol. 33, pp. 214–222. Springer, Heidelberg (1975)
19. Stern, J.: Characterization of some classes of regular events. Theoret. Comput. Sci. 35, 17–42 (1985)
20. Straubing, H.: On finite $\mathcal{J}$-trivial monoids. Semigroup Forum 19, 107–110 (1980)
21. Straubing, H.: On logical description of regular languages. In: Rajsbaum, S. (ed.) LATIN 2002. LNCS, vol. 2286, pp. 528–538. Springer, Heidelberg (2002)
22. Straubing, H., Thérien, D.: Partially ordered finite monoids and a theorem of I. Simon. J. Algebra 119, 393–399 (1988)
23. Volkov, M.V.: Reflexive relations, extensive transformations and piecewise testable languages of a given height. Internat. J. Algebra Comput. 14, 817–827 (2004)

Construction of Tree Automata from Regular Expressions

Dietrich Kuske and Ingmar Meinecke*

Institut für Informatik, Universität Leipzig, Germany
{kuske,meinecke}@informatik.uni-leipzig.de

Abstract. Since recognizable tree languages are closed under the rational operations, every regular tree expression denotes a recognizable tree language. We provide an alternative proof to this fact that results in smaller tree automata. To this aim, we transfer Antimirov's partial derivatives from regular word expressions to regular tree expressions. For an analysis of the size of the resulting automaton as well as for algorithmic improvements, we also transfer the methods of Champarnaud and Ziadi from words to trees.

1 Introduction

One of the most prominent topics in formal language theory is the comparison of different finite descriptions for potentially infinite objects – the languages. The result of Kleene [13] states the equivalence between finite automata and regular expressions for languages of finite words. The transformation of a finite automaton into an equivalent regular expression is a prototypical example of dynamic programming. The converse transformation is of direct practical consequence e.g. in text processing. For this reason, several methods were proposed within the last decades to find more efficient algorithms, see [15,16] for surveys. For teaching purposes, one often uses an inductive construction. The most common construction is the standard or position automaton (Glushkov [9] and McNaughton and Yamada [14]). Brzozowski's construction [3] of a deterministic finite automaton uses derivates of regular expressions. This approach was modified by Antimirov [1] who defined *partial derivatives* to construct a non-deterministic automaton from a regular expression E.

Kleene's theorem was lifted to the setting of trees [17], also cf. [8,7], which are one of the most fundamental concepts in computer science. A regular tree expression defines a language of ordered trees. An inductive construction even produces a tree automaton accepting this language. The number of states of this automaton is exactly the number of iterations in E plus $\|E\|$ where $\|E\|$ is the number of occurrences of symbols from the ranked alphabet in E. In this paper, we define partial derivatives for regular tree expressions and build by their help a non-deterministic finite tree automaton recognizing the language denoted by the regular expression. The concept of partial derivatives will yield a tree automaton with at most $\|E\|$ states and $\|E\|^2$ transitions. The construction of this tree automaton and the correctness proof is combined with algorithmic considerations to build this automaton. We adapt and modify the approach

---

* The second author was supported by the German Research Foundation (DFG).

M. Ito and M. Toyama (Eds.): DLT 2008, LNCS 5257, pp. 491–503, 2008.
© Springer-Verlag Berlin Heidelberg 2008

by Champarnaud and Ziadi [5,6] in the word case who extended work of Berry and Sethi [2]. Here, we use linearizations of regular tree expressions. The main idea is to distinguish occurrences of the same symbol at different positions in the regular expression. By doing so, we can ensure a certain uniqueness of the partial derivatives. As it turns out, the partial derivatives of the original regular expression are just projections of the partial derivatives of the linearized regular expression. This approach results in two main advantages: Firstly, the desired automaton is in fact a quotient of an automaton that stems from the linearized regular expression. This way we also get the upper bound on the number of transitions mentioned above. Secondly, the theoretical results allow for an efficient algorithm working in the syntax-tree of E. We obtain an algorithm with $\mathcal{O}(R \cdot |E|^2)$ space and time complexity where R is the maximal rank of a symbol occuring in the finite ranked alphabet Σ and $|E|$ is the size of the regular expression.

Beside the standard and the partial derivative construction there are other proposals in the literature how to obtain an automaton from a regular expression. Especially, it would be interesting whether the construction of the follow automaton [11,12,4] carries over to the setting of trees. In this paper we consider ranked trees. However, regular expressions were explored for unranked trees in connection with XML. They are used in pattern matching, see e.g. [10]. We wonder whether the concept of partial derivatives can lead to fruitful results and algorithms in this area.

2   Trees, Automata, and Regular Expressions

Throughout this paper, we fix a finite ranked alphabet $\Sigma = (\Sigma_m)_{m \geq 0}$. The set T_Σ of trees over Σ is defined by the syntax

$$t = f(\underbrace{t, t, \ldots, t}_{m \text{ times}})$$

where $f \in \Sigma_m$. A subset $L \subseteq T_\Sigma$ is called a *tree language*.

A *tree automaton* over Σ is a tuple $\mathcal{A} = (Q, \Sigma, F, \Delta)$ where Q is a set of states, $F \subseteq Q$ is the set of final states, and $\Delta = (\Delta_m)_{m \in \mathbb{N}}$ is the set of transitions such that $\Delta_m \subseteq Q \times \Sigma_m \times Q^m$ for every $m \in \mathbb{N}$.[1] Especially, $\Delta_0 \subseteq Q \times \Sigma_0$. A finite tree-automaton (or FTA) is a tree automaton $\mathcal{A}$ with only finitely many states and, thus, only finitely many transitions (note that there are only finitely many m with $\Sigma_m \neq \emptyset$).

As to whether a tree t is accepted by a tree automaton $\mathcal{A} = (Q, \Sigma, F, \Delta)$ is defined inductively along the construction of the tree t: if $t = c \in \Sigma_0$, then t is accepted by $\mathcal{A}$ iff there exists a transition $(q, c) \in \Delta_0$ with $q \in F$. For $f \in \Sigma_m$ with $m > 0$, the tree $f(t_1, \ldots, t_m)$ is acccepted by $\mathcal{A}$ iff there exists a transition $(q, f, q_1, \ldots, q_m) \in \Delta_m$ such that $q \in F$ and, for $1 \leq i \leq m$, the tree t_i is accepted by the tree automaton $(Q, \Sigma, \{q_i\}, \Delta)$. The language $L(\mathcal{A})$ recognized by $\mathcal{A}$ is the set of all trees t that are accepted by $\mathcal{A}$. A tree language L is *recognizable* if there is a FTA $\mathcal{A}$ with $L(\mathcal{A}) = L$.

---

[1] In the term-rewriting terminology employed by [7], Δ is the set of rules $f(q_1(x_1), \ldots, q_m(x_m)) \rightarrow q(f(x_1, \ldots, x_m))$.

We next introduce some constructions of tree languages that extend the rational operations on word languages. Let $f \in \Sigma_m$ and $L, L_1, \ldots, L_m \subseteq T_\Sigma$. Then we put

$$f(L_1, \ldots, L_m) = \{f(t_1, \ldots, t_m) \mid t_i \in L_i \text{ for } i = 1, \ldots, m\}.$$

For $L \subseteq T_\Sigma$ and $c \in \Sigma_0$ we define for every $t \in T_\Sigma$ inductively the non-uniform *substitution* $t[c \leftarrow L]$:

- $c[c \leftarrow L] = L$ and $d[c \leftarrow L] = \{d\}$ for every $d \in \Sigma_0$ with $d \neq c$,
- $f(t_1, \ldots, t_m)[c \leftarrow L] = f(t_1[c \leftarrow L], \ldots, t_m[c \leftarrow L])$.

Then the *c-product* of $L_1, L_2 \subseteq T_\Sigma$ is the language $L_1 \cdot_c L_2 = \bigcup_{t \in L_1} t[c \leftarrow L_2]$. Now the iterated *c*-products are defined for $L \subseteq T_\Sigma$ by

$$L^{0,c} = \{c\} \text{ and } L^{n+1,c} = L^{n,c} \cup L \cdot_c L^{n,c}.$$

The *c-iteration* of L is defined as $L^{*c} = \bigcup_{n \geq 0} L^{n,c}$.

It is well-known that a tree language L is recognizable if and only if it can be denoted by a regular expression. These *regular expressions* are defined by the following syntax

$$E = f(\underbrace{E, E, \ldots, E}_{m \text{ times}}) \mid E + E \mid E \cdot_c E \mid E^{*c}$$

where $f \in \Sigma_m$ and $c \in \Sigma_0$.

The semantics $[\![E]\!]$ of a regular expression E is defined inductively by

$$[\![f(E_1, E_2, \ldots, E_m)]\!] = f([\![E_1]\!], [\![E_2]\!], \ldots, [\![E_m]\!]), \quad [\![E + F]\!] = [\![E]\!] \cup [\![F]\!],$$
$$[\![E \cdot_c F]\!] = [\![E]\!] \cdot_c [\![F]\!], \text{ and} \qquad\qquad [\![E^{*c}]\!] = [\![E]\!]^{*c}.$$

For a set M of regular expressions, we put $[\![M]\!] = \bigcup_{E \in M} [\![E]\!]$.

The set of all regular expressions over the ranked alphabet Σ is denoted by $\mathrm{EXP}(\Sigma)$. Let $|E|_f$ denote the number of occurrences of the letter $f \in \Sigma$ in E. The *alphabetic width* $\|E\|$ of E is the number $\sum_{f \in \Sigma} |E|_f$ of occurrences of symbols from Σ in E. The *size* $|E|$ of E is defined inductively by: $|c| = 1$ for $c \in \Sigma_0$, $|f(E_1, \ldots, E_m)| = \sum_{i=1}^m |E_i| + 1$, $|E + F| = |E \cdot_c F| = |E| + |F| + 1$, and $|E^{*c}| = |E| + 1$. Every regular expression E can be understood as a tree over the ranked alphabet $\Sigma \cup \{+, \cdot_c, {}^{*c} \mid c \in \Sigma_0\}$ where $+$ and $\cdot_c$ have rank 2 and ^{*c} has rank 1. This tree is called the *syntax-tree* t_E of E.

3 A Direct Construction

In this section, we will construct from a regular expression E a tree automaton $\mathcal{A}_E$ that accepts $[\![E]\!]$. The finiteness of this automaton will only be proved later. Our construction is based on partial derivates that we have to define and investigate first.

Let M be a set of regular expressions, F some regular expression, and $c \in \Sigma_0$. Then $M \cdot_c F$ denotes the set

$$M \cdot_c F = \{E \cdot_c F \mid E \in M\}.$$

Similarly, we put for a set $\mathcal{M}$ of m-tuples of regular expressions

$$\mathcal{M} \cdot_c F = \{(E_1 \cdot_c F, E_2 \cdot_c F, \ldots, E_m \cdot_c F) \mid (E_1, E_2, \ldots, E_m) \in \mathcal{M}\}.$$

Let $\Sigma_{\geq 1} = \bigcup_{m \geq 1} \Sigma_m = \Sigma \setminus \Sigma_0$ denote the set of non-constant symbols from the ranked alphabet Σ.

Definition 3.1. *For $g \in \Sigma_{\geq 1}$ and a regular expression E, we define the sets $g^{-1}(E)$ of m-tuples of regular expressions inductively by*

$$- g^{-1}(f(E_1, E_2, \ldots, E_n)) = \begin{cases} \{(E_1, E_2, \ldots, E_n)\} & \text{if } f = g \\ \emptyset & \text{if } f \neq g \end{cases}$$

$$- g^{-1}(E + F) = g^{-1}(E) \cup g^{-1}(F)$$

$$- g^{-1}(E \cdot_c F) = \begin{cases} g^{-1}(E) \cdot_c F & \text{if } c \notin \llbracket E \rrbracket \\ g^{-1}(E) \cdot_c F \cup g^{-1}(F) & \text{otherwise} \end{cases}$$

$$- g^{-1}(E^{*c}) = g^{-1}(E) \cdot_c E^{*c}.$$

Following Antimirov, we define further functions ∂_w for finite words $w \in \Sigma_{\geq 1}^*$ over the alphabet $\Sigma_{\geq 1}$. By ε we denote the empty word.

Definition 3.2. *Let E be a regular expression. Then $\partial_\varepsilon(E) = \{E\}$ and, for $w \in \Sigma_{\geq 1}^*$ and $g \in \Sigma_{\geq 1}$, the set $\partial_{wg}(E)$ consists of all regular expressions F that appear in some tuple from $g^{-1}(E')$ for some $E' \in \partial_w(E)$. For a set of words $W \subseteq \Sigma_{\geq 1}^*$ and a regular expression E, we put $\partial_W(E) = \bigcup_{w \in W} \partial_w(E)$.*

The function ∂_w is called the partial derivative w.r.t. w.

Note that $\partial_{wg}(E) = \partial_g(\partial_w(E)) = \bigcup_{E' \in \partial_w(E)} \partial_g(E')$ for all $w \in \Sigma_{\geq 1}^*$ and $g \in \Sigma_{\geq 1}$. Further note that we consider derivatives with respect to words over the non-constant symbols from Σ and not with regard to trees.

A symbol $f \in \Sigma$ occurs unguarded in E if no ancestor in the syntax tree t_E is labeled by an element of Σ. We will be interested in the number $\langle E \rangle_f$ of unguarded occurrences of f in E that can be computed inductively:

- $\langle f(E_1, \ldots, E_m) \rangle_f = 1$ and $\langle g(E_1, \ldots, E_m) \rangle_f = 0$ for $g \neq f$,
- $\langle E_1 + E_2 \rangle_f = \langle E_1 \cdot_c E_2 \rangle_f = \langle E_1 \rangle_f + \langle E_2 \rangle_f$, and
- $\langle E^{*c} \rangle_f = \langle E \rangle_f$.

Proposition 3.3. *Let E be a regular expression and $g \in \Sigma_{\geq 1}$. Then $|g^{-1}(E)| \leq \langle E \rangle_g$. Especially, if $|E|_g = 0$ then $g^{-1}(E) = \partial_g(E) = \emptyset$.*

Next, we express the semantics of a regular expression $\llbracket E \rrbracket$ in terms of the semantics of the tuples from $g^{-1}(E)$.

Proposition 3.4. *For any regular expression E, we have*

$$\llbracket E \rrbracket = \bigcup \{g(\llbracket G_1 \rrbracket, \ldots, \llbracket G_m \rrbracket) \mid g \in \Sigma_{\geq 1}, (G_1, \ldots, G_m) \in g^{-1}(E)\}$$

$$\cup \{c \in \Sigma_0 \mid c \in \llbracket E \rrbracket\}. \tag{1}$$

Let E be a regular expression and let $Q_E = \partial_{\Sigma_{\geq 1}^*}(E)$. Then we define a set of transitions Δ_E as

$$\{(F, f, (G_1, G_2, \ldots, G_m)) \mid F \in Q_E, f \in \Sigma_m, m \geq 1, (G_1, \ldots, G_m) \in f^{-1}(F)\}$$
$$\cup \{(F, c) \mid F \in Q_E, c \in \Sigma_0, c \in [\![F]\!]\} \, .$$

Furthermore, let $\mathcal{A}_E = (Q_E, \Sigma, \{E\}, \Delta_E)$ denote the tree automaton whose only final state is the regular expression E.

Theorem 3.5. *Let E be a regular expression over the ranked alphabet Σ. Then $\mathcal{A}_E$ is a tree automaton that accepts $[\![E]\!]$.*

Proof. We show by induction on $n \in \mathbb{N}$: for all trees $t = f(s_1, \ldots, s_m)$ of size n and all regular expressions F, the tree automaton $\mathcal{A}_F$ accepts t iff $t \in [\![F]\!]$.

Since there are no trees of size 0, the base case is trivial. So let $t = f(s_1, \ldots, s_m)$ for some $m \geq 0$. For $m = 0$ we have $t = c \in \Sigma_0$. Now c is accepted by $\mathcal{A}_F$ iff there is a transition $(F, c) \in \Delta_F$. But this is the case iff $c \in [\![F]\!]$. Now suppose $m > 0$. Then t is accepted by $\mathcal{A}_F$ iff there exists a transition $(F, f, (G_1, \ldots, G_m)) \in \Delta_F$ such that s_i is accepted by the tree automaton $(Q_F, \Sigma, \{G_i\}, \Delta_F)$ for all $1 \leq i \leq m$. Note that the reachable part of the automaton $(Q_F, \Sigma, \{G_i\}, \Delta_F)$ is the set of states Q_{G_i}. Hence, s_i is accepted by this automaton iff it is accepted by $\mathcal{A}_{G_i}$. By the induction hypothesis, this is equivalent to saying $s_i \in [\![G_i]\!]$. Since this holds for all $1 \leq i \leq m$, we have that t is accepted by $\mathcal{A}_F$ iff there exists $(G_1, \ldots, G_m) \in f^{-1}(F)$ with $s_i \in [\![G_i]\!]$ which is, by Proposition 3.4, equivalent to saying $t \in [\![F]\!]$. □

So far, we did not prove that the tree automaton $\mathcal{A}_E$ has only finitely many states, i.e., that $[\![E]\!]$ is recognizable. Theorem 4.14 will show that the number of states is linear and that the number of transitions is quadratic in the size of E. This will only be achieved after going through the following two constructions.

4   An Indirect Construction Via Linearizations

The idea of the indirect construction is as follows: In a regular expression E, uniquely mark the occurrences of letters from $\Sigma_{\geq 1}$. Then apply our direct construction to the resulting regular expression $\overline{E}$. The projection of this automaton accepts $[\![E]\!]$. As it turns out, a quotient of the automaton one obtains this way is isomorphic to the result of the direct construction.

4.1   Linear Regular Expressions

A regular expression E is *linear* if every letter $f \in \Sigma_{\geq 1}$ occurs at most once in E. Note that $c \in \Sigma_0$ may occur more than once. The following proposition is a consequence of Proposition 3.3.

Proposition 4.1. *Let E be a linear regular expression and $g \in \Sigma_m$ for $m \geq 1$. Then $|g^{-1}(E)| \leq 1$ and therefore $|\partial_g(E)| \leq m$.*

We consider partial derivatives w.r.t. non-empty words for linear regular expressions:

Proposition 4.2. *Let E, F be linear regular expressions over the alphabet Σ such that also $E + F$ and $E \cdot_c F$ are linear. Let $w \in \Sigma_{\geq 1}^*$ and $g \in \Sigma_{\geq 1}$. Then the following hold true:*

$$- \ g^{-1}(\partial_w(E + F)) = \begin{cases} g^{-1}(\partial_w(E)) & \text{if } |E|_g > 0, \\ g^{-1}(\partial_w(F)) & \text{otherwise.} \end{cases}$$

$$- \ g^{-1}(\partial_w(E \cdot_c F)) = \begin{cases} g^{-1}(\partial_w(E)) \cdot_c F & \text{if } |E|_g > 0 \\ \bigcup \{g^{-1}(\partial_v(F)) \mid \exists u \in \Sigma_{\geq 1}^* : w = uv \ \& \ c \in [\![\partial_u(E)]\!]\} \\ \hspace{7cm} \text{otherwise.} \end{cases}$$

- There are suffixes $v_1, \ldots, v_k$ of w such that

$$g^{-1}(\partial_w(E^{*c})) = \bigcup_{1 \leq i \leq k} g^{-1}(\partial_{v_i}(E)) \cdot_c E^{*c}.$$

Proof (Idea). The claim for the sum is shown easily. For the product and the iteration we proceed by induction on $|w|$. For the product, we have to perform an elaborate case distinction. □

Proposition 4.3. *Let E be a linear regular expression, $u, w \in \Sigma_{\geq 1}^*$, and $g \in \Sigma_{\geq 1}$. Then we have:*

1. *$|g^{-1}(\partial_w(E))| \leq 1$,*
2. *if $\partial_{ug}(E) \neq \emptyset$ and $\partial_{wg}(E) \neq \emptyset$, then $\partial_{ug}(E) = \partial_{wg}(E)$,*
3. *if $g^{-1}(\partial_u(E)) \neq \emptyset$ and $g^{-1}(\partial_w(E)) \neq \emptyset$, then $g^{-1}(\partial_u(E)) = g^{-1}(\partial_w(E))$.*

Proof. Note that the second claim is a consequence of the third one. The proof of the first statement can easily be extracted from our proof of the third one.

First consider the case $E = f(E_1, \ldots, E_n)$. Since E is linear, there is at most one i with $|E_i|_g > 0$, if no such i exists, set $i = 1$. Then we have

$$g^{-1}\partial_u(E) = \begin{cases} g^{-1}\partial_{u'}(\{E_1, \ldots, E_n\}) = g^{-1}\partial_{u'}(E_i) & \text{if } u = fu', \\ \{(E_1, \ldots, E_n)\} & \text{if } u = \varepsilon \ \& \ f = g, \\ \emptyset & \text{otherwise} \end{cases}$$

where the first case is due to $|E_j|_g = 0$ for $j \neq i$, and, similarly for $g^{-1}\partial_w(E)$.

Recall that by assumption $g^{-1}\partial_u(E) \neq \emptyset$ and $g^{-1}\partial_w(E) \neq \emptyset$. If $f = g$ is the first letter of $u = fu'$, then $\emptyset \neq g^{-1}\partial_u(E) = g^{-1}\partial_{u'}(E_i) = \emptyset$ since E is linear, a contradiction. Hence either $f \neq g$ is the first letter of u or $f = g$ and u is empty. Since the analogous holds for w, we obtain $u = \varepsilon$ iff $w = \varepsilon$. Now the claim follows immediately from the induction hypothesis.

For $E = E_1 + E_2$ the claim is immediate by the last proposition and the induction hypothesis.

Let $E = E_1 \cdot_c E_2$. If $|E_1|_g > 0$, then $g^{-1}\partial_u(E) = g^{-1}\partial_u(E_1) \cdot_c E_2$ as well as $g^{-1}\partial_w(E) = g^{-1}\partial_w(E_1) \cdot_c E_2$. Since these two sets are non-empty, so are the sets $g^{-1}\partial_u(E_1)$ and $g^{-1}\partial_w(E_1)$. Hence, by the induction hypothesis, the claim follows. Suppose now $|E_1|_g = 0$. Then $g^{-1}\partial_u(E)$ is a finite union of sets of the form $g^{-1}\partial_{u'}(E_2)$ where every u' is a suffix of u. The induction hypothesis implies that any two non-empty of them are equal, i.e., $g^{-1}\partial_u(E) = g^{-1}\partial_{u'}(E_2)$ for some u'. Similarly, $g^{-1}\partial_w(E) = g^{-1}\partial_{w'}(E_2)$ for some word w'. Now the claim follows from the induction hypothesis.

A similar argument can be applied in case $E = F^{*c}$ with $g^{-1}\partial_{u'}(F) \cdot F^{*c}$ in place of $g^{-1}\partial_{u'}(E_1)$.   □

By Propositions 4.2 and 4.3 we conclude

Corollary 4.4. *For a linear regular expression E and $w \in \Sigma_{\geq 1}^+$ we have $\partial_w(E^{*c}) = \partial_u(E) \cdot_c E^{*c}$ for some non-empty suffix u of w.*

Next, we bound the number of partial derivatives of a linear regular expression.

Proposition 4.5. *Let E be a linear regular expression. Then $|\partial_{\Sigma_{\geq 1}^+}(E)| \leq \|E\| - 1$ and $|\partial_{\Sigma_{\geq 1}^*}(E)| \leq \|E\|$.*

Proof. Note that $\partial_{\Sigma_{\geq 1}^*}(E) = \partial_{\Sigma_{\geq 1}^+}(E) \cup \{E\}$. By induction on E we get: For $E = f(E_1, \ldots, E_n)$, $g \in \Sigma_{\geq 1}$, and $u \in \Sigma_{\geq 1}^*$

$$\partial_{gu}(E) = \begin{cases} \partial_u(\{E_1, \ldots, E_n\}) & \text{if } g = f, \\ \emptyset & \text{if } g \neq f. \end{cases}$$

Hence, $|\partial_{\Sigma_{\geq 1}^+}(E)| \leq \sum_{i=1}^n |\partial_{\Sigma_{\geq 1}^*}(E_i)| \leq \sum_{i=1}^n \|E_i\| = \|E\| - 1$. For $E = E_1 + E_2$ we use Proposition 4.2 and the induction hypothesis and obtain the assumption. If $E = E_1 \cdot_c E_2$, then again by Proposition 4.2: $|\partial_{\Sigma_{\geq 1}^+}(E)| \leq |\partial_{\Sigma_{\geq 1}^+}(E_1)| + |\partial_{\Sigma_{\geq 1}^+}(E_2)| \leq \|E_1\| - 1 + \|E_2\| - 1 < \|E\| - 1$. Finally, we conclude by Corollary 4.4 $|\partial_{\Sigma_{\geq 1}^+}(E^{*c})| \leq |\partial_{\Sigma_{\geq 1}^+}(E)| \leq \|E\| - 1 = \|E^{*c}\| - 1$.   □

4.2   The Projection Construction

Recall that Theorem 3.5 provides a possibly infinite tree automaton $\mathcal{A}_E$ that accepts $[\![E]\!]$. Assuming E to be linear, we are now in the position to improve this result:

Corollary 4.6. *Let E be a linear regular expression over the ranked alphabet Σ. Then $\mathcal{A}_E$ is a finite tree automaton with at most $\|E\|$ many states and $\|E\| \cdot |\Sigma|$ many transitions that accepts $[\![E]\!]$.*

Proof. The equality $L(\mathcal{A}_E) = [\![E]\!]$ was shown in Theorem 3.5. Since the set of states of $\mathcal{A}_E$ equals $\partial_{\Sigma_{\geq 1}^*}(E)$, the finite tree automaton has at most $\|E\|$ many states by Proposition 4.5. For $f \in \Sigma_{\geq 1}$ and $D \in Q_E$, there is at most one transition of the form $(D, f, (G_1, \ldots, G_m))$ by Proposition 4.3(1), i.e., there are at most $\|E\| \cdot |\Sigma_{\geq 1}|$ many transitions whose label belongs to $\Sigma_{\geq 1}$. In addition, there can be $|Q_E \times \Sigma_0| \leq \|E\| \cdot |\Sigma_0|$ many transitions of the form (D, c) with $c \in \Sigma_0$.   □

Remark 4.7. We even proved for a linear regular expression E that $\mathcal{A}_E$ is a deterministic top-down automaton which implies the number of transitions given in the corollary.

Note that given two alphabets Γ and Σ with $\Gamma_0 \subseteq \Sigma_0$ and a mapping $\eta : \Gamma \to \Sigma$ with $\eta(\Gamma_m) \subseteq \Sigma_m$ for every $m \in \mathbb{N}$, we can extend η naturally to $\eta : \mathrm{EXP}(\Gamma) \to \mathrm{EXP}(\Sigma)$ by:

- $\eta(f(E_1, \ldots, E_m)) = \eta(f)(\eta(E_1), \ldots, \eta(E_m))$,
- $\eta(E + F) = \eta(E) + \eta(F)$, $\eta(E \cdot_c F) = \eta(E) \cdot_c \eta(F)$, and $\eta(E^{*c}) = (\eta(E))^{*c}$.

Definition 4.8. *Let E be a regular expression over the ranked alphabet Σ. A linear regular expression $\overline{E}$ is a* linearization *of E w.r.t. η over the ranked alphabet $\overline{\Sigma}$ if $\eta : \overline{\Sigma} \to \Sigma$ is a mapping with $\eta(\overline{\Sigma}_m) \subseteq \Sigma_m$ such that $\eta(c) = c$ for every $c \in \overline{\Sigma}_0$ and $\eta(\overline{E}) = E$.*

Note that both the constants from Σ_0 and the operations $\cdot_c$ and ^{*c} remain unchanged. By abuse of notation, we denote also the two natural continuations of η to $\overline{\Sigma}^*$ and to $T_{\overline{\Sigma}}$ by η. The following lemma is easily shown:

Lemma 4.9. *Let E be a regular expression and $\overline{E}$ a linearization of E w.r.t. η. Then $\eta(\llbracket \overline{E} \rrbracket) = \llbracket E \rrbracket$.*

Let E be an arbitrary regular expression. Then one can construct a small finite tree automaton $\overline{\mathcal{A}}_E$ accepting $\llbracket E \rrbracket$ as follows: firstly, construct some linearization $\overline{E}$ of E w.r.t. η (we can assume that every symbol from $\overline{\Sigma}$ appears in $\overline{E}$ and therefore $|\overline{\Sigma}| \leq \|\overline{E}\| = \|E\|$). Secondly, build the finite tree automaton $\mathcal{A}_{\overline{E}}$ which then has at most $\|\overline{E}\| = \|E\|$ many states and $\|\overline{E}\| \cdot |\overline{\Sigma}| \leq \|E\|^2$ many transitions. Thirdly, replace the transitions $(\overline{F}, \overline{f}, (\overline{G}_1, \ldots, \overline{G}_m))$ of this automaton by $(\overline{F}, \eta(\overline{f}), (\overline{G}_1, \ldots, \overline{G}_m))$. Then, by Lemma 4.9, the following is immmediate:

Corollary 4.10. *Let E be a regular expression. Then $\overline{\mathcal{A}}_E$ is a finite tree automaton with at most $\|E\|$ many states and $\|E\|^2$ many transitions that accepts $\llbracket E \rrbracket$.*

4.3 The Quotient Construction

We will now collapse some of the states of the automaton $\overline{\mathcal{A}}_E$. The resulting automaton will turn out to be isomorphic to the automaton $\mathcal{A}_E$ from our first construction.

We define the following equivalence relation $\sim$ on $Q_{\overline{E}}$:

$$\overline{F} \sim \overline{H} : \Longleftrightarrow \eta(\overline{F}) = \eta(\overline{H}).$$

Let $\overline{G}_i, \overline{H}_i \in Q_{\overline{E}}$ with $\overline{G}_i \sim \overline{H}_i$ for $i = 1, \ldots, m$ and $\overline{f}_1, \overline{f}_2 \in \overline{\Sigma}_{\geq 1}$ with $\eta(\overline{f}_1) = \eta(\overline{f}_2) = f$. Then $\overline{f}_1(\overline{G}_1, \ldots, \overline{G}_m) \sim \overline{f}_2(\overline{H}_1, \ldots, \overline{H}_m)$. Hence, $\sim$ is a congruence relation. We denote the congruence class of $\overline{G} \in Q_{\overline{E}}$ by $[\overline{G}]$. Since $\sim$ is a congruence, the following quotient FTA is well-defined: $\widetilde{\mathcal{A}}_E = \left(Q_{\overline{E}}/_\sim, \Sigma, \{[\overline{E}]\}, \Delta'_E \right)$ where

$$\Delta'_E = \{ ([\overline{F}], f, ([\overline{G}_1], \ldots, [\overline{G}_m])) \mid (\overline{F}, f, (\overline{G}_1, \ldots, \overline{G}_m)) \in \overline{\Delta}_E \}.$$

We will show that the FTA $\widetilde{\mathcal{A}}_E$ is isomorphic to $\mathcal{A}_E$ and, thus, in particular accepts the language $[\![E]\!]$. Therefor, we have to clarify that $\eta(\overline{F}) \in Q_E = \partial_{\Sigma_{\geq 1}^*}(E)$ for every $\overline{F} \in Q_{\overline{E}}$. The following fundamental relation between the partial derivatives of E and of $\overline{E}$ is shown for partial derivatives w.r.t. a single letter g by an induction on the construction of E:

Proposition 4.11. *Let E be a regular expression over the ranked alphabet Σ and $\overline{E}$ a linearization of E w.r.t. η. Then we have for every $g \in \Sigma_{\geq 1}$*

$$g^{-1}(E) = \bigcup_{\overline{g} \in \eta^{-1}(g)} \eta(\overline{g}^{-1}(\overline{E})) \text{ and } \qquad \partial_g(E) = \bigcup_{\overline{g} \in \eta^{-1}(g)} \eta(\partial_{\overline{g}}(\overline{E})).$$

Now we lift this result to arbitrary partial derivatives w.r.t. arbitrary words.

Theorem 4.12. *Let E be a regular expression over the ranked alphabet Σ and $\overline{E}$ a linearization of E w.r.t. η. Then we have for every $w \in \Sigma_{\geq 1}^*$*

$$\partial_w(E) = \bigcup_{\overline{w} \in \eta^{-1}(w)} \eta(\partial_{\overline{w}}(\overline{E})).$$

Proof. We proceed by induction on the length of w where the case $w = \varepsilon$ is trivial. By Proposition 4.11, the assumption holds for $|w| = 1$. Now consider $w = ug$ with $u \in \Sigma_{\geq 1}^*$ and $g \in \Sigma_{\geq 1}$. Using the induction hypothesis and Proposition 4.11 we get:

$$\partial_{ug}(E) = \partial_g(\partial_u(E)) = \partial_g\left(\bigcup_{\overline{u} \in \eta^{-1}(u)} \eta(\partial_{\overline{u}}(\overline{E})) \right) = \partial_g \eta \left(\bigcup_{\overline{u} \in \eta^{-1}(u)} \partial_{\overline{u}}(\overline{E}) \right) \quad (\star)$$

Consider the set $\overline{H} = \bigcup\{\partial_{\overline{u}}(\overline{E}) \mid \overline{u} \in \eta^{-1}(u)\}$. Since it consists of finitely many regular expressions, there exists a set of linear regular expressions H' over some ranked alphabet Σ' and a function $\theta : \Sigma' \to \overline{\Sigma}$ such that $\theta(H') = \overline{H}$. In other words, H' consists of linearizations of the regular expressions in $\overline{H}$ w.r.t. θ. Then the regular expressions from H' are also linearizations of the regular expressions in $H = \eta(\overline{H})$ w.r.t. $\alpha = \eta\theta$. Hence we get from Proposition 4.11 and the above

$$\partial_{ug}(E) = \partial_g(H) = \bigcup_{g' \in \alpha^{-1}(g)} \alpha(\partial_{g'}(H'))$$

$$= \bigcup_{\overline{g} \in \eta^{-1}(g)} \eta\left(\bigcup_{g' \in \theta^{-1}(\overline{g})} \theta\partial_{g'}(H') \right) = \bigcup_{\overline{g} \in \eta^{-1}(g)} \eta\left(\partial_{\overline{g}} \left(\bigcup_{\overline{u} \in \eta^{-1}(u)} \partial_{\overline{u}}(\overline{E}) \right) \right)$$

$$= \bigcup_{\overline{g} \in \eta^{-1}(g)} \bigcup_{\overline{u} \in \eta^{-1}(u)} \eta(\partial_{\overline{ug}}(\overline{E})) = \bigcup_{\overline{w} \in \eta^{-1}(ug)} \eta(\partial_{\overline{w}}(\overline{E})).$$

$\square$

Now we can identify the result of the quotient construction.

Theorem 4.13. *The finite tree automaton $\widetilde{\mathcal{A}}_E$ is isomorphic to $\mathcal{A}_E$.*

Proof. The state isomorphism is given by $\varphi : {}^{Q_{\overline{E}}}/_\sim \to Q_E : [\overline{G}] \mapsto \eta(\overline{G})$. Firstly, φ really maps into Q_E. Indeed, $\overline{G} = \partial_{\overline{w}}(\overline{E})$ for some $\overline{w}$. By Theorem 4.12, $\eta(\overline{G}) \in \partial_{\Sigma_{\geq 1}^*}(E) = Q_E$. Injectivity of φ is obvious by the definition of $\sim$. Surjectivity follows from Theorem 4.12. Certainly, $\varphi([\overline{E}]) = E$. Now, suppose $([\overline{F}], f, ([\overline{G}_1], \dots, [\overline{G}_m])) \in \Delta'_{\overline{E}}$. Then there is a $\overline{f}$ such that $(\overline{F}, \overline{f}, (\overline{G}_1, \dots, \overline{G}_m)) \in \Delta_{\overline{E}}$ which means $(\overline{G}_1, \dots, \overline{G}_m) \in \overline{f}^{-1}(\overline{F})$. But due to Proposition 4.11, $(G_1, \dots, G_m) \in f^{-1}(F)$ where $G_i = \eta(\overline{G}_i)$ and $F = \eta(\overline{F})$. Vice versa, if $(G_1, \dots, G_m) \in f^{-1}(F)$, then there is an $\overline{f} \in \Sigma_{\geq 1}(\overline{E})$ with $(\overline{G}_1, \dots \overline{G}_m) \in \overline{f}^{-1}(\overline{F})$. Moreover, we have for $c \in \Sigma_0$:

$$([\overline{F}], c) \in \Delta'_{\overline{E}} \iff c \in [\![\overline{F}]\!] \iff c \in [\![\eta(\overline{F})]\!] \iff (\eta(\overline{F}), c) \in \Delta_E. \qquad \square$$

Now we will show that the FTA $\mathcal{A}_E$ from Theorem 3.5 is finite. The number of transitions of $\mathcal{A}_E$ is obviously bounded from above by $|Q_E| \cdot |\Sigma| \cdot |Q_E|^R \leq \|E\|^{R+1} \cdot |\Sigma|$ where R is the maximal rank appearing in Σ. However, as we will show next, there is a much smaller bound.

Theorem 4.14. *Let E be a regular expression. Then $\mathcal{A}_E$ is a finite tree automaton with at most $\|E\|$ many states and $\|E\|^2$ many transitions that accepts $[\![E]\!]$.*

Proof. The equality $L(\mathcal{A}_E) = [\![E]\!]$ was shown in Theorem 3.5. The numbers of states and transitions of $\mathcal{A}_E$ equal those of $\widetilde{\mathcal{A}}_E$ by Theorem 4.13. Since $\widetilde{\mathcal{A}}_E$ is a quotient of $\overline{\mathcal{A}}_E$, the result follows from the estimates in Corollary 4.10. $\qquad \square$

Compare this to the inductive construction of a finite tree automaton accepting $[\![E]\!]$: For union and c-product, one takes the disjoint union of the argument automata and adds some transitions. For c-iteration, one has to add one new state in order to accept the tree c. Hence, the inductive construction yields a finite tree automaton whose number of states equals $\|E\|$ plus the number of c-iterations applied in the construction. The number of transitions of that automaton is very difficult to analyse.

5   Algorithmic Issues

Due to Theorem 4.13, we can construct the FTA $\widetilde{\mathcal{A}}_E$ to get the automaton $\mathcal{A}_E$. Following this line, Champarnaud and Ziadi [5] gave in the case of words an algorithm with an $\mathcal{O}(\|E\| \cdot |E|^2)$ space and time complexity. By algorithmic refinements they enhanced the algorithm to one with an $\mathcal{O}(|E|^2)$ space and time complexity. We can mainly adapt this algorithm for the construction of the FTA $\mathcal{A}_E$ from a regular tree expression E. Since the algorithm is based on a more detailed analysis of the structure of partial derivatives, we first prove some more facts about them.

5.1   Form of Partial Derivatives

Positions p of a tree are defined as usually as finite words over $\mathbb{N}$ where ε is the position of the root. Positions of an expression E are understood as those in the syntax-tree t_E. Every position p denotes a sub-expression of E.

Theorem 5.1. *Let $E \in \mathrm{EXP}(\Sigma)$ and $D \in \partial_w(E)$. Then there exist positions $p_1, \ldots, p_n$ in the syntax tree t_E of E and constant symbols $c_1, \ldots, c_n \in \Sigma_0$ such that*

- $D = H_1 \cdot_{c_1} H_2 \ldots H_{n-1} \cdot_{c_{n-1}} H_n$ *where H_i is the sub-expression of E at p_i,*
- *the number n is bounded by the number of products $\cdot_c$ and stars $*^c$ appearing in E,*
- *if p_i is a prefix of p_j, then $i \geq j$.*

Proof (Sketch). It follows immediately from Proposition 4.2, Corollary 4.4, and Theorem 4.12 that every partial derivative D is a product of sub-expressions of E and that n is bounded by the number of products and stars in E. The last claim is proven by induction on the length of w. It is trivial for $w = \varepsilon$. For $w = f \in \Sigma_{\geq 1}$ we proceed by induction on the construction of E. For $|w| > 1$ the claim follows easily from the induction hypothesis when considering partial derivatives of a product. $\qquad\square$

5.2 An Algorithm for Computing the Automaton

We only give a sketch of the algorithm. Details of the algorithm for word expressions are given in [5]. The adaption to tree expressions does not cause much trouble.

Firstly, compute the syntax-tree t_E of E. Note that the syntax-tree of a linearization $\overline{E}$ is obtained from t_E by labelling each $g \in \Sigma_{\geq 1}$ additionally with its position. In the sequel, we will mainly work within the syntax-tree.

Computing the States. Recall that the partial derivatives of E are projections of the partial derivatives of the linearization $\overline{E}$, cf. Theorem 4.12. Moreover, due to Proposition 4.3 the nonempty partial derivatives $\partial_{\overline{wg}}(\overline{E})$ depend just on the last symbol $\overline{g}$, i.e., the unique position in the syntax-tree labelled by $\overline{g}$. Now we calculate the partial derivatives of $\overline{E}$. Afterwards we identify partial derivatives that describe the same partial derivative of E.

1. *Computing the Linearized Partial Derivatives.* For every position in the syntax-tree t_E labelled by some $g \in \Sigma_{\geq 1}$ calculate the partial derivatives $\partial_{\ldots \overline{g}}(\overline{E})$ by following the path from the respective position of $\overline{g}$ to the root of t_E. Hereby, we have to collect at every node on this path labeled by a product or a star the factors of the partial derivative, cf. Theorem 5.1. Moreover, if we pass a $\cdot_c$-node from the right, then we have to check whether c is in the semantics of the sub-expression to the left of this node. Note that for $g \in \Sigma_m$ we have up to m partial derivatives. We do not just save the set $\partial_{\ldots \overline{g}}(\overline{E})$ but also the respective tuple, i.e., the ordering of the partial derivatives which stems from the ordering of the sons of the $\overline{g}$-node.
2. *Identification of Partial Derivatives.* This step is done by a lexicographic ordering and an identification of consecutive partial derivatives.

Computing the Transitions (D, c) for $c \in \Sigma_0$. We compute for every sub-expression F of E the set $[\![F]\!] \cap \Sigma_0$ which can be easily done in the syntax-tree. Afterwards we calculate for every partial derivative D the set $[\![D]\!] \cap \Sigma_0$ which can be done using the product structure of D, cf. Theorem 5.1.

Computing the Transitions for $f \in \Sigma_{\geq 1}$. We can calculate for every linearized sub-expression $\overline{F}$ the so-called FIRST-sets, i.e., those $\overline{f} \in \Sigma_{\geq 1}$ such that $\overline{f}$ has an unguarded occurrence in $\overline{F}$. Now we can compute from those sets the respective FIRST-sets for every linearized partial derivative $\overline{D}$. But for every linearized symbol $\overline{f}$ in the FIRST-set we obtain the unique $\overline{f}$-transition from $\overline{D}$ by the unique tuple of the linearized partial derivatives from $\partial_{...\overline{f}}(\overline{E})$. Note that this tuple was obtained already in the computation of states. A projection gives the transitions for f.

Complexity. The algorithm for word expressions has an $\mathcal{O}(\|E\| \cdot |E|^2)$ space and time complexity, cf. [5]. The algorithm for regular tree expressions as sketched above follows exactly the same lines but has to keep track of tuples of partial derivatives instead of singletons as it is the case for words. Hence, the algorithm has an $\mathcal{O}(R \cdot \|E\| \cdot |E|^2)$ space and time complexity where $R \geq 1$ is the maximal rank appearing in the ranked alphabet Σ (and at least 1).

The improvements suggested by Champarnaud and Ziadi, mainly a preprocessing of star sub-expressions of $\overline{E}$ and an improved computation of the FIRST-sets in the syntax-tree, carry over to the above algorithm. Hence, we shall get for such an improved algorithm an $\mathcal{O}(R \cdot |E|^2)$ space and time complexity.

References

1. Antimirov, V.: Partial derivatives of regular expressions and finite automaton constructions. Theoretical Computer Science 155, 291–319 (1996)
2. Berry, G., Sethi, R.: From regular expressions to deterministic automata. Theoretical Computer Science 48, 117–126 (1986)
3. Brzozowski, J.A.: Derivatives of regular expressions. J. Assoc. Comput. Mach. 11, 481–494 (1964)
4. Champarnaud, J.-M., Nicart, F., Ziadi, D.: Computing the follow automaton of an expression. In: Domaratzki, M., Okhotin, A., Salomaa, K., Yu, S. (eds.) CIAA 2004. LNCS, vol. 3317, pp. 90–101. Springer, Heidelberg (2005)
5. Champarnaud, J.-M., Ziadi, D.: From c-continuations to new quadratic algorithms for automaton synthesis. Intern. J. of Algebra and Computation 11(6), 707–735 (2001)
6. Champarnaud, J.-M., Ziadi, D.: Canonical derivatives, partial derivatives and finite automaton constructions. Theoretical Computer Science 289, 137–163 (2002)
7. Comon, H., Dauchet, M., Gilleron, R., Löding, C., Jacquemard, F., Lugiez, D., Tison, S., Tommasi, M.: Tree automata techniques and applications (October 2007) (release October 12, 2007), http://www.grappa.univ-lille3.fr/tata
8. Gécseg, F., Steinby, M.: Tree languages. In: Handbook of Formal Languages, ch. 1, vol. 3, pp. 1–68. Springer, Heidelberg (1997)
9. Glushkov, V.M.: The abstract theory of automata. Russian Mathematical Surveys 16, 1–53 (1961)
10. Hosoya, H., Pierce, B.: Regular expression pattern matching for XML. SIGPLAN Not. 36(3), 67–80 (2001)
11. Hromkovic, J., Seibert, S., Wilke, T.: Translating regular expressions into small ε-free non-deterministic finite automata. J. Comput. System Sci. 62, 565–588 (2001)
12. Ilie, L., Yu, S.: Constructing NFAs by optimal use of positions in regular expressions. In: Apostolico, A., Takeda, M. (eds.) CPM 2002. LNCS, vol. 2373, pp. 279–288. Springer, Heidelberg (2002)

13. Kleene, S.E.: Representations of events in nerve nets and finite automata. In: Shannon, C.E., McCarthy, J. (eds.) Automata Studies, pp. 3–42. Princeton University Press, Princeton (1956)
14. McNaughton, R.F., Yamada, H.: Regular expressions and state graphs for automata. IEEE Transactions on Electronic Computers 9, 39–57 (1960)
15. Sakarovitch, J.: Éléments de théorie des automates. Vuibert (2003)
16. Sakarovitch, J.: The language, the expression, and the (small) automaton. In: Farré, J., Litovsky, I., Schmitz, S. (eds.) CIAA 2005. LNCS, vol. 3845, pp. 15–30. Springer, Heidelberg (2006)
17. Thatcher, J.W., Wright, J.B.: Generalized finite automata theory with application to a decision problem of second-order logic. Math. Systems Theory 2(1), 57–81 (1968)

Balance Properties and Distribution of Squares
in Circular Words

Roberto Mantaci[1], Sabrina Mantaci[2], and Antonio Restivo[2]

[1] LIAFA - Université Denis Diderot - Paris 7
Case 7014
75205 Paris Cedex 13, France
mantaci@liafa.jussieu.fr
[2] Dipartimento di Matematica ed Applicazioni
Università di Palermo Via Archirafi, 34
90123 Palermo, Italy
sabrina, restivo@math.unipa.it

Abstract. We study balance properties of circular words over alphabets of size greater than two. We give some new characterizations of balanced words connected to the Kawasaki-Ising model and to the notion of derivative of a word. Moreover we consider two different generalizations of the notion of balance, and we find some relations between them. Some of our results can be generalised to non periodic infinite words as well.

Introduction

In this paper we deal mostly with the characterization of balanced finite circular words over an alphabet of any size. Informally, a word w is said balanced if for any letter a in the alphabet, and for any two factors u and v of w having the same length, the number of occurrences of a's in u and v are "almost" the same, i.e. they can differ at most by 1.

Throughout the last fifty years, this subject has been widely developed in literature for two-letter alphabets. In fact, for binary alphabets, the infinite non periodic balanced words coincides with the Morse and Hedlund's *Sturmian words* (cf. [9]), defined as the infinite sequences having exactly $n + 1$ distinct factors of length n. Because of their numerous properties, Sturmian words have applications in several fields of research. Such a versatility explains also the existence of many equivalent definitions. Unfortunately, this is not the case when we generalize to alphabets of size greater than two. In fact, each generalization to larger alphabets of the different definitions of Sturmian words generates a different set of words. This is also why a theory of balanced words over general alphabets is a very difficult topic and has not been totally investigated yet.

In this paper we are interested in periodic balanced sequences v^ω. For two-letter alphabets, it is proved that v^ω is balanced if and only if v is a conjugate of one of the *Standard words*, the finite "bricks" of Sturmian words (cf. [2]). We note that the notion of balanced periodic words over an alphabet of

M. Ito and M. Toyama (Eds.): DLT 2008, LNCS 5257, pp. 504–515, 2008.

size larger than two also appears in the statement of the Fraenkel conjecture (cf. [7]). As a direct consequence of a result of Graham, one can prove that balanced sequences on a set of letters having different frequencies must be periodic (cf. [14]). The problem of characterizing balanced words over any alphabet has been developed by Altman, Gaujal and Hordijk [1] in the field of optimal routing in queuing networks. Hubert in [11] proves the existence of balanced periodic words for any alphabet and gives an algorithm for their construction. Moreover, he also proves that for alphabets of size greater than three there exist balanced words that are not obtained by this construction. Vuillon in [14] provides a thorough survey on the topic. The reader can also look at the references therein.

Notice that a way to deal with infinite periodic words, is to represent them as a circular (finite) word. This approach, that we have decided to use in this paper, is also connected with the interest recently devoted to circular words, since they have important applications in some data compression algorithms, such as the ones using the Burrows-Wheeler Transform [12], in combinatorics [3], and in automata theory [6]. We remark that the recent interest to circular words also comes from molecular biology, since mitochondrial DNA, as well as the genome of bacteria, are actually circular sequences of nucleotides [8].

In this paper we consider some new approaches for dealing with circular balanced words. In Section 2, we present one of these approaches, connected with the characterization of balanced sequences given by Cameron and Wu in [4] using the Kawasaki-Ising model. We show that this characterization can be expressed more simply only in terms of combinatorial properties of the word itself, namely by the combinatorial object that we call *square vector*. Furthermore, such characterization is more suitable for generalizations to larger alphabets. In Section 3 we introduce the notion of *derivative of a word*, that allows to give a new characterization of balanced words and that suggest also a new method for their construction knowing the number of occurrences of each letter in the word (when a balanced word with such a letter distribution exists). In Section 4 we apply the notions and the tools that we have introduced in the previous section to the study of two more general notions of balance. The first is the notion of m-balance, formally introduced by Cassaigne, Ferenczi and Zamboni [5] or by Heinis [10]. A word w is called m-balanced if for any letter a and for any two factors of w of the same length, the difference of the numbers of a's in the two factors is at most m. A complementary concept, introduced by Sano, Miyoshi and Kataoka [13] is the notion of m-uniformly distributed words. Informally, a word w is m-uniformly distributed if for any letter a, and for any two factors of the form aua in w having the same number of a's, the lengths of the two factors may differ at most by m. The notion of derivative allows us to find some connections between these two notion of balance. We conclude the paper by noting that several of the results contained herein are easily extended to non periodic infinite words and we highlight some open problems.

1   Definitions and Notations

Let Σ be an alphabet and let Σ^* denote the set of words over Σ. Let $w \in \Sigma^*$ be a word and let w_i denote the i-th character of w. As usual, $|w|$ denotes the length of w, and $|w|_a$ the number of occurrences of the letter a in w; we will sometimes refer to this integer as the a-*length* of the word w.

In what follows, w will be always considered as a circular word, that is, the end of the word is connected with its beginning. Actually, the circular word w is a way to represent the class of conjugates of w, or also the infinite periodic word w^ω. In this paper our focus is on the balance property of circular words. We recall that a word w is said to be *balanced with respect to a letter* a if and only if for any two factors u and u' of w such that $|u| = |u'|$, one has $||u|_a - |u'|_a| \leq 1$. A word is said to be *balanced* if it is balanced with respect to all the letters of the alphabet. For a two-letter alphabet, being balanced is equivalent to being balanced with respect to one letter. It is proved that balanced circular words over two letters alphabets are exactly the conjugacy classes of Standard words (cf. [2], [3]). Notice that balanced circular words correspond to balanced infinite words that are called *periodic* in literature (see [1]).

2   Squares and Kawasaki-Ising Model

Let us recall the definition of the Kawasaki-Ising model, used in [4] in order to characterize circular balanced words over two-letter alphabets. In the general Ising model on a graph G, each vertex i of G is assigned a spin, denoted by $\sigma(i)$ or σ_i, which is either $+1$ (called *up*) or -1 (*down*). An assignment of spins to all the vertices of G is called a *configuration* and is denoted by σ. The number of vertices in the up spin in a configuration σ, denoted by $|\sigma|_+$, is called the *weight* of σ. The Kawasaki-Ising model $KI(k,n)$ on a cycle graph C_n with n vertices consists of the configurations σ such that $|\sigma|_+ = k$ for a given number k.

Here we assume $V(C_n) = \{0, \ldots, n-1\}$ and the vertices are consecutively labelled, then a configuration can be represented as $(\sigma_0, \ldots, \sigma_{n-1})$ (see Fig. 1).

Among all configurations in $KI(k,n)$, one special important class is the class of *regular configurations*. Roughly speaking, they are the configurations that are close to the random ones. In the words model, they correspond to Standard words, the finite version of Sturmian words [9]. One important feature of regular configurations is that they provide some good extremal properties, which have important applications in many fields.

For a configuration $\sigma = (\sigma_0, \ldots, \sigma_{n-1})$ and for an integer $p \in \{1, \ldots, n-1\}$, Cameron and Wu in [4] define the Hamiltonian of order p of σ as:

$$H_p(\sigma) = \frac{1}{(2n)^p} \sum_{i=0}^{n-1} \sigma_i \sigma_{i+p}$$

and then define the Hamiltonian of σ as:

$$H(\sigma) = \sum_{p=1}^{n-1} H_p(\sigma)$$

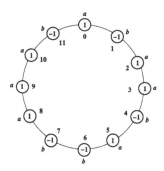

Fig. 1. The configuration $\sigma = (1, -1, 1, 1, -1, 1, -1, -1, 1, 1, 1, -1) \in KI(12, 7)$ and its corresponding word $T(\sigma) = abaababbaaab$

Cameron and Wu also defined a bijection T between the set $\mathbb{W}_{k,n}$ of all circular words on the alphabet $\{a, b\}$ of length n having exactly k occurrences of a and the set of configurations $KI(k, n)$. Such a bijection associates a word w with a configuration σ such that $\sigma_i = 1$ if and only if $w_i = a$ (see again Figure 1). The main result of Cameron and Wu in [4] is that a word w in $\mathbb{W}_{k,n}$ is balanced if and only if $T(w)$ is a configuration of minimum Hamiltonian in $KI(k, n)$. They also note that for two configurations σ and τ, $H(\sigma) < H(\tau)$ if and only if there exists an integer $l \in \{1, \ldots, n - 1\}$ such that

$$H_i(\sigma) = H_i(\tau) \text{ for } 1 \leq i \leq l - 1 \text{ and } H_l(\sigma) < H_l(\tau).$$

In other words, a configuration σ has minimal Hamiltonian if and only if its Hamiltonian vector $\overline{H}(\sigma) = \{H_1(\sigma), H_2(\sigma), \ldots, H_{n-1}(\sigma)\}$ is minimal according to the lexicographic order.

On the other hand, for a circular word $w = w_0 w_1 \ldots w_{n-1}$ of length n over any alphabet and an integer $p \in \{1, \ldots, n - 1\}$, one can define :

$$S_p(w) = |\{j \mid 0 \leq j \leq n - 1 \text{ and } w_j = w_{j+p}\}|.$$

The integer $S_p(w)$ counts the number of "squares" (in the sense of pairs of identical letters) at distance p. The *square vector* S of the word w is then defined as:

$$S(w) = (S_1(w), \ldots, S_{n-1}(w)).$$

For a circular word $w \in \mathbb{W}_{k,n}$, there is a simple relation between the vector $S(w)$ and the Hamiltonian vector $\overline{H}(T(w))$. Indeed, if we note $\sigma = T(w)$, then

$$\sum_{i=0}^{n-1} \sigma_i \sigma_{i+p} = |\{j \mid 0 \leq j \leq n - 1 \text{ and } w_j = w_{j+p}\}| +$$

$$-|\{j \mid 0 \leq j \leq n - 1 \text{ and } w_j \neq w_{j+p}\}|$$
$$= S_p(w) - (n - S_p(w))$$
$$= 2S_p(w) - n$$

Therefore, $S_p(w) = \frac{1}{2}[(2n)^p H_p(\sigma) + n]$ and hence the vector $S(w)$ is lexicographically minimum among all the square vectors of words in $\mathbb{W}_{k,n}$ if and only if the vector $\overline{H}(T(w))$ is lexicographically minimum among all Hamiltonian vectors of configurations in $KI(k,n)$. This remark allows to characterize balanced words only in terms of their square vector.

Proposition 1. *A word w in $\mathbb{W}_{k,n}$ is balanced if and only if its square vector $S(w)$ is lexicographically minimum among the square vectors of all words in $\mathbb{W}_{k,n}$.*

We will show that this characterization of balanced words can be extended to the case of alphabets of any size. Let $\Sigma = \{a_1, a_2, \ldots, a_k\}$ be an alphabet. We recall that by definition a word is balanced if and only if it is balanced with respect to each letter of the alphabet.

For a circular word $w \in \Sigma^*$ and for a letter $a_i \in \Sigma$, we define $\varphi_{a_i}(w)$ the image of w under the morphism φ_{a_i} such that :

$$\varphi_{a_i}(x) = \begin{cases} a_i & \text{if } x = a_i \\ b & \text{otherwise} \end{cases} .$$

The following lemma is straightforward.

Lemma 1. *A word $w \in \Sigma^*$ is balanced if and only if $\varphi_{a_i}(w)$ is balanced (in the binary alphabet $\{a_i, b\}$) for all $a_i \in \Sigma$.*

We recall that the Parikh vector for a word $w \in \Sigma^*$ is a vector P of length $|\Sigma|$ such that for all i, the i-th component P_i is the number of occurrences of the letter a_i in the word w. We note $\mathbb{W}(P)$ the set of all words whose Parikh vector is P. We also recall that when $k > 2$ a balanced word having a given Parikh vector P may not exist. For instance, there is no balanced word having $(3, 2, 1)$ as Parikh vector. In the following theorem we give a characterization of balanced words having a fixed Parikh vector (if it exists), in terms of their square vector.

Theorem 1. *Let P be a Parikh vector such that a balanced word in $\mathbb{W}(P)$ exists. Then $w \in \mathbb{W}(P)$ is balanced if and only if $S(w)$ is the lexicographically minimum among all square vectors of words in $\mathbb{W}(P)$.*

We note that if two words w and w' are obtained one from the other by a combination of the operations of *letter renaming* and *word reversal*, they satisfy the same balance properties and have the same square vectors.

We would also like to remark that the square vector $S(w)$ does not characterize the word w not even up to word reversal and letter renaming, as the following example shows it:

Example 1. Consider the circular words $w = aaabbacb$ and $w' = aabaacbb$. One can verify that both w and w' have the same square vector, i.e. $S(w) = S(w') = (3, 1, 3, 2, 3, 1, 3)$ and they are not equivalent up to word reversal and letter renaming. Also note that the square vector is symmetric, therefore half of it is somewhat redundant.

One could also define the square vector $S(w)_a$ of a word w with respect to a letter a as the integer vector whose p-th component is

$$|\{j \mid 0 \leq j \leq n-1 \text{ and } w_j = w_{j+p} = a\}|.$$

The collection of all square vectors with respect to each of the letters does not characterize the word w in $\mathbb{W}(P)$ either. The two words in the preceding example have the same square vector, both with respect to a and with respect to b, i.e. $S(w)_a = S(w')_a = (2, 1, 2, 0, 2, 1, 2)$ and $S(w)_b = S(w')_b = (1, 0, 1, 1, 1, 0, 1)$.

A natural question is to ask whether this may happen when one takes the minimal square vector in the class $\mathbb{W}(P)$ and, in particular, whether two non-equivalent balanced words may exist in $\mathbb{W}(P)$. It is known that when the alphabet has size 2 or 3, the balanced word having a given Parikh vector, if it exists, is unique (cf. [1]). However, this is false in general and in particular for $|\Sigma| = 5$. The words $abcaebacbad$ and $abcaebacabd$ have the same Parikh vector $(4, 3, 2, 1, 1)$, the same square vector $(0, 1, 4, 2, 3)$ and they are both balanced. It is unknown to us what the answer is when the size of the alphabet is 4.

3 Balance Properties and Distances

By *distance* of two letters w_i and w_j of a word $w = w_0\, w_1\, \ldots\, w_{n-1}$ we mean the difference $i - j$, this difference being computed modulo n.

Definition 1. *Given a circular word w and a letter a, we define derivative word of w with respect to the letter a as the circular sequence of integers (a word on the alphabet $\mathbb{N}$) corresponding to the distances of all pairs of consecutive a's in w. The derivative word of w with respect to a is denoted by $\partial_a(w)$ and clearly has length $|w|_a$.*

Example 2. Consider the word $w = abaabbbabba$. Then $\partial_a(w) = 21431$.

The derivative of w with respect to a letter "contains more information" than the square vector with respect to such letter, indeed it is clearly possible to reconstruct the square vector from the derivative.

We will show that it is possible to characterize balanced words in terms of their derivative. Before doing that, we will introduce a definition.

Definition 2. *We say that a factor v of a word w is an a-chain if $v = aua$, where $u \in \Sigma^*$.*

An a-chain is then just a factor that begins and ends with an a. Notice that the derivative word of w is obtained by subtracting 1 from the lengths of the a-chains aua factors of w such that u has a-length equal to 0.

For the remainder of the paper, it will be convenient to define by analogy the derivative with respect to a of a factor of w that is also an a-chain aua. By doing this, we will commit a little abuse of notation since aua will not be seen in this case as a circular word, that is, the a-chain created from the concatenation of the last letter with the first letter of aua will not be taken into account in the computation of the derivative $\partial_a(aua)$. The advantage resulting from this little abuse is that $\partial_a(aua)$ is then a (circular) factor of $\partial_a(w)$. For instance, the derivative with respect to a of the a-chain $abbaaba$, factor of $w = abaabbbabba$, is 312, which is a factor of the (circular) word $\partial_a(w) = 21431$.

The main result of this section is the following theorem, stating a characterization of balanced words with respect to a fixed letter in terms of their derivatives.

Theorem 2. *Let w be a circular word over an alphabet Σ and $a \in \Sigma$ a letter, then w is balanced with respect to the letter a if and only if there exists an integer d such that $\partial_a(w)$ is a balanced word over the two letters $\{d, d+1\}$.*

Remark 1. The integer d in Theorem 2 is only a function of the Parikh vector of the word w. It is simply the value of the integer division $\lfloor |w|/|w|_a \rfloor$. The number of occurrences of the letter d in $\partial_a(w)$ is $|w|_a - (|w| \bmod |w|_a)$, while the number of occurrences of the letter $d+1$ is $(|w| \bmod |w|_a)$.

Given a Parikh vector P, it is then always possible to compute the derivative words with respect to each letter of the balanced word in $\mathbb{W}(P)$ (if it exists). For each letter a, it suffices to compute $d = \lfloor |w|/|w|_a \rfloor$ and then write the balanced word on the two-letter alphabet $\{d, d+1\}$ containing $|w|_a - (|w| \bmod |w|_a)$ occurrences of the letter d and $(|w| \bmod |w|_a)$ occurrences of the letter $d+1$. It is indeed well known that on an alphabet of size 2, a balanced word for a given distribution of the two letters always exists and is unique (up to letter renaming), cf. [3].

However, when the number of letters of the alphabet is greater than 2, and analogously to multi-variable differential equations theory, "integrating" the system obtained from the values of all partial derivatives (that is, determining a balanced word w whose partial derivatives are given with the method just described) is a very difficult problem.

From the partial derivative with respect to a letter a, one can determine (up to a shift modulo $n = |w|$) the sequence of positions where the occurrences of the letter a are to be placed. If the partial derivative $\partial_a(w)$ is the word $d_1 d_2 \ldots d_l$ and the first occurrence of a is placed at position 0, then the j-th occurrence of a is to be placed at position $\sum_{i=1}^{j-1} d_i$. The difficulty consists in determining appropriate (circular) shifts for the position sequence for each letter in such a way that the obtained shifted sequences are pairwise disjoint and cover the entire integer interval $\{0, 1, \ldots, n-1\}$.

This problem is just a different formalization of the well-known problem of covering the set of natural integers with so-called *Beatty sequences*, that is, with sequences of the form $\{\lfloor \alpha n + \beta \rfloor \mid n \in \mathbb{N}\}$. The position sequences for a letter a obtained using the derivative as described in the previous paragraph are indeed Beatty sequences where

$$\alpha = \frac{|w|}{|w|_a} \text{ and } \beta = -\frac{|w|}{|w|_a}.$$

Interestingly, much more is known about the problem of covering the integers with Beatty sequences with irrational coefficient α (corresponding to infinite non periodic words) than it is when the coefficients are rational (corresponding to infinite periodic words, that is, to circular words, the case that we deal with here) and it appears that, while it is possible to determine the conditions on α_1, β_1, α_2 and β_2 so that the two Beatty sequences $\{\lfloor \alpha_1 n + \beta_1 \rfloor \mid n \in \mathbb{N}\}$ and $\{\lfloor \alpha_2 n + \beta_2 \rfloor \mid n \in \mathbb{N}\}$ are disjoint when the coefficient α_i's are irrational, there is no analogous result when the α_i's are rational. This is somewhat surprising since in this case, because of the periodicity, the problem can be reduced to the covering of the integer interval $\{0, 1, \ldots, n - 1\}$.

From the algorithmic point of view, this means that we do not know any algorithm to compute the appropriate shifts of the position sequence of a given letter that is substantially more efficient than the naive algorithm that tries all possible shifts and verifies if a certain shift is "good". Note that such an algorithm may also need to use backtracking. In fact, after determining an appropriate shift for the position sequence of the b's in such a way that the resulting sequence does not intersect the position sequence of the a's, one may realize that the choice of this shift does not leave any possible choice for an appropriate shift for the position sequence of the c's. The only improvement that we have determined to this algorithm is that shifts obtained as sums of consecutive integers occurring in the derivative with respect to another letter are certainly "bad" (in the sense that the resulting sequences would not be disjoint) and do not need to be tested.

Example 3.

Consider the Parikh vector $P = (7, 7, 5)$.

Therefore, the derivatives of the balanced word in $\mathbb{W}(P)$ (if it exists) with respect to a and to b are equal to the balanced word on the alphabet $\{\lfloor \frac{19}{7} \rfloor, \lfloor \frac{19}{7} \rfloor + 1\}$ (that is, on the alphabet $\{2, 3\}$) having $7 - (19 \bmod 7) = 2$ occurrences of the letter 2 and $(19 \bmod 7) = 5$ occurrences of the letter 3. This is the word $2\,3\,3\,2\,3\,3\,3$.

The derivative with respect to c is the balanced word on $\{\lfloor \frac{19}{5} \rfloor, \lfloor \frac{19}{5} \rfloor + 1\}$ (that is, on the alphabet $\{3, 4\}$) having $5 - (19 \bmod 5) = 1$ occurrences of the letter 3 and $(19 \bmod 5) = 4$ occurrences of the letter 4. This is the word $3\,4\,4\,4\,4$.

If we choose to place the first occurrence of a at position 0, then from the derivative with respect to a we can deduce the following position sequence for a: $0, 2, 5, 8, 10, 13, 16$. Note that this sequence coincides with the Beatty sequence $a_n = \lfloor \frac{19}{7} n - \frac{19}{7} \rfloor$. Since the derivatives with respect to a and b are equal, the position sequence of the b's is the same and hence needs to be appropriately shifted in order to make it disjoint from the sequence of the a's. A shift equal to 1 does that : $1, 3, 6, 9, 11, 14, 17$, however this choice turns out to be bad, as the sequence of positions left free : $4, 7, 12, 15, 18$ is not compatible with the derivative with respect to c (which would have to be $3\,5\,3\,5\,5$). Then we need to try a different shift for the position sequence of b. The shift equal to 2 is certainly bad because 2 is one of the partial sums of integers appearing consecutively in the derivative with respect to a (it is one of the letter occurring in the derivative and hence it is a sum of one term). The same can be said of the shift equal to 3. The shift equal to 4 gives the sequence $4, 6, 9, 12, 14, 17, 1$ leaving free the positions $3, 7, 11, 15, 18$ which are compatible with the derivative with respect to c.

Hence, we have been able to reconstruct a balanced word for the Parikh vector $P = (7, 7, 5)$. This word is:

$$0\ 1\ 2\ 3\ 4\ 5\ 6\ 7\ 8\ 9\ 10\ 11\ 12\ 13\ 14\ 15\ 16\ 17\ 18$$
$$a\ b\ a\ c\ b\ a\ b\ c\ a\ b\ a\ \ c\ \ b\ \ a\ \ b\ \ c\ \ a\ \ b\ \ c$$

4   Other Balance Properties

The following definition, which can be found in Cassaigne, Ferenczi and Zamboni [5] or in Heinis [10], is a natural generalization of the notion of balance.

Definition 3. *Let w be a circular word over Σ and let m be a nonnegative integer. We say that w is m-balanced if for any letter $a \in \Sigma$, and for any pair of factors u and v of w having the same length, we have $||u|_a - |v|_a| \leq m$.*

Obviously, if a word w is m-balanced, then it is also m'-balanced for all $m' > m$.

In [13] Sano, Miyoshi and Kataoka introduced another balance property generalizing the definition of balanced words, which they also called m-balance before realizing that such terminology had already been used in literature. They eventually reserved the term m-balance for the notion they introduce, and create a new term for the already known notion of m-balance (as in Definition 3). In this paper we choose to maintain the term m-balance for the notion in Definition 3 and introduce the term m-*uniform distribution* for the notion of Sano et al. [13].

Definition 4. *For a nonnegative integer m and a word w on Σ, a letter a is m-uniformly distributed in w if it satisfies the following: whenever there exists an a-chain aua in w, any factor u' in w such that $|u'| = |u| + m + 1$ satisfies $|u'|_a \geq |u|_a + 1$. A word w is m-uniformly distributed if each letter is m-uniformly distributed in w.*

As for m-balance, if a word w is m-uniformly distributed, then it is also m'-uniformly distributed for all $m' > m$.

Sano et al. provide a necessary and sufficient condition for a word to be m-uniformly distributed. In our opinion this characterization is clearer and more intuitive than the original definition and we would like to take it as an alternate definition of m-uniformly distributed word.

Definition 5. *A word w on Σ is m-uniformly distributed if (i.) each letter in w appears either at most once or infinitely often and (ii.) for each letter a having at least two occurrences in w, any two a-chains aua and $au'a$ such that $|u|_a = |u'|_a$ satisfy $||u| - |u'|| \leq m$.*

Note that in case of periodic infinite words, this definition is equivalent to condition (ii.) alone.

In a certain sense, the notion of m-balance and the notion of m-uniform distribution are symmetrical to each other. In the former, one compares factors of w having equal length and verifies that the difference of the numbers of occurrences of a letter is less than or equal to m. In the latter, one compares factors of

w having the same number of occurrences of a letter (starting and ending with that letter) and verifies that the difference of the length of the two factors is less than or equal to m.

Both notions of m-balance and of m-uniform distribution are generalizations of the balance property. Indeed, when $m = 1$ both definitions are equivalent to the definition of balanced words. However, as Sano et al. note in their paper, the notion of m-uniform distribution is stronger than the notion of m-balance .

Theorem 3. [Sano, Miyoshi, Kataoka] *If a word w is m-uniformly distributed, then it is also m-balanced. If $m = 1$ then the converse also holds.*

Sano et al. also note that the converse of this theorem does not hold for $m \geq 2$ and provide the example of the word $w = cbcbaca$, which is 2-balanced, but not 2-uniformly distributed. Indeed bcb and $bacacb$ are two b-chains where the internal factors $u = c$ and $u' = acac$ contain no b's and $|u'| - |u| = 3$.

Furthermore, we would like to note that if a word w is m-balanced, this does not imply that w is k-uniformly distributed for some k larger than m. For instance for any integer k, one can consider the word $a^{k+1}bb$, which is 2-balanced, but is not k-uniformly distributed, as it contains a b-chain of length 2 and a b-chain of length $k + 3$.

In the previous section we have given a characterization of balanced words in terms of their derivatives. In the remainder of this section we will present some results relating the two notions of m-balance and m-uniform distribution with properties of the derivative (and its "iterations", which we will define next).

We note that the derivative word with respect to a letter a is obtained by subtracting one from the lengths of the a-chains aua of w with $|u|_a = 0$. By generalization, we give the following

Definition 6. *For an integer $k \geq 1$, for a word w on Σ and for a letter a in Σ, we call k-th (partial) derivative of w with respect to a (noted $\partial_a^k(w)$), the word of length $|w|_a$ obtained by subtracting 1 from the lengths of the a-chains aua of w with $|u|_a = k - 1$.*

For instance, $\partial_a^2(abaabbbabba) = 35743$.

Note that the k-th (partial) derivative of w with respect to a can be easily computed from the (first) derivative with respect to a. Indeed, if $t = |w|_a$ and $\partial_a(w) = i_0 \, i_1 \, \ldots \, i_{t-1}$ then $\partial_a^k(w)$ is the word of length t whose j-th letter is obtained as $\sum_{p=0}^{k-1} i_{j+p}$ (the sums $j + p$ being computed modulo t).

Returning to the same example, we had $\partial_a(abaabbbabba) = 21431$, and indeed : $3 = 2 + 1$; $5 = 1 + 4$; $7 = 4 + 3$; $4 = 3 + 1$; $3 = 1 + 2$.

The following proposition is simply an immediate consequence of the alternate definition of m-uniformly distributed words.

Proposition 2. *A letter a is m-uniformly distributed in the word w if and only if for all k, there exists an integer d_k such that $\partial_a^k(w) \in \{d_k, d_k+1, \ldots, d_k+m\}^*$.*

We recall our characterization of balanced words in terms of properties of the derivative : a circular word w is balanced with respect to the letter a if and only if

1. $\partial_a(w)$ is a word on the two-letter alphabet $\{d, d+1\}$ for a given integer d;
2. $\partial_a(w)$ is balanced.

We would like to study what kind of balance properties are satisfied by a word characterized by weaker conditions than these two. We start by weakening condition 2 and suppose that the word $\partial_a(w)$ is m-balanced.

Proposition 3. *Let w be a circular word over an alphabet Σ and $a \in \Sigma$ a letter. If there exist an integer d such that $\partial_a(w)$ is an m-balanced word over the two letters $\{d, d+1\}$ then the letter a is m-uniformly distributed in w.*

When $m = 1$, the converse of the proposition obviously holds (it just becomes Theorem 2), while it certainly does not hold when $m \geq 2$. For instance, for $m = 2$, there exist 2-uniformly distributed words whose partial derivative respect to a given letter is a word over an alphabet of kind $\{d, d+1, d+2\}$.

We will weaken now condition 1.

Proposition 4. *Let w be a circular word over an alphabet Σ and $a \in \Sigma$ a letter. If there exist an integer d and an integer m such that $\partial_a(w)$ is a balanced word over the $m+1$ letters $\{d, d+1, \ldots, d+m\}$ then the letter a is m'-uniformly distributed in w, with*

$$
m' = \begin{cases} \left(\frac{m+1}{2}\right)^2 & \text{if } m \text{ is odd} \\[2mm] \frac{m^2+2m}{4} & \text{if } m \text{ is even} \end{cases}.
$$

Note that the converse of this proposition does not hold when $m \geq 2$. Indeed it is false that for a 2-uniformly distributed letter a in a word w, the derivative $\partial_a(w)$ is a balanced word on an alphabet $\{d, d+1, d+2\}$. Take for instance the word $w = a * a * a * *a * *$ (where the symbol $*$ stands for any letter different from a). The letter a is 2-uniformly distributed in this word as $\partial_a(w) = 2\,2\,3\,3$, $\partial_a^2(w) = 4\,5\,6\,5$ and $\partial_a^3(w) = 7\,8\,8\,7$ are all words on alphabets Σ such that $\max(\Sigma) - \min(\Sigma) \leq 2$, but $\partial_a(w)$ is not a balanced word, as it contains the factors $2\,2$ and $3\,3$.

5   Conclusions

While our work focuses on balance properties of finite circular words, it is important to note that several of our results apply to infinite words of any kind (periodic or not). With a little tweaking, it is indeed possible to define the derivative of any infinite word (in this case the derivative is infinite as well) and basically extend Theorem 2 and the results of Section 4 to non periodic words.

Before closing this article, we would like to briefly highlight a few open questions and possible developments.

As we stressed in Section 2, it is unknown to us how to characterize Parikh vectors for which a balanced word exists. The algorithmic aspects of this problem boils down to finding an efficient algorithm to generate balanced words given their Parikh vector. Finally, Propositions 3 and 4 define a subclass of m-uniformly distributed words. It would be interesting to find out if the words belonging to such class can be characterized combinatorially.

References

1. Altman, E., Gaujal, B., Hordijk, A.: Balanced sequences and optimal routing. Journal of the ACM 47(4), 752–775 (2000)
2. Berstel, J., Seebold, P.: Sturmian words. In: Lothaire, M. (ed.) Algebraic Combinatorics on Words, ch. 2, pp. 45–110. Cambridge University Press, Cambridge (2002)
3. Borel, J.P., Reutenauer, C.: On Christoffel classes. RAIRO-Theoretical Informatics and Applications 40, 15–28 (2006)
4. Cameron, P.J., Wu, T.: A new characterization of balanced words. In: Proceedings of the 6th International Conference on Words, Marseille, France, pp. 63–71 (2007)
5. Cassaigne, J., Ferenczi, S., Zamboni, L.Q.: Imbalances in Arnoux-Rauzy sequences. Ann. Inst. Fourier (Grenoble) 50(4), 1265–1276 (2000)
6. Castiglione, G., Restivo, A., Sciortino, M.: Circular words and automata minimization. In: Proceedings of the 6th International Conference on Words, Marseille, France, pp. 79–89 (2007)
7. Fraenkel, A.S.: Complementing and exactly covering sequences. J. Combin. Theory, Ser. A 14, 8–20 (1973)
8. Gusfield, D.: Algorithms on Strings, Trees, and Sequences. Cambridge University Press, Cambridge (1997)
9. Hedlund, G.A., Morse, M.: Symbolic dynamics ii. Sturmian trajectories. Amer. J. Math. 62 (1940)
10. Heinis, A.: On low-complexity bi-infinite words and their factors. Journal de théorie des nombres de Bordeaux 13(2), 421–442 (2001)
11. Hubert, P.: Suites équilibrées. Theoretical Computer Science 242, 91–108 (2000)
12. Mantaci, S., Restivo, A., Sciortino, M.: Burrows-Wheeler transform and Sturmian words. Informat. Proc. Lett. 86, 241–246 (2003)
13. Sano, S., Miyoshi, N., Kataoka, R.: m-Balanced words: a generalization of balanced words. Theor. Comput. Sci. 314(1), 97–120 (2004)
14. Vuillon, L.: Balanced words. Bull. Belg. Math. Soc. Simon Stevin 10(5), 787–805 (2003)

MSO Logic for Unambiguous Shared-Memory Systems*

Rémi Morin

Aix-Marseille université — UMR 6166 — CNRS
Laboratoire d'Informatique Fondamentale de Marseille
163, avenue de Luminy, F-13288 Marseille Cedex 9, France

Abstract. Shared-memory systems appear as a generalization of asynchronous cellular automata. In this paper we relate the partial-order semantics of shared-memory systems to Mazurkiewicz trace languages by means of a new refinement construction. We show that a set of labeled partial orders is recognized by some unambiguous shared-memory system if and only if it is definable in monadic second-order logic and media-bounded.

Introduction

Partially ordered multisets (for short, pomsets) are a usual setting to describe the concurrent behaviors of a distributed or parallel system [1,10,12,16,21]. For the communication paradigm based on messages, the standard notation of message sequence charts has been investigated intensively in the past years, see e.g. [5,12,14]. The somewhat simpler paradigm of communication by means of synchronizations is illustrated by asynchronous automata [24] and corresponds to the classical framework of Mazurkiewicz traces [9]. The variant of asynchronous *cellular* automata is a kind of shared-memory systems where each process communicates with a fixed set of neighbors [8]. Similarly to [11] we study here a more general model where the communication connectivity evolves dynamically along executions. As a result the pomsets accepted by this kind of shared-memory systems are no longer Mazurkiewicz traces.

In the deterministic case, we showed in [19] that shared-memory systems recognize precisely the sets of pomsets that are consistent and regular, two notions borrowed from [3]. In this paper we investigate the more general setting of *unambiguous* shared-memory systems. Roughly speaking, a shared-memory system is unambiguous if any behaviour can be executed in only one way. We characterize the class of pomset languages that arise from unambiguous shared-memory systems. For this we introduce the notion of a *media-bounded* set of pomsets. Our main result asserts that a set of pomsets is recognized by some unambiguous shared-memory system if and only if it is media-bounded and definable in monadic second-order logic (Theorem 3.4).

In this work we present a new refinement construction that allows us to represent media-bounded sets of pomsets by Mazurkiewicz trace languages (Theorem 2.11). Moreover this refinement preserves MSO-definability (Theorem 3.2). As a consequence we get a kind of generalization of the main result from [3] that asserts that any regular consistent set of pomsets can be refined onto a regular set of Mazurkiewicz traces. Moreover this result is optimal because only media-bounded sets of pomsets can be refined onto a set of Mazurkiewicz traces (Cor. 2.12).

---

* Supported by the ANR project SOAPDC.

M. Ito and M. Toyama (Eds.): DLT 2008, LNCS 5257, pp. 516–528, 2008.

Preliminaries. A *pomset* (or partially ordered multiset) over an alphabet Σ is a triple $t = (E, \preccurlyeq, \xi)$ where $(E, \preccurlyeq)$ is a finite partial order and ξ is a mapping from E to Σ *without autoconcurrency*: $\xi(x) = \xi(y)$ implies $x \preccurlyeq y$ or $y \preccurlyeq x$ for all $x, y \in E$. A pomset can be seen as an abstraction of an execution of a concurrent system [16,20,21,10]. In this view, the elements e of E are *events* and their label $\xi(e)$ describes the action that is performed in the system by the event $e \in E$. Furthermore, the order $\preccurlyeq$ describes the dependence between events. We denote by $\mathbb{P}(\Sigma)$ the class of all pomsets over Σ.

Let $t = (E, \preccurlyeq, \xi)$ be a pomset and $x, y \in E$. Then y *covers* x (denoted $x \!-\!\!\prec y$) if $x \prec y$ and $x \prec z \preccurlyeq y$ implies $y = z$. An *order extension* of a pomset $t = (E, \preccurlyeq, \xi)$ is a pomset $t' = (E, \preccurlyeq', \xi)$ such that $\preccurlyeq \subseteq \preccurlyeq'$. A *linear extension* of t is an order extension that is linearly ordered. It corresponds to a sequential view of the concurrent execution t. Linear extensions of a pomset t over Σ can naturally be regarded as words over Σ. By $\text{LE}(t) \subseteq \Sigma^*$, we denote the set of linear extensions of a pomset t over Σ. Two isomorphic pomsets admit the same set of linear extensions. Noteworthy the converse property holds [22]: If $\text{LE}(t) = \text{LE}(t')$ then t and t' are two isomorphic pomsets. In the sequel of this paper we do not distinguish between isomorphic pomsets.

An *ideal* of a pomset $t = (E, \preccurlyeq, \xi)$ is a subset $H \subseteq E$ such that $x \in H \wedge y \preccurlyeq x \Rightarrow y \in H$. The restriction $t' = (H, \preccurlyeq \cap (H \times H), \xi \cap (H \times \Sigma))$ is then called a *prefix* of t and we write $t' \leqslant t$. For all $z \in E$, we denote by $\downarrow z$ the ideal of events below z, i.e. $\downarrow z = \{y \in E \mid y \preccurlyeq z\}$. For any set of pomsets $\mathcal{L}$, $\text{Pref}(\mathcal{L})$ denotes the set of prefixes of pomsets from $\mathcal{L}$. The language $\mathcal{L}$ is called *prefix-closed* if $\text{Pref}(\mathcal{L}) = \mathcal{L}$.

1   A Generalization of Asynchronous Cellular Automata

Throughout the paper we fix some finite alphabet Σ. The notion of a shared-memory system studied in this work is based on a set $\mathcal{I}$ of processes together with a distribution $\text{Loc} : \Sigma \rightarrow 2^{\mathcal{I}}$ which assigns to each action $a \in \Sigma$ a subset of processes $\text{Loc}(a) \subseteq \mathcal{I}$. Intuitively each occurrence of action a induces a *synchronized* step of all processes from $\text{Loc}(a)$. For that reason we assume that $\text{Loc}(a)$ is non-empty for all $a \in \Sigma$. The pair (Σ, Loc) is often called a distributed alphabet.

1.1   Shared-Memory Systems

Processes of a shared-memory system can communicate by means of a set $\mathcal{R}$ of shared variables (or registers) taking values from a common set of data $\mathcal{D}$; in particular the initial contents of this shared memory is formalized by a *memory-state* $\chi_{\text{init}} : \mathcal{R} \rightarrow \mathcal{D}$ that associates to each register $r \in \mathcal{R}$ a value $\chi_{\text{init}}(r) \in \mathcal{D}$. Intuitively each action corresponds to the reading of the values of a subset of registers (a guard) and the writing of new values in some other registers. A *valuation* is a partial function $\nu : \mathcal{R} \rightharpoonup \mathcal{D}$; it will correspond to the reading or the writing of some values in a subset of registers. The *domain* $\text{dom}(\nu)$ of a valuation ν is the set of registers r such that $\nu(r)$ is defined. We denote by $\mathcal{V}$ the set of all valuations.

Now each process $i \in \mathcal{I}$ is provided with a set of local states S_i together with an initial local state $\imath_i \in S_i$. A *global state* $s = (s_i)_{i \in \mathcal{I}}$ consists of one local state s_i for each process $i \in \mathcal{I}$ and a *configuration* $q = (\chi, s)$ is a pair made of a memory-state $\chi : \mathcal{R} \to \mathcal{D}$ and a global state s. We let the Cartesian product $Q = \mathcal{D}^{\mathcal{R}} \times \prod_{i \in \mathcal{I}} S_i$ denote the set of all configurations. The initial configuration $\iota = (\chi_{\text{init}}, s)$ corresponds to the initial memory-state χ_{init} and the initial global state $s = (\iota_i)_{i \in \mathcal{I}}$. Given a memory-state $\chi : \mathcal{R} \to \mathcal{D}$ and a subset of registers $R \subseteq \mathcal{R}$, we let $\chi|R$ denote the valuation with domain R such that $\chi|R(r) = \chi(r)$ for all $r \in R$. Given some action a, some process j and some global state $s = (s_i)_{i \in \mathcal{I}}$, we denote by $s|a$ the partial state $(s_i)_{i \in \text{Loc}(a)}$ and by $s|j$ the local state s_j. For each $a \in \Sigma$ we denote by S_a the set of partial states $S_a = \prod_{i \in \text{Loc}(a)} S_i$. A *transition rule* is a quintuple (ν, s, a, ν', s') where $a \in \Sigma$, $\nu, \nu' \in \mathcal{V}$ are two valuations and $s, s' \in S_a$ are two partial states. For convenience we put $\rho = (\nu_\rho, s_\rho, a_\rho, \nu'_\rho, s'_\rho)$, $R_\rho = \text{dom}(\nu_\rho)$ and $W_\rho = \text{dom}(\nu'_\rho)$ for each transition rule ρ.

DEFINITION 1.1. *A shared-memory system (for short, an SMS) over some distributed alphabet (Σ, Loc), some initial memory-state $\chi_{\text{init}} : \mathcal{R} \to \mathcal{D}$, and local states $(S_i, \iota_i)_{i \in \mathcal{I}}$ consists of a subset of transition rules Δ together with a subset $F \subseteq Q$ of final (or accepting) configurations.*

Intuitively action a can occur synchronously on all processes from $\text{Loc}(a)$ in some configuration $q = (\chi, s)$ if there exists a transition rule $\rho \in \Delta$ such that $a_\rho = a$, $\nu_\rho = \chi|R_\rho$, and $s_\rho = s|a$. In that case processes from $\text{Loc}(a)$ may perform a joint move to the new partial state s'_ρ and write the new values $\nu'_\rho(r)$ in registers from W_ρ. The step consisting of all these moves and all these changes is considered atomic.

1.2 Partial Order Semantics of Shared-Memory Systems

Following a classical trend in concurrency theory [16,20,21] we want to describe the concurrent executions of a shared-memory system $\mathcal{S}$ by means of labeled partial orders in such a way that the ordering of events represents the must-happen-before relation between occurrences of actions. Since each process works sequentially, events occurring on the same process must be comparable. Furthermore any two events that change the value of some register should be comparable, that is, we consider Exclusive-Write systems. Now if one event writes a new value in some register read by another event then these two events should be comparable as well; otherwise it would be unclear which value is actually read by the second event. In that way we have characterized which pairs of transition rules may occur concurrently. We formalize this *May-Occur-Concurrently* relation by means of a binary relation $\| \subseteq \Delta \times \Delta$. For any two transitions rules $\rho, \rho' \in \Delta$ we put $\rho\|\rho'$ if $\text{Loc}(a_\rho) \cap \text{Loc}(a_{\rho'}) = \emptyset$, $W_\rho \cap (R_{\rho'} \cup W_{\rho'}) = \emptyset$, and $W_{\rho'} \cap (R_\rho \cup W_\rho) = \emptyset$.

In order to reason about which registers are read by each event and how events change the local states of processes and the values of registers, we make use of the notion of run. Let $t = (E, \preccurlyeq, \xi)$ be a pomset over Σ. A *run* of t over $\mathcal{S}$ is a mapping $\rho : E \to \Delta$ which maps each event e from E to some transition rule $\rho(e) \in \Delta$ such that $a_{\rho(e)} = \xi(e)$. In order to reflect the May-Occur-Concurrently relation, two events are incomparable in t only if their transition rules are independent. This is formalized by Axiom V_1 below. The partial order of events in t results from the transitive closure of the covering relation $\prec\!\!\!\!-$ and can be represented by its Hasse diagram. Since we want the partial order to reflect the *Must-Happen-Before* relation, any edge from the covering relation must represent some dependence between the corresponding transition rules. This is formalized by Axiom V_2 below. As a consequence the run ρ is called *valid* if V_1 and V_2 are satisfied:

V_1: For all events $e_1, e_2 \in E$ with $\rho(e_1) \,\| \rho(e_2)$, we have $e_1 \preccurlyeq e_2$ or $e_2 \preccurlyeq e_1$;
V_2: For all events $e_1, e_2 \in E$ with $e_1 \mathop{-\!\!\!\prec} e_2$, we have $\rho(e_1) \,\| \rho(e_2)$.

We assume now that ρ is a valid run for t. Let $H \subseteq E$ be an ideal of t. The configuration $q_{\rho,H}$ at H corresponds intuitively to a snapshot of the system after all events from H have occurred along the execution of t w.r.t. ρ: The value of each register is the value written by the last event that has modified this value and the local state of each process is the local state reached after the last joint move performed by that process. Formally $q_{\rho,H}$ is the configuration $q_{\rho,H} = (\chi_{\rho,H}, s_{\rho,H})$ defined by the next two conditions:

- For all registers $r \in \mathcal{R}$, we put $\chi_{\rho,H}(r) = \nu'_{\rho(e)}(r)$ if e is the greatest event in H such that $r \in W_{\rho(e)}$, and $\chi_{\rho,H}(r) = \chi_{\text{init}}(r)$ if there is no such event.
- For all $i \in \mathcal{I}$, we put $s_{\rho,H}|i = s'_{\rho(e)}|i$ if e is the greatest event in H such that $i \in \text{Loc}(\xi(e))$, and $s_{\rho,H}|i = \imath_i$ if there is no such event.

Due to V_1 events satisfying $r \in W_{\rho(e)}$ are totally ordered so there exists at most one maximal event satisfying this condition. A similar observation holds for events satisfying $i \in \text{Loc}(\xi(e))$. Therefore $q_{\rho,H}$ is well-defined. Note here that $q_{\rho,\emptyset}$ corresponds to the initial configuration $\imath$. Now we say that a valid run ρ is compatible with $\mathcal{S}$ if the configuration reached after all events below e enables the execution of the transition rule $\rho(e)$. Formally a valid run ρ of t is *compatible with* $\mathcal{S}$ if for all events $e \in E$ the configuration (χ, s) at $\downarrow e \setminus \{e\}$ satisfies $\chi|R_{\rho(e)} = \nu_{\rho(e)}$ and $s|\xi(e) = s_{\rho(e)}$. A pomset that admits a compatible run corresponds to a potential execution of $\mathcal{S}$.

DEFINITION 1.2. *A pomset* $t = (E, \preccurlyeq, \xi)$ *is accepted by* $\mathcal{S}$ *if it admits a compatible run* ρ *such that the configuration* $q_{\rho,E}$ *belongs to* F. *The language* $\mathcal{L}(\mathcal{S}) \subseteq \mathbb{P}(\Sigma)$ *recognized by* $\mathcal{S}$ *collects all pomsets accepted by* $\mathcal{S}$. *The SMS* $\mathcal{S}$ *is called* **unambiguous** *if each* $t \in \text{Pref}(\mathcal{L}(\mathcal{S}))$ *admits a unique compatible run.*

In this paper we focus on *unambiguous* shared-memory systems: Roughly speaking each prefix of a pomset from $\mathcal{L}(\mathcal{S})$ can be executed in only one way. This definition is somewhat more restrictive than the classical notion of unambiguity [2,7,25] because we consider all prefixes of $\mathcal{L}(\mathcal{S})$ instead of $\mathcal{L}(\mathcal{S})$ itself and moreover we assume that shared-memory systems have a single initial state.

EXAMPLE 1.3. Our running example corresponds to a Producer-Consumer system. Its alphabet is $\Sigma = \{p, c\}$ where p represents a production of one item and c a consumption. Its behaviour consists of all *ladders*, that is, pomsets over Σ that consist of a chain of n production events and a chain of n consumption events and such that the kth consumption covers the kth production and no consumption is below any production. An example of a ladder is depicted in Figure 1. We leave it to the reader to find some unambiguous shared-memory system over Σ that accepts precisely the set of all ladders. With no surprise such an SMS needs infinitely many registers.

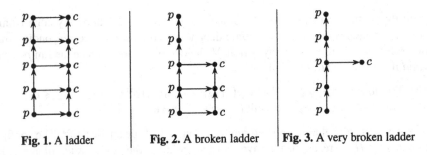

Fig. 1. A ladder   **Fig. 2.** A broken ladder   **Fig. 3.** A very broken ladder

In the rest of this paper we consider finite shared-memory systems, only: The set of registers $\mathcal{R}$, the set of values $\mathcal{D}$ and the set of processes $\mathcal{I}$ and the sets of local states S_i of any SMS are finite.

1.3 Partial Commutations and Asynchronous Cellular Automata

Let us recall some basic definitions from Mazurkiewicz trace theory [9]. The concurrency of a distributed system is often represented by an *independence relation* over the set of actions Σ, that is a binary, symmetric and irreflexive relation $\| \subseteq \Sigma \times \Sigma$. The associated *trace equivalence* is the least congruence $\sim$ over Σ^* such that for all $a, b \in \Sigma$, $a \| b$ implies $ab \sim ba$. A *trace* $[u]$ is the equivalence class of a word $u \in \Sigma^*$. We denote by $\mathbb{M}(\Sigma, \|)$ the set of all traces w.r.t. $(\Sigma, \|)$. Let $u \in \Sigma^*$; then the trace $[u]$ is precisely the set of linear extensions $\mathrm{LE}(t)$ of a unique pomset $t = (E, \preccurlyeq, \xi)$, that is, $[u] = \mathrm{LE}(t)$. Moreover t satisfies the following additional properties:

M_1: For all events $e_1, e_2 \in E$ with $\xi(e_1) \not\| \xi(e_2)$, we have $e_1 \preccurlyeq e_2$ or $e_2 \preccurlyeq e_1$;

M_2: For all events $e_1, e_2 \in E$ with $e_1 \mathbin{-\!\!\prec} e_2$, we have $\xi(e_1) \not\| \xi(e_2)$.

Conversely the linear extensions of a pomset satisfying these two axioms form a trace of $\mathbb{M}(\Sigma, \|)$. Thus one usually identifies $\mathbb{M}(\Sigma, \|)$ with the class of pomsets satisfying M_1 and M_2. Note that these requirements are similar to the criteria of a valid run.

Now an interesting particular case of shared-memory systems from the literature is provided by the notion of an asynchronous cellular automaton [8,25].

DEFINITION 1.4. *Let* $\| \subseteq \Sigma \times \Sigma$ *be some independence relation. An* asynchronous cellular automaton *over* $(\Sigma, \|)$ *(for short, an ACA) is a shared-memory system such that* $\Sigma = \mathcal{I} = \mathcal{R}$ *and for all* $a \in \Sigma$: $\mathrm{Loc}(a) = \{a\}$, $S_a = \{\imath_a\}$ *is a singleton and moreover* $(\nu, \imath_a, a, \nu', \imath_a) \in \Delta$ *implies* $\mathrm{dom}(\nu) = \{b \in \Sigma \mid b \not\| a\}$ *and* $\mathrm{dom}(\nu') = \{a\}$.

Intuitively each action corresponds to a process which owns a register whose value describes its current local state. Moreover these systems are Owner-Write: No process may write in the register of some other process. On the other hand, two processes read the values of each other if they are dependent w.r.t. the given independence relation.

Observe now that the May-Occur-Concurrently relation $\| \subseteq \Delta \times \Delta$ and the given independence relation $\| \subseteq \Sigma \times \Sigma$ match each other: $\rho \| \rho'$ if and only if $a_\rho \| a_{\rho'}$ for any two transitions rules $\rho, \rho' \in \Delta$. As a consequence the pomsets accepted by an ACA according to Definition 1.2 are Mazurkiewicz traces from $\mathbb{M}(\Sigma, \|)$. Actually Definition 1.2 coincides with the usual semantics of an ACA.

Consider now some asynchronous cellular automaton S such that each transition rule occurs in at least one compatible run. Then S is unambiguous if and only if for each $a \in \Sigma$ and each configuration $q = (\chi, s)$ there exists at most one transition rule ρ such that $a_\rho = a$ and $\nu_\rho = \chi|R_\rho$. Thus unambiguity and determinism coincide in the particular case of asynchronous cellular automata.

2 Media-Bounded Pomset Languages

In this section we introduce the notion of a media-bounded set of pomsets. We observe that the language of any (finite) unambiguous shared-memory system is media-bounded (Prop. 2.4). Moreover we show that a set of pomsets is media-bounded if and only if it can be refined onto a set of Mazurkiewicz traces (Cor. 2.12).

2.1 Media-Bound of a Set of Pomsets

In this subsection we fix a set of pomsets $\mathcal{L}$ and define its media-bound.

DEFINITION 2.1. *Let $t = (E, \preccurlyeq, \xi)$ be a pomset over Σ and $e \in E$ be an event of t. Let $a, b \in \Sigma$ be two distinct actions. Then e activates (a, b) in t if $\xi(e) = a$ and there exists a pomset $t^\circ = (E^\circ, \preccurlyeq^\circ, \xi^\circ)$ in $\mathcal{L}$ satisfying the two following conditions:*

– *t is a prefix of t°, and*
– *there is some event $f \in E^\circ \setminus E$ such that $e \relbar\mkern-9mu\prec^\circ f$ and $\xi^\circ(f) = b$.*

Thus an event $e \in E$ labeled by a activates (a, b) in t if t is a prefix of some pomset t° from $\mathcal{L}$ in which e is covered by some event labeled by b. Observe that if e activates (a, b) in t then e activates (a, b) in the prefix $\downarrow_t e$, too. Moreover no event from t activates (a, b) if t is not a prefix of some pomset from $\mathcal{L}$.

EXAMPLE 2.2. We continue Example 1.3 and consider again the set $\mathcal{L}$ of all ladders. We call *broken ladder* any prefix of a ladder. The broken ladder t of Figure 2 consists of 5 production events and 3 consumption events. Among all production events there are three events that are covered by consumption events. These production events do not activate (p, c) in t. The other two production events do activate (p, c) in t since we can complete the pomset t with two consumption events that will cover these two production events and get a pomset from $\mathcal{L}$.

DEFINITION 2.3. *For any pomset t over Σ we let $\alpha_{a,b}(t)$ denote the number of events in t that activate (a, b). The media-bound of $\mathcal{L}$ is the least upper bound $B \in \mathbb{N} \cup \{\infty\}$ of all $\alpha_{a,b}(t)$ where $t \in \mathbb{P}(\Sigma)$ and (a, b) is a pair of distinct actions from Σ. The language $\mathcal{L}$ is called media-bounded if its media-bound is finite.*

It is clear that the set of all ladders from Examples 1.3 and 2.2 is *not* media-bounded. The next result proves that this language is recognized by no unambiguous SMS.

PROPOSITION 2.4. *The language of any unambiguous SMS is media-bounded.*

For latter purposes, we observe also here that any Mazurkiewicz trace language is media-bounded and its media-bound is at most 1.

LEMMA 2.5. *Let* $\| \subseteq \Sigma \times \Sigma$ *be an independence relation. If* $\mathcal{L} \subseteq \mathrm{M}(\Sigma, \|)$ *then the media-bound of* $\mathcal{L}$ *is at most 1.*

2.2   The Notion of Refinement

A characterization of media-bounded sets of pomsets relies on the notion of refinement that we introduce now. Let Σ_1 and Σ_2 be two alphabets and $\pi : \Sigma_1 \to \Sigma_2$ a mapping from Σ_1 to Σ_2. This mapping extends into a map from $\Sigma_1^\star$ to $\Sigma_2^\star$. It extends also in a natural way into a function that maps each pomset $t = (E, \preccurlyeq, \xi)$ over Σ_1 to the structure $\pi(t) = (E, \preccurlyeq, \pi \circ \xi)$. The latter might not be a pomset over Σ_2 in case some autoconcurrency appears in it (see preliminaries). This situation can occur if $\pi(a) = \pi(b)$ for two distinct actions $a, b \in \Sigma$ while there are two events e and f in t that are labelled by a and b and that are not comparable.

Refinements allow to relate sets of pomsets $\mathcal{L}_1$ and $\mathcal{L}_2$ that are identical up to some relabeling. We require that the relabeling $\pi : \Sigma_1 \to \Sigma_2$ induces a bijection from $\mathcal{L}_1$ onto $\mathcal{L}_2$ and a bijection from $\mathrm{Pref}(\mathcal{L}_1)$ onto $\mathrm{Pref}(\mathcal{L}_2)$.

DEFINITION 2.6. *Let* $\mathcal{L}_1$ *and* $\mathcal{L}_2$ *be two sets of pomsets over* Σ_1 *and* Σ_2 *respectively. A mapping* $\pi : \Sigma_1 \to \Sigma_2$ *from* Σ_1 *to* Σ_2 *is a* refinement *from* $\mathcal{L}_2$ *onto* $\mathcal{L}_1$ *if* $\pi(t)$ *is a pomset for each* $t \in \mathcal{L}_1$, $\pi(\mathcal{L}_1) = \mathcal{L}_2$ *and* $\pi : \mathrm{Pref}(\mathcal{L}_1) \to \mathrm{Pref}(\mathcal{L}_2)$ *is one-to-one.*

Note that the requirement $\pi(\mathcal{L}_1) = \mathcal{L}_2$ implies that π induces a map from $\mathcal{L}_1$ onto $\mathcal{L}_2$. It follows that π induces a map from $\mathrm{Pref}(\mathcal{L}_1)$ onto $\mathrm{Pref}(\mathcal{L}_2)$, too. The latter is a bijection since we require also that it should be one-to-one. It follows that the mapping $\pi : \mathcal{L}_1 \to \mathcal{L}_2$ is a bijection, too. The next result shows that refinements preserve media-boundedness.

PROPOSITION 2.7. *Let* $\pi : \Sigma_1 \to \Sigma_2$ *be a refinement from* $\mathcal{L}_2$ *onto* $\mathcal{L}_1$. *If* $\mathcal{L}_1$ *is media-bounded then* $\mathcal{L}_2$ *is media-bounded.*

As observed above in Lemma 2.5, any set of Mazurkiewicz traces is media-bounded. Thus if a set of pomsets can be refined onto a set of Mazurkiewicz traces then it is media-bounded. In the next section we establish the converse property (Cor. 2.12).

2.3   A Characterization of Media-Bounded Sets of Pomsets

Let $\mathcal{L}$ be a set of pomsets over Σ and $B \in \mathbb{N} \cup \{\infty\}$ be its media-bound. For each pomset $t = (E, \preccurlyeq, \xi)$ over Σ such that t admits a unique maximal event $e_{\max}$, we define the *activated number* $\kappa_{a,b}(t) \in \mathbb{N} \cup \{\bot\}$ by induction on the number n of events labelled by a in t. First, in case $n = 0$, we put $\kappa_{a,b}(t) = \bot$. The induction step proceeds as follows. First, if $e_{\max}$ does not activate (a, b) in t then $\kappa_{a,b}(t) = \bot$. We assume now that $e_{\max}$ activates (a, b) in t. In particular we have $\xi(e_{\max}) = a$ and $n \geqslant 1$. Let E' be the subset of events $e \in E \setminus \{e_{\max}\}$ that activate (a, b) in t. Then we put $\kappa_{a,b}(t) = \min(\mathbb{N} \setminus \{\kappa_{a,b}(\downarrow e) \mid e \in E'\})$.

We stress that the activated number $\kappa_{a,b}(t)$ equals $\bot$ if and only if $e_{\max}$ does not activate (a, b) in t. Furthermore for each event $e \in E'$, we have $\kappa_{a,b}(\downarrow e) \neq \bot$ because e activates (a, b) in $\downarrow e$. Note also that $\kappa_{a,b}(t) = \bot$ if $t \notin \mathrm{Pref}(\mathcal{L})$.

LEMMA 2.8. *The media-bound of $\mathcal{L}$ is an upper bound of all $\kappa_{a,b}(t) + 1$ where a and b are two distinct actions of Σ and t is a pomset over Σ that admits a unique maximal event which activates (a, b) in t.*

Thus for all pomsets $t \in \mathbb{P}(\Sigma)$ with a unique maximal event $e_{\max}$ and for all pairs (a, b) of distinct actions from Σ, $\kappa_{a,b}(t) \neq \bot$ implies $\kappa_{a,b}(t) < B$.

In the rest of this section we assume that $\mathcal{L}$ is media-bounded, i.e. $B < \infty$. We denote by X the set of all partial functions from Σ to $\{0, 1, ..., B - 1\}$. For $x \in X$ and $a \in \Sigma$, we write $x(a) = \bot$ to denote that x is undefined for a. We consider the alphabet $\Gamma = \Sigma \times X \times X$ provided with the independence relation $\|$ such that $(a, x, y) \nparallel (b, x', y')$ if $a = b$ or $x(b) = y'(a) \neq \bot$ or $x'(a) = y(b) \neq \bot$. Clearly the binary relation $\|$ is irreflexive and symmetric. Note here that Γ is a finite alphabet because $B < \infty$. We let π_1 denote the first projection from Γ to Σ: We put $\pi_1(a, x, y) = a$. The next definition explains how a pomset t from $\mathcal{L}$ is refined into a Mazurkiewicz trace $\beta(t) \in \mathbb{M}(\Gamma, \|)$.

DEFINITION 2.9. *Let $t = (E, \preccurlyeq, \xi)$ be a pomset over Σ. For each event $e \in E$ labelled by $\xi(e) = c$, we put $\gamma(e) = (c, x, y) \in \Gamma$ where*
- $x(c) = \bot$ *and* $x(b) = \kappa_{c,b}(\downarrow e)$ *for all* $b \in \Sigma \setminus \{c\}$,
- $y(c) = \bot$ *and for all* $a \in \Sigma \setminus \{c\}$:

$$y(a) = \begin{cases} \kappa_{a,c}(\downarrow f) & \text{if } \exists f \in E, f \prec\!\!\!-\, e \wedge \xi(f) = a \\ \bot & \text{otherwise} \end{cases}$$

We write $\beta(t) = (E, \preccurlyeq, \gamma)$.

Recall that t has no autoconcurrency. It follows that $y(a)$ is well-defined because such an event f is unique if it exists. Clearly $\pi_1(\beta(t)) = t$ for all pomsets $t \in \mathbb{P}(\Sigma)$.

EXAMPLE 2.10. We continue Examples 1.3 and 2.2. Since $\Sigma = \{p, c\}$ we can identify a function $x \in X$ or $y \in X$ in a labeling (c, x, y) with an element from $\mathbb{N} \cup \{\bot\}$. Then Figure 4 shows a broken ladder $t \in \mathcal{L}$ and the associated pomset $\beta(t)$.

Now it is easy to check that for each pomset $t \in \mathcal{L}$, the pomset $\beta(t)$ satisfies the two characteristic properties M_1 and M_2 of Mazurkiewicz traces (cf. Subsection 1.3).

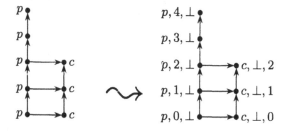

Fig. 4. A broken ladder t and its associated Mazurkiewicz trace $\beta(t)$

THEOREM 2.11. *The first projection $\pi_1 : \Gamma \to \Sigma$ is a refinement from $\mathcal{L}$ onto the subset of Mazurkiewicz traces $\beta(\mathcal{L}) \subseteq \mathbb{M}(\Gamma, \|)$.*

As an immediate consequence of Lemma 2.5, Proposition 2.7 and Theorem 2.11, we get the expected characterization of media-bounded sets of pomsets.

COROLLARY 2.12. *A set of pomsets $\mathcal{L}$ is media-bounded if and only if it admits a refinement $\pi : \mathcal{L}' \to \mathcal{L}$ onto a set of Mazurkiewicz traces $\mathcal{L}'$.*

3   Expressive Power of Unambiguous Shared-Memory Systems

In this section we characterize the class of pomset languages that are recognized by some unambiguous shared-memory system. We show that these languages are exactly the set of pomsets that are media-bounded and definable in Monadic Second-Order (MSO) logic (Theorem 3.4).

3.1   SMS Languages Are MSO-Definable

Formulae of the MSO logic that we consider involve first-order variables $x, y, z...$ for events and second-order variables $X, Y, Z...$ for sets of events. They are built up from the atomic formulae $P_a(x)$ for $a \in \Sigma$ (which stands for "the event x is labeled by the action a"), $x \preccurlyeq y$, and $x \in X$ by means of the boolean connectives $\neg, \vee, \wedge, \to, \leftrightarrow$ and quantifiers $\exists, \forall$ (both for first order and for set variables). We denote by $\mathrm{MSO}(\Sigma)$ the set of all formulae of MSO. Formulae without free variables are called sentences.

The satisfaction relation $\models$ between pomsets and sentences is defined canonically with the understanding that first order variables range over events of E and second order variables over subsets of E. The set of pomsets which satisfy a sentence φ is denoted by $\mathrm{Mod}(\varphi)$. We say that a set of pomsets $\mathcal{L}$ is MSO-*definable* if there exists a sentence φ such that $\mathcal{L} = \mathrm{Mod}(\varphi)$.

It is straightforward but a bit tedious to establish that the language of any SMS is MSO-definable by means of somewhat classical techniques [23,9,11].

LEMMA 3.1. *The language of any shared-memory system is MSO-definable.*

3.2   Main Result

The main technical contribution of this section lies in the following result which shows that the refinement $\pi_1 : \beta(\mathcal{L}) \to \mathcal{L}$ defined in the previous section preserves MSO-definability.

THEOREM 3.2. *Let $\mathcal{L}$ be a media-bounded set of pomsets. If $\mathcal{L}$ is MSO-definable then $\beta(\mathcal{L})$ is MSO-definable, too.*

Proof sketch. Since $\mathcal{L}$ is MSO-definable, the subset $\mathcal{L}' = \{t \in \mathbb{M}(\Gamma, \|) \mid \pi(t) \in \mathcal{L}\}$ is MSO-definable, too. Clearly $\beta(\mathcal{L}) \subseteq \mathcal{L}'$. We claim that there exists some sentence φ such that for all $t \in \mathcal{L}'$, we have $t \in \beta(\mathcal{L})$ iff $t \models \varphi$, that is, we can check whether the relabeling of a trace $t \in \mathbb{M}(\Gamma, \|)$ corresponds to the definition of β. ∎

We adapt now to the present setting the main technical lemma from [19]. A shared-memory system is called *singular* if the set of local states of each process $i \in \mathcal{I}$ is a singleton $S_i = \{\iota_i\}$. Furthermore a singular SMS is called *cellular* if $\Sigma = \mathcal{I}$ and $\mathrm{Loc}(a) = \{a\}$ for each $a \in \Sigma$.

LEMMA 3.3. *Let $\mathcal{L}$ and $\mathcal{L}'$ be some sets of pomsets over Σ and Σ' respectively such that there exists a refinement $\pi : \Sigma' \to \Sigma$ from $\mathcal{L}$ onto $\mathcal{L}'$. If $\mathcal{L}'$ is the language of a cellular unambiguous SMS $\mathcal{S}'$ then there exists a* singular *unambiguous SMS $\mathcal{S}$ that accepts $\mathcal{L}$. Moreover $\mathcal{S}$ and $\mathcal{S}'$ share the same configurations.*

Proof sketch. We consider an SMS $\mathcal{S}$ over Σ with the same processes, local states, registers and data as $\mathcal{S}'$. In particular $\mathcal{S}$ is singular since each process has a single local state. The SMS $\mathcal{S}$ is defined by the two following requirements:

- For each $a \in \Sigma$, we put $\mathrm{Loc}(a) = \pi^{-1}(a)$. Observe here that $\mathrm{Loc}(a) \subseteq \Sigma' = \mathcal{I}$.
- $(\nu, \iota | a, a, \nu', \iota | a) \in \Delta$ if there is $x \in \mathrm{Loc}(a)$ such that $(\nu, \iota | x, x, \nu', \iota | x) \in \Delta'$.

Note that $\mathcal{S}$ and $\mathcal{S}'$ share a common set of configurations Q: If $\mathcal{S}'$ is finite then $\mathcal{S}$ is finite, too. We claim that the shared-memory system $\mathcal{S}$ is unambiguous and recognizes $\mathcal{L}$. ∎

By Proposition 2.4 and Lemma 3.1, the set of pomsets accepted by an unambiguous shared-memory system is media-bounded and MSO-definable. Our main result below asserts the converse property.

THEOREM 3.4. *Let $\mathcal{L}$ be a set of pomsets. The following conditions are equivalent:*
   (i) *$\mathcal{L}$ is recognized by some unambiguous shared-memory system.*
   (ii) *$\mathcal{L}$ is media-bounded and definable in MSO logic.*
   (iii) *there exists a refinement from $\mathcal{L}$ onto a regular set of Mazurkiewicz traces.*

Proof. Let $\mathcal{L}$ be an MSO-definable and media-bounded set of pomsets over Σ. By Theorems 2.11 and 3.2, there exists a refinement $\pi : \Gamma \to \Sigma$ from $\mathcal{L}$ onto an MSO-definable set of Mazurkiewicz traces $\mathcal{L}' \subseteq \mathbb{M}(\Gamma, \|)$. By [23], $\mathcal{L}'$ is a regular set of Mazurkiewicz traces. It follows from [8] that $\mathcal{L}'$ is accepted by some unambiguous asynchronous cellular automaton $\mathcal{S}°$ over $\Gamma, \mathcal{I}, \mathcal{R}$ and $\mathcal{D}$ (Def. 1.4). By Lemma 3.3, $\mathcal{L}$ is recognized by some unambiguous SMS. ∎

3.3   Comparisons with Generalized Asynchronous Cellular Automata

The model of shared-memory systems we have considered in this paper is similar to the notion of *generalized* asynchronous cellular automata investigated in [11]. This model can be identified with a shared-memory system such that $\mathcal{I} = \mathcal{R}$, $S_i = \{\iota_i\}$ for each process $i \in \mathcal{I}$ (that is, each process owns a register whose value describes its current state), $\mathrm{Loc}(a)$ is a singleton for each action a (so processes never synchronize) and moreover $(\nu, \iota_a, a, \nu', \iota_a) \in \Delta$ implies that $\mathrm{dom}(\nu') = \{\mathrm{Loc}(a)\}$ which means that each process writes only in its own register. Then the set of transition rules of each action a can be represented by a mapping $\delta_a : \mathcal{V} \to 2^{\mathcal{D}}$. It is easy to see that these shared-memory systems form another generalization of asynchronous cellular automata.

Let us consider now the set $\mathcal{L}_{vbl}$ of all very broken ladders: A *very broken ladder* is a pomset t that consists of a chain of $n \geqslant 1$ production events and a single consumption event that covers one production event (Fig. 3). It is clear that $\mathcal{L}_{vbl}$ is not

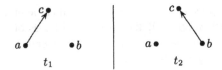

Fig. 5. A non-consistent set of pomsets

media-bounded. Consequently it is not recognized by some unambiguous SMS. Observe here that the notion of of media-bound is different from the prime-bound introduced in [15] (see also [11, Section 6]) because $\mathcal{L}_{vbl}$ is prime-bounded but not media-bounded. Moreover $\mathcal{L}_{vbl}$ is accepted by some (unambiguous) generalized asynchronous cellular automaton *according to any of the semantics studied in* [11]. Thus the acceptance condition adopted in Def. 1.2 differs from the one investigated in [11].

Similarly to [11], it is possible to restrict the logic to existential formulae and get similar results. This means that MSO is equivalent to EMSO for media-bounded sets of pomsets. This follows actually from Theorem 3.2: If $\mathcal{L}$ is media-bounded and MSO-definable then the set of Mazurkiewicz traces $\beta(\mathcal{L})$ is MSO-definable, hence EMSO-definable; since $\mathcal{L} = \pi \circ \beta(\mathcal{L})$, $\mathcal{L}$ is EMSO-definable, too.

Conclusion

As opposed to [19], we have considered in this paper non-deterministic shared-memory systems: A shared-memory system is called *deterministic* if for all actions $a \in \Sigma$ and all reachable configurations $q = (\chi, s)$ there is at most one transition rule $\rho \in \Delta$ such that $a_\rho = a$, $\nu_\rho = \chi|R_\rho$ and $s_\rho = s|a$. Intuitively this means that there is at most one rule that allows an occurrence of action a in each configuration. Any deterministic SMS is unambiguous but the converse fails. Actually determinism reduces the expressive power of shared-memory systems: We showed in [19] that *deterministic* finite shared-memory systems accept precisely the set of pomsets that are regular and consistent, two notions borrowed from [3]. A set of pomsets $\mathcal{L}$ is called consistent if any two distinct prefixes of pomsets from $\mathcal{L}$ have disjoint sets of linear extensions. In particular any set of Mazurkiewicz traces is consistent. As observed in [13] the class of regular consistent sets of pomsets is closely related to the semantics of stably concurrent automata [6]. By Theorem 3.4 any finite non-consistent set of pomsets is accepted by some unambiguous SMS, but no deterministic one. For instance the set of pomsets $\mathcal{L} = \{t_1, t_2\}$ depicted in Figure 5 is recognized by some unambiguous non-deterministic SMS.

Given an MSO-definable and media-bounded set of pomsets $\mathcal{L}$, the new refinement presented in Section 2 yields a Mazurkiewicz trace language $\beta(\mathcal{L})$. We established in Theorem 3.2 that $\beta(\mathcal{L})$ is MSO-definable, too. This theorem is the key ingredient of a new proof of the main result from [3] which asserts that any regular consistent set of pomsets can be refined onto a regular set of Mazurkiewicz traces. This result was proved to be crucial for applications in several frameworks such that concurrent automata and event structures [13,18] and also deterministic shared-memory systems [19]. Since we deal in this paper with non-consistent sets of pomsets, we had to establish a kind of generalization of that powerful result.

Message Sequence Charts (MSCs) are a popular model often used for the documentation of telecommunication protocols. An MSC gives a graphical description of message exchanges between processes in the form of a labeled partial order. For few years several papers have developped the theory of MSCs [5,12,14] at some abstract and somewhat simplified level. In particular all these studies rely on the assumption that channels are reliable. As a consequence the MSC languages accepted by message-passing systems form a consistent set of pomsets. On the other hand Kuske's lemma [14] allows us to refine any regular set of message sequence charts onto a regular set of Mazurkiewicz traces. This rather simple observation turned out to be quite powerful to transfer results from the theory of Mazurkiewicz traces to the framework of MSC languages [4,14,17]. However this connection fails if we consider MSCs with message loss. Still the new refinement construction presented in Section 2 applies in that setting. For that reason we are applying at present this refinement construction in order to develop a language theory for bounded lossy-channel systems.

References

1. Alur, R., Grosu, R.: Shared Variables Interaction Diagrams. In: 16th IEEE Int. Conf. on Automated Software Engineering, pp. 281–288. IEEE Computer Society, Los Alamitos (2001)
2. Arnold, A.: Rational ω-languages are non-ambiguous. TCS 26, 221–223 (1983)
3. Arnold, A.: An extension of the notion of traces and asynchronous automata. RAIRO, Theoretical Informatics and Applications, Gauthiers-Villars 25, 355–393 (1991)
4. Baudru, N., Morin, R.: Safe Implementability of Regular Message Sequence Charts Specifications. In: Proc. of the ACIS 4th Int. Conf. SNDP, pp. 210–217 (2003)
5. Bollig, B., Leucker, M.: Message-passing automata are expressively equivalent to EMSO logic. TCS 358, 150–172 (2006)
6. Bracho, F., Droste, M., Kuske, D.: Representations of computations in concurrent automata by dependence orders. TCS 174, 67–96 (1997)
7. Carton, O., Michel, M.: Unambiguous Büchi automata. TCS 297, 37–81 (2003)
8. Cori, R., Métivier, Y., Zielonka, W.: Asynchronous mappings and asynchronous cellular automata. I&C 106, 159–202 (1993)
9. Diekert, V., Rozenberg, G.: The Book of Traces. World Scientific, Singapore (1995)
10. Diekert, V., Métivier, Y.: Partial Commutation and Traces. In: Rozenberg, G., Salomaa, A. (eds.) Handbook of Formal Languages, vol. 3, pp. 457–534 (1997)
11. Droste, M., Gastin, P., Kuske, D.: Asynchronous cellular automata for pomsets. TCS 247, 1–38 (2000)
12. Henriksen, J.G., Mukund, M., Narayan Kumar, K., Sohoni, M., Thiagarajan, P.S.: A Theory of Regular MSC Languages. I&C 202, 1–38 (2005)
13. Husson, J.-F., Morin, R.: On Recognizable Stable Trace Languages. In: Tiuryn, J. (ed.) FOSSACS 2000. LNCS, vol. 1784, pp. 177–191. Springer, Heidelberg (2000)
14. Kuske, D.: Regular sets of infinite message sequence charts. I&C 187, 80–109 (2003)
15. Kuske, D.: Asynchronous cellular automata and asynchronous automata for pomsets. In: Sangiorgi, D., de Simone, R. (eds.) CONCUR 1998. LNCS, vol. 1466, pp. 517–532. Springer, Heidelberg (1998)
16. Lamport, L.: Time, Clocks, and the Ordering of Events in a Distributed System. Commun. ACM 21, 558–565 (1978)
17. Morin, R.: On Regular Message Sequence Chart Languages and Relationships to Mazurkiewicz Trace Theory. In: Honsell, F., Miculan, M. (eds.) FOSSACS 2001. LNCS, vol. 2030, pp. 332–346. Springer, Heidelberg (2001)

18. Morin, R.: Concurrent Automata vs. Asynchronous Systems. In: Jedrzejowicz, J., Szepietowski, A. (eds.) MFCS 2005. LNCS, vol. 3618, pp. 686–698. Springer, Heidelberg (2005)
19. Morin, R.: Semantics of Deterministic Shared-Memory Systems. In: CONCUR 2008. LNCS, vol. 5201, pp. 36–51. Springer, Heidelberg (2008)
20. Nielsen, M., Plotkin, G., Winskel, G.: Petri nets, events structures and domains, part 1. TCS 13, 85–108 (1981)
21. Pratt, V.: Modelling concurrency with partial orders. International Journal of Parallel Programming 15, 33–71 (1986)
22. Szpilrajn, E.: Sur l'extension de l'ordre partiel. Fund. Math. 16, 386–389 (1930)
23. Thomas, W.: On logical definability of trace languages Technical University of Munich, report. TUM-I9002, 172–182 (1990)
24. Zielonka, W.: Notes on finite asynchronous automata. RAIRO, Theoretical Informatics and Applications 21, 99–135 (1987)
25. Zielonka, W.: Safe executions of recognizable trace languages by asynchronous automata. In: Meyer, A.R., Taitslin, M.A. (eds.) Logic at Botik 1989. LNCS, vol. 363, pp. 278–289. Springer, Heidelberg (1989)

Complexity of Topological Properties of Regular ω-Languages

Victor L. Selivanov[1,*] and Klaus W. Wagner[2]

[1]A.P. Ershov Institute of Informatics Systems, Siberian Division of the Russian
Academy of Sciences
vseliv@nspu.ru
[2]Institut für Informatik, Julius-Maximilians-Universität Würzburg
wagner@informatik.uni-wuerzburg.de

Abstract. We determine the complexity of topological properties of regular ω-languages (i.e., classes of ω-languages closed under inverse continuous functions). We show that they are typically NL-complete (PSPACE-complete) for the deterministic Muller, Mostowski and Büchi automata (respectively, for the nondeterministic Rabin, Muller, Mostowski and Büchi automata). For the deterministic Rabin and Streett automata and for the nondeterministic Streett automata upper and lower complexity bounds for the topological properties are established.

1 Introduction

The study of decidability and complexity questions for properties of regular languages is a central research topic in automata theory. Its importance stems from the fact that finite automata are fundamental to many branches of computer science, e.g., databases, operating systems, verification, and hardware and software design.

For properties of regular languages, many decidablity and complexity results were obtained. However, for properties of regular ω-languages, only a couple of facts about their complexity (recalled below) seems to be known so far.

In this paper we determine the complexity of topological properties of regular ω-languages given by different types of ω-automata. Topological properties are classes of ω-languages which are closed under inverse continuous functions. Defining the *Wadge reducibility* $\leq_\mathrm{w}$ on the Cantor space as the many-one reducibility via continuous functions, the topological properties are the classes of ω-languages which are closed under Wadge reducibility. The classes $\{L' \mid L' \leq_\mathrm{w} L\}$ for ω-languages L are called *elementary* topological properties; every topological property is the union of elementary topological properties. Obviously, there is a bijection between the elementary topological properties and the Wadge degrees.

To explain our results, let us recall some facts from [Wag79] where the Wadge degrees of regular ω-languages (over any alphabet A with at least two symbols) were determined, in particular the following results were established:

---

* Supported by DFG Mercator program and by RFBR grant 07-01-00543a.

M. Ito and M. Toyama (Eds.): DLT 2008, LNCS 5257, pp. 529–542, 2008.

1. The structure $(\mathcal{R}; \leq_w)$ of regular ω-languages under the Wadge reducibility is almost well-ordered with order type ω^ω, i.e., for each ordinal $\alpha < \omega^\omega$ there is a regular ω-language $A_\alpha \in \mathcal{R}$, such that $A_\alpha <_w A_\alpha \oplus \overline{A}_\alpha <_w A_\beta$ for $\alpha < \beta < \omega^\omega$, and any regular set is Wadge-equivalent to one of the sets A_α, $\overline{A}_\alpha$, and $A_\alpha \oplus \overline{A}_\alpha$ where $\alpha < \omega^\omega$ (here $\oplus$ denotes the marked union).

2. The elementary topological properties of regular ω-languages are $\mathcal{R}_\alpha =_{\text{def}}$ $\{L \mid L \leq_w A_\alpha\}$, co-$\mathcal{R}_\alpha =_{\text{def}} \{L \mid L \leq_w \overline{A}_\alpha\}$, and $\mathcal{R}_{\alpha+1} \cap$ co-$\mathcal{R}_{\alpha+1} = \{L \mid L \leq_w A_\alpha \oplus \overline{A}_\alpha\}$, for $\alpha < \omega^\omega$.

3. The Wadge-degrees of regular ω-languages are $\mathcal{R}'_\alpha =_{\text{def}} \{L \mid L \equiv_w A_\alpha\} = \mathcal{R}_\alpha \setminus$ co-$\mathcal{R}_\alpha$, co-$\mathcal{R}'_\alpha =_{\text{def}} \{L \mid L \equiv_w \overline{A}_\alpha\} = $ co-$\mathcal{R}_\alpha \setminus \mathcal{R}_\alpha$, and $\widetilde{\mathcal{R}}_\alpha =_{\text{def}} (\mathcal{R}_{\alpha+1} \cap \text{co-}\mathcal{R}_{\alpha+1}) \setminus (\mathcal{R}_\alpha \cup \text{co-}\mathcal{R}_\alpha)$, for $\alpha < \omega^\omega$.

4. All elementary topological properties of regular ω-languages and all Wadge-degrees of regular ω-languages are decidable (the regular ω-languages given by deterministic Muller automata).

A natural question is to determine the complexity of the classes listed under 2. for different popular types of ω-automata such as deterministic or nondeterministic Büchi, Muller, Rabin, Streett and Mostowski (or parity) automata. To our knowledge, only a couple of results in this direction were established so far.

Theorem 1.   *1.* **[KPB95, WY95]** *For every* $\alpha < \omega^\omega$, *given a deterministic Muller automaton* $\mathcal{M}$, *one can decide in polynomial time whether* $L_\omega(\mathcal{M}) \in \mathcal{R}_\alpha$.

2. **[SVW87]** *The problem of deciding, given a nondeterministic Büchi automaton* $\mathcal{M}$ *with input alphabet* A, *whether* $L_\omega(\mathcal{M}) = A^\omega$, *is PSPACE-complete.*

3. **[SVW87]** *The problem of deciding, given a nondeterministic Büchi automaton* $\mathcal{M}$, *whether* $L_\omega(\mathcal{M}) = \emptyset$, *is NL-complete.*

The Statements 2 and 3 above are related to the classes $\mathcal{R}_\alpha$ because $\mathcal{R}_0$ coincides with $\{\emptyset\} = \{L \mid L \leq_w \emptyset\}$ and the dual class co-$(\mathcal{R}_0)$ for $\mathcal{R}_0$ coincides with $\{A^\omega\} = \{L \mid L \leq_w A^\omega\}$.

We will determine the complexity of all elementary topological properties of regular ω-languages and all Wadge-degrees of regular ω-languages w.r.t. the

automata type	$\mathcal{C}$	deterministic		nondeterministic	
		lower bound	upper bound	lower bound	upper bound
Muller	$= \mathcal{R}_0$	NL	NL	NL	NL
	$\neq \mathcal{R}_0$	NL	NL	PSPACE	PSPACE
Rabin	$= \mathcal{R}_0$	NL	NL	NL	NL
	$\neq \mathcal{R}_0$	P	P^{NP}	PSPACE	PSPACE
Streett	$= \text{co-}\mathcal{R}_0$	NL	NL	P	co-NP
	$\neq \text{co-}\mathcal{R}_0$	P	P^{NP}	PSPACE	EXPSPACE
Mostowski	$= \mathcal{R}_0$	NL	NL	NL	NL
	$\neq \mathcal{R}_0$	NL	NL	PSPACE	PSPACE
Büchi	$= \mathcal{R}_0$	NL	NL	NL	NL
	$\neq \mathcal{R}_0$	NL	NL	PSPACE	PSPACE

mentioned types of ω-automata. Our results are represented in the following table. Let $\mathcal{C}$ be an elementary topological property of regular ω-languages, or a Wadge-degree of regular ω-languages. For deterministic Büchi automata this is restricted to $\mathcal{C} \subseteq \mathcal{R}_\omega$ because they can accept only such regular ω-languages from $\mathcal{R}_\omega$. The lower bounds mean hardness for the complexity class in question.

Due to the page restriction we had to omit most of the proofs in this extended abstract.

2 ω-Languages and Topology

For a set S, let $P(S)$ be the class of subsets of S. For a class $\mathcal{C} \subseteq P(S)$, let co-$\mathcal{C}$ be the dual class $\{\overline{C} \mid C \in \mathcal{C}\}$, and let $\mathrm{BC}(\mathcal{C})$ be the Boolean closure of $\mathcal{C}$.

Fix a finite alphabet A containing more than one symbol. Let A^* and A^ω denote respectively the sets of all words and of all ω-words (i.e. sequences $\alpha :$ $\mathbb{N} \to A$) over A. The empty word is denoted by ε. Let $A^+ = A^* \setminus \{\varepsilon\}$ and $A^{\le\omega} = A^* \cup A^\omega$. For $n \in \mathbb{N}$, let A^n be the set of words of length n. Note that all our results are formulated for arbitrary fixed alphabet A.

The set A^ω carries the Cantor topology with the open sets $W \cdot A^\omega$, where $W \subseteq X^*$. Let $\mathcal{B}$ denote the class of Borel subsets of A^ω, i.e. the least class containing the open sets and closed under complement and countable union. Borel sets are organized in a hierarchy the lowest levels of which are as follows: G and F are the classes of open and closed sets, respectively; G_δ (F_σ) is the class of countable intersections (unions) of open (resp. closed) sets; $G_{\delta\sigma}$ ($F_{\sigma\delta}$) is the class of countable unions (intersections) of G_δ- (resp. of F_σ-) sets, and so on. In the modern notation of hierarchy theory, $\boldsymbol{\Sigma}^0_1 = G$, $\boldsymbol{\Sigma}^0_2 = F_\sigma$, $\boldsymbol{\Sigma}^0_3 = G_{\delta\sigma}$, $\boldsymbol{\Sigma}^0_4 = F_{\sigma\delta\sigma}$ and so on, $\boldsymbol{\Pi}^0_n =_{\mathrm{def}}$ co-$\boldsymbol{\Sigma}^0_n$ is the dual class for $\boldsymbol{\Sigma}^0_n$, and $\boldsymbol{\Delta}^0_n = \boldsymbol{\Sigma}^0_n \cap \boldsymbol{\Pi}^0_n$. The sequence $\{\boldsymbol{\Sigma}^0_{n+1}\}_{n<\omega}$ is known as *the finite Borel hierarchy*. It may be in a natural way extended on all countable ordinals. The resulting sequence called *the Borel hierarchy* exhausts the class $\mathcal{B}$. For any $n > 0$, the class $\boldsymbol{\Sigma}^0_n$ contains $\emptyset, A^\omega$ and is closed under countable unions and finite intersections, while the class $\boldsymbol{\Delta}^0_n$ is closed under complement and finite unions. For any $n > 0$, we have the strict inclusions $\boldsymbol{\Sigma}^0_n \cup \boldsymbol{\Pi}^0_n \subset \mathrm{BC}(\boldsymbol{\Sigma}^0_n) \subset \boldsymbol{\Delta}^0_{n+1}$.

For $L, K \subseteq A^\omega$, L is said to be *Wadge reducible* to K (in symbols $L \le_{\mathrm{w}} K$), if $L = g^{-1}(K)$ for some continuous function $g : A^\omega \to A^\omega$. The Wadge reducibility on $P(A^\omega)$ is a preorder. By $\equiv_{\mathrm{w}}$ we denote the induced equivalence relation which gives rise to the corresponding quotient partial ordering. Following a well established jargon, we call this ordering the structure of Wadge degrees [Wa72, Wa84]. The operation $L \oplus K = \{0 \cdot \xi \cup i \cdot \eta \mid 0 < i < k, \xi \in L, \eta \in K\}$ on subsets of A^ω_k induces the operation of least upper bound in the structures of Wadge degrees. Any level of the Borel hierarchy is closed under the Wadge reducibility in the sense that every set reducible to a set in the level is itself in that level. Moreover, every $\boldsymbol{\Sigma}$-level $\mathcal{C}$ (and also every $\boldsymbol{\Pi}$-level) of the Borel hierarchy has a Wadge complete set C which means that $\mathcal{C} = \{L \mid L \le_{\mathrm{w}} C\}$. For additional information on ω-languages see e.g. [Sta97, Th90, Th96].

3   Finite Automata Accepting ω-Languages

Finite automata may accept ω-languages in different ways. Here we briefly recall some acceptance modes and corresponding facts that will be used later.

By *deterministic pre-automaton* (over A) we mean a triple $\mathcal{M} = (S, A, \delta)$ consisting of a finite non-empty set S of states, an input alphabet A and a transition function $\delta : S \times A \to S$. The transition function is naturally extended to the function $\delta : S \times A^* \to S$ defined by induction $\delta(s, \varepsilon) =_{\text{def}} s$ and $\delta(s, xa) =_{\text{def}} \delta(\delta(s, x), a)$ where $x \in A^*$ and $a \in A$. For input sequences from A^ω define the function $\delta : S \times A^\omega \to S^\omega$ by $\delta(s, \xi)(n) = \delta(s, \xi[n])$ where $\xi[n]$ is the prefix of ξ of length n.

Nondeterministic pre-automata are defined in the same way only now the transition function is of the form $\delta : S \times A \to P(S)$ which is extended to the function $\delta : S \times A^* \to P(S)$ by $\delta(s, \varepsilon) =_{\text{def}} \{s\}$ and $\delta(s, xa) =_{\text{def}} \bigcup_{s' \in \delta(s,x)} \delta(s', a)$ where $x \in A^*$ and $a \in A$. For input sequences from A^ω define the function $\delta : S \times A^\omega \to P(S^\omega)$ by $\delta(s, \xi) =_{\text{def}} \{\eta \mid \eta(0) = s \wedge \forall i(\eta(i+1) \in \delta(\eta(i), \xi(i)))\}$.

Unlike automata on finite words, for automata on ω-words the acceptance conditions were defined in different way by different authors. As a result, there are several notions of automata accepting ω-words (which we generally call ω-automata). For $\eta \in S^\omega$, let $\inf(\eta)$ be the set of all $s \in S$ which occur infinitely often in η.

Let (S, A, δ) be a deterministic pre-automaton and $s_0 \in S$. The quintuple $\mathcal{M} = (S, A, \delta, s_0, \mathcal{F})$ is called a deterministic

- *Büchi automaton* if $\mathcal{F} \subseteq S$; it recognizes the set
  $L_\omega(\mathcal{M}) = \{\xi \in A^\omega \mid \inf(\delta(s_0, \xi)) \cap F \neq \emptyset\}$,
- *Muller automaton* if $\mathcal{F} \subseteq P(S)$; it recognizes the set
  $L_\omega(\mathcal{M}) = \{\xi \in A^\omega \mid \inf(\delta(s_0, \xi)) \in \mathcal{F}\}$,
- *Rabin automaton* if $\mathcal{F} \subseteq P(S)^2$; it recognizes the set
  $L_\omega(\mathcal{M}) = \{\xi \in A^\omega \mid \exists((E, F) \in \mathcal{F})(\inf(\delta(s_0, \xi)) \cap E = \emptyset \wedge \inf(\delta(s_0, \xi)) \cap F \neq \emptyset)\}$,
- *Mostowski automaton* (known also as *Rabin chain automaton* or *parity automaton*) if it is a deterministic Rabin automaton such that $\mathcal{F} = \{(E_1, F_1), (E_2, F_2), \ldots, (E_m, F_m)\}$ satisfies $E_1 \subseteq F_1 \subseteq E_2 \subseteq F_2 \subseteq \cdots \subseteq E_m \subseteq F_m$,
- *Streett automaton* if $\mathcal{F} \subseteq P(S)^2$; it recognizes the set
  $L'_\omega(\mathcal{M}) = \{\xi \in A^\omega \mid \forall((E, F) \in \mathcal{F})(\inf(\delta(s_0, \xi)) \cap E \neq \emptyset \vee \inf(\delta(s_0, \xi)) \cap F = \emptyset)\}$.

Deterministic Streett automata and deterministic Rabin automata are formally the same objects, and $L'_\omega(\mathcal{M}) = A^\omega \setminus L_\omega(\mathcal{M})$ for every deterministic Rabin automaton $\mathcal{M}$.

The nondeterministic versions of the introduced types of automata are defined in the usual way: We start with a nondeterministic pre-automaton and instead of the acceptance condition $H(\inf(\delta(s_0, \xi)))$ we use the acceptance condition $\exists \eta(\eta \in \delta(s_0, \xi) \wedge H(\inf(\eta)))$, i.e. there is an infinite run such that the corresponding sequence of states satisfies the acceptance condition.

Theorem 2. *For any ω-language $L \subseteq A^\omega$ the following statements are equivalent:*

1. *L is recognized by a deterministic Muller (Rabin, Mostowski, Streett) automaton.*
2. *L is recognized by a nondeterministic Büchi (Muller, Rabin, Mostowski, Streett) automaton.*
3. *L is a finite union of sets $U \cdot V^\omega$ where $U \subseteq A^*$ and $V \subseteq A^+$ are regular languages.*

The ω-languages satisfying the assertions above are called *regular ω-languages*. Let $\mathcal{R}$ be the class of regular ω-languages.

Theorem 3. *1. $\mathcal{R} \subset \mathrm{BC}(\Sigma_2^0)$.*
2. [La69, SW74] The deterministic Büchi automata accept exactly the regular Π_2^0-sets.

For the above defined types of automata we introduce the abbreviations B, M, R, P, and S for Büchi, Muller, Rabin, Mostowski (parity), and Streett automata, resp., and D and N stand for deterministic and nondeterministic, resp. In this way, for example, NB is the name for nondeterministic Büchi automata. Let $\mathcal{C}$ be a class of ω-languages, and let T be a type of automata. We consider the

Problem $(\mathcal{C})_T$:
Given: An automaton $\mathcal{M}$ of type T.
Question: Does $\mathcal{M}$ accept an ω-language in $\mathcal{C}$?

Because of the duality of the deterministic Rabin acceptance and the deterministic Streett acceptance we have (where $\equiv_m^{\log}$ denotes the many-one logspace equivalence)

Proposition 1. *If $\mathcal{C}$ is a class of ω-languages then $(\mathcal{C})_{\mathrm{DS}} \equiv_m^{\log} (\text{co-}\mathcal{C})_{\mathrm{DR}}$.*

By Theorem 2 all the introduced classes of ω-automata (besides deterministic Büchi automata) are equivalent in the sense that they recognize the same ω-languages. Moreover, the well known proofs of these equivalences are effective, i.e. from a given automaton of some type one can compute an equivalent automaton of any other type. When one is interested in complexity considerations (as we are here), the computational resources needed for finding the equivalent automaton and its size become important. For types T, T' of ω-automata we write $T \leq_m^P T'$ if there exists a polynomial time computable function f such that, for every automaton $\mathcal{M}$ of type T, the result $f(\mathcal{M})$ is an automaton of type T' which accepts the same ω-language as $\mathcal{M}$. The following relationship to decision problems is obvious:

Proposition 2. *Let T and T' be two types of ω-automata, and let $\mathcal{C}$ be a class of ω-languages. Then $T \leq_m^P T'$ implies $(\mathcal{C})_T \leq_m^P (\mathcal{C})_{T'}$.*

Unfortunately, some of the well known reductions in Theorem 2 do not work in polynomial time. For some cases one can even prove that this is not possible. In

[Sa88] an overview on possibility or impossibility of polynomial time reductions between different types of ω-automata is given (for more recent papers on this see also [Lo99, Ya06]).

Theorem 4. [Sa88] *The following figure represents some results on polynomial time reductions between different types of ω-automata. A solid line means that there exists a polynomial time reduction from the notion below to the notion above. A dotted arc means that polynomial time reduction in this direction is not proved and not disproved. Moreover, there are no further polynomial time reductions between these types of ω-automata which do not already follow from the solid lines and dotted arcs.*

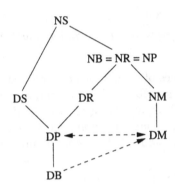

4   Topological Properties of Regular ω-Languages

Topological properties are classes of ω-languages which are closed under Wadge reducibility, i.e., under inverse continuous functions. Theses are just the classes $\{L \mid \exists L'(L' \in \mathcal{C} \wedge L \leq_w L')\}$ where $\mathcal{C} \subseteq P(A^\omega)$. We are interested in topological properties of *regular ω-languages*, these are just the classes $\widehat{\mathcal{C}} =_{\mathrm{def}} \{L \mid \exists L'(L' \in \mathcal{C} \wedge L \leq_w L')\} \cap \mathcal{R}$ where $\mathcal{C} \subseteq \mathcal{R}$. If $[L]_w$ is the $\equiv_w$-equivalence class which includes $L \subseteq A^\omega$ (the *Wadge degree* of L) then we have $\widehat{\mathcal{C}} = \bigcup_{L \in \mathcal{C}} \widehat{[L]_w}$ for every $\mathcal{C} \subseteq \mathcal{R}$. That means: we know all topological properties of regular ω-languages if we know all *elementary topological properties* $\widehat{[L]_w}$ of regular ω-languages L. Furthermore, we know these, if we know all *regular Wadge degrees* $[L]_w \cap \mathcal{R}$. We define the family $\mathcal{T} =_{\mathrm{def}} \{[L]_w \cap \mathcal{R} \mid L \in \mathcal{R}\}$ of all regular Wadge degrees and the family $\widehat{\mathcal{T}} =_{\mathrm{def}} \{\widehat{[L]_w} \mid L \in \mathcal{R}\}$ of all elementary topological properties of regular ω-languages.

These families of classes were completely characterized in [Wag79] by some invariants of deterministic Muller automata. We recall in this section the definitions and results from this paper which we need here. In what follows let $\mathcal{M} = (S, A, \delta, s_0, \mathcal{F})$ be a deterministic Muller automaton.

A subset $S' \subseteq S$ is called a *loop* if there exist an $s \in S$ and $x, z \in A^*$ such that $\delta(s_0, x) = \delta(s, z) = s$ and $\{\delta(s, y) \mid y$ is an initial part of $z\} = S'$. A loop

S_2 is *reachable* from a loop S_1 if there exists an $s \in S_1$ and an $x \in A^*$ such that $\delta(s, x) \in S_2$.

For $m \geq 1$, an m^+*chain* is a sequence $(S_1, S_2, \ldots, S_m)$ of loops such that $S_1 \subset S_2 \subset \cdots \subset S_m$, $S_1, S_3, \cdots \in \mathcal{F}$, and $S_2, S_4, \cdots \in P(S) \setminus \mathcal{F}$. An m^-*chain* is a sequence $(S_1, S_2, \ldots, S_m)$ of loops such that $S_1 \subset S_2 \subset \cdots \subset S_m$, $S_1, S_3, \cdots \in P(S) \setminus \mathcal{F}$, and $S_2, S_4, \cdots \in \mathcal{F}$.

For $m, n \geq 1$, an $(m, n)^+$*superchain* is a sequence $(T_1, T_2, \ldots, T_n)$ such that $T_1, T_3, \ldots$ are m^+chains, $T_2, T_4, \ldots$ are m^-chains, and the loops from T_{i+1} are reachable from the loops of T_i for $i = 1, 2, \ldots, n-1$. An $(m, n)^-$*superchain* is a sequence $(T_1, T_2, \ldots, T_n)$ such that $T_1, T_3, \ldots$ are m^-chains, $T_2, T_4, \ldots$ are m^+chains, and the loops from T_{i+1} are reachable from the loops from T_i for $i = 1, 2, \ldots, n-1$.

Now define the characteristics

$\mathrm{m}^+(\mathcal{M}) =_{\mathrm{def}} \max\{m \mid \text{there exists an } m^+\text{chain in } \mathcal{M}\}$,
$\mathrm{m}^-(\mathcal{M}) =_{\mathrm{def}} \max\{m \mid \text{there exists an } m^-\text{chain in } \mathcal{M}\}$,
$\mathrm{m}(\mathcal{M}) \;\;=_{\mathrm{def}} \max\{\mathrm{m}^+(\mathcal{M}), \mathrm{m}^-(\mathcal{M})\}\}$,
$\mathrm{n}^+(\mathcal{M}) =_{\mathrm{def}} \max\{n \mid \text{there exists an } (\mathrm{m}(\mathcal{M}), n)^+\text{superchain in } \mathcal{M}\}$,
$\mathrm{n}^-(\mathcal{M}) =_{\mathrm{def}} \max\{n \mid \text{there exists an } (\mathrm{m}(\mathcal{M}), n)^-\text{superchain in } \mathcal{M}\}$, and
$\mathrm{n}(\mathcal{M}) \;\;=_{\mathrm{def}} \max\{\mathrm{n}^+(\mathcal{M}), \mathrm{n}^-(\mathcal{M})\}$.

The characteristics $\mathrm{m}^+(\mathcal{M})$, $\mathrm{m}^-(\mathcal{M})$, $\mathrm{n}^+(\mathcal{M})$, and $\mathrm{n}^-(\mathcal{M})$, are invariants of all automata accepting the same language:

Theorem 5. *For deterministic Muller automata $\mathcal{M}, \mathcal{M}'$, if $L_\omega(\mathcal{M}) = L_\omega(\mathcal{M}')$ then $\mathrm{m}^+(\mathcal{M}) = \mathrm{m}^+(\mathcal{M}')$, $\mathrm{m}^-(\mathcal{M}) = \mathrm{m}^-(\mathcal{M}')$, $\mathrm{n}^+(\mathcal{M}) = \mathrm{n}^+(\mathcal{M}')$, and $\mathrm{n}^-(\mathcal{M}) = \mathrm{n}^-(\mathcal{M}')$.*

Theorem 5 justifies the following definition. Let L be an ω-language and let $\mathcal{M}$ be a deterministic Muller automaton such that $L_\omega(\mathcal{M}) = L$. Then $\mathrm{m}^+(L) =_{\mathrm{def}} \mathrm{m}^+(\mathcal{M})$, $\mathrm{m}^-(L) =_{\mathrm{def}} \mathrm{m}^-(\mathcal{M})$, $\mathrm{n}^+(L) =_{\mathrm{def}} \mathrm{n}^+(\mathcal{M})$, and $\mathrm{n}^-(L) =_{\mathrm{def}} \mathrm{n}^-(\mathcal{M})$.

For $m, n \geq 1$, define the classes

$C_m^n =_{\mathrm{def}} \{L \mid \mathrm{m}(L) = m \wedge \mathrm{n}^+(L) = n-1 \wedge \mathrm{n}^-(L) = n\}$,
$D_m^n =_{\mathrm{def}} \{L \mid \mathrm{m}(L) = m \wedge \mathrm{n}^+(L) = n \wedge \mathrm{n}^-(L) = n-1\}$,
$E_m^n =_{\mathrm{def}} \{L \mid \mathrm{m}(L) = m \wedge \mathrm{n}^+(L) = \mathrm{n}^-(L) = n\}$,
$\widehat{C}_m^n =_{\mathrm{def}} \{L \mid \mathrm{m}(L) < m \vee (\mathrm{m}(L) = m \wedge \mathrm{n}^+(L) < n)\}$,
$\widehat{D}_m^n =_{\mathrm{def}} \{L \mid \mathrm{m}(L) < m \vee (\mathrm{m}(L) = m \wedge \mathrm{n}^-(L) < n)\}$, and
$\widehat{E}_m^n =_{\mathrm{def}} \{L \mid \mathrm{m}(L) < m \vee (\mathrm{m}(L) = m \wedge \mathrm{n}(L) \leq n)\}$.

Some important relations between these classes are given by the following theorem.

Theorem 6. *Let $m, n \geq 1$.*

1. $D_m^n = \mathrm{co}\text{-}C_m^n$ and $\widehat{D}_m^n = \mathrm{co}\text{-}\widehat{C}_m^n$.
2. $\widehat{C}_m^n \cup \widehat{D}_m^n \subset \widehat{E}_m^n = \widehat{C}_m^{n+1} \cap \widehat{D}_m^{n+1}$.
3. $\widehat{C}_{m+1}^1 \cap \widehat{D}_{m+1}^1 = \bigcup_{n \geq 1} \widehat{C}_m^n = \bigcup_{n \geq 1} \widehat{D}_m^n = \bigcup_{n \geq 1} \widehat{E}_m^n = \{L \mid \mathrm{m}(L) \leq m\}$.
4. *The classes C_m^n, D_m^n, and E_m^n form a partition of the class of regular ω-languages.*
5. $C_m^n = \widehat{C}_m^n \setminus \widehat{D}_m^n$, $D_m^n = \widehat{D}_m^n \setminus \widehat{C}_m^n$, and $E_m^n = \widehat{E}_m^n \setminus (\widehat{C}_m^n \cup \widehat{D}_m^n)$.

The following theorem shows the topological nature of the classes $\widehat{C}_m^n$, $\widehat{D}_m^n$ and $\widehat{E}_m^n$.

Theorem 7. 1. *For $m, n \geq 1$, there hold $\widehat{C}_m^n = \widehat{\overline{C_m^n}}$, $\widehat{D}_m^n = \widehat{\overline{D_m^n}}$ and $\widehat{E}_m^n = \widehat{\overline{E_m^n}}$.*
Hence these classes are topological properties of regular ω-languages.
2. *$\widehat{C}_1^1 = \{\emptyset\}$ and $\widehat{D}_1^1 = \{A^\omega\}$.*
3. *$\widehat{C}_1^2$ is the class of regular open languages, and $\widehat{D}_1^2$ is the class of regular closed languages.*
4. *$\widehat{C}_2^1$ is the class of regular G_δ-languages, and $\widehat{D}_2^1$ is the class of regular F_σ-languages.*
5. *For $m, n \geq 1$, the classes C_m^n and D_m^n are regular Wadge degrees.*
6. *For $n \geq 1$, the class E_1^n is a regular Wadge degree.*

From this theorem we know that the classes $\widehat{C}_m^n$ and $\widehat{D}_m^n$ for $m, n \geq 1$, and the classes $\widehat{E}_1^n$ for $n \geq 1$ are elementary topological properties of regular ω-languages. So one has to look at the classes $\widehat{E}_m^n$ for $m \geq 2$ and $n \geq 1$, how they split into elementary topological properties of regular ω-languages. For this reason define
$d^+ S =_{\text{def}} \{s \mid s \in S \text{ and an } (m(\mathcal{M}), n(\mathcal{M}))^+\text{superchain can be reached from } s\}$
and $d^- S =_{\text{def}} \{s \mid s \in S \text{ and an } (m(\mathcal{M}), n(\mathcal{M}))^-\text{superchain can be reached from } s\}$. Notice that $d^+ S \neq \emptyset$ implies $s_0 \in d^+ S$, that $d^- S \neq \emptyset$ implies $s_0 \in d^- S$, and that the defining condition $m(\mathcal{M}) = m \wedge n^+(\mathcal{M}) = n^-(\mathcal{M}) = n$ of E_m^n is equivalent to $m(\mathcal{M}) = m \wedge n(\mathcal{M}) = n \wedge d^+ S \cap d^- S \neq \emptyset$.

The *derivation* $d\mathcal{M}$ of a Muller automaton $\mathcal{M} = (S, A, \delta, s_0, \mathcal{F})$ is defined as follows. If $m(\mathcal{M}) = 1$ or $n^+(\mathcal{M}) \neq n^-(\mathcal{M})$ then $d\mathcal{M} =_{\text{def}} \mathcal{M}$. Otherwise $d\mathcal{M}$ is defined as the Muller automaton $d\mathcal{M} =_{\text{def}} ((d^+ S \cap d^- S) \cup \{s^+, s^-\}, A, d\delta, s_0, \mathcal{F} \cap P(d^+ S \cap d^- S))$ where $s^+, s^- \notin d^+ S \cap d^- S$ and

$$
d\delta(s, a) =_{\text{def}} \begin{cases}
\delta(s, a), & \text{if } s, \delta(s, a) \in d^+ S \cap d^- S, \\
s^+, & \text{if } s \in d^+ S \cap d^- S \text{ and } \delta(s, a) \in d^+ S \setminus d^- S, \\
s^-, & \text{if } s \in d^+ S \cap d^- S \text{ and } \delta(s, a) \notin d^+ S, \\
s^+, & \text{if } s = s^+, \\
s^-, & \text{if } s = s^-.
\end{cases}
$$

For $r \geq 1$, define the r-th derivation of $\mathcal{M}$ by $d^0 \mathcal{M} =_{\text{def}} \mathcal{M}$ and $d^{r+1}\mathcal{M} =_{\text{def}} d(d^r \mathcal{M})$.

Theorem 8. *For deterministic Muller automata $\mathcal{M}$ and $\mathcal{M}'$, if $L_\omega(\mathcal{M}) = L_\omega(\mathcal{M}')$ then $L_\omega(d\mathcal{M}) = L_\omega(d\mathcal{M}')$, i.e., the derivation is an invariant of all automata accepting the same language.*

Theorem 8 justifies the following definition. Let L be an ω-language and let $\mathcal{M}$ be a deterministic Muller automaton such that $L_\omega(\mathcal{M}) = L$. Then $d(L) =_{\text{def}} L_\omega(d\mathcal{M})$. For $\mathcal{C} \subseteq \mathcal{R}$ define $d(\mathcal{C}) =_{\text{def}} \{d(L) \mid L \in \mathcal{C}\}$ and $d^{-1}(\mathcal{C}) =_{\text{def}} \{L \mid d(L) \in \mathcal{C}\}$.

Theorem 9. 1. *If $L \in E_m^n$ for $m \geq 2, n \geq 1$ then $d(L) \in C_m^1 \cap D_m^1$.*
2. *If $L \in C_m^n$ or $L \in D_m^n$ for $m, n \geq 1$ or $L \in E_1^n$ for $n \geq 1$ then $d(L) = L$.*

For a class $\mathcal{C} \subseteq \mathcal{R}$ and $m, n \geq 1$ we define $\mathrm{E}_m^n \mathcal{C} =_{\mathrm{def}} \{L \mid L \in \mathrm{E}_m^n \wedge d(L) \in \mathcal{C}\} = \mathrm{E}_m^n \cap d^{-1}(\mathcal{C})$. Now the family $\mathcal{T}$ of all regular Wadge degrees can be characterized as follows.

Theorem 10

$$\mathcal{T} = \{\mathrm{E}_{m_1}^{n_1} \mathrm{E}_{m_2}^{n_2} \dots \mathrm{E}_{m_{r-1}}^{n_{r-1}} \mathrm{C}_{m_r}^{n_r} \mid r \geq 1, m_1 > m_2 > \dots > m_r \geq 1, n_1, n_2, \dots, n_r \geq 1\} \cup$$
$$\{\mathrm{E}_{m_1}^{n_1} \mathrm{E}_{m_2}^{n_2} \dots \mathrm{E}_{m_{r-1}}^{n_{r-1}} \mathrm{D}_{m_r}^{n_r} \mid r \geq 1, m_1 > m_2 > \dots > m_r \geq 1, n_1, n_2, \dots, n_r \geq 1\} \cup$$
$$\{\mathrm{E}_{m_1}^{n_1} \mathrm{E}_{m_2}^{n_2} \dots \mathrm{E}_{m_{r-1}}^{n_{r-1}} \mathrm{E}_1^{n_r} \mid r \geq 1, m_1 > m_2 > \dots > m_{r-1} > 1, n_1, n_2, \dots, n_r \geq 1\}.$$

For our decision algorithms the following theorem will be important.

Theorem 11. *For $m \geq 2$ and $n \geq 1$, if $\mathcal{C} \subseteq \mathrm{C}_m^1 \cap \mathrm{D}_m^1$ then $\widehat{\mathrm{E}_m^n \mathcal{C}} = \widehat{\mathrm{C}}_m^n \cup \widehat{\mathrm{D}}_m^n \cup \mathrm{E}_m^n \widehat{\mathcal{C}}$.*

An interesting relationship between the structure of $\mathcal{T}$ and $\widehat{\mathcal{T}}$, resp., and the ordinal numbers below ω^ω should be mentioned. It is well-known that every non-zero ordinal $\alpha < \omega^\omega$ can be presented in the form $\alpha = n_1 \cdot \omega^{m_1} + n_2 \cdot \omega^{m_2} + \dots + n_r \cdot \omega^{m_r}$ where $r \geq 1$, $m_1 > m_2 > \dots > m_r \geq 0$ and $n_1, n_2, \dots, n_r \geq 1$ (*). This gives a bijection between the ordinals below ω^ω and the classes of type $\mathrm{E}_{m_1}^{n_1} \mathrm{E}_{m_2}^{n_2} \dots \mathrm{E}_{m_{r-1}}^{n_{r-1}} \mathrm{C}_{m_r}^{n_r}$. If α is presented in the form (*) then we define $\mathcal{R}_\alpha =_{\mathrm{def}} \mathrm{E}_{m_1+1}^{n_1} \mathrm{E}_{m_2+1}^{n_2} \dots \mathrm{E}_{m_{r-1}+1}^{n_{r-1}} \mathrm{C}_{m_r+1}^{n_r+1}$. Then co-$\mathcal{R}_\alpha' = \mathrm{E}_{m_1+1}^{n_1} \mathrm{E}_{m_2+1}^{n_2} \dots \mathrm{E}_{m_{r-1}+1}^{n_{r-1}} \mathrm{D}_{m_r+1}^{n_r+1}$. For $\alpha = n_1 \cdot \omega^{m_1} + n_2 \cdot \omega^{m_2} + \dots + n_{r-1} \cdot \omega^{m_{r-1}} + n_r$ where $r \geq 1$, $m_1 > m_2 > \dots > m_{r-1} \geq 1$ and $n_1, n_2, \dots, n_r \geq 1$ we obtain $\widetilde{\mathcal{R}}_{\alpha+1} = \mathrm{E}_{m_1+1}^{n_1} \mathrm{E}_{m_2+1}^{n_2} \dots \mathrm{E}_{m_{r-1}+1}^{n_{r-1}} \mathrm{E}_1^{n_r+1}$ where $\widetilde{\mathcal{R}}_{\alpha+1} =_{\mathrm{def}} (\mathcal{R}_{\alpha+1} \cap \text{co-}\mathcal{R}_{\alpha+1}) \setminus (\mathcal{R}_\alpha \cup \text{co-}\mathcal{R}_\alpha)$ and $\mathcal{R}_\alpha =_{\mathrm{def}} \widetilde{\mathcal{R}}_\alpha'$. Thus we have $\mathcal{T} = \{\mathcal{R}_\alpha', \text{co-}\mathcal{R}_\alpha', \widetilde{\mathcal{R}}_{\alpha+1} \mid \alpha < \omega^\omega\}$ and $\widehat{\mathcal{T}} = \{\mathcal{R}_\alpha, \text{co-}\mathcal{R}_\alpha, \mathcal{R}_{\alpha+1} \cap \text{co-}\mathcal{R}_{\alpha+1} \mid \alpha < \omega^\omega\}$.

We have $\mathcal{R}_\alpha \cup \text{co-}\mathcal{R}_\alpha \subseteq \mathcal{R}_{\alpha+1} \cap \text{co-}\mathcal{R}_{\alpha+1}$ for $\alpha < \omega^\omega$. Hence, $(\mathcal{T}; \leq_w)$ and $(\widehat{\mathcal{T}}; \subseteq)$ have a quasi-linear structure.

5   Upper Bounds for Deterministic Automata

Let $\mathcal{M} = (S, A, \delta, s_0, \mathcal{E})$ be a deterministic ω-automaton of some type, where $\mathcal{E}$ describes an acceptance condition for this type. Obviously, $\mathcal{M}$ is equivalent to the deterministic Muller automaton $\widetilde{\mathcal{M}} = (S, A, \delta, s_0, \{S' \mid S' \text{ satisfies condition } \mathcal{E}\})$, i.e., we have $L_\omega(\mathcal{M}) = L_\omega(\widetilde{\mathcal{M}})$. In fact, deterministic ω-automata of arbitrary types can be considered as succinct presentations of deterministic Muller automata. Hence the definitions of chains, superchains, and the characteristics m^+, m^-, n^+, and n^- apply also to these types of ω-automata. For $X \in \{M, R, S, P, B\}$, let

$$\mathrm{Chain}_{DX} =_{\mathrm{def}} \{(\mathcal{M}, m, s, +) \mid \mathcal{M} \text{ is a deterministic } X\text{-automaton}, m \geq 1,$$
$$\text{and } s \text{ belongs to an } m^+\text{chain of } \mathcal{M}\} \cup$$
$$\{(\mathcal{M}, m, s, -) \mid \mathcal{M} \text{ is a deterministic } X\text{-automaton}, m \geq 1,$$
$$\text{and } s \text{ belongs to an } m^-\text{chain of } \mathcal{M}\},$$

$\text{Super}_{\text{DX}} =_{\text{def}} \{(\mathcal{M}, m, n, s, +) \mid \mathcal{M} \text{ is a deterministic } X\text{-automaton}, m, n \geq 1, \text{and}$
$\text{an } (m, n)^+\text{superchain of } \mathcal{M} \text{ is reachable from } s\} \cup$
$\{(\mathcal{M}, m, n, s, -) \mid \mathcal{M} \text{ is a deterministic } X\text{-automaton}, m, n \geq 1, \text{and}$
$\text{an } (m, n)^-\text{superchain of } \mathcal{M} \text{ is reachable from } s\}$

Proposition 3. *Let $\mathcal{M}$ be a deterministic X-automaton, and let $m, n \geq 1$.*

1. $\text{m}^+(\mathcal{M}) \geq m \iff$ *there exists an $s \in S$ such that $(\mathcal{M}, m, s, +) \in \text{Chain}_{\text{DX}}$.*
2. $\text{m}^-(\mathcal{M}) \geq m \iff$ *there exists an $s \in S$ such that $(\mathcal{M}, m, s, -) \in \text{Chain}_{\text{DX}}$.*
3. $\text{n}^+(\mathcal{M}) \geq n \iff (\mathcal{M}, \text{m}(\mathcal{M}), n, s_0, +) \in \text{Super}_{\text{DX}}$.
4. $\text{n}^-(\mathcal{M}) \geq n \iff (\mathcal{M}, \text{m}(\mathcal{M}), n, s_0, -) \in \text{Super}_{\text{DX}}$.

It turns out that, for deciding the topological degrees, the complexity of Chain plays a central role. Knowing its complexity, the complexity of the topological properties follows in a uniform way. (As to the notation in the following theorem: If $\mathcal{K}$ is the class of all languages accepted by machines of a certain type, and $\mathcal{L}$ is an arbitrary language class, then $\mathcal{K}^{\mathcal{L}}$ is the class of all languages which can be accepted by machines of the given type when using oracles from $\mathcal{L}$. If $\mathcal{L} = \{A\}$ is a singleton then we write $\mathcal{K}^A$ rather than $\mathcal{K}^{\mathcal{L}}$. For a standard textbook on complexity see e.g. [BDG95].)

Lemma 1. *Let $X \in \{M, R, S, P, B\}$.*

1. $\text{Super}_{\text{DX}} \in \text{NL}^{\text{Chain}_{\text{DX}}}$.
2. *There exists an $\text{L}^{\text{NL}^{\text{Chain}_{\text{DX}}}}$-algorithm which, given a deterministic X-automaton $\mathcal{M}$, computes the characteristics $\text{m}^+(\mathcal{M})$, $\text{m}^-(\mathcal{M})$, $\text{n}^+(\mathcal{M})$, and $\text{n}^-(\mathcal{M})$.*
3. *There exists an $\text{L}^{\text{NL}^{\text{Chain}_{\text{DX}}}}$-algorithm which, given a deterministic X-automaton $\mathcal{M}$, computes $\text{d}\mathcal{M}$.*
4. *For every $\mathcal{C} \subseteq \mathcal{R}$, if $(\mathcal{C})_{\text{DX}} \in \text{NL}^{\text{Chain}_{\text{DX}}}$ then $(\text{d}^{-1}\mathcal{C})_{\text{DX}} \in \text{NL}^{\text{Chain}_{\text{DX}}}$.*

Theorem 12. *For $X \in \{M, R, S, P, B\}$ and $\mathcal{C} \in \mathcal{T}$, the problems $(\hat{\mathcal{C}})_{\text{DX}}$ and $(\mathcal{C})_{\text{DX}}$ are in $\text{NL}^{\text{Chain}_{\text{DX}}}$.*

6   Deterministic Muller Automata

In this section, let $\mathcal{M} = (S, A, \delta, s_0, \mathcal{F})$ be a deterministic Muller automaton where $\mathcal{F} = \{S_1, S_2, \ldots, S_r\}$. We define a few problems needed for our algorithm. Let $m, n \geq 1$.

$(\mathcal{M}, i, j) \in \text{Subset} \iff_{\text{def}} S_i \subset S_j$
$(\mathcal{M}, i, j) \in \text{Subseteq} \iff_{\text{def}} S_i \subseteq S_j$
$(\mathcal{M}, s, s') \in \text{Reach} \iff_{\text{def}} \exists x (x \in A^* \wedge \delta(s, x) = s')$
$(\mathcal{M}, i) \in \text{Loop} \iff_{\text{def}} S_i \text{ is a loop of } \mathcal{M}$
$(\mathcal{M}, i, j) \in \text{Between}^+ \iff_{\text{def}} \exists k (S_k \text{ is a loop of } \mathcal{M} \text{ and } S_i \subset S_k \subset S_j)$
$(\mathcal{M}, i, j) \in \text{Between}^- \iff_{\text{def}} (\mathcal{M}, i, j) \notin \text{Between}^+ \wedge \exists S'(S' \text{ is a loop of } \mathcal{M} \wedge S_i \subseteq S' \subseteq S_j)$
$(\mathcal{M}, i) \in \text{Outside}^+ \iff_{\text{def}} \exists k (S_k \text{ is a loop of } \mathcal{M} \text{ and } S_i \subset S_k)$
$(\mathcal{M}, i) \in \text{Outside}^- \iff_{\text{def}} (\mathcal{M}, i) \notin \text{Outside}^+ \wedge \exists S'(S' \text{ is a loop of } \mathcal{M} \text{ and } S_i \subseteq S')$
$(\mathcal{M}, i) \in \text{Inside}^+ \iff_{\text{def}} \exists k (S_k \text{ is a loop of } \mathcal{M} \text{ and } S_k \subset S_i)$
$(\mathcal{M}, i) \in \text{Inside}^- \iff_{\text{def}} (\mathcal{M}, i) \notin \text{Inside}^+ \wedge \exists S'(S' \text{ is a loop of } \mathcal{M} \text{ and } S' \subseteq S_i)$

Lemma 2. *1. The problems* Subset, Subseteq, Reach, Loop, Between$^+$, Between$^-$, Outside$^+$, Outside$^-$, Inside$^+$, *and* Inside$^-$ *are in* NL.
2. Chain$_{\mathrm{DM}} \in$ NL.

Lemma 2 provides the upper bound of the main result of this section.

Theorem 13. *For $\mathcal{C} \in \mathcal{T}$, the problems $(\widehat{\mathcal{C}})_{\mathrm{DM}}$ and $(\mathcal{C})_{\mathrm{DM}}$ are NL-complete.*

7 Deterministic Mostowski and Büchi Automata

Here we can prove the same results as for deterministic Muller automata.

Lemma 3. *The problems* Chain$_{\mathrm{DP}}$ *and* Chain$_{\mathrm{DB}}$ *are in* NL.

To understand the following theorem remember that deterministic Büchi automata can accept just the sets from $\widehat{C}_2^1$, i.e., from C_2^1, C_1^n, D_1^n, and E_1^n for $n \geq 1$.

Theorem 14. *1. For every $\mathcal{C} \in \mathcal{T}$, the problems $(\widehat{\mathcal{C}})_{\mathrm{DP}}$ and $(\mathcal{C})_{\mathrm{DP}}$ are NL-complete.*
2. The problems $(C_2^1)_{\mathrm{DB}}$, $(\widehat{C}_1^n)_{\mathrm{DB}}$, $(C_1^n)_{\mathrm{DB}}$, $(\widehat{D}_1^n)_{\mathrm{DB}}$, $(D_1^n)_{\mathrm{DB}}$, $(\widehat{E}_1^n)_{\mathrm{DB}}$, and $(E_1^n)_{\mathrm{DB}}$ are NL-complete for $n \geq 1$.

8 Deterministic Rabin and Streett Automata

We start with the complexity of chains and superchains. Just by guessing a possible chain or superchain and testing whether it is really one we obtain

Proposition 4. *The problems* Chain$_{\mathrm{DR}}$, Chain$_{\mathrm{DS}}$, Super$_{\mathrm{DR}}$, *and* Super$_{\mathrm{DS}}$ *are in* NP.

From Theorem 12 we obtain immediately that the problems $(\widehat{\mathcal{C}})_{\mathrm{DR}}$ and $(\mathcal{C})_{\mathrm{DR}}$ are in $\mathrm{P^{NP}}$ for all $\mathcal{C} \in \mathcal{T}$. However, in some cases there are better upper bounds in terms of the Boolean hierarchy $\{\mathrm{NP}(n)\}_{n \geq 1}$ over NP (see e.g. [We85]); recall that NP(1) coincides with NP, NP(2) is the class of differences of NP-sets and NP(3) is the class of sets $(A \setminus B) \cup C$ where A, B, C are NP-sets. Unfortunately, in most cases there remains a gap between upper bound and lower bound. We consider Rabin automata first.

Theorem 15. *1. The problem $(C_1^1)_{\mathrm{DR}}$ is NL-complete.*
2. The problem $(D_1^1)_{\mathrm{DR}}$ is P-hard and in co-NP.
3. The problems $(C_m^n)_{\mathrm{DR}}$ and $(D_m^n)_{\mathrm{DR}}$ for $m+n > 2$, and the problems $(E_m^n)_{\mathrm{DR}}$ for $m, n \geq 1$ are P-hard and in NP(2).
4. The problems $(\widehat{C}_m^n)_{\mathrm{DR}}$ and $(\widehat{D}_m^n)_{\mathrm{DR}}$ for $m+n > 2$, and the problems $(\widehat{E}_m^n)_{\mathrm{DR}}$ for $m, n \geq 1$ are P-hard and in co-NP(3).
5. For every $\mathcal{C} \in \mathcal{T} \setminus \bigcup_{m,n \geq 1}\{C_m^n, D_m^n, E_m^n\}$, the problems $(\widehat{\mathcal{C}})_{\mathrm{DR}}$ and $(\mathcal{C})_{\mathrm{DR}}$ are P-hard and in $\mathrm{P^{NP}}$.

Because of Proposition 1 we obtain

Theorem 16.   *1. The problem* $(D_1^1)_{DS}$ *is NL-complete.*
  2. The problem $(C_1^1)_{DS}$ *is P-hard and in co-NP.*
  3. The problems $(C_m^n)_{DS}$ *and* $(D_m^n)_{DS}$ *for* $m + n > 2$, *and the problems* $(E_m^n)_{DS}$
   for $m, n \geq 1$ *are P-hard and in* NP(2).
  4. The problems $(\widehat{C}_m^n)_{DS}$ *and* $(\widehat{D}_m^n)_{DS}$ *for* $m + n > 2$, *and the problems* $(\widehat{E}_m^n)_{DS}$
   for $m, n \geq 1$ *are P-hard and in* co-NP(3).
  5. For every $C \in \mathcal{T} \setminus \bigcup_{m,n \geq 1} \{C_m^n, D_m^n, E_m^n\}$, *the problems* $(\widehat{C})_{DS}$ *and* $(C)_{DS}$ *are*
   P-hard and in P^{NP}.

It should be noticed that we would obtain exact complexity results for deterministic Rabin and Streett automata if we could show that Chain_{DR} (or, equivalently, Chain_{DS}) is in P. By Theorem 12, Theorem 15, and Theorem 16 we obtain

Theorem 17. *Assume* $\text{Chain}_{DR} \in P$.
  1. For all $C \in \mathcal{T} \setminus \{C_1^1\}$, *the problems* $(\widehat{C})_{DR}$ *and* $(C)_{DR}$ *are P-complete.*
  2. For all $C \in \mathcal{T} \setminus \{D_1^1\}$, *the problems* $(\widehat{C})_{DS}$ *and* $(C)_{DS}$ *are P-complete.*

However, we even do not know the complexity of the problem $(D_1^1)_{DR}$, that is the problem of whether every loop of a given deterministic Rabin automaton satisfies the acceptance condition of this automaton. We know that this problem is P-hard and in co-NP, but we do not know whether this problem is in P or co-NP-complete.

9   Nondeterministic Automata

Let $\mathcal{M} = (S, A, \delta, s_0, \mathcal{E})$ be a nondeterministic ω-automaton of some type. A set $S' \subseteq S$ is a *loop* of $\mathcal{M}$ if there are $l \geq 1$, $x \in A^*$, $a_1, \dots, a_l \in A$, and $s_1, \dots, s_l \in S$ such that $\{s_1, \dots, s_l\} = S'$, $s_1 \in \delta(s_0, x)$, $s_{j+1} \in \delta(s_j, a_j)$ for $j = 1, \dots, l-1$, and $s_1 \in \delta(s_l, a_l)$.

Theorem 18. *Let* $T \in \{NR, NM, NP, NB\}$.
  1. The problem $(C_1^1)_T$ *is NL-complete.*
  2. For every $C \in \mathcal{T} \setminus \{C_1^1\}$, *the problems* $(C)_T$ *and* $(\widehat{C})_T$ *are PSPACE-complete.*

This PSPACE-upper bound is established by the use of determinization. Instead one could use the algebraic characterization of the hierarchy of regular ω-languages [PP04, Wi93, Pi97, CP97, CP99, DR06, Ca07]. W.l.o.g. it suffices to show that $(\mathcal{R}_\alpha)_{NB} \in \text{PSPACE}$ for each $\alpha < \omega^\omega$. In the algebraic approach, one associates to any nondeterministic Büchi automaton $\mathcal{M} = (Q, A, \delta, i, F)$ a finite ω-semigroup (S_+, S_ω), a set $P \subseteq S_\omega$ and a morphism $\varphi : (A^+ \cup A^\omega) \to (S_+, S_\omega)$ of ω-semigroups such that $L_\omega(\mathcal{M}) = \varphi^{-1}(P)$ and $L_\omega(\mathcal{M}) \in \mathcal{R}_\alpha$ iff P has no μ_α-alternating tree. The last notion is formulated similar to the corresponding notions in [Wag79, Se98] in terms of two preorders on the so called linked pairs of S_+ (most explicitly the details are written in [Ca07]). It is not hard to see that checking the last condition yields a desired PSPACE-algorithm for $L_\omega(\mathcal{M}) \in \mathcal{R}_\alpha$.

Theorem 19. *1. The problem* $(C_1^1)_{NS}$ *is P-hard and in* co-NP.
2. For every $C \in \mathcal{T} \smallsetminus \{C_1^1\}$ *the problems* $(C)_{NS}$ *and* $(\widehat{C})_{NS}$ *are* PSPACE-*hard and in* EXPSPACE.

References

[BDG95]   Balcazar, J.L., Diaz, J., Gabarro, J.: Structural Complextiy I. Springer, Heidelberg (1995)

[Ca07]   Cabessa, J.: A Game Theoretic Approach to the Algebraic Counterpart of the Wagner Hierarchy. PhD Thesis, Universities of Lausanne and Paris-7 (2007)

[CP97]   Carton, O., Perrin, D.: Chains and superchains for ω-rational sets, automata and semigroups. International Journal of Algebra and Computation 7(7), 673–695 (1997)

[CP99]   Carton, O., Perrin, D.: The Wagner hierarchy of ω-rational sets. International Journal of Algebra and Computation 9(7), 673–695 (1999)

[DR06]   Duparc, J., Riss, M.: The missing link for ω-rational sets, automata, and semigroups. International Journal of Algebra and Computation 16(1), 161–185 (2006)

[KPB95]   Krishnan, S., Puri, A., Brayton, R.: Structural complexity of ω-automata. In: Mosses, P.D., Schwartzbach, M.I., Nielsen, M. (eds.) CAAP 1995, FASE 1995, and TAPSOFT 1995. LNCS, vol. 915, pp. 143–156. Springer, Heidelberg (1995)

[La69]   Landweber, L.H.: Decision problems for ω-automata. Math. Systems Theory 4, 376–384 (1969)

[Lo99]   Löding, C.: Optimal bounds for the transformation of omega-automata. In: Pandu Rangan, C., Raman, V., Ramanujam, R. (eds.) FST TCS 1999. LNCS, vol. 1738, pp. 97–109. Springer, Heidelberg (1999)

[MS72]   Meyer, A., Stockmeyer, L.J.: The equivalence problem for regular expressions with squaring requires exponential time. In: Proc. of the 13th IEEE Symp. on Switching and Automata Theory 1972, pp. 125–129 (1972)

[Pi97]   Pin, J.-E.: Syntactic semigroups. In: Handbook of Formal Languages, pp. 679–746. Springer, Heidelberg (1997)

[PP04]   Perrin, D., Pin, J.-E.: Infinite Words. Pure and Applied Math, vol. 141. Elsevier, Amsterdam (2004)

[Sa88]   Safra, S.: On the complexity of ω-automata. In: Proc. of the 29th IEEE FOCS 1988, pp. 319–327 (1988)

[Se98]   Selivanov, V.L.: Fine hierarchy of regular ω-languages. Theoretical Computer Science 191, 37–59 (1998)

[SVW87]   Sistla, A.P., Vardi, M.Y., Wolper, P.: The complementation problem for Büchi automata with applications to temporal logic. Theoretical Computer Science 49, 217–237 (1987)

[SW74]   Staiger, L., Wagner, K.: Automatentheoretische und automatenfreie Characterisierungen topologischer Klassen regulärer Folgenmengen. Elektronische Informationsverarbeitung und Kybernetik 10, 379–392 (1974)

[Sta97]   Staiger, L.: ω-Languages. In: Handbook of Formal Languages, vol. 3, pp. 339–387. Springer, Heidelberg (1997)

[Th90]   Thomas, W.: Automata on infinite objects. In: Handbook of Theoretical Computer Science, vol. B, pp. 133–191. Elsevier, Amsterdam (1990)

[Th96]   Thomas, W.: Languages, automata and logic. In: Handbook of Formal Languages, vol. 3, pp. 133–191. Springer, Heidelberg (1997)

[Wa72]   Wadge, W.: Degrees of complexity of subsets of the Baire space. Notices AMS 19, 714–715 (1972)

[Wa84]   Wadge, W.: Reducibility and determinateness in the Baire space. PhD thesis, University of California, Berkely (1984)

[Wag79]   Wagner, K.: On ω-regular sets. Information and Control 43, 123–177 (1979)

[Wi93]   Wilke, T.: An algebraic theory for for regular languages of finite and infinite words. Int. J. Alg. Comput. 3, 447–489 (1993)

[We85]   Wechsung, G.: On the Boolean closure of NP. In: Budach, L. (ed.) FCT 1985. LNCS, vol. 199, pp. 485–493. Springer, Heidelberg (1985)

[WY95]   Wilke, T., Yoo, H.: Computing the Wadge degree, the Lipschitz degree, and the Rabin index of a regular language of infinite words in polynomial time. In: Mosses, P.D., Schwartzbach, M.I., Nielsen, M. (eds.) CAAP 1995, FASE 1995, and TAPSOFT 1995. LNCS, vol. 915, pp. 288–302. Springer, Heidelberg (1995)

[Ya06]   Yan, Q.: Lower Bounds for Complementation of omega-Automata Via the Full Automata Technique. In: Bugliesi, M., Preneel, B., Sassone, V., Wegener, I. (eds.) ICALP 2006. LNCS, vol. 4052, pp. 589–600. Springer, Heidelberg (2006)

Author Index

Lecture Notes in Computer Science

Sublibrary 1: Theoretical Computer Science and General Issues

For information about Vols. 1– 4974
please contact your bookseller or Springer